STATISTICAL ANALYSIS

Holt, Rinehart and Winston
QUANTITATIVE METHODS SERIES
ALBERT J. SIMONE, Adviser
University of Cincinnati

ROBERT G. BROWN
Decision Rules for Inventory Management

YA-LUN CHOU
Statistical Analysis: With Business and Economic Applications

STATISTICAL ANALYSIS

With Business and Economic Applications

Ya-lun Chou

St. John's University
Jamaica, New York

HOLT, RINEHART AND WINSTON, INC.
New York Chicago San Francisco Atlanta Dallas
Montreal Toronto London Sydney

Cover quotation: W. J. Youden

Library of Congress Catalog Card Number: 69-11874

03-073095-3

Printed in the United States of America

1 2 3 4 5 6 7 8 9

To My Wife, Pauline

PREFACE

Ever since the publication of my *Applied Business and Economic Statistics* in 1963, many college instructors have written to my publisher and to me, urging that I write another, more comprehensive statistics text which would include some of the more frequently used advanced and modern topics. This volume is the result of my response to such requests.

In preparing this text, I have kept certain guidelines constantly in view. First of all, this book is intended for use in either a one-semester or a two-semester course at the undergraduate or the graduate level in statistics for students majoring in economics and business who have had only freshman mathematics.

Second, I have tried to design this book as an introductory text as well as a terminal text. To this end, the coverage, though not encyclopedic, is more complete than most elementary texts on statistics in the market today. I believe that the topics included are sufficient to enable students to conduct independent research in economics and management science whenever statistical techniques are required. To this end also, I have attempted to obtain a proper balance between the underlying statistical principles and their applications. This reflects my feeling that theory is practical and that practice too far removed from theory easily leads one astray. My intention is to introduce the reader to a way of thinking, as he is led through the basic field of statistics, that will enable him to venture into advanced courses of statistical analysis with minimum difficulty and maximum understanding.

Realizing the limits to the amount of statistics that the average student can learn during a specific period, I have made flexibility my third guideline. If this book is used at the undergraduate level, some of the more advanced topics, such as the analysis of variance, nonparametric statistics, and multiple association analysis, can be eliminated without disturbing continuity. If pressed for time, the instructor may make further cuts from Chapters 10, 11, 15, and 16.

Finally, to enable the student to acquire a "feeling" of usefulness I have introduced many examples in the text and have included numerous problems of great variety at the end of each chapter. I hope that the student will work out some of the problems regularly even if the material presented in the text or in lectures seems absolutely clear to him.

I am grateful to all those users of my first book who have encouraged me to take up this project. I also would like to take this occasion to express my thanks to Dr. Joseph F. Sinzer, Vice-President of Academic Affairs, and Dr. Nathan Becker, Dean of the Business School, both of Pace College, for their loyal and continuous support of my research efforts.

I am also indebted to the Literary Executor of the late Sir Ronald A. Fisher, F.R.S., to Dr. Frank Yates, F.R.S., and to Oliver & Boyd Ltd., Edinburgh, for permission to reprint Tables II and XXXIII from their book *Statistical Tables for Biological, Agricultural and Medical Research.*

Ya-lun Chou

FRESH MEADOWS, NEW YORK
NOVEMBER 1968

CONTENTS

10. Statistical Testing 298

11. Sampling and Sample Designs 350

STATISTICAL ANALYSIS

1

STATISTICAL DECISION PROBLEMS

1.1

The Nature of Statistics

Statistics is a method of decision making in the face of uncertainty, on the basis of numerical data, and at calculated risks. The importance of this decision procedure is obvious when we consider that we live in a world where future events are always fraught with varying degrees of uncertainty.

It may be recognized at the very outset that there are two kinds of uncertainty. One is that uncertainty which arises because of *randomness*. When a coin is tossed, the outcome is random and not at all certain. In a single toss, the outcome may be either a head or a tail. This type of uncertainty, however, is very simple in principle because the *law of randomness* applicable to the situation is *easily* known. On the basis of pure reasoning, we know not only that the tossing of a coin may turn up a head or a tail, but also that both outcomes are equally likely to occur. In other words, the randomness of the population of all possible tosses is such that half of them must be heads and the other half must be tails. So in the case of this type of uncertainty, the principle is simple to handle and the decision easy to make. For instance, if you were asked to participate in a coin-tossing game under the condition that you would win two dollars if the coin turned up a head and that you would lose one dollar if the coin turned up a tail, you would certainly be inclined to accept the offer because you know that heads and tails are equally likely to occur.

Another type of uncertainty springs from the fact that the laws of randomness, or the state of nature, of a given problem situation is unknown. For instance, if the coin is obviously biased, then the head and the tail will not have equal chance to occur. That is, one face will have a greater chance to appear than the other, perhaps in favor of the tail. In this case, where the state of nature is not known, how would you decide if you were offered the same bet as before? If you wish to solve this statistically, you should perform experiments and take observations in order to test the fairness of the coin. Suppose you have tossed the coin ten times and observed that all trials in your experiment turned up tails, you may therefore conclude that the coin is biased and decide to decline the offer. You may know intuitively that your decision is wise but you may not know precisely why. Let us then put this problem into a logical framework and analyze it briefly.

First, the fairness of the coin lies in the distribution of the outcomes of all possible tosses of the coin. We know that if the coin is fair, then half of these outcomes must be heads and half of them must be tails. We are willing to play with such a coin because each toss may be regarded as a random selection from this infinite population and we can expect that in the long run heads will come up roughly 50 percent of the time. In the previous test of the fairness of the coin, the ten trials may be considered as a sample of ten observations taken from this imaginary population of all possible tosses. It is possible, from the viewpoint of chance variations, that a fair coin can give ten consecutive tails, as in the experiment. Thus, even though the present sample is the worst possible for testing the fairness of the coin, it is still not completely inconsistent with the assumption that the coin is fair. Despite this possibility, however, the sample information has led to the conclusion that the coin is not fair and consequently a decision is made to turn down the invitation. This decision, furthermore, can be considered wise. Why? Because, while a sample of ten tails in tossing a fair coin is possible, it is highly improbable. As you shall learn, if you had been sampling from a population that was half heads, the probability of ten tosses being all tails is $1/1024$, or just about one chance in a thousand. It is indeed a very rare event. You must realize, at this point, that your decision not to accept the offer might be wrong since the sample information has not established the complete certainty of the state of nature. However, in view of the extreme rareness of the sample if the coin is perfect, you could afford to *behave* with confidence on the assumption that the coin is indeed biased.

The foregoing example illustrates the procedure of statistical decision making. *Statistical decision* can thus be defined as the process of making a decision on the basis of numerical data and probability considerations. Numerical data are used to estimate the true state of nature, upon which both the selection of action and the desirability of action depend. For various reasons, which will be explained later, the provision of numerical data is often limited to a part of the population. Since partial (sample) information will not eliminate uncertainty altogether, statistical decision is often made in the face of some degree of uncertainty. As a result, there exists, in statistical decision, the risk of wrong action. By appropriate statistical theory and methods, however, risks of wrong decisions can be calculated and controlled. Indeed, a major role of inductive statistics is to design devices for measuring such risks.

1.2 Stages of Statistical Investigations

It is true that certain old techniques have gained significance and some new ones have been introduced because of the application of statistical methods to particular problems. However, problems of the application of statistical method to various fields of study are the same. Therefore, before anything more is said about statistics, it will be beneficial to remember the common problems of its application to any type of investigation. For this purpose we shall list in logical order the stages an investigator goes through in his attempt to answer questions of interest or importance: (1) formulation or definition of the problem; (2) design of the experiment; (3) collection of data; (4) classification, tabulation, and description of results; (5) generalization or final inference. We shall now explain briefly the problem or problems involved in each of these stages.

1. Statistical analysis, as a manufacturing process, begins with raw materials, which are numerical data, and through the process flow the finished products, which consist of whatever useful information or valuable conclusions lie buried among the crude data. In manufacturing an output, the producer must first decide what is to be produced and then set up precise specifications of the output quality. Otherwise he may not know what kind of raw materials and how much are to be fed into the process. Thus in statistics, the first problem is to know exactly what is to be investigated: to formulate the problem or question as precisely as possible. Only then can the investigator decide what the relevant data for the problem are. If he fails to do this, data collected may be entirely irrelevant or may even tend to becloud rather than clarify the problem. It is well to remember that the quality of statistical conclusions depends upon the appropriateness and accuracy of data, which in turn depend upon exactness in formulating the problem. Statistical techniques, no matter how refined and precise, cannot yield useful results for arriving at decisions on problems if they are applied to inappropriate data.

In statistics, the totality of pertinent data that may be collected and observed in a given problem is called a *population* or a *universe*. For instance, if the problem at hand is to discover the percentage of unemployment in the labor force at a particular time in the United States, the population—or universe—in this case will include all observations about the individual workers, as to whether they are employed or unemployed, in the United States at that time. Again, if we wish to determine the average life of a lot of 1000 automobile tires, the population for this problem consists of the different durabilities in terms of, say, number of miles traveled, of all the 1000 tires in the lot. It is interesting to note that a statistical population consists of a collection of observations of one kind or another. These observations constitute statistical data that are collected in accordance with the formulation of a particular statistical problem or with the definition of a specific statistical population.

2. Once the problem has been precisely formulated, the investigator must decide whether to study the population in its entirety or to observe only a part of it. The former procedure is called a *complete enumeration* or a *census;* the latter is referred to as *sampling.* In practice, it is often costly, time-consuming, and even physically impossible to conduct a census, and we have to resort to sampling as the only practical method of making the study. This is what is done, for example, by the Bureau of the Census in its Current Population Survey each month. Among other things, this survey furnishes estimates of employment and unemployment based on data from a sample of 25,000 households distributed throughout the United States.

In sampling we try to make a decision, on the basis of sample data, about the population from which the sample is drawn. Hence the sample must obviously represent the population adequately. To obtain a representative sample is one of the most important components of statistical theory. It involves such questions as these: How large should the sample be? What type of data is to be collected? How are the data to be collected? These questions belong to what is called *sample design* or the *design of the experiment* in statistics. Proper care must be taken in planning and designing an experiment; otherwise we may not be able to reach any valid conclusion. The design of the experiment is the second stage of a statistical investigation.

As will be shown in a later chapter there are many types of sampling design, and each of them has its merits, depending on the type of population and the aim of the survey. However, most of the time we shall deal with random samples. A *random* sample may be defined as one drawn under conditions that each item in the population has an equal chance to be selected. This emphasis is due to the fact that only random samples can be treated with probability, and probability considerations are the basis for making valid inferences.

3. The third stage is to collect the data in accordance with the design of the experiment. From an over-all point of view, this stage is mainly routine in nature even though it is often the most time-consuming and costly component of the whole statistical process. In formulating the problem and in designing the sample, a good deal of judgment and a sound knowledge of the subject matter are required. In contrast, good results in collecting data demand the absence of personal judgment.

4. When data are collected they should be arranged in readable forms. They may be classified in some systematic manner and presented in a table. Or, for the purpose of conveying to the reader the true meaning of facts with greater ease or with more force, data may be presented by graphs or diagrams. Descriptive measures, such as percentages, averages, or standard deviations, are then calculated. A measure derived from the sample data is called a *statistic.* If a measure is computed from observations of the entire population, however, it is referred to as a *parameter.* From a sample of 25,000 households, for instance, we may compute

the number of unemployed as a percentage that may turn out to be, say, 6 percent of all the observations made from the sample. This fourth stage of the process, as can be seen, defines the scope of what is called *descriptive statistics.*

5. When the sample includes the whole population, that is, when an investigation is made by a complete enumeration, the last stage of the survey is the descriptive statistics. By then the characteristics of the whole population would have been described and revealed and a decision of one sort or another on the problem could be readily made. However, when the sample consists of only a part of the universe, the study will not stop with descriptive measures. A sample is often not examined for its own sake. Rather, it is analyzed in the hope that it can throw some light on the true state of the population. In other words, the last stage consists of trying to answer on the basis of the sample statistic the original problem or question formulated, which is always concerned with the population parameter. In this last stage we try to say something more, for instance, than that 6 percent of the workers in the 25,000 households surveyed are unemployed in a given month. We try to use this information as an estimate of the total unemployment situation of all the workers throughout the United States. To do so we are in effect making a generalization about all other workers whom we have not observed. This problem is not so formidable as it may appear. As we shall discover, such generalizations or inferences can indeed be made by appropriate theories and techniques. In fact, the study of the methodology and reasoning used in arriving at such conclusions, called *inductive statistics,* is not only the most valuable but also the most fascinating part of statistics. The value of inductive statistics rests on the fact that it is characteristically a scientific method of investigation. As a process of arguing from the traits of a small sample to those of a larger population, our knowledge is increased if the conclusions reached are true. The fascination of inductive statistics lies, then, in its appeal to our speculative imagination.

Even though sample information is the rule rather than the exception in statistical decision, in order to obtain maximum efficiency, we must, before the selection of the sample, ask many important questions: What type of information is needed? How should we proceed to solve a problem if and when complete information can be made available? How much information is needed for a particular problem? How are decisions to be related to observations? Answers to all these questions are connected with the concept of *statistical definition of the decision problem.*

Thus, whenever there is a decision problem and it is recognized that the uncertainty attached to it can be removed by numerical data, the very first step in the decision procedure is to translate the problem into statistical terms. Statistical definition of a problem involves two distinct but closely related steps: defining the statistical population and describing the state of nature of the defined population. The latter step involves the use of statistical methods to summarize the pertinent facts about the population: the computation of the parameter or parameters necessary for the decision.

The concentration on statistical definition of a problem at the initial stage of a survey is a very rewarding effort, even if the decision will ultimately be made on the basis of a sample. First of all, a precise definition of the population is indispensable in the selection of the sample, since this will avoid the frequent mistake of drawing the sample from the wrong population. Second, thinking in terms of complete information at the beginning will enable us to concentrate on the *design of decision*—the design of a survey which goes beyond the mere collection or presentation of numerical facts—which can lead directly to decisions. Third, too much initial stress on data gathering often leads to converting a decision problem into a data-collection problem. Little attention is paid to the problem of how to use the data when data are obtained. Consequently, numerical information gathered often proves to be of no use for the decision problem in question.

In short, the stress on statistical definition of the problem is the stress on a priori planning in attacking the decision problem. Modern statistics demands that decisions be related to observations before observations are made. This demand calls for a decision maker to specify the possible decisions a priori that are conditional upon the actual observations. Suppose you are faced with a decision problem of jumping from the window or running out the door through the smoke when your house is on fire. You may make the decision a priori, relating the action to be taken to an observation which is to be made, with the toss of a coin. The rule may be this: if the head turns up, jump from the window; if the tail appears, run to the door through the smoke. This illustration may strike you as absurd. Nevertheless, it does bring out the necessary condition of a good decision rule: relate decisions to observations before observations are made. Throughout this book, you will learn many methods of constructing statistical decision rules. Now, we may turn our attention to a discussion of the various aspects of defining decision problems statistically.

1.3 Populations, Elementary Units, and Observations

In statistics, the term *population* is used to mean an aggregate of individual items, whether composed of people or things, that are to be observed in a given problem situation. The individual items in a population are called *elementary units*. To define a population is in a sense to limit the scope of the elementary units. Elementary units possess certain *characteristics*, sometimes referred to as *traits* or *properties*, which may be qualitative or quantitative in nature. For instance, a decision problem may involve observing the effectiveness of a shipment of certain drugs. Here the population includes all the individual units of drugs—the elementary units—in the shipment, and the characteristic of the elementary unit to be observed is its quality, its ability to cure some disease. Another problem may require observing the family income in a given state. Here the population includes all the households in that state, and the trait to be observed is the quantity of family income of the elementary units, the households. The result of observing an elementary unit is called an *observation*.

Since an observation is made on each elementary unit in the process of an investigation, we may consider a population as the totality of all pertinent observations which may be made in a given decision problem.

Note that both the definition of a population and the characteristic of its elementary units to be observed depend upon the nature of the decision problem at hand. If the following list of examples is studied thoughtfully the importance of this concept will be evident.

Example 1 Male Adults' Hats

1. Decision problem: To decide the appropriate quantities of hat production for male adults in accordance with varying sizes.
2. Population: All American male adults.
3. Characteristic: Circumference of each male adult's head.

Example 2 Housing Supply

1. Decision problem: To determine the effective supply of apartment units in New York City.
2. Population: All apartments in New York City.
3. Characteristic: Availability.

Example 3 Acceptance of an Order

1. Decision problem: To decide whether to accept or reject a shipment of television tubes.
2. Population: All television tubes in the shipment.
3. Characteristic: Time to failure.

Example 4 Tax Policy

1. Decision problem: For the government to determine a reasonable corporate profits tax.
2. Population: All corporations in America.
3. Characteristic: Net corporate incomes.

Example 5 Corporation Policy

1. Decision problem: For a firm to set up a pension plan.
2. Population: All employees in the firm.
3. Characteristic: Length of service of employees.

Example 6 Quality of Production Process

1. Decision problem: To determine the quality of steel rods produced by a certain process.
2. Population: All steel rods which could have been produced by that process.
3. Characteristic: Tensile strength, or diameter, or weight, or all three traits.

Example 7 Sales Forecast

1. Decision problem: To estimate future consumers' demand.
2. Population: All retail stores and mail-order houses.
3. Characteristic: Volume of sales.

EXAMPLE 8 TESTING ECONOMIC THEORY

1. Decision problem: To test whether marginal propensity to consume decreases with increase in income.
2. Population: All American families.
3. Characteristic: Family income and consumption expenditures.

EXAMPLE 9 PERSONAL INVESTMENT

1. Decision problem: To decide on investment in common stocks.
2. Population: All common stocks listed in the New York Stock Exchange.
3. Characteristic: Dividend rates.

From the preceding list, it may be found that populations can be classified into two types: infinite and finite. An *infinite population* is one that includes an infinitely large number of elementary units. For instance, the sixth example in the list is such a population, since in this case the population is associated with the process of producing steel rods. Theoretically, such a population includes all the possible steel rods that could be produced if the manufacturing process continued to operate indefinitely under given operating conditions. Another illustration of infinite population is all possible outcomes when a die is thrown continuously and indefinitely. In the case of an infinite population complete information is clearly impossible to secure; consequently decision must be made on the basis of a sample. A population which is not indefinitely large or which contains only a finite number of items is said to be a *finite population.* There are a number of examples of finite population in the above list. The student can easily think of many others.

It is also important to note that an aggregate of elementary units which is considered as the population in a given problem situation may also be considered as a sample in a different problem situation. For instance, in the third example, where the problem is to determine whether to accept or reject a shipment of television tubes, the population is defined to include all the tubes in the shipment. This very shipment, however, can also be considered as a sample drawn from an infinite population of television tubes if the problem is to make a decision on the process of production itself. In other words, the first problem is to determine the quality of the tubes in the shipment, and the second is to make a decision on the quality of the production process.

Another point to be noted is that for a given problem the population may be defined in different ways. For instance, to estimate future consumers' demand we may also define the population as all family units in the United States, and the characteristic to be observed becomes family income. For the same problem (the seventh example), the population includes all retail stores and mail-order houses, and the characteristic to be observed is the volume of sales. The first definition is based on the theory that consumers' demand, among other things, may be considered as a function of income. The second definition is founded on the knowledge that past sales can be considered as a reliable estimate of future purchases. Clearly, both sets of data are useful in solving the same problem. However, as to which definition would yield better results depends upon the investigator's personal judgment. The important implication here is that in any statistical survey, the investigator must have a thorough knowledge in the field of investigation in addition to his skill in statistical techniques. This is true not only in defining statistical populations, but also at all stages of the statistical procedure. A statistician who does not have the theoretical knowledge in a given field should form a partnership with one who does in order to do research in that field fruitfully. Indeed, in many business and economic decision-making research projects, teamwork is the rule rather than the exception.

At this stage, the student may ask, Can the same population be used for decision making in different problem situations? This is an important but relatively easy question. A moment's reflection will reveal that the same population and the characteristic of its elementary units, say, all American families and family incomes, can be used as a basis for decisions concerning (a) the future consumers' demand, (b) the formulation of an individual income tax policy, (c) the standard of living of American people, (d) the pattern of distribution of income, and so forth. Actually, it is the possibility of using the same population to solve different problems that makes it advisable, once the nature of the decision problem is known, to check to see if there exist data, collected by others for other purposes, that can be used to advantage in solving our own problem.

1.4 Measurements of Elementary Units

Statistics deals with numerical data. Characteristics of the elementary units must be expressed in terms of numbers so that they can be fed to statistical methods for refinement and analysis. Traits or characteristics of elementary units are of two kinds: quantitative and qualitative. Quantitative traits can be transformed into numerical data simply by measuring them directly by some units of measurement, such as a foot, an inch, a dollar, a pound, a yard. The results of measuring the quantities of the elementary units form the totality of observations, which are expressed numerically in terms of the units of measurement used. The values which the quantitative observations may assume are called *variates*. The variates together are called a *variable*.

When the characteristics are qualitative they are not directly measurable. The elementary units can be classified only as having or not having a certain quality or property. Nevertheless, in many cases, the qualitative observations of elementary units, which are called *attributes*, may be expressed numerically. This is done by counting the numbers of the elementary units which have a certain attribute, such as males in a human population, or satisfactory or nondefective items in a population of some industrial products. In the process of enumerating, we score 1 for an element which has the attribute and 0 for one which lacks it. We are in effect assigning the qualitative observations numerical values, 0 or 1. With the completion of accounting we would get the total by adding the 1s together, and in so doing, we obtain a numerical result. The ability to express qualitative data quantitatively via the indirect approach to measurement opens up many areas, where direct measurement can never be applied, to statistical treatment.

Very often quantitative data can also be treated qualitatively if the nature of the problem demands. As an example, a collection of ages of all the persons in a state is a variable, but it can also be treated as attributes if we are interested in, say estimating the eligible voters in that state. In this decision problem each person may be characterized by either of the two attributes: voting age or non-voting age. To treat a variable as an attribute is quite common in the work of statistical quality control.

The discussion of measuring elementary units also leads us to the necessity of defining both the elementary unit and the characteristic of the elementary unit to be measured more carefully. In many cases it is simple and easy to identify the elementary unit in the population. Occasionally, however, this proves quite difficult unless the elementary unit is precisely defined. For instance, in decision problems concerning a housing situation, we often need to know the number of rooms in a given area instead of just the housing units. But what is a room? Do we consider a bath, a kitchen, a terrace, or a foyer a room? Confusion in measurement also often arises because of a lack of precise definition of the characteristic. The most obvious illustration is the characteristic of age. How are we to record a person's age? At his last birthday? At his approaching birthday? At his nearest birthday? Also, are we interested in the characteristic age in years? Age to the nearest month? Age to the nearest day? Answers to these questions, of course, depend upon the purpose of the study. They should nevertheless be stated before actual measurements are made.

Before we go on to a new section, let us return to the concept of a variable. A variable is a set of variates which, in turn, are the values of the quantitative observations. When the values of a variable are arranged into some order, from large to small or vice versa, a *statistical series* is formed. A series may be continuous or discrete. A *continuous series* is a variable which can assume any numerical value within a specific range. In such a series successive values may differ by infinitesimal amounts. In other words, a continuous series is one in which the units may be divided into fractions of any size, no matter how small, so that there is a continuous flow of measurements with graduations infinitely minute. Measurements of weight, height, length, time, or temperature represent a continuous series, since, though the scale is graded in units, cases can be found coinciding with any point on the scale. Thus, a ton of steel can be divided into hundredweights which, in turn, can be reduced to pounds, to ounces, or to millionths of an ounce, and so on without theoretical limits. Similarly, the diameters of steel rods may be measured to the nearest 0.001 inch, to the nearest 0.0001 inch, to the nearest 0.00001 inch, or as precisely as desired. Also, the height of a man may be 5 feet, $5\frac{1}{7}$ feet, $6\frac{13}{24}$ feet, or $8\frac{2}{57}$ feet, or any other point on the scale between the shortest and the tallest height possible for a man. Clearly, all these examples are a continuous series because, in each case, the small units of measurements touch each other in a continuous flow of the variates.

The nature of some type of data, on the other hand, is such that the unit of measurement is not divisible. That is, the unit of measurement can be defined only in terms of integers. Consequently, each elementary unit has to be taken and measured in whole units or none. Observations of such data can assume only discontinuous values; that is, the variates vary with definite amounts. For instance, a family may have one, two, three, or eight children; but to have $2\frac{1}{2}$ or $5\frac{1}{4}$ or any other fractional unit is clearly an impossibility. Again, the highest degree of precision in measuring family income is to the nearest cent. It cannot take any value ending in a fraction of one cent. A series in which measurements cannot fall between the units on the scale is called a *discrete series.*

Although the theoretical distinction between continuous and discrete variation is clear and precise, in practical statistical work it is only an approximation. The reason is that even the most precise instruments of measurement can be used only to a finite number of places. Thus, every theoretically continuous series can never be expected to flow continuously with one measurement touching another without any break in actual observations. However, this practical drawback should not invalidate the simplifying assumption of the continuity of quantitative observations in statistical theory. Moreover, as will be explained in Chapter 2 and elsewhere, in the application of statistical methods, it is often helpful to treat discrete series as if they were continuous. What should be kept in mind is that in interpreting and using statistical results, the investigator should always pay attention to the logical distinction between the continuous and the discrete variables.

1.5 Distribution of Observations

With a series we have a large group of numerical observations that form what we call *statistical data.* These observations vary in value from item to item. We are, however, not interested in the individual values separately, since statistics is mainly concerned with information about the whole population. To provide information about a population, it is neither necessary nor efficient to handle the observations individually. Instead, we usually try to summarize the individual observations in one form or another when we study a population. The most comprehensive summary of a complete collection of observations is called the *frequency distribution.* A detailed discussion of the construction and uses of frequency distributions will be found in Chapter 2. Here we wish to use this concept only in relation to the problem of defining the population.

When the resistance of a lot of 500 units of a certain electrical product is measured and the results of the measurement (observations) are arranged into a series in ascending order, we first of all notice the range of the series, which is the difference between the highest and the lowest measurements of the series. Then we may proceed to divide the range into a number of steps, or classes, over

TABLE 1.1 FREQUENCY DISTRIBUTION OF 500 UNITS OF A CERTAIN ELECTRICAL PRODUCT CLASSIFIED ACCORDING TO RESISTANCE

Resistance (ohms)	Frequency
2.7–2.9	2
3.0–3.2	18
3.3–3.5	48
3.6–3.8	97
3.9–4.1	138
4.2–4.4	104
4.5–4.7	69
4.8–5.0	20
5.1–5.3	4

which the observations are distributed. As a result a frequency distribution of the observations emerges. It is so called because it shows the number of observations (frequency) that falls into each of the classes. Each class will evidently have its own frequency, since the number of measures falling in each class will probably differ from that in any other class. A frequency distribution may be presented in tabular form, which is referred to as the *frequency table.* Table 1.1 is such a table.

It may be noted again for stress that this distribution portrays a lot of 500 units of a certain electrical product. It is considered here as a presentation of a population only in the sense that the decision problem at hand is concerned with the quality of this specific lot. If, however, our concern is with the quality of the manufacturing process of this electrical product, this distribution then actually presents a sample for the data covered.

In concluding this section, we may point out that the concept of frequency distribution has been introduced as a summary measure of a statistical population. The value of this concept does not lie in the fact that it can summarize the population when the population is available. Instead, its value is that when we think in terms of the population distribution, even if the population is unavailable, we should be able to know which form of statistical data is needed in a problem situation. In addition, frequency distribution is often needed to summarize data for further analysis. It is for these reasons that a detailed discussion of this concept will be presented later.

1.6

Specifying the Decision-Parameters

As was explained earlier, whenever a problem can be solved with the aid of numerical data, the very first thing to do is to define the problem statistically; that is, define the statistical population and specify the decision-parameter. We have hitherto discussed the various aspects of defining a statistical population; now we shall study the problem of selecting decision-parameters.

A *parameter* is also called a *true value*. It is, as may be recalled, a single value derived by statistical methods in order to describe in a summary fashion the pertinent characteristics about a population. A population may have many characteristics and, therefore, it has just as many parameters. It often has a minimum value, a maximum value, a range, a total value of all the individual values or an aggregate, an average value of all the individual values, and so forth. A *decision-parameter* is a trait or property of a population that is required as a basis for making a decision. Clearly, not all parameters of a population are decision-parameters in a given problem situation. As in the case of defining a population, the selection of the appropriate decision-parameter depends upon the nature of the problem in hand. Different parameters are needed in different problem situations.

The subsequent sections are devoted to the question of selecting decision-parameters. The key to this topic is to ask: What must we know about the population in order to make a decision on the problem in hand? In answering this question, however, we shall not attempt to make a comprehensive description of all possible population parameters and their corresponding values in making various types of decisions. We shall, instead, study a few basic and simple parameters that will reinforce the idea that the selection of the appropriate decision-parameter depends upon the type of decision we wish to make. Furthermore, we shall not be concerned with the detailed computational procedures for the parameters: statistical methods of summarizing population characteristics are the same as those used to compute the corresponding sample statistics, and the latter will be fully treated elsewhere.

1.7 Frequency Distributions

Since we already have some knowledge of the nature of frequency distributions, we shall begin with this concept. First of all, it must be recognized that a frequency distribution consists of a set of parameters, since each class frequency is a trait of the population. Often this whole set of parameters is needed for a decision problem. For illustration, consider a New York department store that has been in business for five years. In its men's shoe department, 4500 pairs of shoes have been sold in the past and now management has decided to carry an inventory of 1000 pairs in the forthcoming year. The problem is to decide how many pairs of shoes in each and every size it must carry. This problem situation clearly points to the fact that a frequency distribution of shoes sold in the past in accordance with size is needed for a decision. Now suppose the management has such a distribution constructed from sales records in the past, and the results are presented in the first two columns in Table 1.2.

In order to be more useful for the present problem, the absolute frequencies must be converted into relative frequencies. This is done by dividing the total

TABLE 1.2 DISTRIBUTION OF SALES OF MEN'S SHOES BY SIZES IN A NEW YORK DEPARTMENT STORE, 1964–1968

Size	Number of Pairs	Relative Frequency
$4\frac{1}{2}$	10	0.2
5	20	0.4
$5\frac{1}{2}$	40	0.8
6	50	1.1
$6\frac{1}{2}$	150	3.3
7	300	6.7
$7\frac{1}{2}$	610	13.6
8	920	20.4
$8\frac{1}{2}$	800	17.8
9	750	16.7
$9\frac{1}{2}$	420	9.3
10	200	4.4
$10\frac{1}{2}$	150	3.3
11	40	0.8
$11\frac{1}{2}$	30	0.7
12	10	0.2
Total	4500	100.0

frequency into the absolute frequency of each class and multiplying it by 100. Thus the relative frequency of a class is the absolute frequency of that class expressed as a percentage of the total frequency. Relative frequencies of the data under consideration are recorded in the third column in the above table. (Note that the sum of the relative frequencies is not precisely 100 because of rounding.) Now, with this information, a decision for the present problem can be readily made: the relative frequency distribution tells the department store manager that he should carry 0.2 percent of his stock in size $4\frac{1}{2}$, 3.3 percent in size $6\frac{1}{2}$, 20.4 percent in size 8, and so on.

In summary, we note that the problem is to decide on the proportion of a store's shoe inventory that should be in various sizes after the total quantity of inventory has been determined. For this problem the population is defined as the sizes of its customers' shoes and the appropriate decision-parameters are the relative frequency distribution of this population. Observe again that both the definition of the statistical population and the selection of the decision-parameter are determined by the problem situation.

1.8 The Aggregate

The *aggregate* of a population is simply the *total* or *sum* of the numerical values of the individual observations in the population. In order to study statistical methods with ease and precision, it is essential to assign symbols to represent different measures and to express the relationships between different quantities by formulas. In general, population parameters are designated by letters of the Greek alphabet, and sample statistics by letters of the English alphabet. Now if we let A, the Greek capital letter alpha, stand for the aggregate and N as the total number of observations in the population, then the first sentence in this paragraph can be written as

$$\mathrm{A} = \text{the sum of the } N \text{ values of the variable}$$

Also, in terms of notations, a population is generally represented by a set of observations of $x_1, x_2, x_3, \cdots, x_n$, wherein each symbol stands for the value of one individual observation. Thus, the previous expression may be changed to

$$\mathrm{A} = x_1 + x_2 + x_3 + \cdots + x_n$$

For convenience, Σ, the Greek capital letter sigma, is used to represent the operation of summing. With this notation, the above equation can be further reduced to

$$\mathrm{A} = \sum_{i=1}^{n} x_i$$

where the letter i is used to designate the serial number, and the right side of the equation reads "the sum of the values of the variable from the first through the nth." In practice, however, it is common to omit the subscript to the x and the references to it above and below the Σ. The formula for the population aggregate can then be finally presented as follows:

$$\mathrm{A} = \Sigma x \tag{1.1}$$

Needless to say, the aggregate is a reasonable property only for a finite population. The aggregate of any infinite population would most probably also be infinite. Now, the question is, Under what conditions should we select the aggregate as a decision-parameter?

The aggregate, as simple as it may be, is actually the key decision-parameter in many decision problems. In the field of economics, Gross National Product (GNP)—total monetary value of all goods and services produced during a year at market prices—is needed to decide, say, on the maximum proportion of the nation's productive facilities that can be transferred to defense programs; the total personal income is helpful to reach a decision on a possible change in income tax legislation; the total quantity of money in circulation is required, among other things, for the Federal Reserve System to determine whether more currency should be issued or whether part of the currency outstanding should be withdrawn. In many business problems, attention is also focused on some aggregates. For example, before deciding to construct an apartment building in a given city, the real-estate firm must know the aggregate amount of the apartment units currently available. Again, information on the aggregate inventory on hand is needed before a decision can be made whether to increase or decrease the present stock. The following specific illustrations should crystallize these general ideas.

Among its other responsibilities, the Federal Reserve System regulates the nation's supply of money. Now, if from the studies of the System's economic research staff, it is considered that the appropriate supply of money should be $150 billion for the next six months, the decision problem is to determine whether the present supply of money should be increased, decreased, or left unchanged. The present supply, which, incidentally, is the sum of federal reserve currency, treasury currency, and demand deposits, may be considered as a population aggregate. Here, A, the aggregate, is the key decision-parameter, since, if the System can determine A, it also decides on the course of action to be taken. For example, if A is $135 billion, the System should increase the supply by $15 billion; if A is $160 billion, it should decrease the supply by $10 billion; If A is $150 billion, it should leave the supply as it is.

It is interesting to note, from this simple illustration, the basic nature of the decision rule: alternative courses of action must be related to observations before observations are made.

1.9 The Proportion

The *proportion* of a population, usually designated as π (the Greek letter pi), is the ratio of the number of the elementary units, x, which possess a certain attribute, to the number of all the elementary units in the population, N. Symbolically,

$$\pi = x/N \tag{1.2}$$

For instance, if a survey is to be conducted on the smoking habits of the students in a certain college and it is found that 250 of the 800 students smoke cigarettes, then $x = 250$, $N = 800$, and

$$\pi = x/N = {}^{250}\!/_{800} = 0.3165, \text{ or } 31.65 \text{ percent}$$

The proportion, though a very simple concept, is one of the most useful parameters in many decision problems. In deciding whether a lot of merchandise ordered should be accepted or rejected, the decision-parameter is usually π, the proportion of defective items. In a network's decision on a course of action concerning a TV program, the proportion is also the key decision-parameter. For instance, the network may decide to continue the program if 12 percent or more of the audience watches it, and the network may drop the program if its rating is less than 0.12, or 12 percent. Again, the federal government may decide to take one type of action if the proportion of unemployment is 0.08 or more; to take another type of action if it is between 0.05 and 0.08; to take still another type of action if it is less than 0.05 but greater than 0.03; to take no action at all if it is 0.03 and less.

The proportion as a decision-parameter, like all decision-parameters, describes adequately the true state of nature of the statistical population. Consequently, it eliminates the uncertainty of the problem situation and aids decision on the proper course of action. Clearly, the results of a decision depend not only on the decision made but also on the true value of the proportion. Unfortunately, it is often impossible to make π available. In the first place, we can count neither the number of elementary units having or lacking a certain attribute nor the total number of elementary units of the population if it is infinite. In the second place, even when dealing with finite populations, the calculation of π is merely academic because observations can be made only by the physical destruction of the elementary units, as in the case of testing the tensile strength of steel rods. In situations such as these, where complete information cannot be made available, our only alternative is to reduce the uncertainty of the true population values by utilizing sample information so that a basis may be furnished for rational action in the face of uncertainty. In any case, it is worth repeating, our focus on appropriate decision-parameters at the initial stage of a survey enables us to answer such questions as these: What kind of information do we need for the problem? How should it be used if and when made available?

1.10 The Arithmetic Mean

Another property of a population used frequently as a decision-parameter is the measure of an "average." In statistics, as we shall see, there are many types of averages. The most important average, however, is the arithmetic mean, which we shall briefly discuss in this section. This particular measure also happens to be the most common average. When people talk about an average they usually refer to this measure. *The arithmetic mean* is simply the sum of the variates in a population divided by the total number of the variates. Thus, let μ (the Greek letter mu) be the population arithmetic mean, $x_1, x_2, x_3, \cdots x_n$, the values of the N observations of the population, the formula for the arithmetic mean becomes

$$\mu = \frac{x_1 + x_2 + x_3 + \cdots x_n}{N}$$

which, by using the summation sign, may be written as

$$\mu = \frac{\Sigma x}{N} \tag{1.3}$$

The arithmetic mean is often the key decision-parameter in statistical problem situations. Indeed, as will be seen, the chapters on inductive statistics in this book are mainly concerned with this concept. It would suffice simply to mention here a few situations where the arithmetic mean describes adequately the true state of the problem and thereby serves as the basis for decision. For instance, in deciding whether or not a shipment of cargo should be accepted, the arithmetic mean may be the appropriate decision-parameter. It may be, say, the mean durability of a shipment of automobile tires, or the mean weight of a lot of drained weight of canned meat, or the mean tensile strength of an order of steel wires. Again, in setting up the price of the products of a firm, management may use the arithmetic mean of the selling prices of all the firms in the industry as the basis for decision. Finally, decisions on changes in wage rates are often made on the basis of some price index—average of price-relatives. As a matter of fact, the wages of more than four million unionized workers in this country are tied by contract to the Consumer Price Index constructed and published by the United States Bureau of Labor Statistics.

It may be observed at this point that the arithmetic mean and the aggregate can sometimes be used interchangeably as a basis for selecting the appropriate course of action. The reason is the mathematical relationship between these two concepts. From the properties of their formulas it is easy to see that

$$\mu = \mathrm{A}/N \quad \text{and} \quad \mathrm{A} = N\mu$$

In a problem situation, however, where either the mean or the aggregate can serve as the decision-parameter, the former is always preferred over the latter. This is so because the aggregate may often be misleading, whereas the arithmetic mean is always clear and exhaustive. For instance, if it is desired to determine the relative efficiencies of two firms producing an identical commodity, we may use either total costs—the aggregates—or the average costs—the arithmetic mean—as the decision-parameter. Clearly both parameters will serve equally well in this problem if the two firms also by chance produce the same amount of outputs. However, the use of total-cost data would easily lead to wrong conclusions if the firms have different levels of output. To avoid this possible confusion, the aggregate should be converted into per unit figures; that is, total costs should be changed into average costs of production before the decision is made.

1.11

The Variance and the Standard Deviation

From our previous discussions on measuring elementary units and frequency distributions, it is easy to see that every population is characterized by some degree of variability. The value of one variate usually differs from all the rest in the same series. Thus, we may find that steel wires produced by the same manufacturing process may differ from each other in tensile strength. Radio tubes in the same shipment may have different durabilities. Prices of the same drug vary from store to store in the same city. Family incomes of all the households in the same nation are seldom the same.

Although populations are almost invariably characterized by some variability, the degree of variation in one may be greater than that in another. When a population has a relatively small variability, we say its elementary units are relatively consistent or uniform. On the other hand, when the elementary units in a population differ greatly from each other in value, they are said to have a high degree of variability. But how can variability, or uniformity—the lack of variability—be measured? Also, why should we be concerned with variability? We shall answer these two questions briefly, but in reverse order.

We are concerned with variability mainly because our decisions in many problem situations depend upon the population variations. In other words, a measure that can describe variation adequately will serve as the decision-parameter. To illustrate, suppose there are available to management two different manufacturing processes with which to produce automobile batteries. Moreover, suppose batteries produced by both methods have the same mean life. Under these circumstances, how can management decide which process is better? There may be several criteria for this decision problem. However, the degree of variability should be considered as the key decision-parameter. The reason is that, other things being equal, the more uniform, or the less variable, the outputs the better the process.

The fact that every manufacturer always strives to produce a product that will be as uniform as possible should not blind us to the possibility that in some situations a given course of action may be taken only if the population has a high degree of variability. Examples in which variability is a criterion for decision may be found in stock speculation, where speculators are interested in buying or selling only securities which have great variations in prices, and in the garment industry, where clothes are produced in different sizes.

How can we define a procedure for obtaining a single number to represent the variation in a population? As a matter of fact, there are several methods for this purpose. However, we shall introduce only variance and its related measure, standard deviation, in this section. Our special attention to these measures is due to their mathematically definable properties and to their unique importance in the theory of sampling.

The *variance* of the population is simply the average of the squared deviations of the individual values from the mean; namely, if we denote variance by σ^2, we have

$$\sigma^2 = \frac{(x_1 - \mu)^2 + (x_2 - \mu)^2 + (x_3 - \mu)^2 + \cdots + (x_n - \mu)^2}{N}$$

$$= \frac{\Sigma(x - \mu)^2}{N} \tag{1.4}$$

Standard deviation is defined as the square root of mean squared deviations from the mean; that is,

$$\sigma = \sqrt{\frac{\Sigma(x - \mu)^2}{N}} \tag{1.5}$$

Both the variance and the standard deviation are measures of variability in a population. These two measures are closely related, as indicated by formulas (1.4) and (1.5). It is easy to remember that the variance is the average squared deviations from the arithmetic mean and that the standard deviation is the square root of the variance. Later, it will be shown that sometimes it is more convenient to use the variance; and on other occasions to think in terms of the standard deviation.

We shall have the opportunity in subsequent chapters to show the importance of variance and standard deviation in the development of statistical methods as well as their proper use and interpretation. For the time being it is necessary to point out two things about these two measures that you must now remember. In the first place, the variance or the standard deviation is a single summary number that describes the variability in a population. The smaller the value of σ^2 or σ, the smaller the variability or the greater the uniformity in the population. Second, in general, the term "variation" means the departure from a norm. Thus, a measure of variability is one that measures the extent to which there are differences between individual observations and some central or average value. Indeed, it should be clear from the present discussion that the definition of variance is based on the differences of the individual items from the arithmetic mean of the population.

Glossary of Formulas

(1.1) $A = \Sigma x$ The aggregate of a population is defined as the sum of values of all individual variates of the population. This formula can be applied only to finite populations.

(1.2) $\pi = x/N$ The proportion of a population is the ratio of the number of individual observations having a certain attribute to the total number of observations of the population.

(1.3) $\mu = \dfrac{\Sigma x}{N}$ The arithmetic mean of a population is considered as the sum of values of all the individual observations in the population divided by the total number of observations in the population.

(1.4) $\sigma^2 = \dfrac{\Sigma(x - \mu)^2}{N}$ The variance of a population is the sum of the squares of the deviations of the individual observations from the arithmetic mean of those observations divided by the number of observations.

(1.5) $\sigma = \sqrt{\dfrac{\Sigma(x - \mu)^2}{N}}$ The standard deviation of a population is simply the square root of the population variance. Note that the variance is the mean square of the deviations and that the standard deviation is the square root of the mean square of the deviations.

Problems

1.1 As you know, in any scientific study it is of vital importance to define precisely the technical terms in the field. You should therefore try to understand every term thoroughly the first time you see it. What are the new terms you have learned in this chapter? What do they mean?

1.2 Most of you are now also taking a course in accounting, and you now know that both courses are concerned with figures. Do you think there is any difference between these two branches of studies? If so, what is it?

1.3 Statistics has been used to study both social and physical sciences. Is there a basic difference in statistical methods as applied in these two different areas? Why or why not?

1.4 In what sense do we say that a basic knowledge in statistics is essential to becoming an efficient citizen in a modern democracy?

1.5 It was stated that there are five distinct stages in a statistical investigation. What are they? Can you regroup the chapter headings in the table of contents in accordance with these stages?

1.6 What is the difference between descriptive statistics and inductive statistics? Can you think of some examples to illustrate the difference?

1.7 "Statistics is always concerned with mass phenomena and never with a single observation." Comment.

1.8 What is the most difficult aspect in decision making? Why does this difficulty arise?

1.9 What do we mean by uncertainty? Describe the two types of uncertainty introduced in the text. Which type of uncertainty is the main concern in decision making? Why?

1.10 What is the meaning of statistical decision? Give some examples of your own to illustrate the process of statistical decision.

1.11 Can we always make the right decision when we use statistical theory and methods? Why or why not?

1.12 Give some illustrations to show the decision rule that decisions are related to observations before observations are made.

1.13 Consider the following problem situations:

a. A 1.5 percent sales tax is proposed as a source of governmental revenue in a small town. The financial adviser to the mayor estimated that unless the annual yield of the tax is at least $15,000 it would not be worthwhile to consider the proposal. Thus it is suggested that the total annual retail sales of the city be determined in order to calculate the yield of the proposed tax.

b. An electrician plans to open a TV service shop in a suburb. To provide a basis for his inventory policy, he feels he must have information about the number and make of television sets owned by the families in the area.

c. In order to provide information on business opportunities in a certain state, the state government wishes to estimate the current population and per capita income in the state.

d. A soap manufacturer wishes to determine whether housewives prefer floating bath soaps over the ordinary type.

For each of the above problem situations:

1. Specify the alternative courses of action, states of nature, and consequences.
2. Indicate the uncertainty attached to the problem and explain how numerical data could eliminate or reduce the uncertainty.
3. Formulate a decision rule.

1.14 What is a statistical definition of the problem?

1.15 Although we generally use sample information as the basis of statistical decision, we must always think in terms of complete information at the initial stage of a survey. What are the reasons for this rule?

1.16 Define in your own words the following terms: elementary unit, observation, variate, attribute, and decision-parameter.

1.17 Differentiate these pairs of concepts and cite some examples of your own for each: continuous and discrete series, and finite and infinite population.

1.18 Define the population, identify the elementary unit, and state the characteristic to be observed in each of the following problem situations:

Problem 1 To determine whether a roulette wheel is biased.

Problem 2 To decide the premium rates of life insurance.

Problem 3 To determine the farm policies of a central government.

Problem 4 To decide on the type of heating that should be used for a new housing project in a small city.

1.19 Give some examples of decision problems where the frequency distribution, the aggregate, the proportion, the arithmetic mean, or the standard deviation may be considered as the key decision-parameter.

1.20 Many problems usually involve two or more decision-parameters. Can you think of any problem that might need, say, the arithmetic mean and the standard deviation for its solution?

1.21 It is physically impossible to make a decision on the basis of complete information in many problems. Can you cite such problems? What are the reasons for your selection?

2

FREQUENCY DISTRIBUTIONS

2.1

Introduction

Statistical decisions are based on numerical data. However, data are needed only when the state of nature is unknown and the decision maker is actually uncertain about the most desirable course of action to be taken. Again, if data are needed, we must further decide whether the decision should be based on complete or partial information. Thus, a decision can be made without any numerical observation, on the basis of some observations, or on all pertinent observations. Which of these three courses should be taken in a given problem situation? The answer to this question depends upon the statistical definition of the problem. Sometimes we may find that, once a problem is expressed in statistical terms, no more uncertainty exists as to which course of action is the wisest. In such a case, no observations would be necessary to reach a decision.

In many situations, however, statistical definitions of problems indicate only what type of information is needed and how it should be used, once obtained, without eliminating or reducing uncertainties associated with the problems. Under these circumstances, data must be collected and used for final decisions. Whether complete information—population data—or partial information—sample data—should be used in a given problem situation depends upon many factors.

In general, population data are preferred if the population is relatively small, if it is not very costly to obtain complete observations, and if observation can be made without destroying the elementary units. However, these conditions are seldom met; consequently, statistical decision, as a rule, is based on sample information. In using sample data, we are mainly concerned with three questions: How large should the sample be? How should sample data be selected? What constitutes a well-designed sample? These questions fall into the scope of sampling design—a topic to be taken up in Chapter 11.

After data have been made available, the next step in statistical analysis is to describe their properties. To describe data is to compute a proportion, a mean, a standard deviation, a correlation, or some other summary measure of the observations comprising a population or a sample. It may be recalled that a descriptive measure for a population is called a parameter and that the analogous summary measure for a sample is a statistic. Hence the term *descriptive statistic* refers to either parameter or statistic. It is important to keep clearly in mind at all times whether a population or a sample is under investigation. To aid in this distinction, as pointed out in Chapter 1, we shall use Greek letters to represent population parameters and Roman letters to represent statistics. We shall, beginning with this chapter, discuss the methods of computing sample statistics. It should be remembered, however, that a statistic is often needed not for its own sake but for learning about the corresponding parameter. In other words, the main purpose of computing a statistic is to furnish a basis for making inferences about a population parameter. This is important because the solution of a decision problem depends upon our knowledge of the population parameter.

The type of data needed depends upon the decision problem to be solved. Appropriate descriptive measures, in turn, depend upon the nature of the data. Data may be gathered simultaneously so that the time element is not involved. This type of data, furthermore, may include only one variable or two or more variables, called, respectively, *univariate, bivariate,* and *multivariate data.* Illustrations of univariate data include the diameters of a collection of steel wires, the heights or ages of a group of people, the wages of a given type of worker, the different uses of land in a certain region, a lot of industrial output classified according to defective and nondefective items, and so on. For this type of data we are mainly interested in describing the distribution of the individual items, their averages, and their tendency to vary from the average.

Income and consumption, heights and weights of college students, test scores and sales of the salesmen of a corporation, and prices and quantities of a commodity bought are illustrations of bivariate data. An example of multivariate data may be the price of beef, the price of pork, and the quantity of beef sold. For data that include two or more variables, description of the relationships between the variables is our main concern.

A sample may contain data that show the magnitude of a variable at successive points of time. Such data are called *time series.* Examples are the population of the United States since 1800, production of cars during the past fifty years, average wholesale prices of the years 1955–1968, and so forth. Here our interest is in describing the various types of change of a variable through time.

Summary measures for bivariate data and time series will be presented in Chapters 16 through 20. In this chapter and in the two that follow we shall be concerned with descriptive statistics for a single variable collected at a given point of time.

As has just been mentioned, the description of a single variable, for which time is not a factor, involves three main measures: the distribution, the average, and the variation of the variable. Computations of the last two measures often need the first as the basis. Hence we shall begin our formal analysis of mass data with the distribution of a variable. At this point, however, it is important to distinguish between a numerical (or quantitative) and a categorical (or qualitative) distribution. A distribution is said to be *numerical* if the data are recorded in magnitudes. A *categorical* distribution, on the other hand, refers to data classified qualitatively. (Can you think of some illustrations for each type of distribution?) The sections immediately following deal with the construction of numerical distributions and their graphic presentation. Problems concerning the construction of categorical distributions will be taken up in section 2.8.

2.2 Organizing Data: The Array and Frequency Table

Statistical data in the raw form may be presented as in Table 2.1, which itemizes the average weekly earnings of 142 messenger boys in New York City. What does this table mean?

Since the figures are arranged more or less at random, to the untrained mind the table means almost nothing. Anyone who is good at scanning a set of figures will be able to discover, after a couple of minutes and with considerable effort, the minimum and maximum values in the table. He may also know that he can obtain an "average" of some kind by summing up the individual values and dividing the sum by the number of items in the series. But this would be all the information he could squeeze out of the table.

To get more information and to get it quickly from numerical data, the data must be organized in some systematic fashion. The simplest device for organizing data systematically is to form an *array*, an arrangement of items according to magnitude. It can be formed either in ascending (from the lowest to the highest values) or descending (from the highest to the lowest values) order. In Table 2.2, the average weekly earnings of 142 messenger boys have been rearranged from Table 2.1 into an array of the ascending order.

Such an array has certain distinct advantages over the data in the raw form. Many characteristics of the variable can be learned easily and quickly from it. First, simply a glance at the array tells us that the range of the average weekly earnings of the messenger boys in the sample is \$3.91. Second, it shows clearly that there is a great concentration of the average weekly earnings near \$67.50.

TABLE 2.1

Average Weekly Earnings of 142 Messenger Boys in New York City, 1968

$68.00	$68.03	$67.79	$67.48	$68.05	$67.05	$ 66.03
66.07	67.09	67.49	67.80	68.06	68.05	68.07
69.50	69.54	69.48	69.46	67.47	67.78	69.00
66.10	66.14	67.81	68.09	67.51	67.12	68.11
67.82	67.53	67.14	67.15	67.20	67.17	66.95
67.47	67.75	67.96	68.99	68.98	68.93	68.85
68.82	68.45	68.30	68.20	68.15	66.21	66.29
67.20	67.18	67.83	67.54	67.85	67.55	68.15
68.17	67.85	68.18	69.94	69.20	69.15	68.95
67.94	67.75	67.44	66.93	66.42	67.74	66.93
66.34	66.37	67.22	67.55	67.85	68.22	68.29
67.88	67.55	67.23	66.40	68.91	67.92	67.73
67.41	66.92	66.40	67.71	67.92	67.91	68.87
68.84	68.80	66.91	69.84	69.45	68.78	67.90
66.87	67.38	67.68	66.84	67.32	67.68	67.88
66.73	66.83	66.81	67.32	67.66	67.86	68.64
69.61	69.65	68.33	67.89	66.53	67.25	66.60
67.57	66.71	67.25	67.58	67.80	68.40	69.73
69.24	68.60	66.80	69.21	66.78	69.12	69.20
68.47	69.17	66.75	67.28	67.61	67.83	
67.29	67.65	67.85				
Total						$9632.74

TABLE 2.2

Array for Average Weekly Earnings of 142 Messenger Boys in New York City, 1968

$66.03	$66.91	$67.32	$67.71	$67.90	$68.30	$69.15
66.07	66.92	67.38	67.73	67.91	68.33	69.15
66.10	66.93	67.41	67.74	67.92	68.40	69.20
66.14	66.93	67.44	67.75	67.92	68.45	69.20
66.21	66.95	67.47	67.75	67.94	68.47	69.21
66.29	67.05	67.47	67.78	67.96	68.60	69.24
66.34	67.09	67.48	67.79	68.00	68.64	69.45
66.37	67.12	67.49	67.80	68.03	68.73	69.46
66.40	67.14	67.51	67.80	68.05	68.78	69.48
66.40	67.15	67.53	67.81	68.05	68.80	69.50
66.42	67.17	67.54	67.82	68.06	68.82	69.54
66.53	67.18	67.55	67.83	68.07	68.84	69.61
66.60	67.20	67.55	67.83	68.09	68.85	69.65
66.71	67.20	67.55	67.85	68.11	68.87	69.73
66.75	67.22	67.57	67.85	68.15	68.91	69.84
66.78	67.23	67.58	67.85	68.15	68.93	69.94
66.80	67.25	67.61	67.85	68.17	68.95	
66.81	67.25	67.65	67.86	68.18	68.98	
66.83	67.28	67.66	67.88	68.20	68.99	
66.84	67.29	67.68	67.88	68.22	69.00	
66.87	67.32	67.68	67.89	68.29	69.12	

Finally, it also reveals roughly the distribution pattern of the series. Although there are a number of gaps among the average weekly earnings, the series does give a roughly continuous appearance. In addition, the array shows that about 85 percent of the items are distributed between $66.50 and $69.50 and more items are below $66.50 than above $69.50.

The array is often a practical and useful device in presenting data which include only a small number of items. Thus, for example, to study the subsidies that a state grants its cities for the operation of public utilities, the leading cities in the state may be arranged in an array according to population. It may also be instructive to array the per capita income of the various states and to consider the reasons for the differences. The array, however, is still a very cumbersome form of data organization, especially when a large sample is involved. Moreover, its usefulness is exhausted after a few types of information have been obtained from it. It is therefore desirable to compress the data into a more compact form and to permit the use of more complex analytical methods to reveal still other obscure relationships in the data. To condense and simplify data without losing the essential details is the purpose of the *frequency distribution*, an arrangement of data which shows the frequency of occurrence of values in each of a relatively small number of size-classes. The tabular presentation of such a summary of data is known as the *frequency table.*

To distribute a mass of raw data over the classes that have already been established we use either a tally sheet or the entry form. (Or, if the data are recorded on punch cards—a widely used procedure in handling large samples—the sorting and counting can be automatically made in a single procedure.) Tallying, as shown in Table 2.3, involves establishing classes and representing each item which falls in each class by one diagonal stroke; the number of items in each class is then counted.

The use of the entry form is illustrated by Table 2.4. In this procedure, the classes are set up horizontally at the top in an ascending order from left to right. The actual items are entered into the appropriate classes. The items that fall into each class are then counted and recorded at the bottom of that class. The total figures in the classes constitute the class frequencies. The entry form is more

TABLE 2.3 TALLY SHEET FOR AVERAGE WEEKLY EARNINGS OF 142 MESSENGER BOYS IN NEW YORK CITY, 1968

Average Weekly Earnings	Tally					Number of Messenger Boys
$66.00–$66.49	𝍸	𝍸	/			11
66.50– 66.99	𝍸	𝍸	𝍸			15
67.00– 67.49	𝍸	𝍸	𝍸	𝍸	////	24
67.50– 67.99	𝍸	𝍸	𝍸	𝍸	𝍸	
	𝍸	𝍸	𝍸			40
68.00– 68.49	𝍸	𝍸	𝍸	𝍸		20
68.50– 68.99	𝍸	𝍸	////			14
69.00– 69.49	𝍸	𝍸	/			11
69.50– 69.99	𝍸	//				7

TABLE 2.4 ENTRY FORM: AVERAGE WEEKLY EARNINGS OF 142 MESSENGER BOYS IN NEW YORK CITY, 1968

\$66.00–\$66.49	\$66.50–\$66.99	\$67.00–\$67.49	\$67.50–\$67.99	\$68.00–\$68.49	\$68.50–\$69.99	\$69.00–\$69.49	\$69.50–\$69.99
66.03	66.53	67.05	67.51	68.00	68.60	69.00	69.50
66.07	66.60	67.09	67.53	68.03	68.64	69.12	69.54
66.10	66.71	67.12	67.54	68.05	68.73	69.15	69.61
66.14	66.75	67.14	67.55	68.05	68.78	69.15	69.65
66.21	66.78	67.15	67.55	68.06	68.80	69.17	69.73
66.29	66.80	67.17	67.55	68.07	68.82	69.20	69.84
66.34	66.81	67.18	67.57	68.09	68.84	69.21	69.94
66.37	66.83	67.20	67.58	68.11	68.85	69.24	
66.40	66.84	67.20	67.61	68.15	68.87	69.45	(7)
66.40	66.87	67.22	67.65	68.15	68.91	69.46	
66.42	66.91	67.23	67.66	68.17	68.93	69.48	
	66.92	67.25	67.68	68.18	68.95	(11)	
(11)	66.93	67.25	67.68	68.20	68.98		
	66.93	67.28	67.71	68.22	68.99		
	66.95	67.29	67.73	68.29			
		67.32	67.74	68.30	(14)		
	(15)	67.32	67.75	68.33			
		67.38	67.75	68.40			
		67.41	67.78	68.45			
		67.44	67.79	68.47			
		67.47	67.80				
		67.47	67.80	(20)			
		67.48	67.81				
		67.49	67.82				
			67.83				
		(24)	67.83				
			67.85				
			67.85				
			67.85				
			67.85				
			67.86				
			67.88				
			67.88				
			67.89				
			67.90				
			67.91				
			67.92				
			67.92				
			67.94				
			67.96				
			(40)				

laborious than tallying; however, it has certain advantages over the latter: (1) it is easy to find the wrong entries by scanning the columns; (2) reclassifications can be readily made with little effort if the original classes are unsatisfactory; (3) it enables us to find out how closely the mid-value of a class agrees with the average of the values of the items in that class.

Note that the construction of both tally sheet and entry form can be accom-

TABLE 2.5 FREQUENCY TABLE: AVERAGE WEEKLY EARNINGS OF 142 MESSENGER BOYS IN NEW YORK CITY, 1968

Average Weekly Earnings (dollars)	Number of Messenger Boys (frequencies)
66.00–66.49	11
66.50–66.99	15
67.00–67.49	24
67.50–67.99	40
68.00–68.49	20
68.50–68.99	14
69.00–69.49	11
69.50–69.99	7
Total	142

plished with much greater ease if an array of the series is available. However, the items can also be arranged—with more effort, of course—from the raw data, because the only thing needed for tallying or for making an entry form is the number of classes, which, as will soon be shown, can be determined once the range and the number of items in the series are known. In practice, an array is never constructed solely for the purpose of making a frequency distribution, since too much time and effort are required for making it.

The results of the tally sheet or the entry form may now be presented in a frequency table, such as Table 2.5.

There are some technical terms for the frequency table which the student should know. Each stated interval, such as 66.00–66.49, is called a *class*. The boundaries of classes are called *class limits*. The lower values of the classes, such as 66.00, 66.50, and so on, are called *lower limits;* the higher values, such as 66.49, 66.99, and so on, are the *upper limits*. The width of a class is called the *class interval*, which is the difference between two successive lower class limits. In Table 2.5, the class interval is \$0.50 or 50¢. The number of values of the series (here, wages) which fall in a class is known as the *class frequency*. In Table 2.5, the total number of classes is 8 and the total number of frequencies is 142. Hence, the table summarizes 142 variates by distributing them into 8 size-classes.

The frequency distribution achieves condensation of data by losing the identity of the individual values. Despite this loss of identity, a great deal has been gained by this condensation. First, it can be seen that all the information revealed by the array can be obtained from the frequency distribution with greater ease. Second, the frequency distribution not only shows clearly the concentration of the individual values; it also brings out the pattern of the tendency for the individual values to vary above and below the concentration. Third, with data formed into a frequency distribution, comparisons between two or more series can be made more readily. As will be shown later in this chapter, such comparison is further facilitated when frequency distributions are presented in graphic forms. Finally, frequency tables are indispensable for speeding up computations of many other descriptive measures that will be taken up in the next two chapters.

2.3

Problems of Constructing Numerical Frequency Distributions

There are many important considerations bearing on the construction of numerical frequency distributions. The first main problem is the determination of an appropriate number of classes. Although there is no precise rule for this decision, we generally try not to have too many or too few classes for a frequency distribution. The basic reason for grouping data is to reveal the pattern of distribution. The use of too many classes tends to produce irregularities in class frequencies (such as 5, 11, 8, 20, 30, 28, and so on) and to obscure the concentration of values. The employment of too few classes oversummarizes the data, and some valuable information is lost in the process. In practice, we try not to have a frequency distribution with less than 5 or more than 15 classes. The precise number of classes to be used for a given variable depends upon personal judgment and other considerations bearing on the construction of frequency distribution to be discussed later in this section.

To determine the approximate number of classes, the beginner may use the *Sturges rule* as a guide:

$$k = 1 + 3.3 \log n \tag{2.1}$$

where

k = the approximate number of classes
n = the total number of observations in the sample
log = the ordinary logarithm to the base of 10

For instance, the series used as an illustration here is composed of 142 observations; we have, then,

$$k = 1 + 3.3 \log 142 = 1 + 3.3(2.15229) \doteq 8$$

This is exactly the number of classes we have for the distribution of the messenger boys' earnings. However, the student must realize that the Sturges formula yields only the approximate number of classes. Thus, when $k = 8$ we may eventually choose a number close to it, such as 7 or 9. Also, the use of this formula may give unreasonable results when the number of observations is either very large or very small. As a final warning, let the student be reminded that the Sturges rule is not a substitute for sound judgment.

The selection of the appropriate number of classes may also be aided by two additional considerations: (1) The number of classes selected should result in a monomodal instead of a bimodal distribution. (A monomodal distribution has only a single peak, with class frequencies continuously increasing to a maximum and then decreasing progressively. A *bimodal* distribution contains two peaks.) (2) The number of classes should enable the observations to cluster around the *class mid-point,* which is the point in the middle of the class. The class mid-point, as will be seen, gains importance because, in a frequency distribution, the individual values in each class are represented by its mid-point.

Once the number of classes is selected, the approximate class interval can then be derived. This is done by dividing the number of classes into the range of the series. The approximate class interval, for convenience in additional computations as well as easy reading, should be changed to some convenient number, like 1, 5, 10, 25, and so on; a number like 7, 13, 29, and so on, should be avoided. For instance, the approximate class interval for the data of the messenger boys' earnings is slightly less than \$0.49 and the actual class interval used is \$0.50. To meet this requirement, it is permissible to write the lower class limit of the lowest class below the lowest value of the series and to write the upper class limit higher in value than the highest value in the series. (Compare Tables 2.2 and 2.5.)

Whenever possible, a frequency table should be formed with equal class intervals: uniformity facilitates comparisons of class frequencies and computations of certain statistical measures. Often, however, it may turn out to be impractical and even undesirable to observe this rule. Sometimes there are large gaps in the data and, consequently, the use of equal class intervals would result in great irregularities—with no frequencies or only a few items in some classes. Moreover, uniform class intervals may show unnecessary details of parts of high-valued items and insufficient information for the distribution of those of low value. Because of these reasons, frequency distributions of varying class intervals are often constructed. Table 2.6 shows a distribution of unequal class intervals. In this instance, class intervals vary from \$1,000.00 to \$500,000.00. The purpose in this case is to show more detailed information of the distribution of tax returns from lower-income groups.

To overcome the shortcomings necessitated by varying class intervals, we compute what is called the *frequency density*. This is done by estimating what the class frequencies would be if a uniform class interval were used. Thus, in column 3 of Table 2.6, frequency densities are computed, assuming the same class interval of \$1,000.00. On the basis of frequency densities the distribution can now be interpreted and graphed and certain measures can be calculated. The introduction of the concept "frequency density" enables us to define more precisely the quantity measured by the size-classes. "Frequency" is a loose description of this quantity. More precisely, it represents the density of the observations—the number of cases—per unit of the class-sizes. This idea will become more apparent when we discuss graphic presentations of frequency distributions later.

After the number of classes and the class intervals are fixed, there remains the important problem of setting up the class limits. The following considerations bear on this problem.

Class limits should be mutually exclusive: there should be no doubt as to the class to which a given item belongs. Thus, such class limits as those shown below are not mutually exclusive, and should never be used:

\$1000–\$1500
1500– 2000
2000– 2500
.
.
.

TABLE 2.6 DISTRIBUTION OF FEDERAL INCOME TAX RETURNS BY SIZE OF INCOME, 1964

Adjusted Gross Income (in dollars)	Number of Returns (in thousands)	Frequency Density, Number of Returns per $1000 of Income (in thousands)
Under 1000	520	520
1000 and under 2000	4274	4274
2000 and under 3000	4231	4231
3000 and under 4000	4995	4995
4000 and under 5000	5365	5365
5000 and under 6000	5474	5474
6000 and under 7000	5315	5315
7000 and under 8000	4801	4801
8000 and under 9000	3858	3858
9000 and under 10,000	3019	3019
10,000 and under 15,000	6593	1318.6
15,000 and under 20,000	1457	291.4
20,000 and under 50,000	1209	40.3
50,000 and under 100,000	159	3.19
100,000 and under 500,000	35	0.0875
500,000 and under 1,000,000	1	0.0020
1,000,000 and more	*a*	0.0000
Total	51,306	

a Less than 500.
Source: Data adapted from the Economic Almanac, 1967–1968, pp. 434–435.

With such a designation of class limits, we would not know whether, say, a $1500 item belongs to the first or the second class.

The selection of proper class limits depends to a great extent on whether the data are continuous or discrete. It may be recalled that a series is continuous if, between any two values in it, no matter how close together they may be, there is always another value. Common examples cited before include heights, weights, time. Also, a series is said to be discrete if the variates can assume only definite values, such as the number of people in a family or the number of fatal automobile accidents occurring daily in the United States. Strictly speaking, business and economic data are seldom continuous. Even a variable that is inherently continuous may appear to be discrete because ordinary measuring units are limited in preciseness. Thus, if measurements for length (a continuous variable) are made by estimating to the nearest thousandth of an inch, the set of measurements may contain such figures as 0.345 and 0.346 in., but no value between the two successive thousandths of an inch. In spite of this, it is standard practice in statistics to treat some variables as continuous if the gaps between the successive values are very small and can therefore be considered, without any consequence, to be nonexistent. For instance, any series expressed in terms of money to the nearest cent may be treated as a continuous series.

Although a number of methods of stating class limits have been used, there are but two basic forms, as illustrated below:

Overlapping Limits

Average Hourly Earnings
\$1.00 and under \$1.25
1.25 and under 1.50
1.50 and under 1.75
1.75 and under 2.00

Nonoverlapping Limits

Average Hourly Earnings
\$1.00–\$1.24
1.25– 1.49
1.50– 1.74
1.75– 1.99
.
.
.

Both designations can be used for continuous data or data treated as continuous. Note that the words "and under" in the overlapping form implies that any earnings less than \$1.25 are cast in the first class whereas an item of \$1.25 belongs to the second class. So interpreted, it can be seen that the two forms for continuous data are actually equivalents.

In dealing with discrete data, the nonoverlapping form of class limits should always be used. Thus, if we were making a study of accidents occurring daily in industrial plants in Detroit, the following classification would be employed:

Number of Industrial Accidents
0– 4
5– 9
10–14
.
.
.

In designing the class limits, consideration must also be given to the location of class mid-points. When observations are cast into various classes, their exact values are lost; computations from frequency distributions are therefore made by treating the series as if all were distributed evenly throughout the intervals. For this reason, the class limits should be so selected as to make the mid-points reasonably representative of the values falling within the class limits. To achieve this, the class limits may be so written as to make the mid-points coincide with the concentrations of the observed values. As an example, consider the cafeteria checks, nearly all of which have values with a multiple of 5 cents. In such a case class limits may be designated as 3–7, 8–12, 13–17, and so on, which would provide mid-points of 5, 10, 15, and so on, as desired.

To locate the mid-points, we use the following formula:

$$m_1 = \frac{L_1 + U_1}{2} \tag{2.2}$$

where

m_1 = mid-point of the first class
L_1 = lower class limit of the first class
U_1 = upper class limit of the first class

When the class intervals are uniform, the mid-points of the successive classes will be separated from the mid-point of the first class and from each other by the width of the class interval. It is important to note that the class interval is found by subtracting the lower class limit of the preceding class from the lower class limit of the next class. It is not always possible to find the interval by subtracting the lower limit from the upper limit of a class. For example, in a classification such as 0–9, 10–19, 20–29, and so on, the class interval is 10 and *not* 9.

Here, in connection with the statement of class limits and the determination of mid-points, attention is called to the distinction between *nominal* class limits and *actual* class limits. *Nominal* limits are simply those which are stated in a frequency table. Thus, if we establish for certain observations reported classes and write 10–14, 15–19, 20–24, and so on, these same limits are then nominal. The same nominal limits are often used to accommodate data whose actual limits are different. The actual limits depend upon whether and how the data are rounded. Here we consider three alternatives.

First, data are not rounded, as in the case of the number of industrial accidents, or prices originally reported in dollars and cents. In such a case, the actual class limits are identical with the nominal class limits as previously stated. The class interval is 5 and the mid-points are 12, 17, 22, and so on.

Second, data are rounded to the nearest whole unit, such as the nearest of a whole dollar, or the nearest of a whole pound. If this be the case, a reported figure of 10 actually lies between 9.5 and 10.5; a reported figure of 14 actually falls between 13.5 and 14.5; and so on. This implies that an item of 14, which would be entered into the first class, may actually be as large as 14.5. Consequently, actual limits are 9.5–14.5, 14.5–19.5, 19.5–24.5, and so on, which give the same class interval and mid-points as in the first case.

Third, age data are sometimes rounded to the last full year. As such, the actual limits become 10 and under 15, 15 and under 20, 20 and under 25, and so on. The class interval is still 5 but the mid-points have become 12.5, 17.5, 22.5, and so on.

It can be seen from the above discussion that the mid-points are always determined on the basis of actual limits. In order to avoid possible confusion, the student is advised to construct frequency tables with actual class limits at all times.

Finally, there is the requirement that, whenever possible, there should be no *open-end class intervals*. A class is said to be open if, instead of assigning definite values to its limits, we write ". . . or more," ". . . or less," "greater than . . . ," or "less than . . . ." An illustration of open-end classes is presented in Table 2.7.

The worst feature of open classes is that they do not tell us how much "under" or "over" a given value may fall. Furthermore, the ranges of open classes cannot be defined. Consequently, we are presented with difficulties in putting them in

TABLE 2.7 DISTRIBUTION OF SPENDING UNITS AND TOTAL MONEY INCOME BEFORE TAXES, 1963

Family Personal Income	Percent of Families	Percent of Income
Under 2,000	11	2
2,000–3,999	18	7
4,000–5,999	20	13
6,000–7,999	18	17
8,000–9,999	12	14
10,000–14,999	13	21
15,000 and over	8	26
All Cases	100	100

Source: U.S. Department of Labor, Survey of Current Business (April 1964), 4.

graphic form and in basing calculations upon the data contained in the table. Nevertheless, there are often situations where open classes are clearly desirable. In dealing with data containing a few values which are much smaller or much larger than the rest, the use of open classes can accommodate a wide range of values without requiring a large number of classes or classes that hide too much of the relevant information. It is advisable, when this is done, that the values in the open classes be given in a footnote to the table. This practice will eliminate some of the difficulties of open classes.

In conclusion, it may be pointed out that the frequency table, like other types of data presentation, is always constructed to serve some specific purpose. The technical requirements outlined above must be supplemented by sound subjective judgment if proper and useful frequency distributions are to be formed.

2.4 Graphic Presentation of Frequency Distributions

A numerical distribution may be presented graphically in two basic forms. One way is to picture it graphically by a histogram (or column diagram). For instance, if the frequency distribution in Table 2.5 is shown in this form, it would give a picture like Figure 2.1. The basic requirements in constructing the histogram are the same as those for other arithmetic graphs. Some special features in the construction of a histogram are enumerated below.

(1) The class frequencies are plotted against the Y axis, and the class intervals are scaled on the X axis. The Y axis must start with zero and it must not have any scale break, even though it is not necessary to show the zero on the X axis.

(2) A space, from half to the full size of the class interval, is left at each end of the X axis.

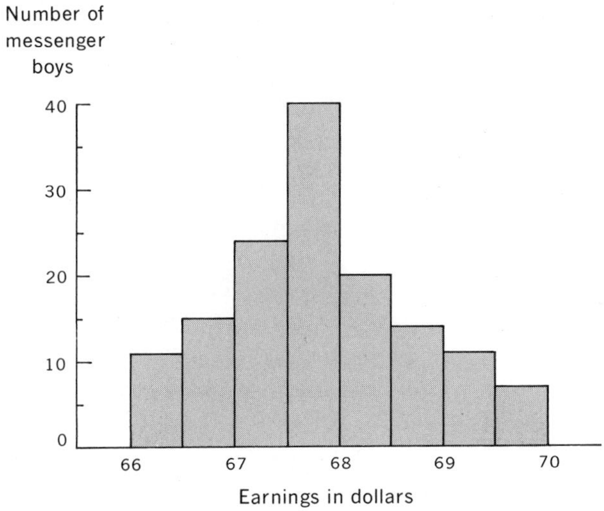

Figure 2.1 Histogram showing 142 messenger boys' average weekly earnings in New York City, 1968. (Source: Table 2.5.)

(3) The X scale designations are usually placed at the class limits. If the data are continuous or treated as such, the upper limit of one class will coincide with the lower limit of the following class. If the data are discrete, the X scale should be labeled in terms of the lower class limits only, or the class frequencies may be indicated by vertical bars separated from each other by a small space and both limits of each class marked. Sometimes, the X scale is labeled by placing the mid-value of each class at the center of the base of the rectangle.

(4) The X scale is equally spaced when the class intervals are uniform. In a varying class interval distribution, the X scale should be adjusted accordingly. For instance, if two class intervals, 100 and 500, are used, then the space on the X axis for the classes of 500 intervals should be five times as wide as those for 100 intervals. In such a case, frequency densities, not the original class frequencies, are plotted.

(5) A frequency distribution with open-end classes can be plotted on a histogram only by eliminating the open-end classes. The information concerning them is given in figures on both sides of the histogram.

(6) Each rectangle, enclosed by the vertical class limit lines on the sides, the horizontal line drawn on the top, and the X axis on the base, represents the frequencies in one class interval.

It should be noted that the area in each bar in Figure 2.1 is equal to the frequency of the corresponding class in Table 2.5. Also, the total area enclosed by the whole diagram is equal to the total frequency. In general, from the construction of a histogram, we may understand the important point that the areas above the various intervals on the X scale are proportional to the frequencies in those intervals of a distribution. From this property of the histogram, it can be seen that the quantity measured along the Y scale called frequency, actually represents

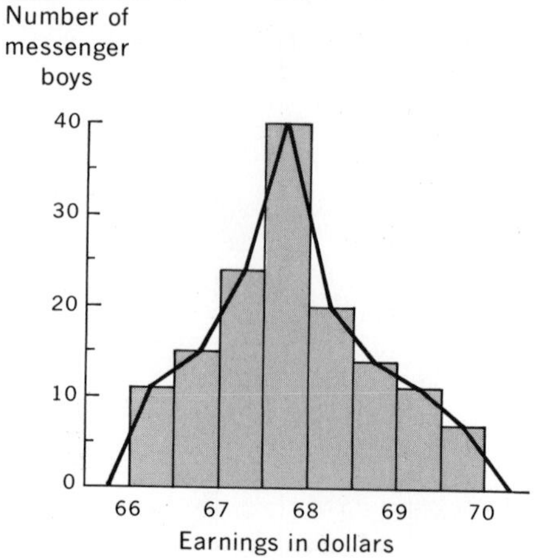

Figure 2.2 Histogram and frequency polygon: 142 messenger boys' average weekly earnings in New York City, 1968. (Source: Table 2.5.)

the density of the observations per unit of the X scale. Thus, in our illustration, instead of labeling the vertical axis as "number of messenger boys," we may more precisely call it "number of messenger boys per \$0.50," \$0.50 being the intervals we have selected for convenience.

Another way of presenting a frequency distribution graphically is to draw a *frequency polygon*. This is done, if a histogram is already available, by simply placing a dot on the mid-point of the top of each bar in the histogram and then connecting these dots by straight lines. This method is illustrated by Figure 2.2. Very often, however, a polygon is constructed without setting up the rectangles. Without the histogram the polygon is obtained by locating the coordinates: the ordinates, which are the class frequencies, and the abscissas, which are the mid-points. These points are then joined by straight lines. (See Figure 2.3.)

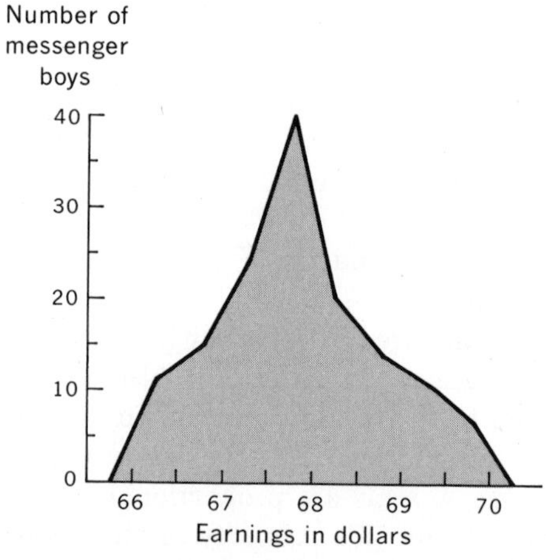

Figure 2.3 Frequency polygon showing 142 messenger boys' average weekly earnings in New York City, 1968. (Source: Table 2.5.)

Although the histogram is a very effective and vivid graphical presentation of frequency distributions, the polygon does not represent the basic data very well. The outstanding shortcoming of the polygon is that the areas under it, because of irregularities of data, are not proportional to the frequencies. One practice, aimed at eliminating some of this drawback, is to close the polygon at the base by extending both ends of the curve to the midpoints of two hypothetical classes at the extremes of the distribution that have zero frequencies. Note that both Figures 2.2 and 2.3 have been drawn in this manner.

There are, nevertheless, at least two important reasons for mentioning the frequency polygon. First, the polygon is especially suitable for comparing different distributions. When several distributions are to be compared on the same graph, it is much clearer to superimpose frequency polygons than to superimpose histograms, especially when all the distributions have the same class limits. For instance, Figure 2.4 shows, by superimposing two smoothed frequency distributions—suggested and derived from polygons—the change in the distribution of income in the United States between 1947 and 1958. It reveals clearly that, while the basic pattern of income distribution in 1958 remained the same as in 1947, there were many more families in high-income brackets and fewer families in the low-income brackets in 1958 than in 1947. With more careful reading of the diagram, it can also be discovered that the number of families and unattached individuals with incomes under $4000 decreased by 8.5 million from 1947 to 1958. Also, the number with incomes greater than $4000 increased by 18.1 million, of which 9.8 million moved up to the bracket between $4000 and $8000 and 8.3 million reached the bracket of $8000 and above.

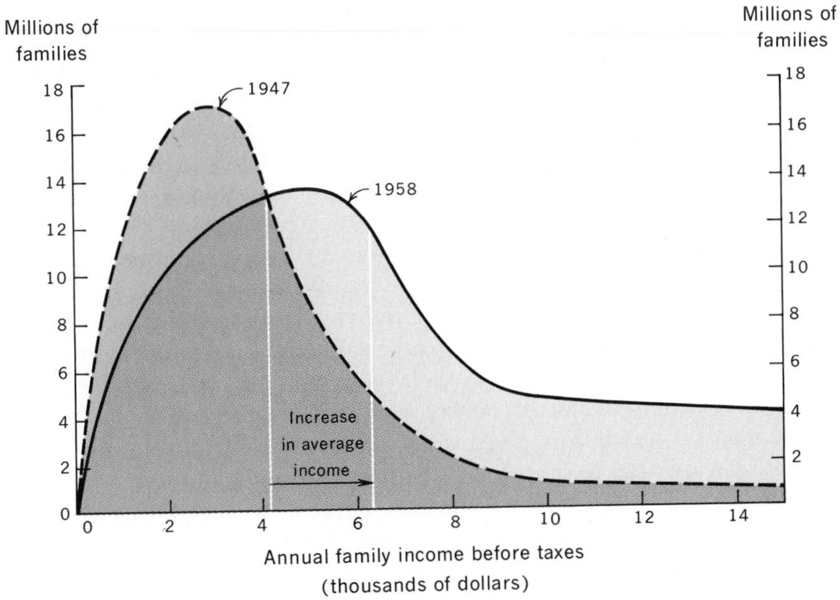

Figure 2.4 Number of families in various income brackets in 1947 and 1958. (From Harold I. Lunde, *Underwriting Prosperity*, Bureau of Economic Studies, Macalester College, 1960.)

2.5 The Smoothing of Curves and Population Models

A second advantage of the frequency polygon is that it suggests the use of a smooth curve as an idealized representation of the population distribution. A sample consists of only a limited number of items and therefore its distribution would be marked with irregularities and breaks. However, if the items in the sample are increased and class intervals are decreased continuously, we would expect the distribution to become smoother and smoother as well as more and more regular, since, in such a process, the accidental irregularities that affect a small number of items would be gradually eliminated. When the sample becomes very large, class intervals will become very narrow and yet each will contain a substantial number of frequencies. At the same time if the vertical scale, which measures frequency, is reduced so as to make the area of the histogram for this extremely large sample equal to the area for the observed sample, then the histogram would practically form a smooth curve. Thus, Figure 2.4, which portrays the distribution income of all the tens of millions of families and independent consuming units in the United States, gives the results of complete smooth and continuous smooth curves.

It was pointed out that the areas under the polygon are not proportional to the frequencies. This is due to the irregularities of the data. In a smooth curve, these irregularities have been eliminated. Thus, the area under the smooth curve between ordinates selected at given points on the X axis is assumed theoretically to be proportional to the frequency of observations between the given values.

The smooth curve gains importance because it is taken to represent the true distribution of the population from which the sample is drawn. The derivation of a smooth curve by enlarging the sample, however, is generally a practical impossibility. What we usually do is to approximate the population distribution on the basis of sample data. This can be done either by smoothing out the sharp points in the frequency polygon with an expert hand or by fitting a smooth curve to the sample data with some mathematical formula. With either method there is a great deal of latitude in curve smoothing. Obviously, two different experts can hardly draw an identical smooth curve to the same data. Differences in the judgment of statisticians may also lead to their using different formulas for the same sample. Thus, a variety of smooth curves which differ considerably but appear to fit the data may be drawn for a given set of observations. One may conceive that some smooth curve fitted to sample data represents a population, but he may not infer from the sample what particular smooth curve is the underlying distribution of the population. The latter is one of the most difficult topics in what is called *analytical statistics*. What may be mentioned here is that, because of the surprising latitude in curve fitting, the smooth curve should always be presented with the histogram.

The smooth curves are alternatively called *population models*. They are so called because they depict the salient features of population distributions. The term "population model" also suggests generalizations of the shapes of population distributions, symmetrical, skewed, **U**-shaped, and so on. These generalizations are of great utility in statistical analysis because they provide simplified methods of describing the basic characteristics of populations. There are additional reasons for our interest in population models. One is that, as introduced in Chapter 1, population distribution is an important parameter for decision making. Another, as will be discussed in later chapters, is that statistical inferences often require our knowledge of population models. A third reason is that a population model, being represented by smooth curves, lends itself more readily to mathematical treatment.

As the reader can easily imagine, population models can assume an almost infinite variety of shapes and sizes. We shall introduce in the following paragraphs some of the population models that are more frequently encountered in business and economic statistics.

Curve *a* in Figure 2.5 is a population model of special interest and importance in statistics. Note that this curve of distribution is bell-shaped, or symmetrical. The largest frequencies, in the center, spread out so that they are smaller and smaller, like a bell, with very small frequencies at both extremes. It is indeed strange that whenever a trait occurring at random is measured, the curve of the resulting distribution often resembles the one shown as Curve *a*. Whether it is the tensile strength of steel bars produced by a given process, the height of men, the intelligence of students, the size of rice, or errors of repeated measurements of a given characteristic that we measure, the curve of distribution always looks like a bell. Because of this the bell-shaped model is commonly called the *curve of normal*

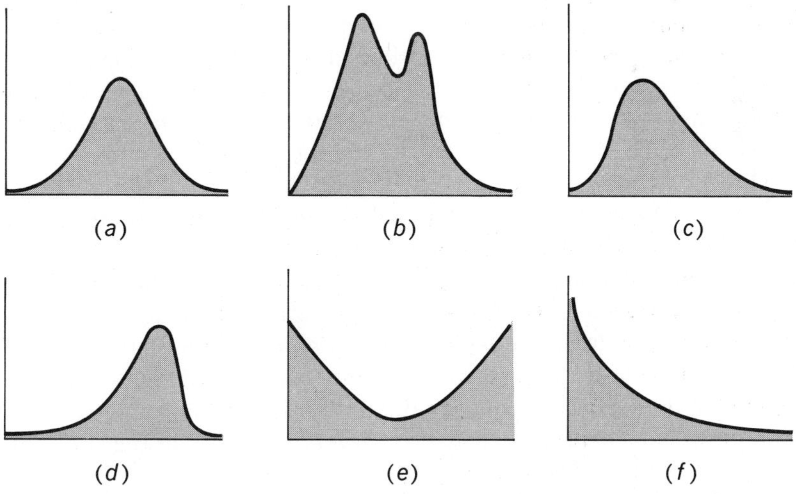

Figure 2.5 Curves representing principal varieties of population models. (a) The normal curve, (b) the bimodal curve, (c) the positively skewed curve, (d) the negatively skewed curve, (e), the **U**-shaped curve, (f) the reversed **J**-shaped curve.

distribution or simply the *normal curve*. We shall have more to say about this curve in Chapter 7. For the time being what may be stressed is that the student should not take the normal curve to be an ordinary curve. As a matter of fact, the normal curve has very precise definition and mathematical properties.

Curve *b* in Figure 2.5 portrays what is known as the bimodal distribution, meaning that it has two peaks. This type of distribution occurs when a population contains elements that can be divided into two different classes differing from each other in the characteristics being measured. Or we may say that the population is not homogeneous. For instance, the distribution for all demand deposit balances in the United States would assume this model. For in such a distribution we would find a prominent peak at a relatively low value for the balances held by consuming units and another distinct peak at a relatively high value for balances held by business firms and other organizations. Also, if we were to study the height of college students, a bimodal distribution would result—one concentration for men and another for women.

Curves *c* and *d* in Figure 2.5 are models for skewed distributions. A skewed distribution has only one peak but it is located at either the lower or the higher end of the curve. It is thus *asymmetrical*. When the longer tail of the curve is on the right, we say the distribution is skewed to the right or *positively skewed*. When the longer tail of the curve is on the left, we say the distribution is *skewed to the left* or *negatively skewed*. The positively skewed model is very common in economic and business data. For instance, the distribution of family income (see Figure 2.4) is skewed to the right, indicating that some families (relatively few in comparison with the over-all group) receive incomes much higher than those received by the majority of families. Also, the distribution of the number of retail stores by amount of sales would be positively skewed since there are a large number of small stores and a few very large ones. An example of the negatively skewed model would be the number of firms distributed according to the ratio of cost of sales to net sales in an industry. The negatively skewed model describes well a population whose variates have an upper limit. For instance, the upper limit of the ratio of sales cost to net sales would be unity, or a 100 percent. A firm cannot stay in business very long if its ratio of sales cost to net sales is unity.

The **U**-shaped curve, as illustrated by Curve *e* in Figure 2.5, describes a distribution that contains predominately low and high values, with intermediate values relatively scarce. The **U**-shaped model is rather rare, but some economic series do conform to it. For instance, the distribution of the nations of the world according to their stages of economic development would reveal bunching at two extremes. Most of the countries are either highly developed or underdeveloped, with only a few at the intermediate stages. Again, if we define the unemployed workers as those who do not work, the frequency distribution of the unemployed according to age groups would emerge as a **U**-shaped curve. This indicates that most of those who do not work are either very young or very old.

A population may also take the **J**-*shaped* or *reversed* **J**-*shaped curve*, such as Curve *f* in Figure 2.5, where the frequencies of occurrence increase or decrease continuously along the horizontal scale. The reversed **J**-shaped curve would be a good approximation for the distribution of corporations classified according to the size of assets, or the distribution of business failures with the frequencies measured by the duration of time in operation. Can you explain why?

In conclusion, it may be said that although the normal distribution holds a very important place in statistical theory, there are business and economic variables that are not normally distributed. For numerous reasons we must have a knowledge of various population models. It is true that most of the time we do not know the true underlying population distributions, but their models can be approximated either by fitting a smooth curve to sample data or by pure deductive reasoning. Our ability to do so greatly facilitates our further development of statistical techniques.

2.6

Cumulative Arrangement of Variates

Up to now our discussion has been confined to the arrangement of numerical data in exclusive classes. For certain purposes it is desirable to arrange data cumulatively. Very often we may wish to answer such questions as these: How many people in the United States are making $6000 or more a year? How many salesmen in the company are selling a given amount of outputs or more per week? When serial bonds secured by equipment are issued, how much of the equipment will be in use at each maturity date of the bonds? How many of the automobile tires produced by a factory can last 30,000 miles or more? How many of them can last only 25,000 miles or less? Answers to this type of question can be readily found if the frequency distributions are arranged cumulatively.

Cumulative frequencies can be formed on either a "less than" or a "more than" basis. Tables 2.8 and 2.9 illustrate these two types of cumulative frequencies.

Table 2.8 shows data cumulated upward and Table 2.9 shows data cumulated

TABLE 2.8 CUMULATIVE FREQUENCY TABLE: AVERAGE WEEKLY EARNINGS OF 142 MESSENGER BOYS IN NEW YORK CITY, 1968

Weekly Earnings	Number of Messenger Boys Earning Less Than a Given Value
Less than $66.50	11
Less than 67.00	26
Less than 67.50	50
Less than 68.00	90
Less than 68.50	110
Less than 69.00	124
Less than 69.50	135
Less than 70.00	142

TABLE 2.9 CUMULATIVE FREQUENCY TABLE: AVERAGE WEEKLY EARNINGS OF 142 MESSENGER BOYS IN NEW YORK CITY, 1968

Weekly Earnings	Number of Messenger Boys Earning More Than a Given Value
$66.00 or more	142
66.50 or more	131
67.00 or more	116
67.50 or more	92
68.00 or more	52
68.50 or more	32
69.00 or more	18
69.50 or more	7

downward. In the first case, frequencies are cumulated from smaller to greater values and therefore are referred to as upper class limits. In the second case, frequencies are accumulated on a "more than" basis. Thus they are obtained by using the lower class limits as references. From these tables, we may readily determine the number of messenger boys earning more than or less than a given value. By reading either table, we can also discover the number of messenger boys who are making more than but less than given values. For instance, can you determine yourself how many messenger boys are making $67.00 or more?

Cumulative tables have many distinct advantages that lead to their wide use. Life tables, for example, are usually arranged in these forms. Their general utility, however, is necessarily limited by the classification system selected to condense the data. Unless a way can be found to enable us to interpolate mathematically, we are then limited to the values of scale actually specified as in Tables 2.8 and 2.9. The way of mathematical interpolation is paved by presenting cumulative frequencies in graphic forms. A graph for a cumulative frequency distribution often takes the form of an *ogive*, or the **S** *curve*. To construct an ogive, the Y axis must be drawn to accommodate the total frequencies. When the "more than" basis is used, frequencies accumulated for each class are plotted at the lower class limits. When the "less than" basis is used, cumulated frequencies are plotted at the upper class limits. The points are then joined by a series of straight lines and the **S** curve may be extended to the zero base of the Y axis. Two ogives for data in Tables 2.8 and 2.9 are plotted in Figure 2.6.

The ogive is particularly adaptable to interpolation. However, it must be realized that the use of straight lines to connect the series of known points is only an approximation to the distribution of the items within any given interval. To be sure, this approximation is generally considered to be sufficiently accurate under most circumstances. Thus, from the "more than" curve in Figure 2.6, we can determine that about 32 boys earned more than $68.50 per week, or about half of the boys (71) earned more than $67.76. Also, from the "less than" curve, we can approximate that about half of the boys earned less than $67.76. Here it is of interest to note that a horizontal line drawn from the point where downward and

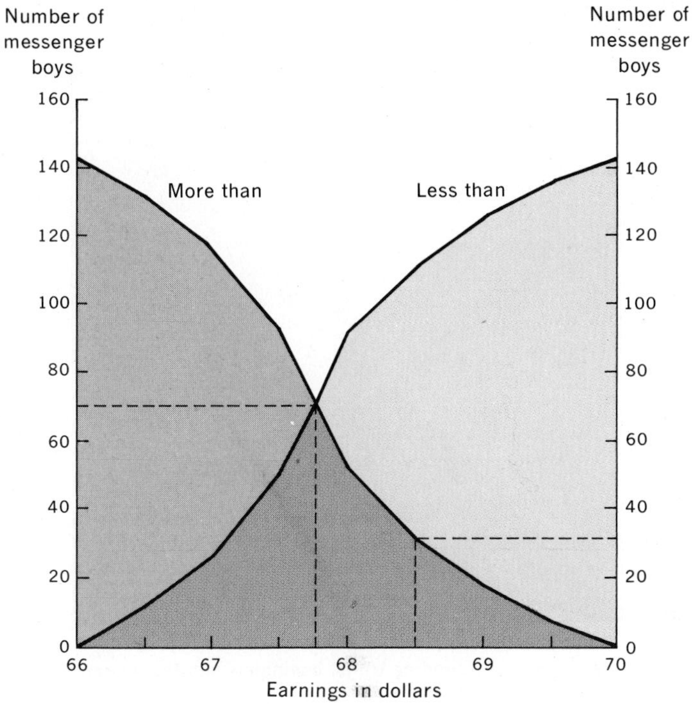

Figure 2.6 The ogive: showing cumulative frequency distributions of 142 messenger boys' average weekly earnings in New York City, 1968. (Source: Tables 2.8 and 2.9.)

upward cumulated frequency curves intersect to the vertical scale would divide the total frequency into two equal parts. If a perpendicular be dropped from this point of intersection to the horizontal scale, the point thus located would be the value of the middle item in the distribution. This value, which is \$67.76 in our illustration, is called the *median*—implying that half of the items in the distribution is above and the other half is below it. The median is a type of average and will be discussed in the next chapter.

From the ogive we can also determine by interpolation the number of observations falling within a given interval. For instance, how many messenger boys earned more than \$67.30 but less than \$69.50? From the "less than" curve, we find that 135 boys earned less than \$69.50 and that 40 boys earned less than \$67.30. Subtracting, we obtain 95 as the desired number. The same number can, of course, be interpolated from the "more than" curve as well. Try to find this out for yourself.

From the above discussion it is obvious that the ogive has eliminated in large part the limitations set by the specific class intervals. For the same distribution, the shape of the curve will be basically the same even though the class intervals and the class frequencies may vary. For this reason, not only do uneven class intervals not distort the ogive, the curve can be constructed directly from the

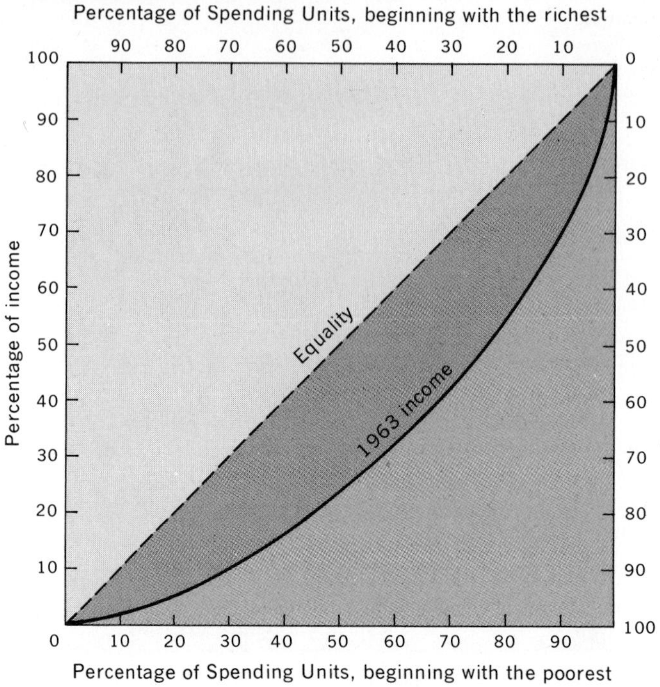

Figure 2.7 Distribution of income before taxes in the United States in 1963. (Source: Table 2.7.)

array. Moreover, different ogives can be compared on the same graph even if their classifications are not the same. The cumulative curve is therefore one of the simplest yet the most useful statistical devices.

A particularly useful arrangement of cumulative frequencies is found in the Lorenz curve used to study income distribution, as shown in Fig. 2.7. This graph shows that if there is absolute equality in income distribution, cumulatively expressed, it would then be represented by the diagonal dashed line at an angle of 45°. That is, if each family received an equal share, 10 percent of the people would receive 10 percent of the income, 20 percent of the people would receive 20 percent of the income, and so on. The solid curve shows the actual distribution of income in 1963. The departure of the curve from the diagonal line measures the degree of deviation of distribution from equality. This particular curve shows that the richest 10 percent of the American people received almost 30 percent of the total national income, whereas the poorest 50 percent received about 22 percent of the total. Note also that nearly half of the total income went to only 25 percent of income-receiving units, leaving only half to be distributed among the remaining 75 percent. The curve therefore shows effectively the degree of inequality in the distribution of income in 1963. It can be seen that the Lorenz curve is also an effective device with which to compare the income distributions at different time periods or under different conditions. For instance, on the same graph, incomes in 1963 can be compared with those of 1968 in order to discover if the degree of inequality has increased or decreased during the interval of time. Or incomes in 1963, before taxes, can be compared with the same year's incomes, after taxes, in order to evaluate the effect of personal taxes upon income distribution.

2.7 Relative Distributions

You may have already noted that in Table 2.7 the frequencies are given in percentages instead of in absolute numbers. Such an arrangement is called the *relative frequency distribution*. In contrast, a frequency expressed in terms of absolute figures is referred to as the *absolute frequency distribution*. Relative frequencies are derived by dividing the absolute class frequencies in each class in turn by the total frequency in the distribution. Table 2.10 is an illustration of relative frequency distributions.

When only relative frequencies are shown in a table, the total number of observations may be stated in a footnote. This is a desirable practice, since it enables anyone to compute the absolute frequencies for the distribution if he so desires.

The area under the histogram or the smooth curve of a percentage frequency distribution equals 100 percent, or 1, if the frequencies are stated in decimal fractions. The density interpretation of the frequency axis is then made in terms of the percentage or the fraction of total frequency per unit of the class axis.

The main utility of the relative frequency distribution is that its graph—the histogram, the polygon, or the smooth curve—lends more readily to the comparison of different distributions, especially if they differ greatly in total numbers of

TABLE 2.10 AVERAGE DAILY EARNINGS OF PHYSICIANS AND STOCKBROKERS IN NEW YORK CITY, 1968

	Number of Persons		Relative Frequencies	
Daily Earnings	*Physicians*	*Stockbrokers*	*Physicians*	*Stockbrokers*
$30 and under $ 36	1	11	0.1	0.3
36 and under 42	24	135	2.4	4.2
42 and under 48	150	247	15.0	7.7
48 and under 54	322	466	32.2	14.5
54 and under 60	185	658	18.5	20.5
60 and under 66	120	596	12.0	18.5
66 and under 72	78	579	7.8	18.0
72 and under 78	66	379	6.6	11.8
78 and under 84	22	115	2.2	3.6
84 and under 90	17	25	1.7	0.8
90 and under 96	14	3	1.4	0.1
96 and under 102	1	—	0.1	—
Total	1000	3214	100.0	100.0

Source: The author.

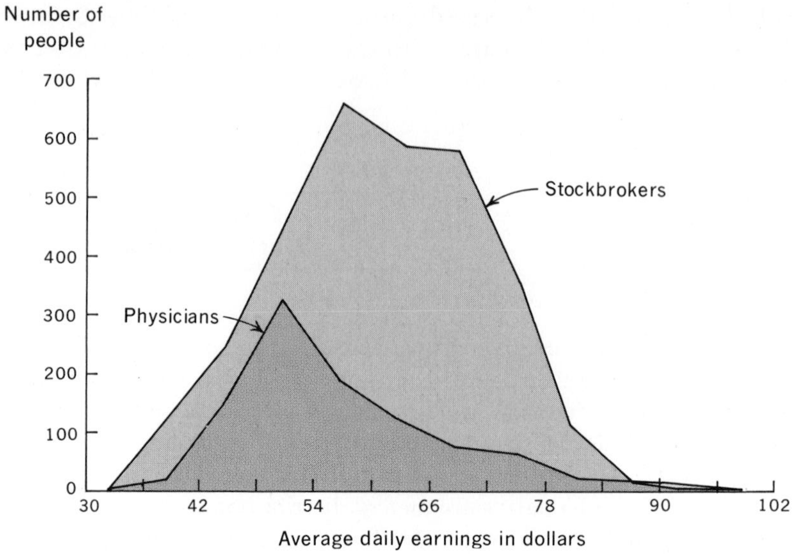

Figure 2.8 Average daily earnings of 1000 physicians and 3214 stockbrokers, 1968. (Source: Table 2.9.)

observations. This point is clearly brought out by comparing Figures 2.8 and 2.9. The former contains the polygons of the absolute distributions in Table 2.10. It is extremely difficult to discover from this graph the similarities or the differences between the two distributions. More serious is the fact that some wrong impressions may even be obtained from it. For instance, it seems to indicate that a

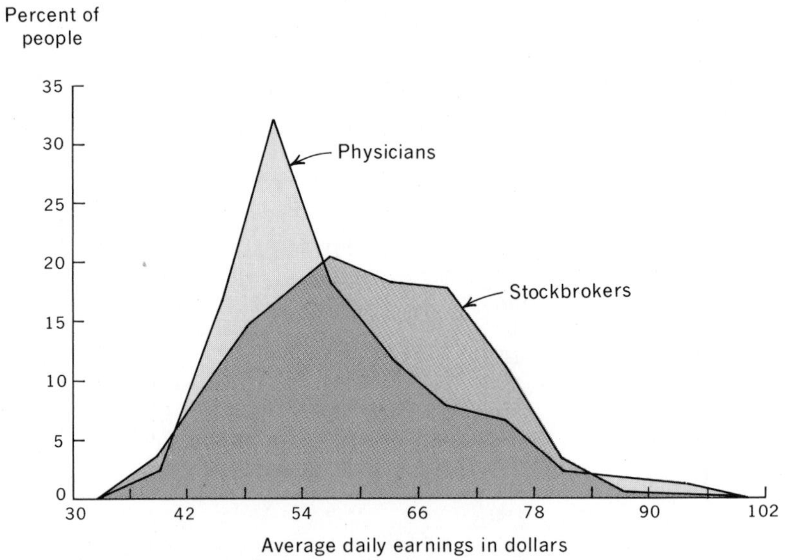

Figure 2.9 Percentage distributions of average daily earnings of 1000 physicians and 3214 stockbrokers, 1968. (Source: Table 2.9.)

greater proportion of the stockbrokers fall within the interval of \$30 and \$60. Actually, the opposite is true.

When, however, reference is made to the polygons of the percentage frequency distributions of the two groups—Figure 2.9—several important conclusions can be drawn easily and accurately. First, both distributions are located at about the same value. Second, the modal earning of the physicians appears at a lower value than does that of the stockbrokers. Third, although the series of physicians is more compactly distributed around the peak, the distribution of stockbrokers resembles a flat top. Finally, a greater percentage of the physicians have their average daily earnings between the interval \$42 and \$58, but a smaller proportion between \$58 and \$84 as compared with the stockbrokers.

The value of comparing relative frequencies becomes even greater when we are dealing with series whose minimum values or maximum values or both differ widely from each other. An exercise for the student who wants to get a really good insight into the utility of relative frequencies is to draw different smooth relative curves and to try to interpret them verbally.

2.8 Categorical Frequency Distributions

A *categorical frequency distribution* shows the number of observations, absolute or relative, that fall into each of the few qualitative classes. Table 2.11 contains such a distribution.

On the one hand, it is relatively simple to construct categorical distributions since we need not be bothered with such mathematical details as class limits, class intervals, mid-points, and so on. On the other hand, it is much more troublesome to determine the number of categories (classes) to be used. As in the case of numerical distributions, the categories must be mutually exclusive and sufficient to accommodate all data. It is, however, very easy to err by selecting overlapping classes for qualitative data. For instance, to study the various kinds of food sold by a grocery chain store, we might adopt some unsatisfactory classes, such as

TABLE 2.11 COSTS OF ACCIDENTAL INJURIES FROM MOTOR VEHICLES IN 1965

Type of Cost	Costs in Millions of Dollars
Wage loss	2,400
Medical expenses	550
Overhead cost of insurance	2,850
Total	5,800

Source: Data adapted from the Economic Almanac, 1967–1968.

"meat and meat products," "vegetables," "frozen foods," and so forth. With such classes, we would have trouble in deciding to which class an item such as "frozen meats" or "frozen vegetables" belongs.

There is another point concerning qualitative distributions that must be mentioned. The graph for the categorical distribution, absolute or relative, is the ordinary horizontal bar chart. Here, the student should not confuse the bar chart with the histogram. The frequency scale on the bar chart, unlike that on the histogram, does not represent density. It has no other significance except the absolute or relative numbers it represents. Also, neither the width nor the areas of the bars have any meaning, as they do in the histogram. In conclusion it may be said that all that the bar chart and the histogram have in common is that both are constructed with bars.

Glossary of Formulas

(**2.1**) $k = 1 + 3.3 \log n$ This formula is known as the Sturges rule. It aids in the determination of the number of classes for numerical frequency distributions. In using this formula remember that it yields only an approximate result. The actual number to be selected also depends upon many other important considerations if an effective frequency table is to be constructed. In the formula, k stands for the approximate number of classes; n, for the number of observations in the sample.

(**2.2**) $m_1 = \dfrac{L_1 + U_1}{2}$ This formula is used to locate the mid-points of the classes in a frequency distribution. In it, m_1 is the mid-point of the first class. L_1 and U_1 are the lower and upper class limits of the first class respectively. When uniform class intervals are used, mid-points of the successive classes are separated from the mid-point of the first class and from each other by the width of the class interval. To apply this formula to a distribution with unequal class intervals, the mid-point of each class should be determined independently. Now, m_1, L_1, and U_1 may be considered as the mid-point of the lower and upper class limits of a given class. In dealing with rounded data, take care to use the actual class limits to locate the mid-points.

Problems

2.1 When is it advisable to use frequency distribution to organize data?

2.2 What are the advantages and disadvantages of the array? Is the array necessary for the construction of a frequency distribution? Why or why not?

2.3 What are the advantages of the entry form over the tally sheet for the construction of frequency tables?

2.4 What are the objections to unequal class intervals and to open classes? Under what circumstances is the use of unequal class intervals or open classes necessary and even desirable?

2.5 The class interval is always the difference between the upper and lower class limits. True?

2.6 The nominal class limits may or may not be the same as the actual class limits. Discuss.

2.7 What important role do mid-points play in a numerical distribution?

2.8 What is the precise definition of the quantity measured by the vertical scale in a histogram?

2.9 In what sense do we say that the histogram is superior to the polygon as the graphic representation of a frequency distribution?

2.10 What important functions does the frequency polygon have?

2.11 To a given set of sample data only one particular smooth curve can be fitted. True?

2.12 Why do we call the smooth curves "population models"?

2.13 Which population model do you consider to be a good approximation of the underlying distribution of each of the following populations of variates? Give reasons.
a. the heights of American soldiers
b. student grades on difficult problems
c. diameters of steel bars produced by a given process
d. annual earnings of American doctors
e. number of occupants living in dwelling units in United States
f. number of corporations filing corporate income tax returns classified by size of assets
g. the ratio of the value of plant and equipment to the net worth of all the firms in the textile industry in the United States
h. the monthly rents of four-room apartments in New York City

2.14 Why is the concept "population model" important in statistical analysis?

2.15 The normal curve is important because it is the most common and the best population model for business and economic data. Right? Wrong? Why?

2.16 For what type of problems do we need to construct cumulative frequency distributions?

2.17 What advantages or disadvantages does the ogive have over the corresponding frequency table?

2.18 A cumulative frequency distribution is always set up by using the upper class limits. Agreed?

2.19 At what X value in the series will the "more than" and the "less than" ogives intersect? What is the name of this value?

2.20 What is a relative frequency distribution? Under what circumstances are relative frequency distributions especially effective in comparing different sets of variates?

2.21 With the aid of the concept of relative frequency, can you explain why there are as many expensive cars in poor neighborhoods as in fashionable suburbs? Also, do you think it is reasonable to expect that the weights of rats may show a greater variation than the weights of elephants?

2.22 If the distribution of income after taxes in the United States in 1963 is plotted from Figure 2.7, would you expect the curve to be closer to the 45° line or otherwise than the curve already plotted there? Explain.

2.23 Give a critical evaluation of Table 2.12, which is a frequency table.

TABLE 2.12 FREQUENCY DISTRIBUTION OF RENTED DWELLINGS IN TORONTO, CANADA, 1968 (CLASSIFIED ON THE BASIS OF RENTAL VALUE)

Monthly Rental	Number of Dwellings in Each Class (in thousands)
Under $100.00	327
$ 100.00–149.99	349
150.00–199.99	521
200.00–299.99	1039
300.00–499.99	1075
500.00–749.99	189
750.00–999.99	24
1000.00 and over	9
Total	3533

2.24 According to the Sturges rule, how many classes would you use to a set of data that contains (1) 50 observations, (2) 100 observations, (3) 1500 observations, or (4) 20,000 observations?

2.25 What are the (1) actual limits, (2) class intervals, and (3) mid-points of the following classifications?

A. DATA REPORTED IN DOLLARS AND CENTS	B. DATA ROUNDED TO NEAREST POUND	C. DATA ROUNDED TO NEAREST TENTH OF AN INCH
$15.00–$24.99	5– 9	2.0–3.9
25.00– 34.99	10–14	4.0–5.9
.	.	.
.	.	.
.	.	.

2.26 Which classification or classifications stated below should be suggested to group the sales of a 5-and-10-cent store? Give reasons.

(1)	(2)	(3)	(4)	(5)
1– 6	3– 7	2.5– 7.5	5–10	0.5– 5.5
7–13	8–12	7.5–12.5	10–15	5.5–10.5
14–19	13–17	12.5–17.5	15–20	10.5–15.5
.	.	.	.	.

Sample A

Average Monthly Wage Rates of 250 New England Textile Factory Workers, 1959

$184.90	$188.13	$190.34	$192.04	$194.42
197.65	199.35	201.05	202.53	204.11
205.47	206.49	206.83	208.00	209.04
209.05	209.55	210.07	210.25	210.74
212.44	211.75	212.27	212.44	212.44
212.95	213.16	213.63	213.63	213.99
214.13	214.50	214.83	214.83	215.34
215.50	216.20	216.35	216.40	216.86
217.11	217.37	217.42	217.54	217.71
218.05	218.27	218.27	218.34	218.56
218.90	219.14	219.60	219.60	219.75
219.75	220.26	220.43	220.95	221.28
221.45	221.50	221.62	221.80	222.30
222.95	222.95	222.95	222.98	223.50
223.83	224.00	224.00	224.32	224.51
224.68	225.00	225.25	225.53	225.55
225.70	225.87	225.92	226.04	226.04
226.27	226.59	226.70	226.72	226.88
227.00	227.06	227.40	228.25	228.25
228.40	228.55	228.93	229.10	229.27
229.45	229.93	229.93	230.12	230.15
230.15	230.50	230.63	230.63	230.75
230.80	230.80	230.80	230.96	231.00
231.00	231.14	231.47	231.50	231.65
231.65	231.87	231.87	231.87	231.90
232.15	232.33	232.50	232.58	232.59
232.67	232.75	232.77	232.77	233.20
233.24	233.35	233.52	233.55	233.86
233.90	234.00	234.20	234.20	234.20
234.20	234.35	234.35	234.70	234.88
235.00	235.21	235.50	235.70	235.70
235.75	235.93	236.07	236.50	236.75
236.80	236.92	237.27	237.30	237.60
237.77	237.80	237.80	238.10	238.28
238.62	238.96	238.96	239.13	239.30
239.30	239.50	239.50	239.50	239.50
239.70	239.98	240.15	240.32	240.65
240.78	240.88	241.00	241.00	241.14
241.14	242.02	242.47	242.54	242.70
243.58	244.06	244.13	244.50	244.50
245.09	245.25	245.80	246.10	246.24
246.78	247.00	247.56	248.14	248.65
249.51	249.80	251.35	252.00	252.24
252.55	252.55	254.25	255.11	255.45
256.64	256.90	257.83	259.70	261.06
262.08	263.00	264.45	264.45	265.65
265.77	267.35	270.75	272.45	274.49
274.83	277.44	278.00	281.68	283.33
285.00	291.00	293.53	295.59	308.15
310.12	314.34	319.22	322.75	330.17

Total of the series is $58,709.31.

Sample B

Average Monthly Wage Rates of 150 Oil Refinery Employees in New Jersey, 1959

$280.05	$280.90	$281.12	$284.49	$286.50
289.94	295.48	297.57	301.95	303.02
304.11	307.48	309.56	311.75	312.83
315.01	315.55	316.00	317.28	318.43
319.23	320.46	321.05	321.44	321.65
322.53	323.80	323.80	324.00	325.90
326.75	326.98	329.18	330.27	331.36
332.45	333.54	335.72	336.81	337.90
340.08	340.08	340.15	341.23	342.26
343.35	344.44	344.44	344.44	345.25
345.78	346.62	347.48	347.78	347.90
348.67	348.80	348.95	349.99	350.95
352.07	352.55	353.00	353.00	354.25
355.64	355.80	356.23	357.52	358.61
359.70	360.39	360.79	361.88	361.88
361.97	363.14	364.27	365.08	365.08
366.24	366.77	366.95	367.33	368.42
369.51	370.60	371.69	372.50	372.50
373.48	373.89	374.96	375.54	375.54
375.54	378.21	379.11	379.25	379.99
380.00	380.00	380.00	380.31	380.55
381.50	381.50	381.50	382.42	384.46
384.77	385.29	386.95	388.04	389.21
390.22	391.31	393.48	394.00	394.95
397.94	398.55	400.06	402.21	403.31
405.48	406.57	407.66	408.53	410.44
411.33	413.21	414.02	415.29	419.38
419.38	419.38	419.47	420.16	420.16
421.07	421.89	424.05	425.10	428.19
429.25	432.36	436.44	437.55	439.82

Total of the series of $54,332.36.

2.27 The two preceding samples were collected by a management consulting firm. Both sets of data are presented as arrays. (Read horizontally from left to right.)
a. Construct a frequency distribution by tallying for Sample A. (Use $20 as the class interval and start with $180.) Name it Table 2P.1.
b. Construct a frequency distribution by the entry form for Sample B. (Use $20 as the class interval and start with $280.) Call it Table 2P.2
c. Construct histograms for data in Tables 2P.1 and 2P.2.
d. Compare the distributions as presented in Tables 2P.1 and 2P.2. (*Hint:* Compute relative frequencies and present them as polygons in the same graph for analysis.)
e. How many workers in Sample A made less than $200.00 per month? more than $250.00 per month? more than $195.25 but less than $215.75? (*Hint:* What additional reorganization of the data is needed to answer these questions?)

2.28 What categories would you establish in order to construct frequency distributions for the following sets of data?
a. Number of policyholders of an insurance company according to the education of the head of the family.
b. A distribution of retail stores in your home town. Also, where the information can be obtained.
c. A distribution of foods sold in a supermarket.
d. Land use in an agricultural region.
e. Number of banks distributed among the several categories established according to the specialized functions they perform.

3

MEASURES OF CENTRAL TENDENCY

3.1

Types of Averages

In the preceding chapter we discussed how sample data could be reduced to compact, comprehensible, and communicable form by the frequency distribution. The frequency distribution is not only a method of organizing data; it is also a descriptive measure. Indeed, it may be considered as a set of descriptive statistics, since each number showing the density of observations in a class is a descriptive statistic. Very often, however, we need a single descriptive statistic that can focus attention more sharply on the nature of data being measured. For instance, when the consumption expenditures of the Russians are compared with those of the Americans, or when the wages of the New England textile workers are compared with those of the oil refinery employees in New Jersey, or when the deposits of commercial banks in New York are contrasted with deposits of commercial banks in London, the use of a single number is clearly more advantageous than the frequency distribution.

A single number used to describe a series must be "representative" of the data measured by it. To be representative, the number must reflect the tendency of the individual items in the series to concentrate at, and be distributed about, certain central values. For this reason, a representative number is considered as a *measure of central tendency*. More commonly, it is known as an *average*.

There are a number of statistical summary measures that are employed as averages. Averages, based on their mathematical properties, can be classified into two main categories: *computed* and *position averages.* The former includes the arithmetic mean, the geometric mean, the harmonic mean, and the quadratic mean; the latter comprises the median and the mode. We shall follow this order for the discussion in this chapter.

3.2 Nature of the Arithmetic Mean

The arithmetic mean, because of ease of computation and long usage, is the best-known and most commonly used average. Sometimes, it is simply referred to as "the mean" or "the average," but appropriate adjectives must be used when references are made to other types of means.

The *arithmetic mean* is the sum of the observations in a sample divided by the number of observations in the sample. Thus, let x_i = values of individual items in the sample, n = the number of items in the sample, and $\bar{x}$ = the arithmetic mean of the sample, and we have

$$\bar{x} = \frac{\Sigma x}{n} \tag{3.1}$$

Applying the formula to the sample data in Table 2.5, we have

$$\bar{x} = \frac{\Sigma x}{n} = \frac{\$9632.74}{142} = \$67.84$$

Before we proceed to a detailed discussion of the computation of the mean, it is interesting to note its mathematical properties.

First of all, the mean is a representative or typical value in the sense that it is the center of gravity—the point of balance. That is, the values of the observations on one side of the mean equal the values of the observations on the other side of the mean. The mean is typical also in that its value may be substituted for the value of each item in the series without changing the total. For instance, if each messenger boy had a weekly earning of \$67.84, the total earnings of all the 142 messenger boys would have remained \$9632.74. Conversely, if \$9632.74 were divided equally among the 142 messenger boys, each would have \$67.84. This property can easily be seen from the formula for the mean. Since $\bar{x} = \Sigma x/n$, we may derive, by multiplying both sides of the equation by n, the result $n\bar{x} = \Sigma x$. This feature often becomes the important consideration for us in deciding whether to use the mean instead of other types of averages that do not possess an analogous algebraic property. The application of this simple property can be illustrated by the use of the mean by a chain-restaurant organization in order to obtain an estimate of the total daily sales of a new restaurant along a certain highway: it is found simply by multiplying the number of cars passing by the new restaurant daily by the mean-sales per car in other outlets operated by the same organization.

Another property of the mean is that the algebraic sum of deviations from the mean (that is, taking due account of algebraic signs) is zero. Let d stand for the deviation of any item from the mean ($d = x - \bar{x}$), and this property can be written symbolically, $\Sigma d = 0$. Consider the following numerical illustration:

TABLE 3.1 DEVIATIONS FROM THE MEAN

	Observations	Deviations
	x	d
	2	−3.5
	3	−2.5
$\bar{x} = \frac{44}{8}$	4	−1.5
	5	−0.5
	6	+0.5
$= 5.5$	7	+1.5
	8	+2.5
	9	+3.5
	$\Sigma x = 44$	$\Sigma d = 0$

Algebraically, this property is derived from the fact that $n\bar{x} = \Sigma x$. Since the sum of variables, each reduced by a constant, is equal to the sum of the variables, minus the constant multiplied by the number of terms in the sum, we can then deduce from the equation that

$$\begin{aligned}(x_1 - \bar{x}) + (x_2 - \bar{x}) + \cdots + (x_n - \bar{x}) \\ = x_1 + x_2 + \cdots + x_n - \bar{x} - \bar{x} \cdots \text{ (}n\text{ times)} \\ = \Sigma x - n\bar{x} = \Sigma x - \Sigma x = 0\end{aligned}$$

As will be indicated shortly, this property, that is, $\Sigma d = 0$, is the theoretical basis for the short-cut method of computing the arithmetic mean.

The third mathematical property of the mean is that the sum of the squared deviations of the items from the mean is less than the squared deviations from any other point in a series. Symbolically, Σd^2 = a minimum. (Readers may wish to verify this by way of some numerical illustrations of their own.) Because of this property the mean usually serves as a basis for measures of dispersion, and it is essential for fitting the curves to observed data. These points will be taken up in subsequent chapters.

3.3 *Proportions as Means*

Qualitative data are usually classified into two proportions: one as having a certain trait, another as not having a certain trait. Proportions are ordinarily thought of as fractions or percentages, but can be thought of as special cases of the arithmetic mean. They can be computed by using formula (3.1) and the properties of the mean applied to them.

Suppose we are to determine the proportion (mean) of the number of eligible voters in a group of citizens: we shall assign a value of 1 to each person qualified to vote and the value 0 to each person not qualified. The sum of the 1s would be Σx and the mean would be obtained by dividing this sum by total number of persons, n, in the group. In dealing with qualitative data, we usually substitute x for Σx and p for $\bar{x}$. Thus, $\bar{x} = \Sigma x/n$ is replaced by $p = x/n$. In the latter equation, p stands for the proportion of the items that have a certain trait in the sample.

3.4

Computing the Arithmetic Mean

Formula (3.1) is suitable only for small samples. If samples are large, such as the sample of messenger boys, this method is too laborious and too subject to error. If thousands or tens of thousands of items are used to compute the arithmetic mean, the problem of simple addition would become next to impossible, even with the aid of an adding machine. An easier and more efficient method is to construct a frequency distribution of the data and then to calculate the arithmetic mean for the distribution.

The arithmetic mean is computed from a frequency distribution under the basic assumption mentioned in the preceding chapter. That is, it is assumed that all the values included within the limits of a class interval are distributed evenly in it, and therefore the mid-point of each class could be used to represent the actual values of the cases. Three methods are available to compute the arithmetic mean from a frequency distribution: the basic or long method, and two short-cut methods. These are discussed successively below.

THE LONG METHOD

This method is illustrated by Table 3.2. It involves four simple steps: First, determine the mid-points. Second, obtain the products of frequency and mid-point fm by multiplying each mid-point by the corresponding class frequency. Third, sum up these products to get the total value of the distribution Σfm. Fourth, divide this total by the number of items in the sample n. These steps may be summarized by the following formula:

$$\bar{x} = \frac{\Sigma fm}{n} \tag{3.2}$$

Note that there is a difference of 1 cent between the two values of $\bar{x}$ for the ungrouped data and frequency distribution after rounding. Before rounding, the difference is still less. Such a difference is usual rather than an exception, since it is only when the variable is continuous and the distribution symmetrical that the value of the arithmetic mean calculated from the ungrouped data is in

TABLE 3.2 COMPUTATION OF ARITHMETIC MEAN OF AVERAGE WEEKLY EARNINGS OF 142 MESSENGER BOYS IN NEW YORK CITY, 1968, BY THE LONG METHOD

Average Weekly Earnings	Mid-points (*m*)	Number of Messenger Boys (*f*)	Products of Frequency and Mid-points (*fm*)
$66.00–$66.49	$66.25	11	728.75
66.50– 66.99	66.75	15	1001.25
67.00– 67.49	67.25	24	1614.00
67.50– 67.99	67.75	40	2710.00
68.00– 68.49	68.25	20	1365.00
68.50– 68.99	68.75	14	962.50
69.00– 69.49	69.25	11	761.75
69.50– 69.99	69.75	7	488.25
Total		142	9631.50

Source: Table 2.5.

$$\bar{x} = \frac{\Sigma fm}{n} = \frac{\$9{,}631.50}{142} = \$67.83$$

close agreement with the arithmetic mean from the frequency distribution. Otherwise, the agreement would be much less close. In any case, however, if the class limits and intervals are properly selected and if there are no irregularities in the distribution because of too small a sample, this difference would not be more than a few percentage points. Also, other things being equal, the narrower the class interval, the greater would be the agreement between the means computed from the ungrouped and the grouped data.

THE SHORT-CUT METHOD IN ORIGINAL UNITS

The short-cut methods are made possible by a unique mathematical property of the arithmetic mean, as has been previously discussed: the algebraic sum of the deviations of the values from the arithmetic mean is zero. From this property it can be reasoned that when any point other than the exact mean is selected, the algebraic sum of the deviations from this point will not be zero. Let us, for instance, choose 4 in the same data used before as the arbitrary point, known technically as the assumed mean, symbolized as A, and compute the deviations of other values in the series from A below:

	Individual Values x	Deviations from the Assumed Mean d
	2	−2
	3	−1
	4	0
$A = 4$	5	+1
	6	+2
	7	+3
	8	+4
	9	+5
	$\Sigma x = 44$	$\Sigma d = +12$

From the total deviations from the assumed mean, Σd, we can obtain the average deviation from the assumed mean by dividing the total deviation by the number of items in the series. Thus

$$\text{average deviation} = \frac{\Sigma d}{n} = \frac{+12}{8} = +1.5$$

Technically, the expression $\Sigma d/n$ is called the correction factor and is identified by the symbol c. When this correction factor is added to the arbitrary point, the assumed mean, the result will be the arithmetic mean:

$$\bar{x} = A + \frac{\Sigma d}{n} = 4 + \frac{12}{8} = 5.5$$

It should be noted that the total deviation in the foregoing illustration is positive because the value of the assumed mean is less than the value of the exact mean. If a value greater than the true mean is selected as the assumed mean, then the total deviation, and hence the correction factor, will be negative. And when the assumed mean is identical with the exact mean, then $\Sigma d = 0$.

This device can readily be applied to grouped data. Steps required in computing the arithmetic mean from a frequency distribution by this technique are as follows:

First, select one of the mid-points as the assumed mean, denoted by A. Even though the mid-point of any class can be selected and will yield the same result, for convenience and ease of computation, it is advisable to select the mid-point of one of the center classes.

Second, compute the deviation from the assumed mean of the items in each class by obtaining the difference between the mid-point of each class interval and the assumed mean. That is, $d = x - A$. As the mid-point of each class is assumed to be the average value of all items in that class, the value of d represents the average deviation of the items in the class from the assumed mean.

Third, obtain the total deviation for all items in each class interval by multiplying the average deviation, d, by the frequency of the class, f. This gives fd.

Fourth, secure the total deviation from the assumed mean of all the items by adding up all the total deviations for all class intervals. This is Σfd.

Fifth, calculate the correction factor, that is, the average deviation about the assumed mean, by dividing Σfd by n, the number of items in the series. This is $\Sigma fd/n$.

Finally, the arithmetic mean is derived when the value of the correction factor is added to the assumed mean.

The formula for the computation is

$$\bar{x} = A + c \quad \text{or} \quad \bar{x} = A + \frac{\Sigma fd}{n} \tag{3.3}$$

where

A = the assumed sample mean
c = correction factor
f = frequency of each class
d = deviation of mid-point of each class from the assumed mean
n = total number of items in the sample

TABLE 3.3 COMPUTATION OF ARITHMETIC MEAN OF AVERAGE WEEKLY EARNINGS OF 142 MESSENGER BOYS IN NEW YORK CITY, 1968, BY THE SHORT-CUT METHOD IN ORIGINAL UNITS

Average Weekly Earnings	Mid-points (m)	Number of Messenger Boys (f)	Deviations from A (d)	fd (dollars)
\$66.00–\$66.49	\$66.25	11	−1.50	−16.5
66.50– 66.99	66.75	15	−1.00	−15.0
67.00– 67.49	67.25	24	−0.50	−12.0
67.50– 67.99	67.75	40	0.00	00.0
68.00– 68.49	68.25	20	+0.50	+10.0
68.50– 68.99	68.75	14	+1.00	+14.0
69.00– 69.49	69.25	11	+1.50	+16.5
69.50– 69.99	69.75	7	+2.00	+14.0
Total		142		+11.0

$$\bar{x} = A + \frac{\Sigma fd}{n} = \$67.75 + \frac{11}{142} = \$67.75 + \$0.07746 = \$67.83$$

Table 3.3 is an illustration of computing the arithmetic mean by this method. This method is called the short-cut method with original units because the deviations, d, are written in the actual magnitudes—in dollars in our illustration—by which the mid-point of each class deviates from the assumed mean.

THE SHORT-CUT METHOD IN CLASS-INTERVAL UNITS

The calculation of the arithmetic mean from a frequency distribution may be further simplified if the distribution has uniform class intervals. When the class intervals are of equal size, the deviation of the mid-point of one class from the next is always constant and is equal to the width of the class interval. The distribution shown in Table 3.3 has a constant difference of \$0.50 between the mid-points, which is equal to the class intervals, such as, for instance, \$66.00–\$66.49. Thus, the deviations of any one class from any other class may be measured in terms of class intervals and for this reason the term "short-cut method in class-interval units" is derived.

The computation of the arithmetic mean by the short-cut method in class-interval units, as demonstrated by Table 3.4, involves the following steps:

First, as before, the mid-point of one of the center classes is chosen as the assumed mean.

Second, deviations from the assumed mean are taken in terms of class intervals. The mid-point of the first class deviates from the assumed mean by -4 class intervals, the second class, by -3 class intervals, the sixth class, by $+1$ class interval, and so on, in our illustration, where the mid-point of the fifth class (\$68.25) has been selected as the assumed mean. Hence only the numbers -4, -3, -2, -1, 0, $+1$, $+2$, $+3$ are recorded in the deviation column. It should be noted that each of these numbers actually represents one class interval. For instance, -4 actually means $-\$2.00$ because the class interval is \$0.50 in

TABLE 3.4 COMPUTATION OF ARITHMETIC MEAN OF AVERAGE WEEKLY EARNINGS OF 142 MESSENGER BOYS IN NEW YORK CITY, 1968, BY THE SHORT-CUT METHOD IN CLASS-INTERVAL UNITS

Average Weekly Earnings	Mid-points (m)	Number of Messenger Boys (f)	Deviation from Assumed Mean (d')	fd'
\$66.00–\$66.49	\$66.25	11	−4	− 44
66.50– 66.99	66.75	15	−3	− 45
67.00– 67.49	67.25	24	−2	− 48
67.50– 67.99	67.75	40	−1	− 40
68.00– 68.49	68.25	20	0	0
68.50– 68.99	68.75	14	+1	+ 14
69.00– 69.49	69.25	11	+2	+ 22
69.50– 69.99	69.75	7	+3	+ 21
Total		142		−120

$$\bar{x} = A + \left(\frac{\Sigma fd'}{n}\right) i = \$68.25 + \left(\frac{-120}{142}\right) 0.50$$
$$= \$68.25 + (-0.42254)$$
$$= \$67.83$$

the illustration. That is, the mid-point of this extreme class is two dollars below the value of the assumed mean, but it is recorded as −4 instead of as −\$2.00. It is this step, that is, deviations are expressed in terms of class-interval units instead of as original units, which makes this method simpler because it eliminates almost all decimal fractions and large totals.

Third, the deviations are then multiplied by their frequencies in order to get the total deviation of all the cases in each class from the assumed mean, fd'.

Fourth, the total deviation of the series from the assumed mean in class-interval units is then obtained by summing up the products of the deviations times their frequencies, $\Sigma fd'$.

Fifth, now we must divide $\Sigma fd'$ by the number of items, n, in order to get the average deviation.

Sixth, since $\Sigma fd'/n$ is the average amount by which the values vary from the assumed mean in class-interval units, this result must be multiplied by the size of the class interval in order to convert the average deviation back to original units. The procedure is $\left(\frac{\Sigma fd'}{n}\right) i$, which is the correction factor for the assumed mean.

Last, when the correction factor is added to the assumed mean, the arithmetic mean results. This method is generalized into a formula as follows:

$$\bar{x} = A + \left(\frac{\Sigma fd'}{n}\right) i \qquad (3.4)$$

where all symbols have been previously defined.

It will be noticed that the results from all three methods are, as they should be, the same. Since the short-cut method in class-interval units is a great time-saver, it is always advisable to use it in computing the arithmetic mean from a frequency distribution, especially when the sample is large.

3.5 Weighted Arithmetic Mean

Theoretically speaking, all arithmetic means are weighted averages. If no specific weights are given to each and every value in the series, each case is assigned an equal weight of 1. In computing the arithmetic mean from grouped data, the class frequencies may be considered as a series of weights for the various mid-points. Thus when the arithmetic mean is computed under the assumption of equal weights, it should be called the *simple arithmetic mean*, as it is often designated.

Sometimes, in computing index numbers or averaging percentages or products, it is necessary to assign proper specific weights to the various items in the calculation of the mean if a correct result is to be secured. When varying weights are used in its computation, the arithmetic mean can be properly spoken of as "weighted." For the purpose of illustration, we assume that in a management consulting firm, 10 junior consultants receive \$60.00 each, 4 seniors receive \$85.00 each, and 1 "specialist" receives \$125.00 per day. It is evident that the different wage rates are of varying importance and that it is necessary to weight these rates with the number of persons receiving the several wages in order to determine the correct average wage payment. The arithmetic mean becomes:

$$\frac{(\$60 \times 10) + (\$85 \times 4) + (\$125 \times 1)}{10 + 4 + 1} = \$71.00$$

The formula for the above may be written as

$$\bar{x}_w = \frac{\Sigma Wx}{\Sigma W} \tag{3.5}$$

where

$\bar{x}_w$ = the weighted arithmetic mean
x = values of the individual items
W = weight, or number of times an item of the series is counted

As a matter of fact, the weighted mean is suggested by the property that $n\bar{x} = \Sigma x$. This point can be explained by the following consideration. In averaging a series of averages, it is necessary to weight each average with the number of cases used to compute that specific average. For the purpose of illustration, suppose there are three firms, in a manufacturing industry, which produce an identical commodity. The average costs and production data of the firms are as shown in the following table:

TABLE 3.5 COST AND PRODUCTION OF A CERTAIN COMMODITY

Firm	Average Cost (x)	Total Production (W)	Total Cost (W_x)
A	$1.50	200,000 units	$ 300,000.00
B	1.00	400,000 units	400,000.00
C	1.05	800,000 units	840,000.00
Total	$3.55	1,400,000 units	$1,540,000.00

It should be clear that the data in column 2 of Table 3.5 are average costs *per unit* of production. If a simple arithmetic mean is computed from these average costs data the result becomes $1.183. It is computed under the assumption that all three firms are of equal size and this average of the averages does not mean the average cost per unit any longer but average cost *per firm*. Therefore the new average, $1.183, is not consistent with the meaning of the original average costs. This point may be demonstrated by the fact that when the total production of the three firms combined (1,400,000 units) is multiplied by the new average ($1.183) a value of $1,656,200 results and this value is $116,200 greater than the total costs of the three firms combined.

Hence, in averaging these average costs they should be weighted by their respective quantities of output, as is done in the computation

$$\bar{x}_w = \frac{\Sigma Wx}{\Sigma W} = \frac{\$1{,}540{,}000}{1{,}400{,}000} = \$1.10$$

Now the value of $\bar{x}_w$ means that the average cost per unit of output in the whole industry is $1.10. This result is now correct because it is consistent with the meaning of the original averages. It will be noticed also that this method of weighting, in fact, has the effect of changing the average costs per unit of each firm back to the original total costs of production of each firm, thereby back to the original total costs in the whole industry.

3.6 The Geometric Mean

The geometric mean may be defined as the nth root of the product of n items or values.

For ungrouped data, the formula is

$$G = \sqrt[n]{(x_1)(x_2)(x_3) \cdots (x_n)} \tag{3.6}$$

where

G = the geometric mean
n = number of items in the sample
x = values of the items

For example, the geometric mean of 2 pounds, 4 pounds, and 8 pounds is

$$G = \sqrt[3]{(2)(4)(8)} = \sqrt[3]{64} = 4 \text{ lb.}$$

For the same series, $\bar{x} = 4.7$ pounds. It is always true that the arithmetic mean is greater than the geometric mean for any series of positive values, unless the items being averaged are of the same value, in which case the two averages are the same.

The computation of the geometric mean is made much easier by reducing the formula to its logarithmic form, as

$$\log G = \frac{\log x_1 + \log x_2 + \log x_3 + \cdots + \log x_n}{n}$$

$$= \frac{\Sigma \log x}{n} \tag{3.7}$$

Thus, the logarithm of the geometric mean is equal to the arithmetic mean of the logarithms of the values. Because of this relationship, to compute the geometric mean from grouped data the same technique for computing the arithmetic mean by the long method may be used, except that the logarithms of the mid-points are used in the computation instead of the actual values of

TABLE 3.6 COMPUTATION OF THE GEOMETRIC MEAN OF WHOLESALE PRICE RELATIVES IN THE UNITED STATES, MARCH, 1936
(1926 = 100)

Price Relatives	Number of Relatives (f)	Mid-points (m)	$\log m$	$f \log m$
20.0– 29.9	6	25	1.3979	8.3874
30.0– 39.9	17	35	1.5441	26.2497
40.0– 49.9	23	45	1.6532	38.0236
50.0– 59.9	54	55	1.7404	93.9816
60.0– 69.9	106	65	1.8129	129.1674
70.0– 79.9	175	75	1.8751	328.1425
80.0– 89.9	163	85	1.9294	314.4922
90.0– 99.9	114	95	1.9777	225.4578
100.0–109.9	72	105	2.0212	145.5264
110.0–119.9	22	115	2.0607	45.3354
120.0–129.9	13	125	2.0969	27.2597
130.0–139.9	3	135	2.1303	6.3909
140.0–149.9	1	145	2.1614	2.1614
Total	769			1390.5760

Source: Leonard Ascher, "Variations in Price Relatives Distributions, January, 1927, to December, 1936," Journal of the American Statistical Association, June, 1937.

$$\log G = \frac{\Sigma f \log m}{n} = \frac{1{,}390.5760}{769} = 1.80829$$

$$G = 64.3$$

the mid-points. The formula is

$$\log G = \frac{f_1 \log m_1 + f_2 \log m_2 + f_3 \log m_3 + \cdots + f_n \log m_n}{n}$$

$$= \frac{\Sigma f \log m}{n} \qquad \textbf{(3.8)}$$

The computation of the geometric mean from grouped data is illustrated by Table 3.6. These steps are involved:

First, obtain the logarithm of the mid-point of each class.

Second, multiply the logarithm of the mid-point by its corresponding class frequency.

Third, get the sum of the products.

Fourth, divide the sum by the number of items.

Last, take the antilogarithm of the result.

3.7 Properties and Uses of the Geometric Mean

It may be noted that the geometric mean is meaningful only for sets of observations that are all positive. In addition, the geometric mean possesses two mathematical properties. The first is its characteristic that the product of the values of series will remain unchanged when the value of the geometric mean is substituted for each individual value. For example, the geometric mean for series 2, 4, 8, 16, and 32 is 8; therefore, we have

$$(2)(4)(8)(16)(32) = 32{,}768 = (8)(8)(8)(8)(8)$$

The other property of the geometric mean is that the sum of the deviations of the logarithms of the original observations above or below the logarithm of the geometric mean are equal; namely, $(\log x - \log G) = 0$. Alternatively, we may say that the value of the geometric mean is such as to balance the ratio deviations of the observations from it. Thus, using the same previous numbers, we can see that

$$(8/2)(8/4) = 8 = (16/8)(32/8)$$

Because of this property, we may say that the geometric mean is typical in the sense that its value balances the ratio deviations of the items from it. Because of this property, too, the geometric mean is especially adapted to average ratios, rates of change, and logarithmically distributed series. In certain special cases of averaging ratios or percentages, such as the computation of the price index, the geometric mean can yield meaningful and logical results that the arithmetic mean cannot do. For example, consider the data in Table 3.7, which shows the

TABLE 3.7 PRICES OF COMMODITIES A AND B IN 1962 AND 1967

	Price		Price Relatives			
			1962 = 100		*1967 = 100*	
Commodity	*1962*	*1967*	*1962*	*1967*	*1962*	*1967*
A	\$ 5	\$10	100	200	50	100
B	20	10	100	50	200	100
Average			100	125	125	100

price changes of two commodities, A and B, from 1962 to 1967. During this period the price of A increased 100 percent and the price of B decreased by 50 percent. What was the relative average change in price?

The arithmetic mean is the wrong answer to that question. As indicated by the computations in Table 3.7, the application of $\bar{x}$ in this case has led to a nonsensical conclusion that the prices in 1967 were, on the average, 25 percent higher than those in 1962, when 1962 was chosen as the base, and that the prices in 1962 were 25 percent higher than those in 1967 when 1967 was chosen as the base. The arithmetic mean of ratios is inconsistent.

A consistent result is reached, however, when the geometric mean is applied in this situation. Since

a. if 1962 is chosen as the base, the prices of 1967 were 100 percent of the prices of 1962,

$$G = \sqrt[2]{200 \times 50} = \sqrt[2]{10{,}000} = 100, \text{ and}$$

b. if 1967 is chosen as the base, the prices of 1962, as they should be, were also 100 percent of the prices in 1967, and

$$G = \sqrt[2]{50 \times 200} = 100$$

This method reduces the large upward moving relative changes to ratios equal to the smaller decreasing relative changes by giving each change equal weight when the nth root of the products of the relatives is extracted. The arithmetic mean has an upward bias because its method overemphasizes the large increasing relatives by treating them as absolute numbers instead of as relative numbers which they really are.

The most useful application of the geometric mean is to average rates of changes, since the average of rates of changes can be measured correctly only by this method. Suppose the population of a certain city had increased from 100,000 to 260,000 during the period 1910–1960. What is the average rate of change per decade? It must be smaller than 32 percent, as the result of the arithmetic mean indicates ($160\% \div 5 = 32\%$), because the growth of population is compounding. Table 3.8 illustrates the correct rate of change obtained by the method of the geometric mean.

TABLE 3.8 COMPUTATION OF THE GEOMETRIC MEAN OF POPULATION CHANGE OF A CERTAIN CITY BY DECADES, 1910–1960

Year	Population	Relatives of Previous Figure (x)	Log of Relatives ($\log x$)
1910	100,000		
1920	120,000	1.200	0.07918
1930	145,000	1.208	0.08207
1940	180,000	1.241	0.09377
1950	220,000	1.222	0.08707
1960	260,000	1.182	0.07262
Total			0.41471

$$\log G = \frac{\Sigma \log x}{n} = \frac{0.41471}{5} = 0.08294$$

$G = 1.21043$, or about 21% per decade

The average rate of increase in population for the last decade is

$$\log G = \frac{\Sigma \log x}{n} = \frac{0.07262}{10} = 0.007262$$

$$G = 1.01687, \text{ or about } 1.7\%$$

These calculations of average rates of change are based upon the assumption of a constant rate of change. When the calculation involves a considerable number of years, it is usually undertaken by using the formula

$$p_n = p_o(1 + r)^n \tag{3.9}$$

where

p_o = the amount at the beginning of the period
p_n = the amount at the end of the period
r = rate of change
n = number of time periods

Solving for r from the above equation, we have

$$\frac{p_n}{p_o} = (1 + r)^n \text{ or}$$

$$\sqrt[n]{\frac{p_n}{p_o}} = (1 + r) \text{ or}$$

$$r = \sqrt[n]{\frac{p_n}{p_o}} - 1$$

This expression is commonly called the compound interest formula because it is used to compute the relationship of compound interest. For example, at what rate of interest would \$1000 double in 10 years? Solving this simple problem by this formula, we have

$$r = \sqrt[10]{\frac{\$2000}{\$1000}} - 1 = \sqrt[10]{2} - 1$$

Solving the above by the use of logarithms, we have

$$\log (r + 1) = \frac{\log 2}{10} = \frac{0.30103}{10} = 0.030103$$

$$r + 1 = 1.0718$$

$$r = 1.0718 - 1 = 0.0718 \text{ or } 7.18\%$$

Observe that in the previous computation logarithms were used; hence the compound interest formula may be rewritten as

$$\frac{\log \left(\frac{p_n}{p_o}\right)}{n} = \log (1 + r)$$

or the computation may be made without computing the natural number of ratios but by subtracting the logarithms so that

$$\frac{\log p_n - \log p_o}{n} = \log (1 + r)$$

Also, in finding the amount at the end of the period, considerable labor may be saved by using logarithms instead of natural numbers. The expression becomes

$$\log p_n = \log p_o + n \log (1 + r) \qquad \textbf{(3.10)}$$

The pattern of the frequency distribution of a series is often an important consideration in selecting the most representative average. If the absolute values of the distribution, plotted on an arithmetic scale, are fairly symmetrical, the arithmetic mean is definitely preferable to the geometric mean. However, if a series is logarithmically distributed, that is, if the distribution is markedly skewed to the right in such a way that, when a logarithmic X scale is used in plotting or when the logarithms of the mid-points are plotted on an arithmetic scale, a symmetrical frequency polygon results, then the geometric mean would appear to be more representative. A logarithmically distributed series is encountered when the series has a definite limit at one extreme and theoretical infinity at the other. A good illustration of this type of distribution is the percentage changes in prices. In theory as well as in practice, the price of a commodity may increase infinitely from a given base, but it can never decrease more than 100 percent. Thus, the percentage changes, as natural numbers, would distribute themselves asymmetrically, with the range of deviations above the arithmetic mean greatly exceeding the range below. However, the logarithms of percentage changes would tend to group themselves symmetrically about the logarithm of the geometric mean. In conclusion, the geometric mean is preferred for a logarithmic distribution because in such a distribution not the absolute deviations but the deviations of ratios about central tendency tend to be symmetrical.

3.8 The Harmonic Mean

If the reciprocal of the value of each item is taken, the arithmetic mean of the reciprocals is computed, and the reciprocal of this mean is taken, the result is known as the harmonic mean. Or, more compactly, the harmonic mean is the reciprocal of the arithmetic mean of the reciprocals of the observations. Symbolically,

$$H = \frac{1}{\dfrac{1/x_1 + 1/x_2 + \cdots + 1/x_n}{n}} = \frac{1}{\dfrac{\Sigma(1/x)}{n}} = \frac{n}{\Sigma(1/x)} \tag{3.11}$$

To make computation easier, formula 3.11 may be rewritten as

$$1/H = \frac{1/x_1 + 1/x_2 + \cdots + 1/x_n}{n} = \frac{\Sigma(1/x)}{n} \tag{3.12}$$

The harmonic mean of the three values 2, 4, and 8 is

$$\frac{1}{H} = \frac{1/2 + 1/4 + 1/8}{3}$$

$$= 7/24$$

$$H = 3.43$$

For the same data, the arithmetic mean is 4.7 and the geometric mean is 4. For any series, the values of which are not all the same and which does not have any value of zero, the harmonic mean is always smaller than both the arithmetic mean and the geometric mean. Thus the harmonic mean is considered to have a downward bias, whereas the arithmetic mean has an upward bias.

The harmonic mean is limited to average time rates under certain conditions and certain types of prices. Nevertheless, within this limited area, the harmonic mean yields consistent results in most cases. The arithmetic mean has often been used when the harmonic mean should have been employed. For instance, during World War I, the estimate of the transport capacity of American ships was made by dividing the round-trip distance by the arithmetic mean speed of the ships to be used. The result, allowing for turn-around time, delays, and so forth, gave an estimate of shipping capacity that proved unattainable. Actually, in such a case the harmonic mean should have been used. The arithmetic average number of hours per mile could have been obtained and multiplied by the number of miles to get the mean number of hours per trip. Since the harmonic mean speed is the reciprocal of the arithmetic average number of hours per mile, such a procedure is mathematically equivalent to dividing the number of miles by the harmonic mean speed.

Because of the absolute necessity of using the harmonic mean in some cases and the confusion between the application of the arithmetic and harmonic averages, the harmonic mean deserves more attention than it receives in most elementary textbooks. The proper use of the harmonic mean depends upon a number of considerations. First of all, we must remember that an average refers to some class of units that must be appropriate to the use that the average is to serve. As an example, suppose two workers, A and B, in a clothing factory can make two and three neckties per hour, respectively, their speeds being recorded in either one of the following two forms:

Form 1.	Worker A:	2 neckties per hour
	Worker B:	3 neckties per hour
Form 2.	Worker A:	30 minutes per necktie
	Worker B:	20 minutes per necktie

In the first form, time is held as a constant and production a variable; that is, the outputs are weighted with *equal time periods*. Under this assumption, the interest is centered around the average output per hour, and it is proper to use the arithmetic mean, which in this case is 2.5 neckties per hour $(2 + 3) \div 2$. If 2.5 neckties are produced, on the average, per hour, how much time is required, on the average, to produce one necktie? The answer is 24 minutes because $60 \div 2.5 = 24$.

When the time rate is expressed as in the second form, however, time is treated as the variable, and production, the constant; that is, time is weighted with *equal amounts of output*. With this change in the assumption, interest has now been shifted to the average time required to complete one unit of product; therefore, the harmonic mean should be employed. Thus,

$$\frac{1}{H} = \frac{1/30 + 1/20}{2} = \frac{5}{120}$$

$$H = 24 \text{ minutes per necktie}$$

or, on the average, 2.5 neckties are produced per hour: $60 \div 24 = 2.5$.

It may be pointed out here that, since workers usually work a fixed number of hours each day, the real interest in time study is the number of hours, rather than the number of outputs, which should be considered as a constant. This being the case, the ratio of form 1—units per hour—should be stated and the arithmetic mean used. If, however, the time rates are recorded as in form 2—length of time per unit of output, where the factor, time, desired to be constant, is made variable—only the harmonic mean gives the correct answer.

Next, the harmonic mean is especially adapted to a situation where the observations are expressed inversely to what is required in the average. That is, for example, when average cost per unit is desired but the data show the number of outputs per amount of cost. This point may be made clear by the following illustration.

Suppose you have spent one dollar for 3 dozen oranges in one shop, another dollar for 4 dozen oranges and still another dollar for 5 dozen oranges in two other different stores. What is the "average price" per dozen of oranges? You may easily make the mistake of computing the arithmetic mean for your data. Since you may reason that the average prices in the three different stores are $33\frac{1}{3}$ cents, 25 cents, and 20 cents respectively, the average price per dozen oranges in the three stores must be $(33\frac{1}{3} + 25 + 20)/3 = 26$ cents. You can see that this is incorrect by the following argument: you have spent only three dollars for 12 dozens of oranges; if the average price is 26 cents you would have spent $(26)(12) = \$3.12$! Or you may note that your data are expressed in the form "so many dozens per dollar," but you wish to know the average price of "so many cents per dozen oranges." These two are inverse expressions. Consequently, the appropriate average can be obtained by the harmonic mean, as is done below:

$$\frac{1}{H} = \frac{1/33\frac{1}{3} + \frac{1}{25} + \frac{1}{20}}{3} = 0.04$$

$$H = 1/0.04 = 25 \text{ cents}$$

You may note that, in this present case, the weighted arithmetic mean also yields accurate results. Here the weights are the quantities of oranges bought.

3.9 The Quadratic Mean

The quadratic mean is defined as the square root of the arithmetic mean of the squares of the observations (root-mean-squares). Symbolically,

$$\text{quadratic mean} = \sqrt{\frac{\Sigma x^2}{n}} \tag{3.13}$$

Practically, this measure is never used to average values. It is used, however, to compute standard deviation and to average deviation. Such uses of the quadratic mean will be discussed in the next chapter.

3.10

Nature and Uses of the Median

Thus far the averages considered are computed averages in the sense that they are computed from all observations in the sample. In contrast, a position average, such as the median or the mode, is found by locating the place of a value in a series. Thus, the values of the individual items in the sample will have no influence upon the value of a position average. We shall first consider the median.

The median, as its name indicates, is the value of the middle item in a series when the items are arranged according to magnitude. Its position in a series is such that it divides the series into two equal parts, with an equal number of items on either side of it. If there are five items in a series, say, \$2, \$4, \$5, \$7, and \$8, the median is the value of the third item, \$5, which is exceeded by as many values as it exceeds in the series. There are two values above and two values below \$5 in the series. If there are six items in a series, say, 3 pounds, 4 pounds, 6 pounds, 7 pounds, 8 pounds, and 10 pounds, then any value between 6 and 7 pounds would divide the series into two equal parts, with three items above and three items below it; therefore, any such value could become the median. However, in practice, we usually assume the median to be halfway between the two central items. In our example, therefore, the median would be 6.5 pounds. The median may have values identical with its own on both sides of it. For instance, in a series of 1, 2, 3, 4, 5, 5, 5, 6, 7, 8, 9, the median is 5. Because of this characteristic, the *median* may now be formally defined as that value which divides a series in such a fashion that at least 50 percent of the items are equal to or less than it and at least 50 percent of the items are equal to or greater than it.

From the above discussion it is clear that the value of the median cannot be located from ungrouped data unless an array is made. The value of the middle item is meaningless as a measure of central tendency in a disorderly series because it may have larger as well as smaller values on both sides of it. If the number of items in a series is odd, the median is the value of the middle item. If the number of items is even, the median is the arithmetic mean of the values of the two middle items. Therefore, the value of the median coincides with that of one of the existing items in an array if there is an odd number of items; the value of the median does not necessarily coincide with one of the items in an array if there is an even number of items.

As a position average, the median is influenced only by the number of items and not by the values of items in the series. As cited earlier, the median of the series of 2, 4, 5, 7, and 8 is 5. The median would still be 5 had we changed the series to 1, 2, 5, 9 and 10.

The typical characteristic of the median is that it divides the distribution into two equal parts. In a sense, the median is also a point of balance: it balances the number of items in the series. Thus, the median is especially meaningful in describing observations that are scored, such as rates, grades, and ranks, rather than counted or measured. However, it is meaningless for completely qualitative data.

Compared with the means, the median has an advantage in that it can be located even when the frequency distribution has open classes. However, it has a disadvantage because, if the full distributions are not shown, it is impossible to determine from the medians of different groups the median for the combined group.

The median has the interesting property that the sum of the absolute deviations of the observations (deviations whose signs are ignored) from the median are less than the sum of the absolute deviations from any other point in the distribution. In symbols, $\Sigma|d| =$ a minimum. What we are interested in pointing out here is that this measure is often selected because of this property. For instance, insurance rates are usually medians because in this field it is much more important to have rates that are on the average as close as possible to claim costs and other expenses than to have rates which are precisely accurate as often as possible but which, at the same time, are seriously off for other risks.

Another simple application of this property is the decision on the optimum location of a plant. For example, suppose a restaurant chain organization has seven restaurants on a certain turnpike:

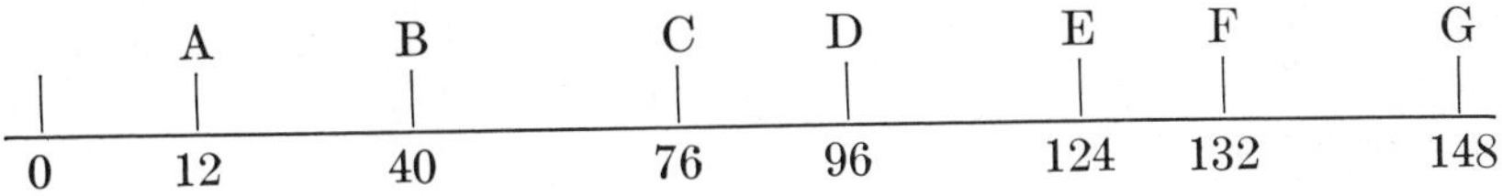

Each restaurant must send each day for fresh food from a central storehouse. To minimize the total of distances from the seven restaurants to the central storehouse, where should it be located? Clearly, it should not be located at the far west or the far east, but somewhere in the center. But where exactly in the center should the location be? Remembering that the sum of absolute deviations from the median is a minimum, it can be easily decided that the central storehouse should be located at Restaurant D or in its vicinity. The median distance is 96 miles and from it the total amount of travel would be minimized, at 276 miles each way. To locate the central storehouse at any other point, the sum of distances will be greater. For instance, at Restaurant C, this sum is 286 miles. The student may check these computations in Table 3.9.

TABLE 3.9 COMPUTATION OF MEDIAN DISTANCE

		Distance	
Restaurant	Location	*From D*	*From C*
A	12	84	64
B	40	56	26
C	76	20	0
D	96	0	20
E	124	28	48
F	132	36	56
G	148	52	72
Total	532	276	286

3.11

Locating the Median from Grouped Data

There are two methods of finding the median from a frequency distribution: by interpolation and by graphic analysis. Both methods will yield identical results, but the value may not be the same as the value obtained from an array.

LOCATING THE MEDIAN BY INTERPOLATION

The method of interpolation is based on the assumption that the data are continuous and that the items in the series are distributed evenly within the class intervals. Here, as for the array, to locate the median the first thing to do is to find the position in the distribution that would divide the series into two equal parts. Using the average weekly earnings of the messenger boys as an illustration again (see Table 3.10), we first locate this position by computing $n/2 = 142/2 = 71$. This indicates that the median is the value on either side of which 71 of these frequencies will fall.

Next, we discover the class in which the point 71 lies. This is accomplished by accumulating the frequencies. It can be seen from Table 3.10 that the cumulative frequency in the fourth class is 90, and we know that it is in this class the space of $n/2$ (which equals 71 in this instance) lies. In other words, the value of the median must be within the range of the fourth class, from $67.50 to $67.99. This class, which contains the median value, is called the *median class*.

TABLE 3.10 LOCATING MEDIAN AVERAGE WEEKLY EARNINGS OF 142 MESSENGER BOYS IN NEW YORK CITY, 1968

Average Weekly Earnings	Number of Messenger Boys	Cumulative Frequencies
$66.00–$66.49	11	11
66.50– 66.99	15	26
67.00– 67.49	24	50
67.50– 67.99	40	90
68.00– 68.49	20	110
68.50– 68.99	14	124
69.00– 69.49	11	135
69.50– 69.99	7	142
Total	142	

Source: Table 5.5.

$n/2 = 71$
$\Sigma f_1 = 50$
$f_{\text{med}} = 40$
$L = \$67.50$
$i = \$\ 0.50$

$$\begin{aligned}\text{Med} &= L + \left(\frac{n/2 - \Sigma f_1}{f_{\text{med}}}\right) i \\ &= \$67.50 + \left(\frac{71 - 50}{40}\right) \$0.50 \\ &= \$67.50 + (0.525)\$0.50 \\ &= \$67.50 + \$0.2625 \\ &= \$67.76\end{aligned}$$

Since in a frequency distribution the values of the individual items are lost except that they are represented by the mid-points of the classes, the exact value of the median can then only be estimated by interpolation from the limits of the median class. But how is the method of interpolation carried out?

Under the assumption that the items are evenly distributed within the class limits, it is clear that if the result of $n/2$ were 50 instead of 71, the median value would be \$67.49, the upper limit of the third class, because it lies at the 50 point. Or if the position of the median were 90, by the same reasoning the median value would be \$67.99. Or if 70 were the median point, it would be equally clear that the median value would be \$67.75; since the point 70 falls at the center of the fourth class, it would be represented by the mid-point value of that class, which is \$67.75.

The actual median value, however, must be greater than the mid-point value of the median class, that is, \$67.75, because the actual position of the median is 71, which is more than midway through the median class. Now, in order to locate the actual median value, we must determine the proportion of the frequencies, which are distributed from the lower class limit to the median point in the median class, to the total frequencies in the median class. This proportion is found by dividing the number of items we are short to make up $n/2$ when we enter the median class by the total number of frequencies in the median class. In our illustration this proportion is $(71 - 50) \div 40 = 0.525$. This means that the position of the median is 52.5 percent through the median class. Obviously, this proportion is in terms of frequencies and must be converted into an actual value in the series by multiplying this proportion by the width of the class interval of the median class. Thus, $0.525 \times \$0.50 = \$0.2625$. When this value is added to the lower class limit of the median class ($\$67.50 + \$0.2625$), we have \$67.76, which is the median.

The preceding discussion of locating the median by interpolation may be summarized as follows:

First, compute the value of $n/2$.

Second, accumulate the frequencies.

Third, find the class whose cumulative frequency is the first to exceed $n/2$. This is the median class.

Fourth, perform the following operations:

a. deduct the cumulative frequency of the class immediately preceding the median class from $n/2$;
b. divide this difference by the number of frequencies in the median class;
c. multiply this quotient by the size of the class interval of the median class; and
d. add this product to the lower class limit of the median class to obtain the value of the median.

These steps may be generalized by the following expression:

$$\text{Med} = L + \left(\frac{n/2 - \Sigma f_1}{f_{\text{med}}}\right) i \tag{3.14}$$

where

- Med = the median
- L = the lower class limit of the median class
- n = total number of items in the sample
- Σf_1 = the sum of all the frequencies accumulated just before entering the median class
- f_{med} = the frequencies in the median class
- i = the size of the class interval of the median class

The value of the median obtained in our illustration is $67.76. It is the value that divides the items in the distribution into two equal parts. It means that half of the 142 messenger boys received average weekly wages less than $67.76 and half of them had wages greater than this amount.

LOCATING THE MEDIAN BY GRAPHIC ANALYSIS

It was mentioned in Chapter 2 that when a frequency distribution has been accumulated on the "more than" and "less than" basis, and when they are properly plotted into ogives, the median can be located by dropping a perpendicular from the intersection point of these curves to the X axis (see Figure 2.6). The median can also be easily obtained from either the "more than" ogive

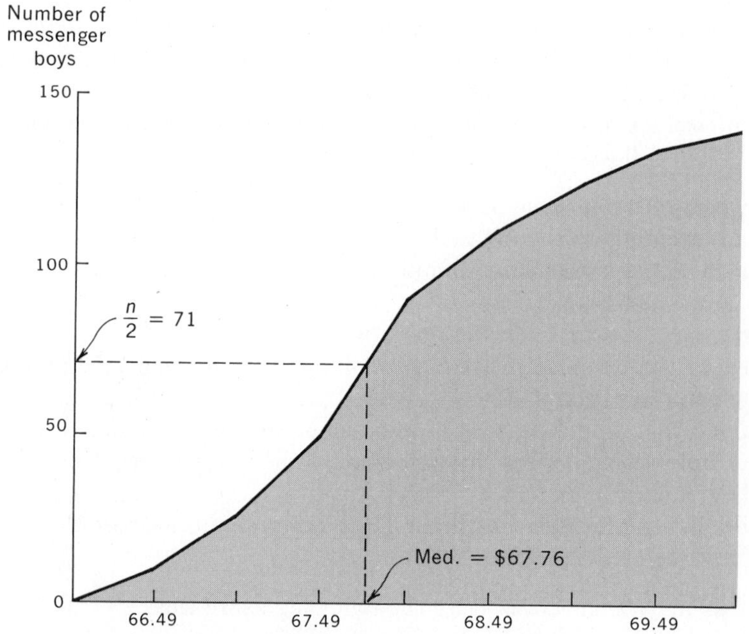

Figure 3.1 Graphic location of median average weekly earnings of 142 messenger boys in New York City, 1968.

or the "less than" ogive alone. To estimate the value of the median from a single cumulative frequency curve, we first locate the point of $n/2$ on the Y scale. Then a horizontal line is drawn from this point to the ogive, and a perpendicular is dropped from the point where this horizontal line cuts the ogive to the X scale. The point where this perpendicular meets the X scale is the median. The application of this method is illustrated by Figure 3.1.

3.12

Nature and Uses of the Mode

The mode is that value of a series which appears more frequently than any other. Thus it is found by discovering the value that appears most frequently. This value can be discovered immediately when the data are arrayed. For example, in the series 1, 2, 3, 4, 4, 5, 6, and 7, the mode is 4. We can therefore consider the mode as typical in the sense that it is the most probable value in a series. The mode differs from other types of averages discussed before in that the mode of a monomodal distribution (for ungrouped data) always coincides with an actual value in the series.

The mode is often the concept in the mind of most people when they speak of "averages." When students say the average grade in their classes is C, for example, they usually mean that this, more than any other, is the grade received by students. The "average consumer" often means that type of consumer who appears more frequently with regard to his pattern of consumption or other quality. The term "typical wage" refers to the modal wage most of the time. The modal size of men's shoes is the typical size bought because more people will buy this size than will buy any other. Thus, we use the mode in preference to other averages if we like to point to the most typical value of the series.

Consider the problem of managing operations of the basement of a department store. Typically, people go shopping in the basement for inexpensive merchandise. Among other things, to make low prices possible, there must be a large turnover per square foot of space. This, in turn, requires that the basement carry sizes and items demanded by the largest number of people possible, such as kitchen utensils, toilet equipment, children's wares, and so on. Similarly, the volume sales of 5- and 10-cent stores, discount houses, grocery chains, counter restaurants, and so forth, are all pitched to the modal customer.

Another interesting application of the mode is the checking of calculations. If a calculation is repeated a number of times, the result accepted is the result that appears the greatest number of times.

Although the mode is a very simple and useful concept, its application presents many troublesome aspects. First, a distribution may reveal that two or more values repeat themselves an equal number of times, and in such a situation there is no logical way of determining which value should be chosen as the mode. Strictly speaking, with a discrete series, any value is called a mode if its number of appearances is not exceeded by either of the adjacent values. As long as the frequencies of the modal values are not equal, however, we might decide to select the value with the highest frequency as the mode for the series. This practice implies the assumption that the population may not contain such irregularities.

Second, in dealing with a continuous series whose values follow each other in a continuous sequence, there would be no mode unless the data are grouped. Also, even for discrete data, we may not find any value that appears more than once, as in the case of the populations of the large cities in the United States.

Finally, the mode is a very unstable value. It changes radically if the method of rounding the data changes. For illustration, let us assume a series of weights with different methods of rounding as follows:

Actual Weight	Rounded to Nearest Pound	Rounded to Last Pound
3.000	3	3
3.500	4	3
3.571	4	3
3.784	4	3
4.500	4	4
4.831	5	4
6.115	6	6
6.115	6	6

From the above, the mode for the series in actual weights is 6.115 pounds; when the data are rounded to the nearest pound, the mode becomes 4 pounds; when the data are rounded to the last pound, still another mode value appears—3 pounds. This erratic behavior gives reason to doubt the significance of the value of the mode determined by this method.

All these difficulties can be overcome totally or in part when the data are arranged into a frequency distribution. In such a distribution, any tendency of the data to cluster around a certain value is more clearly shown and therefore the problem created by the first situation may be solved. Moreover, the erratic nature of the mode obtained from an array due to different methods of rounding can be stabilized to a great extent by determining the mode from the frequency distribution of the same data. This is so because when they are grouped into classes, those items which differ in values only on the right of the decimal points may still fall in the same class irrespective of the method of rounding. Finally, a modal value may appear in grouped data even though there is no repetition of any value in the array, or because the data are continuous. A good illustration of this point would be the series of the population of each large city in the United States, cited earlier; no repetition of any single value can be found from its array, but a modal population appears when the same data are grouped.

Let us now discuss the methods of finding the mode from grouped data.

3.13

Mode for Grouped Data

It should be stated at the very beginning that it is only possible to determine the "true" mode from grouped data by advanced mathematical methods that are clearly beyond the scope of this book. A number of methods are available in elementary statistics, however, by which reasonably accurate approximations of the value of the mode may be obtained. Three such methods will be discussed in this section: (1) estimating the mode by algebraic interpolation within the modal class; (2) locating the mode by geometric interpolation; and (3) determining the mode by the empirical method.

METHOD OF ALGEBRAIC INTERPOLATION

To estimate the mode by the method of algebraic interpolation we first find the modal class by inspecting the frequency distribution. The *modal class* has the highest class frequency. It is clear that the modal class for the distribution of the average weekly earnings of the 142 messenger boys in Table 3.10 is the fourth class, because it includes more cases—40—than any other class.

Next, we must estimate where, within the class interval of the modal class, the value of the mode lies. The mid-point value of the modal class can be used as the modal value if, and only if, the situation is such that there are an infinite number of cases in the distribution and an infinitely small class interval is used. The logic behind this statement is that when the class interval is being reduced, the value of the mode tends to be delimited by a smaller and smaller range of values, and it tends to come closer and closer to the mid-point value of the class of the highest frequency. However, this reduction in the size of the class interval is definitely limited by the number of cases contained in the series. In reality, this ideal situation does not exist; therefore, an approximation of the modal value must be interpolated around the mid-point of the modal class.

To interpolate the modal value within the modal class, we must observe the fact that despite our previous mid-point assumption, the items within a class are not distributed evenly and smoothly. There is usually a tendency for the items to gravitate toward the point of the greatest density. This point of gravity will pull the modal value away from the mid-point of the modal class toward either the lower or the upper class limit. In the distribution of the average weekly earnings of the messenger boys the frequency in the premodal class (the class that precedes the modal class) is 24, while the frequency of the post-modal class (the class which follows the modal class) is 20. This suggests that the point of the greatest concentration is pulled toward the lower limit of the modal class and the modal value will be below the mid-point of that class.

The method of algebraic interpolation for the value of the mode is made by resorting to the formula:

$$\text{Mo} = L + \left(\frac{\Delta_1}{\Delta_1 + \Delta_2}\right) i \tag{3.15}$$

where

L = the lower limit of the modal class
Δ_1 = the difference between the frequency of the modal class and the frequency of the premodal class (signs neglected)
Δ_2 = the difference between the frequency of the modal class and the frequency of the postmodal class (signs neglected)
i = the size of the class interval of the modal class

From Table 3.10, we find

$L = \$67.50$
$\Delta_1 = 40 - 24 = 16$
$\Delta_2 = 40 - 20 = 20$
$i = \$0.50$

Applying the above formula to these data, we have

$$\text{Mo} = \$67.50 + \left(\frac{16}{16 + 20}\right) \$0.50 = \$67.72$$

This mode means that approximately \$67.72, rather than any other amount, was the weekly wage received by more messenger boys in New York City in 1968.

Before turning to the other methods we should consider briefly some of the problems involved in estimating the mode from a frequency distribution. First of all, it may be noticed that the erratic nature of the mode is not totally eliminated by estimating it from the grouped data, since the concentrating point in the distribution depends upon the system of classification. The position of the mode changes if the size of the class interval changes. Thus we are obliged to conclude that the modal value is the most unstable measurement of central tendency among all the averages.

Further, the modal value from a highly skewed distribution is meaningless because it is too close to one end of the data to be very representative of the series.

Also, if a series has two separate and distinct points of concentration—two modal classes—the mode is largely a useless device for measuring central tendency. If, however, a distribution has two distinct modal classes which are adjacent, their combined class interval may be used as the base for interpolating the modal value.

METHOD OF GEOMETRIC INTERPOLATION

Geometric interpolation to find the mode is made from the histogram of the distribution. This method is illustrated in Figure 3.2. Here we first draw a diagonal from the lower limit of the modal class at the point established by the frequency in the premodal class to the opposite point in the rectangle delimited by the modal class. Next, we draw another diagonal from the upper

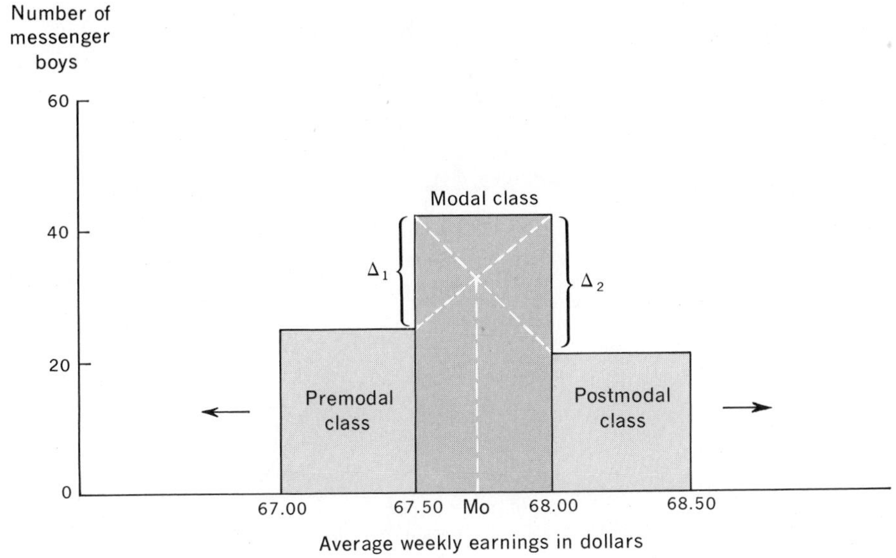

Figure 3.2. Geometric interpolation of the mode, average weekly earnings of 142 messenger boys in New York City, 1968.

limit of the modal class at the point established by the postmodal class frequency to the opposite point of the rectangle described by the modal class. Then, if we drop a perpendicular from the point of intersection of these two diagonals, the modal value is found at the point on the base line where this perpendicular cuts. It should be noted that the two methods of geometric and algebraic interpolation are comparable to and therefore may be expected to yield the same result.

It will be noticed that the value of the mode estimated by the graphic approach is approximately \$67.72, the same as that obtained by the method of algebraic interpolation.

THE EMPIRICAL METHOD

The empirical method is based upon the relationship among the positions of the mean, the median, and the mode in a smooth distribution, such as the three polygons in Figure 3.3.

These three averages are of identical value if they are computed from a symmetrical distribution, as demonstrated by Figure 3.3a. From this figure it can be seen that the modal value corresponds to the maximum point (ordinate) of the polygon. This maximum ordinate, which divides the symmetrical curve into two equal parts, is, therefore, the median value. Because it is a symmetrical distribution, furthermore, the extreme low values and the extreme high values exert equal influence upon the mean; consequently the value of the mean would be at the middle of the range of the distribution that corresponds with the maximum point of the curve.

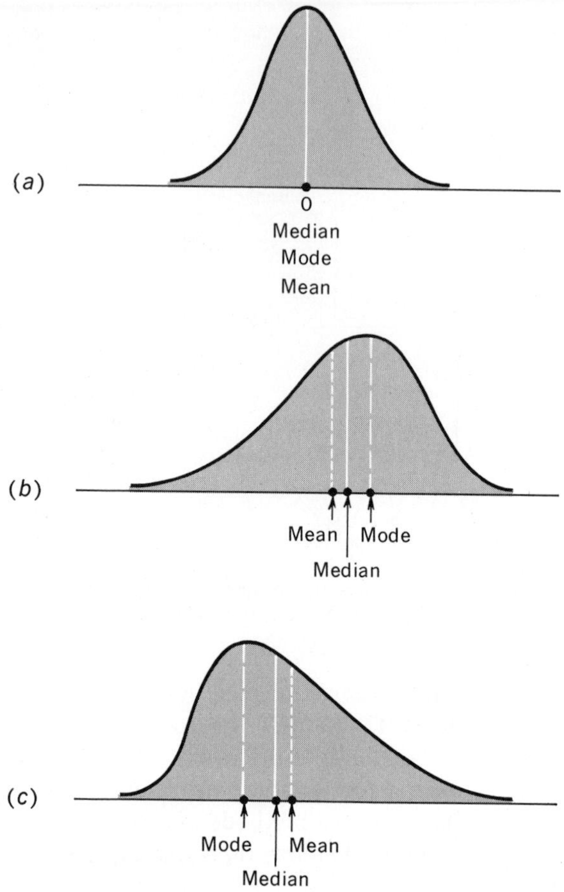

Figure 3.3 Hypothetical frequency polygons showing theoretical position of mode, median, and mean: (a) symmetrical distribution, (b) distribution skewed to the left, (c) distribution skewed to the right.

If, however, the distribution is skewed, either positively or negatively, the mean, the mode, and the median will pull apart. In Figure 3.3b, for instance, the mode is still located at the highest ordinate in the distribution, but the median is pulled to the left of the mode, toward the direction of the low values, so as to divide the area under the curve into two equal parts. The mean, being a value most influenced by the extreme values, will be pulled still further toward the direction of the extreme values than will the median. If we are to consider a distribution skewed to the right, as is that of Figure 3.3c, the same relationship will occur, but in the opposite direction.

Ordinarily, in a moderately skewed distribution, when these average values diverge, the median will be found to be approximately one third as far away from the mean as the mode. Since the mean and the median can be exactly determined, the value of the mode may be estimated from the former two values.

The formula used to describe this relationship is

$$\text{Mo} = \text{Mean} - 3(\text{Mean} - \text{Median}) \tag{3.16}$$

For the average weekly earnings of the messenger boys, the mean is \$67.83 and the median is \$67.76. Substituting these data in the above formula, we have

$$\begin{aligned} \text{Mo} &= \$67.83 - 3(\$67.83 - \$67.76) \\ &= \$67.83 - \$0.21 \\ &= \$67.62 \end{aligned}$$

Observe that the result obtained by this method is different from that secured by algebraic interpolation. This point leads us to two important conclusions concerning the mode. First, the empirical estimate is based upon the assumption that the theoretical relationship of the positions of the mean, the median, and mode usually found in moderately skewed frequency distributions exists. However, this relationship may not exist in a particular frequency distribution. Thus, the uncertainty of the true mode remains. Second, since the elementary methods of estimating the mode discussed in this chapter may not be expected to yield the same result, the student who limits himself to them should use his own judgment as to which of the modal values should be selected to describe the series.

3.14 Summary and Conclusions

Several types of averages have been discussed in this chapter. Except for some specific situations, different averages usually yield different values for the same group of data. Moreover, each average has a unique meaning of its own. Because of these considerations, a very important and often difficult question arises. What average should be used to describe or represent a given series? Perhaps before this question, another fundamental question must be asked: Should an average be used at all in a given situation?

As to whether or not it is appropriate to use a single number to describe a series, we must remember that an average is a measure of central tendency. Hence, unless the data show a clear, single concentration of observations, an average may not be meaningful at all. This evidently precludes the use of any average to typify a bimodal, a U-shaped, or a J-shaped distribution. With this in mind, we may then proceed to determine what average to use by first considering the following queries:

(1) What purpose is the average designed to serve?
(2) Should we permit extreme values in the series to influence the average?
(3) Should the average be used for further computation?

(4) What is the model of the distribution? Is it symmetrical or skewed? If it is positively skewed, is it of the logarithmic type?

(5) How are we to record the observations to be averaged? natural numbers? ratios? rates? averages?

(6) What class of units should be used for the average? Are the observations expressed inversely to what is required in the average?

(7) In what sense do we expect the average to be typical? to balance the individual values? to balance the ratios? to balance the number of items? to typify the most frequent value?

(8) What weights, implicit or explicit, if any, should be used?

With answers to these questions clearly in mind, we can then make a rational selection of averages with reference to the characteristics of the averages and the numerical relationships among them.

The characteristics of various averages are summarized below:

(1) *The arithmetic mean.* (a) It is affected by every observation and is influenced by the absolute magnitudes of the extreme values in the series. (b) As a computed average, it is capable of algebraic manipulation. From its basic equation, $\bar{x} = \Sigma x/n$, when two of the magnitudes are known, the third can also be obtained. That is, $\Sigma x = n\bar{x}$ and $n = \Sigma x/\bar{x}$. (c) The sum of algebraic deviations from the mean is zero, $\Sigma d = 0$, and the sum of the squared deviations from the mean is a minimum, $\Sigma d^2 =$ a minimum. (d) In a sampling sense, the mean is a stable statistic. This concept will be developed in a later chapter. (e) The mean is typical in the sense that it is the center of gravity, balancing the values on either side of it.

(2) *The geometric mean.* (a) It is also affected by all items in the series, but it gives less weight to extremely high values than does the arithmetic mean. (b) It is strictly determined for positive values but cannot be used to average negative values or values with a zero term. (c) It balances the ratio of the values to the geometric mean. (d) It is thus adapted to average rates of change, to ratios between measures, and to ratios of price change. (e) It is also capable of algebraic manipulation. When it is substituted for the values from which it is computed, the same product is secured.

(3) *The harmonic mean.* (a) As a computed average, it is also affected by all observations. However, since the reciprocals are averaged, it gives more weight to the smaller values. This is just the opposite of the mean. (b) It is capable of algebraic manipulation. (c) It is adapted to average time rates and price movements. It is also useful when the observations are expressed inversely to what is required in the average.

(4) *The median.* (a) It is a position average that is affected by the number of items, not by the values of the observations. Thus, it (b) is not influenced by the magnitude of the extreme deviations from it, and (c) it can be located from open-end distributions. (d) The sum of absolute deviations from the median is a minimum. (e) It is typical in the sense that it balances the number of items in the series. (f) It is meaningless for completely qualitative data but meaningful as long as data can be ranked, such as the grades, A, B, C, D, E, F. The median

is the most suitable average to describe observations that are scored rather than computed or measured.

(5) *The mode.* (a) Since it is the point of greatest concentration, the mode is the most typical average for a distribution. Because of this property, (b) the mode is meaningless unless the distribution includes a large number of observations and possesses a distinct central tendency. (c) Like the median, the mode is not influenced by extreme values, is not capable of algebraic manipulation, and can be located from an open-end distribution. (d) Unlike all other averages, the mode is always represented by actual items in the series, except when it is calculated by the empirical method. (e) The mode is the most unstable average and its true value is difficult to determine.

Certain numerical relationships exist among the averages:

(1) For any series, except one whose observations are of identical value, the arithmetic mean is always greater than the geometric mean, which, in turn, is greater than the harmonic mean.

(2) For a symmetrical and unimodal distribution, mean = median = mode.

(3) For a positively skewed distribution, the mean has the largest value, the mode the smallest value, and the median about one third the distance from the mean toward the mode.

(4) For a negatively skewed distribution, the mean is the smallest, the mode the largest, and the median is again about one third the distance from the mean toward the mode.

In conclusion, it may also be pointed out that a complete description of a distribution occasionally calls for two or more of these averages. It is true that presenting two or more averages creates an added burden for the investigator as well as for the consumer of statistics. However, the work this extra burden entails is fully justified if it presents a more complete description of the data than is possible from a single measure.

Glossary of Formulas

(3.1) $\bar{x} = \frac{\Sigma x}{n}$ The arithmetic mean of a sample. It is defined as the sum of all individual observations in the sample divided by the number of observations in the sample.

(3.2) $\bar{x} = \frac{\Sigma fm}{n}$ This formula is used to compute the sample mean from a frequency distribution by the long method. Here, m stands for mid-points, and f refers to the class frequencies.

(3.3) $\bar{x} = A + \frac{\Sigma fd}{n}$ The sample mean may be computed by the short-cut method in original units. It is done by selecting one mid-point as the assumed mean, A, and adding to it a correction factor, $\Sigma fd/n$. Here, $d = x - A$. This method can be applied to both equal and unequal class intervals.

(3.4) $\bar{x} = A + \left(\frac{\Sigma fd'}{n}\right) i$ The mean can also be obtained by the short-cut method in class-interval units if the distribution is of equal class intervals. This formula differs from (3.3) in that here d' instead of d is used. Note that d' is the deviation step. Consequently, the correction factor must be converted back to the original units by multiplying it by the class interval, i.

(3.5) $\bar{x}_w = \frac{\Sigma Wx}{\Sigma W}$ This equation is used to compute the weighted arithmetic mean. W is the weight assigned to the individual observations that are being combined into the average.

(3.6) $G = \sqrt[n]{(x_1)(x_2)(x_3) \cdots (x_n)}$ The general equation for the geometric mean. It is the nth root of the product of the n items in the sample. This formula is seldom used unless n is very small. With larger n, formula (3.7) is usually employed.

(3.7) $\log G = \frac{\Sigma \log x}{n}$; $[G = \text{antilog} (\log G)]$ This is the standard formula for the computation of the geometric mean. Note that this equation transforms the observation into logarithms, and takes the arithmetic mean of the logarithms. Therefore, the antilogarithm of the result must be taken to obtain the geometric mean.

(3.8) $\log G = \frac{\Sigma f \log m}{n}$ This formula is used to compute the geometric mean for grouped data.

(3.9) $p_n = p_o(1 + r)^n$ This equation is commonly known as the compound interest formula. It can be rearranged to solve for r, which is the average rate of growth of a variable.

(3.10) $\log p_n = \log p_o + n \log (1 + r)$ This is the logarithmic form of the previous formula.

(3.11) $H = \frac{1}{\frac{\Sigma(1/x)}{n}} = \frac{n}{\Sigma(1/x)}$ The general formula for computing the harmonic mean. Note here that the natural numbers are transformed into reciprocals and then back to natural numbers after the arithmetic mean of the reciprocals is taken.

(3.12) $1/H = \frac{\Sigma(1/x)}{n}$ This is the reciprocal of formula (3.11). It is used instead of (3.11) in order to make computations easier. It should be noted that the reciprocal of the result must be taken to get the harmonic mean if this formula is used.

(3.13) quadratic mean $= \sqrt{\frac{\Sigma x^2}{n}}$ This formula for the quadratic mean (root-mean-square) is seldom used as a measure of central tendency.

(3.14) $\text{Med} = L + \left(\frac{n/2 - \Sigma f_1}{f_{\text{med}}}\right) i$ This equation is used for the computation of the median from grouped data. The median is the lower class limit, L, of the median class plus the proportion of the distance of the class interval of the median class at which the total frequency is divided into two equal parts.

(3.15) $\text{Mo} = L + \left(\frac{\Delta_1}{\Delta_1 + \Delta_2}\right) i$ This equation is employed to compute the mode from grouped data. The mode is the lower class limit of the modal class, L, plus the term in which Δ_1 is the absolute difference between the frequency of the modal class and the frequency of the premodal class, Δ_2 is the absolute difference between the frequency of the modal class and the frequency of the postmodal class, and i is the class interval. The mode is therefore located by this equation at a point in the modal class that is determined by the relative frequencies of the two classes on either side of the modal class.

(**3.16**) Mo = Mean − 3(Mean − Median) The empirical method of estimating the mode. This equation is recommended when it is desirable to measure the mode but when the data are not sufficient for its computation. This formula is derived by an assumed relationship between the arithmetic mean, the median, and the mode in a smooth distribution.

Problems

3.1 What is the meaning of the concept "average"?

3.2 What are the reasons for using averages?

3.3 Give various interpretations of the arithmetic mean, the geometric mean, the harmonic mean, the median, and the mode.

3.4 Is the arithmetic mean always weighted? Why or why not?

3.5 Why is the arithmetic mean the most widely used average? Do you think it should be?

3.6 The arithmetic mean is often preferred over the median and the mode because the mean can be used for further computation. Why is this property of the mean important?

3.7 Why cannot the geometric mean be used to average observations which have zero or negative values? Do you think this property to be an important disadvantage of the geometric mean? Explain.

3.8 Why is the arithmetic mean greater than the geometric mean, and which, in turn, is greater than the harmonic mean when they are computed from the same series?

3.9 Would it be possible for more than 50 percent of the production workers to receive a wage lower than the mean wage of these workers in the same plant? In connection with your answer, what average would you expect to be used by the public relations man of the plant? by the business agent of the labor union?

3.10 If the average height of the American soldiers were 5 feet and 7 inches, could the most typical height of the American soldiers be greater or smaller than this value? Why or why not?

3.11 If a salesman says he has to call on four prospects "on the average" to make a sale, which average do you suppose he means? Explain.

3.12 Alfred Kinsey and others reported in their *Sexual Behavior of the Human Male* that "nearly all the distribution curves on human sexual behavior are strongly skewed to the right. The means are quite regularly higher than the location of the body of the population would lead one to expect. A few high rating individuals apparently raise the mean, give a distorted picture using the arithmetic mean." What average would be most appropriate under the circumstances? Why?

3.13 Give a specific example of your own for each of the following cases:
a. The median is preferred to the arithmetic mean.
b. The geometric mean would be more satisfactory than the arithmetic mean.
c. The harmonic mean must be used instead of the arithmetic mean.
d. The mode would be preferred to the median.
e. The median would be preferred to the mode.
f. No average would be meaningful.

3.14 Which average—the mean, the mode, or the median—would you use in each of the following situations? Explain your choice.
a. the average number of rooms in apartments in Queens, New York
b. the average annual earnings of Boston lawyers
c. the average temperature for the summer months in your native city

3.15 Indicate (but do not compute) whether you can derive the mean, the mode, or the median from the following distribution. Explain in each case.

SIZE OF ELECTRONICS DISTRIBUTORS IN NORTHEASTERN UNITED STATES

Volume Group	Number of Distributors
under $ 100,000	84
$100,000 under 250,000	144
250,000 under 500,000	98
500,000 under 1,000,000	70
over 1,000,000	54
	450

3.16 A laundry uses two different makes of washing machine. According to its past experience, the following results have been recorded:

Make of Machine	Median Life	Mean Life
A	3500 hours	4000 hours
B	4000 hours	3500 hours

If both makes are of the same price, which make should the laundry purchase from now on? Give reasons.

3.17 A traveling salesman made five trips during the past two months, making the sales tabulated below:

Trip	Number of Days	Value of Sales	Sales per Day
1	3	$ 100.00	$ 33.33
2	5	250.00	50.00
3	7	400.00	57.14
4	4	90.00	22.50
5	6	210.00	35.00
Total	25	$1050.00	$197.97

The sales manager criticized the salesman's performance as not very good since his mean sales are only $39.59 (197.97/5 = 39.59). The salesman replied that the sales manager was unfair in making such a statement for his mean sales were as high as $42.00 (1,050.00/25 = $42.00). (a) What does each average mentioned here mean? (b) Which mean seems to you appropriate in this case?

3.18 The following are the starting monthly salaries of ten college graduates:

$325, $400, $350, $336, $384, $319, $420, $400, $425, and $399

Compute the arithmetic mean, the mode, and the median for the series.

3.19 The following data represent the December, 1967, indices of department store sales in the twelve federal reserve districts:

111, 105, 113, 116, 133, 136, 115, 118, 113, 113, 134, and 118

Compute the geometric mean of these index numbers.

3.20 On a certain trip a motorist bought 10 gallons of gasoline at 29 cents per gallon, 15 gallons at 31 cents per gallon, 20 gallons at 33 cents per gallon, and 15 gallons at 28 cents per gallon. What is the average price per gallon of gasoline paid by the motorist?

3.21 Mr. X went to City B from City A by plane at 250 miles an hour and returned by train at 40 miles an hour. The distance between A and B is 1000 miles. What is his average speed? Check to see if your answer is logical.

3.22 If A can finish a certain operation in 30 minutes, B in 20 minutes, C in 25 minutes, and D in 40 minutes, what is the average time required to finish the operation?

3.23 From Problem 2.27,

a. Compute the arithmetic mean for Sample A by the long method. (Data in Table 2P.1.)

b. Compute the arithmetic mean for the same data again by the short-cut method in original units.

c. Locate the median and the mode for the data in Table 2P.1.

d. Compute the mean, the median, and the mode for the data in Table 2P.2. (For the mean use the short-cut method in class-interval units.)

3.24 Compute the mean, the mode, and the median for data in Table 1.1 in Chapter 1.

3.25 Compute the mean, the mode, and the median for data in Table 1.2 in Chapter 1.

4

MEASURES OF VARIATION AND SKEWNESS

4.1

Other Characteristics of Frequency Distributions

In the preceding chapter we began the description of frequency distributions with averages. An average, as a single significant value adopted to represent the central tendency of a series, is a very useful and powerful measure. However, the use of a single value to describe a distribution conceals many important facts. Decision making often demands the revelation of these concealed characteristics of the distribution. We must now, therefore, develop statistical measures to summarize and describe those concealed characteristics.

In the first place, not all the observations in a series are of the same value as the derived average. Almost without exception, the items included in a distribution always depart from the central value, although the degree of departure varies from one series to another. Moreover, little can be revealed about the dispersion even when several averages are computed for the series. For instance, we cannot tell at all which distribution has a greater or a smaller degree of dispersion from the information given below:

	Distribution A	Distribution B
Mean	15	15
Median	15	12
Mode	15	6

Thus, a measure of the degree of *dispersion* or *variation* is needed in order to give a more complete description of the chief characteristics of a distribution or to make possible effective comparison of two or more distributions.

Second, distribution shapes differ from series to series. Some are symmetrical; others are not. Hence, to describe a distribution we also need a measure of the degree of symmetry or asymmetry, of the balance or lack of balance, on both sides of the central tendency. The descriptive statistic for this characteristic is called the measure of *skewness*.

Finally, there are differences of the degree of peakedness among different distributions. This attribute is known as *kurtosis*. To measure kurtosis is to define the pattern of scatter of observations among the classes near the central value as compared with the scatter of observations near both ends of the distribution. Kurtosis is measured with reference to the peakedness of the normal distribution, which is of "intermediate peakedness"; that is, the normal curve is *mesokurtic*. A distribution with a flat top is called *platykurtic* and one with a pronouncedly peaked top is referred to as *leptokurtic*. (See Figure 4.1.)

Variation is by far the most important characteristic of a distribution: it may be either a basis for decision making or a measure for developing further statistical theory and method. Although skewness is an important characteristic for defining the precise pattern of a distribution, it is rarely calculated in business and economic series. As to kurtosis, we hardly use it at all in elementary statistics. For these reasons, we shall devote the major portion of this chapter to discussing various measures of variation, and the remaining space to skewness. We shall not consider the measurement of kurtosis.

4.2 Significance of Variation

Variety is not only the spice of life but also the essence of statistics. Quantitative data, the raw material for statistical analysis, are always characterized by differences in values among the individual observations. These quantitative differences are as important as the tendency of the items to cluster around a central value in a series. Just as we say that statistics is the science of averages, we can say that all statistical methods are techniques of studying variation. After all, different patterns of frequency distributions are caused by different degrees of variation. Furthermore, many other powerful analytical tools in statistics, such as the correlation study, the testing of hypotheses, the defining of confidence limits of generalizations, the analysis of fluctuations, techniques of production control, and so on, are based on the measures of variation of one kind or another. The significance of the study of variation should thus be clear.

Our immediate concern with the measures of variation is to stress their importance as a supplement to the measures of central tendency in analyzing

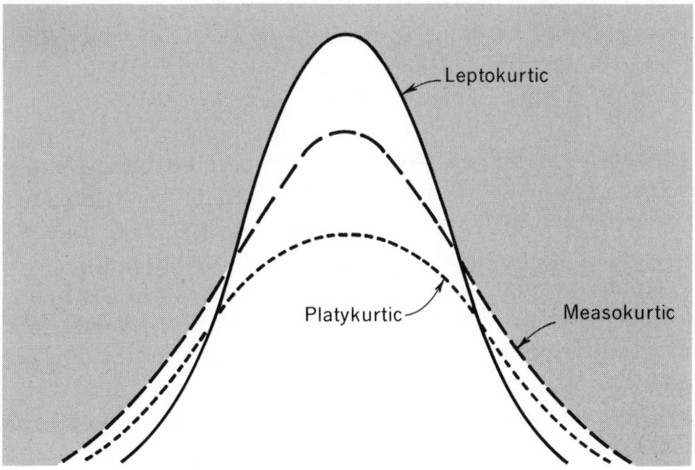

Figure 4.1 Illustration of kurtosis.

frequency distributions. An average gives no information whatsoever for us to judge its representativeness because it conceals the actual distribution of the items in a series. It is therefore of little use unless the degree of variation which occurs about it is given. If it is found that the scatter about the central value is very large—that the items are widely scattered—the average is then of little use as a typical value: it does not represent most of the individual items very well. If, however, the scatter is found to be small—that the items fall very closely together and close to the center—the average then represents the series quite well: it is nearly the same size as the other items in the series. In short, the degree of variation of a series must be measured and given if we are to know how representative of the distribution the average is.

An average, though derived from the individual items of a group, often exhibits a behavior pattern different from that of many individual items. The behavior of the latter is just as important as, if not more than, that of the former. An excellent example is the "general price level," an average of many prices, which is used by economists to reflect the value of money. When this average changes from time to time, we find that many of the individual prices often move in the opposite direction from the course taken by the average. Moreover, this average may remain stable over a period of time, while some of the individual prices may increase, others decrease, and still others remain stable. This over-all stability of the average thus conceals the facts that there is a unique distribution of individual prices at each index (the general price level) and that continuous changes are taking place in the pattern of price distribution through time. Individual price changes in different directions, even if the net change may be zero, thereby yielding a stable average, may indicate serious economic disturbances, such as alternations in the relative positions of different industries and maladjustment in output and pricing. The variation of individual prices, therefore, is also a major concern of economic theorists. Because of public policy we must pay equal attention to the general price level and to the degree of dispersion of prices through time in order to achieve economic stability without serious disturbances in the economic structure.

In production, a small amount of variation or, conversely, a high degree of uniformity of output is often of vital importance. In mass production, the supposedly identical manufactured parts must be interchangeable. Interchangeability requires uniformity in dimensions. It can be easily imagined that the degree of variability may be as important—perhaps more important—as the average value of manufactured items. The variability in the lifetime of light bulbs, for instance, is more crucial than the average life if the bulbs are used in inaccessible places where replacements are extremely difficult to make. Also, the variability in the durability of motorcycle tires used by the police may be more critical than the average. Again, you may have heard the saying that even though the average strength of the links of a chain is great, the variability also contributes to its strength.

An average, as a typical value, may be used to compare two or more series. For instance, we may compare the average annual income of two nations. Such a type of comparison is significant in that it may reveal the relative standards of living in two countries. However, the difference in or the similarity of the material well-being of the countries cannot be told fully by such a comparison alone. The material welfare of a people depends on the size of per capita income as well as on the degree of variation of individual incomes. In other words, we may find that both countries may have the same annual average income, $3,000, but that in one nation the individual incomes vary from $1,000 to $10,000 whereas in the other they vary from $200 to $1,000,000. It is quite likely that the majority of the people in the first nation belong to the middle class, receiving about $3,000 a year, whereas in the second nation there may be many very poor as well as many very rich people with a relatively small number of middle-class citizens. Thus, the wide difference in the social and economic structure—an important factor in determining economic welfare—between the two nations can be revealed only by the measures of variation in individual incomes together with the averages.

Perhaps the significance of variation was most vividly expressed by Darrel Huff when he wrote in his delightful little book entitled *How to Lie with Statistics:* "There is another kind of little figure . . . . It is the one that tells the range of things or their deviation from the average that is given . . . . Place little faith in an average or a graph or a trend when those important figures are missing. Otherwise you are as blind as a man choosing a campsite from a report of mean temperature alone. You might take 61 degrees as a comfortable annual mean, giving you a choice in California between such areas as the inland desert and San Nicolas Island off the south coast. But you can freeze or roast if you ignore the range. For San Nicolas it is 47 to 87 degrees but for the desert it is 15 to 104."

Let us now move on to see how the degree of variation is measured.

Variation may be expressed either in terms of the units of measurement used for the original data or in terms of a relative number—a ratio or a percentage. When absolute variation is being measured, the answer is expressed in the original units. When a measure of relative variation has been computed, the answer is expressed either as an abstract number or as a percentage. In this chapter we shall consider four basic measures of absolute variation: (1) the range, (2) the semi-interquartile range, (3) the average deviation, and (4) the standard deviation. Three measures of relative variation will be discussed in this chapter: (1) coefficient of variation, (2) coefficient of average deviation, and (3) coefficient of quartile deviation.

4.3 The Range

The simplest measure of variation is the range. It is the difference between the highest and the lowest values in a series. If the largest item is 100 and the smallest item is 20, the range is then 100 − 20 = 80. For grouped data, the range is obtained by subtracting the value of the lower class limit of the lowest class from the upper class limit of the highest class. Thus, the range of average weekly earnings of 142 messenger boys in Table 3.2 is \$69.99 − \$66.00 = \$3.99. Very often the ranges for the same data in grouped and ungrouped forms may differ from each other. For instance, the range of the messenger boys' average weekly earnings computed from the ungrouped data in Table 2.1 is \$69.84 − \$66.03 = \$3.81 instead of \$3.99, which is the range for the same data in the form of a frequency distribution. Such a difference is brought about by the fact that the class limits extend slightly beyond the actual data.

Judging from the method of obtaining the range, we may see that the range is not only the simplest but also the crudest measure of dispersion. The range, therefore, has certain defects. Most important of all, the range, being a positional measure and depending on the highest and lowest items alone, may be unduly influenced by one unusual value in the sample. Besides, for the same reason, the range is in no way a measure of scatter of the intervening items relative to the typical value. The fact that most of the items in many distributions tend to concentrate in a few classes near the center indicates that narrower and weighted measures should be employed to describe the distribution. Finally, the range is highly sensitive to the size of the sample. The range tends to change, though not proportionately, in the same direction as the size of the sample varies. This is so because an increase in the number of items in a sample cannot possibly decrease the range already determined. When the number is increased it is possible that some item may have a greater value than the maximum and some item a smaller value than the minimum values in the previous sample. For this reason the range cannot be interpreted properly without the number of observations. This disadvantage is especially serious in comparing the variability of different distributions.

The range, despite its numerous shortcomings, may be used quite fruitfully as a measure of dispersion for many purposes. It is perhaps most useful in a situation where one desires to know only the extent of the extreme dispersion under "normal" conditions. If either the largest or the smallest item is unusual, the range reveals nothing about the "normal" distribution of the items. Thus, stock market reports are frequently stated in terms of their ranges, by quoting high and low prices of stocks over a period of time. When no exceptional movements of stock prices occur, the quoted range may measure the "normal" variation. When, however, exceptional movements are present, the range reveals the effects of temporary disturbing conditions in the market. Consequently, a comparison of price ranges of stocks through time may show the effects of exceptional factors upon different stocks as well as the fluctuations of stock prices caused by ordinary trading conditions. Sometimes the range may have certain distinct advantages in measuring the dispersion of a symmetrical and continuous series; in such a case, the mean can be approximated by taking the average of the two extreme values. For instance, the difference between the heights of the shortest and the tallest may be used to measure the variation of heights of the male students of a certain college and the mean of these two values will give an average, often called the *mid-range*, which would generally agree with the figure obtained by averaging the heights of all the boys. As a matter of fact, it is the practice of meteorologists to derive the daily mean temperature by averaging only the maximum and minimum temperatures instead of using all the twenty-four readings of the day.

The range can also be used to advantage when the sample is small, especially when the same sampling operation is repeated often and an average of each successive result is utilized, as occurs in statistical quality control. In quality control, since the same sample size is used repeatedly, comparisons between ranges are not affected by differences in sample size. Furthermore, the range reveals almost as much about the variability as do the values of all the individual observations when the sample size is small, such as in the case of statistical quality control.

Finally, the range, being easy to compute and a common way of describing dispersion, is often used in engineering and medical reports.

4.4 The Interquartile Range and the Quartile Deviation

The range just discussed expresses the extreme variability of the observations. It is therefore subject to the chance of erratic extreme items; also, it fails to take account of the scatter within the range. From this there is reason to believe that if the dispersion of the extreme items is discarded, the limited range thus established might be more instructive. For this purpose there has been developed a measure called the *interquartile range*, the range which includes the middle 50 percent of the distribution. That is, one quarter of the observations at the lower end and another quarter of the observations at the upper end of the distribution are excluded in computing the interquartile range

To obtain the interquartile range, we begin by finding the first and third quartiles and then subtract the first quartile from the third. The process of locating the first and third quartiles parallels that of locating the median. As a matter of fact, the median is identical with the second quartile, which is found at the point indicated by $n/2$. Similarly, the first quartile, as the value in the series at or below which one fourth of the items lies, is at the point indicated by $n/4$. Also, the third quartile, as the value of the variable at or below which three fourths of the observations fall, is found at the point indicated by $(3n/4)$. Thus, to find the quartile values, as with the median, a certain degree of hairsplitting is required for precision. Formulas for locating the quartiles from grouped data can be written as

$$Q_1 = L + \left(\frac{n/4 - \Sigma f_1}{f_{Q_1}}\right) i \tag{4.1}$$

where

Q_1 = the first quartile
L = the lower class limit of the Q_1 class—the first class whose cumulative frequency exceeds $n/4$ items
Σf_1 = the sum of all frequencies in the classes preceding the Q_1 class
f_{Q_1} = the frequency in the Q_1 class
i = the class interval of the Q_1 class

$$Q_2 = \text{the median}$$

$$Q_3 = L + \left(\frac{3n/4 - \Sigma f_1}{f_{Q_3}}\right) i \tag{4.2}$$

where

Q_3 = the third quartile
L = the lower class limit of the Q_3 class—the first class whose cumulative frequency exceeds $(3n)/4$ items
Σf_1 = the sum of all frequencies in the classes preceding the Q_3 class
f_{Q_3} = the frequency in the Q_3 class
i = the class interval of the Q_3 class

To get quartile values from an array, we simply locate the rank of items by $(n + 1)/4$, $(n + 1)/2$, or $(3n + 1)/4$ and read off the value for Q_1, Q_2, or Q_3. If fractional values occur, interpolation is made between the two values corresponding to the two items between which a fraction falls. For example, if there are 20 items in the sample, the position of Q_1 is $(20 + 1)/4 = 5.25$ spaces. This indicates that the value of Q_1 is that of the fifth item, plus 25 percent of the difference between the values of the fifth and sixth items. Thus, if the value of the fifth item is \$10 and that of the sixth, \$12 then $Q_1 = \$10 + (\$12 - \$10)(0.25) = \$10.50$.

Now, let us find the interquartile range for the average weekly earnings of the 142 messenger boys in New York City by using formulas (4.1) and (4.2).

$$Q_1 = L + \left(\frac{n/4 - \Sigma f_1}{f_{Q_1}}\right) i$$

$$= \$67.00 + \left(\frac{142/4 - 26}{24}\right) \$0.50$$

$$= \$67.20$$

$$Q_3 = L + \left(\frac{\frac{3n}{4} - \Sigma f_1}{f_{Q_3}}\right) i$$

$$= \$68.41$$

And thus the interquartile range in the present case is $Q_3 - Q_1 = \$68.41 - \$67.20 = \$1.21$. This range is a measure of the variability in earnings of the messenger boys. It indicates that the central 50 percent of the average weekly earnings of the 142 messenger boys varies within the value of $1.21.

Very often the interquartile range is reduced to the form of the *semi-interquartile range* or *quartile deviation* by dividing it by 2. Thus, let QD be the semi-interquartile range; in symbols we have

$$QD = \frac{Q_3 - Q_1}{2} \tag{4.3}$$

QD is a very convenient expression because the dispersion can now be associated with and stated in terms of the median. As for the present case, the median is $67.76, the interquartile range $1.21, and the semi-interquartile range 60 cents. Were the distribution of the messenger boys' weekly earnings normal, we could then say that exactly 50 percent of the cases fall within the range of $\$67.76 \pm \$0.60$; that is, the median plus *and* minus the semi-interquartile range. The logic behind this conclusion is not difficult to follow. In a normal distribution, the first and third quartiles are equidistant from the median. It follows, therefore, that the interquartile range and the range of the median plus and minus the semi-interquartile range are the same distance. Moreover, since exactly 50 percent of the observations fall within the interquartile range, the range of the median $\pm QD$ would also include exactly 50 percent of the items.

In reality, however, one seldom finds a series in business and economic data that is perfectly symmetrical. Nearly all distributions of social series are asymmetrical. In an asymmetrical distribution, Q_1 and Q_3 are no longer equidistant from the median; if it is positively skewed, Q_1 is closer to the median than Q_3; if it is negatively skewed, the opposite is true. Thus, in a moderately skewed distribution, the median $\pm QD$ yields values close to, instead of exactly at, the values of Q_3 and Q_1. As a result, the median $\pm QD$ for a skewed series includes only approximately 50 percent of the observations.

Another statistical measure, $K \pm QD$, will include exactly 50 percent of the items of an asymmetrical distribution. K is obtained by the following expression:

$$K = Q_1 + QD \tag{4.4}$$

Applying this measure to the data of average weekly earnings of the messenger boys, we secured this result:

$$K = \$67.20 + \$0.60 = \$67.80$$

It may be noted that if the distribution is perfectly symmetrical, the value of K is the same as those of the median, the mode, and the mean. The difference between the values of the K and the median in this case, therefore, indicates that the distribution of the average weekly earnings of the messenger boys in our example is skewed.

As pointed out earlier, the interquartile range or the semi-interquartile range is a measure that excludes the highest 25 percent and the lowest 25 percent, giving a range within which the central 50 percent of the observation falls, and unlike the crude range, is not affected by the extreme values. The proper interpretation of the semi-interquartile range is that if it is very small it then describes high uniformity or small variation of the central items. Because of this, QD can be used to compare the degrees of variation in different distributions.

This measure, however, has an outstanding shortcoming: it does not make use of the values of all the observations; consequently QD is not affected by the different distribution patterns of those items below the first and above the third quartiles. In addition, QD is only a distance on a scale and is not measured from an average. If we really desire to measure variation in the sense of showing the scatter around an average, we must include the deviation of each and every item from an average in the measurement. Furthermore, the quartile deviation, being a positional measure, is not amenable to algebraic manipulation; therefore, it is impossible to find the quartile deviation for combined groups. Another drawback of this measure is that it is not often useful for statistical inference.

4.5 The Average Deviation

The *average deviation* or the *mean deviation*, as it is sometimes called, is simply the arithmetic mean of the deviations from an average value, either mean or median. Theoretically, because the sum of deviations of the items from the median is a minimum when signs are ignored, it is desirable to use the median as the point of reference when the average deviation is employed as a measure of variation. In practice, however, the mean is more frequently used. In any case, the average used must be clearly stated in a given problem so that no confusion in meaning may arise.

Computing the average deviation from a list of values involves the following steps:

(1) Compute the deviation d of each observation x from the arithmetic mean $\bar{x}$: symbolically, $d = x - \bar{x}$.

(2) Obtain the sum of the deviations from the mean and ignore the signs of the deviations, $\Sigma|d|$. (The symbol $|d|$ indicates that signs are to be ignored.)

(3) Divide the sum of the deviations from the mean by the number of observations in the series. The result so obtained is the average deviation.

Letting AD stand for the average deviation, the general formula becomes

$$AD = \frac{\Sigma|x - \bar{x}|}{n} = \frac{\Sigma|d|}{n} \tag{4.5}$$

For a large sample with a mean having decimal fractions, the task of computing the average deviation by this formula becomes time-consuming and very tedious. Under such conditions it is more convenient to obtain the average deviation from a frequency distribution. Deviations may be measured from the mean and multiplied by class frequencies. This precedure is summarized by the following expression, in which the symbols have all been previously defined:

$$AD = \frac{\Sigma f|m - \bar{x}|}{n} = \frac{\Sigma f|d|}{n} \tag{4.6}$$

The value of AD is always interpreted in connection with an average. If the distribution is symmetrical, the average (mean or median) $\pm AD$ is the range that will include 57.5 percent of the items in the series. If it is moderately skewed, then we may expect approximately 57.5 percent of the items to fall within this range. Hence, if AD is relatively small the distribution is highly compact or uniform, since more than half of the cases are concentrated within a small range around the mean.

The average deviation is useful in dealing with small samples with no elaborate analysis required. It is rarely used for large samples and grouped data. The limited employment of the average deviation is due to its mathematical peculiarity. The student may have already noted that if the average deviation is to be calculated the plus and minus signs of the deviations must be ignored. If the signs of the deviations are not ignored, the net sum of the deviations will be zero if the reference point is the mean, or approximately zero if the reference point is the median. This procedure, however, is algebraically illogical. Perhaps the most important reason for studying the average deviation is that it will help the student to understand another superior measure of dispersion called the *standard deviation.*

However, the serious drawbacks of the average deviation should not blind us to its practical utility. Because of its simplicity in meaning and computation, it is especially effective in reports presented to the general public or to groups not familiar with statistical methods. It may also be mentioned in passing that the National Bureau of Economic Research has found that the average deviation is the most practical measure of variation to use in forecasting business cycles.

4.6 Nature and Uses of the Standard Deviation

The standard deviation is by far the most important statistic among all the measures of variation. It may be considered as a special form of the average deviation from the mean. The standard deviation differs from the average deviation in that it is computed by using the squared deviations from the mean whereas the average deviation is derived by employing the absolute deviations. Obtaining the standard deviation involves five simple steps:

(1) Take the deviation of each observation from the mean, $(x - \bar{x})$, denoted as d.

(2) Square the deviations, $(x - \bar{x})^2$ or d^2.

(3) Sum up the deviations. This sum may be considered as the *variation* and is symbolized as $\Sigma(x - \bar{x})^2$, or Σd^2.

(4) Obtain the mean of the squared deviations, $\Sigma d^2/n$. This value is called the *sample variance* and is denoted by s^2.

(5) Extract the square root of the variance, $\sqrt{\Sigma d^2/n}$. The result thus derived is the *sample standard deviation* and is represented by the symbol s.

Thus the standard deviation is the square root of the variance which, in turn, is the mean of the squared deviations from the mean. These two measures may be written fully as follows:

$$s^2 = \frac{\Sigma(x - \bar{x})^2}{n} \tag{4.7}$$

$$s = \sqrt{\frac{\Sigma(x - \bar{x})^2}{n}} \tag{4.8}$$

To use these formulas, it is necessary to carry $\bar{x}$ to a sufficient number of decimal places in order to obtain greater accuracy. To illustrate the application of these formulas, we shall use two hypothetical samples, both of which contain the results of the measurements of the resistance of a certain electrical product produced by two different companies, A and B.

Product of Company A			Product of Company B		
Resistance (ohms)	$(x - \bar{x})$	$(x - \bar{x})^2$	*Resistance (ohms)*	$(x - \bar{x})$	$(x - \bar{x})^2$
3.5	−0.53	0.2809	2.7	−1.33	1.7689
3.6	−0.43	0.1894	2.8	−1.23	1.5129
3.7	−0.33	0.1098	2.9	−1.13	1.2769
3.8	−0.23	0.0529	3.4	−0.63	0.3969
3.9	−0.13	0.0169	4.0	−0.03	0.0009
4.0	−0.03	0.0009	4.5	+0.47	0.2209
4.2	+0.17	0.0289	4.7	+0.67	0.4489
4.3	+0.27	0.0729	4.8	+0.77	0.5929
4.4	+0.37	0.1369	5.2	+1.17	1.3689
4.9	+0.87	0.7569	5.3	+1.27	1.6129
40.3 ohms		1.6455	40.3 ohms		9.2010

Company A:

$$\begin{aligned} \bar{x} &= 4.03 \text{ ohms} \\ s^2 &= \Sigma(x - \bar{x})^2/n \\ &= 1.6455/10 \\ &= 0.16455 \end{aligned}$$

$$\begin{aligned} s &= \sqrt{\Sigma(x - \bar{x})^2/n} \\ &= \sqrt{0.16455} \\ &= 0.405 \text{ ohm} \end{aligned}$$

Company B:

$$\begin{aligned} \bar{x} &= 4.03 \text{ ohms} \\ s^2 &= \Sigma(x - \bar{x})^2/n \\ &= 9.2010/10 \\ &= 0.92010 \end{aligned}$$

$$\begin{aligned} s &= \sqrt{\Sigma(x - \bar{x})^2/n} \\ &= \sqrt{0.92010} \\ &= 0.959 \text{ ohm} \end{aligned}$$

Formulas (4.7) and (4.8) are introduced first because they give the definitions of the variance and standard deviation. Their application, however, is unduly time-consuming, since the mean must be subtracted from each observation. It becomes especially laborious when $\bar{x}$ contains a large number of digits or decimal places. As indicated before, if $\bar{x}$ is rounded in order to save time and labor, the variance and standard deviation will sacrifice accuracy. For these reasons, we wish to develop more efficient methods of computation. This is done by taking advantage of the relationship that

$$\begin{aligned} \Sigma(x - \bar{x})^2 &= \Sigma(x^2 - 2\bar{x}x + \bar{x}^2) \\ &= \Sigma x^2 - 2\bar{x}\Sigma x + \Sigma\bar{x}^2 \\ &= \Sigma x^2 - 2\bar{x}\Sigma x + \bar{x}\Sigma x \\ &= \Sigma x^2 - \bar{x}\Sigma x \\ &= \Sigma x^2 - (\Sigma x)^2/n \end{aligned}$$

From this relationship, we can then convert formulas (4.7) and (4.8) into the following forms:

$$s^2 = \frac{\Sigma(x - \bar{x})^2}{n} = \frac{\Sigma x^2 - (\Sigma x)^2/n}{n} = \frac{\Sigma x^2}{n} - \left(\frac{\Sigma x}{n}\right)^2 \qquad \textbf{(4.9)}$$

$$s = \sqrt{\frac{\Sigma x^2}{n} - \left(\frac{\Sigma x}{n}\right)^2} \qquad \textbf{(4.10)}$$

It is interesting to note that when formula (4.9) or (4.10) is employed, deviations are taken from an assumed mean of zero. Thus, the algebraic sum of the deviations is, in effect, identical with the sum of the original values, Σx. But the sum of square deviations from any value other than the actual mean must be greater than that from the actual mean. As a result, the variance computed from an assumed mean of zero must be greater than the variance from the actual mean: $\Sigma x^2/n$ is greater than $\Sigma(x - \bar{x})^2/n$. Therefore to compensate for this error, from the variance computed from the assumed mean of zero we must subtract a correction factor, which is equal to the square of the mean, $(\Sigma x/n)^2$. When this is done, the variance and the standard deviation computed by adopting formulas (4.9) and (4.10) should be identical with those derived from formulas (4.7) and (4.8). This is illustrated by the calculations shown below.

Product of Company A	
Resistance (ohms)	
x	x^2
3.5	12.25
3.6	12.96
3.7	13.69
3.8	14.44
3.9	15.21
4.0	16.00
4.2	17.64
4.3	18.49
4.4	19.36
4.9	24.01
40.3	164.05

$$\begin{aligned} s^2 &= \Sigma x^2/n - (\Sigma x/n)^2 \\ &= 164.05/10 - (40.3/10)^2 \\ &= 0.164 \end{aligned}$$

$$\begin{aligned} s &= \sqrt{0.164} \\ &= 0.405 \text{ ohm} \end{aligned}$$

The standard deviation, though the most important and the most widely used measure, has the least obvious interpretation among all the measures of dispersion. Both its importance and its utility are derived from the fact that not only is it the most reliable summary descriptive measure of dispersion; it also plays a vital role in statistical inference.

As a descriptive measure, the standard deviation is essentially a mean. From the process of its computation, it can be seen that it is the root-mean-square of the deviations about the arithmetic mean. More precisely, as noted in the previous chapter, the quadratic mean is the root-mean-square average of a set of observations; hence we may say that the standard deviation is the quadratic mean of the deviations. The standard deviation is a kind of mean of algebraic deviations that give positive values. These positive values increase when the variability increases and are in the same units as the original observations. This observation furnishes us with a proper interpretation of the standard deviation. Returning to the previous illustrations of the electrical products produced by Company A and Company B, we note that both products have an identical mean resistance of 4.03 ohms; but the standard deviation for B's product is 0.959 ohm, which is much greater than A's 0.405 ohm. From these results, we can conclude that A's product is superior to B's: A's product is much more uniform in quality and more dependable in resistance than B's. To argue in the reverse, we may point out that although B's product has the same mean resistance as A's, the former shows much greater variability with some items of very high and others of very low resistance.

The standard deviation, as the student may have already sensed, is also useful in judging the representativeness of the mean. A small s means a high degree of uniformity of the observations as well as homogeneity of a series; a large s means just the opposite. Thus, if we have two or more comparable series with identical or nearly identical means, it is the distribution with the smallest standard deviation that has the most representative mean.

The analytical function of the standard deviation in statistical inference rests on its relation to the normal distribution. A normal distribution is completely defined by its mean, μ, and its standard deviation, σ. (Note that μ and σ are referred to here because, when we speak of the normal distribution, we are thinking in terms of a specific type of population model. A sample, unless it is as large as the population itself, can never assume the exact shape of the normal distribution.) That is, the mean plus and minus a given number of standard deviation will include a precise proportion of the number of observations in the normal distribution. (See Figure 4.2.) To make a decision on the basis of a sample, we must allow for sampling variability. This is so because a sample statistic can never be expected to be the same as the corresponding population parameter. Moreover, if we were to collect a large number of samples of the same kind from the same population, we would expect the same statistic computed to vary from sample to sample. Frequently, however, the variability of a sample statistic can be approximated by a normal distribution which, as has already been noted, is completely defined by its mean and standard deviation. It is because of this that the standard deviation plays a vital role in statistical inference. This point will become clear to the student in the following chapters.

As for the distribution of a sample, as long as it resembles a normal distribution—only moderately skewed and not of the leptokurtic or platykurtic type—the standard deviation is a good estimate of the percentages of observations included by given distances from the mean. Hence the following relation-

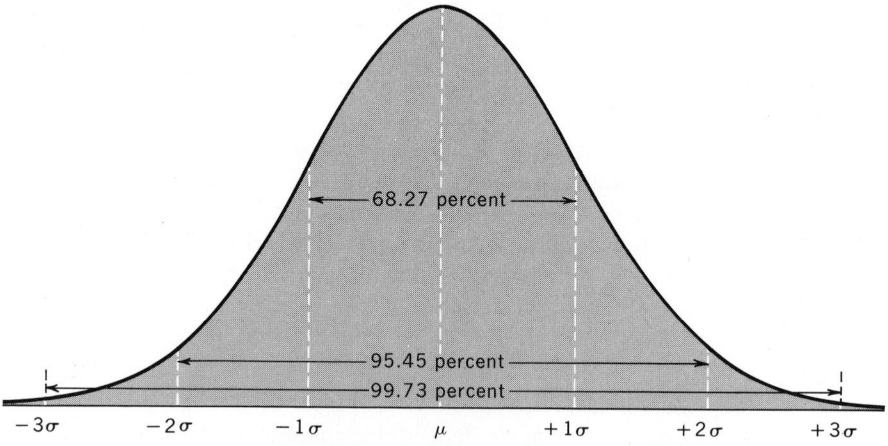

Figure 4.2 Percentage of items included under normal curve for selected values of σ.

ships may be expected to hold approximately:

$\bar{x} \pm s$ includes about 68 percent of the items
$\bar{x} \pm 2s$ includes about 95 percent of the items
$\bar{x} \pm 3s$ includes about 99 percent of the items

The property of the standard deviation that it is an accurate gauge of the proportions of items falling within given distances from the mean leads us to two more important applications of the standard deviation. The first is to use it as a basis for prediction. For instance, suppose a laboratory test finds that the average burning life of a sample of 100 television picture tubes is 500 hours and the standard deviation of burning lives is 50 hours. Assuming that the burning lives of all television picture tubes which could have been produced by the same process are normally distributed and that its mean and standard deviation are about the same as found in the sample, we may then predict that practically all burning lives would be in the range of 350–650 hours; 95 percent would be in the range of 400–600 hours; 68 percent would be in the range of 450–550 hours. It is interesting to note that in this example we have used the sample mean and sample standard deviation as the estimates of the population mean and population standard deviation, respectively. The reliability of the prediction depends upon the accuracy of these estimates, which, in turn, and among other things, depends upon the size of the sample. A sample of 100 in this case should be considered as sufficient. We shall have opportunity in the subsequent chapters to discuss the problems of estimation, the determination of sample size, and other matters of statistical inference.

The other use of the standard deviation is to consider it as a standard unit of measurement, measuring the deviation of a given observation from the mean. That is, we consider an x value as being a given number of standard deviations, denoted by z, from (above or below) the mean. To obtain the z value, which is often called the *standard measure*, or the *standardized value*, is to convert the difference between the mean and an individual observation into units of the standard deviation. Thus, for the population we have

$$z = \frac{x - \mu}{\sigma} \tag{4.11}$$

and for the sample,

$$z = \frac{x - \bar{x}}{s} \tag{4.12}$$

The standardized value will be used frequently in later chapters to determine the probability of an item of a specified value in a normal population. We wish to point out here that the z value can be employed to compare two individual items, even when they belong to different distributions. To illustrate, suppose a student scored 90 in an intelligence test when he entered a college, the mean for all entering students being 100 and the standard deviation 25. Then

$$z = \frac{90 - 100}{25} = -0.4$$

This indicates that the student in question is not very bright because his intelligence is 0.4 standard deviation below the average. Now, suppose that, by the end of the freshman year, this same student achieved an average in his examinations of 79 points while for all the students $\mu = 70$ and $\sigma = 10$. Then

$$z = \frac{79 - 70}{10} = +0.9$$

This reveals that his actual achievement is 0.9 standard deviation above the average. Evidently, the individual being considered is a better student than the intelligence test indicated.

There are still other important uses of the standard deviation: fitting the normal curve to a frequency distribution and analyzing the coefficient of correlation discussed in later chapters. In any case, the student must have realized by now the value of the standard deviation. It is therefore vital for him to have a thorough understanding of this concept, and he should review this section before he proceeds.

4.7 Computing the Standard Deviation from Grouped Data

Very often the standard deviation is computed at the same time as the mean, and a similar short-cut method, applied for computing the mean, can be used for the standard deviation. The basic formula for computing the standard deviation from grouped data is written as

$$s = \sqrt{\frac{\Sigma fd^2}{n}} \tag{4.13}$$

where $d = (m - \bar{x})$. Equation (4.13) is analogous to equation (4.8). A more efficient method uses a formula analogous to equation (4.10), as

$$s = \sqrt{\frac{\Sigma fm^2}{n} - \left(\frac{\Sigma fm}{n}\right)^2} \tag{4.14}$$

Better still, we can compute the standard deviation by the short method. Here, as in the case of the arithmetic mean, we work with deviations from an assumed mean, A. These deviations may be stated in original units or in class-interval units. The formula for the short method in original units is

$$s = \sqrt{\frac{\Sigma fd^2}{n} - \left(\frac{\Sigma fd}{n}\right)^2} \tag{4.15}$$

where $d = (m - A)$.

The formula for the short method in class-interval units is identical with equation (4.15) but for two exceptions. First, d' instead of d is used, where $d' = (m - A)/i = d/i$. Second, because of the use of d', the result must be multiplied by i, the class interval, in order to convert it back into original units. Thus we have

$$s = \left(\sqrt{\frac{\Sigma f(d')^2}{n} - \left(\frac{\Sigma fd'}{n}\right)^2}\right) i \qquad (4.16)$$

Formulas (4.13), (4.14), (4.15), and (4.16) can be expected to yield the same results when they are applied to the same distribution with the same grouping. In general, however, the standard deviation tends to be slightly greater when calculated from grouped data than when calculated from ungrouped data. Also, as in the case of the mean, the narrower the class interval, the closer would the standard deviation computed from grouped data be in agreement with that derived from ungrouped data. For analytical purposes, the class interval should be less than one fourth of the standard deviation. For descriptive purposes, the error introduced by intervals as large as half of the standard deviation is usually acceptable.

As to the selection of formulas in practice, we may note that equation (4.13) is a definitional formula and its use is too laborious. Formulas (4.14) and (4.15) can be used for both equal and unequal class intervals, but the latter is usually preferred because it is less time-consuming than the former. In dealing with equal class intervals, however, equation (4.16) is the most efficient formula. Tables 4.1 and 4.2 are illustrations of the application of Formulas (4.14) and (4.16), respectively.

TABLE 4.1 COMPUTATION OF STANDARD DEVIATION OF AVERAGE WEEKLY EARNINGS OF 142 MESSENGER BOYS IN NEW YORK CITY, 1968, BY THE EFFICIENT METHOD

Average Weekly Earnings	Mid-points m	f	fm	fm^2
\$66.00–\$66.49	\$66.25	11	728.75	48,279.69
66.50– 66.99	66.75	15	1001.25	66,833.44
67.00– 67.49	67.25	24	1614.00	108,541.50
67.50– 67.99	67.75	40	2710.00	183,602.50
68.00– 68.49	68.25	20	1365.00	43,161.25
68.50– 68.99	68.75	14	962.50	66,171.88
69.00– 69.49	69.25	11	761.75	52,751.19
69.50– 69.99	69.75	7	488.26	34,055.44
Total		142	9631.50	653,396.89
		n	Σfm	Σfm^2

Source: Table 2.5.

$$s = \sqrt{\frac{\Sigma fm^2}{n} - \left(\frac{\Sigma fm}{n}\right)^2}$$
$$= \sqrt{653{,}396.89/142 - (9631.50/142)^2}$$
$$= \sqrt{4601.3865 - 4600.5698}$$
$$= \sqrt{0.8167}$$
$$= \$0.90$$

TABLE 4.2 COMPUTATION OF STANDARD DEVIATION OF AVERAGE WEEKLY EARNINGS OF 142 MESSENGER BOYS IN NEW YORK CITY, 1968, BY SHORT METHOD IN CLASS-INTERVAL UNITS

Average Weekly Earnings	f	d'	fd'	$f(d')^2$
\$66.00–\$66.49	11	−3	−33	99
66.50– 66.99	15	−2	−30	60
67.00– 67.49	24	−1	−24	24
67.50– 67.99	40	0	0	0
68.00– 68.49	20	+1	+20	20
68.50– 68.99	14	+2	+28	56
69.00– 69.49	11	+3	+33	99
69.50– 69.99	7	+4	+28	112
Total	142		+22	470

$$s = \left(\sqrt{\Sigma f(d')^2/n - (\Sigma fd'/n)^2}\right)i = \sqrt{470/142 - (22/142)^2}\,(0.50)$$
$$= \sqrt{3.30986 - 0.02400}\,(0.50) = (1.81)(0.50) = \$0.90$$

Here for the first time in this book, the detailed steps required to compute a major statistic are not given. It is time for the student to learn how to set up the computation table for applying a formula to a given set of data. This is actually not a difficult skill in applied statistics, but it is an important one. Generally speaking, the computation table must contain columns that will yield the necessary quantities to be fed into the formula. The relationships among the variables in the formula, in turn, suggest the columns required. Each column, except those that contain the original data, may be considered as a distinct step in computing a given measure. These rather general statements should become clear to the student when he studies Tables 4.1 and 4.2 carefully, together with formulas (4.14) and (4.16). He will find it even more instructive to write out the steps required to compute the standard deviation by all the four formulas introduced in this section.

4.8 Characteristics of Measures of Absolute Variation

To recapitulate, let us restate the chief characteristic of the various measures of variation discussed so far in order to gain a better understanding of their significance.

THE RANGE

(1) The range is easy to compute and simple in meaning. (2) It is based on two extreme values and is therefore a highly unstable measure. Its value may change greatly if a single figure is added or withdrawn. (3) This measure is a

distance along the scale, which includes 100 percent of the cases but gives no indication of the pattern of distribution between the two extreme values. (4) It is useful as a rough measure of the degree of variation and a good measure for presenting variations to the general public.

THE QUARTILE DEVIATION OR SEMIQUARTILE RANGE

(1) The quartile deviation is also easily calculated and readily understood. (2) It is a distance along the scale, which includes the middle 50 percent of the items within its range. Thus, it is not a measure of dispersion from any specific average. (3) The measure is determined by the number of cases rather than by their values and therefore is not suitable for further algebraic manipulation. (4) The semiquartile range is affected neither by the distribution of the items between Q_1 and Q_3 nor by that outside these quartiles. Thus it is not an accurate measure of dispersion. (5) As a rough measure of dispersion, however, the quartile deviation is superior to the crude range and is especially useful in the case of open-end distributions.

THE AVERAGE DEVIATION

(1) The AD is determined by the value of every item in the series. (2) But it is less affected by the extreme deviations than is the standard deviation. (3) It may be computed from either the arithmetic mean or the median, though the average deviation from the latter is a minimum. (4) In a normal distribution, 57.5 percent of the items in the series fall within the range of the mean or median $\pm AD$. If the distribution is moderately skewed, this range will include approximately 57.5 percent of the cases. (5) Mathematically, the AD is neither as logical nor as convenient as the standard deviation. Consequently, its use is limited and it is overshadowed as a measure of variation by the superior standard deviation. Meanwhile, it should be recognized that easy computation and obvious interpretation have made the average deviation an efficient practical measure under certain circumstances.

THE STANDARD DEVIATION

(1) The standard deviation is affected by the value of every item in the series. (2) It places greater stress upon extremes than does the average deviation because all values are squared in computing the standard deviation. (3) Its definite mathematical property makes it perfectly adaptable to further algebraic computation. (4) It is always computed about the mean, for the sum of the squared deviations from this point is a minimum. (5) The standard deviation is usually used to define areas under the normal curve. Under a normal curve, 68.27 percent of the area (or cases) will be included within the range of the mean, plus and minus one standard deviation; 95.45 percent of the area will be included if two standard deviations are measured off from both sides of the mean; and 99.73 percent of the area will be included if three standard

deviations are measured off. In a moderately skewed distribution these percentages are approximations. (6) Because of its inherent properties, the standard deviation has great practical usefulness in sampling, correlation, statistical inference, and other advanced statistical measurements. (7) One drawback of the standard deviation is its lack of obvious interpretation. To say that it is the root-mean-square of the deviations about the arithmetic mean or the quadratic mean of the deviations will mean nothing to the statistically unsophisticated public.

4.9 Relative Variations

It is quite valid to compare the standard deviations if we wish to compare the dispersions of two or more series that have the same or nearly the same mean and that are expressed in the same unit. There are situations, however, where different distributions may have different means or are stated in different units. Thus, it becomes impossible to compare the standard deviations. For instance, we may find the standard deviation for the annual incomes of physicians to be \$1500, while for college professors the standard deviation is \$1000. Does this prove that the variation in incomes of the physicians is 150 percent as large as the variation in the incomes of the professors? The answer, of course, depends upon the mean incomes of the two groups of people. If the mean income of the physicians is \$20,000 and that of the professors is \$12,000, the *relative* degree of dispersion of the professors' incomes is actually greater than that of the physicians'. Clearly, the difference between two absolute values of variation around different means furnishes no comparable basis for comparison in itself.

Again, suppose we wish to compare the degrees of uniformity in the productivity and mental ability of the same group of workers. For this problem, as to productivity, we may find the mean to be 95 percent of a certain standard in production and the standard deviation to be 25 percent; as to the mental test, the mean number of points scored may be 45 while the standard deviation is 10 points. Here a direct comparison between the two standard deviations will not make any sense. It is impossible to say whether 25 percent is a greater or smaller spread than 10 points.

Thus two or more series with averages of different values or different units of measurement cannot be compared by using their respective absolute measures of dispersion. Under these circumstances, to achieve comparability the measures of absolute variation must be changed into relative forms. The procedure is to express a given measure of absolute variation as a percentage of the average around which the deviations are taken. The result becomes a relative number; therefore the name *measure of relative variation.* Such a measure of relative variability of a given distribution can then be compared readily with similar measures obtained from other series irrespective of the sizes of averages and the nature of the original units of the distributions.

The most commonly used measure of relative variation is the one developed by Karl Pearson called the *coefficient of variation*. It is a ratio of the standard deviation to the mean and is represented by the symbol V. Thus,

$$V = \frac{s}{\bar{x}} \tag{4.17}$$

for the physicians, $V = 1500/20{,}000 = 0.075$, or 7.5%
for the professors, $V = 1000/12{,}000 = 0.0831$, or 8.3%

These results reveal that the incomes received by the physicians are more uniform than those of the professors in our illustration.

Again,

for productivity, $V = {}^{25}\!/_{95} = 0.263$, or 26%
for the mental test, $V = {}^{10}\!/_{45} = 0.222$, or 22%

The greater variability in productivity indicates that productivity also depends on such factors as diligence, interest, strength, concentration, and so forth, in addition to the mental ability of the workers.

Another commonly used relative measure of dispersion is the *coefficient of average deviation*. This measure is represented by the letters V_{ad}. It is expressed in terms of a ratio of the average deviation to the mean:

$$V_{ad} = \frac{AD}{\bar{x}} \tag{4.18}$$

A third measure is the *coefficient of quartile deviation*. It is especially appropriate when the distribution has open ends or when there are extreme values or when it is desired to make only a quick estimate of dispersion and the measure of dispersion is not needed for further computation or other purposes. The coefficient of quartile deviation is symbolized by V_q and is calculated by the formula

$$V_q = \frac{Q_3 - Q_1}{Q_3 + Q_1} \tag{4.19}$$

As in the case of V, both V_{ad} and V_q can be expressed in either decimal or percentage form.

To summarize the comparison of the variability of two or more distributions, absolute measures of variation are comparable only when the distributions have the same or nearly the same mean and when they are expressed in terms of the same units; otherwise, measures of relative variation should be used. For the purpose of comparison, furthermore, the same measure must be used for all the distributions. In other words, the students should be careful not to compare V of one distribution with V_{ad} or V_q of another distribution. Nor is it valid to compare the standard deviation of one with the AD or QD of another series.

4.10

Measure of Skewness

To summarize and describe a frequency distribution, besides measures of central tendency and measures of concentration or dispersion about the central tendency, another measure—one which indicates the degree of skewness—is needed. Skewness, as was mentioned at the beginning of this chapter, is lack of symmetry in a frequency distribution. Two distributions may have the same mean and the same standard deviation; yet they may differ in the degree of skewness or asymmetry. The concept of skewness gains importance from the fact that statistical theory is often based upon the assumption of the normal distribution. A measure of skewness is therefore necessary in order to guard us against the consequence of this assumption.

Several measures of skewness are available, but we shall introduce only the two currently in use.

The first is the *Pearsonian measure of skewness.* This measure is based on the nature of and relationship among the mean, the mode, and the median. The mean is influenced by the size of the extreme values, the median is influenced only by their positions, and the mode is not influenced by them at all. These three averages are identical in a symmetrical distribution, but the mean moves away from the mode when the observations are asymmetrical. Consequently, the distance between the mean and the mode could be used to measure skewness; that is, skewness = mean − mode. The greater this distance, whether positive or negative, the more asymmetrical is the distribution. Such a measure, however, has two outstanding defects in application. First of all, being an absolute measure, the result is expressed in terms of the original unit of the distribution. Second, the same absolute amount of skewness has different significance for different series with different degrees of variation. Thus, this absolute measure of skewness must be expressed in relative terms in order to provide a valid basis for comparison. The obvious procedure to accomplish this is to eliminate the influence of the dispersion, and this is achieved by dividing the absolute skewness by the standard deviation. This result is called the *Pearsonian coefficient of skewness* and may be symbolized by Sk_p. The formula is

$$Sk_p = \frac{\bar{x} - \text{Mode}}{s} \tag{4.20}$$

The use of equation (4.20), however, involves another difficulty which grows out of the fact that the modal value of most distributions obtained by elementary methods is only an approximation, but the location of the median can be more satisfactorily found. As has been shown, in moderately skewed distributions, the mode $= \bar{x} - 3(\bar{x} - \text{Med})$. Thus, we can remove the mode from the foregoing formula for skewness by substituting the median in it because of this relationship. Therefore, the new expression becomes

$$Sk_p = \frac{\bar{x} - [\bar{x} - 3(\bar{x} - \text{Med})]}{s}$$

$$= \frac{3(\bar{x} - \text{Med})}{s} \tag{4.21}$$

Applying this formula to the average weekly earnings of the 142 messenger boys for which $\bar{x}$ = \$67.83, Med = \$67.76, and s = \$0.90, we have

$$Sk_p = \frac{3(67.83 - 67.76)}{0.90} = \frac{3(0.07)}{0.90} = +0.233$$

According to the Pearsonian measure, skewness will be zero for a symmetrical distribution because in such a case $\bar{x}$ = Mode = Med. If the distribution is skewed to the right, the mean is greater than the mode as well as the median: the skewness is positive. For a distribution skewed to the left the result of this measure will be negative, for the mean now is smaller than the mode and median. Furthermore, theoretically this measure varies within the limits of ± 3; however, in reality, it is only upon rare occasions that the values exceed the limits of ± 1. Thus, the distribution of the messenger boys' earnings just computed may be considered as mildly skewed.

Another measure of skewness is *Bowley's measure of skewness*, computed from the quartile values. It has been seen that if the distribution is symmetrical, the first and third quartiles are equidistant from the median. In other words, in a symmetrical distribution, $(Q_3 - \text{Med}) = (\text{Med} - Q_1)$. In asymmetrical distributions, the first difference becomes larger relative to the second in the case of skewness to the right, and vice versa in the case of skewness to the left. Clearly the difference between the values of $(Q_3 - \text{Med})$ and $(\text{Med} - Q_1)$ is a possible basis for measuring skewness. From this reasoning, Bowley has suggested the following relative measure of skewness:

$$Sk_q = \frac{(Q_3 - \text{Med}) - (\text{Med} - Q_1)}{Q_3 - Q_1}$$

$$= \frac{Q_3 + Q_1 - 2\text{ Med}}{Q_3 - Q_1} \tag{4.22}$$

As in the Pearsonian measure of skewness, this measure will be zero for a symmetrical distribution, positive for right skewness, and negative for left skewness. Furthermore, according to this measure, skewness falls within the limits of ± 1.

In conclusion, a few interesting points concerning skewness may be mentioned. The J-shaped and reverse J curves are illustrations of extreme left and right skewness, respectively. It is quite common to encounter positively skewed distributions in economic and business data, particularly in production and price series which can be only as small as zero but can be infinitely large. It is believed that positive skewness is produced by multiplicative forces. For instance, income distribution is usually positively skewed because it is affected by a number of factors, such as education, race, sex, "lucky breaks." Negatively skewed distributions are quite rare, and it is often difficult to furnish a rational explanation for their existence.

Glossary of Formulas

(**4.1**) $Q_1 = L + \left(\frac{n/4 - \Sigma f_1}{f_{Q_1}}\right) i$ Q_1 is the first quartile. It is the value at or below which one fourth of the items in the series lies. It is at the point indicated by $n/4$.

(**4.2**) $Q_3 = L + \left(\frac{3n/4 - \Sigma f_1}{f_{Q_3}}\right) i$ Q_3 is the third quartile. It is the value at or below which three fourths of the distribution lies. It is at the point indicated by $3n/4$.

(**4.3**) $QD = \frac{Q_3 - Q_1}{2}$ The quartile deviation, or semi-interquartile range, is half the distance between the first and the third quartiles. For a normal distribution, median $\pm QD$ includes 50 percent of the items in the series.

(**4.4**) $K = Q_1 + QD$ K is the point half way between the first and the third quartiles. $K \pm QD$ will include 50 percent of the observations in an asymmetrical distribution.

(**4.5**) $AD = \frac{\Sigma|d|}{n}$ The average deviation, or the mean deviation, is the arithmetic average of the absolute deviations of the observations about the mean of the same observations. The two vertical lines on either side of d are used in the formula; that is, only the absolute size of d is of interest and the sign is ignored. AD can also be computed from the median, in which case it would be defined as the arithmetic average of the absolute deviations about the median of the observations.

(**4.6**) $AD = \frac{\Sigma f|d|}{n}$ This equation is used to compute AD from grouped data.

(**4.7**) $s^2 = \frac{\Sigma(x - \bar{x})^2}{n}$ The variance of series of observations is defined as the arithmetic average of the squares of the deviations of the observations from the arithmetic mean of the observations.

(**4.8**) $s = \sqrt{\frac{\Sigma(x - \bar{x})^2}{n}}$ The standard deviation of series of observations is the square root of the variance of those observations. It is the square root of the mean square of the deviations: the "root-mean-square" deviation. Thus it is the quadratic mean of the deviations.

(**4.9**) $s^2 = \frac{\Sigma x^2}{n} - \left(\frac{\Sigma x}{n}\right)^2$ This is a more efficient formula for computing the sample variance from ungrouped data.

(**4.10**) $s = \sqrt{\frac{\Sigma x^2}{n} - \left(\frac{\Sigma x}{n}\right)^2}$ This is a more efficient formula for computing the standard deviation from ungrouped data.

(**4.11**) $z = \dfrac{x - \mu}{\sigma}$ z is the x value expressed as the deviation from the mean of the distribution measured in standard deviations of the distribution.

(**4.12**) $z = \dfrac{x - \bar{x}}{s}$ This equation is identical with (4.11) except for one difference: equation (4.11) refers to the population, whereas (4.12) refers to the sample.

(**4.13**) $s = \sqrt{\dfrac{\Sigma fd^2}{n}}$ This is the definitional equation for computing the standard deviation from grouped data. It is analogous to equation (4.8), where $d = (m - \bar{x})$.

(**4.14**) $s = \sqrt{\dfrac{\Sigma fm^2}{n} - \left(\dfrac{\Sigma fm}{n}\right)^2}$ This is a more efficient method of computing the standard deviation by the long method. It is analogous to equation (4.10).

(**4.15**) $s = \sqrt{\dfrac{\Sigma fd^2}{n} - \left(\dfrac{\Sigma fd}{n}\right)^2}$ This is the short method in original units of computing the standard deviation from grouped data. In the equation $d = m - A$, where A is the assumed mean. This equation can be applied to a distribution grouped with equal or unequal class-intervals.

(**4.16**) $s = \left(\sqrt{\dfrac{\Sigma f(d')^2}{n} - \left(\dfrac{\Sigma fd'}{n}\right)^2}\right) i$ This is the short method in class-interval units of computing the standard deviation from grouped data. In the equation, $d' = (m - A)/i$. This is the most efficient method of computing s, but it should be used only when the distribution is grouped with equal class intervals.

(**4.17**) $V = \dfrac{s}{\bar{x}}$ The coefficient of variation is defined as the ratio of the standard deviation to the mean of the same series of observations. It may be used to compare two or more different series with different means and/or of different units.

(**4.18**) $V_{ad} = \dfrac{AD}{\bar{x}}$ The coefficient of average deviation is the ratio of the average deviation to the mean. As in the case of AD, this measure can also be computed by referring to the median.

(**4.19**) $V_q = \dfrac{Q_3 - Q_1}{Q_3 + Q_1}$ The coefficient of the quartile deviation is symbolized by V_q. This measure is especially useful when the distribution has open ends or extreme values.

(**4.20**) $Sk_p = \dfrac{\bar{x} - \text{Mode}}{s}$ This is the Pearsonian coefficient of skewness. The direction of skewness is influenced by the relative values of the mean and the mode. When the mean is larger, Sk_p takes a positive sign. When the mode is larger, Sk_p takes a negative sign. Dividing the difference between the mean and the mode by the standard deviation is to express the measure in relative rather than in absolute terms.

(**4.21**) $Sk_p = \dfrac{3(\bar{x} - \text{Med})}{s}$ This equation is identical with equation (4.20), except that here the modal value is estimated by the empirical method.

(**4.22**) $Sk_q = \dfrac{Q_3 + Q_1 - 2\,\text{Med}}{Q_3 - Q_1}$ This is Bowley's measure of skewness. It is based on the relative distances between the first and second quartiles and the second and the third quartiles. When the second quartile, or the median, is closer to the first quartile than it is to the third quartile, the distribution is positively skewed, and vice versa.

Problems

4.1 Comment fully on the following statements:

a. "The average has its limitations, but provided they are recognized, there is no single statistical quantity more valuable than the average."—L. H. C. Tippett, *Statistics*, 1943.
b. "Variation seems inevitable in nature · · · Whether one is attempting to control a dimension of a part which is to go into a precision assembly, the resistance of a relay, the acidity of a solution used for dyeing textiles, the weight of the contents of a container, or any other quality of a manufactured product, it is certain that the quality will vary."—E. L. Grant, *Statistical Quality Control*, 1952.
c. Before the representativeness of an average can be assessed, an appropriate measure of variation for the data must be known.

4.2 Despite its numerous disadvantages, the range is one of the two most commonly used measures of variation. What are the disadvantages of the range? What are the reasons for its common use?

4.3 Compared with the range and the standard deviation, respectively, what are the advantages and disadvantages of the semi-interquartile range?

4.4 The average deviation is not a particularly good measure of variation; yet at times it may be more useful than the standard deviation. Explain.

4.5 What are the characteristics of the standard deviation? How should it be interpreted?

4.6 State clearly, step by step, how to compute the standard deviation from the grouped data by using formulas (4.15) and (4.16).

4.7 The following table contains wheat yields on 150 farms:

Bushels of Wheat	Number of Farms
6.5 but under 8.5	5
8.5 but under 10.5	10
10.5 but under 12.5	13
12.5 but under 14.5	22
14.5 but under 16.5	26
16.5 but under 18.5	35
18.5 but under 20.5	23
20.5 but under 22.5	21
	150

a. Apply appropriate methods to the above data to compute
1) the range
2) the range within which the middle 50 percent of the farms fall
3) the average deviation
4) the standard deviation
b. Which measure computed in (a) would you use, if your purpose were
1) to report the variation to the general public?
2) to make prediction or inferences?

4.8 In order to initiate a production bonus plan, the management wishes to know what proportion of the workers should be made eligible for the bonus. What measure would you suggest if

a. only the top 2.5 percent are to receive the bonus?
b. only the top 25 percent are to receive the bonus?
c. only the top 50 percent are to receive the bonus?

4.9 What is the relationship between the mean, the mode, and the median in

a. a normal distribution?
b. a positively skewed distribution?
c. a negatively skewed distribution?

4.10 Why is the coefficient of skewness zero for a symmetrical distribution?

4.11 Consider the following distributions:

	Distribution A	Distribution B
Mean	100	90
Median	90	80
Standard deviation	10	10

a. Distribution A has the same degree of variation as Distribution B. Do you agree?
b. Both distributions have the same degree of skewness. True?

4.12 Experience shows that if an automatic can-filling machine is set to pour a given amount into each can, the mean fill is identical with the setting but the amount of fill in each can varies above and below the setting. A cannery desires to specify the contents as 15 ounces of drained weight. In order to measure the variability of the machine, a sample of 200 filled cans was observed. The results show that the distribution can be approximated by the normal distribution with $\bar{x} = 15$ ounces and $s = 0.08$ ounce.

a. Within what range of amounts of fill would all the cans filled by the machine practically fall?
b. Within what range of drained weight would 95 percent of the cans contain?
c. If it is decided that the can will contain at least 15 ounces of drained weight and, at the same time, avoid excessive overfill, what amount should the machine be set to fill? Why?
d. In your answer to the preceding question are your predictions about the mean fill and the standard deviation reasonable? Defend your reply.

4.13 A necktie factory employs 250 girls. The mean number of neckties per worker per day is 35, with a standard deviation of 4 neckties. The distribution is approximately normal.

a. What is the total number of neckties produced by all the workers each day?
b. How many of the girls can produce 40 neckties or more each day?
c. How many girls can produce only 30 or less each day?
d. The girls work eight hours a day and are paid on a piece rate of \$0.50 per necktie. The legal minimum wage is \$1.60 an hour. Are there any girls whose earnings are below the minimum wage?
e. How many girls are making more than \$18.00 per day?
f. How many girls are making less than \$15.00 per day?

4.14 If the mean height of American males is 68 inches and the standard deviation is 2.5 inches, and the corresponding figures for American females are 64 and 2.5 inches, which shows greater variability? How do you know?

4.15 Two workers on the same job show the following results over a long period of time?

	Worker A	Worker B
Mean time of completing the job (minutes)	30	25
Standard deviation (minutes)	6	4

a. Which worker appears to be more consistent in the time he requires to complete the job? Explain.
b. Which worker appears to be faster in completing the job? Explain.

4.16 a. Compute the standard deviation from data contained in Table 2P.1 by formula (4.15).
b. Compute the standard deviation from data contained in Table 2P.2 by formula (4.16).
c. Which distribution is more uniform?
d. Which distribution has a greater degree of skewness?

4.17 Compute the following measures with data in Table 1.1: the range, the quartile deviation, the average deviation, the variance, the standard deviation, the coefficient of variation, and the Pearsonian coefficient of skewness.

4.18 Compute the same statistics as instructed in the preceding problem for data in Table 1.2.

5

PROBABILITY THEORY

5.1

Probability and Risks in Decision

The last three chapters were concerned with descriptive statistics—measures computed to describe the characteristics of numerical data. Descriptive statistics can be used directly for decision making if they are population parameters. However, if they are sample statistics, another step must be taken before decision can be made. This step involves the generalization about parameters on the basis of statistics. Such a process is called statistical inference, which comprises two types of problems: estimation and testing hypothesis. In *estimation*, we use a sample statistic to estimate the corresponding parameter. In *statistical testing*, we try to determine whether or not a sample statistic is consistent with a statement made about the corresponding population parameter—that is, consistent with a given hypothesis.

The result of estimating a parameter by a statistic is necessarily uncertain. A statistic, derived from a limited number of observations, may always be expected to differ from the true value of the population. Thus, in estimation, we are always confronted with the risk of being in error. To use an estimate with confidence, then, we must know about the size of error. We must be able to answer the question, How closely does a sample statistic approach the corresponding parameter? The answer to such a question can be given with complete certainty only by investigating the entire population itself—a task that is sometimes impossible and usually impractical. When the parameter is not known, the most that can be done is to discover the approximate limits of the difference between a parameter and a statistic. This is done by expressing the risk of an error as a probability and giving it a numerical value. With measures of probability, the degree of uncertainty attached to general conclusions can be ascertained, and consequently sample studies can be utilized to reach the best possible approximations of the attributes of the population.

Precisely, a *statistical hypothesis* is some assertion about a population characteristic which may, or may not, be capable of verification. For instance, to return to the coin game mentioned in Chapter 1, recall that you decided the coin was biased by 10 trials because all the tosses turned up tails. In that case you rejected the hypothesis that the coin was fair and perfect. But suppose the coin was indeed unbiased: you would have rejected a true hypothesis.

Again suppose you are very fond of unshelled peanuts, but you are repelled whenever you find wormy meats in them. One day, passing a nut store, you see a sign on its window that says: "There are no wormy meats in our unshelled peanuts." You go in and open ten shells and find no wormy meats. You decide that the shop's claim is true and buy a bag. Before you finish it on your way home, however, you discover wormy meat in one shell. You feel disappointed because through your sampling technique you made the wrong decision by concluding that the shop's claim was true. Technically, we say that you accepted a false hypothesis.

These two examples illustrate that in testing a hypothesis we are subject to two classes of error: (1) the rejection of a true hypothesis and (2) the acceptance of a false hypothesis. To commit the first type of error leads a person to make the wrong decision of not doing what should be done; to commit the second type of error leads him to make the wrong decision of doing what should not be done. The important lesson to be learned here is that it is almost impossible to avoid both types of error in decision making. Statistics is mainly concerned with establishing decision rules by evaluating the risks involved in a particular decision rule. The evaluation of risks is made by probability considerations. The student may also be reminded that statistics cannot aid a person to make the right decision every time; it can only assist him to control the probability of being wrong at specified levels. For instance, if we establish a decision rule for a given type of problem, with the probability of the risk of being wrong at 0.01, then in the long run our decisions would be right 99 out of 100 times.

Probability theory, in addition to being the foundation of the classical decision procedures of estimation and testing, is an indispensable tool for all kinds of formal studies that involve uncertainty. Indeed, since its humble beginnings at the gambling tables in seventeenth-century France, probability theory has been developed and employed to treat and solve many weighty problems. It is involved with the observation of the life span of a radioactive atom, of the phenotypes of the offspring, and of the crossing of two species of plants. It is concerned with the construction of econometric models, with managerial decisions on planning and control, with the occurrence of accidents of all kinds, and with random disturbances in an electrical mechanism. It is used in discussions about the sex of an unborn baby, the position of a particle under diffusion, and the number of double stars in a region of the heaven.

The universality of the phenomenon of uncertainty thus gives importance to the consideration of probability theory. We shall introduce in this chapter various interpretations of the concept of probability and the basic rules for calculating probability values. In the chapters that follow, we shall discuss applications of probability theory.

5.2 Sample Space and Sample Points

The concept of probability is employed not only for various types of scientific investigations, but also for many problems in everyday life. We often hear and make such statements as these: "It is likely to rain tonight." "My chance of passing the course of statistics is better than even." "The odds against candidate A to win are 3 to 2." And so on. In each case, the statement refers to a situation which is uncertain, but the speaker nevertheless expresses some degree of confidence that his prediction will be verified. Furthermore, the degree of confidence may be considered to range from zero, for an impossible outcome, to one, for an outcome that is certain to happen. We shall see that probability provides a mathematical foundation for such assertions. Now, let us develop the concept of probability formally in terms of "sample space," "sample points," and "events," all of which are associated with what we call a *random experiment.*

Suppose we are interested in finding out what is going to happen if a certain act is performed. We can proceed to perform the act one or more times under the same specified conditions. Such a *process* is called a *random,* a *chance,* or a *stochastic process* and it has a number of unique features. First, each performance is called a *trial.* All trials conducted under the same conditions form an *experiment.* Different types or different numbers of trials are therefore different experiments. Note that an experiment may consist of a single trial. The result of a trial is called an *outcome,* a *sample point,* or an *elementary event.* The collection, or the set, of all possible outcomes (that is, the sample points) of an experiment constitutes a *sample space.* Let us introduce a few examples to clarify these concepts.

Consider the flipping of a coin. Here, we have a single-trial experiment whose sample space consists of two sample points, a head and a tail. Let S be the sample space and H and T be the elementary events; we may write

$$S = \{H,T\}$$

In future discussions, we need to know the number of sample points in a sample space. It would be convenient for us to use a symbol, such as $n(S)$, to denote it. Thus, for the experiment of a single coin, we have $n(S) = 2$.

An ordinary die is tossed. For this experiment, we have $n(S) = 6$, and

$$S = \{1,2,3,4,5,6\}$$

This sample space can also be represented by six points on a one-dimensional space as in Figure 5.1 below.

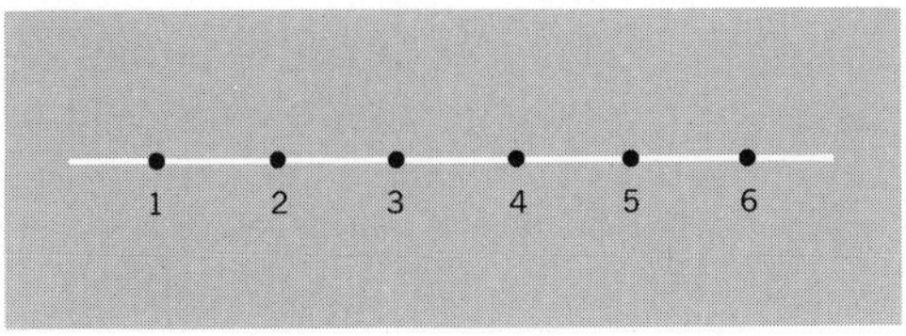

Figure 5.1 Sample space of one die.

A coin is tossed twice, or two coins are tossed together. Now, we have an experiment with two trials and four possible outcomes. Namely,

$$S = \{(HH),(HT),(TH),(TT)\}$$

An alternative way of showing this sample space is the *Venn diagram*, see Figure 5.2.

A coin and a die are tossed together. For this experiment, there are twelve outcomes in the sample space as listed below:

$$S = \{(H1),(H2), \cdots ,(H6),(T1),(T2), \cdots ,(T6)\}$$

Another way of showing this sample space is by a two-dimensional space, as in Figure 5.3, where each point is an ordered pair (x,y), such as $(H1)$, $(T4)$,

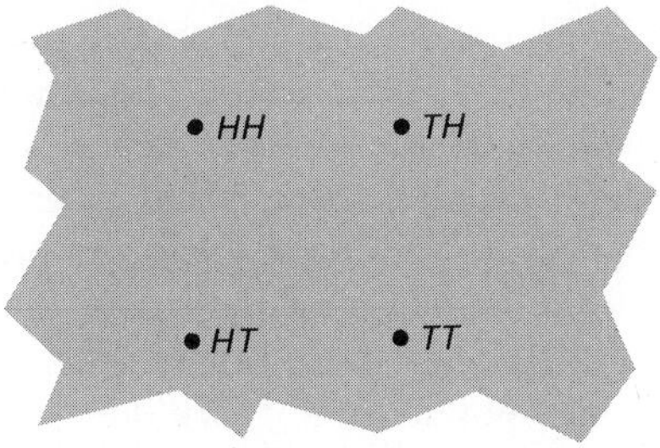

Figure 5.2 Sample space of two coins.

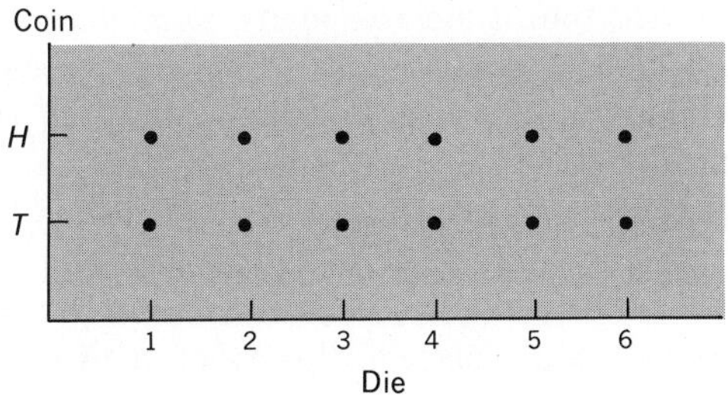

Figure 5.3 Sample space of a coin and a die.

and so on. We should note that sample points may be single elementary events, as in the first two examples, or joint elementary events, as in the last two examples.

Suppose we receive a shipment of 100 radio tubes, each of which is either defective (bad) or nondefective (good). If we draw 4 tubes at random from the shipment, what constitutes the sample space?

Solution of this problem requires an understanding of the phrase *random selection*. An item is selected at random from a collection of items—the population—if the selection procedure is such that each item in the collection is equally likely to be drawn. Random selection is one of the most important types of sampling experiment, and it may be conducted with or without replacement. When selection is made *with replacement*, each item is drawn and returned to the collection after observation before another item is selected. Thus the same item may be chosen more than once. However, at each selection, each and every item in the population has the same chance of being chosen. When random selection is made *without replacement*, we may have either of two procedures. First, the items are drawn one at a time; in this case, at each selection all the items in the population at the time have an equal chance to be chosen. Second, a number of items, called a sample, may be drawn simultaneously; for this method, all samples of the same size have the same chance to be drawn.

Returning to our example, we shall see that there are many ways to express the sample space, depending upon the purpose of our analysis. One method of determining the sample space is by a tree diagram. For each trial, there are two possible outcomes: nondefective (G), and defective (B). This experiment consists of four trials, and the trials may be made with or without replacement; therefore there are sixteen sample points as indicated by Figure 5.4.

The above example leads to another important property of sample points. In some cases, sample points in a given sample space are equally likely; in other cases, they are not. We can see intuitively that in each of the first four examples, the sample points are equally likely to occur; that is, each outcome has the same chance to occur as another. However, for the last example, the sixteen outcomes are equally likely only if the numbers of defective and nondefective items are the same in the shipment. In reality, however, we are quite certain there are fewer defective items than nondefective items. For instance, if there are only 20 defective tubes in the shipment, then we would expect the sample point, say, $GGGG$, to have a much greater chance than the sample point, say, $BBBB$, to occur, since there are four times as many nondefective tubes as defective ones. But how much greater chance does $GGGG$ have to appear than $BBBB$? This is one of the many interesting questions that the theory of probability attempts to answer.

As the last example of sample space, let us consider this: A bag contains five identical chips, numbered 1, 2, 3, 4, and 5. The chips are stirred and two chips are taken from the bag blindly, one after the other without replacement. How can we describe the sample space of this experiment in terms of the numbers on the chips?

To answer this question we shall let X be the number on the first chip and Y

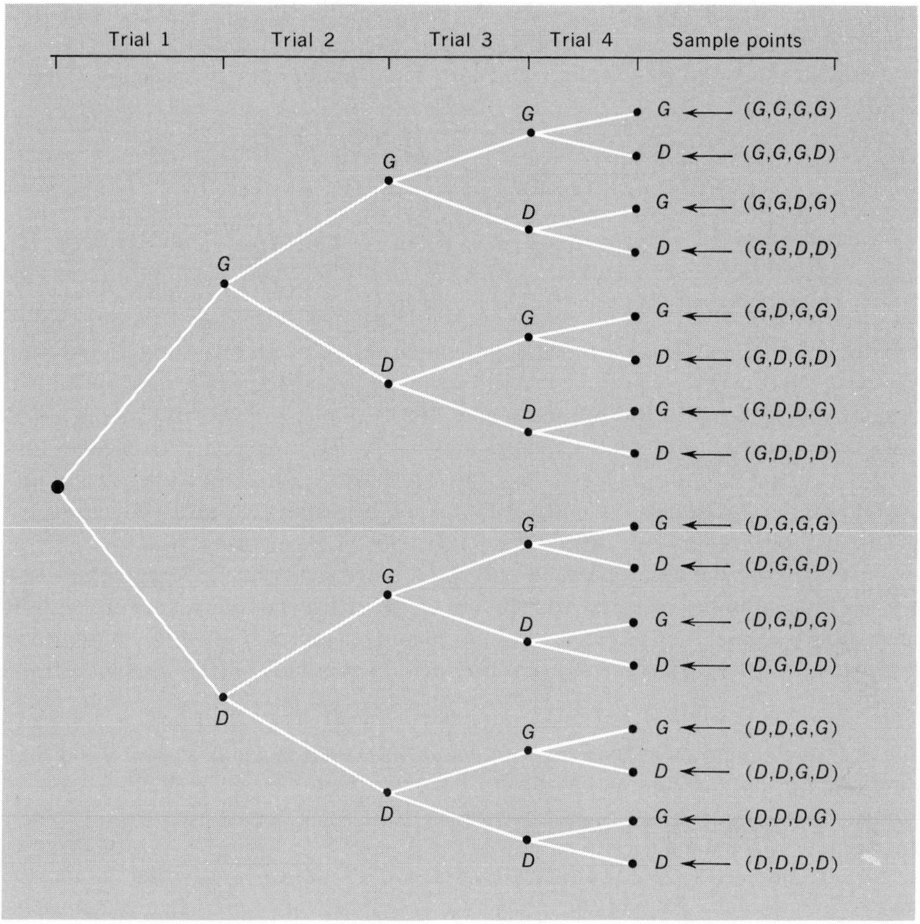

Figure 5.4 Sample points of four random draws obtained from 100 radio tubes via tree diagram.

be the number on the second chip. Both X and Y can have any one of the five numbers. Each possible outcome of the experiment is an ordered pair of numbers, (x,y). Furthermore, since drawings are made without replacement, no sample

TABLE 5.1 SAMPLE SPACE OF TWO NUMBERED CHIPS

x: Number on first chip	y: Number on second chip				
	1	*2*	*3*	*4*	*5*
1	(—)	(1,2)	(1,3)	(1,4)	(1,5)
2	(2,1)	(—)	(2,3)	(2,4)	(2,5)
3	(3,1)	(3,2)	(—)	(3,4)	(3,5)
4	(4,1)	(4,2)	(4,3)	(—)	(4,5)
5	(5,1)	(5,2)	(5,3)	(5,4)	(—)

point in the sample space can have the same value: that is, $x \neq y$. Then, the whole sample space can be described by way of a table (Table 5.1).

Do you think the 20 elementary events in the above sample space are equally likely?

5.3

Events

We have just seen that a sample space is a set S which contains all possible outcomes of a random experiment, and that any trial within the experiment results in one outcome which corresponds to exactly one element in the sample space. For example, the toss of a perfect coin yields two possible outcomes, a head and a tail. However, any particular toss of the coin will produce either a head or a tail, not both. Thus, we say the different sample points are *mutually exclusive* in the sense that no two elementary events can occur simultaneously at a single trial.

We have so far been concerned only with finite sample spaces; namely, those with a finite number of elementary events. In a finite sample space, any subset of sample points (that is, any combination of elementary events), is called a *compound event*, or simply *event*. Events include particularly the following three subsets:

(1) An elementary event is also an event because it is a subset which contains only one sample point in S.

(2) The sample space S is an event in the sense that it is a subset which contains all the elementary events in S.

(3) A subset which contains no sample points, called *the null set*, is also an event. It refers to an event which can never happen—an impossible event.

We shall denote an event by E and the number of sample points in the event set by $m(E)$. If we are concerned with more than one event, we denote them as $E_1, E_2, \cdots$, or as $A, B, \cdots$. Now, a few examples of events are in order.

In the sample space of tossing two coins, the event "appearance of two heads" is the subset E_1 whose element is the sample point (HH) in S. Namely,

$$E_1 = \{(HH)\} \quad \text{and} \quad m(E) = 1$$

In the same sample space, the event "tails on the first coin" is

$$E_2 = \{(TH),(TT)\} \quad \text{and} \quad m(E) = 2$$

See Figure 5.5.

In the sampling experiment which involves the selection of 4 radio tubes from a shipment of 100, the event "at most 1 bad tube" is the subset which contains either no or one defective tube. From Figure 5.4, we can see that this event has five sample points; that is,

$$E = \{(GGGG),(GGGB),(GGBG),(GBGG),(BGGG)\}$$

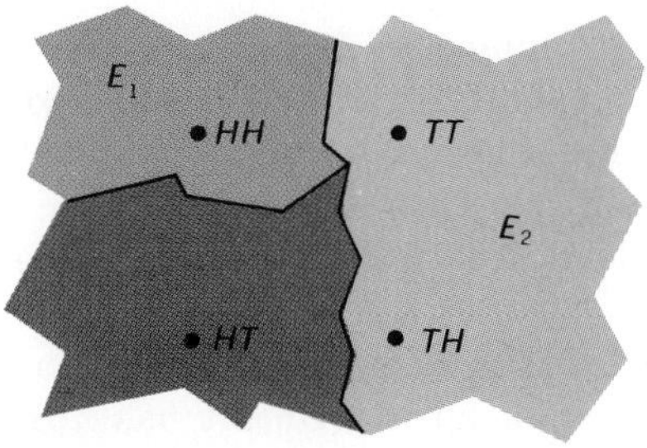

Figure 5.5 The events $E_1 = \{HH\}$ and $E_2 = \{TT,TH\}$ in the two-coins experiment.

Two dice are tossed. Let E_1 be the event "the sum of spots on the dice is greater than 12"; E_2 be the event "the sum of spots on the dice is divisible by 3"; and E_3 be the event "the sum is greater than or equal to 2 and is less than or equal to 12." Then we have,

$$E_1 = \{\phi\}$$
$$E_2 = \{(1,2),(1,5),(2,1),(2,4),(3,3),(4,2),(5,1),(6,6)\}$$
$$E_3 = \{S\}$$

But what constitutes the sample space S in this experiment?

Note that $E_1 = \phi$ refers to the fact that E_1 is the null set, the set with no element in it.

In concluding this section, let us observe two important aspects of "events." First, an event is said to have occurred if at least one of its elements has occurred. Second, only events have probabilities. The theory of probability is concerned with assigning weights (probabilities) to elementary events, and calculating event probabilities from them. While all the rules of probability calculus are based upon the operations of ratios, there are three different schools of thought on the interpretation, or the definition, of probability: the *classical*, the *relative frequency*, and the *subjective* theories. We shall now introduce them in that order.

5.4

Classical Definition of Probability

The classical approach to probability is the earliest. This school of thought assumes that all the possible outcomes of an experiment are mutually exclusive and equally likely. The words "equally likely" convey the notion of "equally probable" to the classicists; that is, to them, each outcome of an experiment has the same chance of appearing as any other and, therefore, can be assigned the same weight (probability) for its occurrence as any other. Thus, if there are n possible outcomes of an experiment, each sample point is assigned a number of $1/n$ as its probability. For example, in the tossing of a coin, the probability of a head is equal to the probability of a tail, and is equal to $\frac{1}{2}$. Again, when an ordinary and well-balanced die is thrown, each of the six possible outcomes has $\frac{1}{6}$ as its weight. Each card drawn at random from a well-shuffled bridge deck has the same chance to be drawn, that is, 1 chance in 52, or $\frac{1}{52}$.

The above remarks lead to the rather obvious classical definition of probability. If the sample space of an experiment has $n(S)$ mutually exclusive and equally likely outcomes, and if an event, as defined on this sample space, has $m(E)$ elements, then the probability of this event, denoted as $P(E)$, is the ratio of $m(E)$ to $n(S)$. That is,

$$P(E) = \frac{m(E)}{n(S)} \tag{5.1}$$

For example, a die is tossed. Let A be the event that an odd number will occur, B be the event that the number will be greater than 4, and C be the event that either a 1 or a 2 will occur. Then the probabilities of these events, by (5.1), are

$$P(A) = \frac{m(A)}{n(S)} = \frac{3}{6} = \frac{1}{2}$$

$$P(B) = \frac{m(B)}{n(S)} = \frac{2}{6} = \frac{1}{3}$$

$$P(C) = \frac{m(C)}{n(S)} = \frac{2}{6} = \frac{1}{3}$$

We draw a card at random from a bridge deck and we have these probabilities:

$$\begin{aligned} P(\text{an ace}) &= \tfrac{4}{52} = \tfrac{1}{13} \\ P(\text{a heart}) &= \tfrac{13}{52} = \tfrac{1}{4} \\ P(\text{a black card}) &= \tfrac{26}{52} = \tfrac{1}{2} \\ P(\text{an ace or a king}) &= \tfrac{8}{52} = \tfrac{2}{13} \end{aligned}$$

Note that the last event refers to either an ace or a king. This means that any one of the 4 aces or any one of the 4 kings is considered as favorable. We have, therefore, for this subset 8 sample points.

We should point out that the classical theory, under the assumption of equally likely outcomes, depends upon a priori analysis. Thus, we meet no difficulty if we are concerned with a well-balanced coin, an unbiased die, an honest roulette wheel, or any other experiment whose outcomes are symmetrical and can be deduced by logic. But what about a coin that is unbalanced? A loaded die? A crooked roulette wheel? In each of these cases, the classical approach of assigning equal probability would offer us nothing but confusion. Fortunately, when a priori reasoning fails due to the lack of symmetry, we can consult the relative frequency theory of probability.

5.5

Relative Frequency Theory of Probability

The relative frequency theoreticians agree that the only valid procedure for determining event probabilities is through repetitive experiments. For example, when a coin is tossed, what is the probability that the coin will turn up heads? The relative frequency theorist would approach the problem by actually tossing the coin, say, 100 times under the same conditions, and then calculating the proportion of times the coin fell heads up. Suppose the coin falls heads 45 times out of 100, then the ratio $45/100$ is used as the *estimate* of the probability of heads, $P(H)$, of this particular coin. A moment's reflection will show that even if the coin were perfect, we might not have exactly 50 heads out of 100 tosses. In other words, we cannot expect to obtain the true probability from repeated experiments. However, if the coin were perfect, the estimate would approach the true ratio (probability) as the number of trials increased.

The above discussion leads us to two definitions of probability in terms of relative frequency:

(1) If an experiment is performed n times under the same conditions and there are m outcomes, $m \leq n$, favoring an event, then an *estimate* of the probability of that event is the ratio m/n.

(2) The estimate of the probability of an event m/n approaches a limit, the true probability of the event, when n approaches infinity; that is,

$$P(E) = \lim_{n \to \infty} \frac{m}{n}$$

Clearly, in reality we can never obtain the probability of an event as given by the above limit. In practice, we can only try to have a close estimate of $P(E)$ based on a large n. For convenience and application, we shall treat the estimate of $P(E)$ as if it were actually $P(E)$ and write the working *relative frequency definition of probability* as

$$P(E) = \frac{m}{n} \tag{5.2}$$

To define $P(E)$ as a limit as n approaches infinity, however, does emphasize that probability involves a long-run concept. This means that when we toss a balanced die six times, it is almost impossible for each of the six numbers to appear exactly once. If, however, we toss the die over and over again, for a large number of times, we can expect, in the long run or on the average, each of the six faces of the die to appear about $1/6$ of the time. It is exactly in this sense that we say the probability of getting any one of the numbers on a die in a random toss is $1/6$.

The relative frequency theory is also called the *objective* or *empirical* definition of probability. It is so called because, according to the relative frequency theorists, the probability of an event is determined objectively by repetitive empirical observations.

5.6 Personalistic View of Probability

An objectivist is quite at home in talking about probability in connection with the toss of a coin or with the manufacture of a mass-produced output. He can readily think of the number of automobile tires produced vis-a-vis the probability of one defective tire as the long-run ratio of the number of defective tires to the total number of tires produced. He would, however, be helpless at unique events—events which occur just once or which cannot be subjected to repetitive experiments. Thus, he would not care to talk about the probability of a Columbus discovering America, or of a Senator Edward Kennedy becoming the Democratic candidate for President in 1972. As a result, a large class of problems is beyond the reach of the objectivists. This limitation of both the relative frequency theory and the classical assumption of symmetry theory has prompted the birth of the personalistic view of probability. While the relative frequency theory is still the most popular definition of probability, the personalistic position is steadily gaining strength.

The *personalistic* or *subjective* theorist regards probability as a measure of personal confidence in a particular proposition, such as a belief that Red China will be admitted to the United Nations no later than the mid-1970's. A subjectivist would assign a weight between zero and one to an event, according to his degree of belief for its possible occurrence. For example, if he is twice as confident of the occurrence of event A as he is of event B, according to his judgment he would decide that $P(A) = 2/3$ and $P(B) = 1/3$, if A and B are the two possible propositions in a situation. In general, the probabilities he assigns to various events for a given proposition are weighted averages, adding up to unity.

The subjective point of view grants that different reasonable individuals may differ in their degrees of confidence, even when offered the same evidence. Consequently, personal probabilities for the same event may differ in the eyes of different decision makers. Furthermore, the subjectivist can apply probabilities to all problems that a classicist or an objectivist studies, and to many more. When the condition of symmetry is met, the personalist would obviously acknowledge a priori reasoning, and when the amount of data is large, the subjectivist would usually get the same answers as the objectivist. But in addition, a personalist would take such propositions as Columbus, Kennedy, and Red China in his stride.

While there exist three different positions on the interpretation of probability, there is hardly any disagreement on the foundation of probability at the mathematical level. Each school defines probability as a ratio, or a proportion. Thus, once weights are assigned to sample points, or probability of events are determined, by any one of the three procedures, the same probability calculus can be applied. We also realize that each interpretation has its own merits, and we shall use whichever approach is convenient and appropriate for the problem under consideration.

5.7

Axioms of Probability

Probability calculus has *set theory* as its foundation and is developed from three axioms, or postulates, of event probabilities. In this section we shall discuss these axioms with reference to finite sample space.

The first axiom is *positiveness.* The probability of an event is nonnegative; it is either zero or positive. When $P(E) = 0$, we say E is the null set with no sample point in it. Clearly, the null set refers to an impossible event; for example, the probability that a human being can live forever is zero. The axiom of positiveness may be denoted as $P(E) \geq 0$.

The second axiom is *certainty.* This postulate states that the probability of a whole sample space is 1; that is, $P(S) = 1$. Alternatively, we say that the probability of an event which will definitely occur is 1; for example, the probability of one individual's eventual death is 1.

Note that the first two axioms indicate that the probability of any event ranges from 0 to 1. We have, therefore, $0 \leq P(E) \leq 1$.

The third axiom is called *unions.* Suppose E is a composite result, say, of two simple events e_1 and e_2, what is the probability of the occurrence of E? Since on each trial of an experiment exactly one simple event (or sample point) can occur, we must then have $P(E) = P(e_1) + P(e_2)$. This is the consequence of the axiom of unions, which may be stated as: The probability of a composite event E is the sum of the probabilities of the simple events of which E is composed.

The axiom of unions permits us to concentrate our attention on the elementary events when we assign probabilities. As soon as we know the weights of the sample points, we know that the probability of any event defined on the sample space is the sum of probabilities of these sample points, whose union is the event set.

From these three axioms we can now deduce the basic probability calculus rules—rules to use in computing the probabilities of various types of events.

5.8

Mutually Exclusive Events

Two events, A and B, are said to be *mutually exclusive,* or *disjoint,* if A and B do not contain any sample points in common. Thus, disjoint events cannot occur simultaneously. (See Figure 5.6.) The probability of either A or B occurring is the sum of their separate

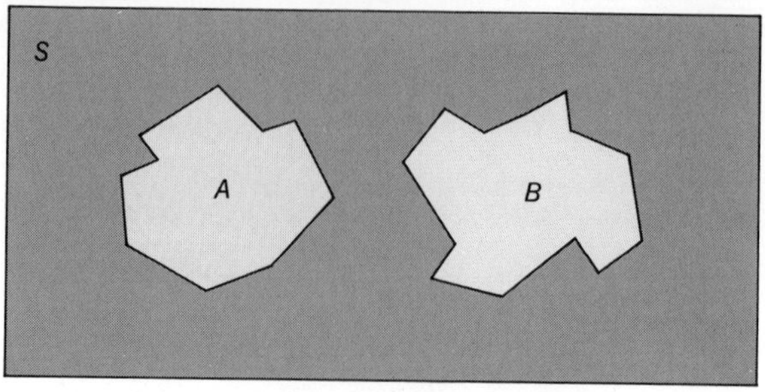

Figure 5.6 Mutually exclusive events.

probabilities:

$$P(A \text{ or } B) = P(A \cup B) = P(A) + P(B) \tag{5.3}$$

where $A \cup B$ reads "A union B."
Equation (5.3) is sometimes called the *additive rule for mutually exclusive events.*

If a card is drawn at random from a bridge deck, what is the probability that (1) either an ace or a queen will appear? (2) a heart or a club? (3) an honor card or a 3?

To solve this problem, note that there are 52 cards (that is, 52 equally likely outcomes) in the deck, made up of four suits: spades, hearts, diamonds, and clubs, each with 13 cards; the honor cards are aces, kings, queens, jacks, and tens. Thus,

$$\begin{aligned} &(1) \quad P(\text{an ace or a king}) = 4/52 + 4/52 = 2/13 \\ &(2) \quad P(\text{a heart or a club}) = 13/52 + 13/52 = 1/2 \\ &(3) \quad P(\text{an honor or a 3}) = 20/52 + 4/52 = 6/13 \end{aligned}$$

The law of addition for disjoint events can be generalized to any number of events. For example, n events are said to be mutually exclusive if no two of them have any sample points in common. Thus, if $E_1, E_2, \cdots, E_n$ are n disjoint events, then,

$$P(E_1 \cup E_2 \cup \cdots \cup E_n) = P(E_1) + P(E_2) + \cdots + P(E_n) \tag{5.4}$$

A coin and a die are tossed. Let A be the event "$H5$ or $H6$," B be the event "$T4$, $T5$, or $T6$," and C be the event "$H3$." The probability of any one of these events to occur is then,

$$P(A \cup B \cup C) = P(A) + P(B) + P(C) = 2/12 + 3/12 + 1/12 = 1/2$$

(Note that there are 12 sample points in the preceding experiment and that each point is assigned a weight of $1/12$.)

If the events $E_1, E_2, \cdots, E_n$ are disjoint and exhaustive—that is, their union is the sample space—then we say that the n events form a *partition* of S into n subsets, and $P(E_1 \cup E_2 \cup \cdots \cup E_n) = P(S) = 1$. Partition of a sample space is shown by Figure 5.7.

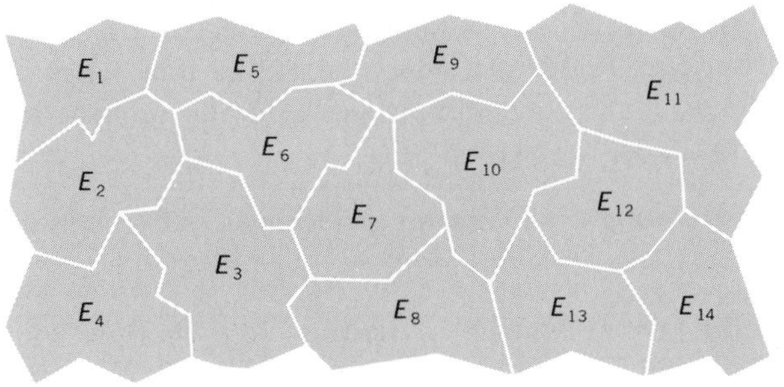

Figure 5.7 Partition of sample space.

5.9 Overlapping Events

Two events, A and B, are said to be *overlapping* or *joint* if they have sample points in common. Those sample points which belong to both A and B form a subset which is called the *intersection* of A and B, denoted by $A \cap B$ (read "A intersection B") or by AB (read "A multiplied by B"). We know that, in the case of disjoint events, $P(A \cup B)$ is the sum of probabilities of the points in the set $A \cup B$. That is, $P(A) + P(B)$ is the sum of probabilities of the points in A and the sum of probabilities of the points in B. Now, for overlapping events, the quantity $P(A) + P(B)$ includes the probabilities of the sample points in the intersection set twice. Therefore, if we subtract $P(A \cap B)$—the probability of the intersection—once, we will have the sum of probabilities of all points in AB, each taken just once. The logic of this procedure is shown by Figure 5.8. We finally conclude that, if A and B are joint events, then the probability of obtaining either A or B is

$$P(A \cup B) = P(A) + P(B) - P(A \cap B) \tag{5.5}$$

For example, if a card is drawn at random from a bridge deck, what is the

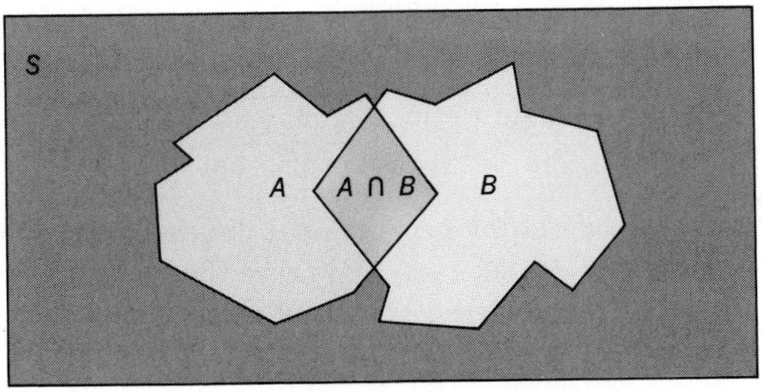

Figure 5.8 Overlapping events.

probability that either a spade or a king will occur? Clearly a spade and a king are not mutually exclusive events, since one of the four kings is a spade. It is also clear that there is only one sample point in the intersection set, the king of spades. Thus, by (5.5), we have

$$\begin{aligned} P(\text{a spade or a king}) &= P(\text{spade}) + P(\text{king}) - P(\text{spade and king}) \\ &= 13/52 + 4/52 - 1/52 = 4/13 \end{aligned}$$

A coin and a die are thrown. Let A be the event that the coin turns up tails and B be the event that a 3 or a 4 shows up on the die. What is the probability that either A or B will appear? In this example, as shown by Figure 5.9, there are two sample points, $T3$ and $T4$, in the intersection. Thus,

$$P(A \cup B) = 1/2 + 2/6 - 2/12 = 2/3$$

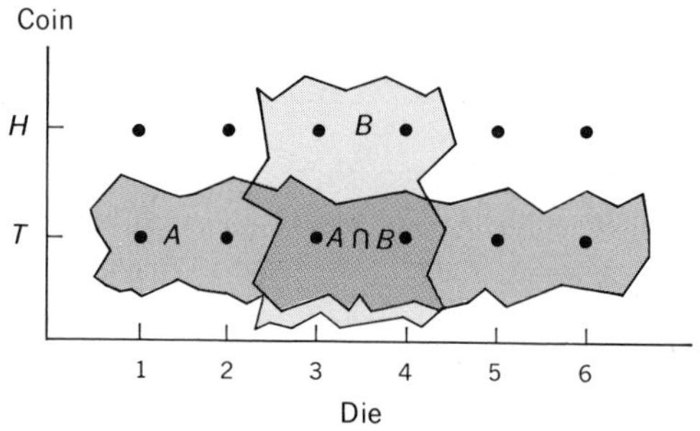

Figure 5.9 The event that the coin turns up tails and that the die shows 3 or 4 in the coin-die experiment.

Let us consider another application of the additive rule for overlapping events. In a lot of 1200 golf balls, 40 have imperfect covers, 32 cannot bounce, and 12 have both defects. If a ball is selected at random from the lot, what is the probability that the ball is defective?

The probability for the balls to have imperfect covers is $40/1200$, and the probability for the balls not to bounce is $32/1200$. Here we have two joint events, and a ball that has both defects is counted as a defective twice, once for its imperfect cover and once for its inability to bounce. We are told there are 12 balls which have both defects. Consequently,

$$\begin{aligned} P(\text{defective}) &= P(\text{imperfect cover}) + P(\text{inability to bounce}) \\ &\quad - P(\text{imperfect cover and inability to bounce}) \\ &= 40/1200 + 32/1200 - 12/1200 = 0.05 \end{aligned}$$

The additive rule for overlapping events can be extended to cover any number of joint events. In particular, if A, B, and C are three joint events, as suggested by Figure 5.10, the probability that any one of these events will occur is

$$\begin{aligned} P(A \cup B \cup C) = P(A) + P(B) + P(C) - P(A \cap B) - P(A \cap C) \\ - P(B \cap C) + P(A \cap B \cap C) \end{aligned} \qquad \textbf{(5.6)}$$

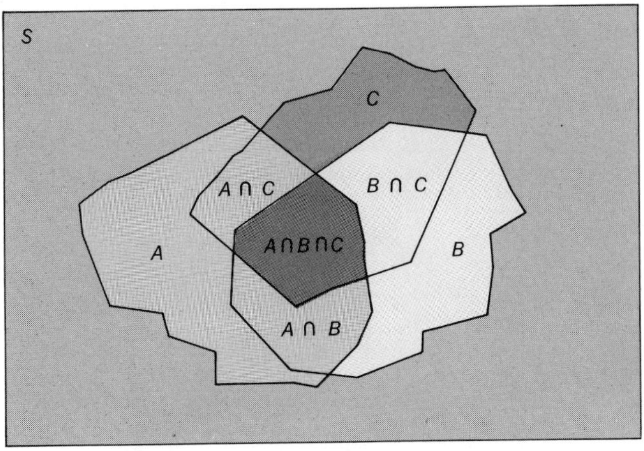

Figure 5.10 Three joint events: A, B, and C.

5.10 Complementary Events

Two events, E and $\bar{E}$, are said to be *complementary events* if the second is a subset which contains all the elementary events in the sample space that are not in the first; that is, complementary events are mutually exclusive and their union is the sample space S. (See Figure 5.11.) It follows, then, that

$$P(\bar{E}) = P(S) - P(E) = 1 - P(E) \tag{5.7}$$

which in particular implies that the null set is the complement of S.

What is the probability of obtaining a 1, 2, 3, 4, or 5 when an ordinary die is tossed? In a problem like this, it is much more convenient and compact to obtain $P(E)$ by first obtaining $P(\bar{E})$, which is $P(6)$ in this case, and then applying

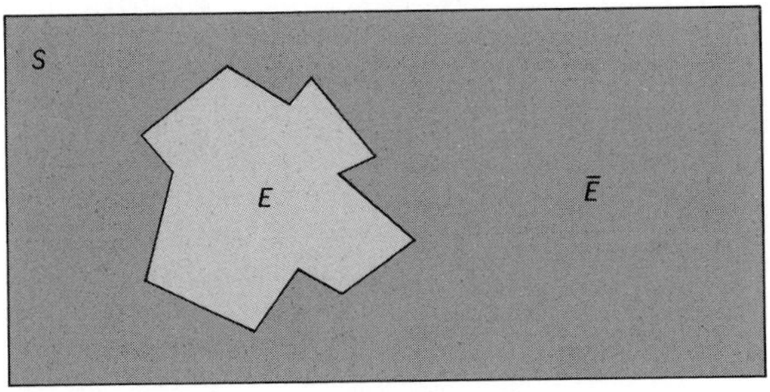

Figure 5.11 Complementary events.

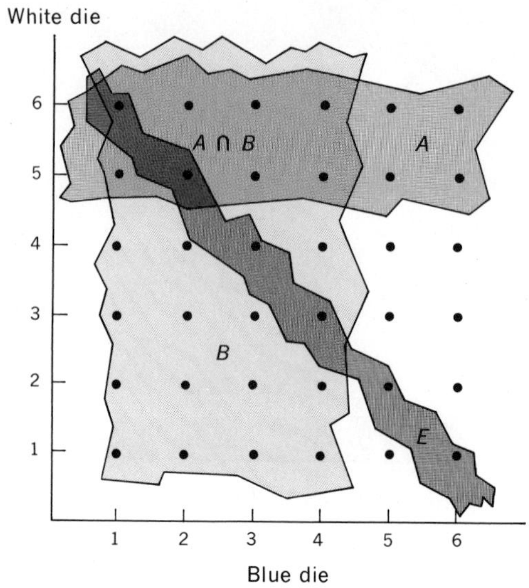

Figure 5.12 Sample space for two dice with E: the sum is 7; A: $W \geq 5$; and B: $b \leq 4$.

the *law of complementation*. Hence,

$$P(1,2,3,4 \text{ or } 5) = 1 - P(6) = 1 - 1/6 = 5/6$$

What is the alternative way to calculate this probability?

Consider one more example. If two ordinary dice are tossed, what is the probability that the sum of the points on the dice will not be 7? In this experiment, there are 36 possible outcomes, and 6 of these correspond to the event that the sum is 7 as shown by Figure 5.12. Assuming equally likely outcomes and denoting the event "7" by E, then we have

$$P(\bar{E}) = 1 - P(E) = 1 - 6/36 = 5/6$$

5.11 Independent Events

Two or more events are said to be *joint* or *compound* if they occur in unison or in sequence. If, for example, a head is followed by a tail in two tosses of a coin, we say that the occurrences of the head and the tail are compound events. Furthermore, joint events are said to be *independent* if the outcome of one does not affect, and is not affected by, the other. That is, if events are independent, the occurrence or nonoccurrence of any event in one trial will not in any way affect the probability of any other event in any other trial. For example, the probability of heads in a toss of a coin is $1/2$, and the probability of getting another head is still $1/2$ in another trial. Thus, the two events, H and H, and the two trials are independent. If A and B are independent, then the probability that A and B will occur in unison, denoted by $P(A \cap B)$, is the product of their separate probabilities. Namely,

$$P(A \cap B) = P(AB) = P(A)P(B) \tag{5.8}$$

As an example, a white die, w, and a blue die, b, are tossed. What is the probability that $w \geq 5$ and $b \leq 4$?

For solving this problem, we note that the sample space consists of the 36 possible combinations of the white and the blue die, each with 6 possible outcomes, and that the event under consideration requires that the two conditions must be satisfied simultaneously. If A is the event that $w \geq 5$ and B is the event that $b \leq 4$, then what is required is to know the number of sample points that both sets have in common; that is, the intersection, $A \cap B$. In Figure 5.12, we can see that there are 8 points in the intersection set. We have then $P(A \cap B) = 8/36$. By counting we find that there are 12 sample points in A and 24 sample points in B. As a result, $P(A) = 12/36$ and $P(B) = 24/36$. Thus,

$$P(A \cap B) = (12/36)(24/36) = 8/36$$

which verifies that the multiplicative rule gives the correct answer.

The above example reveals two important aspects of the multiplicative rule for independent events. First, if two events are independent, they must satisfy (5.8). Second, if A and B are two independent events with nonzero probabilities, then there must be at least one common elementary event in the intersection set $A \cap B$.

A coin and a die are tossed. Show that the event "head on the coin" and the event "the die shows even" are independent.

To solve this problem, we may let A be the first event defined on the sample space for outcomes of the coin S_1, and B be the second event defined on the sample space for outcomes of the die S_2, and then proceed to find the probability of the joint outcomes of A and B to see if the multiplicative law for independent events is satisfied. From the statement of the problem, we have

$$P(A \cap B) = \frac{m(A)}{n(S_1)} \times \frac{m(B)}{n(S_2)}$$

We know that $P(A) = 1/2$, since

$$A = \{H\} \qquad \text{and} \qquad S_1 = \{H,T\}$$

Similarly, $P(B) = 3/6 = 1/2$, since

$$B = \{2,4,6\} \qquad \text{and} \qquad S_2 = \{1,2,3,4,5,6\}$$

Now, we note that

$$A \cap B = \{(H2),(H4),(H6)\} \qquad \text{and}$$
$$S = \{(H1),(H2), \cdots ,(H6),(T1), \cdots ,(T6)\}$$

Therefore,

$$P(A \cap B) = 3/12 = 1/4 = P(A)P(B) = (1/2)(1/2)$$

This result verifies that A and B are independent. It is interesting to observe that if two events are independent, their joint probability is the product of their separate probabilities defined on their respective sample spaces.

When there are more than two independent events, the probability of their simultaneous occurrence is the product of the separate probabilities. For example, if A, B, and C are three independent events, we then have

$$P(A \cap B \cap C) = P(A)P(B)P(C) \tag{5.9}$$

The probability of a 1 to occur on a certain die is estimated to be $\frac{1}{5}$. Assuming this is the true ratio, what is the probability to obtain three 1s in three tosses of this die? ANSWER: $P(\text{three 1s}) = (\frac{1}{5})(\frac{1}{5})(\frac{1}{5}) = \frac{1}{125}$

5.12 Conditional Probability

Often when we are concerned with event probabilities we already have some information with respect to an experiment. The availability of such information in effect reduces the original sample space to one of its subsets. That is, we are dealing with probabilities of a section rather than of the whole sample space. Clearly, the probability of an event about which we had some information would be different from a situation where we had no information. For example, the probability that a card drawn at random from a bridge deck will be an ace is greater if we know that the card, already selected, is an honor card than if we lack this information. Again, a student selected from those students who had achieved honor grades in mathematics and logic will have a greater chance to receive an A in a course on statistics than one drawn from all students now taking statistics. Also, we can easily appreciate the fact that the probability that a household selected at random from Greater Los Angeles will have an annual income of over $25,000 is different from that for one chosen randomly from, say, Beverly Hills.

In each of the above examples, attention is focused on the probability of an event in a subset, or subpopulation, of the original sample space; the probability of an event in the subset is greater than it is in the original sample space. Each such subset is a reduced sample space and is specified by new conditions (information) beyond the initial conditions which yielded the original sample space. Probabilities associated with events defined on the subpopulations are called *conditional probabilities.* We shall now introduce the law governing conditional probabilities by an example.

Consider the experiment of tossing a coin and a die together. Given that the coin falls heads, find the probability that the die falls even.

We know that the original sample space for the above experiment consists of 12 equally likely outcomes. Among them, the coin falls heads 6 times. Ignoring the other 6 outcomes associated with tails, we have a reduced sample space S' defined as

$$S' = \{(H1),(H2),(H3),(H4),(H5),(H6)\}$$

The elementary events in S' are equally likely and, therefore, equal weights may be assigned as their probabilities, making them add up to unity. Thus, we have for each sample point a probability of $1/6$. Next, we observe that the event "the die falls even" consists of 3 points in this reduced sample space; that is, $(H2)$, $(H4)$, and $(H6)$. Consequently, the probability for this event to occur in S' is $3/6 = 1/2$. Or, the number, $1/2$, is the conditional probability that the die falls even, given that the coin falls heads.

In the above example, though we have used independent events, we must not obscure the usefulness of conditional probability. We must note here that it is not due to coincidence that the conditional probability is the same as unconditional probability in our example of the coin and the die. This point is made clear in our next example.

Let us study the concept of conditional probability further by considering the events "die falls even," denoted by A, and "coin falls heads," denoted by B, with reference to the whole sample space of the coin–die experiment. Clearly

$$\begin{aligned} A &= \{(H2),(H4),(H6),(T2),(T4),(T6)\}, \qquad \text{and} \\ B &= \{(H1),(H2),(H3),(H4),(H5),(H6)\} \end{aligned}$$

A moment's reflection will show that if we want to determine the probability of A given B, we must know the elements that belong to A and B; that is, the set $A \cap B$. From sets A and B given above, we see that

$$A \cap B = \{(H2),(H4),(H6)\}$$

We know that there are 12 equally likely sample points in the original sample space, and, therefore,

$$\begin{aligned} P(A) &= 6/12, \\ P(B) &= 6/12, \text{ and} \\ P(A \cap B) &= 3/12 \end{aligned}$$

Since we wish to find the probability of A given B, the event set B becomes then the reduced sample space, and $A \cap B$ becomes the event set in question. Thus, for this example, the probability that the die falls even given that the coin falls heads is

$$\frac{P(A \cap B)}{P(B)} = \frac{3/12}{6/12} = \frac{1}{2}$$

as obtained before.

From the preceding discussion, we may state the *law of conditional probability* as follows: The probability of A given B, denoted as $P(A|B)$, where the vertical line is read "given," is defined by the expression,

$$P(A|B) = \frac{P(A \cap B)}{P(B)}, \quad \text{if } P(B) > 0 \tag{5.10}$$

It can be shown, along the same line of reasoning, that the probability of B given A is

$$P(B|A) = \frac{P(B \cap A)}{P(A)}, \quad \text{if } P(A) > 0 \tag{5.11}$$

For instance, in the coin–die experiment above, the probability that the coin falls heads, B, given that the die falls even, A, is

$$P(B|A) = \frac{P(B \cap A)}{P(A)} = \frac{3/12}{6/12} = \frac{1}{2}$$

which turns out to be the same as the probability that the die falls even given that the coin falls heads. But are $P(A|B)$ and $P(B|A)$ equal to each other all the time?

An urn contains four white balls, numbered 1, 2, 3, and 4, and six black balls, numbered 5, 6, 7, 8, 9, and 10. A ball, selected at random from the urn, is white; what is the probability that the ball chosen is numbered 1? For this problem, we have

$$P(\text{white ball}) = 4/10 = 0.4$$
$$P(1 \text{ and white}) = 1/10 = 0.1$$

Therefore,

$$P(\text{numbered } 1|\text{white ball}) = 0.1/0.4 = 0.25$$

A card drawn at random from a bridge deck and turns out to be an honor card. What is the probability that it is also an ace? To answer this question, we recall that there are 20 honor cards, 4 of which are aces, in the deck of 52 cards. Hence,

$$P(\text{ace}|\text{honor card}) = \frac{P(\text{ace} \cap \text{honor card})}{P(\text{honor card})}$$

$$= \frac{4/52}{20/52} = \frac{1}{5}$$

In concluding this section, we shall make a few observations in connection with the definition of conditional probability.

First, all event probabilities are associated with some sample space. In effect, $P(A) = P(A|S)$, where S is omitted and understood. The conditional sign is employed only when there is some subset of S, S', which contains all outcomes of an experiment. In such a case, we have, in particular, $P(A|A) = 1$.

Second, the probabilities in the ratio on the right side of (5.10), or of (5.11), are event probabilities in the original sample space. The same result can be obtained if we first convert to the reduced sample space, because probabilities assigned to events defined on the reduced sample space are such that they are proportional to those assigned to the events defined on the original sample space. In other words, the total probability of events defined on the reduced sample space add up to unity. This is the reason that the conditional probability of an event is greater than it is in the original sample space. We can see that additional information is a valuable thing.

Third, the probability of the reduced sample is greater than zero because otherwise the conditional probability is not defined.

Fourth, $P(A|B)$ is seldom $P(A \cap B)$. The former may also differ from $P(A)$. On the other hand, if A and B are independent events and neither has zero probability, then $P(A|B) = P(A)$ and $P(B|A) = P(B)$. For instance, in the coin–die experiment, $P(A|B) = P(A) = 1/2$. In other words, the information that the coin falls heads does not alter the probability that the die falls even.

5.13 Dependent Events

Compound events are said to be *dependent* if the occurrence or nonoccurrence of one event in any one trial affects the probability of other events in other trials. For example, two cards are drawn randomly in succession, without replacement, from a bridge deck. The probability that the first selection will be a red card is $26/52$; the probability that the second card will be red, however, depends upon the outcome of the first draw. If it is a red card, the probability of drawing a red card the second time is $25/51$. However, if the first selection produces a black card, then the probability of drawing a red card on the second selection is $26/51$. Thus, in the case of dependent events, the probability of any event is conditional to, or depends upon, the occurrence or nonoccurrence of other events. From definitions of conditional probabilities, we can see that if A and B are dependent events, then,

$$P(A \cap B) = P(B)P(A|B) \tag{5.12}$$

and

$$P(B \cap A) = P(A)P(B|A) \tag{5.13}$$

Order is of no significance in the intersection set, since $A \cap B = B \cap A$. This property of intersection yields the following result:

$$P(A \cap B) = P(B)P(A|B) = P(B \cap A) = P(A)P(B|A) \tag{5.14}$$

which is sometimes called the *multiplicative law for dependent events.*

The manager of a firm estimates that if he puts an ad in a certain magazine, the ad will be read by half of its subscribers, and 1 percent of those who read the ad will buy the product advertised. Assuming these estimates are accurate, then we can say,

$$\begin{aligned} P(\text{read ad and buy product}) &= P(\text{read ad})P(\text{buy product}|\text{read ad}) \\ &= (0.5)(0.01) = 0.005 \end{aligned}$$

In a lot of 100 manufactured items, 15 are defective. Suppose 2 items are drawn at random without replacement; what is the probability that both are defective? Letting D denote defectives, we have

$$P(D \cap D) = P(D)P(D|D) = (15/100)(14/99) = 21/990$$

An urn contains three white and seven black balls; two balls are drawn, without replacement, at random. We desire to set up the sample space of this experiment and assign probabilities to the sample points. Evidently, the sample space of this experiment is as follows:

$$S = \{(WW),(WB),(BW),(BB)\}$$

where, for instance, (BW) means the first ball is black *and* the second ball is white. Random selection here ensures that all balls at any drawing are equally likely to be chosen. Thus, we have the following results:

$$\begin{aligned}
P(WW) &= P(W)P(W|W) = (3/10)(2/9) = 6/90 \\
P(WB) &= P(W)P(B|W) = (3/10)(7/9) = 21/90 \\
P(BW) &= P(B)P(W|B) = (7/10)(3/9) = 21/90 \\
P(BB) &= P(B)P(B|B) = (7/10)(6/9) = 42/90
\end{aligned}$$

The preceding example illustrates the application of the multiplicative rule to assign weights in a sample space. The sum of weights assigned in any sample space, as in the above illustration, must add up to unity.

The multiplicative law for dependent events can be extended to cover more than two events. For example, the probability of the simultaneous, or joint, occurrence of three dependent events, A, B, and C, is

$$P(A \cap B \cap C) = P(A)P(B|A)P(C|A \cap B) \tag{5.15}$$

If three cards are drawn from a deck of bridge cards in succession without replacement, what is the probability that all are aces? Denote aces by A, and by (5.15), we obtain

$$P(AAA) = P(A)P(A|A)P(A|AA) = (4/52)(3/51)(2/50)$$

Again, if five random selections are made from a deck of bridge cards simultaneously, what is the probability that all five cards are hearts? The answer, by an analogy of (5.15), is:

$$P(5 \text{ hearts}) = (13/52)(12/51)(11/50)(10/49)(9/48)$$

5.14 The Elimination Theorem

Situations often arise where we are interested in finding the probability of an *ultimate* event which is dependent on the occurrence or nonoccurrence of events in the intermediate stages of an experiment. The principle which aids us to determine such a probability is called the *elimination theorem*. This theorem turns out to be a combination of the additive and multiplicative rules. Let us develop this theorem by way of an example.

According to the judgment of a political scientist, the probability for the United States to have a Democratic administration in the 1970s is 0.55, and the

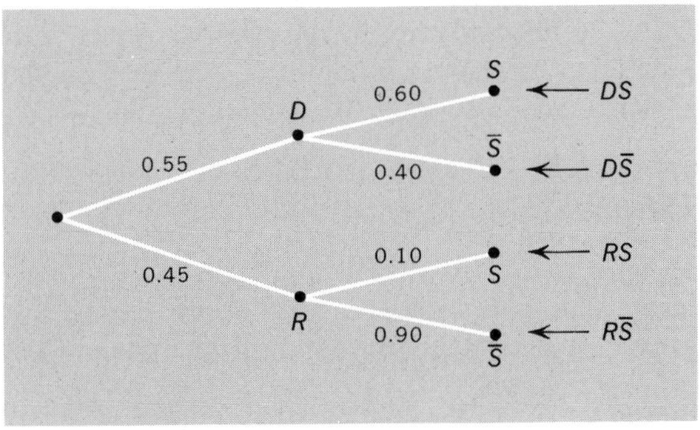

Figure 5.13 Tree diagram showing probability of socialized medicine.

corresponding probability for a Republican administration is 0.45. Moreover, under a Democratic administration, the probability for legislation of socialized medicine is 0.6, and the corresponding probability under a Republican administration is only 0.1. What we would like to know *now* is the probability of having socialized medicine in the 1970s.

We shall try to determine the probability of the ultimate event—socialized medicine—with the aid of a tree diagram, as shown in Figure 5.13, which is a digest of the information given in the problem above. In Figure 5.13,

$$
\begin{aligned}
D &= \text{Democratic administration} \\
R &= \text{Republican administration} \\
S &= \text{Socialized medicine legislation, and} \\
\bar{S} &= \text{No socialized medicine legislation}
\end{aligned}
$$

We can see that the probability of having a Democratic administration and socialized medicine, and the probability of having a Democratic administration and no socialized medicine, are, respectively

$$
\begin{aligned}
P(DS) &= P(D)P(S|D) = (0.55)(0.60) = 0.330 \\
P(D\bar{S}) &= P(D)P(\bar{S}|D) = (0.55)(0.40) = 0.220
\end{aligned}
$$

Likewise,

$$
\begin{aligned}
P(RS) &= P(R)P(S|R) = 0.045 \\
P(R\bar{S}) &= P(R)P(\bar{S}|R) = 0.405
\end{aligned}
$$

The foregoing four outcomes are mutually exclusive, and the first and the third represent the event of socialized medicine legislation in the 1970s. Thus, either the first or the third outcome is considered favorable. Clearly, the desired probability is the sum of the separate probabilities of these two favorable outcomes. That is,

$$P(S) = P(DS) + P(RS) = 0.330 + 0.045 = 0.375$$

This result indicates that if the political scientist's estimates are accurate, the prospects for us to have socialized medicine in the 1970s are not very bright.

We shall now formally generalize the *elimination theorem* as this: Let H_1,

$H_2, \cdots,$ and H_n be n mutually exclusive events, whose union is the sample space, each of which has a positive probability; then for an event E, which must occur with one of these mutually exclusive events, we have

$$\begin{aligned} P(E) &= P(H_1)P(E|H_1) + P(H_2)P(E|H_2) + \cdots + P(H_n)P(E|H_n) \\ &= \Sigma P(H_i)P(E|H_i) \end{aligned} \tag{5.16}$$

It is interesting to note that (5.16) refers to a situation where E is an arbitrary event generated from the sample space, and where $P(E|H_i)$ is the probability of E given the appearance of H_i. The elimination theorem is sometimes called the *theorem of total probability.*

An ordinary die is tossed once. If a 1 appears, a ball is drawn from Urn I. If a 2 or a 3 appears, a ball is drawn from Urn II; otherwise, a ball is drawn from Urn III. Urn I contains five white, three green, and two red balls. Urn II contains one white, six green, and three red balls. Urn III contains three white, one green, and six red balls. We wish to determine (a) the probability that a red ball is chosen, and (b) the probability that Urn II is selected, given that the ball is red.

For this experimental design, the following probabilities are implied:

$$\begin{array}{llll} P(\text{I}) = 1/6 & P(W|\text{I}) = 5/10 & P(G|\text{I}) = 3/10 & P(R|\text{I}) = 2/10 \\ P(\text{II}) = 2/6 & P(W|\text{II}) = 1/10 & P(G|\text{II}) = 6/10 & P(R|\text{II}) = 3/10 \\ P(\text{III}) = 3/6 & P(W|\text{III}) = 3/10 & P(G|\text{III}) = 1/10 & P(R|\text{III}) = 6/10 \end{array}$$

From these probabilities and by the elimination theorem, we have

$$\begin{aligned} \text{(a)}\quad P(R) &= P(\text{I})P(R|\text{I}) + P(\text{II})P(R|\text{II}) + P(\text{III})P(R|\text{III}) \\ &= (1/6)(2/10) + (2/6)(3/10) + (3/6)(6/10) \\ &= 13/30 \end{aligned}$$

and by the rule of conditional probability, we have

$$\text{(b)}\quad P(\text{II}|R) = \frac{P(\text{II} \cap R)}{P(R)} = \frac{P(\text{II})P(R|\text{II})}{P(R)} = \frac{(2/6)(3/10)}{13/30} = \frac{3}{13}$$

Can the reader explain why the probability of selecting Urn II, given a red ball, is much less than the probability of choosing a red ball?

5.15 Bayes' Theorem

In the last example, we have posed this question: If the random selection turns out to be a red ball, what is the probability that it comes from Urn II? Such a question seems to entail some kind of reasoning that is inverse or backward. It is inverse, in a sense, to a more natural way of asking the question, namely: If the toss of the die indicates Urn II, what is the probability of drawing a red ball from it? We shall see now, however, it is exactly upon this sort of backward reasoning that a very useful probability rule, called *Bayes' Theorem,* has been built. This theorem is developed by way of an example in the next few paragraphs.

A coin is tossed. If it turns up heads, two balls will be drawn from Urn I; otherwise, two balls will be drawn from Urn II. Urn I contains two black and six white balls. Urn II contains seven black balls and one white ball. In both cases, the selections are to be made with replacement. What is the probability that (a) Urn I is used, and (b) that Urn II is used, given that both balls drawn are black?

To solve this problem, let us first denote the event that Urn I is used by H_1 and the event that Urn II is used by H_2. According to the conditions of this experiment, we must have

$$P(H_1) = 1/2 \quad \text{and} \quad P(H_2) = 1/2$$

These values are called *a priori,* or *prior, probabilities* because they exist before we gain any information from the experiment itself. Here, H_1 and H_2 are mutually exclusive and exhaustive. Now, if we let E be the event that "both balls are black," then,

$$E = \{(H_1|BB),(H_2|BB)\}$$

What we seek to find are the conditional probabilities $P(H_1|E)$, which is $P(H_1 \cap E)/P(E)$, and $P(H_2|E)$, which is $P(H_2 \cap E)/P(E)$. To obtain these probabilities, we need, first of all, the probabilities of the intersection sets $H_1 \cap E$ and $H_2 \cap E$. From information given in the problem, we must have

$$P(H_1 \cap E) = P(\mathrm{I} \cap BB) = P(\mathrm{I})P(BB|\mathrm{I}) = (1/2)(2/8)(2/8) = 4/128$$
$$P(H_2 \cap E) = P(\mathrm{II} \cap BB) = P(\mathrm{II})P(BB|\mathrm{II}) = (1/2)(7/8)(7/8) = 49/128$$

In those compound probabilities, the terms $P(BB|\mathrm{I})$ and $P(BB|\mathrm{II})$, or, in general $P(E|H_i)$, are called *likelihoods* because they indicate how likely the event E under consideration is to occur *given* each and every one of H_i: urns in our example.

Next, we need the probability of the event E, the denominator of the ratios for our desired conditional probabilities. By the elimination rule, we see that

$$P(E) = P(H_1 \cap E) + P(H_2 \cap E)$$
$$= 4/128 + 49/128 = 53/128$$

Finally, we have

$$P(H_1|E) = \frac{P(H_1 \cap E)}{P(E)} = \frac{4/128}{53/128} = \frac{4}{53}$$

$$P(H_2|E) = \frac{P(H_2 \cap E)}{P(E)} = \frac{49/128}{53/128} = \frac{49}{53}$$

These two values are called *a posterior,* or *posterior, probabilities* because they are determined after the results of the experiment are known; that is, at least after we know the colors of the balls after they have been drawn. It is interesting to note how the evidence provided by the outcome of two black balls is revealed in the high posterior probability of using Urn II, where black balls predominate.

The procedure used for finding posterior probabilities from prior probabilities (which often represent a decision maker's belief in the original propositions, or hypotheses, H_i) and likelihoods (which are probabilities of an event conditional to the outcomes of an experiment) is called *Bayesian rule.* From our understanding of the foregoing example, we may now state this theorem as follows: If $H_1, H_2, \cdots$, and H_n are mutually exclusive events whose union is the sample space of an experiment, and if E is an arbitrary event defined on this sample space such that $P(E) > 0$, *Bayes' Theorem* says that the probability of H_i, given E, is

$$P(H_i|E) = \frac{P(H_i \cap E)}{P(H_1 \cap E) + P(H_2 \cap E) + \cdots + P(H_n \cap E)}$$

$$= \frac{P(H_i)P(E|H_i)}{\Sigma P(H_i)P(E|H_i)} \tag{5.17}$$

where the denominator is $P(E)$ as defined by (5.16).

A manufacturing firm produces steel pipes in three plants with daily production volumes of 500, 1000, and 2000 units, respectively. According to past experience, it is known that the fraction of defective outputs produced by the three plants are, respectively, 0.005, 0.008, and 0.010. If a pipe is selected at random from a day's total production and found to be defective, from which plant does that pipe come?

According to Bayes' Theorem, we have from the problem the following events:

H_1: production volume of the first plant, which is 500 units per day
H_2: production volume of the second plant, which is 1000 units per day
H_3: production volume of the third plant, which is 2000 units per day
E: a defective item

From these events, we see that $P(H_i|E)$ is the probability that the item is produced by the ith plant, *given* that the item is defective. Also, $P(H_i \cap E)$ is the probability that the items are produced by the ith plant and are defective. Information in the problem gives the following probabilities in connection with the random selection of a pipe from a day's total production:

(a) *Prior probabilities:*

$$P(H_1) = \frac{500}{500 + 1{,}000 + 2{,}000} = \frac{1}{7}$$

$$P(H_2) = \frac{1000}{3500} = \frac{2}{7}$$

$$P(H_3) = \frac{2000}{3500} = \frac{4}{7}$$

(b) *Likelihoods:*

$$P(E|H_1) = 0.005$$
$$P(E|H_2) = 0.008$$
$$P(E|H_3) = 0.010$$

(c) *Joint probabilities:*

$$\begin{aligned} P(H_1 \cap E) &= P(H_1)P(E|H_1) = (1/7)(0.005) = 0.005/7 \\ P(H_2 \cap E) &= P(H_2)P(E|H_2) = (2/7)(0.008) = 0.016/7 \\ P(H_3 \cap E) &= P(H_3)P(E|H_3) = (4/7)(0.010) = 0.040/7 \\ \text{Sum} &= P(E) = \Sigma P(H_i)P(E|H_i) = 0.061/7 \end{aligned}$$

(d) *Posterior probabilities:*

$$P(H_1|E) = \frac{P(H_1 \cap E)}{P(E)} = \frac{0.005/7}{0.061/7} = \frac{5}{61}$$

$$P(H_2|E) = \frac{P(H_2 \cap E)}{P(E)} = \frac{0.016/7}{0.061/7} = \frac{16}{61}$$

$$P(H_3|E) = \frac{P(H_3 \cap E)}{P(E)} = \frac{0.040/7}{0.061/7} = \frac{40}{61}$$

Since $P(H_3|E)$ is by far the greatest posterior probability, it is then most probable that the defective item has been drawn from the outputs of the third plant. As a check of the above calculations, the sum of all the posterior probabilities must be unity.

The preceding example illustrates the main application of the Bayesian rule in decision making. That is, a decision maker is often confronted with a set of mutually exclusive and exhaustive hypotheses, $H_1, H_2, \cdots,$ in connection with a phenomenon which is subject to test by experiment. Before the performance of an experiment, the decision maker may find it difficult to assign prior probabilities associated with the set of hypotheses. He is then often forced to assign prior probabilities in such a way that the weights assigned are proportional to the "degree of belief" he has in various hypotheses. Thus, different decision makers may assign different prior probabilities to the same set of hypotheses. For instance, to judge whether a coin is fair, one may initially assign equal probabilities to the occurrences of heads and tails. Another person, somehow believing that the coin is biased in favor of heads, may be inclined to give, say, $2/3$, as the prior probability of heads and $1/3$ as that of tails. Such initial differences, however, do not hinder the utility of the Bayesian theorem, since its very aim is to modify the prior probabilites, $P(H_i)$, by experiment. After the experiment is performed, we would replace $P(H_i)$ by $P(H_i|E)$. Furthermore, we may conduct a new experiment by using posterior probabilities of the preceding experiment as prior probabilities. As we proceed with repeated experiments, evidence accumulates and modifies the initial prior probabilities and, thereby, modifies the intensity of a decision maker's belief in various hypotheses. Repeated estimates will soon produce such low posterior probabilities for some hypotheses that they can be eliminated from further consideration. In other words, the more evidence we accumulate, the less important are the prior probabilities. The only restriction on the application of the Bayesian rule is that all hypotheses must be tenable in a given situation and that none is assigned a prior probability of 0 or 1.

5.16

Permutation and Combination

By this time, the reader is familiar with the fact that the probability of an event E is a ratio of the number of those outcomes in E to the total number of outcomes of an experiment. As we have seen, such numbers are easily obtained by the method of actual listing, if the total number of sample points is small. When an experiment involves a large sample space, it would be next to impossible to list all the sample points. Consequently, some efficient method of counting, or of finding, the number of ways of doing things is desirable. To this end, we shall in this section introduce the concepts of "permutation" and "combination," both of which have as their basis an elementary rule called *the fundamental principle.*

THE FUNDAMENTAL PRINCIPLE

The fundamental principle is concerned with such questions as this: If we have three letters, A, B, and C, in how many different orders can we arrange them? One way to answer this question is to list all the possible arrangements in a tree diagram and then count them. See Figure 5.14.

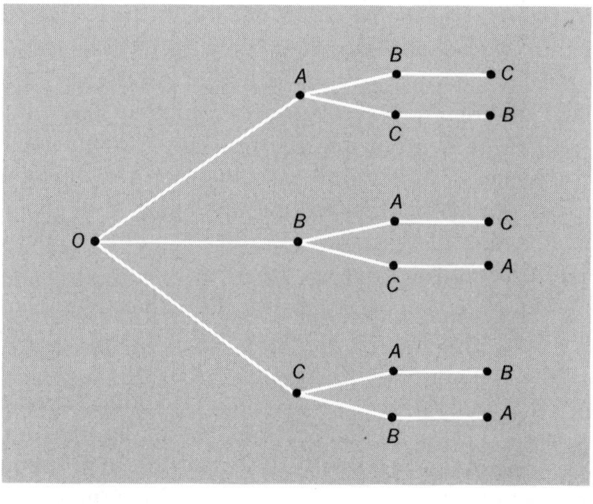

Figure 5.14 Possible ways of ordering three letters.

Figure 5.14 denotes the initial point, the origin, O. We see that, in the first place, there are three possible ways to arrange the three letters. Then, for each of the letters in the first place, there are two possible ways to arrange the two letters other than the one in the first place. After the second places are taken, for each of these two positions, there is only one possible way to place the last letter other than the two in the first and the second places. Thus, there are all together $(3)(2)(1) = 6$ possible arrangements, taken three letters at a time. It is important to note that the tree diagram takes *order* into consideration: a change in the order yields a different arrangement. For instance, ABC and ACB count as two distinct arrangements of the three letters.

The foregoing example suggest a definition of the *fundamental principle* as follows: If an experiment can have exactly n_1 distinct outcomes; if for each of these n_1 outcomes, a second experiment can yield n_2 distinct outcomes; if for each of these n_2 outcomes, a third experiment can yield n_3 distinct outcomes; and so on for k experiments; then the combined experiments, from the first to the kth, can have exactly, $(n_1)(n_2)(n_3) \cdots (n_k)$ outcomes.

Thus, if a coin and a die are tossed together, there are six ways the die can fall, and for each of these the coin has two ways to fall. Consequently, the die and the coin can fall in $(6)(2) = 12$ distinct ways.

Recall that the 12 outcomes of the die–coin experiment are equally likely. From this observation, we move to an important generalization about the property of the fundamental principle. If n_1 outcomes of the first experiment, n_2 outcomes of the second experiment, and so on to the n_k outcomes of the kth experiment, are all equally likely, then the $(n_1)(n_2) \cdots (n_k)$ elementary outcomes of the compound experiment, that is, the k experiments together, are also equally likely.

PERMUTATIONS

An arrangement of a number of objects in a definite order is sometimes called a *permutation*. To facilitate our discussion on counting procedures, we need an understanding of the factorial sign. The notation "$n!$" is read "n factorial" and is the product of all positive integers from 1 to n. That is,

$$n! = n(n-1)(n-2) \cdots (3)(2)(1) = n(n-1)! \tag{5.18}$$

In particular,

$$\begin{aligned} 1! &= 1 \\ 2! &= (2)(1!) = 2 \\ 3! &= (3)(2)(1) = (3)(2!) = 6 \\ 4! &= (4)(3!) = (4)(3)(2)(1) = 24 \end{aligned}$$

We also define that

$$0! = 1$$

Note that

$$\begin{aligned} 30! &= 30(29!) \\ 100! &= 100(99!) = (100)(99)(98)(97)(96)(95!) \end{aligned}$$

or, in general,

$$n! = r!(r + 1)(r + 2) \cdots (n - 1)n \qquad (5.19)$$

where $r < n$. For example,

$$\frac{10! - 9!}{11!} = \frac{9!(10) - 9!}{9!(10)(11)} = \frac{9!(10 - 1)}{9!(10)(11)} = \frac{9}{110}$$

Now, by the fundamental principle and the factorial sign, we may state: The *number of permutations* of *n distinct objects*, taken all together at a time, denoted as ${}_nP_n$, is $n!$ That is,

$${}_nP_n = n! \qquad (5.20)$$

For example, there are five flags of different colors. How many different messages can be sent by using all the five flags at a time? Answer:

$${}_5P_5 = 5! = (5)(4)(3)(2)(1) = 120$$

Formula (5.20) must be modified if the n objects are not distinct. For instance, if k of the n objects are alike, how many permutations can we have when n things are taken at a time? To answer this, let us assume that the modified number of permutations is x and proceed to calculate x. We know that if all the n objects are distinct, there are $n!$ permutations. Now, with k objects alike, the number of distinct arrangements is x. For each of these x arrangements, there would be $k!$ possible arrangements if the k objects were not alike. By the fundamental principle, we must have $xk! = n!$ so that $x = n!/k!$

Employing the same line of reasoning, we can see that if a set of n objects of which k_1 are alike, k_2 are alike, and so on to s kinds of like elements, each kind of element is different from those of any other kind, then the number of permutations taken all together is

$$\binom{n}{k_1, k_2, \ldots, k_s} = \frac{n!}{k_1!k_2! \cdots k_s!} \qquad (5.21)$$

where $k_1 + k_2 + \cdots + k_s = n$.

Consider the word "statistics." Here, we have 10 letters in all, of which there are 3 *s*s, 3 *t*s, 2 *i*s, 1 *a*, and 1 *c*. Thus, $n = 10$, $k_1 = 3$, $k_2 = 3$, $k_3 = 2$, $k_4 = 1$, and $k_5 = 1$. The total number of distinct permutations, using all ten letters of the word "statistics," is, by (5.21),

$$\binom{10}{3,3,2,1,1,} = \frac{10!}{3!3!2!1!1!} = 50{,}400$$

In many situations, we may wish to determine the number of permutations of n different objects in which some, not necessarily all, of the objects are used. For instance, consider a football squad of 40 players. How many different teams can the coach field when two teams composed of the same men but assigned to different positions are treated as different teams? What is required here is the number of permutations of 40 different players, taken 11 at a time. By the fundamental principle, we know that there are 40 possibilities to fill the first position, 39 choices for the second position, and so on until there remains but one position to be determined. At this point, 10 players have already been assigned; there are hence 30 choices for the last position. As a result, the number of teams that the coach can field is

$$(40)(39)(38) \cdots (31)(30)$$

This product can also be expressed in terms of the factorial sign, since

$$(40)(39) \cdots (30) = \frac{(40)(39) \cdots (30)(29) \cdots (2)(1)}{(29)(28) \cdots (2)(1)} = \frac{40!}{29!}$$

The above example indicates that the number of permutations of n things, taken r at a time, $r < n$, denoted by ${}_nP_r$, is

$${}_nP_r = \frac{n!}{(n-r)!} \tag{5.22}$$

Four signal flags are to be run up, one above the other, on a flagpole. How many different messages can be run up if there are available six flags, each of a different color?

By (5.22), the answer is

$${}_6P_4 = \frac{6!}{(6-4)} = \frac{6!}{2!} = \frac{2!(3)(4)(5)(6)}{2!} = 360$$

COMBINATIONS

A permutation requires that the objects assume a definite order. Occasions may arise, however, in which order is of no significance in an arrangement of objects. For example, the ordered set of five cards $AKKAK$ is different from that of $KKKAA$, but to a poker player, they represent the same hand, a full house. He is interested only in what the cards are rather than in the order of the cards. For the case where order is of no importance, we need what is called *combination* for the purpose of accounting.

A *combination* is a set of r things taken from n objects, $r \leq n$, regardless of the order in which the r things are selected. The number of combinations of n objects taken r at a time is denoted by ${}_nC_r$, C_r^n, or $\binom{n}{r}$. To illustrate, suppose seven directors of a company wish to elect three of their group to attend a conference sponsored by the United States Chamber of Commerce. How many different groups of three representatives can be selected? This is a problem of combination, since it makes no difference who is to be chosen first, second, or third. If, however, the delegation is going to have a chairman, a vice-chairman, and a secretary, order assumes importance and the problem is one of permutation. There are then ${}_7P_3 = 210$ ways of choosing the three representatives from the seven directors. If the three representatives are to have the same standing, any three directors, who form $3! = 6$ permutations, represent but one combination. From this, we can deduce that the number of combinations of n things taken r at a time must be the corresponding number of permutations divided by r! That is,

$$ {}_nC_r = \binom{n}{r} = \frac{{}_nP_r}{r!} = \frac{n!}{r!(n-r)!} \tag{5.23} $$

which indicates that $\binom{n}{r}$ is equal to ${}_nP_r$ when r objects are alike and $(n - r)$ objects are alike, r being different from $(n - r)$. Now, for the problem of selecting three representatives of the same standing from seven, we have,

$$ \binom{7}{3} = \frac{7!}{3!(7-3)!} = 35 \text{ ways} $$

As another example of combinations, let us determine the number of committees that can be formed from 5 American, 4 Russian, and 3 French delegates at the United Nations if each committee is to comprise 3 American, 2 Russian, and 1 French delegate. To determine this number, we note that we can select 3 from 5 Americans in $\binom{5}{3}$ ways, 2 from 4 Russians in $\binom{4}{2}$ ways, and 1 from 3 French in $\binom{3}{1}$ ways. By the fundamental principle, we have then

$$ \binom{5}{3}\binom{4}{2}\binom{3}{1} = \left(\frac{5!}{3!2!}\right)\left(\frac{4!}{2!2!}\right)\left(\frac{3!}{2!1!}\right) = 180 \text{ committees} $$

The formula for combination suggests the symmetry between $\binom{n}{r}$ and $\binom{n}{n-r}$. That is, if we select r from n objects, we automatically select a group of $(n - r)$ from n, simply by leaving them behind. Therefore,

$$ \binom{n}{r} = \frac{n!}{r!(n-r)!} = \binom{n}{n-r} \tag{5.24} $$

To make the symmetry formula valid for $n = r$, we define $\binom{n}{0} = 1$, just as we have defined $0! = 1$.

If, from a shipment of 100 radio tubes, each of which is either defective or nondefective, 4 tubes are drawn at random without replacement, how many possible samples can be selected?

This is a problem of combination, since the order of the elements that enter the sample is of no consequence. Also, we have, for this problem, $n = 100$ and $r = 4$. Hence, we can have

$$ \binom{100}{4} = \binom{100}{96} = \frac{100!}{4!96!} = 3{,}921{,}225 \text{ samples} $$

which are equally likely.

In ending this section, it is important to observe two aspects about combination. First, the number of combinations of n things taken r at a time corresponds to random selection without replacement. Second, all the possible combinations yielded by $\binom{n}{r}$ are equally likely because they represent the random selection of r things from the universal set of n things.

5.17 Probability on Large Sample Space

When the number of possible outcomes of an experiment is large, the procedure of combination can be used to calculate probabilities for sample spaces whose elements are equally likely, as illustrated by examples studied in this section.

A box contains 15 electronic tubes, 5 of which are defective. If two tubes are drawn together, at random, from the box, what is the probability that (a) both are good? (b) one is good and one is bad? (c) at least one is good?

To solve this problem, we see that for all the three cases, the sample space is the same. Namely, it contains $\binom{15}{2}$ equally likely sample points, with the same probability of $1\Big/\binom{15}{2}$ for each outcome. Among these elementary events, there are $\binom{10}{2}$ elements in the event set that two are good tubes. So,

$$\text{(a)}\quad P(\text{both tubes are good}) = \frac{\binom{10}{2}}{\binom{15}{2}} = \frac{9}{21}$$

Next, among the $\binom{15}{2}$ combinations, some satisfy the description "one is good and one is bad." There are $\binom{10}{1}$ ways to select one good tube and there are $\binom{5}{1}$ ways of selecting one bad tube. Thus, by the fundamental principle, there are $\binom{10}{1}\binom{5}{1}$ ways to select one good and one bad tube. Consequently,

$$\text{(b)}\quad P(\text{one good and one bad tube}) = \frac{\binom{10}{1}\binom{5}{1}}{\binom{15}{2}} = \frac{10}{21}$$

Finally, the event "at least one is good" is the complement of the event "none is good." Thus,

$$\text{(c)}\quad P(\text{at least one is good}) = 1 - P(\text{none is good})$$
$$= 1 - \frac{\binom{5}{2}}{\binom{15}{2}} = \frac{19}{21}$$

There are 100 instructors at a certain college, 70 men and 30 women. Five instructors are selected at random to evaluate the college's admission procedures. What is the probability that the sample consists of (a) all men? (b) exactly 3 men?

For this example, the number of equally likely samples, assuming selections are made without replacement, is

$$\binom{100}{5} = 75{,}287{,}520$$

A sample consisting of men only can be selected in $\binom{70}{5} = 12{,}103{,}014$ ways. A sample consisting of exactly 3 men and 2 women can be drawn in $\binom{70}{3}\binom{30}{2} =$ 23,811,900 ways. With these values, the desired probabilities become

$$\text{(a)}\qquad P(5\text{ men}) = \frac{\binom{70}{5}}{\binom{100}{5}} = \frac{12{,}103{,}014}{75{,}287{,}502} = 0.1608$$

$$\text{(b)}\quad P(3\text{ men and 2 women}) = \frac{\binom{70}{3}\binom{30}{2}}{\binom{100}{5}} = 0.3163$$

Cards are dealt face up from a well-shuffled bridge deck, one at a time, until the first king appears. What is the probability that the first king appears (a) at the third card? (b) at the rth card? (c) at the rth card or sooner?

For this experiment, there may be a number of ways to show the sample space. One way which is convenient for our present discussion is as follows. There are 4 kings, K, and 48 non-king, $\bar{K}$, in the deck. The sample space then may be thought to consist of all the possible arrangements of 4 Ks and 48 $\bar{K}$s in 52 numbered positions. The total number of arrangements, by the modified formula for permutations, is $(52!)/(4!48!)$.

Now, the event that the first king appears at the third card is one (in the sample space) in which the first three symbols of each of its elements must be $\bar{K} - \bar{K} - K$, since the first king is at the third place. The number of sample points in this event set is therefore all possible arrangements which can be made with the remaining 3 Ks and 46 $\bar{K}$s; that is, (49!)/(3!46!). Therefore, we have,

$$\text{(a)} \quad P(\text{1st king at the 3d card}) = \frac{\binom{49}{3}}{\binom{52}{4}} = 0.0313$$

Next, if the first king appears at the rth card, then the remaining 3 Ks and $(48 - (r - 1))$ $\bar{K}$s can form $\binom{52-r}{3}$ arrangements. Hence,

$$\text{(b)} \quad P(\text{1st king at the } r\text{th card}) = \frac{\binom{52-r}{3}}{\binom{52}{4}}$$

$$r = 1, 2, 3, \cdots, 49$$

Finally, if we let E be the event "first king at the rth card or earlier," then $\bar{E}$ is the event "4 kings after the rth card." The number of sample points belonging to $\bar{E}$ is the number of arrangements that can be made in the remaining $52 - r$ cards taken 4 Ks and $48 - r$ $\bar{K}$s, since the first r symbols of every member in this event are all non-kings. This number is $(52 - r)!/4!(48 - r)!$ Thus,

$$\text{(c)} \quad P(E) = 1 - P(\bar{E}) = 1 - \frac{\binom{52-r}{4}}{\binom{52}{4}}$$

If $r = 10$, for example,

$$P(E) = 1 - \frac{\binom{52-10}{4}}{\binom{52}{4}} = 1 - \frac{111{,}930}{589{,}225} = 0.81$$

which reveals that it is highly likely that a king will appear at the 10th card or sooner.

As the last example of computing probabilities from large sample space, let us consider this: If five cards are drawn at random simultaneously from a well-shuffled bridge deck, what is the probability that a flush hand is selected?

Recall that a bridge deck consists of 52 cards and thus the sample space of this experiment is $\binom{52}{5}$. Also, in the deck there are four suits, each with 13 cards, and a flush hand is one in which all 5 cards are of the same suit. Hence, there are $\binom{13}{5}$ ways to select a flush hand from a given suit. Successful outcomes, however, can come from any one of the four suits. Therefore, the total number of sample points which satisfy the description "a flush hand" is $4\binom{13}{5}$. From these considerations, we see that

$$P(\text{flush hand}) = \frac{4\binom{13}{5}}{\binom{52}{5}} = 0.00198$$

Such a result indicates that a flush hand is a rare event, since it occurs less than once in every 500 deals of five cards each time.

5.18

Randomness, Probability, and Predictability

Probability has been introduced in this chapter in connection with random experiment. "Randomness" is one of those words whose meaning everybody knows but few can define. Surprisingly, one often tries in vain to find a precise definition of randomness in most standard texts. The word "random" was first introduced in this book with reference to "random sample" in Chapter 1. As you may recall, a sample is random if every item in the population has an equal chance of being included in the sample. Note that probability has crept into this definition under the guise of chance. This observation gives us the clue that randomness may be defined in terms of probability. But before we formally define it, it may be helpful to gain an intuitive understanding of the term by way of a simple demonstration.

Number ten tags 0, 1, 2, 3, 4, 5, 6, 7, 8, 9 and place them in a bag. Mix them thoroughly. Draw one and record the number. Return it to the bag. Mix the contents thoroughly. Draw another and record the number. Return it and repeat the process over and over again. This process was taken by the author for 80 trials, with the actual sequence of the results as follows:

1455	5252	4428	8190	0477	9115	3336	6947	5064	2227
0883	9905	1116	4377	7787	3269	4123	3074	5538	9994

The results produced in this manner are called *random numbers* or *random digits*. These random digits form what is known as a *random variable*. The process by which the random sequence is produced is a random process. The population is one in which each number has a probability of 0.1.

Randomness is thus produced by a process that must be random itself. The same random process is capable of producing different sequences from the same population. Achievement of randomness in the process, as we shall see, requires deliberate design and careful planning. Furthermore, the randomness of a process is often judged by the sequence it produces.

A *variable* is *random* if it can take any of a number of values, each of which has a certain probability. It may be either discrete or continuous. If it takes a finite number of values, their probabilities are simply their relative frequencies. With a continuous random variable, we speak only of the probability that its value will lie within or at such and such intervals.

Moreover, the actual sequence of a random variable is usually marked by irregularities of the individual occurrences, but when they are arranged in an orderly way, such as a frequency distribution, they will form a population with distinct attributes. This implies that large numbers generate regularity and stability and therefore make possible the prediction of mass behavior in terms of probability. For example, suppose we did not know the actual figures on the numbered tags in the bag in the previous experiment. Then we observe that from the actual sequence of 80 random draws, we obtained the number 5 nine times. We may tentatively establish the relative frequency of getting 5 as $9/80$, or 11.25 percent. With this as a basis, we can predict the future outcomes of the appearance of 5. Thus, if we call 5 every time before the draw is made, we would expect to be right, in the long run, about 11.25 percent of the time. However, the accuracy of our prediction in this case may not be very high because the relative frequency was established with only 80 trials. Had we made, say, as many as 10,000 trials, the relative frequency would decrease to its mathematical probability, which is 10 percent of the time.

The above illustration indicates that the law of randomness operates in unison with the law of large numbers. That is, the reliability of predicting the outcomes of an event increases with the increase in the number of random observations used to determine the relative frequency of that event. Thus, the relative frequencies of death of persons of various ages, or from fire, accidents of a certain kind, and so on, which have been determined on the basis of large numbers of past occurrences, have enabled the insurance companies to predict, with high degrees of accuracy, the future outcomes in these various fields.

That randomness makes possible the prediction of mass behavior also implies the unpredictability of a single outcome, the behavior of which is irregular and not associated with probability. Before a draw is made, for instance, only one number can be produced and any one of the ten numbers in the bag could appear. As the insurance companies often state, "We don't know who will die, but we know how many."

Let us turn to another aspect of randomness and predictability. The digits produced by our simple experiment are not only random but independent. Recall that an *event* is said to be *independent* if the probability of its outcome does not influence, and is not influenced by, the outcome of other events. The independence of the digits is ensured by the procedure of replacement: the probability of each outcome at each trial remains unaltered.

Sequences of random and independent variables often appear in "runs." Statistically, a *run* is defined as a succession of individual letters (or numbers) which is followed and preceded by a different letter (or number) or by no letter (or number) at all. For instance, the sequence of the 80 random digits has the following 55 runs:

1–4–555–2–44–2–88–1–9–00–4–
77–9–11–5–333–66–9–4–7–5–0–6
–4–222–7–0–88–3–99–0–5–111–6–
4–2–7777–8–7–3–2–6–9–4–1–2
–33–0–7–4–55–3–8–999–4

This illustrates once again that randomness produces irregularities instead of uniformity but that it produces stability in relative frequencies in the long run. The fact that a random sequence tends to cluster is not too difficult to understand. If we are to have 10 random draws from the bag and wish to produce a completely uniform sequence as 0, 1, 2, 3, 4, 5, 6, 7, 8, 9, the probability of success, as you have already learned, is 1/10,000,000,000—almost an impossibility! As a matter of fact, if a sequence has too few runs we may suspect the randomness of the process that produces it. It is exactly for this reason that we mentioned earlier that the randomness of a process can be evaluated by the sequence it produces.

The clustering property of a random and independent variable often leads to the erroneous application of the law of chance. The *maturity of chance* theory in gambling, for example, is based on the misunderstanding of the property of probability because of clusters observed in a random sequence. According to this theory, one should favor bets on outcomes that have turned up less frequently than others in the past. This theory is fallacious because it fails to recognize that the probability of occurrence of an independent random event is unaffected by the outcomes of other events. If, in a dice game, 7 has appeared 10 times in a row, what is the probability that a 7 will occur again in the next toss of a pair of dice? It is still $1/6$. It is indeed very unlikely that ten 7s will occur in a row before the game starts. But once 10 tosses are made, the probability of the next toss is independent of the first 10 tosses and of subsequent tosses. The probability of a 7 appearing again will neither increase nor decrease. In other words, there is no reason to expect that there must be an excess of other numbers to compensate for the excess of 7s. When more tosses are made, the ratio will approach $1/6$. As Tippet has said, the law of large numbers works by its "swamping" effect rather than by compensation.

But does clustering afford an obvious method for prediction? In tossing coins, for instance, we might predict that the next toss will be the same as the last in order to take advantage of clustering in a random sequence. Would our chances of being right increase? Certainly not! Such a method can yield correct forecasts only half the time if the coin is perfect. This is, of course, the same percentage of accurate prediction by any other method of forecasting, such as predicting heads and tails alternatively. If in 100 tosses there are 10 tails too many, the same ratio cannot be expected to remain unchanged when more tosses are made. This difference, amounting to $1/10$ of 100 tosses, appears to be large, but if we continue to 10,000 tosses, and the difference remains at 10, it then represents only $1/1000$ of the number of tosses. The longer we continue to increase the number of tosses, the smaller this ratio becomes. In other words, in coin tossing, we expect the proportion of heads and that of tails to be approximately equal in the long run; but it is inaccurate to expect that in the long run there must be as many heads as tails. Thus it is easy to see why in the long run the equality between the proportion of heads and that of tails does not require that the one that has appeared less frequently should tend to catch up with the other.

Let us conclude this chapter by reminding the reader that probability is a unique property of the population, not associated with any particular outcome of an event. Also, it should be remembered that randomness dictates that, although it is possible to predict mass behavior, individual events are unpredictable. Predictability of mass behavior is based upon the fact that large numbers generate regularity and stability. The law of large numbers, however, applies only to random sequences in which the probability of each possible outcome remains constant. Finally, we must observe that when we apply the law of large numbers to predict chance events in practice, the probabilities themselves are often unknown. For instance, suppose you have tossed a coin, which is believed to be perfect, 10 times and all the occurrences are tails; you would then have reason to suspect that the subsequent tosses would not swamp the first 10 tails. That is, you would have made the inference from a sample of 10 that the coin was not fair—the population proportion was not fifty-fifty.

Glossary of Formulas

(**5.1**) $P(E) = \frac{m(E)}{n(S)}$ Under the assumptions of mutually exclusive and equally likely outcomes in the sample space, the classical interpretation views the probability of an event as a ratio of the points in that event set to the total number points in the sample space.

(**5.2**) $P(E) = \frac{m}{n}$ The relative frequency theory insists that event-probabilities can be established only by repeated experiments. According to this school of thought the probability of an event is the ratio of the number of favorable outcomes for the event to the total number of trials.

(**5.3**) $P(A \cup B) = P(A) + P(B)$ This is the additive rule for mutually exclusive events. This rule states that if A and B are mutually exclusive, then the probability that either A or B will occur is the sum of their separate probabilities.

(**5.4**) $P(E_1 \cup E_2 \cup \cdots \cup E_n) = P(E_1) + P(E_2) + \cdots + P(E_n)$ This is the extension of the additive rule to n disjoint events. From this extension, we see that $P(E_1 \cup E_2 \cup \cdots \cup E_n) = P(S) = 1$. This states that if the n events are mutually exclusive and exhaustive, then the n events are said to form a partition of the sample space S into n subsets.

(**5.5**) $P(A \cup B) = P(A) + P(B) - P(A \cap B)$ The additive rule for two overlapping, or joint, events states that if A and B are not mutually exclusive, then the probability that either A or B will occur is the sum of their separate probabilities less the probability that both A and B will occur. If $P(A \cap B)$ is not deducted from $P(A) + P(B)$, then we would have counted the intersection AB twice.

(**5.6**) $P(A \cup B \cup C) = P(A) + P(B) + P(C) - P(A \cap B) - P(A \cap C) - P(B \cap C) + P(A \cap B \cap C)$ This is the additive rule for three joint events. It is interesting to know the necessity of the last term in this formula.

(**5.7**) $P(\bar{E}) = P(S) - P(E) = 1 - P(E)$ For the law of complementation, we observe that $P(E) + P(\bar{E}) = P(S)$ and $P(E) = 1 - P(\bar{E})$. This rule enables us to compute $P(E)$ via $P(\bar{E})$ or vice versa.

(**5.8**) $P(A \cap B) = P(A)P(B)$ This rule states that if A and B are independent events, then the probability that both A and B will occur simultaneously is the product of their separate probabilities. The product rule is valid only if AB is not an empty set, that is, neither $P(A)$ nor $P(B)$ should have zero probability.

(**5.9**) $P(A \cap B \cap C) = P(A)P(B)P(C)$ The product rule for three independent events also requires that none of the three events should have zero probability.

(**5.10**) $P(A|B) = \dfrac{P(A \cap B)}{P(B)}$ and (**5.11**) $P(B|A) = \dfrac{P(B \cap A)}{P(A)}$

These two formulas define what is called conditional probability. In general $P(A|B) \neq P(B|A)$. $P(A|B)$ is the probability of the occurrence of A given that B has occurred, and B is the reduced sample space. And $P(B|A)$ is interpreted in the same fashion. If A and B are mutually exclusive, then $P(AB) = 0$, and $P(A|B) = P(B|A) = 0$. Thus, for these formulas to be meaningful, the reduced sample spaces must not be empty sets. That is, $P(B) > 0$ for (5.10) and $P(A) > 0$ for (5.11).

(**5.12**) $P(A \cap B) = P(B)P(A|B)$ (**5.13**) $P(B \cap A) = P(A)P(B|A)$

(**5.14**) $P(A \cap B) = P(B)P(A|B) = P(B \cap A) = P(A)P(B|A)$

These equations represent the joint or compound probabilities of the simultaneous occurrences of two dependent events, A and B. The logic of the third expression in this group is that order is not significant in intersection sets; namely, $AB = BA$. These product rules, as it can be seen, are simple transformations from the formulas for conditional probabilities.

(**5.15**) $P(A \cap B \cap C) = P(A)P(B|A)P(C|A \cap B)$ This is the product rule for three dependent events, A, B, and C.

(**5.16**) $P(E) = P(H_1)P(E|H_1) + P(H_2)P(E|H_2) + \cdots + P(H_n)P(E|H_n) = \Sigma P(H_i)P(E|H_i)$
This is the expression for the elimination rule or the theorem of total probability. It represents the probability of an event E which must occur with one of n mutually exclusive and exhaustive events, H_i, none of which has zero probability.

(**5.17**) $P(H_i|E) = \dfrac{P(H_i)P(E|H_i)}{\Sigma P(H_i)P(E|H_i)}$ This is the Bayes' Theorem. In this expression, H_i are mutually exclusive events whose union is S; E is an arbitrary event defined on S such that $P(E) > 0$. Note that (a) $P(H_i)$ are prior probabilities which are often arbitrary and subjective; (b) $P(E|H_i)$ are called likelihoods in the sense that each represents how likely E is to occur *given* the appearance of a particular hypothesis (one of the original mutually exclusive events); (c) $P(H_i|E)$ are posterior probabilities which are calculated from the prior probabilities and likelihoods for the purpose of replacing the prior probabilities.

(**5.18**) $n! = n(n-1)(n-2) \cdots (3)(2)(1) = n(n-1)!$ and (**5.19**) $n! = r!(r+1)(r+2) \cdots (n-1)n$. The first equation defines the "n-factorial" sign. The second is derived from the first and is employed to manipulate equations expressed in terms of factorial signs. Note that in the last equation, $r \leq n$.

(**5.20**) ${}_nP_n = n!$ The number of permutations of n things, taking all of these things at a time, by the fundamental principle, is $n!$ In a permutation, order is of significance.

(**5.21**) $\dbinom{n}{k_1, k_2, \cdots, k_s} = \dfrac{n!}{k_1!k_2! \cdots k_s!}$ This equation determines the number of permutations of n objects taken all at a time under the condition that, of the set of n objects, k_1 elements are alike, k_2 elements are alike, $\cdots$, and k_s elements are alike. Note that $k_1 + k_2 + \cdots + k_s = n$.

(5.22) $_nP_r = \dfrac{n!}{(n-r)!}$ The number of permutations of n objects taken r at a time, $r < n$, is defined by this equation.

(5.23) $\dbinom{n}{r} = \dfrac{n!}{r!(n-r)!}$ This formula represents the number of combinations of n things taken r at a time, $r \leq n$. In a combination, order is of no significance. It is interesting to note that $\dbinom{n}{r}$ combinations are equal to the number of permutations of a set of n things taken all at a time when r of its elements are alike and $(n - r)$ of its elements are alike. Furthermore, $\dbinom{n}{r}$ combinations are equally likely and they represent the number of outcomes selected at random without replacement.

(5.24) $\dbinom{n}{r} = \dfrac{n!}{r!(n-r)!} = \dbinom{n}{n-r}$ This expresses the property of symmetry of the combination procedure. Note that $\dbinom{n}{n} = \dbinom{n}{0} = 1$ just as $0! = 1$ by definition.

Problems

5.1 Define the following terms in your own words: (a) random experiment, (b) sample space, (c) sample points, (d) permutation, (e) combination.

5.2 Differentiate the following pairs of concepts:
a. elementary events and events
b. independent events and dependent events
c. mutually exclusive events and overlapping events
d. a priori probability and empirical probability
e. sample space and reduced sample space

5.3 What are the three schools of thought on the interpretation of "probability?" How does each school define probability?

5.4 If probability is interpreted in three different ways, why can the same probability theorems be applied to the calculations of probabilities in all cases?

5.5 Although it is possible to predict mass behavior in terms of probability, probability offers us no aid in the prediction of a particular individual outcome. Comment.

5.6 Because of the clustering tendency of random sequences, we can increase our accuracy in prediction by taking this tendency into account. True?

5.7 What is the maturity chance theory? What is wrong with it?

5.8 "Odds" are a common expression of "probabilities." Can you explain how these two concepts are related according to your understanding of them?

5.9 Give a precise verbal interpretation of each of the following statements:
a. The probability that a train will leave on time from Pennsylvania Station is 0.95.
b. The probability that a student who enters college will graduate is 0.43.
c. The odds are 2:1 in favor of Tom's passing the test in statistics.
d. The probability that the Russians will have a manned space ship on the moon during the next decade is 0.99.
e. At the initial stage of a political campaign between Smith and Jones, the odds are 2:3 against Smith.

5.10 Three perfect coins are tossed. Show the sample space of this experiment by the set notation.

5.11 Suppose you plan to make a survey of families having three children. You desire to record the sex of each child, in order of its birth. Show all the sample points of this experiment by a tree diagram.

5.12 A box contains a penny, a nickel, a dime, and a quarter. Two coins are drawn together from the box at random. Give the sample space via a tree diagram.

5.13 Three radio tubes are selected at random, with replacement, from a lot of 25 such tubes for testing. The test shows that each tube selected is either defective or nondefective. Represent the sample space of this experiment by a Venn diagram.

5.14 The numbers 1, 2, 3, and 4 are written separately on four identical chips, and the chips are put into a box and thoroughly stirred. Two chips are then drawn from the box at random with replacement. Give the sample space of this experiment in the form of a table.

5.15 Find the sample space of the preceding problem, assuming selections are made without replacement.

5.16 Two dice are tossed. Determine the number of sample points in each of the following events defined on the sample space of this experiment:
a. The sum is 7.
b. The sum is 11.
c. The sum is less than 5.
d. The sum is equal to and less than 12.

5.17 Explain whether or not the following events defined on the sample space of the two-dice experiment are mutually exclusive?
a. Either a 7 or a 11 turns up.
b. The number on one die is less than 3 and the number on the other is greater than 4.

5.18 An experiment consists of tossing a coin and a die together. Give a verbal description of the complement of each of the following events:
a. The coin shows heads and the die shows even.
b. The coin shows tails and the die turns up a number greater than 4.

5.19 A random sample of 4 is drawn from a lot of auto tires with the following events defined on the sample space of this experiment:
E_1: At least 1 is defective.
E_2: At least 3 are defective.
E_3: Exactly 1 is defective.
E_4: Exactly 2 are defective.
E_5: At most 1 is defective.
E_6: There are more defectives than nondefectives in the sample.
Give a verbal description of the intersection of each of the following pairs of events:
a. E_1 and E_2
b. E_1 and E_3
c. E_2 and E_3
d. E_1 and E_4
e. E_5 and E_6

5.20 A card is drawn from a well-shuffled bridge deck. What is the probability that it is (a) an ace? (b) a king of diamonds? (c) a black card?

5.21 In a lot of 100 manufactured items, 7 are defective. An item is drawn at random from the lot; what is the probability that it is defective?

5.22 What is the probability that a sum of 7 or 11 will occur if a pair of fair dice are tossed?

5.23 Three contractors, A, B, and C, are bidding for the construction of a new school. Some expert in the construction industry believes that A has exactly half the chance that B has; B, in turn, is $\frac{2}{3}$ as likely as C to win the contract. What is the probability for each to win the contract if the expert's estimates are accurate?

5.24 A card is selected at random from a bridge deck. Find the probability or odds that the card selected is an honor card.

5.25 One student is chosen from a school where there are 500 boys and 200 girls. What is the probability that the student selected is a boy?

5.26 A person holds a ticket in a lottery that sells 150 tickets and that offers one first prize, two second prizes, and three third prizes. What is the probability that he will win (a) the first prize? (b) the second prize? (c) the third prize? (d) a prize?

5.27 Two dice are thrown. What is the probability of not getting a double?

5.28 During a political race, the probability of A to win is $\frac{2}{5}$ and that of B is $\frac{1}{3}$. What is the probability of either A or B to win if they are in the same race?

5.29 A card is chosen at random from a bridge deck. What is the probability that (a) either an ace or a king will appear? (b) either an ace or a spade will appear?

5.30 If a die is thrown, what is the probability that either an even number or a number divisible by three will occur?

5.31 A clear and a red die are tossed. What is the probability that $c \leq 3$ or $r \leq 2$?

5.32 Experience shows that bolts produced by a certain process are too large 10 percent of the time and are too small 5 percent of the time. If a prospective buyer selects a bolt at random from a lot of 500 such bolts, what is the probability that it will be neither too long nor too short?

5.33 There are 100 students at a certain university who study at least one of the two languages, Chinese and Russian. Furthermore, it is known that 50 students enrolled for Chinese and 70 enrolled for Russian. What is the probability that a student selected at random studies Chinese or Russian?

5.34 Suppose there are 120 students, with other conditions remaining as in the preceding problem, what is the probability that a student chosen at random studies both Chinese and Russian?

5.35 Two coins are thrown. What is the probability that this experiment will result in two tails?

5.36 One bag contains four white and two black balls, and another contains three balls of each color. A ball is drawn at random from each bag; what is the probability that one is white and the other is black?

5.37 Two coins and a die are tossed together. What is the probability that both coins fall heads and the die shows a number less than 3?

5.38 What is the probability that a machine that produces 8 defective items in every 1000 will produce 5 defective items in a row?

5.39 A factory produces a mechanism which consists of three independently manufactured parts. It is known that 1 percent of part one, 4 percent of part two, and 2 percent of part three produced are defective. What is the probability that a complete mechanism is not defective?

5.40 An investor buys five stocks. The probabilities for prices of these stocks to increase are, respectively, 0.7, 0.8, 0.6, 0.5, and 0.3. What is the probability that the prices of all the five stocks will increase?

5.41 A speculator buys "futures" in wheat and sells "short" in soybeans simultaneously. The probabilities that both commodities will fall in prices are 0.2 and 0.9, respectively. What is the probability that he will gain in both markets?

5.42 A clear die and a blue die are thrown. If it is found that $c + b < 4$, what is the probability that $c = 1$?

5.43 Two dice are tossed. What is the probability that one die shows odd, given that the other die shows 6?

5.44 Referring to the two-dice experiment again, what is the probability of getting a sum of 7, if it is known that at least one die turned up 5?

5.45 Given that 10 percent of a college's faculty are women and that 2 percent of all women faculty are blue-eyed, what is the probability that a faculty member selected at random is blue-eyed if we know that a women is selected?

5.46 A retailer receives a lot of 25 radio tubes. Unknown to him, 5 of these tubes are defective. He picks 2 tubes randomly in succession without replacement. What is the probability that (a) the first is good and the second is defective? (b) both are defective?

5.47 A buyer will accept a shipment of 20 steel rods if a sample of 3, picked at random without replacement, contains no defectives. What is the probability that he will accept it if it contains 4 defectives?

5.48 Other things being equal, an individual's income increases with his age. If the probability that an individual is over forty-five and earns more than $10,000 a year is 0.15, and if the probability that an individual makes more than this figure is 0.25, what is the probability that an individual is over forty-five if his income exceeds $20,000?

5.49 A company is considering a site for a new plant. The probability that the site is satisfactory for raw material supply and finished product considerations is 0.7. Given the probability that the site is acceptable from the product-market aspect is 0.9, what is the probability that the site is acceptable from the material supply point of view?

5.50 Two slates of candidates are competing for the positions on the Board of Directors of a corporation. The probabilities that the first and the second slates will win are 0.6 and 0.4, respectively. Furthermore, if the first slate wins, the probability of introducing a new product is 0.8, and the corresponding probability if the second slate wins is 0.3. What is the probability that the new product will be introduced?

5.51 Six bridge decks and four pinochle decks, all having identical designs and constructions, are resting on a table. One deck is chosen at random and a card is then selected from it blindly. If the card drawn is a king of hearts, what is the probability that the card came from a bridge deck? From a pinochle deck? (Note that a pinochle deck contains 48 cards; two each of 9, 10, jack, queen, king, and ace in each of four suits: spades, hearts, diamonds, and clubs.)

5.52 For the problem above, suppose there are 5 bridge and 5 pinochle decks and that other conditions of the experiment remain unchanged. If the card drawn turns out to be a 9 of spades, what is the probability that it came from a bridge deck? from a pinochle deck?

5.53 A factory produces a certain type of outputs by three machines. The respective daily production figures are

Machine I: 3000 units
Machine II: 2500 units
Machine III: 4500 units

Past experience shows that 1 percent of the outputs produced by Machine I is defective. The corresponding fraction of defectives for the other two machines are, respectively, 1.2 percent and 2 percent. An item is drawn at random from the day's production run and is found to be defective. What is the probability that it comes from the outputs of (a) Machine I? (b) Machine II? (c) Machine III?

5.54 "Mr. Smith's gardener is not dependable; the probability that he will forget to water the rosebush is $\frac{2}{3}$. The rosebush is in questionable condition anyhow; if watered, the probability for its withering is $\frac{1}{2}$; if not watered, the probability for its withering is $\frac{3}{4}$. Upon returning Smith finds that the rosebush has withered. *What is the probability that the gardner did not water the rosebush?*" (Hans Reichenback, *The Theory of Probability*, University of California Press, 1949.)

5.55 Four urns contain colored balls as shown by the following table. An urn is selected at random and then a ball is drawn from it. The ball chosen turned out to be black.
a. What is the probability that Urn III is selected before the ball is drawn?
b. What is the probability, after the ball has been selected, that Urn III was used?

	Color			
Urn	*Red*	*Blue*	*Black*	*White*
I	5	6	3	4
II	2	2	2	2
III	3	1	1	2
IV	1	5	1	3

5.56 An automobile insurance company has insured 35,000 class *A* drivers (good risks), 50,000 class *B* drivers (median risks), and 15,000 class *C* drivers (bad risks). The probability that a class *A*, *B*, or *C* driver will have one or more accidents during any year is 0.01, 0.04, or 0.15. The company sells Mrs. Jones an insurance policy and within a year she has an accident. What is the probability that she is (a) a class *A* driver? (b) a class *B* driver? (c) a class *C* driver?

5.57 There are two balls in an urn; it is understood that there are three possibilities: both balls are white, or both balls are black, or one is white and one is black. A ball is drawn and turns out to be white. The ball is returned to the urn and a ball is drawn from it a second time. Again the ball turns out to be white. What is the probability that the urn contains two white balls? two black balls? one white and one black ball?

5.58 How many automobile license plates can be made by using two letters followed by a four-digit number?

5.59 Find the number of license plates for the previous problem if repetition of a letter is not allowed.

5.60 How many different ways can a line be formed by 100 people?

5.61 How many different hands of 5 cards each can be formed from the 52 cards in a bridge deck?

5.62 How many different deals are there in a bridge game?

5.63 The director of the examination division of the Federal Internal Revenue Service has been given a list of 120,000 tax returns for examination. Because of staff shortage, only $\frac{1}{2}$ of these returns can be reviewed during the next two months, $\frac{1}{4}$ of the returns can be referred to further study, and the rest must be filed for future action. In how many different ways can these 120,000 tax returns be arranged?

5.64 How many distinct arrangements can be made from the letters of the word "Mississippi," taken all together?

5.65 If S is a set such that $n(S) = h$, what is the number of ordered subsets of elements of S each containing k elements?

5.66 A consultation firm sends out teams of three men on a certain type of jobs. It has a total force of ten consultants. The first man selected is the foreman for the team and the second man assigned is the assistant foreman of the team. How many different teams as such can be formed?

5.67 A survey of economists is conducted to determine the appropriate policies for economic growth. Each economist is asked to rank five policies from a list of ten policies suggested by the investigator. How many differently ranked ballots are possible from each economist?

5.68 How many samples of 2 can be selected from a shipment of 10 electronic tubes?

5.69 A manufacturer receives a lot of 1000 metal parts, 10 percent of which are defective. The following acceptance decision rule is established: (1) a random sample of 30 parts is to be selected; (2) if more than 3 parts in the sample are defective, the entire lot is rejected; (3) otherwise, the lot is accepted.
a. How many different samples of 30 are possible under this decision rule?
b. How many of the possible samples will lead to the rejection of the lot?

5.70 A small boy plays with 12 marbles and 3 boxes. If he puts the marbles into the three boxes at random, what is the probability that he puts 3 marbles in one box, 4 marbles in another, and 5 marbles in the last box?

5.71 In a lot of 50 radio tubes, 5 are defective. If a random sample of 5 is selected without replacement, what is the probability that the sample is composed of (a) 5 nondefectives? (b) 3 nondefectives and 2 defectives? (c) 4 defectives and 1 nondefective?

5.72 Five cards are dealt from a well-shuffled bridge deck. Find the probability that
a. four are kings
b. four are kings and one is an ace
c. one is an ace, one is a king, one is a queen, one is a jack, and one is a 10
d. three are jacks and two are queens
e. three are of one kind of suit and two are of another (same) kind of suit
f. all five are clubs

5.73 A box contains 10 red, 6 black, and 4 white marbles. Three marbles are selected at random without replacement. What is the probability that
a. all are red?
b. one is of each color?
c. three are black or three are white?
d. two are white and one is black or two are black and one is white?

5.74 A box containing ten parts is received from a supplier. In the past, 80 percent of all such boxes had no defectives, 12 percent had one defective part, and 8 percent had two defective parts. Three parts are selected at random from the box just received and one is found to be defective.
a. Before the sample is taken, what is the probability of receiving a box with two defectives?
b. Given the sample result, what is the probability that the box originally contains two defective parts?

5.75 A lot of 15 radio-tape recorders is received by a retailing shop. In the past, it is known that 70 percent of such lots contain no defectives, 15 percent contain 1 defective recorder, 10 percent contain 2 defective recorders, and 5 percent contain 3 defective recorders. 4 recorders are picked out at random together for observation, and it is found that 2 are defective.
a. Before the sample was selected, what was the probability of receiving a box containing 3 defectives?
b. With the sample evidence, what is the probability that the box just received contains 3 defectives?

5.76 For problem 5.69, what is the probability of accepting the shipment according to the decision rule stated before?

5.77 Two investment opportunities are open to a prospective investor. If opportunity *A* turns out to be successful, a profit of $25,000 will result, and the probability of *A*'s success is estimated as 0.7; if *A* turns out to be a failure there will be a loss of $2500. If opportunity *B* succeeds, a profit of $100,000 will materialize; but if it fails there will be a loss of $80,000, and the probability for *B* to fail is 0.6. Which investment opportunity should the investor take if the decision criterion is to maximize profits?

6

PROBABILITY DISTRIBUTIONS

6.1

Random Variables and Their Probability Functions

We have discussed a number of experiments, such as the flip of a coin, the toss of a die, the selection of an item from a day's production run, the drawing of a ball from an urn or of a card from a bridge deck, and so forth. We call these types of experiments *random experiments*, since they may yield different outcomes in repeated performances even under the same conditions. We also have developed various probability theories to calculate probabilities of different kinds of events defined on some sample space. Quite often, in the application of probability theory, we are interested not only in the probability that a particular event will occur, but also in the distribution of probabilities over the whole range of possible outcomes of an experiment. In other words, we are interested in what are called *random variables* and their *probability functions*.

The concept of a random variable is a simple one. Imagine that a simple number could be assigned to each and every elementary event in a sample space. Then various sample points could be paired with various values of a variable. We would then have what is called a *random variable*. For instance, if we assign a value of 0 to heads and a value of 1 to tails, then we have a random variable: the toss of a coin which produces a value of 0 or of 1. Again, in the toss of a die, the number of spots 1, 2, 3, 4, 5, or 6—that may turn up is a random variable.

TABLE 6.1

Sample Points	Number of Heads	Probability
TTTT	0	1/16
TTTH	1	1/16
TTHT	1	1/16
THTT	1	1/16
HTTT	1	1/16
HHTT	2	1/16
THHT	2	1/16
HHTT	2	1/16
HTTH	2	1/16
HTHT	2	1/16
THTH	2	1/16
HHHT	3	1/16
HHTH	3	1/16
HTHH	3	1/16
THHH	3	1/16
HHHH	4	1/16

In general, we denote random variables by capital letters, such as U, V, W, X, Y, and Z, and denote the individual values of random variables by the corresponding lower-case letters. For example, if we name as X a random variable that has, say, five values, then these values can be symbolized as $x_1, x_2, \cdots, x_5$.

Next, we note, any value of a random variable is an event, since there is always some elementary event, or a set of elementary events, assigned that value. Also, since the value of a random variable is an event, it must have a probability. When we have all the possible values that a random variable can assume and the probability of each of these values, we have what is called a *probability function*, or *probability distribution*, or simply the *distribution* of a random variable.

Consider, for example, the toss of four perfect coins together, in which we are interested in the number of heads. This number may be 0, 1, 2, 3, or 4. But what is the distribution of the total probability of 1 over the range of these numbers? To answer this question, let us observe Table 6.1. The appropriate sample points for the sample space are given in the first column; the second column shows the number of heads corresponding to the sample points; and the third column contains the probabilities for the sample points. From the information in the last two columns, we can compute the probability for each possible number of heads resulting from this experiment. For instance, the probability of exactly 1 head is the sum of the probabilities for the sample points $TTTH$, $TTHT$, $THTT$, $HTTT$; the same holds true for other probabilities. Now, if we let X represent the number of heads, the values that X can assume and their associated probabilities must be as shown in Table 6.2.

Table 6.2 contains the probability function, or probability distribution, of X because it is a set of ordered pairs, each of this form: number of heads, probability of that number.

TABLE 6.2 PROBABILITY DISTRIBUTION FOR THE NUMBER OF HEADS FROM TOSSING 4 COINS

VALUES OF X, x_i	0	1	2	3	4
PROBABILITY $P(X = x)$	$1/16$	$4/16$	$6/16$	$4/16$	$1/16$

From our previous discussion we may formally state the following definitions:

(1) A *random variable* is one whose value is a number determined by the outcome of a random experiment.

(2) If X is a random variable which can assume values $x_1, x_2, \cdots, x_k$ with associated probabilities $f(x_1), f(x_2), \cdots, f(x_k)$, then the set whose elements are the ordered pairs $(x_i, f(x_i))$, $i = 1, 2, \cdots, k$, is called the *probability function*, or *distribution*, of X.

From the above definition of probability distribution, we see that this concept enables us to find the probability that the outcome of an experiment is one to which a particular number, x_i, has been assigned. For this reason we can state the probability function of a random variable in a more obvious and practical way; that is, if we use the notation $f(x)$ (read f at x) for the probability that the random variable X assumes the value x, then we write

$$f(x) = P(X = x) \tag{6.1}$$

as the probability function of X. For example, in the four-coin experiment

$$f(0) = P(X = 0) = 1/16$$
$$f(1) = P(X = 1) = 4/16$$
and so on

Equation (6.1) can also be extended to calculate probabilities in the form of inequalities and intervals. To illustrate and use data in Table 6.2 again, we have,

$$P(X > 4) = 0$$
$$P(X \geq 4) = 1/16$$
$$P(X \leq 2) = f(x_0) + f(x_1) + f(x_2) = 11/16$$
$$P(1 \leq X \leq 3) = P(1) + P(2) + P(3) = 7/8$$
and so on

A probability function may be represented by a table, such as Table 6.2, or by an equation. To illustrate how a probability function can be expressed in terms of an equation, let us consider first the toss of a perfect coin. If we let 0 stand for heads and 1 for tails, and if we let Y represent the variable number on each side of the coin, we can then express the toss of a coin as

$$g(y) = P(Y = y) = 1/2,$$
$$y = 0, 1$$

Next, consider the throw of an ordinary die. Let Z represent the variable number on each of the six faces of the die; we can then write

$$h(z) = P(Z = z) = 1/6$$
$$z = 1, 2, \cdots, 6$$

Finally, for a less obvious example, let us take up the experiment of the four coins discussed before. For this experiment, the values of $f(x) = P(X = x)$ are

$$1(1/16), 4(1/16), \cdots, 1(1/16)$$

The numbers 1, 4, 6, 4, and 1 are called *binomial coefficients* and can be obtained from $\binom{4}{x}$ for $x = 0, 1, 2, 3, 4$. As a result, as we shall explain in detail in Chapter 7, the general mathematical statement for all values of this probability function becomes

$$f(x) = P(X = x) = \binom{4}{x}(1/2)^x(1/2)^{4-x}$$
$$x = 0, 1, 2, 3, 4$$

For example,

$$P(X = 0) = \binom{4}{0}(1/2)^0(1/2)^4 = 1/16$$

As a result of the fact that we always focus our attention on the *values* of a random variable, in our notation of a random variable we make no reference to the "independent variable"—the outcomes in a sample space—unless the outcomes are, by chance, the values of the random variable itself, such as the selection of an item from an urn of numbered chips or the toss of a die. Thus, the graphic representation of a random variable is often made without showing the underlying experiment, and the values of the random variable—the dependent variable—are put on the horizontal scale. The vertical axis shows probabilities. In effect, the graph of a random variable turns out to be the graph of its probability function; see, for example, Figure 6.1. For this reason, many authors incline to define a random variable by explicitly stating its probability function, for instance: A

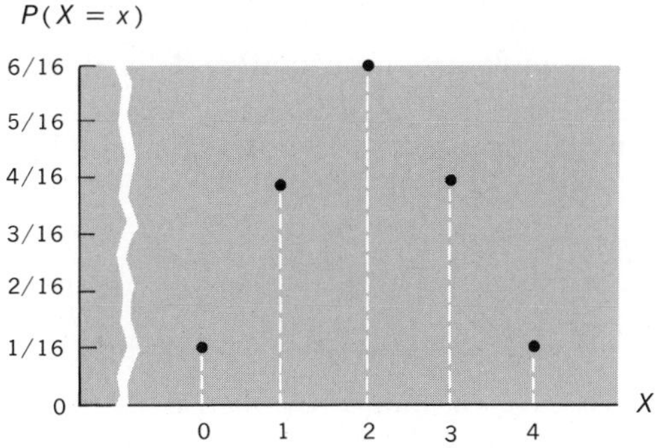

Figure 6.1 Probability distribution of the number of heads in the four-coin experiment.

random variable X is a numerical-valued function on the outcome of a random experiment, each of the number events having a probability $f(x) = P(X = x)$.

To conclude this section, let us summarize the steps in constructing a probability function:

(1) List all possible sample points with their corresponding probabilities for an experiment.

(2) List the value of the random variable under consideration that corresponds to each sample point.

(3) List all possible values of the random variable, $x_1, x_2, \cdots, x_n$ with their associated probabilities, $f(x_1), f(x_2), \cdots, f(x_n)$.

(4) $P(X = x_i) = f(x_i)$ is the sum of probabilities of all sample points corresponding to x_i.

(5) Thus, the set of ordered pairs $\{(x_i), f(x_i)\}$ is the probability function of a random variable; it is usually given by a table or by a general mathematical expression.

6.2 Cumulative Distribution Functions

Quite often, in connection with random variables, we need to ask questions such as this: What is the probability that a random variable assumes a value less than or equal to a particular prescribed number? Answers are given by the cumulative distribution function, CDF, defined as follows: If X is a random variable and x is a real number, then the *cumulative distribution function*, denoted as $F(x)$, which shows the probability that X assumes values less than or equal to x, is

$$F(x) = P(X \leq x) = \Sigma f(x) \tag{6.2}$$

Thus, in defining the CDF of a random variable, probabilities are assigned to all events which are half intervals, such as the event $X \leq x$, where x could be any real number. It is in this sense that CDF is a function of the boundary x. A distribution function is, of course, immediately defined from the corresponding probability function.

Consider a random variable, such as the toss of a coin, having values 0 and 1, each with a probability ½. The CDF then is

$$F(x) = P(X \leq x) = \begin{cases} 0, & \text{if } x < 0 \\ \tfrac{1}{2}, & \text{if } 0 < x < 1 \\ 1, & \text{if } x \geq 1 \end{cases}$$

The graph for the above CDF, as shown by Figure 6.2, has jumps in an amount of ½ at each of the values of 0 and 1.

Table 6.3 shows the possible numbers of heads where four coins are tossed, together with their probabilities $f(x)$ and the cumulative probabilities $F(x)$, as

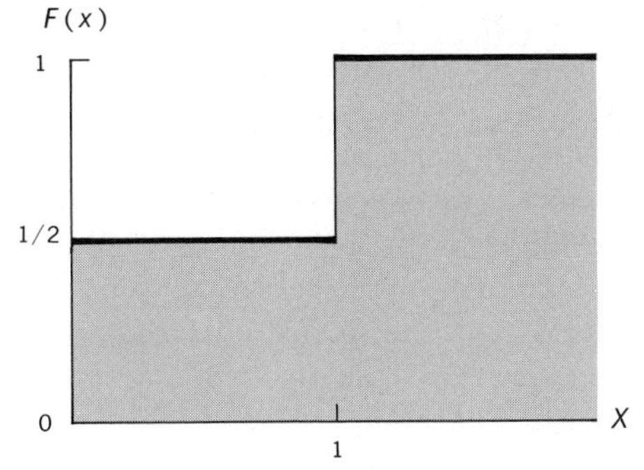

Figure 6.2 The cumulative distribution function of a single coin.

TABLE 6.3 NUMBER OF HEADS, PROBABILITY, AND DISTRIBUTION FUNCTIONS FOR 4 COINS

x	$f(x)$	$F(x)$
0	$1/16$	$1/16$
1	$4/16$	$5/16$
2	$6/16$	$11/16$
3	$4/16$	$15/16$
4	$1/16$	$16/16$

given by (6.2). The CDF has a value for every real number x, not just those listed in the table. For instance,

$$F(1.5) = P(X \leq 1.5) = f(0) + f(1) = F(1) = 5/16$$

In general if x is any number greater than or equal to a and less than b, then $F(x) = F(a)$. This reveals that the cumulative probability between two numbers a and b, $a < b$, is the same as that of its lower bound in a discrete distribution. For example, if a die is tossed—and we let $a = 2$ and $b = 4$—there will be no point between a and b that has a positive probability. The graph of the CDF of four coins is shown by Figure 6.3. Again, we see that jumps occur at $x = 0, 1, 2, 3,$ and 4.

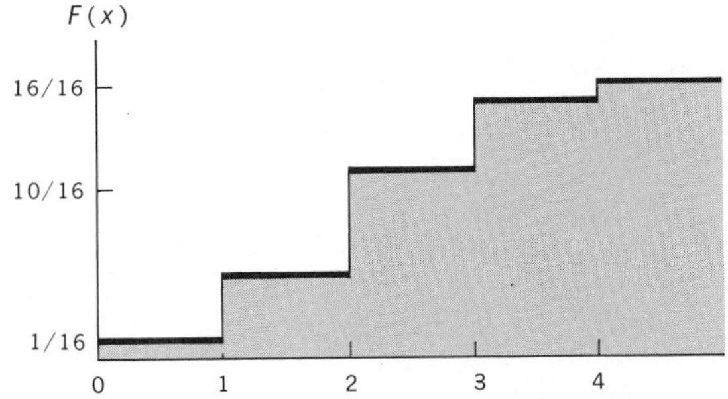

Figure 6.3 The cumulative distribution function of the number of heads in tossing four coins.

The CDF, as a consequence of being a probability, must satisfy certain conditions. First, due to the fact that probability is a number between 0 and 1, we must require that $0 \leq F(x) \leq 1$. Second, since any interval is an event and an event cannot have negative probability, we must have $F(b) > F(a)$, if $a < b$. This condition implies that $F(x)$ is a nondecreasing function, as we move from left to right on the horizontal scale of the graph for a CDF. Finally, we must have $F(\infty) = P(X \leq \infty) = 1$ and $F(-\infty) = P(X \leq -\infty) = 0$.

In concluding this section, we must also note that

$$P(X > x) = 1 - P(X \leq x) \tag{6.3}$$

6.3 Expectation

The two important characteristics of the distribution of a random variable, just as for sample frequency distributions, are its central tendency and its variability. In this section we shall introduce the concept of expectation, which is a measure of central tendency of a random variable.

Expectation is called by many other names, such as the *mathematical expectation*, the *expected value*, or simply the *mean*, of a random variable. This concept is closely related to the familiar notion of the arithmetic mean. All these names will be used interchangeably throughout our discussion.

The mathematical expectation of a random variable is the sum of the products that are obtained by multiplying all possible values of the random variable by their corresponding probabilities. Thus, let X be a random variable with probability function as follows:

VALUES OF X, x	x_1	x_2 · · ·	x_k
PROBABILITY, $f(x)$	$f(x_1)$	$f(x_2)$ · · ·	$f(x_k)$

The *mathematical expectation* of X, denoted by $E(X)$ or by μ_x, is defined to be

$$E(X) = \mu_x = x_1 f(x_1) + x_2 f(x_2) + \cdots + x_k f(x_k) = \Sigma x_i f(x_i) \tag{6.4}$$

A perfect die is tossed; what is the expectation as to the number of dots on the top face? To answer this question, we may recall that this random variable has possible values of 1, 2, 3, 4, 5, and 6, each with a probability of $\frac{1}{6}$. Call this variable X, and by (6.4) we have

$$E(X) = \mu_x = 1(\tfrac{1}{6}) + 2(\tfrac{1}{6}) + 3(\tfrac{1}{6}) + 4(\tfrac{1}{6}) + 5(\tfrac{1}{6}) + 6(\tfrac{1}{6}) = 3.5$$

Four coins are tossed; what is the expectation of the number of heads? The answer for this problem is obtained with the aid of Table 6.4. We denote the random variable as Y.

TABLE 6.4 COMPUTATION OF EXPECTATION OF THE NUMBER OF HEADS WITH FOUR COINS

y_i	$f(y_i)$	$y_i f(y_i)$
0	$1/16$	0
1	$4/16$	$4/16$
2	$6/16$	$12/16$
3	$4/16$	$12/16$
4	$1/16$	$4/16$
	Sum	$32/16$

$$E(Y) = \mu_y = 32/16 = 2$$

Thus, if we toss four coins together each time, over and over again, on the average we shall obtain two heads per toss.

The probability that a house of a certain type will be burned down by fire in any twelve-month period is 0.005. An insurance company offers to sell the owner of such a house a \$20,000 one-year term fire insurance policy for a premium of \$150. What is the company's expected gain?

The "gain," G, for the company is a random variable with possible values of \$150, if the house does not have a fire accident, and $-$\$19,850, if the house is burned down during the year covered by the policy. The probability function of G is then

VALUES OF G, g	150	$-19{,}850$
PROBABILITY, $f(g)$	0.995	0.005

With the above information, we see that

$$E(G) = \mu_G = (150)(0.995) + (-19{,}850)(0.005) = \$50$$

The expected gain for an insurance must be positive in order to enable the company to pay for administrative costs and to build up reserves for paying its beneficiaries and policyholders. However, in all cases of games of chance where games are run for profits, the expected value is negative as shown by the next example.

A lottery sells 10,000 tickets at \$1 per ticket; a prize of \$5000 will be given to the winner of the first draw. Suppose you have bought a ticket; how much should you expect to win?.

Here, the random variable "win," W, has two possible values: \$4999 and $-$\$1. Their respective probabilities are $9999/10{,}000$ and $1/10{,}000$. Thus

$$E(W) = (4999)\left(\frac{1}{10{,}000}\right) + (-1)\left(\frac{9999}{10{,}000}\right) = -\$0.50$$

This amount, a *minus* 50 cents, is the amount you expect to win (lose) each time, if you play this game over and over again.

6.4

Properties of Expectation

If X is a random variable whose probability distribution is known, often we find other random variables which are defined in terms of the same probability distribution as X. That is, each of the probability *functions* of the *random variable* X has an expectation which can be computed from X without discovering the probability function of the related variable first. A few useful laws that govern properties of expectation make this possible. We shall introduce them now via a simple random variable X with the following probability function:

VALUE OF X, x	-1	0	2
PROBABILITY, $f(x)$	0.3	0.1	0.6

For X, we have

$$\mu_x = (-1)(0.3) + (0)(0.1) + (2)(0.6) = 0.9$$

First, if we add a constant, c, to a random variable X, then the expectation of the new random variable $(X + c)$ has the same expectation as X plus the constant. That is $E(X + c) = E(X) + c$. For instance, if we add, say, 3, to our illustrative random variable X, we have

$$\mu_{x+3} = (-1 + 3)(0.3) + (0 + 3)(0.1) + (2 + 3)(0.6) = 3.9 = E(X) + 3$$

This relation is just what we would expect, since if X is increased by c for all its values, it will be increased by c on the average.

Next, analogous to the effect of adding a constant, multiplication by a constant gives $E(cX) = cE(X)$. That is, if we multiply each value of X by a constant c, its average value will be increased by c times. Consider, for instance, the expectation of the variable $(3X)$ as our illustrative problem. Clearly

$$\mu_{3x} = (-3)(0.3) + (0)(0.1) + (6)(0.6) = 2.7 = 3E(X)$$

Finally, the first two properties indicate that by multiplying a constant and adding a constant to every number of X, we multiply the mean of X by a constant and add a constant to the result. For example,

$$E(3X + 3) = (-3 + 3)(0.3) + (0 + 3)(0.1) + (6 + 3)(0.6) = 5.7 = 3E(X) + 3$$

To generalize the foregoing discussion, we introduce this intuitively obvious theorem: If X is a random variable, then for any numerical constant a and b,

$$E(aX + b) = aE(X) + b \tag{6.5}$$

The preceding property can also be stated in another way. Let X be a random variable and let $Z = aX + b$, then $E(Z) = E(aX + b) = aE(X) + b$.

From formula (6.5), we can appreciate that if X is a random variable with mean $E(X) = \mu$, then $E(X - \mu) = 0$. That is, if we deduct from each value of X by the mean of X, then the mean of the random variable, $(X - \mu)$, is zero.

Also, a random variable which is capable of assuming only a single value, say, c, is called a *constant random variable*. The expectation for such a random variable is c. Why?

To wit: If X is a random variable with probability distribution $f(x)$, can we compute the expectation of a function of X by substituting in the formula for the function the possible values of X, multiplying the results by the probabilities of each and every value of X and adding the products?

6.5 Variance

An expectation is a parameter which describes the central tendency of a random variable, or it tells us where the center of mass of the probability distribution of a random variable is located. However, two distributions with the same expectation and, therefore, same probabilities centered at the same point, may still be quite different. For instance, if X takes on values of -1, 0, and 1, with probabilities $\frac{1}{5}$, $\frac{3}{5}$, $\frac{1}{5}$, respectively, then $E(X) = 0$. Another variable, Y, takes on values, say, -10, 0, and 10, with probabilities $\frac{1}{5}$, $\frac{3}{5}$, $\frac{1}{5}$, then $E(Y) = 0$. But X and Y are obviously different, since Y's values have a much greater spread or variability from one trial of an experiment to another trial (see Figure 6.4). Thus, in addition to expectation, which furnishes us a quick picture of the long-run average when an experiment is performed over and over again, we must also know the extent to which outcomes of an experiment are dispersed, away from the concentration of probabilities in the long run, in order to have a complete description of a probability distribution. The reader is now aware that there are many measures of variability, but we shall introduce only the two most important and commonly used devices—the *variance* and its related measure, the *standard deviation*.

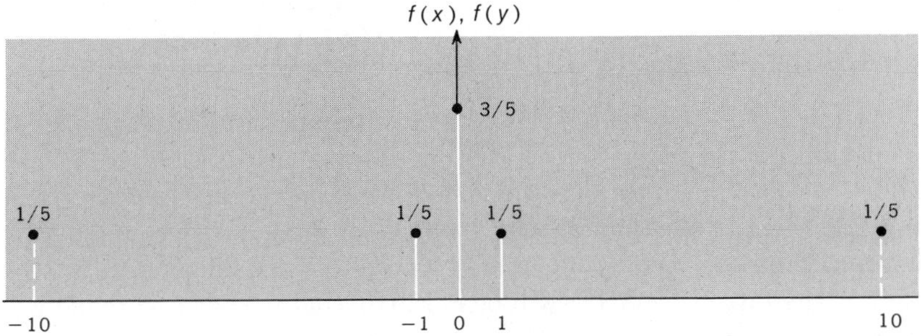

Figure 6.4 Probability distributions of X and Y with identical expectation but different variances.

If X is a random variable with expectation $E(X) = \mu$, the variance of X, denoted by $V(X)$ or σ_x^2, is defined as follows:

$$V(X) = E[(X - \mu)^2] = \Sigma(x_i - \mu)^2 f(x_i) \tag{6.6}$$

Thus, variance of X is the mean squared deviation of X from its mean. Note, variance is a weighted mean of all the squared deviations, each weighted by its probability.

Formula (6.6) is the definitional equation of variance. For actual computation it is more convenient to use its equivalent

$$V(X) = E(X^2) - [E(X)]^2 = E(X^2) - \mu^2 \tag{6.7}$$

which dispenses with the difference $X - \mu$.

Consider again, for example, the toss of an ordinary die. The expectation of variate number spots on the die, X, calculated before, is 3.5, or $7/2$. Now, we wish to determine the variance of X. First, we find

$$E(X^2) = 1^2(1/6) + 2^2(1/6) + 3^2(1/6) + 4^2(1/6) + 5^2(1/6) + 6^2(1/6) = 91/6$$

Next, by (6.7), we have

$$V(X) = E(X^2) - \mu^2 = 91/6 - (7/2)^2 = 35/12$$

For random variables X and Y as represented by Figure 6.4, we have $E(X) = E(Y) = 0$, and, by (6.6),

$$V(X) = -1^2(1/5) + 0^2(3/5) + 1^2(1/5) - 0^2 = 0.4, \quad \text{and}$$
$$V(Y) = -10^2(1/5) + 0^2(3/5) + 10^2(1/5) - 0^2 = 40$$

Note that $V(Y)$ is many times greater than $V(X)$. As a result, we say that a mean of zero is much more representative of X than it is of Y. Alternatively, we say that the total probability of X is concentrated within a narrow range, from -1 to 1; while the total probability of Y is distributed over a wide range, from -10 to 10.

6.6 Properties of Variance

The computation of variance is quite time-consuming if the values of the random variable are at all large. To save time and effort, we rely upon a procedure called the *linear transformation*. It involves shifting the origin of the domain of the values of a variable—changing the scale, so to speak. We note, first of all, that if a random variable is always equal to a constant, c, then its expectation will be c. Furthermore, such a variable will never differ from its expectation and, therefore, the variation about this expectation will be zero. Second, if we add a constant to a random variable X, then $E(X)$ will be increased by the same constant. However, the difference $X - E(X)$ remains unchanged, and therefore $V(X)$ is unaffected by adding the constant to X. Third, if we multiply each value of a random variable by a positive number, say, a, then we also multiply its expectation by a, so that the quantity $X - E(X)$ will be a times as large as before. The square $(X - E(X))^2$ will evidently be a^2 times as large as before; that is, if X is multiplied by a, or $Z = aX$, then $(Z - E(Z))^2 = (aX - aE(X))^2 = (a(X - E(X))^2 = a^2(X - E(X))^2$. Consequently, if a random variable is multiplied by a constant, a, the variance of the variable is multiplied by a^2. In particular, if a random variable is multiplied by $c = -1$, we have $V(-X) = V(X)$.

The preceding properties of variance can be summarized by the *linear transformation theorem*, which states: If X is a random variable with variance $\sigma_x{}^2$, and if a and b are numbers, then

$$V(aX + b) = \sigma^2{}_{aX+b} = a^2\sigma_x{}^2 \tag{6.8}$$

Let the probability distribution of X be

x	5050	5100	5150
$P(X = x)$	0.1	0.4	0.5

Determine $V(X)$. To simplify the solution, we introduce a new random variable, Y, defined as

$$Y = \frac{X - 5100}{50}$$

whose probability function is, clearly,

y	-1	0	1
$P(Y = y)$	0.1	0.4	0.5

For Y, we have

$$E(Y) = (-1)(0.1) + (0)(0.4) + (1)(0.5) = 0.4$$
$$E(Y^2) = (-1^2)(0.1) + (0^2)(0.4) + (1^2)(0.5) = 0.6$$

and, therefore, the variance of Y is

$$\sigma_y{}^2 = E(Y^2) - \mu_y{}^2 = 0.6 - (0.4)^2 = 0.44$$

Now, we may find $\sigma_x{}^2$ on the basis of $\sigma_y{}^2$ by recalling that, from the definition of Y, we must have

$$X = 50Y + 5100$$

Therefore, by (6.8),

$$\sigma_x{}^2 = \sigma^2{}_{(50Y+5100)} = \sigma^2{}_{50Y} = (50)^2\sigma_y{}^2 = 2500\sigma_y{}^2 = 2500(0.44) = 1100$$

The reader should go through the above example by recalling various properties of variance at every step.

For applied statistics, it is more convenient to have a measure of variability in the original units instead of the square of the original units. This leads to the introduction of the *standard deviation,* which is the positive square root of the variance. To state formally, we say that if X is a random variable with $V(X) = \sigma_x^2$, then the standard deviation of X, denoted as σ_x, is

$$\sigma_x = \sqrt{V(X)} = \sqrt{\sigma_x^2} \tag{6.9}$$

In corresponding properties of the variance, we see that the following relations holds for the standard deviation:

$$\sigma_{cx} = |c|\sigma_x \tag{6.10}$$

$$\sigma_{x+c} = \sigma_x \tag{6.11}$$

Can you give a verbal interpretation to each of these two equations?

6.7 Standard Random Variables

As we shall see in future chapters, it is much more compact and convenient to deal with standard random variables. A *standard random variable* is one with zero expectation and with unity variance. Any random variable can be standardized by deducting from it its expectation and dividing the result by its standard deviation. That is, if we let X be a random variable with mean μ_x and standard deviation σ_x, then its standardized form, denoted as Z, is

$$Z = \frac{X - \mu_x}{\sigma_x} \tag{6.12}$$

The expectation of a standardized variable is always 0, since

$$E(Z) = E\left(\frac{X - \mu_x}{\sigma_x}\right) = \frac{E(X) - E(X)}{\sigma_x} = 0 \tag{6.13}$$

where $E(X)$ and σ_x are constants over all the possible values that X can assume.

The variance of a standardized variable is always unity, since

$$V(Z) = E(Z^2) - (E(Z))^2 = E\left(\frac{X - E(X)}{\sigma_x}\right)^2 = \frac{E(X - \mu_x)^2}{\sigma_x^2} = \frac{\sigma_x^2}{\sigma_x^2} = 1 \tag{6.14}$$

which indicates, of course, σ_z is also unity.

Clearly, from our previous discussions on functions of random variables, the Z transformation does not in any way change the pattern of the original probability function. The probability of any Z value is simply the probability of the corresponding value of the original random variable. We also note that each X value corresponds to one, and only one, Z value. The Z transformation has the effects of changing the scale of the original variable and of reducing the individual deviations of the original variable from the mean to standard deviation units.

6.8 The Chebyshev Inequality

Up to now, we have discussed the expectation, the variance, and the standard deviation for probability distributions. We have also pointed out that an expectation measures the concentration of probability values and that a standard deviation measures the dispersion of probability values. Thus, the standard deviation can be employed to furnish information about the way probability accumulates in intervals centered around the mean as their distances widen. Intuitively, we see that a small standard deviation indicates that large deviations from the expectation are improbable. But can we be more precise about the relationship between the measure of central tendency and the measure of variation? Can we, for example, answer such questions as these: (1) What is the percentage of total probability that lies in a specified interval centered at the mean? (2) How wide an interval about the mean is required so that, say, 95 percent of the total probability is included in the interval?

There exists a remarkable theorem called the *Chebyshev* (sometimes spelled Tchebysheff or Tchebichev) *inequality* which enables us to answer just such questions about probability distributions as those raised above. This inequality theorem states that if X is a random variable with mean μ and finite standard deviation σ, then

$$P(|X - \mu| > h\sigma) \leq \frac{1}{h^2} \tag{6.15}$$

This shows that the probability that X assumes a value outside the closed interval from $\mu - h\sigma$ to $\mu + h\sigma$ is never more than $1/h^2$. For instance, the probability for any value of a random variable to deviate from its mean by two standard deviations, 2σ, is at most $\frac{1}{4}$. The probability for any value of a random variable to differ from its mean by 3σ must be less than $\frac{1}{9}$. And so on.

An alternative way to express the Chebyshev inequality is

$$P(|X - \mu| \leq h\sigma) \geq 1 - \frac{1}{h^2} \tag{6.16}$$

which says that at least the fraction $1 - (1/h^2)$ of the total probability of a random variable lies within h standard deviations from the mean. We note that (6.15) and (6.16) are equivalent statements. If the probability assigned to values of X outside the interval $\mu \pm h\sigma$ is at most $1/h^2$, then the probability assigned to values of X within a distance of $h\sigma$ of the mean must be at least

$1 - (1/h^2)$. The utility of Chebyshev inequality is that it holds for any probability distribution.

Let us demonstrate the validity of the Chebyshev inequality by way of an example. Consider a random variable X with the following probability function:

x	1	2	3	4	5
$P(X = x)$	0.1	0.1	0.2	0.3	0.3

What is the probability that is associated with values of X at (a) 2σ and (b) 3σ from its expectation?

For X here, we have

$$\mu = 3.6 \text{ and } \sigma = \sqrt{1.64} = 1.28062$$

For these values, we find that

$$\text{(a)}\quad \begin{aligned} P(|X - \mu| \leq 2\sigma) &= P((\mu - 2\sigma) \leq X \leq (\mu + 2\sigma)) \\ &= P(1.03876 \leq X \leq 6.16124) \\ &= P(2) + P(3) + P(4) + P(5) \\ &= 0.9 \end{aligned}$$

which is much greater than $1 - \frac{1}{4} = \frac{3}{4}$.

$$\text{(b)}\quad \begin{aligned} P(|X - \mu| \leq 3\sigma) &= P(-0.24186 \leq X \leq 7.44186) \\ &= P(1) + P(2) + P(3) + P(4) + P(5) \\ &= 1 \end{aligned}$$

which is also greater than $\frac{8}{9}$ as the Chebyshev inequality dictates.

The graph of the probability function of X together with its expectation, identified by a star $*$, and intervals extending to 2σ and 3σ to the left and right of the expectation, is shown by Figure 6.5.

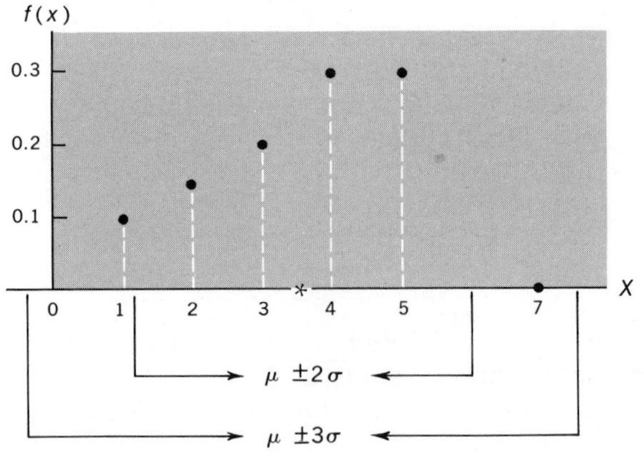

Figure 6.5 Graphic illustration of the application of the Chebyshev inequality.

6.9

Joint Probability Distributions

More than one random variable may be generated from the same experiment. For instance, if we wish to study the relationship between the diameters and the tensile strength of steel wires, we may let the measures of diameter and tensile strength be two random variables whose values are determined by outcomes of the experiment of measuring those properties of a steel wire selected at random from the population, say, of a day's production run. Or, family income, liquid assets, and consumption expenditure can be considered three random variables whose values are the results of observations made of a family chosen at random from households, say, in a given city. Needless to say, four or more random variables may be associated with outcomes of an experiment in some situations.

In the discussion that follows, we shall introduce the notion of the joint probability function of two random variables but shall extend our comments on the expectation and variance of a joint probability function to more than two variables.

To study the relationship between two variables, we are naturally interested in knowing the probability that one variable takes on as a particular value, given the value of the other variable; namely, the joint probability distribution of the two variables under consideration. We are also, just as in the case of a single variable, interested in characteristics of the joint probability distribution: its expectation and variance. We shall now develop these concepts of joint probability distribution via an example.

Four coins are tossed. If we let X be the number of heads and Y be the number of runs, then we have a sample space as presented in Table 6.5. Since all sample points in this sample space are equally likely, each has a probability of $1/16$. From this, we see that the probability distributions of X and Y, respectively, are

PROBABILITY DISTRIBUTION OF X

x	0	1	2	3	4
$P(X = x)$	$1/16$	$4/16$	$6/16$	$4/16$	$1/16$

PROBABILITY DISTRIBUTION OF Y

y	1	2	3	4
$P(Y = y)$	$2/16$	$6/16$	$6/16$	$2/16$

To find the joint probability distribution of X and Y, we ask the question: What is the probability that X and Y jointly take on the values, say, (0, 1)?

TABLE 6.5 SAMPLE POINTS, NUMBER OF HEADS, AND NUMBER OF RUNS OF FOUR COINS

Sample Point	Number of Heads X	Number of Runs Y
$HHHH$	4	1
$HHHT$	3	2
$HHTH$	3	3
$HTHH$	3	3
$THHH$	3	2
$HHTT$	2	2
$HTHT$	2	4
$HTTH$	2	3
$THTH$	2	4
$TTHH$	2	2
$THHT$	2	3
$HTTT$	1	2
$THTT$	1	3
$TTHT$	1	3
$TTTH$	1	2
$TTTT$	0	1

Similar questions can be raised for all other possible ordered pairs of values of (x_i,y_j).

Answers to such questions can be deduced from listings in the original sample space and the sample points that correspond to the joint values of X and Y. For instance, from Table 6.5, we see that when $X = 0$, $Y = 1$, and these are the only possible pair of values among the total of 16 pairs. Thus, we conclude $P(X = 0,\ Y = 1) = 1/16$. The probability when $X = 0$ and $Y =$ any value other than 1 is zero. Similarly, for $X = 1$, $Y = 2$ twice and $Y = 3$ twice. Thus, $P(X = 1,\ Y = 2) = 2/16$ and $P(X = 1,\ Y = 3) = 2/16$. Also, if $X = 1$ and $Y =$ any value other than 2 or 3, the probability is zero. When probabilities of all possible joint events, $P(X = x_i,\ Y = y_j)$, are determined ($i, j = 1, 2, 3, 4$ in our example), then we have a joint probability distribution of X and Y, and these results may be presented in a two-way table as in Table 6.6.

TABLE 6.6 JOINT PROBABILITY DISTRIBUTION OF NUMBER OF HEADS AND NUMBER OF RUNS FOR FOUR COINS

		X: Number of Heads					
		0	1	2	3	4	Row Total
Y: Number of Runs	1	1/16	0	0	0	1/16	2/16
	2	0	2/16	2/16	2/16	0	6/16
	3	0	2/16	2/16	2/16	0	6/16
	4	0	0	2/16	0	0	2/16
Column Total		1/16	4/16	6/16	4/16	1/16	16/16

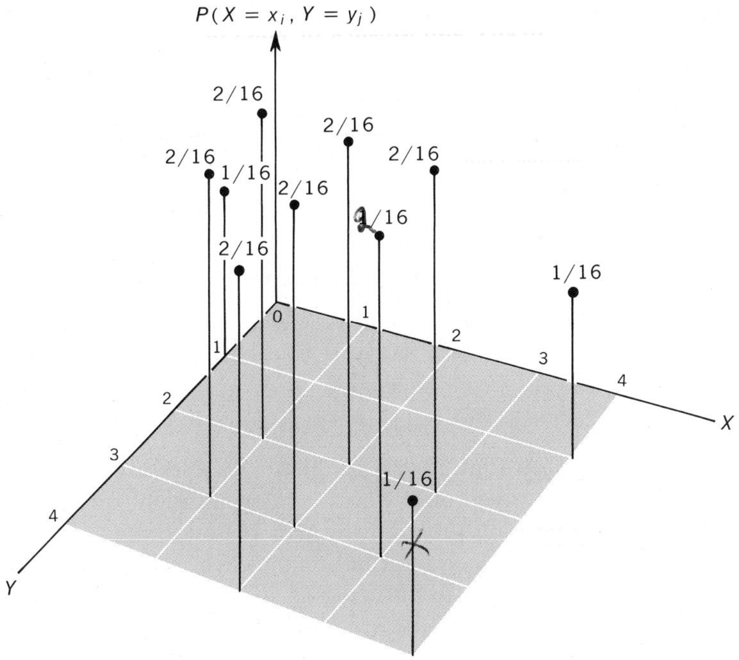

Figure 6.6 Joint distribution of number of heads and number of runs for four coins.

Observe that the column total is $f(x)$ and the row total is $f(y)$ in Table 6.6. Individual probability distributions of X and Y are sometimes called *marginal distributions* because they are found at the margins of a joint probability table.

The graph of a joint probability function of two variables is constructed in a three-dimensional space. Data in Table 6.6 are shown by Figure 6.6.

To generalize the concept of *joint probability functions,* we state: Let X be a random variable with possible values x_i, $i = 1, 2, \cdots, n$, and Y be a random variable with possible values y_j, $j = 1, 2, 3, \cdots, n$, then the probability that X takes on the value of x_i and Y takes on the value of y_j for each ordered pair (x_i,y_j), is the value of the joint probability function of X and Y at (x_i,y_j), or,

$$P(x_i,y_j) = P(X = x_i \text{ and } Y = y_j) \tag{6.17}$$

Note that the relationship between two variables may be either dependent or independent. Dependence implies the existence of real relation between two variables and therefore it is possible to predict the value of one when a particular value of the other is known. Independence implies the lack of relationship between two variables. This distinction can be made easily from the joint probability function of the two variables.

Two random variables are said to be *statistically independent* if each ordered pair of values, (x_i,y_j), as defined in (6.17) satisfies the product rule for independent events. Thus, if X and Y are independent random variables, then

$$P(X = x_i \text{ and } Y = y_j) = P(X = x_i)P(Y = y_i) = P(x_i)P(y_j) \tag{6.18}$$

which, in effect, says that for each ordered pair the joint probability is equivalent to the product of the respective marginal probabilities.

In our illustrative example for joint probability distribution, X and Y are clearly dependent, not independent, since a number of combinations of the values of X and Y have zero probabilities. Consider, for example, the ordered pair $(X = 1, Y = 1)$. For this point, we find, from Table 6.6.

$$P(X = 1 \text{ and } Y = 1) = 0$$

but from the marginal probability distributions, we have

$$P(X = 1) = 4/16$$
$$P(Y = 1) = 2/16$$

and, therefore,

$$P(X = 1)P(Y = 1) = (2/16)(4/16) = 1/32$$

It is interesting to note that if two random variables are dependent, their joint distribution can be derived only when there is sufficient information about the sample space so that joint probabilities can be determined directly, or by analytical deduction. Without such information, however, we must rely on empirical estimates based on relative frequencies of joint events. See, for example, Problem 6.46. If two random variables are independent, on the other hand, then their joint distribution can be directly synthesized from their marginal distributions. See, for example, Problem 6.43.

6.10 Characteristics of Joint Probability Distribution

Just as in the case of a single probability distribution, the behavior of a joint probability distribution is summarized by its expected value and variance. Calculations of the values for a joint probability distribution depend upon the functional relationship between the variables involved. For example, two random variables may be related to each other in various forms of functions, such as, $g(X, Y) = X + Y$, $g(X, Y) = X - Y$, $g(X, Y) = XY$, $g(X,Y) = X/Y$, $g(X,Y) = X$, $g(X,Y) = Y$, and so forth. Among the possible functions mentioned, we are particularly interested in the *additive model*, $X + Y$, and the *product model*, XY. (Of course, the *difference model*, $X - Y$, has the same logical foundation as the additive model.) We shall, in the next few pages, discuss the expectation of these two models and the variance of the additive model.

If X and Y are two random variables with expectations $E(X)$ and $E(Y)$, then the expectation of a sum (difference) of X and Y is the sum (difference) of their respective expectations. Symbolically,

$$E(X + Y) = E(X) + E(Y) \tag{6.19}$$

and

$$E(X - Y) = E(X) - E(Y) \tag{6.20}$$

which is called *additive (difference) law of expectation.*

When we think of $E(X)$ as, say, the long-run average "payoff" from playing a game of X, and $E(Y)$ as the long-run average "payoff" from playing another game, Y, it seems intuitively plausible that $E(X + Y)$ may be thought of as the total average "payoff" from playing both games in the long run. Clearly, the total expected long-run "payoff" from playing both games will equal the average "payoffs" from the two sources separately, be it a sum or a difference. Note that the additive law of expectation holds for both dependent and independent variables.

Considering, for example, the four-coins experiment, we can show that the expected number of heads is

$$E(X) = 2$$

and the expected number of runs is

$$E(Y) = 2.5$$

Now, from Table 6.6, we see that

$$\begin{aligned} E(X + Y) = {} & (0 + 1)(1/16) + (1 + 2)(2/16) + (1 + 3)(2/16) \\ & + (2 + 2)(2/16) + (2 + 3)(2/16) + (3 + 2)(2/16) \\ & + (3 + 3)(2/16) + (4 + 1)(1/16) + (2 + 4)(2/16) = 72/16 = 4.5 \end{aligned}$$

which is equal to $E(X) + E(Y)$, as the additive law of expectation dictates.

The reader may try to verify that $E(X - Y) = -0.5$ for the same example. The additive law of expectation can be extended to any number of random variables. Precisely, if $X_1, X_2, \cdots, X_n$ are any n random variables, which possess expectations, then,

$$\begin{aligned} E(X_1 + X_2 + \cdots + X_n) &= E(X_1) + E(X_2) + \cdots + E(X_n) \\ &= \Sigma E(X_i) \end{aligned} \tag{6.21}$$

By analogy of the additive law of expectation, one may expect to have an additive law of variance. Unfortunately, such a law of variance can hold only if the random variables are statistically independent. To derive a general expression for $V(X + Y)$, let us use the notations $V = X - E(X)$ and $W = Y - E(Y)$, so that $V(X) = E(V^2)$ and $V(Y) = E(W^2)$. Then, the variance of $X + Y$ must be

$$\begin{aligned} V(X + Y) = \sigma^2_{x+y} &= E[(V + W)^2] = E(V^2 + 2VW + W^2) \\ &= E(V^2) + E(W^2) + 2E(VW) = \sigma_x^2 + \sigma_y^2 + 2E(X - \mu_x)(Y - \mu_y) \end{aligned}$$

where $E(X - \mu_x)(Y - \mu_y)$ is called the *covariance* of X and Y and is generally denoted as cov (X,Y). Thus we may write

$$\sigma^2_{x+y} = \sigma_x^2 + \sigma_y^2 + 2 \text{ cov } (X,Y) \tag{6.22}$$

It may be pointed out that cov (X,Y) is a measure of the tendency of X and Y to vary together. It reveals the degree and type of dependence between X and Y. Thus, we can appreciate the fact that if there is no relationship (dependence) between the two variables, that is, if they are statistically independent, the covariance of X and Y must be zero, even though the covariance of X and Y could be zero when they are dependent variables. Then, if X and Y are independent random variables, (6.22) would be reduced to

$$\sigma^2_{x+y} = \sigma_x^2 + \sigma_y^2 \tag{6.23}$$

Similarly,

$$\sigma^2_{x-y} = \sigma_x^2 + \sigma_y^2 \tag{6.24}$$

Thus, we see that only independent random variables have an additive property in variance.

Just as the additive law of expectation can be extended, so can the additive property of variance, to any finite number of random variables. If $X_1, X_2 \cdot \cdot \cdot , X_n$ are statistically independent random variables, then

$$V(X_1 + X_2 + \cdot \cdot \cdot + X_n) = V(\Sigma X_i) = \Sigma V(X_i) \tag{6.25}$$

While the expectation of a sum is the sum of expectations, the expectation of a product is not always the product of expectations. Consider, for example, the joint probability distribution of X and Y below:

		X: −1	X: 2	Row Total
Y	0	0.5	0	0.5
	1	0	0.5	0.5
Column Total		0.5	0.5	1.0

For this distribution, we see that

$$E(X) = (-1)(0.5) + (2)(0.5) = 0.5$$
$$E(Y) = (0)(0.5) + (1)(0.5) = 0.5$$
$$E(XY) = (-1 \times 0)(0.5) + (-1 \times 1)(0) + (2 \times 0)(0) + (2 \times 1)(0.5) = 1$$

Thus,

$$E(XY) = 1 \neq E(X)E(Y) = 0.25$$

It happens that X and Y in our present illustration are not independent variables. It must be noted that dependence does not necessarily preclude the equality between $E(XY)$ and the product $E(X)E(Y)$. However, if X and Y are independent random variables with $E(X)$ and $E(Y)$, then it is always true that

$$E(XY) = E(X)E(Y) \tag{6.26}$$

Consider, for example, the joint probability distribution that follows:

Values of X

		1	2	3	Row Total
Values of Y	0	0.04	0.20	0.16	0.4
	1	0.06	0.30	0.24	0.6
Column Total		0.10	0.50	0.40	1.0

Here, X and Y are clearly statistically independent since each cell in the above table is the product of the corresponding row and column totals. As independent variables, they must satisfy (6.26), as calculations below indicate:

$$E(X) = (1)(0.1) + (2)(0.5) + (3)(0.4) = 2.3$$
$$E(Y) = 0(0.4) + 1(0.6) = 0.6$$
$$E(XY) = (1 \times 0)(0.04) + (1 \times 1)(0.06) + (2 \times 0)(0.2) + (2 \times 1)(0.3) + (3 \times 0)(0.16) + (3 \times 1)(0.24) = 1.38$$

and

$$E(XY) = 1.38 = E(X)(E(Y) = (2.3)(0.6)$$

We note that in this example X and Y are independent, and as a result predication of the value of one variable with the value of another cannot be made. For instance, if $X = 1$, we find $Y = 0$ or 1. Or, if $Y = 0$, then $X = 1$, 2, or 3. However, if we had two dependent variables we would improve the precision of predicting Y from X or X from Y. Thus, in our last example, if $X = -1$, we would find that $Y = 0$; also, if $Y = 1$, then $X = 2$. These, incidentally, are the only possible values for the last joint probability distribution. In a later chapter we shall employ this predictive property to our advantage for association analysis.

In concluding this section, we should point out that the additive model of joint probability distribution is most important for our later discussion. Nearly all special probability models to be introduced in this text possess an additivity property. Recall that if $X_1, X_2, \cdots, X_n$ are any random variables, their sum is a random variable and the expectation of the sum is the sum of their separate expectations. Furthermore, if these random variables are independent, the variance of the sum is also the sum of their separate variances.

6.11

Continuous Random Variables

Up to this point, we have limited our discussion to *discrete random variables*. A *discrete random variable* can take on only a finite number of values. Also, its distribution function, $F(x)$, as has already been pointed out, is one which increases only in finite jumps and which is constant between jumps. As such, we say that $F(x)$ for a discrete random variable is a *step function.*

Often we encounter random variables, such as time or length, which can assume as values any real numbers in an interval or a union of such intervals. Variables of this kind are called *continuous random variables.*

The probability that a continuous random variable assumes any single particular value is zero, since there are infinite numbers of real numbers within the intervals over which X is defined. Consequently, a continous random variable cannot be described by the probability function for discrete random variables. Instead, events in continuous cases are defined in terms of subintervals, and a random variable is characterized by what is called *probability density function,* or simply *density function.*

If X is a continuous random variable which can assume values between $-\infty$ and $+\infty$, then its density function, which shows the probability that X will fall in the interval (a,b), $a < b$, is

$$f(x) = P(a < X < b) = \int_a^b f(x)dx \qquad \textbf{(6.27)}$$

where $\int_a^b f(x)dx$, called a *definite integral,* means the area under the curve of $f(x)$ between the two points, a and b.

The *cumulative distribution function,* CDF, $F(x)$, of a continuous random variable, similar to the discrete case, shows the probability that the random variable assumes values of less than or equal to a particular value. Thus, if X is a random variable as defined above, we have as the CDF of X,

$$F(x) = P(X \leq x) \qquad \textbf{(6.28)}$$

Note from the preceding two definitions that the $f(x)$ and $F(x)$ of a continuous distribution is given by

$$P(a < X < b) = F(b) - F(a) \qquad \textbf{(6.29)}$$

which may be interpreted in this fashion: $P(a < X < b)$ is given by the area under the curve of the density function enclosed by the ordinates elected at $X = a$ and $X = b$ as shown by Figure 6.7.

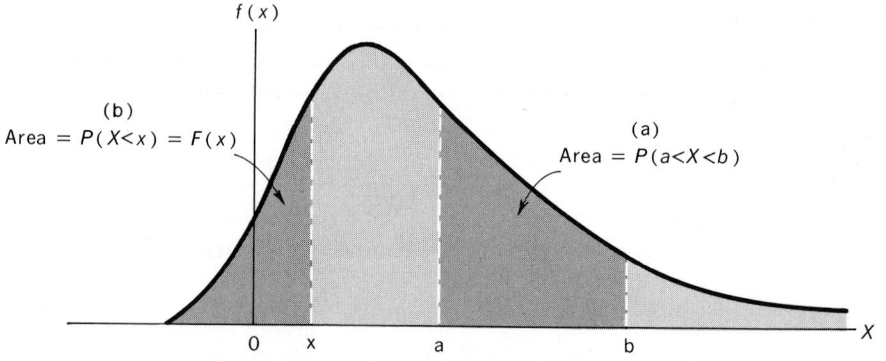

Figure 6.7 Probability interpreted as area (a), and the distribution function as area (b), for a continuous distribution.

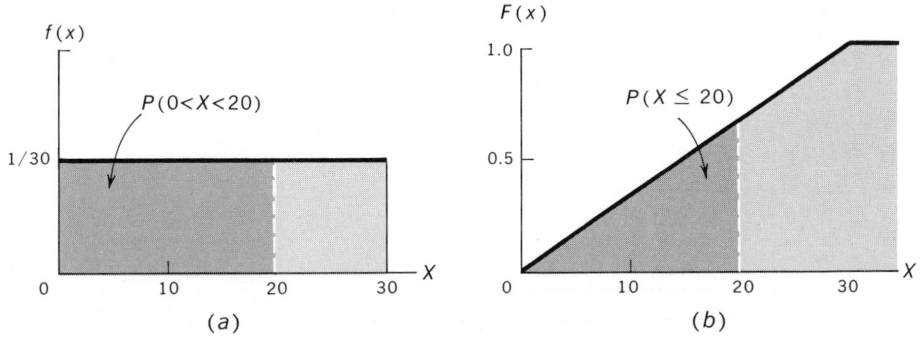

Figure 6.8 Area probability graphs, in terms of (a) density and (b) cumulative functions, for the subway example.

Also note that the total area under the density curve, as total probability of the distribution, must be 1. That is $F(+\infty) = P(-\infty < X < +\infty) = 1$.

Consider, as the first illustration of continuous random variables, this example: Buses in a certain city run every half hour between 10 P.M. and 6 A.M. What is the probability that a man arriving at a bus stop at a random time will have to wait for 20 minutes or less?

Here, the random variable time, T, is a number between 0 and 30 (in minutes). It seems plausible for us to assign to each subinterval of one minute a probability proportional to the length of the interval, T, from 0 to 30. By this rule, each unit interval of one minute would be assigned a weight of 1/30, since the entire interval has a total probability of 1. With these considerations, we have then

$$P(0 < X < 20) = \frac{20 - 0}{30 - 0} = \frac{2}{3}$$

or

$$P(X \leq 20) = F(20) - F(0) = {}^{20}\!/_{30} - 0 = {}^{2}\!/_{3}$$

The solutions of this problem are illustrated by the graphs of density and cumulative functions of T in Figure 6.8. Observe that the graph of the distribution function of a continuous random variable is also continuous. It has no jumps and hence there are no jumps of probability at any value of the random variable.

A discrete probability distribution is like the distribution of discrete mass, which is a collection of point masses at a finite number of points along the horizontal axis. Similarly, a continuous probability distribution is analogous to the distribution of a continuous mass, which has no mass at any point but has mass in subintervals.

In many situations, it is much more convenient to calculate continuous probabilities from the cumulative distribution function of a random variable than from its density. For example, let X be a random variable with the following CDF:

$$F(x) = \begin{cases} 0, \text{ if } x < 0 \\ x^2, \text{ if } 0 < x < 1 \\ 1, \text{ if } x > 1 \end{cases}$$

From this, we can compute probability of any event which is a subinterval of (0,1). For instance,

$$P(X \leq 0.6) = P(0 < X < 0.6) = (0.6)^2 - (0)^2 = 0.36$$
$$P(0.1 \leq X \leq 0.6) = F(0.6) - F(0.1) = F(0.6) - F(0.1) = 0.36 - 0.01 = 0.35$$
$$P(X \geq 0.5) = 1 - F(X < 0.5) = 1 - 0.25 = 0.75$$
$$P(-1.5 < X < 2.7) = F(2.7) - F(-1.5) = 1 - 0 = 1$$
$$P(X < 0.5 \text{ or } X > 0.8) = P(X < 0.5) + P(X > 0.8) = 0.25 + 0.36 = 0.61$$

The expected values and variance of a continuous random variable may be considered as limits of the corresponding expressions for a discrete random variable as the latter approach continuity. Similar to the discrete case, we can think of the continuous probability distribution as a continuous mass distribution. As such, the mean can be interpreted as the center of gravity and variance as the radius of gyration.

In our future discussion of special probability distributions, discrete or continuous, the reader will find that expectations and variances are given without advanced mathematical notations. Furthermore, he will find that probabilities for each of these special distributions have already been tabulated in detail. Thus, as we continue our exploration of the exciting world of probability, the scope of our discussions will be widened but its depth will not be intensified. The reader, in other words, will not encounter more difficult computations than what he has already experienced.

6.12 The Nature of Probability Distributions

We have in this and the preceding chapters introduced the basic elements of probability theory. Probability theory, as we pointed out earlier, is an indispensable tool for decision making whenever uncertainty is encountered. A theory, or a model, developed by a scientific investigation, is mainly an explanation of the behavior of a certain phenomenon. A model is said to be *deterministic* if it enables us to say that certain states or results are certain to follow, given certain initial conditions. Thus, a deterministic model is a cause and effect explanation—certain conditions cause the subsequent state. However, in many situations, such a clear-cut relationship can never be established because of the universality of uncertainty. Given uncertainty, we can have only probabilistic models. A *probabilistic* model enables us to say only that certain states will *probably* result, given certain initial conditions. That is, given such and such conditions, a probabilistic model enables us to deduce a probability distribution for possible subsequent states.

A probability distribution is clearly a population distribution since it gives us probabilities associated with *all* possible values of a random variable. In effect, we are saying that a population is a random variable. A population, as defined in Chapter 1, is a collection of observations made of all elementary elements under a given problem situation. For example, the wages of all secretaries in Chicago constitute a population if our objective is to study certain characteristics of wages of secretaries in that city. Now, if a random selection is made from secretaries in that city, wages which the selected secretary can have are values of a random variable "wage," W. Clearly, values which W can assume must be identical with the population of all possible wages of secretaries in Chicago. Thus, we see our previous definition of "population" is consistent with the meaning of a random variable.

If a population can be considered as a random variable, then we can certainly think of a probability distribution of a random variable as the probability distribution of a population. But, what can we gain by this association? The answer is an obvious one. When we can say that a population is a random variable, a population becomes more than just a collection of observations, since we would also have probabilities for all possible results from making observations.

Often we consider a probability distribution as a theoretical distribution because it is deduced, a priori, from a set of conditions—conditions for performing a random experiment. Also, we may say that a random variable is a theoretical population because its probability or density function is the result of logical deduction. These comments lead us naturally to the conclusion that a random variable may be thought of as the theoretical counterpart of a population as traditionally defined. Thus, one of the big problems confronted by a researcher is to select a "standard" probability distribution to describe the behavior of his data.

We shall introduce, in the following chapter, some special probability distributions which could be adapted to most economic and business variables.

Glossary of Formulas

(6.1) $f(x) = P(X = x)$ The probability function of a discrete random variable X is defined as the probability that X assumes a particular value, x.

(6.2) $F(x) = P(X \leq x) = \Sigma f(x)$ The cumulative distribution function, CDF, of a random variable X shows the probability that X assumes values less than or equal to x. This probability is the sum of probabilities for values of X which are less than or equal to x.

(6.3) $P(X > x) = 1 - P(X \leq x)$ Since the total probability of a distribution is 1, and because of the definition of CDF above, the probability that X assumes values greater than x must be the difference between 1 and $\Sigma f(x)$.

(6.4) $E(X) = \Sigma x_i f(x_i)$ The mathematical expectation of a random variable is a weighted mean, with probabilities as weights. The expectation of a random variable is thus defined as the sum of the products of all possible values of X and their associated probabilities.

(6.5) $E(aX + b) = aE(X) + b$ This expression states the properties of expectation. If each value of a random variable is multiplied by a constant, its expectation is multiplied by the constant. If a constant is added to each value of a random variable, then the expectation increases by the constant.

(**6.6**) $V(X) = \Sigma(x_i - \mu_x)^2 f(x_i)$ and (**6.7**) $V(X) = E(X^2) - \mu_x^2$ The first equation defines the variance of a random variable X as its mean squared deviation of X from its mean. The second, (6.7), is the equivalent of (6.6) but more convenient for computation.

(**6.8**) $V(aX + b) = \sigma^2_{aX+b} = a^2\sigma_x^2$ This equation indicates the linear transformation property of the variance. If a constant is added to each value of a random variable, its variance remains unchanged. If a constant is multiplied by each value of a random variable, its variance is multiplied by the square of the constant.

(**6.9**) $\sigma_x = \sqrt{\sigma_x^2}$ The standard deviation of a random variable is the positive square root of its variance.

(**6.10**) $\sigma_{cX} = |c|\sigma_x$ and (**6.11**) $\sigma_{x+c} = \sigma_x$ These formulas state the properties of the standard deviation. (6.10) says that if each value of X is multiplied by a constant, then its standard deviation is multiplied by the absolute value of the constant. (6.11) indicates that if a constant is added to each value of X, the scale of X has changed but its standard deviation remains unchanged.

(**6.12**) $Z = (X - \mu_x)/\sigma_x$ (**6.13**) $E(Z) = 0$ (**6.14**) $V(Z) = 1$ If X is any random variable with μ_x and σ_x, then it can be transformed into standardized variable Z by (6.12). A standard random variable is one whose expectation is zero and whose variance is unity.

(**6.15**) $P(|X - \mu| > h\sigma) \leq 1/h^2$ and (**6.16**) $P(|X - \mu| \leq h\sigma) \geq 1 - (1/h^2)$ These are the two alternative expressions of the Chebyshev inequality. If X is a random variable and possesses finite mean and variance, then the probability for X to deviate from its mean by $h\sigma$ is less than $1/h^2$, or the probability is at least $1 - (1/h^2)$ that the value of X will fall into the interval of $h\sigma$ from its mean.

(**6.17**) $P(x_i, y_j) = P(X = x_i \text{ and } Y = y_j)$ The joint probability function of X and Y is defined as the probability of the ordered pair (x_i, y_j).

(**6.18**) $P(X = x_i \text{ and } Y = y_j) = P(X = x_i)P(Y = y_j) = P(x_i)P(y_j)$ This expression says that X and Y are statistically independent if, and only if, for each and every pair, (x_i,y_j), the product rule for independent events is satisfied.

(**6.19**) $E(E + Y) = E(X) + E(Y)$ and (**6.20**) $E(X - Y) = E(X) - E(Y)$ The sum (difference) of two random variables is the sum (difference) of their expectations.

(**6.21**) $E(X_1 + X_2 + \cdot \cdot \cdot + X_n) = \Sigma E(X_i)$ The additive property of expectation holds for any number of variables.

(**6.22**) $\sigma^2_{x+y} = \sigma_x^2 + \sigma_y^2 + 2 \text{ cov } (X,Y)$ The variance of a sum is the sum of variances only if the covariance of X and Y is zero, i.e., if X and Y are independent random variables.

(**6.23**) $\sigma^2_{x+y} = \sigma_x^2 + \sigma_y^2$ (**6.24**) $\sigma^2_{x-y} = \sigma_x^2 + \sigma_y^2$

(**6.25**) $V(X_1 + X_2 + \cdot \cdot \cdot + X_n) = \Sigma V(X_i)$

These laws of variance hold only if the random variables under consideration are independent.

(**6.26**) $E(XY) = E(X)E(Y)$ If X and Y are independent, they must satisfy this rule. However, this rule *may* hold if X and Y are dependent.

(**6.27**) $f(x) = P(a < X < b) = \int_a^b f(x)\, dx$ This defines the density function of a continuous random variable. If X is continuous, $P(X = x_i) = 0$, there is only positive probabilities on events defined as intervals.

(**6.28**) $F(x) = P(X \leq x)$ This is the cumulative distribution of a continuous random variable which can take on any real value between $-\infty$ and ∞.

(**6.29**) $P(a < X < b) = F(b) - F(b)$ This expression gives the relationship between the density and distribution functions of a continuous random variable.

Problems

6.1 Let 0 and 1 be the appearances of tails and heads, respectively. Find the probability distribution of the number generated by the toss of a fair coin.

6.2 Two coins are tossed. Find the probability distribution of the number of tails.

6.3 Three coins are tossed. Find and graph the probability function of the number of heads, X.

6.4 Show that the number of tails, Y, has the same probability distribution as the number of heads in tossing these coins.

6.5 Show that X and Y in the preceding two problems are different random variables.

6.6 Three identical marbles, numbered 1, 2 and 3, are placed in a bag. If two marbles are drawn at random with replacement, what is the probability function of the random variable Z, where Z is the sum of the numbers of the marbles?

6.7 Let X be the random variable showing the number of girls in families with four children. What is the probability distribution of X if the births of boys and girls are equally likely?

6.8 Four balls are drawn at random without replacement from an urn which contains six white and four black balls. What are the probability and distribution functions of the number of black balls?

6.9 Sketch the graphs of $f(x)$ and $F(x)$ in the previous problem.

6.10 From a lot of 12 radios containing 5 defectives, a random sample of 3 is drawn without replacement. Find the probability and distribution functions of the number of defectives in the sample.

6.11 A box of articles, 9 good and 1 defective, is tested, one article at a time. If X is the number of the test in which the defective article is located, what are the possible values of X and the corresponding probability function?

6.12 Give a general expression of the probability function in problem 6.8.

6.13 Two dice are tossed. Let X be the total number of dots on the sides turning up. Give a general expression for its $f(x)$.

6.14 Four coins are tossed. Let X be the number of heads and Y be the number of tails. Find the probability distribution of $Z = X - Y$.

6.15 Four coins are tossed. Let X = number of heads and Y = number of tails. Find the distributions of (a) $X + Y$ and (b) XY.

6.16 Three balls are drawn without replacement from a bag containing five white and two black balls. Check if the expected number of black balls is 6/7.

6.17 Determine the expectation of the random variable X given in Problem 6.3.

6.18 Determine $E(X)$ for X in Problem 6.8.

6.19 What is a fair price to pay for entering a game in which one can win \$50 and \$10 with probabilities of 0.2 and 0.5, respectively?

6.20 In a given business venture a man can make a profit of \$1000 or suffer a loss of \$500. The probability of a profit is 0.6. What is the expected profit (or loss) in that venture?

6.21 The probability known is 0.99 that a 30-year-old man will survive one year. An insurance company offers to sell such a man a \$10,000 one-year term life insurance policy at a premium of \$110. What is the company's expected gain?

6.22 In a lottery, 1000 tickets are sold at 25¢ each. There are five cash prizes of \$25, \$20, \$10, \$5, and \$1 respectively. What is the expected net gain for a purchaser of two tickets?

6.23 A random sample of three persons is selected without replacement from a group of four men and three women, to make arrangements for a conference. What is the expected number of men in the sample?

6.24 A roulette wheel has 38 equally spaced openings with numbers 00, 0, 1, 2, · · · , 36. A gambler may bet, say, \$1 on any number. The roulette wheel is spun and a small ball is dropped onto it while it is spinning. If the ball comes to rest on the number on which the gambler has bet, he receives \$35 in addition to his bet of \$1, but otherwise he loses his \$1 What is his expected gain?

6.25 A gambler who has \$700 plays the single-die game with the following system. At the first toss of the die, he bets \$100 on even numbers and quits if he wins. If he loses, he bets \$200 on even numbers at the second toss, and quits if he wins. If he loses again, he bets his final \$400 on even numbers at the third toss. Is the game fair?

6.26 The probability that a man aged 55 will live another year is 0.98. How large a premium should the insurance company charge him for a \$10,000 term life insurance policy for a year if the premium should include a profit and administrative cost of \$50?

6.27 A perfect coin is tossed until the first head is obtained. If X is the number of tosses required to obtain the first head, what is the expected value of X?

6.28 If all permutations of four measurements, 0, 1, 2, and 3, are equally likely, what is the probability function and the expectation of X, where X is the number of turning points?

6.29 Calculate (a) $E(Y)$, (b) $E(Y^2)$, (c) $E(Y - u_y^2)$, and (d) $E(Y^3)$ for the following probability distribution:

y	-20	-10	30
$P(Y = y)$	$3/10$	$2/10$	$5/10$

6.30 A random variable X has the following probability function:

x	-1	0	1
$f(x)$	$1/5$	$3/10$	$1/2$

Find probability distributions for $2X$, $X + 1$, $2X + 1$, X^2 and $(X - 0.3)^2$ and then compute their expectations.

6.31 Let X be the number of tails in a toss of three coins. Compute $E(X)$, $V(X)$, and σ_x.

6.32 Determine variances for random variables in Problems 6.3, 6.6, 6.8, and 6.28.

6.33 If the variance of a random variable is 0.8, what is the variance of the random variable $5X$? $2X$? $X/2$?

6.34 Let X be a random variable with the probability function as follows:

x	10025	10050	10075
$f(x)$	0.2	0.3	0.5

Find σ_x^2.

6.35 Let Y be a random variable with probability function as follows:

y	0.0016	0.0032	0.0064
$f(y)$	$3/5$	$3/10$	$1/10$

Find $E(Y)$, $V(Y)$, and σ_Y.

6.36 A random variable X takes on the value 1 with probability p and the value of 0 with probability $1 - p$. Show that $E(X) = p$ and $V(X) = p(1 - p)$.

6.37 Recall that $\sigma_x^2 = E(X^2) - [E(X)]^2$ and the information that $1 + 2 + \cdots + (n - 1) + n = \frac{n(n + 1)}{2}$ and $1^2 + 2^2 + \cdots + (n - 1)^2 + n^2 = \frac{n(n + 1)(2n + 1)}{6}$. Show that the expectation and variance of X that can take on values $1, 2, \cdots n$, each with $1/n$ as its probability are $u_x = \frac{n + 1}{2}$, $\sigma_x^2 = \frac{n^2 - 1}{12}$.

6.38 Assuming that $\mu_x = 4$ and $\sigma_x = 2$, answer the following:

a. At least how much of the probability lies within the range 0 and 8?

b. What is the minimum value of $P(-2 \leq X \leq 10)$?

c. What is the minimum value of $P(|X| \geq h)$?

d. What value of h guarantees $P(|X| \leq h) \geq 0.96$?

6.39 Why is the Chebyshev theorem useless if $h \leq 1$?

6.40 Using Chebyshev's theorem, determine what value of h guarantees that at least 90 and 99 percent of probabilities, respectively, are within $h\sigma$ of the mean?

6.41 A random variable has a mean value of 5 and a variance of 3.

a. What is the greatest value of $P(|X - 5| < 3)$?
b. What value of h guarantees that $P(|X - 5| \leq h) \leq 0.99$?
c. What is the least value of $P(-1 < X < 11)$?
d. What is the least value of $P(|X - 5| \leq 7.5)$?

6.42 An urn contains four identical chips numbered, one on each side, (1, 1), (2, −1), (3, 1) and (4, −1). Let the chip selected be (x,y). Find $E(X)$, $E(Y)$, $E(X + Y)$, $E(X - Y)$.

6.43 A competitive game of skill is played by teams of two players, each of whom can score 0, 1, 2, and 3 points. The score of team is the sum of the points gained by the two players. For one of such teams, the probability distributions of the two players, respectively, are as follows:

Probability function of scores of the first player

x	0	1	2	3
$f(x)$	0.1	0.3	0.3	0.2

Probability function of scores of the second player

y	0	1	2	3
$g(y)$	0.3	0.1	0.4	0.2

Assuming independence of X and Y, derive (a) the joint probability function of X and Y, (b) the probability function of the total score of the two players, T.

6.44 For Problems 6.43 (a) and (b), compute $E(X)$, $E(Y)$, $E(T)$, and check that $E(T) = E(X) + E(Y)$. Also, show that $E(XY) = E(X)E(Y)$.

6.45 Another team of two for the game, as discussed in Problem 6.43, has a probability function of total score as follows:

T	0	1	2	3	4	5	6
$h(t)$	0.05	0.15	0.20	0.25	0.20	0.10	0.05

Let teams in Problems 6.43 and 6.45 be teams I and II, respectively. Compute the probability of a tie; the probability that Team I wins; and the probability that Team II wins. (*Note:* high total score wins.)

6.46 In an oligarchic industry, the two leading firms control 50 and 30 percent of the total market, respectively. If a random sample of two buyers is selected for observation, what is the joint probability distribution of the number of buyers patronizing each firm in the sample?

6.47 For the last problem, compute $E(X)$, $E(Y)$, $E(X + Y)$ and $E(XY)$. Show that $E(X + Y) = E(X) + E(Y)$ and that $E(XY) \neq E(X)E(Y)$.

6.48 Compute $E(X)$, $E(Y)$, $E(XY)$ for the following distribution Show that $E(XY) = E(X)E(Y)$ for this distribution. Does your result indicate that X and Y are independent? Explain

Y \ X	0	1	2	3	Row Total
1	1/8	0	0	1/8	2/8
2	0	2/8	2/8	0	4/8
3	0	1/8	1/8	0	2/8
Column Total	1/8	3/8	3/8	1/8	8/8

The following joint probability distribution is used for Problems 6.49 through 6.54:

		X: 1	X: 2	X: 7	Row Total
Y	0	0.4	0.2	0.4	0.6
	1	0.1	0.3	0.6	0.4
Column Total		0.5	0.5	1.0	

6.49 Show that X and Y are dependent.

6.50 What would be your prediction of the value of Y if any value of X is given?

6.51 Compute the following probabilities:
a. $P(Y = 0|X = 1)$
b. $P(Y = 1|X = 1)$
c. $P(Y = 0|X = 2)$
d. $P(Y = 1|X = 2)$

6.52 On the basis of probability obtained in the preceding problem, what value of Y would be predicted given that $X = 1$? $X = 2$?

6.53 With the system of prediction in the answer just given, what is the expected percentage of correct predictions of values of Y, given X? What conclusion can you make by comparing the answer to that of Problem 6.50?

6.54 If roles of X and Y are reversed, can you increase the accuracy of predicting X, given Y, than you could without knowledge of Y? Be specific and thorough.

6.55 A child plays with a pair of scissors and a piece of string 9 inches long. He cuts the string in two. What is the probability that the shorter piece is at most 3 inches long?

6.56 If a clock is stopped at a random time, what is the probability that the shorter hand will stop at (a) exactly numeral 2? (b) between 2 and 6?

6.57 What is the probability that, coming at a random time upon a traffic signal which is green 40 seconds and red 20 seconds, one will find the signal green? What is the random variable here?

6.58 The distribution function of a continuous random variable X is

$$F(x) = \begin{cases} cx, & \text{if } 0 \le x < 1 \\ 0, & \text{if } x \le 0 \\ 1, & \text{if } x \ge 1 \end{cases}$$

(a) For this distribution to be continuous, what value should the constant, c, assume?
(b) Compute $P(X = 0.5)$.
(c) Compute $P(|X| < 0.5)$.

6.59 Sketch the graph for the following density function of random variable Y:

$$f(y) = \begin{cases} 1/3, & \text{if } 0 \le y < 1 \\ 2/3, & \text{if } 1 < y \le 2 \\ 0, & \text{otherwise} \end{cases}$$

6.60 Consider the density function of a variable Z as follows:

$$h(z) = Z/48 + 1/16, \quad (2,8)$$

(a) Sketch the graph for Z.
(b) Compute $P(2 < Z < 6)$.
(c) Compute $P(Z \ge 5)$.
(d) Compute $P(|X - 5| < 1/2)$.

7

SPECIAL PROBABILITY MODELS

7.1

Introduction

We have previously mentioned that one of the basic tasks of a scientific investigator is the search for a probability distribution, or a probability model, to describe data on hand. A *probability model* is a mathematical expression derived from a set of assumptions for the twofold purpose of studying the results of a random experiment and of predicting future results of the experiment when performed repeatedly.

We shall, in this chapter, introduce some special probability models which can be employed to approximate a large set of population distributions and which may serve as a beginning for our future discussion on statistical inferences. Some of these are applicable to discrete, and others to continuous, random variables. Whenever possible, for the purpose of calculating probabilities, we shall actually derive the formula from a set of assumptions that characterize a large class of experiments. We shall also rely extensively on prepared tables so that the need for numerical computations is reduced. At this stage, it is more important for the reader to identify a problem situation with the correct model than to memorize the detailed mathematical expression.

7.2 The Bernoulli Model

We begin our study of probability models with the Bernoulli distribution, not only because it is the simplest model but also because it furnishes a basis for the derivation of the binomial probability model, which is associated with a number of frequently occurring situations.

The Bernoulli model applies to a random variable which can assume only two values. For simplicity, let the two values be 1 and 0, with p and q as their respective probabilities; then the probability function of a Bernoulli variable is simply

x_i	$f(x_i)$
1	p
0	q
Sum	1

For this distribution, we note that $\Sigma x_i f(x_i) = p$ and $\Sigma x^2 f(x_i) = E(X^2) = p$. Consequently, we have for the Bernoulli variable,

$$E(X) = \mu = p \tag{7.1}$$

and

$$V(X) = \sigma^2 = E(X^2) - \mu^2 = p - p^2 = p(1 - p) = pq \tag{7.2}$$

Suppose that 60 percent of the employees of a company favor a proposed pension plan. An employee is selected at random, and we let $X = 1$ if he is in favor, and $X = 0$ if he is against, the plan; then the expectation and variance of X become

$$\mu = p = 0.60$$

and

$$\sigma^2 = pq = 0.60(1 - 0.60) = 0.24$$

The Bernoulli model is appropriate in any situation where one is interested in an experiment which results in an event, E, or its opposite, $\bar{E}$, such as success and failure, yes or no propositions, male or female subjects, defective or nondefective items, and so forth.

7.3 The Binomial Model

The binomial probability model may be considered as a series of repeated, independent Bernoulli trials. More precisely, the binomial model refers to a sequence of events which possess the following properties:

(1) A simple experiment is repeated a number of times where the outcomes are independent.

(2) Outcomes of each trial can be classified into two mutually exclusive categories, arbitrarily called "successes" and "failures."

(3) The probability of success in a single trial, denoted by p, remains the same for all trials. The probability of failure in a single trial, denoted by q, is equal to $(1 - p)$.

(4) In a given trial the focus is on whether or not the *successful* outcomes have occurred.

(5) The experiment is performed under the same conditions for a fixed number of trials, say, n.

These conditions are satisfied if we toss a coin, say, ten times, and we define success in terms of the number of heads that turn up; if we select five items at random with replacement from a lot, say, of one hundred units of incoming merchandise and we define success as the number of defective items in the sample; and so on.

A random variable generated under the foregoing conditions is called a *binomial variable;* it is *discrete* and has $n + 1$ possible values. For example, if a coin is tossed twice, then the possible number of heads are 0, 1, and 2; that is, $(2 + 1)$ terms.

This model is useful to answer questions such as this: If we conduct an experiment under the stated conditions n times, what is the probability of obtaining exactly x successes? More specifically, suppose six dice are tossed together, or one die is tossed six times, what is the probability of obtaining exactly two aces?

To answer this question, we note that the outcome of each die is independent from those of all others. Moreover, the outcome of each die may be either an ace or any one of the other five numbers. That is, this experiment satisfies all premises of a binomial model: we have $p = 1/6$, $q = 5/6$, and $n = 6$. Now, any sequence which contains two aces and four other numbers can be considered as a successful outcome. The probability for any such sequence, by the product rule, is p^2q^4, or $(1/6)^2(5/6)^4$. For instance, the probability for a sequence in which the first and the last outcomes are aces, and the remaining four are not, is

$$(1/6)(5/6)(5/6)(5/6)(5/6)(1/6) = (1/6)^2(5/6)^4$$

Furthermore, all sequences that contain exactly two aces each are mutually exclusive events and, therefore, the probability of success, by the law of addition, is the sum of the individual probabilities of all these sequences; that is,

$$(1/6)^2(5/6)^4 + (1/6)^2(5/6)^4 + \cdots$$

where there are as many identical terms as there are ways of selecting two things at a time from six objects, $\binom{6}{2}$—a two-category permutation, $\frac{6!}{2!(4!)}$. These considerations lead us to conclude that the probability of obtaining exactly two aces in six tosses of a die, with the probability of success in each trial being $\frac{1}{6}$, is

$$P(2;\, n = 6,\, p = \tfrac{1}{6}) = \binom{6}{2}\left(\frac{1}{6}\right)^2\left(\frac{5}{6}\right)^4 = \frac{9{,}375}{46{,}656}$$

To generalize: The *binomial probability function* is the probability of obtaining exactly x successes in n independent trials of an experiment, with p as the probability of success for each trial; it is given by the expression

$$b(x;n,p) = \binom{n}{x} p^x q^{n-x} \tag{7.3}$$

The binomial model derives its name from the fact that (7.3) is a term of the "binomial" $(q + p)^n$.

In a lot of 12 television tubes 3 are defective. If a random sample of 3 is drawn from the lot with replacement, what is the probability that (1) exactly 1 is defective, and (2) none or 1 is defective?

Sampling with replacement ensures independence of successive selections, and therefore the binomial model applies. As stated in the problem, $p = \frac{3}{12} = 0.25$. Thus

(1) $b(1;3,0.25) = \binom{3}{1}(0.25)(0.75)^2 = 0.42$

(2) $b(0;3,0.25) + b(1;3,0.25) = \binom{3}{0}(0.25)^0(0.75)^3 + \binom{3}{1}(0.25)(0.75)^2 = 0.84$

The *distribution function* of a *binomial variable*, as that of any discrete random variable, gives the probability of obtaining r successes or less in n trials, with $r \leq n$, and is obtained by adding the individual probabilities for all binomial values equal to or less than r. Namely,

$$B(r;n,p) = P(X \leq r) = b(0;n,p) + b(1;n,p) + \cdots + b(r;n,p) = \sum_{x=0}^{r} b(x;n,p) \tag{7.4}$$

Extensive tables for the binomial distribution function are available. A shorter table of the binomial cumulative probabilities appears as Table A-III in Appendix C. This table, as shown by the following examples, can be used to obtain cumulative as well as individual binomial probabilities.

A coin is tossed ten times; what is the probability of getting 8 heads or more? Here, we have a binomial model with $n = 10$ and $p = \frac{1}{2} = 0.5$. The desired probability is that of obtaining 8, 9, or 10 heads. What we need then is

$$\sum_{x=8}^{10} b(x;10,0.5) = 1 - \sum_{x=0}^{7} b(7;10,0.5)$$

$$= 1 - 0.94531 \qquad \text{from Table A-III}$$

$$= 0.05469$$

In a 20-question, 4-answer, multiple-choice examination, what is the probability of getting (1) exactly 7 correct answers, or (2) 7 or more correct answers, if a student answers randomly? To solve this problem, let us assume that all questions are answered, that all questions are independent, and that the answer is either right or wrong. Under these premises, we would then have a binomial model with $p = 1/4$ and $n = 20$. Then

$$\begin{aligned}(1)\quad b(7;20,0.25) &= B(7;20,0.25) - B(6;20,0.25)\\ &= 0.89819 - 0.78578 \qquad \text{from Table A-III}\\ &= 0.11241\end{aligned}$$

$$\begin{aligned}(2)\quad \sum_{x=7}^{20} b(x;20,0.25) &= 1 - \sum_{x=0}^{6} b(x;20,0.25)\\ &= 1 - 0.78578 \qquad \text{from Table A-III}\\ &= 0.21422\end{aligned}$$

The production output of a certain process is 90 percent perfect; the items which comprise the defective remainder cannot be detected without destroying them through inspection. If a random sample of 15 is drawn from a certain day's production run—which is more than 100,000 units—what is the probability that the sample will contain 12 or less perfect units?

We now have a case where sampling with replacement is impossible because sampling is destructive. However, the population is extremely large compared with the size of the sample; as a result, the probability of getting a perfect item remains approximately at the same value of 0.90 at each selection and, therefore, the binomial model can be employed as an approximation. Here, we have a binomial model with $n = 15$ and $p = 0.90$. We wish to evaluate $\sum_{x=0}^{12} b(x;15,0.90)$. However, because our binomial table gives ps up to only 0.50, probabilities for ps greater than 0.50 cannot be read from it directly. To use the table when $p > 0.50$, we must make some transformation.

In evaluating the sum

$$P(X \leq r) = B(r;n,p) = \sum_{x=0}^{n} b(x;n,p)$$

we see that the probability of r or fewer successes is the same as the probability of $n - r$ or more failures. Consequently, we can write

$$B(r;n,p) = \sum_{x=n-r}^{n} b(x;n,q) = 1 - \sum_{0}^{n-r-1} b(x;n,q) = 1 - B(n - r - 1;n,q) \qquad \textbf{(7.5)}$$

This indicates that when $p > 0.50$, we can find the cumulative binominal probabilities from the table by interchanging the roles of p and q and, thereby, the roles of successes and failures.

For our present problem, to obtain 12 or fewer perfect items is the same as to get 3 or more defective ones. From (7.5), it follows that

$$\begin{aligned}\sum_{x=0}^{12} b(x;15,0.90) &= \sum_{x=3}^{15} b(x;15,0.10)\\ &= 1 - \sum_{x=0}^{2} b(x;15,0.10)\\ &= 1 - 0.81594 \qquad \text{from Table A-III}\\ &= 0.18406\end{aligned}$$

A binomial variable X, the number of occurrences of successes in n independent trials, may be considered as a sum

$$X = X_1 + X_2 + \cdots + X_n$$

where each X_i is a Bernoulli variable with expectation p. Thus, the expectation of the binomial variable may be thought of as the sum of expectations of the n Bernoulli variables by the additive property of expectation. Namely,

$$E(X) = E(X_1) + E(X_2) + \cdots + E(X_n) = \underbrace{p + p + \cdots + p_n}_{n \text{ times}} = np \tag{7.6}$$

Furthermore, the Bernoulli variables are identical and independent. As such, their variances are also additive. That is, the binomial variable, as the sum of n Bernoulli variables, must have as its variance the sum of the variances of the n Bernoulli variables; namely,

$$V(X) = V(X_1) + V(X_2) + \cdots + V(X_n) = \underbrace{pq + pq + \cdots + pq}_{n \text{ products}} = npq \tag{7.7}$$

According to the past experience of a mail-order house, the probability of receiving more than $250,000 in new orders on any day is a constant at 0.1, and the probability of not receiving more than this amount is 0.9. If the next five days are considered, then the number of days on which more than $250,000 in new orders are received is a binomial variable with probability and distribution functions as follows:

No. of days	0	1	2	3	4	5
$b(x)$	0.5905	0.3280	0.0729	0.0081	0.0004	0.0000*
$B(x)$	0.5905	0.9185	0.9914	0.9995	0.9999	0.9999†

* *$b(5) = 0.00001$.* † *This result is due to rounding. Had we carried the probability values to five decimal places, $P(X \geq 5)$ would equal unity.*

For this distribution,

$$E(X) = np = (5)(0.10) = 0.5 \text{ days}$$

which means that on 0.5 days in a five-day period, or on one day in a ten-day period, we would expect to have more than $250,000 in new orders. The variance

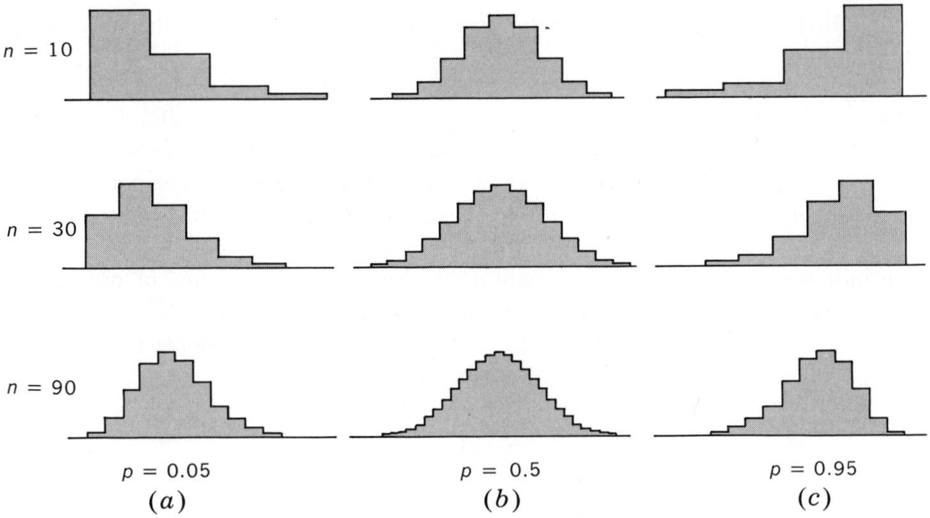

Figure 7.1 Histograms of standardized binomial distributions.

of this distribution is

$$V(X) = npq = (5)(0.10)(0.90) = 0.45$$

Now, if the normal staff in the shipping department can handle up to \$250,000 in new orders per day without delay, what is the probability that shipping will be delayed during the next five days? Delays will occur if, and only if, new orders on any day exceed \$250,000. From $B(x)$ above, we find that the probability that new orders will be equal to or less than this amount on all five days is 0.5905, and thus the desired probability is $1 - 0.5905 = 0.4095$.

Before concluding our discussion on the binomial model, we should note that the binomial distribution is symmetrical if $p = 0.5$, and skewed if $p \neq 0.5$. When $p < 0.5$, it is skewed to the right; when $p > 0.5$, it is skewed to the left. The interchange of p and q in any binomial distribution yields its mirror image. The skewness of the binomial distribution, irrespective of the size of p, becomes less pronounced as n increases. All these properties are revealed by Figure 7.1.

7.4 The Multinomial Model

The binomial theorem, which deals with independent trials with two outcomes, is but a special case of the multinomial law, which is concerned with independent trials with more than two outcomes. As we shall show, the extension of the binomial probability to multinomial probability is simply a process of replacing the binomial coefficients by the multinomial coefficients.

Suppose we have an experiment which is capable of yielding k possible mutually exclusive results, $e_1, e_2, \cdots, e_k$, with respective probabilities $p_1, p_2, \cdots, p_k$ which add up to unity. If we were to repeat this experiment n times *independently*, what is the probability of obtaining exactly x_1 occurrences of e_1; x_2 occurrences of $e_2, \cdots$; and x_k occurrences of e_k; given that $x_1 + x_2 + \cdots + x_k = n$? To answer this question, first we note that the experiments are independent. Thus, the probability of any stated sequence of outcomes is equal to the product of their separate, unconditional probabilities; that is, $p_1^{x_1}p_2^{x_2} \cdots p_k^{x_k}$. Furthermore, the number of distinct sequences yielding the stated number of results of each kind is equal to

$$\binom{n}{x_1,x_2, \cdots ,x_k} = \frac{n!}{x_1!x_2! \cdots x_k!}$$

which is the number of ways to permute n things taken all at a time when x_1 are alike, x_2 are alike, $\cdots$, and x_k are alike. Since all these sequences of outcomes are mutually exclusive and have the same probability of occurrence, the total probability of $(x_1,x_2, \cdots ,x_k; p_1,p_2, \cdots ,p_k)$ is then the sum of all the sequence probabilities which have $\binom{n}{x_1,x_2, \cdots ,x_k}$ terms. Thus, we may define the probability function of a multinomial variable as

$$m(x_1,x_2, \cdots ,x_k; p_1,p_2, \cdots ,p_k) = \binom{n}{x_1,x_2, \cdots ,x_k} p_1^{x_1}p_2^{x_2} \cdots p_k^{x_k} \quad (7.8)$$

where each of the xs can take on any of the values $0, 1, 2, \cdots, n$, subject to the condition that $x_1 + x_2 + \cdots + x_k = n$. The name "multinomial" is derived from the fact that (7.8) is a term of the expansion of the "multinomial" $(p_1 + p_2 + \cdots p_k)^n$.

Fifteen dice are tossed together; what is the probability that each of the odd numbers will appear exactly twice and that each of the even numbers will appear exactly three times? There are six outcomes in a single trial, each with a probability of $1/6$. Now, if we denote x_1 as the number of times that an ace appears, x_2 as the number of times that a two appears, and so on, then we have the desired probability as

$$m(2,3,2,3,2,3; 1/6,1/6,1/6,1/6,1/6,1/6)$$

which is equal to

$$\binom{15}{2,3,2,3,2,3} (1/6)^2(1/6)^3(1/6)^2(1/6)^3(1/6)^2(1/6)^3 = 0.0016$$

Sometimes in a repeated trial process, we may be interested in the probability of the exact numbers of occurrences for only a few, instead of all, of the possible outcomes. For example, suppose that failures of a certain electronic mechanism may be due to six different causes, $e_1, \cdots, e_6$, with respective probabilities of 0.4, 0.3, 0.1, 0.1, 0.05, and 0.05 occurring. These causes are known to be independent, and simultaneous defects are a remote possibility. If failures are observed six times at random, what is the probability that each contains two instances of e_1 and e_2?

To attack this problem, we first note that among the six possible outcomes we are interested in only the probability of x_1 occurrences of e_1 and x_2 occurrences of e_2. The number of occurrences of each of the four other outcomes are not required. Therefore, we actually have a new experiment which has only three outcomes, e_1, e_2, and e_3', with the last representing the occurrence of any one of the four other outcomes in the original experiment. The corresponding probabilities now become p_1, p_2, and p_3', with $p_3' = p_3 + p_4 + p_5 + p_6$. With these designations, we see that $x_3' = n - (x_1 + x_2)$. Similar to (7.8), we have

$$m(x_1,x_2,x_3';\, p_1,p_2,p_3') = \binom{n}{x_1,x_2,x_3'} p_1^{x_1} p_2^{x_2} p_3'^{x_3'}$$

for our present problem, which can be generalized for k outcomes with any number of particular outcomes specified. Turning to the numerical calculations for our present example, we have $p_3' = 0.3$ and $x_3' = 2$. Thus,

$$m(2,2,2;\, 0.4,0.3,0.3) = \binom{6}{2,2,2} (0.4)^2(0.3)^2(0.3)^2 = 0.15552$$

In the multinomial model, the number of occurrences of each possible outcome in n independent trials may be considered as an individual variable, and its expectation and variance are defined as follows:

$$E(X_i) = np_i \tag{7.9}$$
$$V(X_i) = np_i(1 - p_i) \tag{7.10}$$

For the example just given, we have

$$\begin{aligned} E(X_1) &= np_1 = 6(0.4) = 0.24, \\ E(X_2) &= E(X_3') = 6(0.3) = 1.80, \\ V(X_1) &= np_1(1 - p_1) = 6(0.4)(0.6) = 1.44, \quad \text{and} \\ V(X_2) &= V(X_3') = 6(0.3)(0.7) = 1.26 \end{aligned}$$

In concluding this section, we may note that computation of multinomial probabilities can become time-comsuming and tedious when n is at all large. Tabulations of these probabilities are cumbersome and impractical since they must include many possible ps, xs and ns. In many situations where the multinomial model is encountered, we resort to its approximation by other probability models. We have introduced the multinomial law here primarily for the purpose of reference.

7.5 The Hypergeometric Model

The *hypergeometric probability model* applies to a sampling from a finite population, without replacement, whose elements can be classified into two categories. Consider a population of N items, k of which possess a certain characteristic and $N - k$ of which do not possess such a characteristic. Obviously, if a random selection is made from such a population, the result must be either one of the ks (successes) or one of the $N - k$s (failures). It is also clear that if n random selections are made without replacement from the population, each subsequent draw is dependent and the probability of success changes on each draw. Under these conditions, we may wish to find the probability of obtaining exactly x items which are of type k, or successes, in a random sample of size n. The number of successes in n dependent trials is called a *hypergeometric variable* and is generated with three fixed numbers: N, n, and k. Let us now proceed to derive the probability function for this random variable.

First, recall that the total number of ways to draw n items from N things is $\binom{N}{n}$. Next, note that to draw exactly x items of a certain characteristic is the same as to draw $n - x$ items which do not possess the characteristic from a total of $N - k$ items available. Thus, the total number of ways to draw x ks and $n - x$ non-ks is $\binom{k}{x}\binom{N-k}{n-x}$. Finally, if the process of sampling is random, then all items which are still available after any number of selections have been made are equally likely to be chosen. As a result, we can assign equal probabilities to each of the $\binom{N}{n}$ points. These arguments lead us to conclude that the *probability function* of a *hypergeometric variable*, the probability that x of the n items are of type k, is

$$h(N,n,k,x) = \frac{\binom{k}{x}\binom{N-k}{n-x}}{\binom{N}{n}} \tag{7.11}$$

From (7.11), we obtain the *distribution function* of X as follows:

$$H(N,n,k,r) = P(X \leq r) = \sum_{x=\max(0,n-(N-k))}^{r} h(N,n,k,x) \tag{7.12}$$

Table A-IV in Appendix C gives values for both $h(x)$ and $H(x)$ for $N = 10$. (An extensive table for these values is *Table of the Hypergeometric Probability Distribution* by G. J. Lieberman and D. B. Owens, published by Stanford University Press in 1961.)

To derive the expectation of a hypergeometric distribution, note that the hypergeometric variable X may also be considered as a sum, just as in the binomial case, of n variables X_i, except that in the hypergeometric model, $X_1, \cdots,$ X_2 are dependent. But, because the additive property of expectation does not require independence of X_i, it is still true that $E(X) = E(X_1) + \cdots + E(X_n)$, where each $E(X_i)$ is the probability of X at the ith trial, (k/N), if it is not known what has happened at preceding or subsequent trials. Thus, the expectation of the hypergeometric variable X becomes

$$E(X) = n\left(\frac{k}{N}\right) \tag{7.13}$$

The variance is not additive for dependent variables. It can be shown, however, that in the hypergeometric case, the variance is given by the expression below:

$$V(X) = \left(\frac{N-n}{N-1}\right) n \left(\frac{k}{N}\right)\left(\frac{N-k}{N}\right) \tag{7.14}$$

We note that in the binomial case, where sampling is made with replacement, $V(X) = npq = n(k/N)((N-k)/N)$. When sampling without replacement is assumed for the hypergeometric model, a factor $(N-n)/(N-1)$ is introduced.

In a lot of 50 baby chickens, 6 are females. If a random sample of 3 is drawn without replacement, then X, the number of female chickens in the sample, is a hypergeometric variable and its probability function can be evaluated by the general term

$$h(50,3,6,x) = \frac{\binom{6}{x}\binom{44}{3-x}}{\binom{50}{3}}$$

Table 7.1 contains calculations of these terms, together with their cumulative sums.

TABLE 7.1 PROBABILITY AND DISTRIBUTION FUNCTIONS OF THE NUMBER OF FEMALE CHICKENS WITH $N = 50$, $n = 3$ AND $k = 6$

x	$3-x$	$\binom{6}{x}$	$\binom{44}{3-x}$	$\binom{6}{x}\binom{44}{3-x}$	$h(x)$	$H(x)$
0	3	1	13,244	13,244	0.6757	0.6757
1	2	6	946	5,676	0.2896	0.9653
2	1	15	44	660	0.0337	0.9990
3	0	20	1	20	0.0010	1.0000
—	—	—	—	19,600*	1.0000	—

* $\binom{50}{3} = 19{,}600$.

For this model, we have

$$E(X) = n\left(\frac{k}{N}\right) = 3\left(\frac{6}{50}\right) = 0.36 \text{ female chickens,}$$

and

$$V(X) = \left(\frac{N-n}{N-1}\right) n \left(\frac{k}{N}\right)\left(\frac{N-k}{N}\right) = \left(\frac{50-3}{50-1}\right) 3 \left(\frac{6}{50}\right)\left(\frac{50-6}{50}\right) = 0.3039$$

Whenever sampling is destructive, the hypergeometric model should be adopted rather than the binomial model. Fortunately, a comprehensive hypergeometric probability table, as mentioned before, has been made available. In using such a table, one should observe a very interesting property of the hypergeometric distribution: when k and n are interchanged, there will be no effect on the probabilities. That is,

$$h(N,n,k,x) = h(N,k,n,x) \tag{7.15}$$

and

$$H(N,n,k,r) = H(N,k,n,r) \tag{7.16}$$

Consider this example: if 4 chips are drawn together at random from an urn containing 6 black and 4 white chips, what is the probability that 2 or fewer black chips are drawn?

We have here a hypergeometric model with $N = 10$, $k = 6$, and $n = 4$. The desired probability is

$$\begin{aligned} P(X \le 2) &= H(10,4,6,2) \\ &= H(10,6,4,2) && \text{by (7.16)} \\ &= 0.547619 && \text{from Table A-IV} \end{aligned}$$

Now if 5 chips are drawn without replacement from the same urn, what is the probability that (1) all are black? (2) 4 or more are black?

ANSWER: (1)

$$\begin{aligned} P(\text{All five are black}) &= h(10,5,6,5) \\ &= h(10,6,5,5) && \text{by (7.15)} \\ &= 0.023810 && \text{from Table A-IV} \end{aligned}$$

(2)

$$\begin{aligned} P(X \ge 4) &= 1 - P(X \le 3) \\ &= 1 - H(10,5,6,3) \\ &= 1 - H(10,6,5,3) && \text{by (7.16)} \\ &= 1 - 0.738095 && \text{from Table A-IV} \\ &= 0.261905 \end{aligned}$$

7.6

The Discrete Uniform Model

We have encountered the discrete uniform probability model before, though it was not given such a name. For instance, the probability distribution of the number of spots on a die generated by tossing the die is uniform because the probability associated with each and every outcome is a constant $\frac{1}{6}$. The probability function of this model can be written as

$$f(x;6) = \tfrac{1}{6}, \quad x = 1, 2, \cdots, 6$$

In general, the uniform distribution applies to an experiment that can terminate in n mutually exclusive and equally likely ways. Consider, for example: n numbers, one each on n identical chips, are placed in a box and mixed thoroughly, and then one chip is drawn from the box. The mathematical statement for this experiment is clearly

$$u(x;n) = 1/n, \quad x = 1, 2, \cdots, n \tag{7.17}$$

which gives the probability that x assumes for any specific value as $1/n$; it is therefore called the *discrete uniform*, or *rectangular*, *probability model.*

One important aspect of the uniform distribution is that it furnishes a practical means of drawing a random sample from a finite population. Recall that sampling from an infinite population is random if all observations in the sample are made from the same population and if these observations are drawn independently. In the case of a finite population, a sample is said to be random if the sampling procedure is such that all possible samples of the same size have the same probability to be chosen. This distinction exists of course if sampling is made without replacement. When replacement is employed, a finite population can be in effect treated as an infinite one, because then we can never exhaust observations, and each observation is made independent of all the rest. It is interesting to note that both types of samples—one drawn under the condition that all the items in the population have equal probability to be included in the sample, an unconditional random sample, and one drawn under the condition that all samples of the same size have the same probability to be selected, a conditional random sample—are called *simple random samples*, and they are the only kinds of samples we shall consider when we discuss the theory of sampling in the next chapter. Now, let us explain how to use the uniform probability model for selecting a random sample from a finite population.

Consider a population of ten elements designated as 1, 2, $\cdots$, 10. We wish to draw a sample of size three from it. To do this, we first write the ten numbers on ten chips and put them in a box. Next, we select one chip at a time without replacement until three chips have been drawn. By this procedure, we see that the probabilities associated with all selections are uniform. For the first choice, we have $u(x,10) = \frac{1}{10}$. For the second choice, we have $u(x,9) = \frac{1}{9}$. For the third choice, we have $u(x,8) = \frac{1}{8}$. Finally, let us evaluate the probability of drawing a sample which contains any three items, say, 1, 3, and 7. Such a sample can be obtained by drawing first any one of the three numbers with a probability of $\frac{3}{10}$, by drawing second either of the two remaining numbers with a probability of $\frac{2}{9}$, and by drawing third the remaining number with a probability of $\frac{1}{8}$. By the product rule for dependent events, we have the probability of getting this sample as follows:

$$(\tfrac{3}{10})(\tfrac{2}{9})(\tfrac{1}{8}) = 3!7!/10! = 1/(10!/3!7!) = 1\Big/\binom{10}{3}$$

where $\binom{10}{3}$ is the number of all possible samples of size three from a population of ten items. From similar reasoning, each and every one of the possible samples has the same probability of $1\Big/\binom{10}{3}$ of being selected. As a result, the method has yielded a random sample.

As a matter of fact, the preceding arguments constitute a verification of the statement that all combinations are equally likely. We may now generalize that if a sample of size n is drawn from a population with N observations, then there are $\binom{N}{n}$ possible samples, each of which has a probability of $1\Big/\binom{N}{n}$ of occurring. This means that the distribution of samples itself is uniform, with a probability function

$$u(x;r) = 1\Big/\binom{N}{n} \tag{7.18}$$

where $r = \binom{N}{n}$, the number of possible samples of size n and x, stands for the r possible samples. Here x is a variable of n-dimensions since each sample contains n items. It seems that what we need is an "N-sided" die to generate a uniform population of N items. Such a die is clearly impractical to construct. Fortunately, instead, we can use a table of random digits for the same purpose.

A table of random digits is assumed to represent a large random sample whose values are those of a random variable with sample space

$$S = \{0,1,2,3,4,5,6,7,8,9\}$$

and uniform distribution. That is,

$$u(x;10) = 1/10,\ x = 0,\ 1,\ 2,\ \cdots,\ 9$$

To construct a table of random numbers with this underlying uniform distribution of decimal digits, we may number 10 identical chips from 0 consecutively to 9 and place them in a box. One chip is chosen at random and replaced, another chip is chosen at random and replaced, ad infinitum. The number on each chip is recorded in a list before it is returned to the box. When numbers in pairs are used, random numbers from 00 to 99 are obtained. By combining three digits at a time, random numbers from 000 to 999 are secured. And so on. Random numbers can also be generated by a ten-sided die or be programmed by electric computers. Table A-XVI in Appendix C is a short table of random digits. Much more comprehensive tables of random digits are available.

Simple random samples can be obtained by using the table of random digits for sampling with and without replacement. To illustrate, suppose we wish to draw a random sample of 100 items from a population of 9000 items. First, we prepare a list of the individual items in the population, such as households in a town, employees of a corporation, retail prices, or the like, and assign each item on the list a number from 0001 to 9000. Then we may proceed to obtain the sample in the following fashion.

(1) Consult the table in a random manner: open the table and place it in front of you; close your eyes and point your finger to a spot on the table. The figure under your finger becomes the starting point. Then you can read the table vertically, horizontally, or diagonally, that is, in any systematic way once you have made the random start.

(2) Read only the four-digit numbers in the table, since our population consists of 9000 members. If a number is greater than 9000, skip it. If the sample is to be taken without replacement, also skip a number when it appears the second time. Proceed in this way until you have obtained 100 numbers. (If N were between 100 and 999, you would use the first three columns of the table; if N were between 1000 and 9999, the first four columns; if N were between 10,000 and 99,999, the first five columns; and so on.)

(3) Take from the list those 100 items that have the corresponding recorded numbers; they form the desired sample.

With the aid of a table of random digits, simple random sampling becomes an efficient procedure if the population is not large and if locating the sampling units is relatively easy and inexpensive. This procedure could also be a practical one for large populations whose elements are concentrated within a small area. For instance, the study of all the expense vouchers of a large corporation, the investigation of the attitudes of the employees toward a certain management policy in a big company, the survey of students' study habits at a given university, and so on, can easily be made by a simple random sample.

7.7 The Poisson Model

There are events which do not occur as outcomes of a definite number of trials of an experiment but which occur at random points of time or space—see Figure 7.2, where each point represents an occurrence of the event. For events of this kind, we are interested in only the number of occurrences of the event; not in its nonoccurrences. We may wish to know the number of trains arriving at a certain station during a certain hour, but obviously there is no sense in investigating the number of trains not arriving at the station during that hour. Or, we may need to discover the number of printer's errors on each page of a book, but there is no need to determine the number of

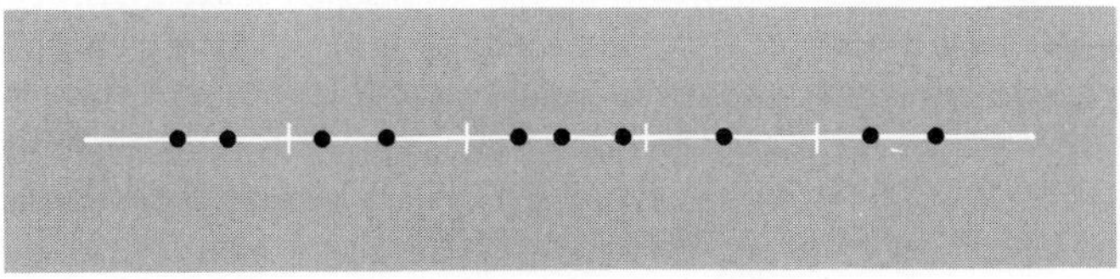

Figure 7.2 Random points in time or space.

correct words on each page. There is a large class of events of this variety. The event may be the number of occurrences of accidents, errors, breakdowns, or other calamities which appear randomly and independently over a continuum of time. The event may be the demand for service upon the cashier or the saleswoman of a department store, upon the stock clerk of a factory, upon the tollbooth of a bridge or tunnel, upon the cargo-handling facility of a port, upon the trunk lines of a telephone exchange, or upon the maintenance man of a machine shop, during a day or some other specific time unit. The event may be the number of defects on the surface of a table, or glassware, or a piece of goods. The event may be the emission of a radioactive particle, and so on.

For each of the above random variables, the following postulates seem appropriate:

First, the probability of n occurrences of the event per unit of time or space, t, is exactly the same, no matter where the interval t starts. That is, the mean rate that the event occurs per unit of time or space is the same, regardless of how the infinite number of unit lengths or unit areas is selected on the t axis in order to determine the average, λ, denoted as Greek lambda.

Second, events which occur in one time interval or region of space are independent of those occurring in any other time interval or region of space, no matter how the interval is selected.

Third, since there exists a constant and positive λ, it is reasonable to conclude that for any small interval of width, h,

(a) the probability that exactly one event will occur in h is approximately proportional to the interval h, that is, λh, in the sense that this probability is equal to $\lambda h + o_1(h)$ and $\underset{h \to 0}{\text{Limit}} \dfrac{o_1(h)}{h} = 0$;

(b) the probability that more than one event will occur in the interval h is a negligible quantity $o_2(h)$ in the sense that $\underset{h \to 0}{\text{Limit}} \dfrac{o_2(h)}{h} = 0$; and

(c) the probability that exactly 0 events occur in the interval h is approximately equal to $1 - \lambda h$ in the sense that it is equal to $1 - \lambda h + o_3(h)$ and $\underset{h \to 0}{\text{Limit}} \dfrac{o_3(n)}{h} = 0$.

We are concerned, under the above conditions, with a random variable X, the number of occurrences of an event in an interval of time, or in an area of space, t. Here X is called a Poisson variable, named after the French mathematician S. D. Poisson (1781–1840), which can assume possible values $0, 1, 2, \cdots$. The Poisson probability function defines the probability of exactly x occurrences, or successes, appearing per unit of time or space, given the average number of occurrences in the given interval, λt, and it is written as

$$p(x;\lambda t) = e^{-\lambda t} \frac{(\lambda t)^x}{x!} \tag{7.19}$$

where p stands for the functional notation of the Poisson distribution, e is a constant and is approximately equal to 2.71828, and λ is the constant proportionality introduced in part (a) of the third Poisson postulate.

It is interesting to note that expression (7.19) is the general term of the Poisson series,

$$e^{-\lambda t} + \lambda t e^{-\lambda t} + \frac{(\lambda t)^2 e^{-\lambda t}}{(1)(2)} + \frac{(\lambda t)^3 e^{-\lambda t}}{(1)(2)(3)} + \cdots$$

which says that the probability of one occurrence is λt times the probability of none; that the probability of two occurrences is $\lambda t/2$ times the probability of one; that the probability of three occurrences is $\lambda t/3$ times the probability of two; and so on. Now, if we let $P_t(n)$ represent the probability of n occurrences of an event per time interval or per area space, then from (7.19), and the Poisson series, we can see that

$$\left.\begin{aligned} P_t(0) &= e^{-\lambda t} \\ P_t(1) &= \lambda t P_t(0) \\ P_t(2) &= (\lambda t/2) P_t(1) \\ P_t(3) &= (\lambda t/3) P_t(2) \\ &\cdots \\ P_t(k) &= (\lambda t/k) P_t(k-1) \end{aligned}\right\} \qquad \textbf{(7.20)}$$

Suppose that calls go through a telephone exchange at a mean rate of four calls per minute, at random. The probability that a call will occur at any instant is infinitely small, but the number of calls during a period of time, say, an hour, is extremely large. Under these conditions, the number of calls per minute is a Poisson variable, with $\lambda t = 4$ and the following probability model:

$$p(x;4) = e^{-4} \frac{4^x}{x!}$$

The individual terms of this probability distribution can be evaluated by (7.20) and are presented by Table 7.2 below.

The distribution function of the Poisson variable, as usual, is obtained by adding the proper number of Poisson probabilities, which gives the probability

TABLE 7.2 PROBABILITY DISTRIBUTION OF THE NUMBER OF CALLS WITH $\lambda t = 4$ PER MINUTE

x	$P(X = x)$	x	$P(X = x)$
0	$P_t(0) = 0.0183$*	8	$P_t(8) = 0.0298$
1	$P_t(1) = 0.0733$	9	$P_t(9) = 0.0123$
2	$P_t(2) = 0.1465$	10	$P_t(10) = 0.0053$
3	$P_t(3) = 0.1954$	11	$P_t(11) = 0.0019$
4	$P_t(4) = 0.1954$	12	$P_t(12) = 0.0006$
5	$P_t(5) = 0.1563$		.
6	$P_t(6) = 0.1042$		.
7	$P_t(7) = 0.0595$		.

*$P_t(0) = e^{-4} = 0.0183$ by Table A-V in Appendix C.

of r or less occurrences. Its mathematical statement is

$$P(r;\lambda t) = \sum_{x=0}^{r} p(x;\lambda t) \tag{7.21}$$

It can be shown that for the Poisson model, the expectation and the variance are identical and equal to λt, namely,

$$E(X) = V(X) = \lambda t \tag{7.22}$$

Due to the fact that $E(X) = \mu = \lambda t$, we can rewrite, for the sake of convenience and compactness, the probability and distribution functions of the Poisson variable as follows:

$$p(x;\mu) = e^{-\mu} \frac{\mu^x}{x!} \tag{7.23}$$

and

$$P(r;\mu) = \sum_{x=0}^{r} p(x;\mu) \tag{7.24}$$

Table A-V in Appendix C gives cumulative Poisson probabilities for selected values of μ. (A fairly extensive table, *Poisson's Exponential Binomial Limits*, tabulated by E. C. Molina, has been published by D. Van Nostrand Company. This table gives $P(r;\mu)$ to at least six decimal places for $\mu = 0.001(0.001)0.02$ $(0.01)0.30(0.1)15.0(1)100$.)

Now, if we say $t = 1$, then we have $\mu = \lambda$, the average number of occurrences per unit interval. This leads us to the observation that any size on the t axis can be used as the basic unit for specifying μ. Once μ is specified, the distribution of probabilities among all possible number of occurrences in an interval of any other size can also be determined. This is so because, with such an interpretation of λ and λt, whenever the basic interval is changed, μ is changed proportionately. Thus, doubling the specified basic unit also doubles the expected number of occurrences per time unit.

To illustrate the preceding arguments, let us use the telephone-call example mentioned above and ask: What is the probability that no more than two calls will occur during the next 2 minutes?

To answer this question, we note that the expected value was given as four calls per minute. Since, for the Poisson variable, expectation is proportional to the size of the interval t, we have now $\mu = 8$ calls per 2 minutes; thus the desired probability is

$$P(2;8) = 0.1375 \qquad \text{from Table A-V}$$

For the same distribution, what is the probability that no calls will occur in any interval of 30 seconds?

Now, $\mu = 4/2 = 2$ per 30 seconds and

$$p(0;2) = P_t(0) = e^{-2} = 0.1353 \qquad \text{from Table A-V}$$

The foregoing property of the Poisson distribution should be familiar to us;

recall that for independent random variables, such as the Poisson variables, additivity holds for both expectation and variance.

A manufacturer of woolen piece goods claims that the average number of flaws in his products is one per 2 square yards. A sample square yard of his product, selected at random, shows three flaws. What is the probability of obtaining three or more flaws in any 1 square yard if the manufacturer's claim is valid?

Here, we have a Poisson variable with $\mu = 0.5$ per square yard, and the desired probability is

$$\begin{aligned}\sum_{x=3}^{\infty} p(x;0.5) &= 1 - \sum_{x=0}^{2} p(x;0.5) \\ &= 1 - P(2;0.5) \\ &= 1 - 0.98561 \qquad \text{from Table A-V} \\ &= 0.01439\end{aligned}$$

In view of this result, what can we conclude about the manufacturer's claim?

A certain state in the United States has, on the average, twelve traffic deaths every three months. What is the probability (a) that there are more than four traffic deaths, and (b) exactly four deaths, in any one month?

Note that $\mu = 4$ per month. Thus

$$\begin{aligned}\text{(a)} \quad \sum_{x=5}^{\infty} p(x;4) &= 1 - \sum_{x=0}^{4} p(x;4) \\ &= 1 - P(4;4) \\ &= 1 - 0.62884 \qquad \text{from Table A-V} \\ &= 0.37116\end{aligned}$$

$$\begin{aligned}\text{(b)} \quad p(4;4) &= P(4;4) - P(3;4) \\ &= 0.62884 - 0.43347 \qquad \text{from Table A-V} \\ &= 0.19537\end{aligned}$$

The Poisson distribution is positively skewed but its skewness decreases continuously with continuous increase in the size of its expectation; see Figure 7.3. Thus, a Poisson distribution is completely defined by its μ, which, however, cannot be determined theoretically. Its value must be estimated empirically. The method of estimating μ involves, first, the selection of an interval of width t for events of interest occurring in time or space. Then, the observation of a large number of intervals, N, of width t, is required. Next, if for integer $x = 0, 1, 2, \cdots$, we let N_x denote the number of intervals in which exactly x successes have occurred, then the total number of successes observed in the N intervals is

$$T = 0(N_0) + 1(N_1) + 2(N_2) + \cdots + x(N_x) + \cdots$$

where

$$N = N_0 + N_1 + \cdots + N_x + \cdots$$

Finally, the ratio T/N is used as an estimate of μ, that is,

$$\hat{\mu} = \text{estimate of } \mu = \frac{T}{N} \tag{7.25}$$

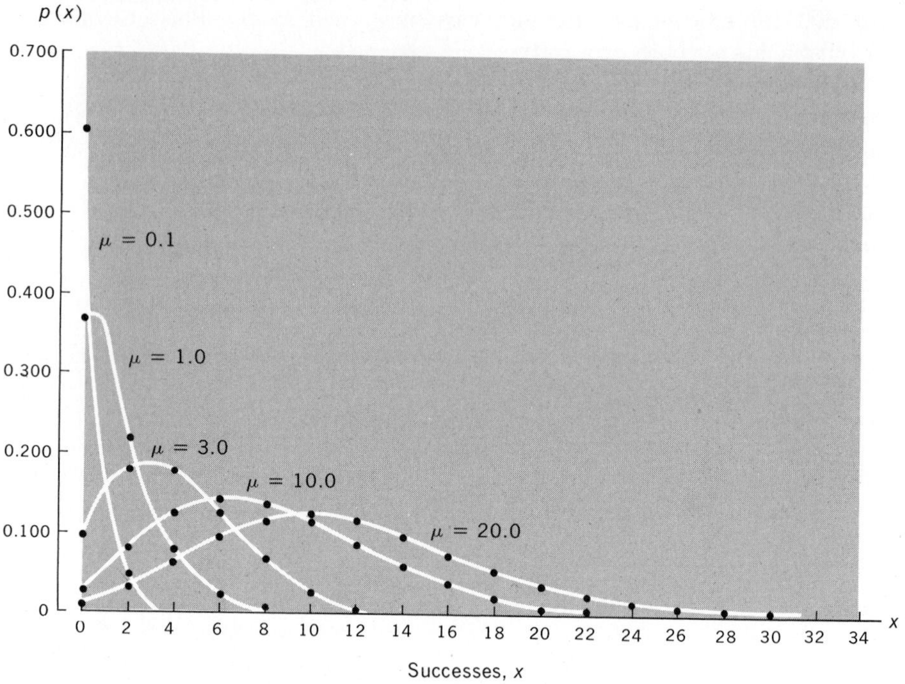

Figure 7.3 Poisson distributions for $\mu = 0.1, 1.0, 3.0, 10.0$, and 20.0.

Let us use, as an example of estimating the Poisson expectation, the data on the vacancies in the U.S. Supreme Court, either by death or by resignation of members, reported by W. A. Wallis in "The Poisson Distribution and the Supreme Court," *Journal of the American Statistical Association*, Vol. 31, 1936, pp. 376–380. These data are contained in the first two columns in Table 7.3.

For data in Table 7.3, there are 96 fixed intervals of one year each which can be thought of as repeating a random experiment 96 times. Thus, we have $N = 96$.

TABLE 7.3

x: Number of vacancies in the year	N_x: Observed number of years with x vacancies	$p(x;0.5)$: Probability of x vacancies	$96p(x;0.5)$: Expected number of years in which x vacancies occur
0	59	0.6065	58.22
1	27	0.3033	29.12
2	9	0.0758	7.28
3	1	0.0126	1.21
over 3	0	0.0018	0.17
	96	1.0000	96.00

In these 96 intervals, the total number of vacancies is

$$T = 0(N_0) + N_1 + 2N_2 + 3N_3 = 0 + 27 + 2(9) + 3(1) = 48$$

With the values of N and T, we have the estimate of μ for this Poisson variable as

$$\hat{\mu} = \frac{T}{N} = \frac{48}{96} = 0.5$$

For the model under consideration, the probability function becomes

$$p(x;0.5) = e^{-0.5}\frac{(0.5)^x}{x!}$$

The individual terms of this function are given in column three of Table 7.3. The last column of this table contains what may be considered as the theoretical values of the number of years in which x vacancies occur. Observe how close the expected numbers are to the actual numbers in this case.

7.8 Relations among the Special Discrete Models

So far we have been concerned with some special discrete probability models. Before taking up continuous cases, it is important for us to realize that all the discrete models introduced are closely related to each other. We have already shown that the binomial variable may actually be considered as the sum of n identical Bernoulli variables. We shall proceed in this section to observe the relations among the binomial, the hypergeometric, and the Poisson models.

First, let us establish the relationship between the hypergeometric and the binomial distributions. Suppose that 30,000 of the 100,000 members of a union are in favor of a strike; if 10 members are selected at random, without replacement, then the probability that exactly x members in the sample are in favor of a strike is given by the hypergeometric law

$$h(100{,}000;10{,}30{,}000,x) = \frac{\binom{30{,}000}{x}\binom{100{,}000 - 30{,}000}{10 - x}}{\binom{100{,}000}{10}}$$

If, however, the sample of 10 is drawn at random with replacement, and the probability of each independent trial remains constant at 30,000/100,000, or 0.3, then the binomial model is appropriate for the number of members in favor of a strike in the sample, and the probability function becomes

$$b(x;10,0.3) = \binom{10}{x}(0.3)^x(0.7)^{10-x}$$

Now, the ratio of those in favor of a strike to the total number of union members will change only very slightly when a sample as small as 10 is drawn without replacement from a population as large as 100,000. In other words, when n is very small and N is very large, the probability of success will not change significantly from one draw to another, and successive draws can in effect be considered as independent. Thus, we would expect the binomial probability to approximate very closely the hypergeometric probability. So, in our numerical example, we would expect $b(x;10,0.3)$ to yield very close approximations to $h(100{,}000,10,30{,}000,x)$. In general, it can be shown that if n, and therefore x, is very small compared to both k and $N - k$, then

$$\frac{\binom{k}{x}\binom{N-k}{n-x}}{\binom{N}{n}} \doteq \binom{n}{x}\left(\frac{k}{N}\right)^x\left(\frac{N-k}{N}\right)^{n-x} \tag{7.26}$$

which says: If a very small sample is taken from a very large population, then nonreplacement (the hypergeometric model) and replacement (the binomial model) give approximately identical results.

The practical utility of (7.26) is that, when conditions for its validity are met, we can obtain a good approximation of a hypergeometric probability by the more easily computed binomial probability. This approximation improves, given n, as N increases. When N becomes infinite, the hypergeometric and binomial probabilities become identical. Because of this, we say that the binomial model is the limit of the hypergeometric model as N approaches infinity.

While the binomial can be used to approximate the hypergeometric probabilities, the binomial, in turn, can be approximated by the Poisson distribution. It can be shown that if n is large and p is small, then the binomial and Poisson models are related by the following equation:

$$\binom{n}{x} p^x q^{n-x} \doteq e^{-np}\frac{(np)^x}{x!} \tag{7.27}$$

where the individual binomial terms are replaced by corresponding Poisson terms with $\mu = np$.

When we find it convenient to approximate the binomial by the Poisson, we must be sure to know what is meant by large n and small p. As a rule of thumb, if $n > 100$ and $p < 0.01$, the binomial and Poisson probabilities agree approximately (to three decimal places) for every value of the random variable. Furthermore, for those x values for which $((x - np)^2/n) < 0.01$, the two models agree to within 1 percent. Finally, if p is sufficiently small, satisfactory approximations of the binomial by the Poisson can be made even when n is as small as 10. Given a small p, the larger n is, the better the results. Given n, the smaller the p is, the more satisfactory the approximations.

Table 7.4 gives the binomial probabilities for $x = 0, 1, 2, 3, 4$, and 5 with $n = 10$ and $p = 0.1$, and with $n = 20$ and $p = 0.05$, respectively. In each case the Poisson approximations are the same since $\mu = 10(0.1) = 20(0.05) = 1$.

TABLE 7.4 BINOMIAL DISTRIBUTIONS AND THEIR POISSON APPROXIMATIONS

x	0	1	2	3	4	5 · · ·
$b(x;10,0.10)$	0.349	0.387	0.194	0.057	0.011	0.0015
$b(x;20,0.05)$	0.358	0.377	0.187	0.060	0.013	0.0022
$p(x;1)$	0.368	0.368	0.184	0.061	0.015	0.0031

The above numerical illustrations reveal that Poisson approximations of the binomial probabilities are quite good even when n is as small as 10 or 20, and that for $n = 20$ the approximations are better than for $n = 10$. However, in practice, it would make no sense to approximate binomial probabilities by the Poisson values when the former can be read directly from tables. Speed is gained without sacrificing accuracy only when n is large and p is sufficiently small, and when the binomial probabilities cannot be obtained from published tables.

Suppose that the probability that a newborn baby will die of a certain disease is 0.00002, what is the probability that out of 100,000 newborn babies (a) 4 or more, (b) exactly 4, will die of this disease?

The probabilities required are clearly binomial.

(a) The probability that 4 or more will die of this disease is

$$\sum_{x=4}^{100,000} \binom{100,000}{x} (0.00002)^x (0.99998)^{100,000-x}$$

Rather than trying to evaluate this seemingly formidable expression, we would be well-advised to approximate it by the Poisson law, with

$$\mu = np = 100,000(0.00002) = 2$$

So,

$$\begin{aligned}\sum_{x=4}^{100,000} p(x;2) &= 1 - \sum_{x=0}^{3} p(x;2) \\ &= 1 - P(3;2) \\ &= 1 - 0.85712 \qquad \text{from Table A-V} \\ &= 0.14288\end{aligned}$$

(b) The probability of exactly 4 deaths is

$$\begin{aligned}P(4;2) - P(3;2) &= 0.94735 - 0.85712 \qquad \text{from Table A-V} \\ &= 0.09023\end{aligned}$$

In concluding this section, let us observe that just as the hypergeometric model takes the binomial model as its limit, so the binomial model takes the Poisson model as its limit. As such, the Poisson can also be employed to approximate the hypergeometric probability via the binomial distribution. We shall see, however, that all these probabilities can be approximated by a single probability model—the normal probability law—under appropriate conditions.

7.9

The Continuous Uniform Distribution

We begin our discussion of continuous models with a simple and familiar case. A random variable whose value can lie only in a certain finite interval for finite numbers a and b, for example, $X = \{\text{real number } x\colon a < x < b\}$, has a *uniform* or *rectangular distribution* if its probability density function is constant over the interval from a to b. Thus, for a continuous uniform model, the probability of any event which is a subinterval of (a,b) would be just the length of that interval in (a,b); but the probability of an event which has no points in common with (a,b) is zero. The density function of a continuous random variable X can therefore be written as

$$\begin{aligned} r(x) &= \frac{1}{b-a}, \quad \text{if } a < x < b \\ &= 0, \quad \text{otherwise} \end{aligned} \tag{7.28}$$

It can be shown that by integrating (7.28), we obtain the distribution function of the uniform model as

$$\begin{aligned} R(x) &= 0 && \text{if } x \le a \\ &= \frac{x-a}{b-a} && \text{if } a < x < b \\ &= 1 && \text{if } x \ge b \end{aligned} \tag{7.29}$$

Figure 7.4 gives the density function of the continuous rectangular distribution.

The rectangular model has a and b as its parameters. Its expectation and variance, respectively, are

$$E(X) = \frac{(a+b)}{2} \tag{7.30}$$

and

$$V(X) = \frac{(b-a)^2}{12} \tag{7.31}$$

Note that when a random variable is uniformly distributed over the interval

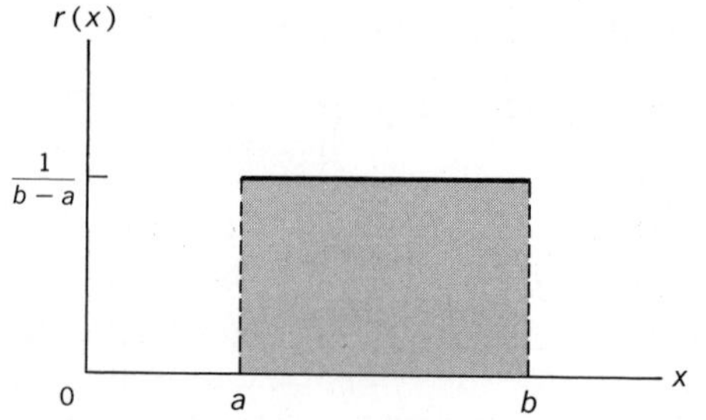

Figure 7.4 Density of a uniform distribution.

from $X = a$ to $X = b$, its density is symmetrical about the center of the interval $(a + b)/2$, and thus this value is both the mean and the median of the distribution. Furthermore, a uniform model is quite analogous to a uniform distribution of mass along a line, such as a uniform thread. As such, the variance of a uniform model $(b - a)^2/12$ is a measure of the moment of inertia about the center of gravity of a uniform thread of mass with length $(b - a)$, where, in the case or probability, mass $= 1$.

Subway trains on a certain line run every half hour between midnight and six in the morning. What is the probability that a man entering the station at a random time during this period will have to wait at least 20 minutes?

The random variable $T =$ Time to the next train, under the assumption of random time of the man's arrival at the station, is uniformly distributed on $0 \leq T \leq 30$. The probability that he has to wait at least 20 minutes is

$$P(T \geq 20) = P(20 \leq T \leq 30) = {}^{10}\!/_{30} = {}^{1}\!/_{3}$$

Furthermore, it may be checked that for T here, the mean wait is 15 minutes and $V(T) = 75$.

7.10 The Exponential Model

The exponential probability model can be derived in a number of ways. One way which can be readily understood is to think of the exponential model as generated from the Poisson process. For example, if the arrivals of automobiles at a tollbooth follow the Poisson law, then the distribution of the length of time between successive arrivals is an exponential variable. Alternatively, the exponential model arises in response to the question: If a sequence of events occurs in time according to the Poisson law, at the rate of μ events per unit of time, how long a time need one wait to observe the first occurrence of an event?

The answer to the foregoing question suggests a method of constructing the exponential model from the Poisson distribution. Now, let us denote X as the variable, time, between events and proceed to determine the distribution function of X by evaluating the event $X > x$ for any specific time interval x. The event $X > x$ implies that it is an event which has not yet occurred in the time interval $(0,x)$. By the Poisson law, this probability is seen to be

$$P(X > x) = P(\text{no occurrence in } (0,x)) = e^{-\mu x}\frac{(\mu x)^0}{0!} = e^{-\mu x}$$

As a result, the CDF of X—the cumulative distribution function of the exponential variable—becomes

$$X(x) = 1 - P(X > x) = 1 - e^{-\mu x} \tag{7.32}$$

where μ is the average number of occurrences per unit of time or space in the Poisson distribution, x is the time interval between occurrences, and e is the base of natural logarithms which is approximately equal to 2.71828. A brief table of exponential functions are given in Table A-VI in Appendix C.

The density function of the exponential variable, as can be shown by differentiating $X(x)$, is

$$x(x) = \mu e^{-\mu x} \tag{7.33}$$

Very often, the exponential model is considered as a special case of what is called the *gamma distribution*, and the exponential density and distribution functions are written in following forms:

$$x(x) = \frac{1}{\beta} e^{-x/\beta} \tag{7.34}$$

and

$$X(x) = 1 - e^{-x/\beta} \tag{7.35}$$

where β (Greek Beta) $= 1/\mu$.

The expectation and the variance of the exponential distribution are found to be, respectively,

$$E(X) = \beta = \frac{1}{\mu} \tag{7.36}$$

and

$$V(X) = \beta^2 = \frac{1}{\mu^2} \tag{7.37}$$

Thus, in the Poisson case expectation and variance are equal, and in the exponential case expectation and standard deviation are equal. It is also interesting to note that the mean of the exponential model is the reciprocal of the mean of the Poisson model. This is a natural result, since the exponential variable refers to time between successive Poisson occurrences. In this connection, we see that the mean of the exponential model must be interpreted as the average time interval between Poisson occurrences, or the expected time until the first occurrence of the event.

Ex The distribution of lengths of time during which a certain make of computer operates effectively, that is, hours of effective operation before the first breakdown, is exponential with an expectation $\beta = 360$ hours. What is the probability that the computer will operate effectively for 180 hours or less? More than 720 hours?

As stated in the problem, the density function of X, the exponential variable lengths of time, is

$$x(x) = \frac{1}{360} e^{-x/360}, \quad (0 \leq x \leq \infty)$$

whose graph is shown by Figure 7.5.

The exponential distribution function in this case is

$$X(x) = 1 - e^{-x/360}$$

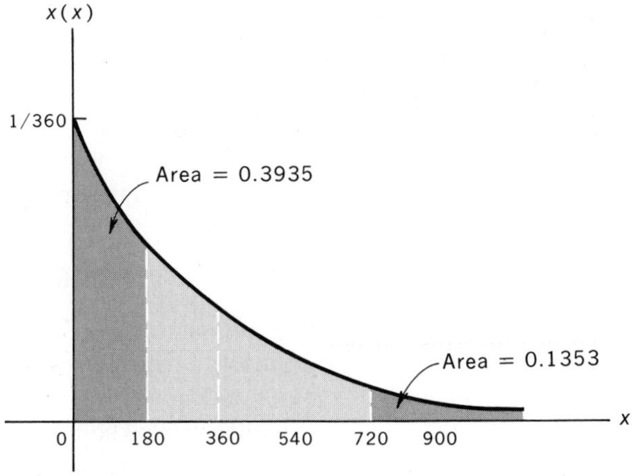

Figure 7.5 Exponential distribution with $\beta = 360$.

Now, at points of 180 and 720 hours (see Figure 7.5), we have

$$\begin{aligned} X(180) &= 1 - e^{-180/360} \\ &= 1 - e^{-0.5} \\ &= 0.3935 \qquad \text{from Table A-VI} \end{aligned}$$

and

$$\begin{aligned} X(720) &= 1 - e^{-720/360} \\ &= 1 - e^{-2} \\ &= 0.8647 \qquad \text{from Table A-VI} \end{aligned}$$

Thus, the probability that the computer will operate effectively less than 108 hours is

$$P(X \leq 180) = X(180) = 0.3935$$

and the probability that it will operate effectively more than 720 hours is

$$P(X \geq 720) = 1 - P(X < 720) = 1 - X(720) = 1 - 0.8647 = 0.1353$$

Workers join a waiting line in front of a workshop to receive tools with an average time between arrivals of 2 minutes. According to the Poisson law, (a) what is the probability that 4 minutes will elapse in which no worker arrives? (b) what is the probability that in a 4-minute interval at least two workers arrive?

In this example, Poisson probabilities are required, but the mean of a corresponding exponential distribution has been given as $\beta = 2$ minutes. From this, we must first obtain the average number of Poisson occurrences per unit of time. By (7.36), we see that $\beta = 1/\mu$, so we have $\mu = 1/\beta$ and, for our problem,

$\mu = 1/2 = 0.5$ arrivals per minute. Thus,

(a) $P(\text{no arrivals in } (0,4)) = p(0;0.5) = e^{-0.5(4)} \dfrac{((0.5)(4)^0}{0!} = e^{-2} = 0.13534$

from Table A-V

(b) $P(\text{at least two arrivals in } (0,4)) = P(\text{two or more arrivals})$

$= 1 - P(X < 1) = 1 - P(1;(0.5)(4)) = 1 - P(1;2) = 1 - 0.40601$
$= 0.59399$ from Table A-V

The exponential distribution is of great importance in applied probability theory because it can be used to describe a variety of random situations. In management science, one often encounters such operational problems as determining the number of tollgates required at highway entrances, the number of telephone lines for serving a given area, the number of helpers at a cafeteria, and so forth. In each of these situations, there is a common feature—the existence of a timewise variable demand for service which can be satisfied by a specific number of servers. This timewise variable demand is found to obey the exponential law. Furthermore, recent studies also reveal that the exponential model describes well the lengths of waiting time, and such numerically valued random phenomena as the life of an electronic tube, the time intervals between breakdowns of electrical mechanism or between accidents of many kinds, and so on. We shall have another opportunity to consider applications of this model when we take up operation research problems in a later chapter.

7.11 The General Normal Model

The most frequently used probability model in economic and business decisions is the *normal distribution,* which may be stated in the general or standard form. A random variable is said to have a general normal distribution if it is continuous, if the constants μ (with a value between $-\infty$ and ∞) and σ (with a value greater than zero) exist, and if its density function is given by the following expression:

$$n(x) = n(\mu,\sigma) = \frac{1}{\sigma\sqrt{2\pi}} e^{-(x-\mu)^2/2\sigma^2} \tag{7.38}$$

where μ and σ are, respectively, the mean and the standard deviation of the normal variable, $e \doteq 2.71828$, and $\pi \doteq 3.14159$. The reader should not worry about his inability to comprehend this seemingly formidable formula, since he will learn immediately that calculation of normal probabilities can be made easily from published tables. He is, however, advised to learn about the geometric properties of the graph of the normal density function as shown by Figure 7.6 and as discussed in the following paragraphs.

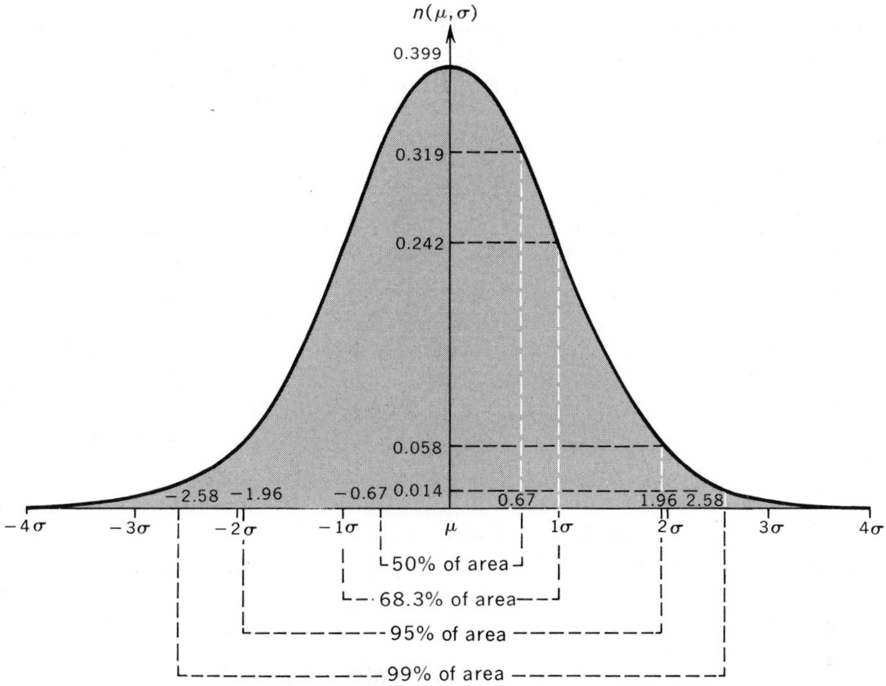

Figure 7.6 Graph of the general normal density function.

First, a normal distribution has as its parameters μ and σ (or σ^2) in the sense that the area under its density curve (normal probability) is completely defined by the distance between μ and a specified value of σ. As such, the working part of the normal density is the exponent $-(x-\mu)^2/2\sigma^2$ because it contains a particular value of the normal variable, X, and the parameters μ and σ^2 of the distribution. The greater the deviation is of a particular x from μ, the greater is the numerator of this exponent. However, this deviation is a squared quantity; as such, two different values of X showing the same absolute deviation from μ have the same probability. This reflects the fact that the normal distribution is symmetrical with respect to μ. That is, $n(\mu + x) = n(\mu - x)$. For example,

$$P(\mu - 0.67\sigma < X < \mu) = P(\mu < X < \mu + 0.67\sigma) \doteq 0.2500, \text{ or}$$
$$P(\mu \pm 0.67\sigma) \doteq 0.5000$$

$$P(\mu - \sigma < X < \mu) = P(\mu < X < \mu + \sigma) \doteq 0.3415, \text{ or}$$
$$P(\mu \pm \sigma) \doteq 0.6830$$

$$P(\mu - 1.96\sigma < X < \mu) = P(\mu < X < \mu + 1.96\sigma) \doteq 4750, \text{ or}$$
$$P(\mu \pm 1.96\sigma) \doteq 0.9500$$

$$P(\mu - 2.58\sigma < X < \mu) = P(\mu < X < \mu + 2.58\sigma) \doteq 0.4950, \text{ or}$$
$$P(\mu \pm 2.58\sigma) \doteq 0.9900$$

and so on.

Second, the fact that the exponent $-(x - \mu)^2/2\sigma^2$ is negative as a whole indicates that the greater the deviation of X from μ, the smaller will be the probability density of X. That is, both tails of the normal distribution experience decreasing density since the further X is from μ, the lower the height of the density function curve becomes. In this connection, we see that, when the value of X is identical with μ, the exponent is zero and, therefore, the density is reduced to $1/\sigma\sqrt{2\pi}$, the largest value of the normal density. Thus, the normal distribution is unimodal, with the modal value at $X = \mu$.

Third, the curve for $n(\mu,\sigma)$ has two points of inflection, located at a distance of σ from either side of the mean μ.

Fourth, the normal distribution has an infinite range so that its density curve never touches the X axis. As a consequence, any interval of numbers will have nonnegative probability. However, the probability of large deviations from μ is negligibly small, as can be seen from the fact that the density curve falls off quickly and that more than 99 percent of its area is enclosed by $\mu \pm 3\sigma$. This property enables us to use the normal distribution to approximate other distributions for which the true range is finite.

Fifth, a change in the value of μ displaces the whole normal distribution, while a change in the value of σ merely alters its relative position with reference to a fixed scale. These facts indicate that the normal distribution is really a family of distributions. See Figure 7.7.

Sixth, a linear change of scale for a normal distribution results in a new normal distribution. That is, if X is a normal variable, then $Y = a + bX$, for $a \neq 0$, is also a normal variable.

Seventh, if $X_1, X_2, \cdots, X_n$ are independent (or merely uncorrelated) normal variables, then their sum, S, is also a normal variable. Furthermore, because of independence, the additivity property holds for both the expectation and the variance here. That is, the expectation of S is the sum of the expectations

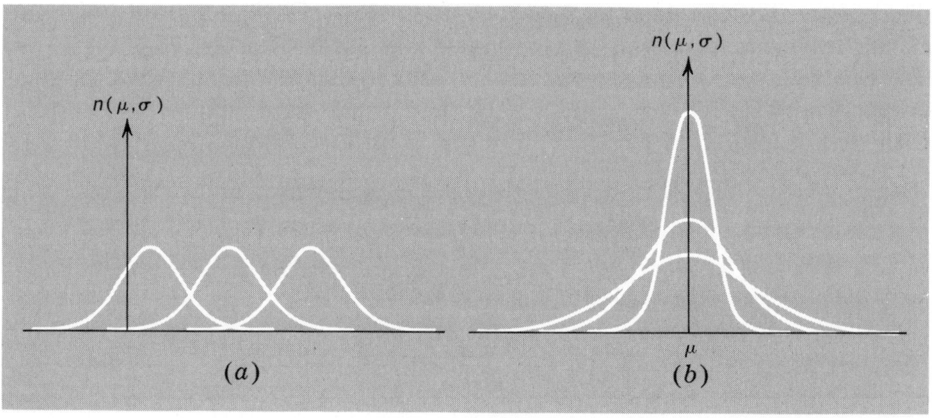

Figure 7.7 (a) Normal distributions with the same standard deviation, different means; (b) normal distributions with the same mean, different standard deviations.

of the n normal variables. Also, the variance of S is the sum of the variances of the n normal variables.

Finally, if X is normally distributed with μ and σ, then $Z = (X - \mu)/\sigma$ is also normally distributed. This transformation from X to Z, as it may be recalled, is named the Z *transformation* and has the effect of reducing X to units in terms of standard deviations. In other words, given a value of X, the corresponding Z value tells us how far away and in what direction X is from its mean, μ, in terms of its standard deviation, σ. For example, $Z = 1.5$ means that the particular value of X is 1.5σ to the right of μ. Similarly, $Z = -2$ means that the particular X value is 2σ to the left of μ. This property of a normal variable, as we shall see immediately, enables us to compute normal probabilities, irrespective of values of μ and σ, from a single probability table for the standard normal distribution.

The normal distribution was first discovered by Demoivre as the limiting form of the binomial model in 1733. It was also known to Laplace no later than 1774, but through a historical error, it has been credited to Gauss, who first made reference to it in 1809. Throughout the eighteenth and nineteenth centuries, various efforts were made to establish the normal model as the underlying law ruling all continuous random variables—thus the name "normal." These efforts failed because of false premises. The normal model has, nevertheless, become the most important probability model in statistical analysis. There is a number of reasons why this is true.

First, many continuous random variables, such as heights of adult males, intelligences of school children, diameters of auto tires of a certain make, tensile strengths of steel wires produced by a certain process, and the like, are normally distributed because of what is known, or presumed to be true, of measurements themselves. In this connection, it is particularly interesting to note that "errors" of repeated measurements of a given dimension are hypothesized to follow the normal probability law. Any measurement or observation is assumed to represent a true magnitude *plus* an error. Each error has a magnitude itself, resulting from a vast collection of factors operative at the moment. Each factor has but a tiny effect on the size and direction of the error; furthermore, errors of measurement work independently with equal force to push the observed measurement up or down, and therefore cancel out in the long run. Thus, we think of errors of measurements as reflections of chance variations which are normally distributed with zero expectation.

Second, as we shall see in Chapter 8, the normal distribution serves as a good approximation of many discrete distributions, such as the binomial or the Poisson model, whenever the exact discrete probability is laborious to obtain or impossible to calculate accurately.

Third, in theoretical statistics, many problems can be solved only under the assumption of a normal population. In applied work, as well, we often find that methods developed under the normal probability law yield satisfactory results, even when the assumption of a normal population is not fully met, despite the fact that the problem can have a formal solution only if such a premise is hypothesized.

Finally, and most important of all, as we shall find in the following chapter, distributions of many sample statistics computed from large samples approach the normal distribution as a limit. As a result, our work on statistical inferences is made easier.

Let us now turn to study the standard normal distribution, which enables us to compute probabilities for any general normal distribution simply by looking at published normal probability tables.

7.12 The Standard Normal Model

As we discussed in Chapter 6, it is often easier to work with standardized probability distributions. This is especially true when we have a density function, such as for the normal variable, involving indefinite integrals which cannot be reduced to elementary functions. For the sake of speed and efficiency in practical work, the concept of the standard normal distribution is indeed indispensable.

A *normal distribution* is said to be of the *standard form* if its mean is zero and its variance, therefore the *standard deviation*, is unity. From the density function of the general normal distribution, we can see that the probability density of the standard normal distribution must be

$$n(x) = n(0,1) = \frac{1}{\sqrt{2\pi}} e^{-x^2/2} \qquad (7.39)$$

The cumulative distribution function corresponding to the standard density, $n(0,1)$, as does any other CDF, gives the probability that the standard normal variable assumes a value equal to or less than x, and can be obtained by integrating (7.39); namely,

$$N(x) = N(0,1) = \int_{-\infty}^{x} n(u)du = P(X \leq x) \qquad (7.40)$$

where $\int_{-\infty}^{x}$ is an integral sign and means the area under the standard normal density curve from $-\infty$ to x.

Graphs of the standard density and distribution functions are shown in Figure 7.8. From these graphs, it can be seen that the standard normal distribution has the same geometric properties as a general normal distribution, with the exception that the former has zero mean instead of any real number and unity variance instead of any positive number.

Any general normal distribution, $n(\mu,\sigma)$, can be converted into the standard normal distribution, $n(0,1)$, by the Z transformation; that is, if X is any normal

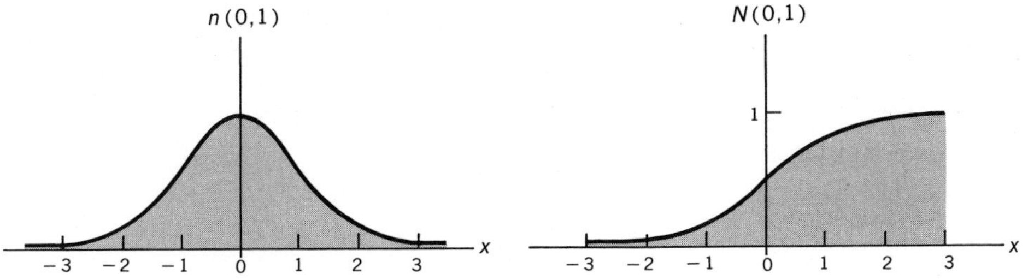

Figure 7.8 Standard normal density and distribution functions.

variable, then its distribution function can be expressed as follows:

$$N(\mu,\sigma) = P(X \le x) = P(\mu + \sigma Z \le x) = P\left(Z \le \frac{x-\mu}{\sigma}\right) = N\left(\frac{x-\mu}{\sigma}\right) \quad \textbf{(7.41)}$$

Thus, given $n(\mu,\sigma)$ for any real numbers a and b, with $a < b$, we would have

$$P(a \le X \le b) = N(b) - N(a) = N\left(\frac{b-\mu}{\sigma}\right) - N\left(\frac{a-\mu}{\sigma}\right) \quad \textbf{(7.42)}$$

As a result, we can compute probabilities for any normal model if we have available an extensive table for $n(0,1)$ or for $N(0,1)$. Furthermore, since the normal distribution is symmetrical about its mean, that is, $n(-x) = n(x)$ or $N(-x) = 1 - N(x)$, it is sufficient to know only the positive values of the normal function in order to know all its values. A table of $N(0,1)$ is given as Table A-VII in Appendix C. We shall now see how to compute normal probabilities by employing expression (7.42) and Table A-VII.

An aluminum factory produces, among other things, a certain type of aluminum-alloy channels. According to past experience, it is known that the stiffness, measured in effective *EI psi*, is normally distributed with $\mu = 2{,}425$ *psi* and $\sigma = 115$ *psi*. This distribution is portrated by Figure 7.9(a).

(1) What is the probability that an aluminum-alloy channel selected at random from this process will have a value between 2250 and 2425 *psi*?

ANSWER:

$$\begin{aligned} P(2250 < X < 2425) &= N(2425) - N(2250) \\ &= N\left(\frac{2425 - 2425}{115}\right) - N\left(\frac{2250 - 2425}{115}\right) \\ &= N(0) - N(-1.52) \\ &= N(0) - (1 - N(1.52)) \\ &= 0.5000 - (1 - 0.9357) \qquad \text{from Table A-VII} \\ &= 0.4357 \end{aligned}$$

See Figure 7.9(b).

(2) What is the probability that the channel drawn will have a value between 2250 and 2500 *psi*?

ANSWER: $$\begin{aligned} P(2250 < X < 2500) &= N\left(\frac{2500 - 2425}{115}\right) - N\left(\frac{2250 - 2425}{115}\right) \\ &= N(0.65) - N(-1.52) \\ &= 0.7422 - 0.0643 \qquad \text{from Table A-VII} \\ &= 0.6779 \end{aligned}$$

See Figure 7.9(b).

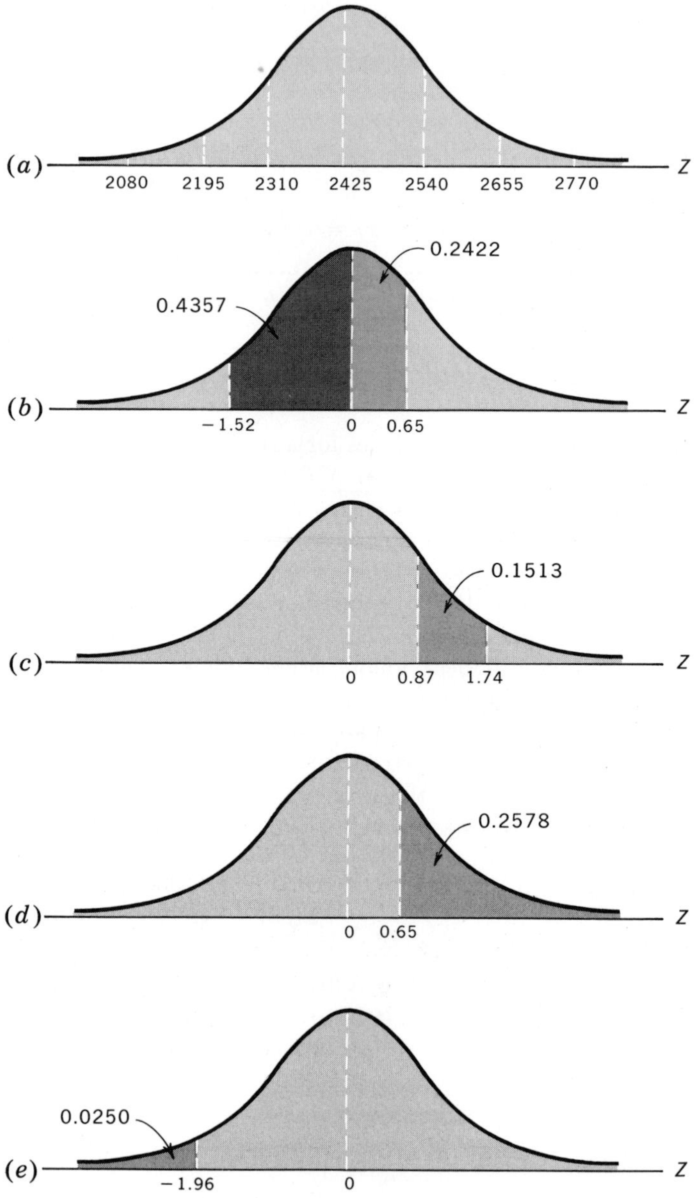

Figure 7.9 Graphic normal probabilities for the aluminum illustrative problem.

(3) What is the probability that the channel drawn will have a value between 2525 and 2625 *psi*?

ANSWER:
$$\begin{aligned} P(2525 < X < 2625) &= N(1.74) - N(0.87) \\ &= 0.9591 - 0.8078 \qquad \text{from Table A-VII} \\ &= 0.1513 \end{aligned}$$

See Figure 7.9(c).

(4) What is the probability that the channel drawn will have a value greater than 2500 *psi*?

ANSWER:
$$\begin{aligned} P(X > 2500) &= 1 - N(0.65) \\ &= 1 - 0.7422 \qquad \text{from Table A-VII} \\ &= 0.2578 \end{aligned}$$

See Figure 7.9(d).

(5) What is the probability that the channel drawn will have a value less than 2200 *psi*?

ANSWER:
$$\begin{aligned} P(X < 2200) &= N\left(\frac{2200 - 2425}{115}\right) \\ &= N(-1.96) \\ &= 0.0250 \qquad \text{from Table A-VII} \end{aligned}$$

See Figure 7.9(e).

If X is a normal variable with $\mu = 80$ and $\sigma = 30$, then

(1)
$$\begin{aligned} P(X < 100) &= N\left(\frac{100 - 80}{30}\right) \\ &= N(0.67) \\ &= 0.7486 \qquad \text{from Table A-VII} \end{aligned}$$

(2)
$$\begin{aligned} P(X > 100) &= 1 - P(X < 100) \\ &= 1 - N(0.67) \\ &= 1 - 0.7486 \qquad \text{from Table A-VII} \\ &= 0.2514 \end{aligned}$$

(3)
$$\begin{aligned} P(100 < X < 150) &= N\left(\frac{150 - 80}{30}\right) - N\left(\frac{100 - 80}{30}\right) \\ &= N(2.33) - N(0.67) \\ &= 0.2415 \qquad \text{from Table A-VII} \end{aligned}$$

(4)
$$\begin{aligned} P(|X - 100| > 25) &= 1 - P(|X - 100| < 25) \\ &= 1 - P(-25 < X - 100 < 25) \\ &= 1 - P(75 < X < 125) \\ &= 1 - (P(X < 125) - P(X < 75)) \\ &= 1 - (N(1.5) - N(-0.17)) \\ &= 0.5007 \qquad \text{from Table A-VII} \end{aligned}$$

An output consists of four components which are produced independently.

The over-all length of this product is the sum, S, of the lengths of the four components, X_1, X_2, X_3 and X_4, all normal random variables with means and variances as follows:

$$\begin{aligned} \mu_1 &= 2 \text{ inches}, & \sigma_1^2 &= 0.010 \\ \mu_2 &= 1 \text{ inch}, & \sigma_2^2 &= 0.006 \\ \mu_3 &= 0.5 \text{ inch}, & \sigma_3^2 &= 0.004 \\ \mu_4 &= 1.5 \text{ inches}, & \sigma_4^2 &= 0.011 \end{aligned}$$

What is the probability that the over-all length for S will meet the specification of 5 ± 0.10 inches?

Here, the desired probability is $P(4.9 < S < 5.1)$. Since lengths of the four components are independently and normally distributed, S must be normally distributed also and we must have

$$\mu_s = \mu_1 + \mu_2 + \mu_3 + \mu_4 = 5 \text{ inches},$$

and

$$\sigma_s = \sqrt{\sigma_1^2 + \sigma_2^2 + \sigma_3^2 + \sigma_4^2} = \sqrt{0.031} = 0.176 \text{ inches}$$

Consequently,

$$\begin{aligned} P(4.9 < S < 5.1) &= N(0.57) - N(-0.57) \\ &= 0.4314 \qquad \text{from Table A-VII} \end{aligned}$$

The preceding result indicates that, on the average, only 43 percent of outputs can be expected to meet the predesignated specifications. This is a very low probability for problems of this kind; the management should either increase the precision of producing the components (reducing variability in production), or relax the standards if relaxation does not impair the output's utility.

It is estimated that a certain machine, on the average, can replace 10,000 man hours, and chances are 50:50 that the saving in terms of mean time could be more than 10,500 hours or less than 9500 hours per week. Now, assuming that a specific normal distribution will fit these estimates, what is the probability that the machine will actually replace less than 9000 man hours per week?

To solve this problem, we must first find the value of the standard deviation for this distribution. To do so, we note that the central 50 percent of the area of a normal distribution is enclosed by $\mu \pm 0.67\sigma$. For the problem, we have

$$\frac{10{,}500 - 9500}{2} = 0.67\sigma$$

That is, if we shift in either the positive or negative direction by 500 hours from the mean, we shift 0.67σ away from the mean. So,

$$500 = 0.67\sigma$$

and

$$\sigma = \frac{500}{0.67} = 746 \text{ hours}$$

With this, the desired probability is seen to be

$$\begin{aligned} P(X < 9000) &= N\left(\frac{9000 - 10{,}000}{764}\right) \\ &\doteq N(-1.31) \\ &= 0.0951 \qquad \text{from Table A-VII} \end{aligned}$$

What does this result mean?

7.13 Summary and Conclusions

Discussion of probability models is an essential step in statistical inference, since we are required to select a model to describe the behavior of the data at hand. We have in this chapter investigated some important special probability models which are known to characterize a large class of random experiments. While in most cases we have tried to derive formulas under different sets of assumptions, we have emphasized throughout that the student should learn to associate the correct model with the problem at hand instead of trying to memorize mathematical statements. As the reader can now agree, calculations of probabilities for various models are quite simple because extensive tables are available.

We began with what is called the Bernoulli model with reference to a random variable which can assume only two values. Then we introduced the binomial distribution as the sum of n independent Bernoulli trials, with the probability of success for each trial remaining as a constant, p. In the binomial case, the two numbers n and p determine the probabilities and may be called parameters of the distribution. The binomial model is appropriate when we think of sampling from an infinite population or sampling with replacement from a finite population, where the elements of the population, infinite or finite, can be classified into two mutually exclusive categories.

We also treated the binomial model as a special case of the multinomial model, which describes the result of n independent trials of an experiment which can be classified into k mutually exclusive classes. We have mentioned the multinomial model merely for the purpose of reference; we shall not use it in our future work.

In general, the hypergeometric model is appropriate when a random sample of size n is drawn without replacement from a population of N items, k of which have a certain characteristic. In contrast to the binomial model, the hypergeometric probability of success changes from one selection to another because successive draws are dependent. It is intuitively clear, however, that if N is large relative to n, the change in k/N could be small enough to be of no consequence. To an extreme, as N becomes infinite, k/N would become a fixed ratio and the hypergeometric probabilities would approach the binomial probabilities with

$p = k/N$. In practice, we are satisfied to approximate hypergeometric probabilities by the binomial values when the sample size is small in relation to the size of the population.

The discrete uniform model refers to a random variable generated by an experiment which can terminate in n mutually exclusive and equally likely ways. One important application of this model arises when we seek a simple random sample from a finite population. When a sample of size n is drawn from a population with N items, on each selection the distribution is uniform and the probability of drawing any specified set of n elements is $1 \Big/ \binom{N}{n}$. In practice, requirements of simple random samples, unconditional and conditional, are satisfied by using a table of random numbers constructed with an underlying uniform distribution of the ten digits from 0 to 9.

The Poisson model is useful in certain situations in which some kind of event occurs repeatedly at random during a time interval or an area of space. A Poisson variable may be considered as the number of occurrences of an event in a time interval or an area of space with possible values 0, 1, 2, $\cdots$. The Poisson model can also be considered as the limit of the binomial when n is large and p is small. To approximate the binomial probabilities by the Poisson values, we use $\mu = np$ and $\mu/n = p$. In this connection, situations may arise in which the hypergeometric is approximated by the binomial, which, in turn, is approximated by the Poisson probability.

As the first continuous model, we introduced the continuous uniform, or rectangular, distribution. If the density function of a random variable is constant over some region of values and zero elsewhere, the variable is said to be uniformly distributed over that region. The density of a uniform model is symmetrical about the center of the interval on which it is defined; hence the central number is both the mean and median of the distribution.

Next, we took up the exponential model, which is also called the negative exponential or Laplace distribution. This model has its origin in the Poisson process. For example, if the number of incoming telephone calls per unit of time obeys the Poisson law, then the distribution of lengths of time between successive telephone calls is exponential. The density function of the exponential variable gives us the probability of the length of time one has to wait to observe the first occurrence of a Poisson event. In the case of Poisson, the mean and the variance are equal; in the case of exponential, the mean and the standard deviation are equal. The exponential model is especially useful in describing the queuing phenomena, to be discussed in Chapter 22.

The last probability model we discussed is the normal distribution, which has been given in both the general and standard forms. This model is by far the most important among all the special probability distributions we have so far investigated. It is important for a number of reasons. First, many random phenomena are known to follow the normal probability law. Second, this model gives a reasonable enough approximation to many other distributions. Third, many other continuous distributions, such as the t, the chi-square, and the F distributions (to be discussed in the future) are derived from the normal distribution. Finally, as we shall see in the following chapter, the distribution of many a sample statistic approaches the normal model as a limit when the size of the sample becomes sufficiently large. Thus, as we proceed with our analytical work, we should have the normal probability model in mind almost all the time.

Glossary of Formulas

(7.1) $\mu = p$ and **(7.2)** $\sigma^2 = pq$

These are the mean and the variance of the Bernoulli model. A random variable is said to be of the Bernoulli type if it can assume only two values: E, with p as its probability of occurrence, and $\bar{E}$, with q as its probability of occurrence. Here, $p + q = 1$.

(7.3) $b(x;n,p) = \binom{n}{x} p^x q^{n-x}$ **(7.4)** $B(r;n,p) = \sum_{x=0}^{r} b(x;n,p)$

(7.5) $B(r;n,p) = 1 - B(n - r - 1;n,q)$ **(7.6)** $E(X) = np$ **(7.7)** $V(X) = npq$

This group of equations is concerned with the binomial probability model. The first is the binomial probability function, which gives the probability of obtaining exactly x successes in n independent trials of an experiment whose outcomes can be classified into two mutually exclusive categories, successes and failures, with probability of success for any single trial as p. The second gives the cumulative probabilities of a binomial variable. The third states the fact that, to obtain cumulative binomial probabilities from published tables, p and q can be interchanged by interchanging the roles of successes and failures. This is valid because the probability of r or fewer successes is the same as the probability of $n - r$ or more failures. The last two expressions here are the expectation and variance, respectively, of a binomial distribution. They are derived by considering a binomial variable as the sum of n independent Bernoulli variables.

(7.8) $m(x_1, \cdots, x_k; p_1, \cdots, p_k) = \binom{n}{x_1, \cdots, x_k} p_1^{x_1} p_2^{x_2} \cdots p_k^{x_k}$ The multinomial probability function gives the probability of obtaining exactly x_1 occurrences of the first kind of outcome, $\cdots$, and exactly x_k occurrences of the kth kind of outcome, when n independent trials are made of an experiment which is capable of yielding k possible mutually exclusive outcomes. Here, p_i refers to the probability of success of the ith outcome in each trial, and $p_1 + \cdots + p_n = 1$. Also, each of the xs can take on any of the values 0, 1, $\cdots$, n, but $x_1 + \cdots + x_k = n$. It is possible to modify this expression to cover a situation where one's interest might be in the probability of the exact number of occurrences of a few, instead of all, the occurrences of all possible outcomes.

(7.9) $E(X_i) = np_i$ and **(7.10)** $V(X_i) = np_i(1 - p_i)$

The number of possible occurrences of each possible outcome in the multinomial model for n independent trials may be treated as an individual variable whose expectation and variance are given as these two equations.

(7.11) $$h(N,n,k,x) = \frac{\binom{k}{x}\binom{N-k}{n-x}}{\binom{N}{n}}$$

(7.12) $$H(N,n,k,r) = P(X \leq r) = \sum_{x=\max(0,n-(N-k))}^{r} h(N,n,k,x)$$

(7.13) $E(X) = n\left(\frac{k}{N}\right)$ (7.14) $V(X) = \left(\frac{N-n}{N-1}\right) n \left(\frac{k}{N}\right)\left(\frac{N-k}{N}\right)$

(7.15) $h(N,n,k,x) = h(N,k,n,x)$ (7.16) $H(N,n,k,r) = H(N,k,n,r)$

The hypergeometric model refers to a random sampling process made without replacement from a population which is finite with N items, k of which possess a certain trait and $N - k$ of which do not possess such a trait. Now, if a conditional random sample is drawn, (7.11) gives the probability that exactly x items are of type k, and (7.12) defines the cumulative probability function of this hypergeometric variable. The expectation and the variance of this model are given by (7.13) and (7.14), respectively. Equations (7.15) and (7.16) state the unique property of a hypergeometric distribution that when k and n are interchanged there will be no effect on probabilities.

(7.17) $u(x;n) = \frac{1}{n}$ This is the probability function of a discrete uniform variable. This model is appropriate when an experiment can terminate in n mutually exclusive and equally likely outcomes. An important aspect of this model is that it furnishes a practical means of selecting a random sample from a finite population.

(7.18) $u(x;r) = 1\bigg/\binom{N}{n}$ Under the assumption that a sample of size n is drawn from a uniformly distributed population with N elements, then the distribution of all possible sample of size n are also uniformly distributed. Here, r is equal to $\binom{N}{n}$, the number of all possible samples of size n.

(7.19) $p(x;\lambda t) = e^{-\lambda t}\frac{(\lambda t)^x}{x!}$

(7.20)
$$
\begin{aligned}
P_t(0) &= e^{-\lambda t}\\
P_t(1) &= \lambda t P_t(0)\\
P_t(2) &= (\lambda t/2)P_t(1)\\
P_t(3) &= (\lambda t/3)P_t(2)\\
&\cdots\\
P_t(k) &= (\lambda t/k)P_t(k-1)
\end{aligned}
$$

(7.21) $P(r;\lambda t) = \sum_{x=0}^{r} p(x;\lambda t)$

(7.22) $E(X) = V(X) = \lambda t$

(7.23) $p(x;\mu) = e^{-\mu}\frac{\mu^x}{x!}$

(7.24) $P(r;\mu) = \sum_{x=0}^{r} p(x;\mu)$

(7.25) $\hat{\mu} = \frac{T}{N}$

Equations (7.19) through (7.25) are connected with the Poisson probability model. A Poisson variable is the number of occurrences of an event in an interval of time, or an area of space, t, and it can take on values of 0, 1, 2, · · · . The Poisson probability function (7.19) defines the probability of success, exactly x occurrences of an event, during a period of time or in a space, given the average number of occurrences in a given interval, λt. (7.20) gives the individual probabilities of a Poisson variable. It reveals that the probability of nonoccurrences is $e^{-\lambda t}$, the probability of one occurrence is λt times the probability of none, and so on. Formula (7.21) is the Poisson distribution function. Equation (7.22) indicates that the expectation and the variance of a Poisson distribution are identical. Because $\lambda t = \mu$ here, the probability and distribution functions of a Poisson variable can also be written more compactly as (7.23) and (7.24). The last expression in this group furnishes a means of estimating a Poisson expectation, where T refers to the total number of successes observed in N intervals and N is the total number of observed intervals.

(7.26) $\dfrac{\binom{k}{x}\binom{N-k}{n-x}}{\binom{N}{n}} \doteq \binom{n}{x}\left(\frac{k}{N}\right)^x\left(\frac{N-k}{N}\right)^{n-x}$ This expression shows that when n is very small relative to N, sampling without replacement (hypergeometric) and sampling with replacement (binomial) give approximately the same probabilities. Thus, under these conditions the binomial can be employed to approximate the hypergeometric probabilities.

(7.27) $\binom{n}{x} p^x q^{n-x} \doteq e^{-np} \frac{(np)^x}{x!}$ When n is large and p is small, the binomial approaches the Poisson values. Thus, when these conditions are met, the Poisson may be employed to approximate the binomial probabilities.

(7.28) $r(x) = \frac{1}{b-a}$, if $a < x < b$
$= 0$, otherwise

(7.29) $R(x) = 0$ if $x \leq a$
$= \frac{x-a}{b-a}$ if $a < x < b$
$= 1$ if $x \geq b$

(7.30) $E(X) = (a+b)/2$
(7.31) $V(X) = (b-a)^2/12$

This group of formulas define the continuous uniform or rectangular probability model. A random variable $X = \{\text{real number } x: a < x < b\}$ is said to have a uniform distribution if its probability density function is constant over the interval (a,b), as indicated by (7.28). The last three equations give, respectively, the cumulative function, the expectation, and the variance of a continuous uniform variable.

(7.32) $X(x) = 1 - e^{-\mu x}$
(7.33) $x(x) = \mu e^{-\mu x}$
(7.34) $x(x) = (1/\beta)e^{-x/\beta}$
(7.35) $X(x) = 1 - e^{-x/\beta}$
(7.36) $E(X) = \beta = 1/\mu$
(7.37) $V(X) = \beta^2 = 1/\mu^2$

The first two equations in this group are the cumulative distribution and the density function of an exponential variable derived from a Poisson process. The exponential variable is then considered as the time interval between Poisson occurrences. Note that here μ is the Poisson average number of occurrences per unit of time or space. The exponential model may also be considered a special case of the gamma distribution and its density and distribution functions may then be given as (7.34) and (7.35). Note now that $\beta = 1/\mu$. Note also that the expectation of the exponential distribution is the reciprocal of the mean of the Poisson model and, as such, it should be interpreted as the expected time until the first occurrence of a Poisson event. Finally, the expectation and the standard deviation of the exponential model are identical.

(7.38) $n(x) = n(\mu,\sigma) = \frac{1}{\sigma\sqrt{2\pi}} e^{-(x-\mu)/2\sigma^2}$. This is the density function of a general normal probability model. A random variable is said to be normal if its mean μ, a constant which may be any real number, and its variance or standard deviation, a constant which is any positive number, exist and if its density function is given by this mathematical expression.

(7.39) $n(x) = n(0,1) = \frac{1}{\sqrt{2\pi}} e^{-x^2/2}$

(7.40) $N(x) = N(0,1) = \int_{-\infty}^{x} n(u)du = P(X \leq x)$

These are the density and distribution functions of the standard normal distribution. This model, as does any standardized random variable, has zero mean and unit variance. Any general normal distribution can be converted into this form by the familiar Z transformation. This possibility makes calculations of normal probabilities a relatively easy task.

(7.41) $N(\mu,\sigma) = P(X \leq x) = P(\mu + \sigma Z \leq x) = P\left(Z \leq \frac{x-\mu}{\sigma}\right) = N\left(\frac{x-\mu}{\sigma}\right)$ By way of the Z transformation, the distribution function of any general normal distribution can be expressed as given here.

(7.42) $P(a \leq X \leq b) = N(b) - N(a) = N\left(\frac{b-\mu}{\sigma}\right) - N\left(\frac{a-\mu}{\sigma}\right)$ This expression follows readily from (7.41). Given $n(\mu,\sigma)$ and any real numbers a and b, with $a < b$, then the probability of any value of $n(\mu,\sigma)$ which would fall within the interval (a,b) can be directly obtained from a table of standard normal probabilities.

Problems

7.1 What is a probability model?

7.2 Under what conditions is the binomial probability model appropriate?

7.3 Cite some business and economic problems in which you know the binomial model can be applied.

7.4 Can you think of some business and economic situations in which the multinomial probability is required?

7.5 Under what conditions does the hypergeometric probability model arise?

7.6 What is the difference between the binomial and the hypergeometric probability laws? Under what conditions can you obtain satisfactory approximations of hypergeometric probabilities by the binomial probability law? In doing this, what expression should you use in place of the binomial parameter p?

7.7 How is the discrete uniform or rectangular probability distribution defined? What is its major application?

7.8 How does the Poisson probability model differ from the binomial and hypergeometric models?

7.9 How does the binomial distribution approach the Poisson distribution as a limit?

7.10 How is the continuous uniform distribution defined? Do you think it might have a practical use in describing business and economic data? If so, what?

7.11 How is the exponential model related to the Poisson model?

7.12 Can you explain why the expectation and the variance are identical in the Poisson case, and the expectation and the standard deviation are identical in the exponential case? Why is the expectation of the exponential distribution the reciprocal of the expectation of the Poisson distribution?

7.13 Can you explain the reason or reasons for the particular shape of a normal distribution?

7.14 Why does the normal distribution hold the most honorable position in probability theory?

7.15 Why do you think that the binomial, the hypergeometric, and the Poisson distributions might approach the normal?

7.16 The light at a traffic intersection is green for 30 seconds at a time, yellow for 5 seconds, and red for 25 seconds, respectively. Assuming that traffic conditions induce random variations in automobile arrival times, so that "making the green light" is a chance event, answer the following:

a. What probability model is appropriate here to successes in 5 and 25 trials?
b. What is the probability of success?
c. What are the mean and the expectation of this model?
d. Use Table A-III to write down 5 and 25 independent trials, respectively.

7.17 Five cards are drawn at random with replacement from a bridge deck. Let X be the number of hearts in the five cards drawn; what is the mean of X? the variance? the standard deviation?

7.18 Suppose it is known that 40 percent of the graduates from a high school go on to college. What is the probability that 15 in an unconditional sample of 30 will go on to college?

7.19 If 100 perfect coins are tossed, (a) what is the probability that exactly 50 heads turn up? (b) What is the probability that the number of heads will not deviate from 50 by more than 10?

7.20 The output of a certain manufacturing process is 1 percent defective.
a. In a random sample of 4 items, taken with replacement, what is the probability that none is defective? that at most 1 is defective?
b. What is the mean number of defectives in a lot of 50 items?

7.21 Ten percent of the radios produced by Company A are defective. If a random sample of 5 items is drawn from a certain day's production run together (which is more than 100,000 units), what is the probability that there will be (a) 0 defectives, (b) 5 defectives, and (c) at least 3 defectives, in the sample?

7.22 A certain make of computer has a probability of 0.05 to give wrong results due to electrical or mechanical failure. To improve the accuracy of calculations, the operator feeds the same data into five such computers, and accepts as accurate the answer that the majority of the computers give. How does this procedure reduce the probability of an error due to a faulty machine?

7.23 If A, B, and C match coins with the understanding that the odd man wins, what is the probability function that A wins in four matches? Determine the probability that (a) A wins exactly twice, (b) B loses exactly twice, and (c) A either wins twice or loses twice.

7.24 If a student is late for his classes 10 percent of the time, what is the probability that he will be late 15 times or more in 100 times? Exactly 15 times?

7.25 The probability for A to make a profit on any business deal is 0.70. What is the probability that he will make a profit exactly 7 times in 10 successive deals?

7.26 Write down the expression without evaluation of this statement: The probability of obtaining exactly 51,000 heads in 100,000 tosses of a perfect coin.

7.27 Twelve dice are tossed together; what is the probability that each number will come up twice?

7.28 A single missile of a certain variety has a probability of 0.25 of shooting down a jet bomber, a probability of 0.25 of damaging it, and a probability of 0.5 of missing it. Also, two damaging shots are capable of downing the plane. If four such missiles are fired, what is the probability of shooting down a jet bomber?

7.29 In a large Canadian factory, it is known that 50 percent of the workers are Canadians, 30 percent are Americans, and 20 percent are South Americans. If a committee of 5 persons is selected, each worker being selected independently, what is the probability that the committee will include 3 Canadians, 1 American, and 1 South American? What is the probability that it will contain 4 Canadians?

7.30 Thirteen cards are drawn at random from a bridge deck; what is the probability that the selection contains five spades and four clubs?

7.31 A bag contains eight oranges and four apples. If four units are drawn from the bag blindly together, (a) what is the probability model for the number of oranges among the four items drawn? (b) Give the probability function of this model.

7.32 A box contains seven white and three black marbles. If three marbles are selected from the box at random without replacement, what are the expectation and the variance of white marbles?

7.33 How would the thirteen spades be distributed in a bridge hand under the assumption of randomness?

7.34 A box contains six good and four bad eggs. If four eggs are drawn at random together, what is the probability of drawing two or fewer good eggs? More than three good eggs?

7.35 The names of four men and six women are written on slips of paper and placed in a hat. Four pieces of paper are drawn from the hat without replacement. What is the probability that the names of two men and two women are drawn?

7.36 A certain make of automobile tire has a flat due to external causes, on the average, once every 2500 miles. Assuming that the occurrences follow the Poisson probability law, determine (a) the probability in a given 500-mile run that more than one flat occurs; (b) the probability that no flat will occur in a trip of 5000 miles.

7.37 A Poisson variable X has a variance of 2.5. Determine $p(4;\mu)$.

7.38 If $P(X = 2) = 0.1839$ for a Poisson variable X, what is the expectation of X?

7.39 A book contains 100 misprints distributed randomly throughout its 100 pages. Assuming a Poisson distribution, determine (a) what is its mean? (b) What is the probability that a page observed at random contains at least two misprints?

7.40 Automobiles arrive at a tollbooth at random at the rate of 300 cars per hour. Calculate the probability that (a) one car arrives during a given 1-minute period; (b) at least two cars arrive during a given 1-minute period; (c) no cars arrive during a period of T minutes.

7.41 Flaws in a certain kind of woolen piece goods occur at random, with an average of one per 100 square feet. What is the probability that a roll 50 by 10 feet will have no flaws? At most one flaw? (*Hint:* $\mu = (50)(10)/100 = 5$.)

7.42 A taxi company has, on the average, 8 flat tires per week. During the past week, 16 flat tires occurred. If the Poisson model is appropriate here, what is the probability of having 16 or more flat tires per week? What conclusion can you draw from this result?

7.43 Accidents on a certain highway occur, at random, at a rate of 20 per month. What is the probability that at least 1 accident occurs in a given half-month interval?

7.44 It is known that the mean number of defects per sheet of carpet of a certain make is two. What is the probability that any sheet of carpet will contain more than two defects?

7.45 Suppose that cars pass a certain traffic intersection at a rate of 30 cars per hour.
a. What is the probability that during a 2-minute interval no car will pass the intersection, or at least 2 cars will pass it?
b. If you observed the number of cars passing the intersection during each of thirty 2-minute intervals, would you be surprised if you found twenty or more of these intervals had the property that either none or at least 2 cars had passed the intersection during that time?

7.46 One thousand television sets are inspected as they come off the production line, and the number of defects per set are recorded below. Estimate the average number of defects per television set for the production process.

x	N_x
0	955
1	42
2	2
3	1
4	0
	1,000

7.47 A sample of 100 is taken without replacement from a student body of 10,000 at a certain university, and it is found that there are 3 foreign students in the sample. What approximate probability would result if there are 500 foreign students at the university?

7.48 A life insurance company has found that the probability is 0.00001 that a person in the 50 to 60 age group will die from a rare disease during a period of one year. If the company has 100,000 policyholders in this age group, what is the approximate probability that the company must pay off more than four claims because of death from this cause?

7.49 An intercontinental ballistic missile has 10,000 parts. The probability that each part will succeed is 0.99998, and all parts work independently. The failure of any part can make the flight a failure. What is the probability of successful flight? (*Hint:* $\mu = np = 0.2$.)

7.50 A company will accept a shipment of 500 articles if, in a conditional random sample of 50, there is at most 1 defective item. If the shipment is 10 percent defective, what is the approximate probability that the shipment will be accepted?

7.51 What is the Poisson approximation of the preceding probability? Do you think the Poisson is better than the binomial approximation in this case? Explain.

7.52 The circumference of a roulette wheel is divided into 37 arcs of equal lengths, numbered from 0 to 36. A ball is placed in the wheel, and the wheel is spun. After the wheel comes to rest, the position of the ball is observed. What is the probability that the ball will be in the arc with a number (a) from 10 to 20 inclusive? (b) that is even? (c) 1?

7.53 Buses on a certain line run every 10 minutes during rush hours. What is the probability that a man arriving at the bus stand at a random time will have to wait at least 8 minutes?

7.54 Let X be uniformly distributed on the interval $0 \leq X \leq 4$.
a. Determine and sketch $r(x)$.
b. Compute $P(|X| > 4/3)$.

7.55 The life of a certain make of radio tube is exponentially distributed with a mean life of 1000 hours.
a. What is the probability that a tube will last less than 1000 hours?
b. More than 1200 hours?
c. If 3 such tubes are taken at random, what is the probability that one will last less than 800 hours, another between 500 and 1200 hours, and still another more than 1200 hours?

7.56 Customers arriving at a certain shop follow the Poisson law, with a mean rate of one customer per minute.
a. What is the probability that 5 or more minutes have passed since the last customer arrived?
b. What is the probability that 5 or more minutes have passed since the next to the last customer arrived?

7.57 A certain digit computer, which operates 24 hours a day, suffers breakdowns at a rate of 0.1 per hour. Suppose the computer has operated satisfactorily for 10 hours, what is the probability that it will continue to operate satisfactorily during the next 10 hours? (*Hint:* An interesting property of the exponential model is that the distribution of future life remains the same as the initial distribution of life. For such computers, an old one is just as good as a new one.)

7.58 From 8 to 9 a.m., workers arrive at a machine-tool shop according to the Poisson law, with an average time between arrivals of 2 minutes.
a. What is the probability that 5 minutes will elapse with no workers arriving?
b. What is the probability that during a 5-minute interval at most 3 workers arrive?

7.59 The distribution, in months of business life, of a certain type of enterprise is exponential with $1/\mu = 70$. What is the probability that a new firm will fail (a) ten months after its birth? (b) twenty months after its birth? What do those answers indicate?

7.60 The random variable X is normally distributed, with $\mu = 10$ and $\sigma^2 = 4$. Determine $P(|X - \mu| > 3)$.

7.61 Let X be a normal variable with mean 100 and variance 49, and compute the following:

a. $P(X = 120)$
b. $P(X > 100)$
c. $P(|X - 95| > 5)$
d. $P(|X - 100| < 10)$

7.62 Random variable Y is normally distributed with mean 65 and standard deviation 6. Compute the following:
a. $P(Y = 36)$
b. $P(Y > 60)$
c. $P(Y < 55)$
d. $P(|Y - 65| < 9)$

7.63 Given that X is normally distributed with mean 10 and $P(X > 12) = 0.1587$, what is the probability that X will fall in the interval (9,11)?

7.64 Inside diameters of steel pipes produced by a certain process are normally distributed with $\mu = 10$ inches and $\sigma = 0.10$ inches. Pipes with inside diameters beyond the range of 10.05 ± 0.12 inches are considered defectives.
a. What is the probability that a pipe produced will be defective?
b. If the process is adjusted so that the inside diameters will have $\mu = 10.10$ and $\sigma = 0.10$ inches, what is the probability that a pipe produced will be defective?
c. If the process is adjusted so that the inside diameters will have $\mu = 10.05$ and $\sigma = 0.06$ inches, what is the probability that a pipe produced will be defective?

7.65 If W is a normal variable and if $P(W < 10) = 0.8413$, what would be its expectation and variance?

7.66 Suppose that heights of American adult males are normally distributed with $\mu = 70$ inches and $\sigma^2 = 9$ inches, how long should mattresses be made in order to accommodate at least 99 percent of them?

7.67 There are two procedures for getting fighter planes ready to take off. Procedure A requires a mean time of 27 minutes with a standard deviation of 5 minutes; for procedure B, $\mu = 30$ and $\sigma = 2$. Which procedure should be used if the available time is 30 minutes? 34 minutes?

7.68 Three independent normal variables are: X_1, with $\mu_1 = 20$ and $\sigma_1^2 = 5$; X_2, with $\mu_2 = 16$ and $\sigma_2^2 = 9$; and X_3, with $\mu_3 = 25$ and $\sigma_3^2 = 7$. Let a new variable, Y, be defined as

$$Y = \frac{X_1 + X_2}{2} - X_3$$

a. How is Y distributed?
b. What are $E(Y)$ and $V(Y)$?

7.69 If $Y = X_1 - X_2$, where both X_1 and X_2 are normal variables, and $\mu_1 = 27$, $\mu_2 = 25$, $\sigma_1 = 3$, and $\sigma_2 = 4$, what is the probability that a random observation of Y will exceed a value of 5? (*Hint:* obtain μ_Y and σ_Y first.)

7.70 The life of electronic tubes of a certain kind is normally distributed with $\mu = 120$ hours and $\sigma = 7$ hours. Six such tubes are used in a circuit. If these tubes alone determine the effective operation of the circuit, and if any one of the tubes fails, the circuit becomes inoperative.
a. What is the probability that the circuit will remain operative after 130 hours?
b. What is the probability that the circuit will remain operative less than 100 hours?

8

THEORY OF SAMPLING

8.1 Introduction

The formulation of decision procedures depends upon our knowledge of the consequences that may result from different actions taken in a given situation and upon the prevailing state of nature at the time of implementing a decision. Here, "state of nature" refers to population models or chance phenomena. Logical analysis of chance phenomena demands ideal probability models, such as those introduced in the preceding chapter. Often, however, we find that the precise properties of the model, or the state of nature, are not known. In order to make reasonable decisions, we must know the model at least approximately. One way of approximating the features of a probability or a population model is by sampling the population directly.

If we take a sample by a random process, then sample statistics are random variables which we can use to estimate the corresponding population parameters. Statistical methods which enable us to infer from limited data (samples) the expected long-run behaviors (populations) are called *inductive statistics* or *statistical inferences.* In the process of forming inferences, we can make errors concerning the state of nature. We must be able to appraise these errors in order to have a measure of confidence in our inductive conclusions. Because these errors are the result of random or chance variations, they can be evaluated only in terms of probabilities. Now, a sample statistic, being a random variable, must have a probability distribution of its own. The distribution of a statistic, commonly called the *sampling distribution* of a statistic, has well-defined properties, as does any probability model. It is from these properties of the distribution of a statistic that we can calculate risks (errors due to chance) involved in making generalizations about populations on the basis of samples.

The purpose of this chapter is to introduce procedures for gathering evidence and for translating data into probabilistic terms amenable to inferences about states of nature. We shall be concerned with the construction of probability distributions for sample statistics and the evaluation of their properties; to this end, our discussion begins with a review of sums of random variables and the Chebyshev inequality. These concepts lead naturally to the consideration of the law of large numbers and the central limit theorem, which—as we shall soon see—form the basis of sampling theory for large samples.

8.2 Sums of Random Variables

Sums of random variables are of interest to us for two reasons. First, often we may find that a random variable under study is inherently an additive combination of ingredients. For example, the binomial variable with parameters n and p is merely the sum of n independent random variables having Bernoulli distributions with a common parameter p. Second, many sample statistics are computed with sums of random variables. For instance, the sample mean is the sum of sample values multiplied by a constant, $1/n$—the reciprocal of the sample size. Furthermore, the sample variance is a constant times the sum of squared deviations.

Now, we may recall that if $X_1, X_2, \cdots, X_n$ are n independent random variables, each having finite mean and variance, then the additivity property holds for both the expectation and the variance of their sum, S. Namely, the expectation of S is the sum of the individual expectations, and the variance of S is the sum of the individual variances. Furthermore, it can be shown that if the X_is are not only independent but also have identical distributions, then $E(S) = n\mu$ and $V(S) = n\sigma^2$.

Computations of probabilities involving a sum often requires that the entire distribution of the sum be known in addition to the expectation and the variance values of the sum. In a simple discrete case, such as the binomial which is a sum of n Bernoulli variables, the sum can be evaluated directly with great ease. However, in most cases, it is quite difficult to derive the distribution of a sum. Fortunately, we find that many of the frequently employed probability models possess the additivity property. For example, the sum of two or more binomial variables is again a binomial variable. Similar statements can also be made, as discussed in the last chapter, for the Poisson and the normal probability models. In each such case, the distribution of a sum is the same as the individual distributions.

8.3

The Chebyshev Inequality

The Chebyshev inequality was introduced in Chapter 6 for the purpose of showing the relationship among the expectation, the variance, and the probability assignments of a probability distribution. We pointed out then that this inequality holds for any probability distribution, provided that its mean and variance exist. Moreover, this inequality can be stated in two ways. First, it asserts that the probability that a random variable X will deviate absolutely from its mean by $h\sigma$ is always less than or equal to $1/h^2$. That is,

$$P(|X - \mu| \geq h\sigma) \leq \frac{1}{h^2}$$

Note that, no matter what pattern the probability distribution assumes, the probability that X will have a value outside the closed interval from $\mu - h\sigma$ to $\mu + h\sigma$ can be no more than $1/h^2$.

The second version of the Chebyshev inequality states that at least $1 - 1/h^2$ of the probability associated with any random variable will lie within h standard deviations of the mean. Namely,

$$P(|X - \mu| < h\sigma) \geq 1 - \frac{1}{h^2}$$

This shows, for example, that the probability that X will assume a value with four standard deviations from the mean is at least 0.9375; similarly, the probability that an observed value of X will lie within ten standard deviations of the mean is 0.99.

The significance of the Chebyshev inequality is a theoretical one. Its generality makes it highly useful in certain discourses of a theoretical nature. One such discourse is the law of large numbers, introduced next.

8.4

Law of Large Numbers

A set of n observations $X_1, X_2, \cdots$ is said to constitute a random sample drawn from a random variable X if X_is are independent variables and each of them is identically distributed as X, a population with mean μ and variance σ^2. Now, if we let

$$S = X_1 + X_2 + \cdots + X_n$$

be the sum of the n observations, and

$$\bar{X} = \frac{1}{n}(X_1 + X_2 + \cdots + X_n)$$

be the sample mean, we would have $E(S) = n\mu$ and $V(S) = n\sigma^2$. As it shall be shown later, we would also have $E(\bar{X}) = \mu$ and $V(\bar{X}) = \sigma^2/n$.

From the last quantity $V(\bar{X}) = \sigma^2/n$, we deduce the amazing fact that the variance of the sample mean approaches zero as n approaches infinity. By the Chebyshev inequality, we know that if a random variable has a small variance, then the probability will be close to 1 that the value of an observation from the random variable will be approximately equal to the mean of the random variable. Now, since the sample mean itself is a random variable, by taking a large enough sample we can make the probability of the sample mean as close to the population mean as we wish—that is, close to 1. Or, if the sample size is large enough, then the probability that the sample mean will differ from the population mean by more than an arbitrarily prescribed positive difference is close to zero. These statements are alternative ways of expressing what is called the *law of large numbers*. This law states: For any arbitrarily prescribed difference $\delta > 0$, then

$$P(|\bar{X} - \mu| < \delta) \to 1 \quad \text{as } n \to \infty \tag{8.1}$$

or

$$P(|\bar{X} - \mu| > \delta) \to 0 \quad \text{as } n \to \infty \tag{8.2}$$

To observe the validity of the above statements, let us replace X by $\bar{X}$, σ^2 by $V(\bar{X}) = \sigma^2/n$, and μ by $\mu = E(\bar{X})$ in the Chebyshev inequality, to find

$$P(|\bar{X} - \mu| < \delta) \geq 1 - \frac{\sigma^2}{n\delta^2}$$

and

$$P(|\bar{X} - \mu| \geq \delta) < \frac{\sigma^2}{n\delta^2}$$

Since σ and δ are positive quantities, the relation $\sigma^2/n\delta^2$ in the preceding inequalities approaches zero as n approach infinity. This proves that the results in equations (8.1) and (8.2) are valid.

The generality that $\bar{X}$ approaches μ in probability is called the *Khintchine's theorem* for the law of large numbers, which was published in 1929. The first general formulation of the law of large numbers, known as the *Bernoulli law* of large numbers, was made by Jacob Bernoulli, and published posthumously in 1713. The Bernoulli theorem states that if S represents the number of successes in n independent identical Bernoulli trials, with probability p of success in each trial, and if $\delta > 0$, then

$$P\left(\left|\frac{S}{n} - p\right| < \delta\right) \to 1 \quad \text{as } n \to \infty \tag{8.3}$$

and

$$P\left(\left|\frac{S}{n} - p\right| > \delta\right) \to 0 \quad \text{as } n \to \infty \tag{8.4}$$

Here, $S/n = \bar{X}$ and represents the frequency of the number of successful outcomes in relation to the total number of trials. Intuitively, the Bernoulli law states that if n is sufficiently large and a fixed integer, then the relative frequency would have a high probability of being close to p. A question that naturally arises is: In order to obtain a "good" estimate of p, how large should n be? That is, we are interested in estimating the size of n in order that the observed frequency will be within a specific distance, δ, from p (whatever the value of p) at a given high level of probability, α (read as Greek letter alpha). For this task, the Chebyshev inequality is again a handy tool. The problem, once values of α and δ are arbitrarily specified, is to find an integer n so that

$$P\left(\left|\frac{S}{n} - p\right| \leq \delta\right) \geq \alpha \qquad \text{for all } p \text{ in } 0 \leq p \leq 1$$

can be satisfied. To find a crude lower bound for this expression by the Chebyshev inequality, we note that, first of all, $\alpha = 1 - \sigma^2/n\delta^2$. Second, $V(S/n) = pq/n \leq 1/4n$ for all p in the closed interval (0,1). This is due to the fact that $pq = p(1 - p) = p - p^2 = \frac{1}{4} - (\frac{1}{2} - p)^2 \leq \frac{1}{4}$. With this information, we can rewrite the foregoing expression as

$$P\left(\left|\frac{S}{n} - p\right| \leq \delta\right) \geq 1 - \frac{1}{4n\delta^2}$$

and it is satisfied if

$$n \geq \frac{1}{4\delta^2(1 - \alpha)} \tag{8.5}$$

For example, if $\delta = 0.1$ and $\alpha = 0.95$, then $n \geq 500$, since

$$n \geq \frac{1}{(4)0.1^2(1 - 0.95)} = 500$$

So, we see that if $n \geq 500$, then $P(|S/n - p| < 0.1) \geq 0.95$. This means that if we take a random sample of at least 500 items, then the probability is at least 0.95 that the observed relative frequency of success will differ from the true but unknown p by less than 0.1.

Our discussion so far has been concerned with what is called the *weak law of large numbers* in contrast to what is called the *strong law of large numbers.* While a consideration of the latter is beyond the scope of this text, it may be interesting to mention just this: The *weak* law of large numbers states that if n is a sufficiently large and fixed integer, then the resulting sample mean or relative frequency is highly likely to be close to μ or to p. It does not say, however, that the sample mean or relative frequency will stay near μ or p when n is increased. Nor does it say that $\bar{X}$ or S/n becomes stabilized at μ or p. These are, however, the exact statements of the *strong* law of large numbers. For example, the strong law may state that $P(\lim_{n\to\infty} \bar{X} = \mu) = 1$. This is certainly a much stronger mathematical statement, but its role in statistical analysis is somewhat obscure.

In closing this section, we may point out that the law of large numbers is the philosophical justification for all attempts to estimate probability empirically. It justifies, for instance, the relative frequency theory of probability. Furthermore, it assures a high level of probability that a sample average is close to the unknown but constant population mean if n is large. This clearly implies that the accuracy of estimates will always increase with increases in the sample size. Obviously, much of analytical statistics could not exist without a theorem of this kind.

8.5 The Central Limit Theorem

We have just shown the validity of the law of large numbers by mathematical expression of this fact: that by averaging an increasingly large number of observations of the value of a quantity we can obtain increasingly more accurate measurements of the expectation of that quantity. It must be noted now, however, that the center of the distribution of a sum S moves off to infinity, and the variance of a sum becomes larger and larger as n increases since, as we know, $E(S) = n\mu$ and $V(S) = n\sigma^2$. It is therefore quite necessary for us to learn something about the pattern of the distribution of a sum. Under certain mild restrictions, the distribution of the sum of a large number of independent random variables, whatever distributions the summands possess, has approximately a normal distribution. This is the essence of what is called the *central limit theorem*, CLT, a name given it by G. Polya in 1920. The utility of this theorem is that it requires virtually no conditions on distribution patterns of the individual random variables being summed. As a result it furnishes a practical method of computing approximate probability values associated with sums of arbitrarily distributed independent random variables. No wonder this theorem is considered the most remarkable theoretical formulation of all probability laws from both the practical and the theoretical viewpoints.

The central limit theorem, first introduced by Demoivre during the early eighteenth century, has been expressed in many forms. One important version treats it as a consequence of sums of identically distributed independent random variables. According to this version, if S is the sum of a large number of identical and independent random variables, each with mean μ and variance σ^2, then the variable $(s - n\mu)/\sigma\sqrt{n}$ approaches the standard normal distribution when n becomes infinite.

The computational significance of CLT is that, for large n, we can express the

$$P(S \leq s) = N\left(\frac{s-\mu}{\sigma/\sqrt{n}}\right)$$

distribution function of S in terms of $N(0,1)$ as follows:

$$\text{OK} \rightarrow P(S \leq s) \doteq N\left(\frac{s-n\mu}{\sigma\sqrt{n}}\right)$$

$$\text{if } m=1 \quad P(S \leq s) = N\left(\frac{s - \mu_{pop}}{\sigma_{pop}}\right) \tag{8.6}$$

where s is a specified value of S. As to the question of how large n should be in order for (8.6) to hold, there is no standard answer. This depends upon the closeness of approximation required and the actual distribution forms of X_i. If the summands are normally distributed, then (8.6) provides exact probabilities no matter how small n is. If nothing is known about the distribution patterns of X_i, the rule of thumb is that n must be 25 or greater in order to have satisfactory approximations. Of course, approximations improve when n becomes larger. The tendency of a sum toward normality is illustrated by Figure 8.1, in which, distributions of the sums of 1, 2, 4, and 8 dice are drawn against a scale of normal probabilities. This figure shows, in addition to the evident tendency toward normality as n increases, that both the mean and the variance of the sum of identically distributed random variables increase with the increase in n.

Now let us apply the central limit theorem in an example. A food-processing factory produces canned ground beef with a mean of 5 ounces and a variance of 0.3 ounces² per can. A box of 60 cans has a mean weight of $(60)(5) = 300$ ounces and a variance of $(60)(0.3) = 18$ ounces². By CLT, the weight of a box, denoted by W, would be approximately normally distributed. With this information, we

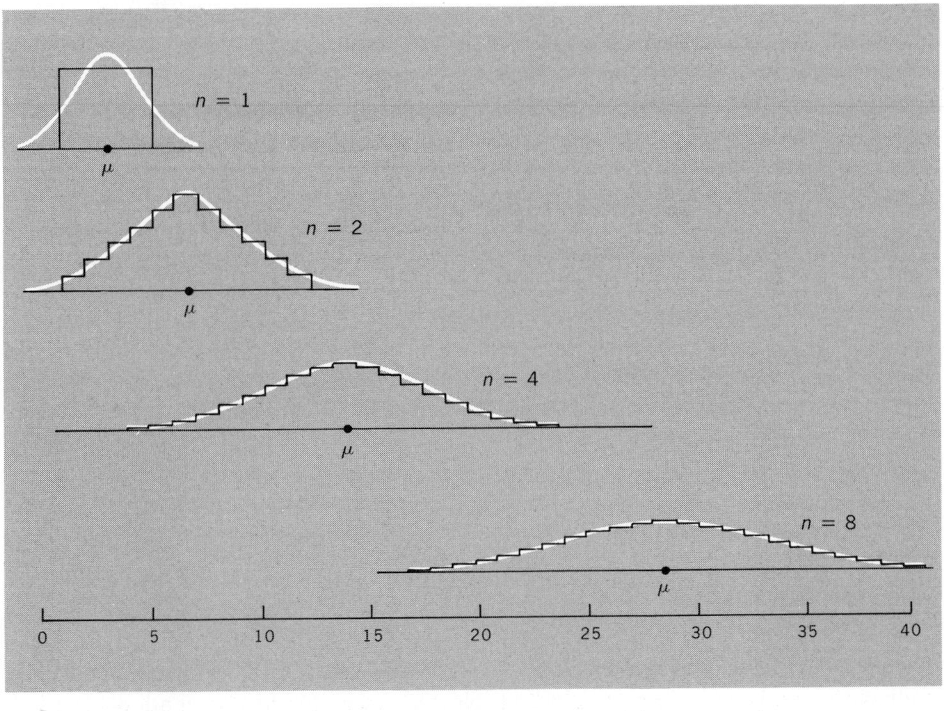

Figure 8.1 Distributions of sums of points on 1, 2, 4, and 8 dice.

would be able to determine probabilities for such various events concerning W as the following:

$$P(W < 290 \text{ ounces}) \doteq N\left(\frac{290 - 300}{\sqrt{18}}\right) = N(-2.36) = 0.0091$$

$$P(W > 310 \text{ ounces}) \doteq 1 - N\left(\frac{310 - 300}{\sqrt{18}}\right) = 1 - 0.9909 = 0.0091$$

$$P(290 < W < 310) \doteq N\left(\frac{310 - 300}{\sqrt{18}}\right) - N\left(\frac{290 - 300}{\sqrt{18}}\right) = 0.9818$$

and so forth.

In the last section we discussed how the Chebyshev inequality could be employed to determine n so that the observed relative frequency would differ from the true proportion by a distance of δ. The estimation of n may now be improved by our new knowledge of the central limit theorem. We shall show in the next chapter that, in accordance with the central limit theorem, the requirement $P(|S/n - p| < 0.1) > 0.95$ is satisfied if $n \geq 100$. Recall that to satisfy this same requirement by the Chebyshev inequality, n must be 500 or greater. This improvement is due to the fact that we know now S/n is approximately normally distributed. Here again we see the value of additional knowledge.

8.6

Sample Observations as Random Variables

Sampling refers to a process of selecting a sample of n items from a population or random variable. There are, as shown in Chapter 11, many alternative sampling procedures. For the time being, it is sufficient to point out that these procedures yield either what are called *judgment* samples or *random* samples. Here we are interested in only a special type of random sample—the simple random samples introduced in the previous chapter. In this section we shall discuss the consequences of random observations and thus provide a basis for our later arguments on the derivation of sampling distributions and their related properties.

An observation made from a random variable is, again, a random variable. An individual observed value is random in the sense that it can assume any of the possible values of a random variable or population. As a random variable, the probability distribution of a sample observation is identical with that of the population of measurements—the random variable under consideration. Imagine an urn which contains eight balls, two with number 1, two with number 3, and four with number 5. Now, if a ball is selected from this urn at random, the ball can have any one of three values—1, 3, or 5—each with a specific probability as follows:

Number on the ball	Probability
1	0.25
3	0.25
5	0.50
Sum	1.00

This is evidently identical with the population probability distribution.

Let us demonstrate the above conclusion in another way. Suppose an unconditional simple random sample of two balls is drawn from the urn, then the possible samples of two balls and their associated probabilities become

First ball	Second ball	Probability
1	1	(0.25)(0.25) = 0.0625
1	3	(0.25)(0.25) = 0.0625
1	5	(0.25)(0.50) = 0.1250
3	1	(0.25)(0.25) = 0.0625
3	3	(0.25)(0.25) = 0.0625
3	5	(0.25)(0.50) = 0.1250
5	1	(0.50)(0.25) = 0.1250
5	3	(0.50)(0.25) = 0.1250
5	5	(0.50)(0.50) = 0.2500
		Sum 1.0000

where the first number drawn is a random variable X_1 and the second number drawn is a random variable X_2. The first assumes the value of 1 in the first three samples, the value of 3 in the second three samples, and the value of 5 in the last three samples. The probability that X_1 will assume the value of 1 is, therefore, the sum of probabilities of the first three samples, and so on. The distribution of X_1 is thus

Value	Probability
1	0.2500
3	0.2500
5	0.5000
Sum	1.0000

Likewise, the probability distribution of X_2, the second number drawn, is identical with that of X_1 and is identical with the distribution of the parent population.

To generalize: When a simple random sample of size n is drawn from a finite population with replacement, or from an infinite population conditionally or unconditionally, we would have n identically distributed random variables, $X_1, X_2, \cdots, X_n$, all possessing the same distribution as the parent population. Also, if the population is large, though finite, the sample observations, X_i, can in practice be treated as if they were identical and independent. From these findings we may finally define a *random sample* as one whose n observations can be described by the mathematical model of n independent random variables, each a replica of the population from which the sample is selected.

We note now that while sample observations are identically distributed when considered as random variables and are denoted by $X_1, \cdots, X_n$, the actual observations of a specific sample which has been drawn are all different and are denoted by $x_1, x_2, \cdots, x_n$. Thus, the term "random sample" refers to the observations thought of as random variables, not to a particular set of actual realizations. The latter may or may not be random, dependent on whether the method employed to obtain a sample ensures the independence of observations. However, if X_is are identically distributed, then a specific collection of sample observations can be thought of as the particular values of a random variable X; that is, $X: x_1, x_2, \cdots, x_n$. In this connection, for a random sample of n items, we may treat the x_is as realizations of X_is; namely,

$$X_1 = x_1, X_2 = x_2, \cdots, X_n = x_n$$

Consequently, if the sample is random, then sample observations thought of as independent random variables or as a particular set of sample values are equivalent and can be used interchangeably.

The thought that sample observations are random variables leads us to the conclusion that any quantity computed from sample observations must also be a random variable. As a random variable, the computed quantity, or statistic, has a probability distribution of its own, called a *sampling distribution*, or the *probability distribution of a statistic*, which can be established directly from the parent population distribution. For example, in the previous illustration, the sum of numbers on the two balls drawn, $X_1 + X_2$, is a statistic and, according to all the possible samples listed before, the possible sums and their associated probabilities are as follows:

$X_1 + X_2$	Probability
2	0.0625 = 0.0625
4	2(0.0625) = 0.0250
6	0.0625 + 2(0.1250) = 0.3125
8	2(0.1250) = 0.2500
10	0.2500 = 0.2500
Sum	1.0000

This is called the *sampling distribution of the sum with* $n = 2$ *drawn from a population, with replacement, of eight balls, two with number* 1, *two with number* 3, *and four with number* 5. We may, and we usually do, call it simply the *distribution of a sum*. The full name, however, does reveal some important features of the probability distribution of a statistic. First, a sampling distribution is generated from a population distribution, known or assumed. Second, the same population may generate an infinite number of sampling distributions for the same statistic, each for a special sample size n. Finally, a population may generate sampling distributions for two or more different statistics. For example, a random sample of n may be taken from a shipment of steel wires of a certain make, and then a distribution for tensile strength, another for diameter, and still another for length, can be derived. Our interest in this chapter, however, is merely concerned with univariate sampling distributions. We shall now introduce sampling distributions for a number of the most frequently encountered statistics; others will be taken up in later chapters.

8.7 Sampling Distribution of the Number of Successes

If a random sample of size n is taken from a population whose elements belong to two mutually exclusive categories—one containing elements which possess a certain trait and the other containing elements which do not possess the trait—then the sampling distribution of the number of successes is a binomial distribution if sampling is made with replacement; and it is the hypergeometric distribution if sampling is made without replacement. Here, for our discussion, we shall assume that an unconditional random sample is taken and therefore the binomial probability is appropriate.

The sampling distribution of the number of successes, being a binomial probability model, would have as its mean $\mu = np$ and as its variance $\sigma^2 = npq$. The standard deviation of a sampling distribution is usually called the *standard error* in the sense that it measures the sampling variability due to chance or to random forces. In this case, the standard error, denoted by σ, is $\sigma = \sqrt{npq}$.

Since we have thoroughly treated the binomial model in the preceding chapter, we shall not include any illustrations of the distribution of the number of successes here. Instead, we shall move directly to the application of the central limit theorem to this sampling distribution. If X is the number of successes in a random sample of size n, the variable $(x - np)/\sqrt{npq}$ approaches the standard normal distribution as n approaches infinity. Consequently, the distribution function of the number of successes can be expressed as

$$P(X \leq x) \doteq N\left(\frac{x - np}{\sqrt{npq}}\right) \tag{8.7}$$

The left side of this equation, as revealed by Figure 8.2, is a step function which has jumps at nonnegative integers only. The right side of it, however, is continuous and increases as x increases from $-\infty$ to ∞. Yet the values of these functions are approximately equal at each argument for large n, with the continuous function passing through about the middle of each step of the discrete function. These approximations are good with $np > 5$ when $p \leq \frac{1}{2}$, and with $nq > 5$ when $p > \frac{1}{2}$. Normal approximations become exact binomial probabilities when n becomes infinite.

EX A coin is tossed 10,000 times; what is the probability that heads will come up less than 4900 times? more than 5200 times? between 4900 and 5200 times?

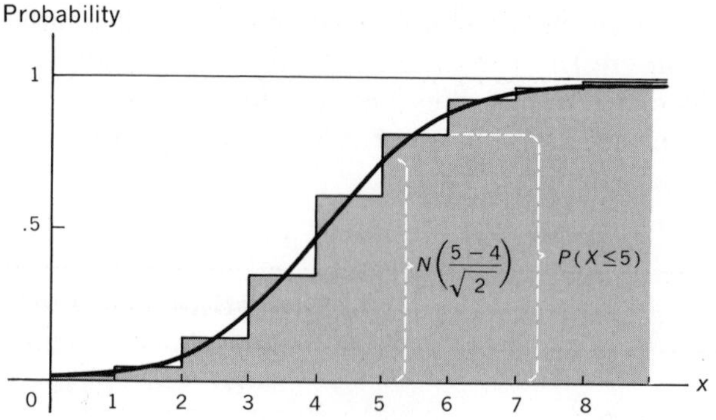

Figure 8.2 Normal distribution function fitted to the binomial distribution function with $n = 8$ and $p = 0.5$.

The distribution of the number of heads with $n = 10{,}000$ taken from an infinite binomial population is approximately normal, with $\mu = np = 5000$ and $\sigma = \sqrt{npq} = 50$. Hence

$$P(X \leq 4900) = N\left(\frac{4900 - 5000}{50}\right) = N(-2) = 0.0228$$

$$P(X \geq 5200) = 1 - N\left(\frac{5200 - 5000}{50}\right) = 1 - N(4) = 0$$

$$P(4900 \leq X \leq 5200) = N(4) - N(-2) = 0.9772$$

Satisfactory normal approximations of the distribution of the number of successes can be made even when n is relatively small if we introduce what is called the *continuity correction factor*, CCF, whose value is a constant of ½. From Figure 8.2 it can be seen that a better approximation to the value of say, 5, can be made by taking the ordinate of the continuous curve half-unit to the right of 5. Consequently, we write, in general,

$$P(X \leq x) \doteq N\left(\frac{x + \frac{1}{2} - \mu}{\sqrt{npq}}\right) \tag{8.8}$$

Consider, for example, a distribution of the number of successes with $n = 25$ and $p = 0.2$. Here, we have $\mu = np = (25)(0.2) = 5$ and

$$\sigma = \sqrt{npq} = \sqrt{(25)(0.2)(0.8)} = 2$$

Then,

$$P(X \leq 5) \doteq N\left(\frac{5 + \frac{1}{2} - 5}{2}\right) = N(0.25) = 0.5987$$

$$P(X = 5) = P(X \leq 5) - P(X \leq 4) \doteq N\left(\frac{5 + \frac{1}{2} - 5}{2}\right) - N\left(\frac{4 + \frac{1}{2} - 5}{2}\right)$$
$$= N(0.25) - N(-0.25) = 0.1974$$

and

$$P(X \geq 5) = 1 - P(X \leq 4) \doteq 1 - N\left(\frac{4 + \frac{1}{2} - 5}{2}\right) = 0.5987$$

These values approximate the exact binomial probabilities of 0.6167, 0.1634, and 0.6167, respectively. Errors of normal approximations in these calculations are therefore 0.018, 0.034, and 0.018. It may also be noted here that for the same size of n, the approximations are better when p is closer to 0.5. Needless to say that, given p, the larger n is, the better the approximation is.

8.8

Sampling Distribution of a Proportion

Recall that a population proportion is defined as $\pi = x/N$, where x is the number of elements which possess a certain trait and N is the total number of items in the population. Recall also that a sample proportion is defined as $p = x/n$, where x is the number items in the sample which possess a certain trait and n is the sample size. Thus, a proportion may be considered as a proportion of successes, and it is obtained by dividing the number of successes by the sample size n. Because of this, we see that if a random sample of n is obtained with replacement, then the sampling distribution of p obeys the binomial probability law and its probability function, in terms of our new symbols, can be written as

$$P\left(\frac{x}{n} \leq p\right) = \binom{n}{x} \pi^x (1 - \pi)^{n-x} \tag{8.9}$$

The expected value and variance of the distribution of p, obtained by dividing n into the corresponding measures for the number of successes, are respectively,

$$E(p) = \frac{n\pi}{n} = \pi \quad \text{and} \tag{8.10}$$

$$V(p) = \frac{n\pi(1 - \pi)}{n^2} = \frac{\pi(1 - \pi)}{n} \tag{8.11}$$

The standard error of p, denoted by σ_p, which measures chance variations of sample proportions from sample to sample, is

$$\sigma_p = \sqrt{V(p)} = \sqrt{\frac{\pi(1 - \pi)}{n}} \tag{8.12}$$

If sampling is made without replacement, then the sampling distribution of p obeys the hypergeometric probability law. In this case, $E(p)$ is still identical with π, but the standard error must be adjusted by a *finite population correction factor*, FPCF, which is equal to $\sqrt{(N - n)/(N - 1)}$; namely,

$$\sigma_p = \sqrt{\frac{\pi(1 - \pi)}{n}} \sqrt{\frac{N - n}{N - 1}} \tag{8.13}$$

It may be noted that if the sample is small relative to the population—say, the former is less than 5 percent of the latter—FPCF would approach unity. Consequently, the introduction of FPCF would produce a practically insignificant difference. This is to say that if a small sample is taken from a large population, we can treat sample observations as if they were independent variables and σ_p can be computed without FPCF.

Now, it may be observed that the distribution of p for an unconditional random sample approaches the normal distribution when n becomes infinite. For a conditional random sample, the distribution of p approaches the binomial model, under conditions specified in the last section, which, in turn, approaches the normal model as a limit. In short, the distribution of p obeys the central limit theorem. Precisely, the variable $(p - \pi)/\sigma_p$ approaches $N(0,1)$ when n becomes infinite. Thus, we state that the distribution function of p with large n is

$$P\left(\frac{x}{n} \leq p\right) \doteq N\left(\frac{p - \pi}{\sigma_p}\right) \tag{8.14}$$

When n is rather small, as in the case of the distribution of the number of successes, satisfactory normal approximations can be obtained by introducing a continuity correction factor of $1/2n$. (Note: $1/2n$ is employed here instead of $\frac{1}{2}$ because a proportion of successes is the number of successes divided by n.) With the CCF, equation (8.14) becomes

$$P\left(\frac{x}{n} \leq p\right) \doteq N\left(\frac{p + \frac{1}{2n} - \pi}{\sigma_p}\right) \tag{8.15}$$

We shall now use a simple example to show how a distribution of a proportion is derived, and to verify that the expectation of p is identical with π and that the standard error of p is as defined by equation (8.13).

A class in advanced statistics consists of four female and six male students. A random sample of five students is drawn from this class without replacement. We wish to evaluate the sampling distribution of the proportion of female students in a sample of five.

In this case, $\pi = 4/10 = 0.4$. The number of all possible samples of five from a population of ten is

$$\binom{10}{5} = \frac{10!}{5!5!} = 252$$

Furthermore, since there are only four female students in the class, there can then be only five possible sample proportions as listed below:

$$p_0 = \frac{0 \text{ females}}{5} = 0.0,$$

$$p_1 = \frac{1 \text{ female}}{5} = 0.2,$$

$$p_2 = \frac{2 \text{ females}}{5} = 0.4,$$

$$p_3 = \frac{3 \text{ females}}{5} = 0.6, \text{ and}$$

$$p_4 = \frac{4 \text{ females}}{5} = 0.8$$

To determine the probability for each of these possible sample results, we note that this distribution of p obeys the hypergeometric law, and its probability values can be obtained directly from Table A-IV. To use this table, we have only to use x/n in the place of x, that is, $h(N,n,k,x) = h(N,n,k,x/n)$. For this problem, we have

$$\begin{aligned}
P(p = 0.0) &= h(10,5,4,0/5) = 0.023810 \\
P(p = 0.2) &= h(10,5,4,1/5) = 0.238095 \\
P(p = 0.4) &= h(10,5,4,2/5) = 0.476190 \\
P(p = 0.6) &= h(10,5,4,3/5) = 0.238095 \\
P(p = 0.8) &= h(10,5,4,4/5) = 0.023810 \\
\text{Sum} & \qquad\qquad\qquad\quad\; 1.000000
\end{aligned}$$

which is the sampling distribution of p with $n = 5$ drawn from a population of ten students with $\pi = 0.4$. The graph of this distribution of p is given by Figure 8.3.

To calculate the expected value and standard error for the above distribution we need the number of samples which will yield each of the possible sample

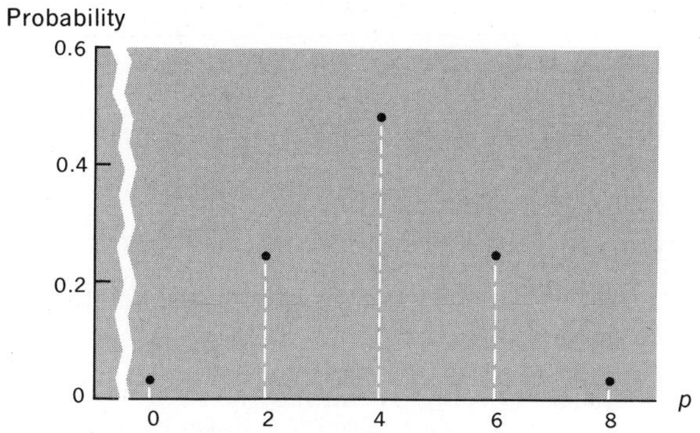

Figure 8.3 Sampling distribution of p with a conditional random sample of 4 from a population with $n = 10$ and $\pi = 0.4$.

TABLE 8.1 COMPUTATIONS OF THE EXPECTATION AND THE STANDARD ERROR OF THE DISTRIBUTION OF p WITH $n = 5$ AND $\pi = 0.4$

p	f	$p - \pi$	$f(p - \pi)$	$f(p - \pi)^2$
0.0	6	−0.4	−2.4	0.96
0.2	60	−0.2	−12.0	2.40
0.4	120	0.0	0.0	0.00
0.6	60	+0.2	+12.0	2.40
0.8	6	+0.4	+2.4	0.96
	252			6.72

proportions. This again is an easy matter since the number of samples which yields a given sample proportion is the product of the total number of all possible samples and the probability of obtaining that sample proportion. Thus,

For $p = 0.0$,	there are	252(0.023810) =	6 samples
For $p = 0.2$,	" "	252(0.238095) =	60 "
For $p = 0.4$,	" "	252(0.476190) =	120 "
For $p = 0.6$,	" "	252(0.238095) =	60 "
For $p = 0.8$,	" "	252(0.023810) =	6 "
Sum			252 samples

Now, with the aid of Table 8.1, we find that

$$E(p) = \Sigma p_i f(p_i) = (0.0)(6/252) + (0.2)(60/252) + \cdots + (0.8)(6/252) = 0.4$$

which, as promised, is identical with the population proportion π.

To compute the standard error directly from the standard formula, we have

$$\sigma_p = \sqrt{\frac{\Sigma f(p - \pi)^2}{\Sigma f}} = \sqrt{\frac{6.72}{525}} = 0.1634$$

Let us now calculate the standard error by the theoretical formula to see if it yields the same result. If it does, we know that the figure for the standard error holds.

$$\sigma_p = \sqrt{\frac{\pi(1 - \pi)}{n}}\sqrt{\frac{N - n}{N - 1}} = \sqrt{\frac{0.4(0.6)}{5}}\sqrt{\frac{10 - 5}{10 - 1}} = 0.1634$$

which is identical with the previous value.

As a second example, let us assume that in a shipment of 20 electronic tubes 6 are defective. If a random sample of 500 is drawn with replacement from this population, how should we go about evaluating the distribution of defective proportion?

Clearly, in this case, it is highly impractical for us to list all the possible sample proportions and their associated probabilities. Fortunately, this is an unnecessary task. All that we need to know are the expectation and the standard error of the distribution of any statistic in order to know about the whole sampling distribution. For the distribution of p with $n = 500$ on hand, we have

$$E(p) = \pi = \frac{6}{20} = 0.3, \text{ and}$$

$$\sigma_p = \sqrt{\frac{0.3(0.7)}{500}} = 0.0205$$

Furthermore, the sample is large enough for this distribution of p to be satisfactorily approximated by the normal distribution. With this information, we can easily estimate the probability for any possible sample proportion for the distribution of p under consideration. For example, what is the probability that the sample proportion will be less than $^{150}/_{500}$? equal to $^{145}/_{500}$? between these two values? greater than $^{165}/_{500}$?

To compute the first probability, we note that $^{150}/_{500} = 0.30$ and that the continuity correction factor here is $\frac{1}{2}(500) = 0.001$. Thus,

$$P\left(\frac{x}{n} \leq 0.3\right) \doteq N\left(\frac{0.3 + 0.001 - 0.3}{0.0205}\right) \doteq N(0.05) = 0.5199$$

Next, the probability for p to be equal to $^{145}/_{500}$ is the same as between $^{145}/_{500}$ and $^{144}/_{500}$, or between 0.290 and 0.288. Therefore,

$$P\left(0.288 \leq \frac{x}{n} \leq 0.290\right) \doteq N\left(\frac{0.290 + 0.001 - 0.3}{0.0205}\right) - N\left(\frac{0.288 + 0.001 - 0.3}{0.0205}\right)$$

$$\doteq N(-0.44) - N(-0.54) = 0.3300 - 0.2946 = 0.0354$$

Now, note that the probability for p to fall between $^{145}/_{500}$ and $^{150}/_{500}$ should be considered as between $^{144}/_{500}$ and $^{150}/_{500}$, or between 0.288 and 0.300. Why? We have then,

$$P\left(0.288 \leq \frac{x}{n} \leq 0.300\right) \doteq N\left(\frac{0.3 + 0.001 - 0.3}{0.0205}\right) - N\left(\frac{0.288 + 0.001 - 0.3}{0.0205}\right)$$

$$= N(0.05) - N(-0.54) = 0.2253$$

Finally, the probability that the sample proportion is equal to or greater than $^{165}/_{500}$ is the probability of one minus that of $^{164}/_{500} = 0.328$. Hence,

$$P\left(\frac{x}{n} \geq \frac{165}{500}\right) \doteq 1 - N\left(\frac{0.328 + 0.001 - 0.3}{0.0205}\right) 1 - N(1.41)$$

$$= 1 - 0.9207 = 0.0793$$

As another example, consider a conditional random sample of 20 drawn from a lot of 100 radio tubes, 80 of which are nondefective. What is the probability that the sample proportion of nondefective tubes will be less than 0.7?

Here, we have

$$E(p) = \pi = 0.8, \text{ CCF} = \frac{1}{2(20)} = 0.025, \text{ and}$$

$$\sigma_p = \sqrt{\frac{0.8(0.2)}{20}}\sqrt{\frac{100-20}{100-1}} = 0.08055$$

Consequently,

$$P\left(\frac{x}{n} \leq 0.7\right) \doteq N\left(\frac{0.7 + 0.025 - 0.8}{0.08055}\right) = N(-0.93) = 0.1762$$

8.9

Sampling Distribution of the Mean

If a random sample of size n is drawn from a population with mean μ and variance σ^2, then the sample observations are independent and identically distributed random variables, as we have already explained. Furthermore, the sample mean, computed as

$$\bar{x} = \frac{1}{n}(x_1 + x_2 + \cdots + x_n)$$

is also a random variable, and the expected value and the variance of the sampling distribution of $\bar{x}$ can be derived simply. First, we note that

$$\begin{aligned} E(\bar{x}) &= E\left(\frac{1}{n}(x_1 + x_2 + \cdots + x_n)\right) = \frac{1}{n}(E(x_1) + E(x_2) + \cdots + E(x_n)) \\ &= \frac{1}{n}(n\mu) = \mu \end{aligned} \tag{8.16}$$

That is, the expectation of the sample mean is the population mean.

Next, since sample observations are considered independent random variables, the additivity property holds for the variance. Namely, the variance of the sum is the sum of the variances. Furthermore, since $V(x_i) = \sigma^2$, the population variance, we have

$$V(\bar{x}) = V\left(\frac{1}{n}\Sigma x_i\right) = \frac{1}{n^2}(V(x_1) + V(x_2) + \cdots + V(x_n)) = \frac{\sigma^2}{n} \tag{8.17}$$

Note that in this derivation we have employed the theorem that the variance of a constant times a variable is equal to the square of the constant times the variable.

The standard error of the mean, which measures the chance variability in sample means, is

$$\sigma_{\bar{x}} = \sqrt{V(\bar{x})} = \frac{\sigma}{\sqrt{n}} \tag{8.18}$$

which reveals that $\sigma_{\bar{x}}$ is smaller than σ. Moreover, it indicates that when $n \to \infty$,

$\sigma_{\bar{x}} \to 0$. Thus, the larger the sample is, the less the fluctuation of sample means is from sample to sample.

If samples are taken from a finite population without replacement, as in the previous cases, a FPCF must be introduced to compute the standard error of the mean. Namely,

$$\sigma_{\bar{x}} = \frac{\sigma}{\sqrt{n}} \sqrt{\frac{N - n}{N - 1}} \tag{8.19}$$

When the parent population is normal, then the sampling distribution of $\bar{x}$ is also normal, no matter how small the sample size is. However, if the population distribution is anything but normal, then the variable $(\bar{x} - \mu)/\sigma_{\bar{x}}$ approaches the standard normal distribution when n becomes infinite. That the distribution of $\bar{x}$ approaches the normal probability model when n increases is illustrated by Figure 8.4. In this figure, the population distribution is given by $n = 1$ and $\mu = 0$. The density functions of $\bar{x}$ for $n = 3$, 6, 10, 50, and 100 are plotted against this population model in the same figure. This figure shows vividly the tendency for the distribution of $\bar{x}$ to approach normality rapidly when n is increased, even though the population is far from being normal itself.

By the CLT stated above, for large n, the distribution function of $\bar{x}$ can then be expressed as follows:

$$P\left(\frac{\Sigma x}{n} \leq \bar{x}\right) \doteq N\left(\frac{\bar{x} - \mu}{\sigma_{\bar{x}}}\right) \tag{8.20}$$

Let us now give an example of a complete evaluation of the distribution of the

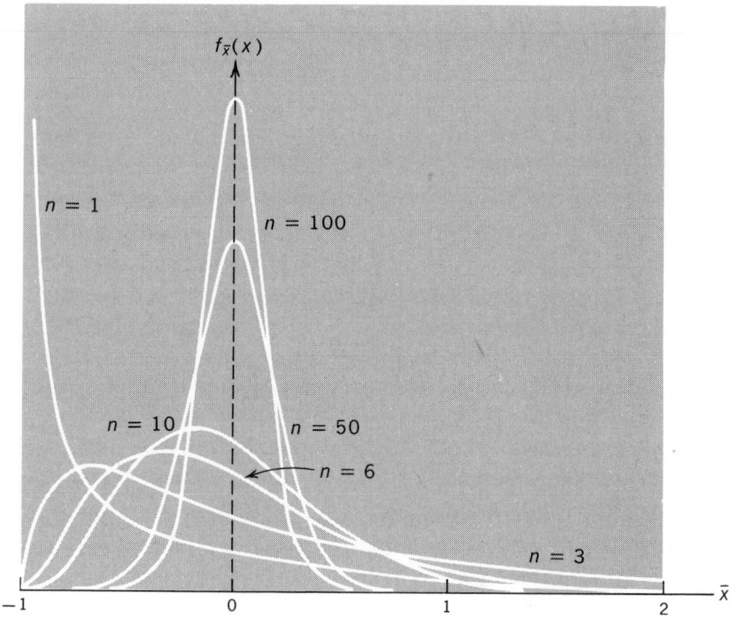

Figure 8.4 Illustration of the effect of sample size on the distribution of $\bar{x}$.

mean. A conditional random sample of size two is drawn from a population which consists of five even numbers, 2, 4, 6, 8, and 10, written on identical chips. Under these conditions, we note first that the mean and the variance of the population are, respectively, 6 and 8. Next, we observe that there are $\binom{5}{2} = 10$ equally likely samples, which are

(2,4), (2,6), (2,8), (2,10), (4,6), (4,8), (4,10), (6,8), (6,10), (8,10).

Clearly, the sample means range consecutively from 3 to 9, with the following probabilities:

	Possible Values of $\bar{x}$	Probability
	3	$1/10 = 0.1$
	4	$1/10 = 0.1$
	5	$2/10 = 0.2$
	6	$2/10 = 0.2$
	7	$2/10 = 0.2$
	8	$1/10 = 0.1$
	9	$1/10 = 0.1$
Sum		$10/10 = 0.1$

For this distribution of $\bar{x}$, the expected value is

$$E(\bar{x}) = \Sigma x_i f(x_i) = 3(0.1) + 4(0.1) + 5(0.2) + \cdots + 9(0.1) = 6$$

which is identical with the population mean stated before. Also, the standard error here is

$$\begin{aligned} \sigma_{\bar{x}} &= \sqrt{\sum_i (\bar{x} - E(\bar{x}))^2 f(\bar{x}_i)} \\ &= \sqrt{(3-6)^2(0.1) + (4-6)^2(0.1) + \cdots + (9-6)^2(0.1)} = \sqrt{3} \end{aligned}$$

To see whether this result agrees with the theoretical formula for $\sigma_{\bar{x}}$, we compute

$$\sigma_{\bar{x}} = \sqrt{\frac{\sigma^2}{n}}\sqrt{\frac{N-n}{N-1}} = \sqrt{\frac{8}{2}}\sqrt{\frac{5-2}{5-1}} = \sqrt{3}$$

Thus we see that the standard error of the mean is related to the population standard deviation as specified by (8.19).

Now, let us consider another example. Steel wires produced by a certain factory have a mean tensile strength of 500 pounds and a standard deviation of 20 pounds. If a random sample of 100 wires is drawn from the production line during a certain week, with a total output of over 100,000 units, what is the probability that the sample mean will differ from 500 pounds by 4 pounds? What is the probability that the sample mean will be less than 496 pounds? more than 504 pounds?

For this case, we have

$$E(\bar{x}) = \mu = 500 \text{ pounds} \quad \text{and} \quad \sigma_{\bar{x}} = \frac{\sigma}{\sqrt{n}} = \frac{20}{\sqrt{100}} = 2$$

Why hasn't the FPCF been used to compute the standard error here? With these values, the desired probabilities become

$$P(|\bar{x} - 500| \geq 4) \doteq 2N\left(\frac{-4}{2}\right) = 2(0.0228) = 0.0456,$$

$$P(\bar{x} \leq 496) \doteq N\left(\frac{496 - 500}{2}\right) = N(-2) = 0.0228, \text{ and}$$

$$P(\bar{x} \geq 504) \doteq 1 - N\left(\frac{504 - 500}{2}\right) = 0.0228$$

It was reported by a certain state that on 125 farms the recorded average wheat yield was 16 bushels per acre, with a standard deviation of 5 bushels. If a random sample of 36 farms is selected without replacement, what is the probability that the sample mean will be less than, or equal to, 14.5 bushels?

ANSWER: $$P(\bar{x} \leq 14.5) \doteq N\left(\frac{14.5 - 16}{\frac{5}{\sqrt{36}}\sqrt{\frac{125 - 36}{125 - 1}}}\right) = N(-2.12) = 0.0170$$

If the recorded average yield is true, then the event of a sample with $n = 36$ to have a mean less than 14.5 bushels per acre would be a very rare event.

8.10 Distribution of the Difference of Two Proportions

When two random samples drawn from two binomial variables are to be compared, it is possible to work only with the *proportion* of successes, not with the *number* of successes, unless both samples are of the same size. For example, during a presidential election, a sample of 100 voters is taken from one state and 40 are found to be in favor of Candidate A; another sample of 150 voters is taken from a second state and 50 are found to be in favor of Candidate A. Clearly, these two sets of figures cannot be evaluated unless they are reduced to proportions. Specifically, what we need here is a probability model of the difference of two proportions.

Two independent random samples are drawn. The first is of size n_1 from a binomial population with π_1, and the second is of size n_2 from a binomial population with π_2. Then the difference of the two sample proportions, $\Delta p = p_1 - p_2$, is again a sample statistic whose probability distribution has as its expected value

$$E(\Delta p) = \Delta\pi = \pi_1 - \pi_2, \tag{8.21}$$

and has as its standard error

$$\sigma_{\Delta p} = \sqrt{\frac{\pi_1(1 - \pi_1)}{n_1} + \frac{\pi_2(1 - \pi_2)}{n_2}} \tag{8.22}$$

Furthermore, if n_1 and n_2 are sufficiently large, then the sampling distribution of Δp is approximately normal. That is, by the CLT, we have

$$P(p_1 - p_2 \leq \Delta p) \doteq N\left(\frac{\Delta p - \Delta\pi}{\sigma_{\Delta p}}\right) \tag{8.23}$$

A game of "heads or tails" is played by two friends, A and B. In this game, each tosses a coin fifty times; one wins the game if he tosses five more heads than the other. What is the probability that A will win the game?

If we assume that the coins used are fair, the sampling distribution of Δp will have

$$E(p) = \pi_1 - \pi_2 = 0.5 - 0.5 = 0, \text{ and}$$

$$\sigma_{\Delta p} = \sqrt{\frac{0.5(0.5)}{50} + \frac{0.5(0.5)}{50}} = 0.1$$

Now, for A to win he must toss five more heads than B in fifty tosses, or A must have $5/50 = 0.1$ more heads than B. Thus, by normal approximation,

$$P(\Delta p \geq 0.1) = 1 - N\left(\frac{0.1 - 0}{0.1}\right) = 1 - 0.8413 = 0.1587$$

It is believed that 10 percent of the batteries produced by Company A are defective, and that 5 percent of those produced by Company B are defective. A random sample of 250 units is drawn from the production line of Company A and 20 are found to be defective; a random sample of 300 units is taken from the production line of Company B and 18 are found to be defective. What is the probability of obtaining such a difference or less in sample proportions if the belief about the population parameters is true?

If the belief were true, then the distribution of Δp would be defined by

$$\Delta\pi = 0.1 - 0.05 = 0.05, \text{ and}$$

$$\sigma_{\Delta p} = \sqrt{\frac{0.1(0.9)}{250} + \frac{0.05(0.95)}{300}} = 0.0228$$

The desired probability, if we note that $\Delta p = 0.08 - 0.06 = 0.02$, becomes

$$P(\Delta p \leq 0.02) \doteq N\left(\frac{0.02 - 0.05}{0.0228}\right) \doteq N(-1.32) = 0.0934$$

In view of this rather small probability, it is highly doubtful that $\pi_1 = 0.1$ and $\pi_2 = 0.05$. Alternatively, this result indicates that it is highly likely that either $\pi_1 < 0.1$ or $\pi_2 > 0.05$.

8.11

Distribution of the Difference of Two Means

In many fields of scientific investigation, we often desire to compare the means of two random variables, such as the effect of two conditions or treatments or production methods. Again, the comparison is made on the basis of two independent random samples: one with size n_1 drawn from one population with μ_1; another with size n_2 drawn from the other population with μ_2. Now, if $\bar{x}_1$ and $\bar{x}_2$ are the two sample means, then we can evaluate the possible difference between μ_1 and μ_2, $\Delta\mu = \mu_1 - \mu_2$, by the difference of the sample means $\Delta\bar{x} = \bar{x}_1 - \bar{x}_2$. The problem here, therefore, is one of determining the properties of the sampling distribution of $\Delta\bar{x}$.

Let X_1, with μ_1 and σ_1^2, and X_2, with μ_2 and σ_2^2, be the two parent populations; then it is easy to see that the expected value of $\Delta\bar{x}$ must be

$$E(\Delta\bar{x}) = \Delta\mu = \mu_1 - \mu_2 \tag{8.24}$$

and the standard error of $\Delta\bar{x}$ is

$$\sigma_{\Delta\bar{x}} = \sqrt{\frac{\sigma_1^2}{n_1} + \frac{\sigma_2^2}{n_2}} \tag{8.25}$$

We know that the sample mean is normally distributed when the sample size becomes infinite. There is reason to believe, then, that if $\bar{x}_1$ and $\bar{x}_2$ are independent and normally distributed, their difference, $\Delta\bar{x}$, must also be normally distributed when both n_1 and n_2 approach infinity. Thus, for sufficiently large integers of n_1 and n_2, the distribution function of $\Delta\bar{x}$ can be expressed as

$$P(\bar{x}_1 - \bar{x}_2 \leq \Delta\bar{x}) \doteq N\left(\frac{\Delta\bar{x} - \Delta\mu}{\sigma_{\Delta\bar{x}}}\right) \tag{8.26}$$

It may be noted here, as in the case of the distribution of the mean, that if X_1 and X_2 are normal variables, then $\Delta\bar{x}$ is also a normal variable however small n_1 and n_2 are.

Spot welds of aluminum of high shearing strength produced by Company A possess a mean of 4500 pounds and a variance of 40,000 pounds². Those produced by Company B possess a mean of 4000 pounds and a variance of 90,000 pounds². If 50 welds of Brand A and 100 welds of Brand B are selected at random and tested, what is the probability that the sample mean strength of A will be at least 600 pounds more than B? At least 500 pounds more than B?

Since, for the distribution of $\Delta\bar{x}$ in this case,

$$\Delta\mu = 4500 - 4000 = 500 \text{ pounds, and}$$

$$\sigma_{\Delta\bar{x}} = \sqrt{\frac{40{,}000}{50} + \frac{90{,}000}{100}} = 41.23 \text{ pounds,}$$

the desired probabilities become

$$P(\Delta\bar{x} \geq 600) \doteq 1 - N\left(\frac{600 - 500}{41.23}\right) = 1 - N(2.43) = 0.0075$$

$$P(\Delta\bar{x} \geq 500) = 1 - N\left(\frac{500 - 500}{41.23}\right) = 1 - N(0) = 1 - 0.5000 = 0.5000$$

Steel pipes produced by a certain process have a mean diameter of 5 inches and a standard deviation of 0.1 inches; what is the probability that two lots, each with 25 pipes, will differ in mean diameter by (a) 0.01 inches or more? (b) by 0.005 inches or more? (c) by 0.005 inches or less?

Here, the distribution of $\Delta\bar{x}$ has

$$\Delta\mu = 5 - 5 = 0, \text{ and}$$

$$\sigma_{\Delta\bar{x}} = \sqrt{\frac{(0.1)^2}{25} + \frac{(0.1)^2}{25}} = 0.0282 \text{ inches}$$

Consequently,

$$\text{(a)} \quad P(|\Delta\bar{x}| \geq 0.01) \doteq 2N\left(\frac{-0.01}{0.0282}\right) = 2N(-0.35) = 0.7264$$

$$\text{(b)} \quad P(|\Delta\bar{x}| \geq 0.005) \doteq 2N\left(\frac{-0.005}{0.0282}\right) = 2N(0.18) = 0.8572$$

$$\text{(c)} \quad P(|\Delta\bar{x}| \leq 0.005) \doteq N\left(\frac{0.005}{0.0282}\right) - N\left(\frac{-0.005}{0.0282}\right) = 0.5714 - 0.4286 = 0.1428$$

The above calculations indicate that, given population parameters, the smaller the difference is between two sample means, the greater the probability is for it to occur. Similar statement can be made for the distribution of Δp.

8.12

Sampling Distribution of Variance

When a random sample of size n is drawn from a population, the sample variance is computed as the sum of squared deviations about the sample mean. That is,

$$s^2 = \frac{1}{n}\Sigma(x_i - \bar{x})^2$$

The probability distribution of the variance so calculated, as it can be shown, has the following expectation:

$$E(s^2) = \frac{n-1}{n}\sigma^2 \tag{8.27}$$

This indicates that, on the average, s^2 is smaller than the population variance σ^2 by a factor of $(n - 1)/n$. In general, as we shall see in the next chapter, a statistic whose expectation is different from the population parameter being estimated is called a "biased" estimate of that parameter. Thus, s^2 is a biased estimate of σ^2. This bias can be corrected by computing the sample variance with a division constant of $n - 1$ instead of n. We shall denote the sample variance by using $n - 1$ as $\hat{s}^2$ and

$$\hat{s}^2 = \frac{1}{n-1}\Sigma(x_i - \bar{x})^2, \text{ or}$$

$$(n-1)\hat{s}^2 = \Sigma(x_i - \bar{x})^2$$

where $n - 1$ is called "$n - 1$ degrees of freedom," a term explained in a later chapter.

To derive the expectation of the sampling distribution of $\hat{s}^2$, we note that, by the linear combination theorem, the variance remains unchanged when the origin is changed. Thus, subtracting a constant μ from x_i and $\bar{x}$, expanding the square and collecting like terms, from the last equation, we have

$$(n-1)\hat{s}^2 = \Sigma(x_i - \mu)^2 - n(\bar{x} - \mu)^2$$

Now, taking expectation on both sides of this expression, we get

$$(n-1)E(\hat{s}^2) = \Sigma E((x_i - \mu)^2) - nE((\bar{x} - \mu)^2) = \Sigma\sigma^2 - n\left(\frac{\sigma^2}{n}\right) = (n-1)\sigma^2$$

Namely,

$$E(\hat{s}^2) = \sigma^2 \tag{8.28}$$

Now, it may be noted that for large samples, s^2 and $\hat{s}^2$ are practically equal and thus s^2 is said to be "asymptotically unbiased." Throughout our discussion in this section we shall be concerned with large samples and therefore assume that $E(s^2) = E(\hat{s}^2) = \sigma^2$.

The calculation of the standard error of s^2, denoted as σ_{s^2}, is much more involved. It can be shown, however, that if the parent population is normally distributed and possesses the fourth moment, μ_4, then

$$\sigma_{s^2} = \sigma^2\sqrt{\frac{2}{n-1}} \tag{8.29}$$

In this expression we have introduced a concept—*moment*. In general, the term "moments" is employed to describe the properties of a distribution, empirical or theoretical. Moments are calculated from the origin or the expectation of a random variable. The first moment about the origin of a random variable, for example, is the mean of the random variable. When moments are computed about the mathematical expectation, the first moment is $\mu_1 = E(X - \mu) = 0$; the second moment is the variance, $\mu_2 = \sigma^2 = E(X - \mu)^2$; the third moment measures the skewness and is defined as $\mu_3 = E(X - \mu)^3$; and the fourth moment is a measure of kurtosis and is defined as $\mu_4 = E(X - \mu)^4$. This last moment has been used to define σ_{s^2} in (8.29).

Ex The variance of weights for truck tires produced by a certain manufacturer is normally distributed, with a mean variance of 10 pounds². If random samples of 100 tires each are taken, what is the percentage of sample variances that will be greater than 10.5? less than 9.75?

To solve this problem, we note that the distribution of s^2 in this case is approximately normal, with

$$E(s^2) = 10 \text{ pounds}^2\text{, and}$$

$$\sigma_{s^2} = \sigma^2 \sqrt{\frac{2}{n-1}} = 10 \sqrt{\frac{2}{99}} = 1.414$$

Thus

$$P(s^2 \geq 10.5) \doteq 1 - N\left(\frac{10.5 - 10}{1.414}\right) = 1 - N(0.35) = 1 - 0.6368 = 0.3632\text{, and}$$

$$P(s^2 \leq 9.75) = N\left(\frac{9.75 - 10}{1.414}\right) = 0.4286$$

A random sample of 200 units is drawn from a population whose variance is 100. What is the probability that the sample variance will differ from the population variance by 2 or more?

The desired probability is

$$P(|s^2 - 100| \geq 2) = 2N\left(\frac{-2}{100\sqrt{2/199}}\right) = 2N(-0.2) = 0.8414$$

This result indicates that 84 percent of the sample variances will fall outside the range of 98 to 102.

8.13 Distribution of Standard Deviation

All comments made on the sampling distribution of s^2 can apply to the sampling distribution of s—sample standard deviation—with one important exception: that is, while the expectation s^2, computed with the division constant of $n - 1$ is identical with σ^2, the expectation of s is only nearly equal to σ when the sample size is greater than 100. Assuming a normal population and a sample size greater than 100, the distribution of s has

$$E(s) \doteq \sigma \quad \text{and} \tag{8.30}$$

$$\sigma_s = \frac{\sigma}{\sqrt{2(n-1)}} \tag{8.31}$$

Ex If a random sample of 200 men is drawn from the American armed forces, in which the standard deviation for heights is 2.5 inches, what is the probability for the sample standard deviation to be less than 2.4 inches? greater than 2.7 inches?

The distribution of s in this case is approximately normal with

$$E(s) \doteq 2.5, \quad \text{and} \quad \sigma_s = \frac{2.5}{\sqrt{2(200-1)}} = 0.125$$

Therefore,

$$P(s < 2.4) \doteq N\left(\frac{2.4 - 2.5}{0.125}\right) = 0.2119, \quad \text{and}$$

$$P(s > 2.7) \doteq 1 - N\left(\frac{2.7 - 2.5}{0.125}\right) = 0.0548$$

8.14 Distributions of Median and Related Measures

If the parent population distribution is normal, and if the sample size is equal to or greater than 30, the distribution of the median is approximately normal with expectation and standard error as follows:

$$E(\text{Med}) = \mu, \quad \text{and} \tag{8.32}$$

$$\sigma_{\text{med}} = \frac{1.2533\sigma}{\sqrt{n}} \tag{8.33}$$

We note that, while the expectation of the median is the same as the expectation of the mean, the standard error of the median is greater than the standard error of the mean by a multiplier of 1.2533.

Recall that the median is a position measure which divides a variable into two equal parts. Recall also that a variable may be divided into four equal parts by three *quartile* measures. The first quartile, Q_1, is a statistic whose value is equal to or less than 25 percent of the values of a variable. The second quartile, Q_2, is identical with the median. The third quartile, Q_3, is a measure which is equal to or greater than 75 percent of the values of a variable. Furthermore, for ungrouped data, we know that $Q_1 = n/4$, $Q_2 = 2n/4$, and $Q_3 = 3n/4$, where n is the sample size, as usual.

Distributions of Q_1 and Q_3 have the same standard error, which is defined as

$$\sigma_{Q_1} = \sigma_{Q_3} = \frac{1.3626\sigma}{\sqrt{n}} \tag{8.34}$$

Furthermore, if the parent population is normal or approximately so and the sample size is greater than 30, distributions of Q_1 and Q_2 are nearly normal, with expectations very nearly equal to the first and second quartiles of the population, respectively.

Measures which divide a set of data into ten equal parts are called *deciles*. Clearly there are nine decile measures, with $D_1 = n/10$, $D_2 = 2n/10$, $D_3 = 3n/10$, $\cdots$, $D_9 = 9n/10$. Remarks made on the distribution of the median apply equally well to distributions of deciles. Expectations of deciles are very nearly equal to their corresponding population parameters. Standard errors of deciles are as follows:

$$\sigma_{D_1} = \sigma_{D_9} = \frac{1.7094\sigma}{\sqrt{n}} \tag{8.35}$$

$$\sigma_{D_2} = \sigma_{D_8} = \frac{1.4288\sigma}{\sqrt{n}} \tag{8.36}$$

$$\sigma_{D_3} = \sigma_{D_7} = \frac{1.3180\sigma}{\sqrt{n}}, \text{ and} \tag{8.37}$$

$$\sigma_{D_4} = \sigma_{D_6} = \frac{1.2680\sigma}{\sqrt{n}} \tag{8.38}$$

Note that the standard error of D_5 is obviously the same as that of the median. Note also how the multiplier factor decreases as the decile value comes closer to the median.

8.15 Significance of Sampling Distributions

The value of a sample statistic computed from a random sample of a given size drawn from a specific population depends upon which elements are included in the sample. A different sample will, usually, yield a different value of the statistic. This observation implies that a sample statistic is expected to be different from its corresponding population parameter, which is often an unknown constant; that is, measuring only a part instead of the entire population results in a margin of error, known as the sampling error—the difference between the value of a statistic and that of a population parameter. It is important to note, in addition, that a possible difference, although referred to as an error, might occur due to chance, not to an error that has been made or will be made because of sampling. Furthermore, we must understand that sampling errors are not literally "errors" as the layman understands the word. It is perhaps more appropriate to call them "sampling variations" or "chance deviations," since their existence is due to the characteristic of chance inherent in random sampling. One basic value of sampling distributions is the assistance that they give us in revealing the patterns of sampling errors and their magnitudes in terms of standard errors—that is, standard deviations of sampling distributions.

A sampling distribution shows that a statistic of a random sample may take on any of a set of values; but these values do not have the same probability of occurrence. It is indeed remarkable that the closer a statistic is to its corresponding parameter, the greater the probability is for the former to happen. As we shall see, this property enables us to evaluate the validity of statistical inferences.

It may now be observed that, in deducing a sampling distribution, we must first make an assumption about the appropriate population parameter. Inasmuch as any value can be assumed for a parameter, depending upon our knowledge or guess of the population, there is no theoretical limit to the number of sampling distributions for the same sample size that can be taken from the same population. There is a sampling distribution for each assumed value of a parameter.

Furthermore, as we have already mentioned, given the assumed value of a parameter, there is a different sampling distribution of a statistic for each specific sample size.

Finally, under the same assumptions about a population and the same sample size, the sampling distribution of one statistic differs from that of another statistic. For example, the pattern of the distribution of $\bar{x}$ will differ from that of s^2, even though both measures are computed from the same sample.

From previous observations, it is easy to appreciate why we should always bear in mind the assumption of the population, the name of the statistic, and the sample size when we think of a sampling distribution.

It will also be useful for us to remember that our discussion in this chapter might appropriately be called "theory of sampling for large samples." In short, when a sampling distribution with large n meets the requirements of the central limit theorem, we have applied the large sampling method. As to how large n should be in order for the CLT to apply depends upon the particular statistic. For the mean, the median, and other position measures, it is sufficient if $n \geq 30$. For sample proportion, it is required that $n(\pi) \geq 5$ if $\pi < 0.5$, and that $n(1 - \pi) \geq 5$ if $\pi > 0.5$. In the case of sample variance, and also of standard deviation, the sample size must be greater than 100. Needless to say, when large sampling theory holds, statistical inferences are made with the normal probability model.

It is also interesting as well as important to note that when sample size is large enough for the CLT to operate, standard errors can be calculated with sample statistics in place of the corresponding population parameters when the latter are unknown; for instance, s can be used for σ, p can substitute for π, and so on. In other words, if the sample size is large enough, the sampling theory for large samples holds whether a standard error is computed with the known population parameter required or its sample estimate.

When samples are small and population parameters required to compute standard errors are unknown, we face what is called the "small sampling method" or "exact sampling theory." With exact sampling theory, each statistic has a particular probability distribution other than the normal probability model, unless the population is itself normal. These probability models will be taken up in Chapter 12.

We shall now turn, with the knowledge gained in the present chapter, to a discussion of the traditional decision procedures of estimation and testing of hypotheses for the case of large samples.

Glossary of Formulas

(8.1) $P(|\bar{X} - \mu| < \delta) \to 1$ as $n \to \infty$ and **(8.2)** $P(|\bar{X} - \mu| > \delta) \to 0$ as $n \to \infty$

These equations are alternative ways of stating the law of large numbers in terms of the sample mean. The former states that the probability for the sample mean to differ from the corresponding population mean absolutely by less than a constant δ approaches 1 as the sample size approaches infinity. The second expression says that the probability for the sample mean to differ from the corresponding population mean absolutely by greater than a constant of δ approaches 0 as n becomes infinite. The implication of the law of large numbers is that we can always make the sample mean be as close to the population mean as we wish by taking a large enough sample.

(8.3) $P(|(S/n) - p| < \delta) \to 1$ as $n \to \infty$ and **(8.4)** $P(|(S/n) - p| > \delta) \to 0$ as $n \to \infty$

These expressions are referred to as the Bernoulli law of large numbers. Here, S is the number of successes, n, the number of trials, and p, the probability of success in each trial. The ratio S/n is the relative frequency of successes. Similar interpretations for the preceding group of equations can be applied here. The Bernoulli law clearly justifies the relative frequency definition of probability.

(8.5) $n \geq \dfrac{1}{4\delta^2(1 - \alpha)}$ This formula gives the size of n required in order for the sample relative frequency to differ from p (which may be thought of as the corresponding population or true proportion) by a value of δ for a specified level of probability, α, or greater. This relationship is derived from the Chebyshev inequality.

(8.6) $P(S \leq s) \doteq N\left(\dfrac{s - n\mu}{\sigma\sqrt{n}}\right)$ This is one of the possible statements of the powerful central limit theorem. It says that if S is the sum of a large number of identically distributed independent random variables, each with mean μ and variance σ^2, then the variable $(s - n\mu)/\sigma\sqrt{n}$ obeys the standard normal probability law. Note that $n\mu = E(S)$ and s is a particular value of the sum. Also, in general, this expression holds if $n \geq 25$.

(8.7) $P(X \leq x) \doteq N\left(\dfrac{x - np}{\sqrt{npq}}\right)$ This says that the distribution function of the number of successes can be approximated by the standard normal distribution for large n because of the central limit theorem. Satisfactory normal approximations of binomial probabilities require that $np \geq 5$ or that $nq \geq 5$. Note also $np = \mu$ and $\sqrt{npq} = \sigma$ here.

(8.8) $P(X \leq x) \doteq N\left(\dfrac{x + \frac{1}{2} - \mu}{\sqrt{npq}}\right)$ To improve normal approximations of binomial—the number of successes—probabilities when n is relatively small, it is necessary to introduce a continuity correction factor of $\frac{1}{2}$.

(8.9) $P((x/n) \leq p) = \dbinom{n}{x} \pi^x(1 - \pi)^{n-x}$

(8.10) $E(p) = \pi$

(8.11) $V(p) = \dfrac{\pi(1 - \pi)}{n}$

(8.12) $\sigma_p = \sqrt{\dfrac{\pi(1 - \pi)}{n}}$

(8.13) $\sigma_p = \sqrt{\dfrac{\pi(1 - \pi)}{n}}\sqrt{\dfrac{N - n}{N - 1}}$

(8.14) $P((x/n) \leq p) \doteq N\left(\dfrac{p - \pi}{\sigma_p}\right)$

(8.15) $P((x/n) \leq p) \doteq N\left(\dfrac{p + \dfrac{1}{2n} - \pi}{\sigma_p}\right)$

This group of equations define the properties of the distribution of a proportion p, which is number of successes divided by n; that is, $p = x/n$. If the population proportion is π and sampling is made with replacement, the distribution of the proportion of successes obeys the binomial probability law and its distribution function is given as (8.9). The next three equations give the expectation, variance, and standard error of a proportion respectively. When sampling is made without replacement, the distribution of p follows the hypergeometric probability law. In this case the expectation of p is still identical with the population proportion π but the standard error must be adjusted by a finite population correction factor as given by (8.13). Equation (8.14) states the fact that the distribution of p obeys the central limit theorem. As in the case of the distribution of the number of successes, satisfactory normal approximations of proportion probabilities for relatively small samples, if a continuity correction factor of $\frac{1}{2n}$ is introduced, are expressed by the last equation in this group.

(8.16) $E(\bar{x}) = \mu$ **(8.17)** $V(\bar{x}) = \sigma^2/n$ **(8.18)** $\sigma_{\bar{x}} = \sigma/\sqrt{n}$

(8.19) $\sigma_{\bar{x}} = \sigma/\sqrt{n}\ (\sqrt{(N-n)/N-1})$ **(8.20)** $P((\Sigma x/n) \leq \bar{x}) \doteq N\left(\frac{\bar{x}-\mu}{\sigma_{\bar{x}}}\right)$

Here we have properties of the sampling distribution of the mean. If the parent population possesses a mean μ and variance σ^2, then the expectation of $\bar{x}$ is identical with the population mean and its variance is related to the population variance as expressed by (8.17). The standard error of the mean is defined by (8.18) if the population is infinite or if sampling is made with replacement from a finite population. When a conditional random sample is drawn from a finite population, $\sigma_{\bar{x}}$ must be determined, as in previous cases, by a finite population correction factor as given in (8.19). If n is sufficiently large, then the variable $(\bar{x} - \mu)/\sigma_{\bar{x}}$ is approximately normally distributed and, therefore, the distribution function of $\bar{x}$ is given as (8.20).

(8.21) $E(\Delta p) = \pi_1 - \pi_2$ **(8.22)** $\sigma_{\Delta p} = \sqrt{\frac{\pi_1(1-\pi_1)}{n_1} + \frac{\pi_2(1-\pi_2)}{n_2}}$

(8.23) $P(p_1 - p_2 \leq \Delta p) \doteq N\left(\frac{\Delta p - \Delta\pi}{\sigma_{\Delta p}}\right)$

If a random sample of n_1 is drawn from a population with π_1, and another sample of n_2 is drawn independently from a population with π_2, then the distribution of the difference between two sample proportions, $p_1 - p_2 = \Delta p$, would have its expectation and standard error defined by (8.21) and (8.22). Furthermore, if n_1 and n_2 are sufficiently large, then the distribution of Δp is approximately normal as revealed by (8.23).

(8.24) $E(\Delta\bar{x}) = \Delta\mu = \mu_1 - \mu_2$ **(8.25)** $\sigma_{\Delta\bar{x}} = \sqrt{\frac{\sigma_1^2}{n_1} + \frac{\sigma_2^2}{n_2}}$

(8.26) $P(\bar{x}_1 - \bar{x}_2 \leq \Delta\bar{x}) \doteq N\left(\frac{\Delta\bar{x} - \Delta\mu}{\sigma_{\Delta\bar{x}}}\right)$

The sampling distribution of the difference of two means, $\Delta\bar{x}$, has as its expectation the difference between the population means, and has as its standard error the square root of the sum of variances of $\bar{x}_1$ and $\bar{x}_2$. If the parent populations are normal, the distribution of $\Delta\bar{x}$ is also normal irrespective of the sizes of n_1 and n_2. If the parent populations are not normal, the distribution of $\Delta\bar{x}$ approaches the normal distribution when n_1 and n_2 become infinite. Thus, for large sample sizes, the distribution function of $\Delta\bar{x}$ can be stated as that of $N(0,1)$ as in (8.26).

(8.27) $E(s^2) = \frac{n-1}{n}\sigma^2$ **(8.28)** $E(\hat{s}^2) = \sigma^2$ **(8.29)** $\sigma_{s^2} = \sigma^2\sqrt{\frac{2}{n-1}}$

The distribution of sample variance s^2, computed with the division constant of $1/n$, has an expectation, as stated by (8.27), smaller than the population variance by a factor of $(n-1)/n$.

However, the sample variance $\hat{s}^2$, obtained by the division constant of $1/(n-1)$, has an expectation identical with the population variance. Expression (8.29) is the standard error of the distribution of s^2 or $\hat{s}^2$, derived under the assumption that the fourth moment of the population exists. It is also assumed that the population is normally distributed. The distributon of s^2 is approximately normal if $n \geq 100$.

(8.30) $E(s) \doteq \sigma$ **(8.31)** $\sigma_s = \dfrac{\sigma}{\sqrt{2(n-1)}}$

If the parent population is normal and if sample size is greater than 100, then the distribution of sample standard deviation is approximately normal, with an expectation very nearly equal to population standard deviation and with a standard error as given by (8.31).

(8.32) $E(\text{Med}) = \mu$ and **(8.33)** $\sigma_{\text{med}} = \dfrac{1.2533\sigma}{\sqrt{n}}$

The distribution of the median has the population mean as its expectation. Its standard error is greater than that for the mean by a multiplier of 1.2533.

(8.34) $\sigma_{Q_1} = \sigma_{Q_2} = \dfrac{1.3626\sigma}{\sqrt{n}}$ Standard errors of the first and the third quartiles are identical.

Recall also that the second quartile is identical with the median.

(8.35) $\sigma_{D_1} = \sigma_{D_9} = \dfrac{1.7094\sigma}{\sqrt{n}}$ **(8.36)** $\sigma_{D_2} = \sigma_{D_8} = \dfrac{1.4288\sigma}{\sqrt{n}}$

(8.37) $\sigma_{D_3} = \sigma_{D_7} = \dfrac{1.3180\sigma}{\sqrt{n}}$ **(8.38)** $\sigma_{D_4} = \sigma_{D_6} = \dfrac{1.2680\sigma}{\sqrt{n}}$

These formulas are standard errors for distributions of deciles. Note that $D_5 = Q_2 =$ Med. Note also, how the constant multiplier increases as the decile measure moves away from the fifth decile, or median, whose standard error has a constant multiplier of 1.2533. If the parent population is normal or approximately so, and if the sample size is greater than 30, then the distribution of each decile is approximately normal, with an expectation nearly equal to its corresponding population decile. The same observations can also be made for distributions of quartiles.

Problems

8.1 Why is the distribution of a sum important for the discussion on sampling theory?

8.2 What are the important properties of the distribution of a sum?

8.3 State the law of large numbers in your own words.

8.4 Comment on the similarities of the two alternative statements of the law of large numbers.

8.5 What is the significance of the law of large numbers?

8.6 What role does the Chebyshev inequality play in the discussion of the law of large numbers?

8.7 What does the central limit theorem say? What is its importance?

8.8 In what sense do we say that a random observation is a random variable? Use a numerical example to support your answer.

8.9 Suppose your class of statistics contains 100 students, numbered from 1 to 100. Select a simple random sample of 10 by using the table of random digits (A-XVI).

8.10 Show that the first number selected in the preceding problem comes from a uniform distribution with $f(x;100) = 1/100$, $x = 1,2,3, \ldots, 100$.

8.11 What is a sampling distribution according to your understanding of the term?

8.12 Can you explain why a sampling distribution can be derived from the process of deduction?

8.13 What do we mean by sampling errors? How are their magnitudes measured?

8.14 Explain in at least two different ways the fact that the larger the sample, the more reliable is the sample statistic as an estimate of the corresponding population parameter.

8.15 A certain type of ground-to-ground missile has a probability of 0.25 to hit the target. Ten such missiles are fired.
a. Write the exact expression that exactly five will hit the target.
b. Write the exact expression that at least five will hit the target.
c. Use the Chebyshev inequality to estimate the number of shots required in order to have a probability of at least 0.5 that the average number of hits will differ from 0.5 by less than 0.25.

8.16 If you wish to estimate the proportion of business executives who have graduate degrees, and to have the estimate be correct within 2 percent with a probability of 0.95 or better, how large a sample should be taken if you (a) are confident that the true proportion is less than 0.2, and (b) have no knowledge at all about the true proportion?

8.17 A random sample is to be taken to estimate the proportion of defectives, p, of a certain production process. Determine the sample size so that the probability is at least (a) 0.95, (b) 0.99, that the observed proportion will differ from the true proportion by less than (i) 1 percent and (ii) 10 percent.

8.18 A die is tossed sixty times.
a. What is the mean of the sum of points?
b. What is the variance of the sum of points?
c. What is the probability that the total number of points will be less than 200?

8.19 A delivery truck is loaded with three different kinds of cartons containing food products. The weight of each kind of carton is a random variable. There are 10 cartons of one kind, 40 of another kind, and 80 of a third kind. Mean weights and standard deviations for these cartons, respectively, are 100, 10, 50; and 5, 25, $\sqrt{20}$—all in pounds. What is the probability that the load on the truck is more than 5100 pounds?

8.20 The reliability of a rocket is the probability p that an attempted launching of the rocket will be successful. It has been established that the reliability of a certain type of rocket is 0.9. A modification of this rocket is made and is being evaluated. Which set of evidence below throws more doubt on the hypothesis that the modified design is only 90 percent reliable?
a. 121 modified rockets tested, 116 performed successfully.
b. 81 modified rockets tested, 78 performed successfully.

8.21 Which of the following sets of evidence throws more doubt on the hypothesis that more voters are in favor of Candidate A than in favor of Candidate B in a presidential election? (a) Of 10,000 voters sampled, 5100 are in favor of A. (b) Of 1000 voters sampled, 510 are in favor of A.

8.22 An urn contains eight numbered balls, three with the number 1 and five with the number 2. Two balls are drawn at random without replacement from the urn.
a. List all the possible samples and their associated probabilities.
b. Derive the sampling distribution of the difference between the two numbers on the balls drawn.
c. What comments are in order from your results in (a) and (b)?

8.23 Three independent observations are made from a normally distributed random variable X with $\mu = 0$ and $\sigma^2 = 1$. The largest of these three observations is a statistic—a function of these observations. The probability distribution of this statistic is

$$\begin{aligned} P(\text{largest observation} \le x) &= P(\text{all observations} \le x) \\ &= P(\text{first observation} \le x) \end{aligned}$$
$$P(\text{second observation} \le x)$$
$$P(\text{third observation} \le x) = N(x)N(x)N(x) = (N(x))^3$$

What is the probability that the largest observation exceeds a value of 2?

8.24 A random sample of 4 is drawn from an infinite population with $p = 0.5$. Derive the sampling distribution of the number of successes (a) in terms of exact probabilities, and (b) in terms of normal approximations.

8.25 If a random sample of 100 is taken from the production line of a process in which

20 percent of the items are defective, what is the probability that 15 or less in the sample are defective?

8.26 Plot the distribution of the number of successes with $n = 20$ and $p = 0.4$ against its normal approximations. In your judgment, are the normal approximations good enough for practical situations?

8.27 Let $n = 3600$ and $p = \frac{1}{3}$, and calculate the probability for the number of successes between $3600p - 20$ and $3600p + 20$.

8.28 Half-inch iron screws manufactured by a certain factory are occasionally not slotted. This occurs at random, and the compound probability of its happening and escaping inspection is 0.02. In a shipment of 2500 such screws, what is the probability that (a) 64 or more will lack slots? (b) 36 or less? (c) between 36 and 64?

8.29 Let the population be ten balls in an urn, five of which are black and five are white. A random sample of six is drawn without replacement; evaluate the sampling distribution of the proportion of white balls. (*Note:* The evaluation of a sampling distribution involves the listing of all sample values with associated probabilities, and calculations of its expected value and standard error.)

8.30 Forty percent of the registered voters in a given state are Republicans. From this population a sample of five is drawn at random. Evaluate the sampling distribution of p, proportion of Republicans. (*Hint:* Since the sample is so very small relative to the population, independence may be assumed here.)

8.31 Let the population be the same as in the preceding problem, and let a random sample of 100 be drawn. Would you be surprised to find that there are 50 Republicans in the sample?

8.32 In a national election, 55 percent of the voters are in favor of a certain candidate; what is the probability that, in a sample of 100, the result will not show a majority in favor of the candidate?

8.33 If an unconditional random sample of 180 items is drawn from a population with $\pi = \frac{1}{3}$, what is the probability that the sample proportion satisfies $50/180 < p < 70/180$?

8.34 On college faculties in New York City, $\frac{1}{6}$ are women. If a random sample of 180 is drawn from this population, determine a. $P(p \geq 0.3)$, and b. $P(0.1 < p < 0.25)$.

8.35 The variance of the distribution of p is found to be 0.0025 with $n = 100$; what is π?

8.36 Forty-six percent of union members are against trading with Red China. What is the probability that a poll would show a majority of this trait if the poll consisted of (a) 100 union members? (b) 1000 union members?

8.37 It is known that 5 percent of the radio tubes produced by a certain manufacturer are defective. If the manufacturer sends out 1000 lots, each containing 100 tubes, in how many of these lots can we expect to have (a) fewer than 90 good tubes? (b) 98 or more good tubes?

8.38 In a random sample of 1000 households selected in West Virginia in 1965, 96 of the heads of families were found to be unemployed. Was this sample result consistent with the assumption that 10 percent of the labor force there in that year were unemployed? Explain.

8.39 Evaluate the sampling distribution of $\bar{x}$ with $n = 2$ from a population of 10 digits, 0, 1, 2, 3, 4, 5, 6, 7, 8, and 9. Assume that sampling is made with replacement.

8.40 Rework the preceding problem under the assumption that a conditional sample of 2 is drawn.

8.41 The distribution of average weekly wages of 3000 secretaries in a Canadian city is found to have $\mu = \$68.00$ and $\sigma = \$3.00$. If 250 random samples, all of the same size of 25, are drawn from this population without replacement, what would be the mean and standard error of the distribution of $\bar{x}$ formed by these 250 samples?

8.42 In the preceding problem, how many samples in the distribution can we expect to have means between \$66.80 and \$68.30?

8.43 A lot of 1000 fried chicken dinners has a mean weight of 12 ounces and a standard deviation of 0.6 ounces. What is the probability that in a random sample of 100 from this population the total weight will be (a) less than 1190 ounces? (b) more than 1195 ounces? (c) between 1190 and 1195 ounces? (*Hint:* That the combined weight will be between 1190 and 1195 ounces is the same as that the sample mean will be between 11.90 and 11.95 ounces.)

8.44 Packages to be delivered by a factory have a mean weight of 300 pounds and a standard deviation of $\sqrt{2500}$ pounds. What is the probability that 25 packages taken at random and loaded on a truck will exceed the specified capacity of the truck, known to be 8200 pounds?

8.45 A liberal arts college has 100 faculty members, 60 of whom have doctoral degrees. Two samples, with $n_1 = n_2 = 30$, are drawn from this faculty independently, with replacement, and whether or not they possess doctoral degrees is noted. What is the probability that the two samples will differ by 8 or more with doctoral degrees?

8.46 Rework the preceding problem under the assumption of nonreplacement.

8.47 Similar products are produced by two manufacturers, A and B. A's output contains 7 percent defectives, and B's contains 5 percent defectives. If an unconditional random sample of 2000 is drawn from each manufacturer's product, what is the probability that the two samples will reveal a difference in proportional defectives of 0.01 percent or more?

8.48 A certain make of ball bearings has a mean weight of 0.5 ounces and a standard deviation of 0.02 ounces. Two unconditional random samples are taken independently from a certain day's production, with $n_1 = 500$ and $n_2 = 800$. What is the probability for the two sample means to differ by more than 0.002 ounces? By less than 0.001 ounces?

8.49 Two different makes of TV picture tubes A and B, possess the following parameters: $\mu_A = 1400$ hours, $\sigma_A{}^2 = 40{,}000$ hours², $\mu_B = 1200$ hours, and $\sigma_B{}^2 = 10{,}000$ hours². A random sample of 125 tubes is drawn from each make; determine the probability that (a) Make A will have a mean life at least 160 hours longer than B; (b) Make A will have a mean life at least 250 hours longer than B.

8.50 The annual salary of professors in a certain city averages $12,000, with a standard deviation of $1,000. In the same city, the annual salary of physicians averages $15,000 with a standard deviation of $1,500. A random sample of 100 is taken from each population; what is the probability for the sample means to differ by less than 5,000? more than 6,000?

8.51 Weights of American soldiers are normally distributed, with a variance of 100 pounds². If a random sample of 200 is drawn from this population, what are the expectation and standard error of the sampling distribution of s^2?

8.52 For the preceding problem, what is the probability that s^2 will be greater than 101? less than σ^2? different from σ^2 by 1.5?

8.53 Tensile strengths of steel wires are normally distributed with a standard deviation of five pounds. Many samples of 100 each are drawn at random, and standard deviations are computed for these samples. Find the expected value and the standard error of the distribution of s.

8.54 For the last problem, what percentage of samples would have standard deviations greater than 5.71 pounds? less than 4.148 pounds?

8.55 A normal population has $\mu = 100$ and $\sigma = 5$. If a random sample of 250 is drawn from this population, is it likely for this sample to have a median of 91 or less?

8.56 If the first quartile of a population with $\sigma = 10$ is 11, what is the probability for Q_1 with $n = 100$ to be less than 10?

8.57 If the third quartile of random variable, with $\sigma^2 = 81$, is 75, what is the probability for Q_3 with $n = 81$ to be greater than 77?

8.58 If 100 students are selected at random from colleges in California and interviewed, and 51 are found to be in favor of open-housing policy, what is the probability for this sample proportion to differ from the true proportion by more than 1 percent?

8.59 The Transit subway Authority in New York City bought 100 light bulbs of Brand A and another 100 of Brand B. On testing these bulbs, it found that $\bar{x}_A = 1{,}300$ hours, $s_A = 90$ hours, $\bar{x}_B = 1{,}250$ hours, and $s_B = 100$ hours. What is the probability that the difference between the two corresponding population means is greater than 40 hours?

8.60 The average IQ of 350 school children observed at random in India is 105. The sample standard deviation for this sample is 5. In view of this information, is it reasonable to assume that the mean IQ for all children in India is greater than 110? Why or why not?

9

ESTIMATION

9.1

Statistical Inferences

With a knowledge of probability theory and sampling distributions, we are prepared to discuss traditional decision procedures for scientific investigations.

In conducting a scientific investigation, the scientist often knows, or is willing to assume, that the population or the random variable X from which he samples is of a certain functional form whose parameter he attempts to evaluate. For example, from theoretical considerations or previous studies, it might be known that the IQ of school children possesses a normal function, $n(x;\sigma,\mu)$, where the mean IQ, μ, is the unknown parameter of interest. In general, when we say that a random variable X is distributed as $f(x;\theta)$, we mean that θ (theta) is the unknown parameter in the frequency function, $f(x)$.

Statistical inferences refer to methods by which one tries to select a random sample from a population and, on the findings of sample observations, attempts (1) to ascertain the value of the unknown parameter θ, or (2) to decide if θ or some function of θ is the same as some preconceived value of θ, say, θ_0. The first of these decision procedures is called *estimation of* θ, and the second is known as *testing a hypothesis of* θ.

In this chapter, we shall be concerned with methods of statistical estimation, and in the next, with methods of hypotheses testing. With respect to estimating a parameter, there are two types of estimates to be considered: point estimation and interval estimation. We shall now introduce them in that order.

9.2 Point Estimation

The procedure in point estimation is to select a random sample of n observations, $x_1, x_2, \cdots, x_n$, from a population $f(x;\theta)$; and then to use some preconceived method to arrive, from these observations, at a number, say, $\hat{\theta}$ (read theta hat), which we accept as an *estimator* of θ. Note that the estimator, $\hat{\theta}$, is a single point on the real number scale and thus the name *point estimation*.

It is important to observe that an estimator $\hat{\theta}$ is a function of n independent sample values. Considering this fact, we should select a function, say, $\hat{\theta} = g(x_1,x_2, \cdots ,x_n)$, which will give us the best estimate of θ. Such a function, unfortunately, is often difficult and sometimes impossible to determine in practice. The next best thing we can do is to insist that the estimator be a "good" one. A good estimator, as common sense dictates, is one which is "close" to the parameter being estimated. More precisely, the quality of an estimator is to be evaluated in terms of unbiasedness, consistency, efficiency, and sufficiency. Let us comment on these desirable properties of an estimator before we move on to see how they can be incorporated into an estimator.

UNBIASEDNESS

An estimator is said to be unbiased if its expected value is identical with the population parameter being estimated. That is, if $\hat{\theta}$ is an unbiased estimate of θ, then we must have $E(\hat{\theta}) = \theta$.

Thus, we see that the sample mean $\bar{x}$, the sample proportion p, the difference between two sample means $\Delta\bar{x}$, and the difference between two sample proportions Δp, are all unbiased estimators, since in each case the expectation of the estimator is equal to the corresponding parameter being estimated. However, the sample variance, as has already been noted, computed with the division constant $1/n$, is a biased estimate of σ^2 because

$$E(s^2) = \frac{n-1}{n}\sigma^2 = \sigma^2 - \frac{\sigma^2}{n}$$

The bias here is the quantity $-\sigma^2/n$. Clearly this negative bias will vanish when n becomes infinite, and therefore s^2 becomes an unbiased estimate of σ^2. Furthermore, as it has also been noted, when the sample variance is computed with a division constant of $1/(n - 1)$ it would be an unbiased estimate of σ^2 however small n is.

The criterion of unbiasedness assumes significance when many investigations produce estimates of the same parameter. These estimates, if unbiased, can be combined to produce a still better estimate of the parameter.

Bias in estimation, however, is not neccessarily undesirable; it may turn out to be an asset in some situations. For example, if the parent population is normally distributed, then it can be shown that a sample variance computed with $1/n$ is a slightly better estimator than one computed with $1/(n-1)$, when they are evaluated on the basis of their mean square deviations about σ^2. Or, more generally, suppose we wish to estimate θ for $f(x;\theta)$, where it may be known that, say, $\theta \leq 0$, on the basis of a sample, and suppose that we insist upon an unbiased estimate; then it may be that some values of $\hat{\theta}$ are positive. If we do arrive at a positive estimate of θ under these conditions, it would be obviously better to set $\hat{\theta} = 0$. Of course, if many estimates are to be obtained for a parameter, it is advisable to report all unbiased estimates even though some of them are impossible results.

It is also of interest to note that many estimators are what we call "asymptotically unbiased" in the sense that the biases reduce to practically insignificant values when n becomes sufficiently large. The estimator s^2 is an example. Usually we would want bias to disappear in large samples. If a bias still exists with large samples, it is customary to remove the bias with a correction factor.

What counts, actually, is that an estimator must be close to the parameter being estimated. Unbiasedness should not be considered as an absolutely desirable property. For example, it may happen that an unbiased estimator is less desirable than a biased estimator if the former has a greater variability than the latter and, as a consequence, the expected value of the latter is closer than that of the former to the parameter being estimated. This being the case, the biased estimator may be preferred.

CONSISTENCY

An estimator $\hat{\theta}$ is said to be consistent for θ if the limit of the probability that $|\hat{\theta} - \theta| < \delta$ is unity when the sample size approaches infinity. In other words, an estimator is said to be consistent if the probability for it to approach the parameter being estimated is 1 as n approaches infinity. This is clearly true for $\bar{x}$ or p, or s^2 as a consequence of the law of large numbers.

Recall that $E(x) = E(\text{Med}) = \mu$; but are sample mean and sample median equally satisfactory to estimate μ? The answer is yes if $f(x;\mu)$ is symmetrical and if consistency is the only criterion under consideration. If, however, the population distribution is skewed so that the population mean and median are not identical, we would expect the sample mean to come closer to the population mean and the sample median to come closer to the population median when sample size is increased. Thus when n approaches infinity or approaches the size of the population, the sample mean will approach the population mean in value, and the sample median will approach the population median (which is different from the population mean) in value. So we see that the sample mean is, but the median is not, a consistent estimate of μ when the population distribution is not symmetrical. Furthermore, the sample mean is an unbiased estimator of μ no matter what form the population distribution assumes, while the sample median is an unbiased estimate of μ only if the population distribution is symmetrical. We conclude, therefore, that the sample mean is better than the sample median as an estimate of μ in terms of both unbiasedness and consistency.

Consistency seems to be a desirable property for an estimator to possess when large samples are employed. When a small sample is used, consistency is of little importance unless the limit of probability defining consistency is reached even with a relatively small size of the sample.

EFFICIENCY

Given that estimators are unbiased or at least consistent, then an estimator $\hat{\theta}_1$ is said to be more efficient than another estimator $\hat{\theta}_2$ for θ if the variance of the first is less than the variance of the second. This criterion seems to be an intuitively clear concept. Obviously, the smaller the variance of an estimator is, the more concentrated is the distribution of the estimator around the parameter being estimated and, therefore, the better this estimator is.

The best illustration of relative efficiency in estimators is the estimation of μ by sample mean and by sample median. If the population is symmetrically distributed, then both the sample mean and the sample median are consistent and unbiased estimators of μ. Yet, we can claim that the sample mean is better than the sample median as an estimate of μ. This claim is made in terms of efficiency. We know that $V(\bar{x}) = \sigma^2/n$ and that $V(\text{Med}) = (1.2533\sigma/\sqrt{n})^2 = 1.57076\sigma^2/n$. That is, $V(\bar{x}) < V(\text{Med})$. As a matter of fact, $V(\bar{x}) \doteq 0.64V(\text{Med})$. Thus, we conclude that the sample mean is more efficient than the sample median as an estimator of μ.

When we say that one estimator is more efficient than another, it seems reasonable to think of an estimator that is the *most* efficient for the parameter being estimated. The most efficient estimator, evident from our understanding of relative efficiency, must be one which possesses a variance as small in value as possible. That is, we think of the most efficient estimator as a *minimum variance estimator.* An estimator $\hat{\theta}$ is said to be a minimum variance estimator of θ if $E(\hat{\theta} - E(\hat{\theta}))^2 < E(\hat{\theta}^* - E(\hat{\theta}^*))^2$, where $\hat{\theta}^*$ is any estimator of θ other than $\hat{\theta}$.

The lower bound of a minimum variance can be found by what is called the Cramer-Rao inequality, according to which, for example, if $\hat{\theta}$ is an estimate of μ, then the variance of $\hat{\theta}$ cannot be less than σ^2/n—$V(\hat{\theta}) \geq \sigma^2/n$. Here, $\hat{\theta}$ could be the sample mean, the sample median, or some other statistic. Whatever it is, its variance has a lower bound of σ^2/n. But we know that $V(\bar{x}) = \sigma^2/n$. Thus, in addition to being unbiased and consistent, $\bar{x}$ is a minimum variance estimate of μ.

SUFFICIENCY

A rigid mathematical statement of this last desirable property, sufficiency in estimation, is rather involved. Fortunately, we find that this concept entails an accurate intuitive meaning. An estimator is said to be sufficient if it conveys as much information as possible about the parameter which is contained in the sample, so that little additional information will be supplied by any other estimator. This rather obvious yet vague statement can be made clearer perhaps by way of an illustration.

Consider, for example, a normal population with a unit variance but an unknown mean, μ. If we wish to estimate this unknown mean, we may take a random sample of, say, two observations. These observations, X_1 and X_2, are assumed to be independent random variables, and each is distributed identically with the population. Now, when we investigate the collection (X_1,X_2), we investigate all the information this sampling experiment yields. Furthermore, the sample set (X_1,X_2), obtained from the parent random variable whose mean we seek to estimate, furnishes all the information that is available for us to compute various estimators.

Now, suppose, we select the following two functions of (X_1,X_2) for evaluation:

$$\hat{\theta}_1 = \frac{X_1 + X_2}{2}, \text{ and}$$

$$\hat{\theta}_2 = \frac{X_1 - X_2}{2}$$

We see that the set $(\hat{\theta}_1,\hat{\theta}_2)$ contains the same information as (X_1,X_2) since one is functionally related to the other. Next we note that in $(\hat{\theta}_1,\hat{\theta}_2)$, $\hat{\theta}_2$ is normally distributed with an expectation of zero and a variance of 0.5. Clearly, the distribution of $\hat{\theta}_2$ does not involve μ, and therefore it does not yield any useful information on the parameter we seek to estimate. Again, in $(\hat{\theta}_1,\hat{\theta}_2)$, $\hat{\theta}_1$ is normally distributed with an expectation identical with μ and a variance of 0.5. Clearly, $\hat{\theta}_1$ is a sufficient estimator of μ since it provides all the information on μ we can obtain from the sample data, and since any other estimator, such as θ_2, cannot yield any more information for the task of estimation at hand.

The significance of sufficiency lies in the fact that if a sufficient estimator exists it is absolutely unnecessary to consider any other nonsufficient estimators; a sufficient estimator ensures that all information that a sample can furnish with respect to the estimation of a parameter is being utilized. It may now be mentioned that $\bar{x}$, p, $\Delta\bar{x}$, and Δp are sufficient estimators of the corresponding parameters μ, π, $\Delta\mu$, and $\Delta\pi$, respectively.

Having decided upon the properties desirable for an estimator to possess, we must now devise methods of estimating functions that may provide estimators satisfying these properties. Among the many important methods, we shall consider only two in this text: the *maximum likelihood function* and the *least square method.* The first is our immediate concern; the latter we shall take up when we treat time series and association analysis.

9.3

Maximum Likelihood Function

We shall develop the concept of the maximum likelihood function by an example. Suppose, after taking a sample of 10 students from colleges and universities, we find that 3 of them are against the admission of Red China to the United Nations. We would like to know, given this sample evidence, how likely is it that the population proportion will be 0.1? 0.2? · · · 0.9? 1.0?

To answer this question, we note that the sample proportion in this case is a binomial variable. If the true proportion π is, say, 0.1, then we have

$$P(p = 0.3) = \binom{10}{3}(0.1)^3(0.9)^7 = 0.0574$$

If the true proportion π is, say, 0.2, then we have

$$P(p = 0.3) = \binom{10}{3}(0.2)^3(0.8) = 0.2013$$

Similarly, we can compute the probability for the sample proportion to be 0.3 given any other possible values of π specified in the query. Let us tabulate these probabilities as follows:

π	0.0	0.1	0.2	0.3	0.4	0.5	0.6	0.7	0.8	0.9	1.0
$P(p = 0.3)$	0.0	.0547	.2013	.2668	.2150	.1172	.0425	.0090	.0008	.0000	0.0

Now, if we think that the true proportion can assume any value between 0 and 1 instead of just those given above, we are thinking of the function $L(x) = P(p = 0.3|\pi = x)$, which is defined for all xs satisfying the inequality $0 \le x \le 1$. $L(x)$ is called the *likelihood function* for π and observed result $p = 0.3$ with $n = 10$. This function clearly gives the conditional probability of $p = 0.3$, given that $\pi = x$. We can graph this function by plotting the values shown in the preceding table and connecting them by a smooth curve. In doing so we obtain what is called the *likelihood curve* for π given $p = 0.3$ with $n = 10$, as shown by Figure 9.1.

We ask: What value of π seems to make the sample outcome observed most likely? The standard answer is: The value of π must be one that makes the proba-

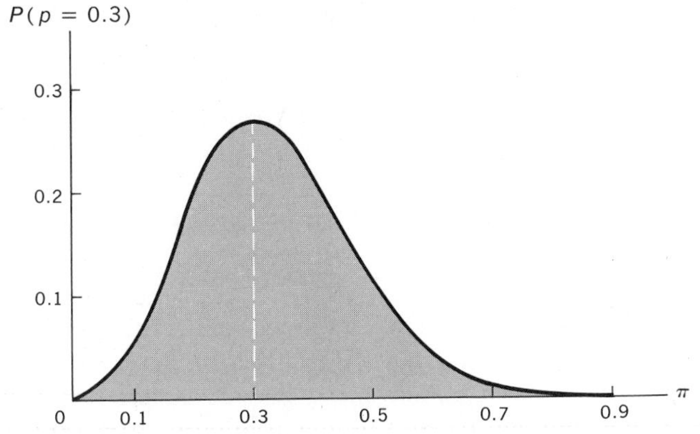

Figure 9.1 Likelihood curve for π given $p = 0.3$ and $n = 10$.

bility $P(p = 0.3)$ as high as possible. Looking at Figure 9.1, we see that this value is $\pi = 0.3$. There is then reason to conclude: In our sample of 10 we found that 3 were against the admission of Red China to the United Nations. Among all values of π, the value of $\pi = 0.3$ seems to make this sample result most likely. We therefore shall estimate π to be 0.3. Having done this, we have made what is called a *maximum likelihood estimate* for π.

To generalize the foregoing discussion, suppose that the population is distributed as $f(x;\theta)$, where θ is the parameter we attempt to estimate. Although θ is unknown, we may well know a set S of numbers to which θ must belong. That is, S is the collection of possible values of the parameter θ. If x_0 is an observed value of the variable X (the population), we can consider the function

$$L(y) = P(X = x_0|\theta = y)$$

as the *likelihood function for θ, given* $X = x_0$. Furthermore, if

$$L(\hat{\theta}) = \max\,\{L(y): y \in S\}$$

then we say that $\hat{\theta}$ is the *maximum likelihood estimate of* θ.

An alternative approach to the likelihood function is often useful. If E is any fixed event, the relationship defined by

$$L(y) = P(X \in E|\theta = y)$$

is the *likelihood function for θ, given the event E*. Here, $L(y)$ is the probability that the event E occurs when y is the true value of the parameter. Furthermore,

$$L(y) = \sum_{x \in E} f(x;y)$$

when the probability function for X is $f(x;\theta)$.

In our presentation of the maximum likelihood function, we begin with a sample result, then proceed to all possible values of the parameter, and finally proceed to the population and find the parameter. The process seems to be straight and simple. These impressions, however, are due to our simple illustration. As a matter of fact, mathematical methods devised by R. A. Fisher, which enable us to find estimators based on maximum likelihood reasoning, are actually beyond the reach of most readers of this text. But the reader may be assured that most statistics introduced in the preceding chapter are obtained by maximum likelihood methods. Let us now mention the results of the maximum likelihood method for a few situations, without proof.

First, if a random sample of size n is drawn from a binomial population, then the MLE of π is $p = x/n$. We know that $E(p) = \pi$; p is therefore an unbiased estimator of π. Furthermore, since $\sigma_p = \sqrt{\pi(1 - \pi)/n}$, we see that σ_p approaches 0 as n approaches infinity. Thus, p is also a consistent estimator of π. Generally, we add a "hat" to the parameter to indicate its estimator. Here, we have $\hat{\pi} = p$.

Second, the MLE of μ is sample mean $\bar{x} = \Sigma x/n$. $\hat{\mu} = \bar{x}$. We have already observed that $\bar{x}$ is an unbiased, consistent, efficient, and sufficient estimator of μ.

Third, the MLE of the difference between two population proportions is the difference between two independent random sample proportions; that is, $\Delta\hat{\pi} = p_1 - p_2 = x_1/n_1 - x_2/n_2$. We note that Δp possesses all the four desirable properties of an estimator. Can you explain?

Fourth, the MLE of the difference between two population means is the difference between two independent sample means. Namely, $\Delta\hat{\mu} = \bar{x}_1 - \bar{x}_2 = \Sigma x_1/n_1 - \Sigma x_2/n_2$. Is $\Delta\bar{x}$ an unbiased, consistent, efficient, and sufficient estimator for $\Delta\mu$?

Fifth, for a normal population the MLE of σ^2 is the sample variance; that is, $\hat{\sigma}^2 = (x - \bar{x}^2)/n = s^2$. It has already been pointed out that s^2 is a biased estimate of σ^2 and that this bias can be corrected by introducing a division constant of $1/(n - 1)$, or this bias disappears when n becomes infinite. In statistical inference, σ^2 or σ may be known from previous studies. Also, due to the fact that if a constant is added to or subtracted from every value of a variable, the distribution of the variable changes only its scale (not its variance), the value of σ known previously can be used even though μ has increased or decreased. And when σ is needed for computation purposes but is unknown, s can be used in its place, as mentioned in the previous chapter, without sacrificing much accuracy if n is large. But is s the MLE of σ?

In concluding our discussion on point estimation, let us observe that the maximum likelihood method provides estimators that are consistent, efficient, and sufficient. It does not, however, provide unbiased estimators in many situations.

9.4 Interval Estimation

As we saw, the procedure in point estimation is to select a function of a sample random variable that will represent "best" the parameter being estimated. Its merit is that it provides an exact value for the parameter under investigation. This merit, however, is also the very deficiency of a point estimate. Being a single point on the real number scale, a point estimate does not tell us how close the estimator is to the parameter being estimated. Moreover, in scientific investigations, it is usually not necessary to know the exact value of a parameter. For example, in ascertaining average family income in a certain city, a knowledge of the *exact* average family income is probably unnecessary; a value within, say, \$50.00 of the true mean will suffice. It is desirable, of course, to have some degree of confidence that the value obtained is within a certain range. A point estimate does not provide such confidence and, therefore, the use of an interval estimate is suggested.

Interval estimation refers to the estimation of a parameter by a random interval, called the *confidence interval,* whose endpoints, L and U with $L < U$, are functions of the observed random variables such that the probability that the inequality $L < \theta < U$ is satisfied in terms of a predetermined number, $1 - \alpha$.

This formulation may be stated as follows:

$$P(|\hat{\theta} - \theta| < kSE(\hat{\theta})) = 1 - \alpha$$

where $SE(\hat{\theta})$ stands for the standard error of the estimator $\hat{\theta}$, $1 - \alpha$ is called the *confidence coefficient*, and k is the *confidence multiplier* corresponding to the value of $1 - \alpha$. We may now note that the inequality in the above expression can be written as

$$|\theta - \hat{\theta}| < kSE(\hat{\theta})$$

or

$$-kSE(\hat{\theta}) < \theta - \hat{\theta} < +kSE(\hat{\theta})$$

Adding $\hat{\theta}$ to all members of this extended inequality, we obtain

$$\hat{\theta} - kSE(\hat{\theta}) < \theta < \hat{\theta} + kSE(\hat{\theta})$$

With this result, we can express the general statement of a confidence interval for estimating θ as this:

$$P[\hat{\theta} - kSE(\hat{\theta}) < \theta < \hat{\theta} + kSE(\hat{\theta})] = 1 - \alpha \tag{9.1}$$

Here,

$\hat{\theta} - kSE(\hat{\theta}) = L$, the *lower confidence limit*, and
$\hat{\theta} + kSE(\hat{\theta}) = U$, the *upper confidence limit*

For obvious reasons, we call L and U confidence limits, which are random end-points of a random interval. The adjective "random" employed here is indicated by P; the P refers to the thing that is random inside the brackets of (9.1), and that is the estimator $\hat{\theta}$.

When the parent population is normal, or when the central limit theorem applies, the sampling distribution of the estimator $\hat{\theta}$ is normal. As a result, the confidence multiplier k turns out to be the standard deviation value of a normal distribution. The most frequently used confidence coefficients and their corresponding confidence multipliers for the normal case are given below:

$1 - \alpha$	k
0.90	1.64
0.95	1.96
0.9545	2.00
0.98	2.33
0.99	2.58

Thus, if it is decided that $1 - \alpha = 0.95$, then we can write

$$P(\theta - 1.96SE(\hat{\theta}) < \theta < \theta + 1.96SE(\hat{\theta})) = 0.95$$

which is commonly called a 95 percent confidence interval.

To give a proper interpretation to a confidence interval, we note from (9.1) that a confidence interval gives a probability of $1 - \alpha$ that the parameter being estimated will fall into the interval bounded by L and U. If the same estimating function is employed for various samples of the same size, then each sample statistic, $\hat{\theta}$, will give a different interval and on the average $1 - \alpha$ of all these intervals will cover the parameter θ. In practice, when we construct only one confidence interval to estimate the parameter, the parameter may or may not fall within the interval. Nevertheless, we have a *level of confidence* of $1 - \alpha$ that it will. More obviously, if we have constructed a 99 percent confidence interval, we say that the probability (confidence) is 0.99 that the interval will include the parameter being estimated. That is, if we construct such an interval again and again for many times, in the long run 99 percent of these intervals will cover the unknown parameter and 1 percent of them will not. Now, since 99 percent of the intervals will cover the parameter, we can certainly afford to *behave* as if each of our intervals will cover the parameter.

The preceding arguments are demonstrated in Figure 9.2. This figure shows that the distribution of the estimator $\hat{\theta}$ is normal. Each circle under the density curve is thought of as a possible value of the estimator, or a group of estimators of equal values. The horizontal scale measures possible values of $\hat{\theta}$, and the vertical scale measures probabilities associated with these values. Now, if we wish to construct, say, a 95.45 percent confidence interval, then $k = 2$. Clearly, if $\hat{\theta}$ assumes a value represented by any circle within the limits $E(\hat{\theta}) \pm 2SE(\hat{\theta})$, then the interval $\hat{\theta} \pm 2SE(\hat{\theta})$ will cover θ, which is equal to $E(\hat{\theta})$. (See, for example, the random interval identified as A.) But, we know that $\hat{\theta}$ would fall within the interval $E(\hat{\theta}) \pm 2SE(\hat{\theta})$ 95.45 percent of the time. Now, if $\hat{\theta}$ happens to be one of those extreme values which are outside the interval of two standard errors from the expectation, then $\hat{\theta} \pm 2SE(\hat{\theta})$ will not cover θ. (See, for example, the random interval identified as B.) But $\hat{\theta}$ deviates from its expectation by 2 standard errors or more only 4.55 percent of the time. Thus, we see that we have a confidence level of 0.9545 in estimating θ with a 95.45 percent confidence interval. Similar demonstrations and statements can be made for confidence intervals with other confidence coefficients by using Figure 9.2 in the same fashion.

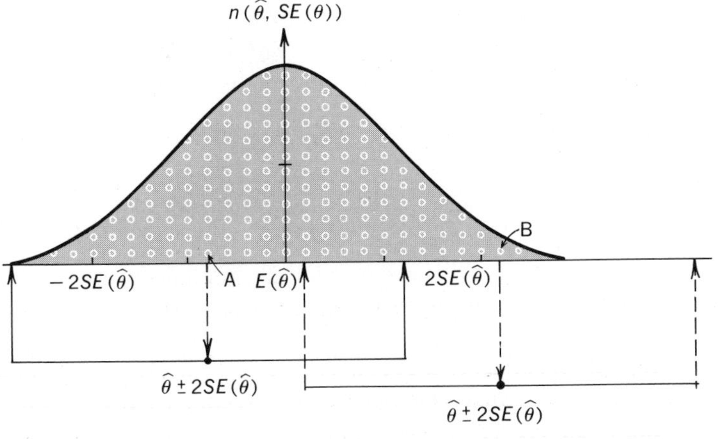

Figure 9.2 Illustration of 95.45 percent confidence intervals.

In interval estimation we naturally wish to have a shorter rather than a longer random interval, since the shorter the interval the more precise is our estimate. However, we must note that the actual width of a confidence interval is dictated by the confidence coefficient and the sample size. Given n, therefore $SE(\hat{\theta})$, the shorter the interval the smaller is $1 - \alpha$, the level of confidence. If we insist upon a given width of interval and a given high degree of confidence, the size of the sample must be increased so that $SE(\hat{\theta})$ is reduced.

9.5 Confidence Limits for μ

We know that the sampling distribution of $\bar{x}$ is normal if the parent population is normal or if the size of the sample is greater than 30. We know also that $E(x) = \mu$, and $\sigma_{\bar{x}} = \sigma/\sqrt{n}$. An interval estimate of μ involves the construction of confidence limits for μ on the basis of $\bar{x}$ as follows:

$$P(\bar{x} - k\sigma_{\bar{x}} < \mu < \bar{x} + k\sigma_{\bar{x}}) = 1 - \alpha$$

or, more compactly,

$$\bar{x} \pm k\sigma_{\bar{x}} \tag{9.2}$$

Thus the procedure follows the general statement on interval estimation made in the previous section: First, obtain a random sample. Second, compute the sample mean. Third, compute the random endpoints with the sample mean as implied by (9.2). Let us now cite an illustration.

The Chamber of Commerce in Miami Beach wishes to estimate the mean expenditures per tourist per visit in that city. For this purpose, a random sample of, say, 100 tourists has been selected for investigation, and it has been found that $\bar{x} = \$800.00$. Now, suppose the standard deviation of expenditures for all tourists is known to be \$120.00, then we have $\sigma_{\bar{x}} = 120/\sqrt{100} = \$12.00$. Finally, suppose the Chamber wishes to construct a 95 percent confidence interval, then the confidence multiplier $k = 1.96$, and with data at hand, we have

$$\text{Lower confidence limit} = L = \bar{x} - 1.96\sigma_{\bar{x}} = 800 - 1.96(12) = \$776.48$$

and

$$\text{Upper confidence limit} = U = \bar{x} + 1.96\sigma_{\bar{x}} = 800 + 1.96(12) = \$823.52$$

So we state: The mean expenditures of tourists per person per visit at Miami Beach falls in the interval from \$776.48 to \$823.52, with a level of confidence at 0.95.

Now, we may observe that there is an error δ involved in the interval estimate of μ. This *error* may be defined as $\delta = |\bar{x} - \mu|$, which is clearly half of the length of a given confidence interval. Furthermore, this error must be less than $k\sigma_{\bar{x}}$ at a confidence level of $1 - \alpha$. That is, if we have constructed, say, a 95 percent confidence interval for μ, we are 95 percent sure that the error $\delta < 1.96\sigma_{\bar{x}}$. Or, we say that the probability is 0.95 that $\delta < 1.96\sigma_{\bar{x}}$.

The last interpretation of error in interval estimate leads us to consider what is called the *risk* of an interval estimate. Such a risk is defined as the probability that the error of estimation will be equal to or greater than $k\sigma_{\bar{x}}$; that is, $P(\delta \geq k\sigma_{\bar{x}})$. For instance, if $1 - \alpha = 0.95$, then the risk is $1 - 0.95 = 0.5$ or α that the error will be equal to or greater than $1.96\sigma_{\bar{x}}$.

An interesting consequence of the discussion on the relationship between error and risk is that it gives us a clue to the determination of a proper sample size, so that a specified width of interval at a given level of confidence can be satisfied. To see why this is true, let us note, for confidence limits for μ, that

$$\delta \leq k\left(\frac{\sigma}{\sqrt{n}}\right)$$

and upon solving this equation for n, we have

$$n \geq \frac{k^2\sigma^2}{\delta^2} \tag{9.3}$$

In our previous example, the error of estimation was

$$\delta = 1.96(12) = \$23.52$$

at a confidence level of 0.95. Now, if the Chamber of Commerce would like to increase the level of confidence to 0.99 without increasing the error of estimation, the sample size would have to be enlarged from 100 to

$$n \geq \frac{k^2\sigma^2}{\delta^2} = \frac{(2.58)^2(120)^2}{(23.52)^2} \doteq 171$$

On the other hand, if the Chamber wishes to reduce the error of estimation to, say, \$20.00, with the level of confidence remaining at 0.95, then the required sample size becomes

$$n \geq \frac{(1.96)^2(120)^2}{(20)^2} \doteq 136$$

As an exercise, the reader might try to find the required sample size in this illustration if it is specified that $1 - \alpha = 0.99$ and $\delta = \$20.00$.

When we are confronted with a completely new situation where the population variance is unknown, we may use the sample variance s^2 from a pilot study as an estimate of σ^2 in determining the desired sample size. Suppose, for example, a new drug for the reduction of temperature has been perfected and a sample variance $s^2 = \hat{\sigma}^2 = 0.1$ degrees has been found. The manufacturer wishes to have an interval estimate of the mean effectiveness (average reduction in temperature) and he is willing to assume an error of 0.1 degrees at $1 - \alpha = 0.99$. How large a sample is required to satisfy these conditions?

Using $\hat{\sigma}^2$ to replace σ^2 in equation (9.3), the solution becomes

$$n \geq \frac{k^2\hat{\sigma}^2}{\delta^2} = \frac{(2.58)^2(0.1)}{(0.1)^2} \doteq 66$$

When population variance is not known, the standard error can be computed with sample variance when $n > 30$. Thus the approximate confidence limits for μ are

$$\bar{x} \pm k\hat{\sigma}_{\bar{x}} = \bar{x} \pm k\frac{\hat{\sigma}}{\sqrt{n}} \tag{9.4}$$

Suppose our drug manufacturer has tried the new drug on 66 patients who had high temperatures and has found that, on the average, the reduction in temperature was 2.5 degrees, with a standard deviation of $\sqrt{0.09}$ degrees; then the approximate 99 percent confidence interval for μ is

$$2.5 \pm 2.58\left(\sqrt{\frac{0.09}{66}}\right) \doteq 2.5 \pm 0.11$$

That is,

$$2.4048 < \mu < 2.5952$$

Note that the error of estimation here is 0.0952, which is approximately 0.1 as specified before.

If a random sample is drawn from a finite population without replacement, and if $n \geq 0.05N$, then the finite population correction factor of $\sqrt{(N-n)/(N-1)}$ should be introduced to calculate $\sigma_{\bar{x}}$ or $\hat{\sigma}_{\bar{x}}$. All formulas in this section have been designed under the assumption of either infinite population or sampling with replacement from a finite population.

9.6

Confidence Limits for π

The sampling distribution of p is asymptotically normal with $E(p) = \pi$ and $\sigma_p = \sqrt{(\pi)(1-\pi)/n}$. In this case a confidence interval may be constructed for π in the same way as that described in the previous section. Corresponding to a confidence coefficient $1 - \alpha$, the approximate confidence limits for π, assuming n is large, are

$$p \pm k\hat{\sigma}_p \tag{9.5}$$

where $\hat{\sigma}_p$ is the unbiased estimate of σ_p defined as

$$\hat{\sigma}_p = \sqrt{\frac{p(1-p)}{n-1}}$$

or the biased estimate of σ_p defined as

$$\hat{\sigma}_p = \sqrt{\frac{p(1-p)}{n}}$$

These two estimates differ but very slightly when n is large. We shall use the biased estimate because it is easier to work with.

We use the estimate of the standard error because the computation of σ_p requires π, the unknown parameter which we try to estimate. We may be assured, however, that when n is large, p is a very good point estimate of π.

The parameter π is undefined for an infinite population. Why? Thus, when we talk about a proportion we are thinking of a finite population. Therefore, when sampling is made without replacement, the estimate of σ_p must be calculated with a finite population correction factor as follows:

$$\hat{\sigma}_p = \sqrt{\frac{p(1-p)}{n}}\sqrt{\frac{N-n}{N-1}}$$

Confidence limits defined by (9.5) require a large sample size. Intuitively, we can see that the more the sample proportion deviates from 0.5, the larger n should be. W. G. Cochran has given us working rules on this point as presented in Table 9.1. According to this table, for example, if $p = 0.1$, the sample size should be at least 600 to use normal approximation.

TABLE 9.1

If p equals	Use normal approximation only if n is at least equal to
0.5	30
0.4 or 0.6	50
0.3 or 0.7	80
0.2 or 0.8	200
0.1 or 0.9	600
0.05 or 0.95	1400

Source: Reprinted from W. G. Cochran, Sampling Techniques (New York: John Wiley & Sons, Inc., 1953), p. 41.

An unconditional random sample of $n = 500$ manufacturing firms has been selected; 80 of these firms report that they are planning to increase investment expenditures during the next two years. What is the 99 percent confidence interval for π?

We have for this problem, $p = 80/500 = 0.16$ and

$$\hat{\sigma}_p = \sqrt{\frac{(0.16)(0.84)}{500}} = 0.0163$$

Consequently, the desired confidence limits are

$$p \pm k\hat{\sigma}_p = 0.16 \pm 2.58(0.0163) = 0.16 \pm 0.042$$

Or, the 99 percent confidence interval is

$$0.118 < \pi < 0.202$$

As an exercise, the reader might construct a 95 percent confidence interval for the previous illustration under the assumption that a conditional random sample was taken from a population of 2500 manufacturing firms.

The error of estimating π, as in the case of the mean, is the difference between the sample proportion and the population proportion; namely, $\delta = p - \pi$. Furthermore, the probability is $1 - \alpha$ that this error is less than $k\sigma_p$. With this concept of maximum error, we can again determine the sample size required in order to satisfy a given degree of precision in estimating π. Now, since

$$\delta = k\sigma_p = k\sqrt{\frac{\pi(1-\pi)}{n}}$$

upon solving for n, we obtain

$$n \geq \pi(1-\pi)\left(\frac{k}{\delta}\right)^2 \tag{9.6}$$

An unfortunate aspect of this result is that we need the value of π, a value we wish to estimate, in order to determine n. A practical solution to this difficulty is to use any information we have, such as, say, a sample proportion obtained in the past, as an estimate of π. When no information is available on π, we may assume $\pi = \frac{1}{2}$. The logic for this assumption is that

$$\begin{aligned}\pi(1-\pi) &= \pi - \pi^2 \\ &= \tfrac{1}{4} - (\tfrac{1}{4} - \pi + \pi^2) \\ &= \tfrac{1}{4} - (\tfrac{1}{2} - \pi)^2\end{aligned}$$

which shows that the maximum value $\pi(1-\pi)$ can have is $\frac{1}{4}$ at $\pi = \frac{1}{2}$.

Suppose that the producer of a new TV program wishes to ascertain the proportion of all viewers watching TV who are tuned in to his show. He is willing to assume an error of ± 5 percent at a confidence level of 0.95; how large a sample should be taken?

By (9.6) and assuming $\pi = \frac{1}{2}$, we find

$$n \geq \left(\frac{1}{2}\right)\left(\frac{1}{2}\right)\left(\frac{1.96}{0.05}\right)^2 \doteq 359$$

This result indicates that if we take a random sample of 359 or larger, the probability is 0.95 that the sample proportion will differ from the true value by 5 percent or less.

Now, suppose the producer knows that in the past programs similar to his new one shared approximately 25 percent of the TV audience; he may then use 0.25 as the value of π, and the sample size required becomes

$$n \geq (0.25)(0.75)\left(\frac{1.96}{0.05}\right)^2 \doteq 288$$

Thus, we see once again the value of information about past experience. In this case, possessing some information of π, we can achieve the same degree of precision with a much smaller sample. This advantage is especially important when sampling is expensive.

Confidence limits for π can be improved, especially when n is relatively small, in two ways. First, since the distribution is discrete, a continuity correction factor of $\frac{1}{2n}$ may be introduced for p to determine a confidence interval for π as follows:

$$\left(p - \frac{1}{2n}\right) - k\hat{\sigma}_p < \pi < \left(p + \frac{1}{2n}\right) + k\hat{\sigma}_p \tag{9.7}$$

which, when n is quite large, reduces to (9.5).

The second procedure is to obtain what is called the *quadratic* confidence interval for π. This is done by solving the inequality

$$|\pi - p| < k\sqrt{\frac{\pi(1 - \pi)}{n}}$$

for π, yielding the following confidence limits for π:

$$\frac{1}{1 + k^2/n}\left(p + \frac{k^2}{2n} \pm k\sqrt{\frac{p(1 - p)}{n} + \frac{k^2}{4n^2}}\right) \tag{9.8}$$

This rather involved formula, when the sample is very large, reduces to the simpler form of (9.5). Furthermore, these and the last limits are still approximate, since the relationship between k and $1 - \alpha$ is determined under the assumption that p is normally distributed, which is only approximately so. However, when n is relatively small, (9.7) and (9.8) do yield more satisfactory approximate confidence limits for π. As an exercise, the reader might work out the example of the manufacturing firms, using both these formulas; he will see that they yield practically the same results as those obtained before.

9.7

Confidence Limits for $\Delta\mu$

When two independent random samples of $n_1 > 30$ and $n_2 > 30$ are taken, then the distribution of the difference between the two sample means, $\Delta\bar{x}$, is approximately normal with $E(\Delta\bar{x}) = \Delta\mu$ and $\sigma_{\Delta\bar{x}} = \sqrt{\sigma_1^2/n_1 + \sigma_2^2/n_2}$. Given the confidence coefficient, the approximate confidence limits for $\Delta\mu$ may be expressed as follows:

$$\Delta\bar{x} \pm k\sigma_{\Delta\bar{x}} \tag{9.9}$$

An automobile manufacturing company has recorded observations of two different makes of batteries, both of which are being used in its cars. Forty observations of Make A showed a mean life of 32 months. Forty-five recordings of Make B showed a mean life of 30 months. Past experience indicates that the standard deviations for both makes of batteries are the same—4 months. What are the confidence limits for the true difference in mean life between the two types of batteries with $1 - \alpha = 0.95$?

For this problem, we have $\Delta\bar{x} = \bar{x}_A - \bar{x}_B = 32 - 30 = 2$ and

$$\sigma_{\Delta\bar{x}} = \sqrt{\frac{4^2}{40} + \frac{4^2}{45}} = 0.87$$

Substituting these data into (9.9), we find

$$2 \pm 1.96(0.87) = 2 \pm 1.7$$

Or the 95 percent confidence interval in this case is

$$0.3 < \Delta\mu < 3.7$$

Two cities are separated by a river. The Chamber of Commerce in City A claims that the average family income of that city is \$500 more than that in City B. The Chamber of Commerce in City B disputes this, and a statistician is hired to settle this argument. He attacks this problem by estimating the true difference between the average family incomes in the two cities. One random sample is drawn from each city and the following results are obtained:

Sample from City A	Sample from City B
$n_1 = 100$	$n_2 = 120$
$\bar{x}_1 = \$5900$	$\bar{x}_2 = \$5800$
$s_1^2 = 9050$	$s_2^2 = 8700$

Since population variances are unknown, the statistician uses sample variances to compute an estimate of the standard error of $\Delta\bar{x}$. This is valid since sample sizes in this case are quite large. And

$$\hat{\sigma}_{\Delta\bar{x}} = \sqrt{\frac{s_1^2}{n_1} + \frac{s_2^2}{n_2}} = \sqrt{\frac{9050}{100} + \frac{8700}{120}} = 12.7671$$

He proceeds then to construct a 99 percent confidence interval for $\Delta\mu$ as follows:

$$\Delta\bar{x} - 2.58\hat{\sigma}_{\Delta\bar{x}} < \Delta\mu < \Delta\bar{x} + 2.58\hat{\sigma}_{\Delta\bar{x}}$$

which, in terms of sample values, turns out to be:

$$100 - 2.58(12.7671) < \Delta\mu < 100 + 2.58(12.7671)$$

or,

$$\$62.20 < \Delta\mu < \$137.80$$

This result indicates that while the average family income in City A may be higher than that in City B, the true difference is far below \$500 as the Chamber of Commerce in City A claimed. Furthermore, we can state with a confidence level of 0.99 that the mean incomes in the two cities differ only from a little over \$62 to a little less than \$138.

9.8

Confidence Limits for $\Delta\pi$

The sampling distribution of Δp, given that p_1 and p_2 are determined from independent random samples and that n_1 and n_2 are sufficiently large as specified by values in Table 9.1, is approximately normal with $E(\Delta p) = \Delta\pi$ and $\sigma_{\Delta p} = \sqrt{\pi_1(1 - \pi_1)/n_1 + \pi_2(1 - \pi_2)/n_2}$. No new knowledge is required for the construction of confidence limits for $\Delta\pi$. However, we are faced with the same difficulty now as that in the case of interval estimation for π; that is, we need the values of π_1 and π_2 (which are unknown) for the calculation of $\sigma_{\Delta p}$. Inasmuch as $\Delta\pi$ is unknown, we are forced to used p_1 and p_2 as point estimators for π_1 and π_2, respectively, and find the estimator for $\sigma_{\Delta p}$ as follows:

$$\hat{\sigma}_{\Delta p} = \sqrt{\frac{p_1(1 - p_1)}{n_1} + \frac{p_2(1 - p_2)}{n_2}}$$

With the estimate for $\sigma_{\Delta p}$ available, if we assume large n_1 and n_2, the approximate confidence limits for $\Delta\pi$ corresponding to a confidence coefficient $1 - \alpha$ are

$$\Delta p \pm k\hat{\sigma}_{\Delta p} \tag{9.10}$$

Prior to a gubernatorial election in a certain state, two random samples, one for male voters with $n_1 = 150$ and one for female voters with $n_2 = 120$, are taken. It is found that 70 men in the first sample are in favor of Candidate X and 60 women in the second sample are in favor of the same candidate. What are the confidence limits for $\Delta\pi$, the true difference in voting preferences between men and women in that state, given that $1 - \alpha = 0.95$?

For the problem at hand, we note that

$$p_1 = 70/150 = 0.47,$$

$$p_2 = 60/120 = 0.50,$$

$$\Delta p = 0.47 - 0.50 = -0.03, \text{ and}$$

$$\hat{\sigma}_{\Delta p} = \sqrt{\frac{(0.47)(0.53)}{150} + \frac{(0.50)(0.50)}{120}} = 0.062$$

With these data, we have as the desired confidence limits

$$-0.03 \pm 1.96(0.062) = -0.03 \pm 0.12152$$

Or, the 95 percent confidence interval for $\Delta\pi$ in this case is

$$-0.15152 < \Delta\pi < 0.09152$$

Had we computed Δp as $p_2 - p_1 = 0.50 - 0.47 = 0.03$, then the desired confidence interval would be

$$-0.09152 < \Delta\pi < 0.15152$$

The negative value for the lower confidence limit here should not be a surprise to us since the error of estimate for this problem is $1.96(0.062) = 0.12152$, which is much larger than the observed difference between sample proportions. The interesting point to note is that the negative value for the lower confidence limit in conjunction with the positive upper limit implies that $\Delta\pi$ may very well be zero. In other words, the sample result does not indicate that men's voting preferences are different from women's in the state. It might be that there is a real difference, that is, $\Delta\pi \neq 0$, but the samples may be too small to bring out this difference. In any event, our proper inference from the above confidence interval should be this: Given the stated conditions of the sampling experiment and the confidence coefficient of 0.95, a difference of 3 percent between two sample percentages does not support the contention that the voting preferences are different between men and women in the said state.

9.9 Confidence Limits for σ^2

Under the assumptions that the parent population is normally distributed and that $n \geq 100$, the sampling distribution of sample variance s^2 is approximately normally distributed with $E(s^2) = \sigma^2$ and $\sigma_{s^2} = \sqrt{2\sigma^2/(n-1)}$. With these assumptions, we can then write

$$P\left(|s^2 - \sigma^2| < k\sqrt{\frac{2\sigma^2}{n-1}}\right) \doteq 1 - \alpha$$

Solving the inequality inside the parentheses, we obtain an equivalent inequality as the confidence interval estimate of σ^2:

$$\frac{s^2}{1 + k\sqrt{2/(n-1)}} < \sigma^2 < \frac{s^2}{1 - k\sqrt{2/(n-1)}}$$

From this inequality, we may express the confidence limits for σ^2 corresponding to the confidence coefficient $1 - \alpha$ as follows:

$$\frac{s^2}{1 \pm k\sqrt{2/(n-1)}} \tag{9.11}$$

It is important to note that in this case

$$L = \frac{s^2}{1 + k\sqrt{2/(n-1)}}$$

and

$$U = \frac{s^2}{1 - k\sqrt{2/(n-1)}}$$

It is also important to note that L and U are not equidistant from s^2.

Ex A random sample of 250 units of canned tomatoes is drawn and the variance in the weights of the cans is found to be 5 ounces. Assuming the weight is normally distributed, then an approximate 99 confidence interval for σ^2 would have the limits

$$\frac{5}{1 \pm 2.58\sqrt{2/249}}$$

Or, we have

$$4.065 < \sigma^2 < 6.494$$

9.10

Confidence Limits for σ

Approximate confidence limits for σ can be obtained from those for σ^2 if the latter are available. This is done simply by taking square roots of the limits for σ^2. For example, the 99 percent confidence interval for σ corresponding to that for σ^2 in the last illustration is

$$\sqrt{4.065} < \sigma < \sqrt{6.494}$$

When we wish to construct confidence limits for σ directly, we note that if the parent population is normal and if $n \geq 100$, then the distribution of s is approximately normal with an expectation very nearly equal to σ and a standard error defined as $\sigma_s = \sigma/\sqrt{2(n-1)}$. To obtain an interval estimate for σ under these assumption, we can state that

$$P\left(|s - \sigma| < k\frac{\sigma}{\sqrt{2(n-1)}}\right) = 1 - \alpha$$

The inequality inside the parentheses is equivalent to

$$\frac{s}{1 + k/\sqrt{2(n-1)}} < \sigma < \frac{s}{1 - k/\sqrt{2(n-1)}}$$

where the extreme members are the confidence limits for σ corresponding to the confidence coefficient $1 - \alpha$. Thus,

$$\frac{s}{1 \pm k/\sqrt{2(n-1)}} \tag{9.12}$$

Ex The standard deviation of tensile strengths of 200 steel wires selected at random was found to be 10 pounds. Find the 95 percent confidence limits for σ.

By (9.12) just given, we find the desired limits to be

$$\frac{10}{1 \pm \dfrac{1.96}{\sqrt{2(199)}}} = \frac{10}{1 \pm 0.098}$$

That is, we have a 95 percent confidence interval estimate for σ here as

$$9.1 < \sigma < 11.1$$

It can be shown that the limits given by (9.12) are approximately equal to

$$s \pm k\left(\frac{\hat{\sigma}}{\sqrt{2n}}\right) \tag{9.13}$$

from which we see that the approximate error in estimating σ is $\delta = s - \sigma$, which cannot be greater than $k(\hat{\sigma}/\sqrt{2n})$. That is,

$$\delta \leq k(\hat{\sigma}/\sqrt{2n})$$

Solving this for n, we have

$$n \leq \left(\frac{1}{2}\right)\frac{k^2\hat{\sigma}^2}{\delta^2} \tag{9.14}$$

For example, how large a sample of steel wires (see previous example) must be drawn in order to be 99 percent sure that the true population standard deviation will not differ from the sample standard deviation by, say, 5 percent or less? We have, by (9.14)

$$n \leq \left(\frac{1}{2}\right)\frac{(2.58)^2(10)^2}{(0.5)^2} \doteq 1320$$

Note, in this calculation, $\delta = 0.5$, which is 5 percent of 10. In any event, to meet the specified precision and confidence level, the sample size must be 1320 or larger.

9.11

Summary and Conclusions

In this chapter we have been concerned with the traditional decision procedure of estimation. Estimation is the statistical method of making inferences about parameter values on the basis of sample statistics. An estimate of a parameter given by a single point derived from sample observations is called a point estimate. An estimator is said to be good if it possesses the properties of unbiasedness, consistency, efficiency, and sufficiency. The maximum likelihood method provides estimators which are consistent, efficient, and sufficient; but it does not always provide estimators that are unbiased. It can be shown that all estimators introduced so far in this text are maximum likelihood estimators.

An estimate of a parameter given by a random interval whose endpoints are functions of sample observations is called an interval estimation. Interval estimates indicate the precision or accuracy of an estimate, and are therefore preferable to point estimates. Each interval estimate gives a probability—a confidence level—that the confidence limits constructed will cover the parameter being estimated.

In interval estimation, the error of estimation, the level of confidence, and the sample size are closely related. The error here is defined as the difference between the estimator and the parameter being estimated. With a probability of $1 - \alpha$, this error cannot be greater than the product of the confidence multiplier and the standard error of the estimator. The probability that the error of estimation is equal to or greater than this product is thought of as the *risk* in estimation; that is, the probability that the confidence interval will not cover the parameter being estimated. Given the sample size, the more precise or shorter the confidence interval is, the lower must be the level of confidence; or if the confidence coefficient is increased, the less precise (or the greater the error of) estimation becomes. In general, if the goal is to have a small error and a high level of confidence, the sample size must be increased. It is possible to determine the minimum size of sample required to satisfy predetermined precision and confidence. Usually, this is done by equating the error of estimation to the maximum error, $(\hat{\theta} - \theta) = \delta = kSE(\hat{\theta})$, and solving for n, which is always involved with the calculation of the standard error of an estimator under investigation.

Our discussion of estimation so far has been made under the assumption that the sampling distribution of an estimator is normally distributed in accordance with the central limit theorem. Insofar as many sampling distributions are only approximately normal, confidence limits constructed with normal confidence multipliers possess only approximate values. Bearing this in mind, we must also say that such approximate confidence limits are quite satisfactory for estimating parameters in many types of investigations.

In our previous discussion, we have also arbitrarily chosen the confidence coefficients as 0.90, 0.95, 0.9545, 0.98, and 0.99; these are the most frequently used values. Another confidence coefficient commonly employed in the past, but one used only in military applications today, is 0.50. The 50 percent confidence interval for θ is given by $\hat{\theta} \pm 0.6745SE(\hat{\theta})$, since 50 percent of the area under the density curve of the standard normal distribution lies between -0.6745 and 0.6745. The quantity $0.6745SE(\hat{\theta})$ is known as the *probable error* of estimate. It means that if a sample statistic is used to estimate the corresponding parameter, the probability is 0.50 that the error of estimate will be less than $0.6745SE(\hat{\theta})$.

Finally, it may be noted that although our treatment covers estimation for only μ, π, $\Delta\mu$, $\Delta\pi$, σ^2, and σ, the same general procedure of interval estimation can be applied to other parameters—such as the median, the mean of binomial variable, and other position measures—providing n is large enough for the central limit theorem to operate. These few parameters have been singled out for evaluation because they are by far the most important decision parameters insofar as univariate data are concerned. We shall continue to be concerned with them as we proceed to take up the traditional decision procedure of hypotheses testing in the following chapter.

Glossary of Formulas

(9.1) $P(\hat{\theta} - kSE(\hat{\theta}) < \theta < \hat{\theta} + kSE(\hat{\theta})) = 1 - \alpha$ This is the general statement for interval estimation. A confidence interval is constructed with an estimator in such a way that the probability is $1 - \alpha$ (the confidence level or coefficient) that the inequality $L < \theta < U$ is satisfied. Here, $L = \hat{\theta} - kSE(\hat{\theta})$, is the lower confidence limit, and $U = \hat{\theta} + kSE(\hat{\theta})$, the upper confidence limit. In this statement also, k is called the confidence multiplier corresponding to a given value of $1 - \alpha$, and SE($\hat{\theta}$) is the standard error of the estimator $\hat{\theta}$. When the distribution of $\hat{\theta}$ is normal, the relationship between $1 - \alpha$ and k is established by the standard normal distribution.

(9.2) $\bar{x} \pm k\sigma_{\bar{x}}$ These are confidence limits for μ constructed with the sample mean. Note, $\sigma_{\bar{x}} = \sigma/\sqrt{n}$.

(9.3) $n \geq (k^2\sigma^2)/\delta^2$ This expression helps us to determine the minimum size of sample required so that a prescribed level of confidence and error of estimation can be satisfied for the estimation of μ. Here, $\delta = \bar{x} - \mu$, which is half of the confidence interval specified.

(9.4) $\bar{x} \pm k(\hat{\sigma}/\sqrt{n})$ Confidence limits for μ when the standard error of the mean is computed with the estimator of the population standard deviation, $s = \hat{\sigma}$.

(9.5) $p \pm k\hat{\sigma}_p$ Confidence limits for π. In this expression $\hat{\sigma}_p = \sqrt{p(1 - p)/n}$ if sampling with replacement is made from a finite population, but when sampling is made without replacement $\hat{\sigma}_p = \sqrt{p(1 - p)/n}\sqrt{(N - n)/(N - 1)}$.

(9.6) $n \geq \pi(1 - \pi)(k/\delta)^2$ To determine the minimum size of n so that a given degree of precision $\delta = p - \pi$ is satisfied at a given level of confidence $1 - \alpha$ for estimating π, this formula may be employed. Here, it is commonly assumed that $\pi = 0.5$ or any other value of π we judge to be appropriate in a given situation.

(9.7) $$\left(p - \frac{1}{2n}\right) - k\hat{\sigma}_p < \pi < \left(p + \frac{1}{2n}\right) + k\hat{\sigma}_p$$

(9.8) $$\frac{1}{1 + k^2/n}\left(p + \frac{k^2}{2n} \pm k\sqrt{\frac{p(1 - p)}{n} + \frac{k^2}{4n^2}}\right)$$

Better approximate confidence limits for π can be constructed with these two expressions when n is relatively small. In (9.7) a continuity correction factor of $1/2n$ is introduced. Expression (9.8) is what is called the quadratic confidence limits for π. It may be noted that if n is quite large, both expressions here are reduced to the form of (9.5).

(9.9) $\Delta\bar{x} \pm k\sigma_{\Delta\bar{x}}$ Confidence limits for $\Delta\mu$. Note that $\sigma_{\Delta\bar{x}} = \sqrt{(\sigma_1^2/n_1) + (\sigma_2^2/n_2)}$. Also, $s^2 = \hat{\sigma}^2$ can be employed here for σ^2 when necessary.

(9.10) $\Delta p \pm k\hat{\sigma}_{\Delta p}$ Confidence limits for $\Delta\pi$. Here, we have

$$\hat{\sigma}_{\Delta p} = \sqrt{p_1(1 - p_1)/n_1 + p_2(1 - p_2)/n_2}.$$

(9.11) $$\frac{s^2}{1 \pm k\sqrt{2/(n - 1)}}$$ Confidence limits for σ^2. It is important to remember here that $L = s^2/(1 + k\sqrt{2/(n - 1)})$ and that $U = s^2/(1 - k\sqrt{2/(n - 1)})$.

(9.12) $$\frac{s}{1 \pm k/\sqrt{2(n - 1)}}$$ Confidence limits for σ. This expression is equivalent to

$$s/(1 + k/\sqrt{2(n - 1)}) < \sigma < s/(1 - k/\sqrt{2(n - 1)}).$$

(9.13) $s \pm k(\hat{\sigma}/\sqrt{2n})$ These confidence limits are approximately the same as (9.12), especially if n is large. Here, evidently, $\hat{\sigma} = s$.

(9.14) $n \geq \left(\frac{1}{2}\right)\frac{k^2\hat{\sigma}^2}{\delta^2}$ This expression gives the minimum size of n required in order to satisfy prescribed precision δ for the confidence coefficient for estimating σ.

Problems

9.1 What do we mean by estimation? How does point estimation differ from interval estimation?

9.2 What are the desirable properties of a point estimate?

9.3 In what way do we say that an interval estimate is better than a point estimate?

9.4 How is the confidence coefficient related to the precision of estimation?

9.5 How is the confidence coefficient related to the risk of estimation?

9.6 Is the precision of estimation different from the error of estimation? If so, how do they differ?

9.7 How does the sample size affect the degree of estimating precision, given the confidence coefficient?

9.8 How does the sample size affect the confidence coefficient, given the degrees of precision?

9.9 "Theoretically speaking, it is possible to have an estimate which is identical with the parameter being estimated. In practice, however, such an estimate is often unnecessary as well as physically impossible." Comment.

9.10 What are the maximum likelihood estimators for μ, π, σ^2, $\Delta\mu$, and $\Delta\pi$, respectively? Evaluate these MLE in terms of the four desirable properties for the parameters being estimated.

9.11 In what sense do we consider estimation as a procedure of decision making?

9.12 A shipment contains a very large number of radio tubes of which a proportion π is defective. A random sample of 10 tubes contains 2 defectives. Plot the likelihood function for π and give the maximum likelihood estimate for π.

9.13 Let π be the proportion of people in a certain city who watch a certain TV program on Friday night. In a random sample of 20 people from this city, 6 are viewers of this program. Graph the maximum likelihood function for π and find a maximum likelihood estimate for π.

9.14 An urn contains n balls numbered 1, 2, 3, . . . , n where n is unknown but the standard deviation of this population is known to be 3. Two balls are drawn at random with replacement; the numbers on them are 7 and 11. How would you go about arguing that $\bar{x} = (7 + 11) = 9$ is the maximum likelihood estimator for μ in this case?

9.15 The standard deviation of average weekly wage rates for elevator operators is known to be \$2.79. An unconditional random sample of 100 elevator operators yields a mean average weekly wage rate of \$87.45. Construct (a) 95 percent and (b) 99 percent confidence intervals for the true mean average weekly wages rates of elevator operators.

9.16 Suppose there are 5000 elevator operators in a large city, and suppose a sample mean of \$87.45 is obtained from a conditional random sample of $n = 100$, what are the (a) 95 percent and (b) 99 percent confidence intervals for μ now?

9.17 Measurements of the diameters of a conditional random sample of 50 from a lot of 200 ball bearings show as a mean $\bar{x} = 1.05$ inches, and as a standard deviation $s = 0.04$ inches.

a. What are the 95 percent confidence limits for the mean diameter of the 200 ball bearings?
b. At what level of confidence could we say that the mean diameter of all 200 ball bearings is 1.05 ± 0.005?

9.18 A conditional random sample of 45 grades taken from a class in statistics of 221

students shows a mean of 70 points and a standard deviation of 9 points.
a. What are the 98 percent confidence limits for the mean of all 221 grades?
b. With what level of confidence could we say that the mean of all 221 grades is 70 ± 1.5?

9.19 In measuring assembly time for a certain mechanism, the engineer estimates that the standard deviation is 0.05 seconds. How many times should the assembly be made in order to be (a) 95 percent and (b) 99 percent certain that the error will not exceed 0.01 seconds?

9.20 A department store wishes to estimate, with a confidence coefficient of 0.98 and a maximum error of \$5.00, the true mean dollar value of purchases per month by its customers with charge accounts. How large a sample should the store take from its records to satisfy the specifications if the standard deviation is known to be \$15.00?

9.21 A sample of 35 wires selected at random from a very large shipment of steel wires shows a mean tensile strength of 1500 pounds and a standard deviation of 20 pounds.
a. If this sample mean is used to estimate the mean tensile strength of the whole shipment, what can be said for the possible error with a probability of 0.95?
b. If the error estimated in (a) is to be reduced by half without reducing the level of confidence, what size sample is required?
c. If the estimate requires that the error estimated in (a) be reduced by half and the probability be 0.99 that the reduced error is a maximum, how large should the sample be?

9.22 In a random sample of 25 college instructors in the United States, 16 are found to hold Ph.D. degrees.
a. Construct an approximate 95 confidence interval for the proportion of all U. S. college instructors holding Ph.D. degrees by using equation (9.5).
b. Do the same by using equation (9.8).
c. Comment on the difference in results for (a) and (b).

9.23 Ten out of 400 articles produced by a certain process are found to have defects. Construct a 99 percent confidence interval for the probability that a single article produced in the process is defective.

9.24 A sample poll of 100 members of the AFL and CIO chosen at random reveals that 55 are against trade with Red China.
a. Find the 95 percent confidence limits for the proportion of all the union members against trade with Red China.
b. What is the minimum size of sample required in order to be 95 percent sure that at least 50 percent of the union members are against trade with Red China?

9.25 During a certain week, it was observed and recorded by a small department store that 5750 of the 12,500 persons who entered the store made at least one purchase. Treating this as a random sample of all potential customers, find the 99 percent confidence limits for the actual proportion of persons entering the store who will make at least one purchase.

9.26 A very close vote between the two major presidential candidates is forecast for a national election. What is the least number of voters that must be polled in order to be (a) 95 percent and (b) 99 percent certain of a decision in favor of either one of the two candidates?

9.27 The auditor of a large bank wishes to estimate the proportion of monthly statements for the bank's demand depositors that have errors of various kinds, and he specifies a confidence coefficient of 0.99 and a maximum error of estimation of 0.25 percent (or 0.0025).
a. What is the minimum sample size required if there is no information available on the true proportion of the monthly statements that are in error?
b. What is the minimum sample size required if the auditor, according to his past experience, believes the true proportion is near 0.01?

9.28 The same graduate examination was given to 50 seniors of University A and to 50 seniors of University B. The mean and variance of grades made by students of A are, respectively, 73.6 and 10; the sample mean and variance of grades made by students of B are, respectively, 72.4 and 8. Construct a 90 percent confidence interval for the true difference in average grades between the seniors of Universities A and B.

9.29 A random sample of 200 Brand A radio batteries show a mean life of 140 hours and a standard deviation of 10 hours. A random sample of 120 Brand B radio batteries yield a mean life of 125 hours and a standard deviation of 8 hours. Give (a) 95 percent and (b) 99 percent confidence limits for the difference of the mean lifetimes of the populations of Brands A and B.

9.30 Two randomly selected groups, of 50 students each, of a secretarial school are taught shorthand by two different systems and then tested for performance in taking dictation. It is found that the first group averaged 120 words per minute with a standard deviation of 11 words, while the second group averaged 110 words per minute with a standard deviation of 10 words. Construct a 99 percent confidence interval for $\Delta\mu$.

9.31 The mean and the standard deviation of the maximum loads supported by 100 cables produced by Alpha Company are found to be 20 tons and 1.1 tons, respectively. The mean and the standard deviation of the maximum loads supported by 60 cables produced by Beta Company are 16 tons and 0.8 tons, respectively. Find (a) 95 percent and (b) 98 percent confidence limits for the difference in mean maximum loads between the two brands of cables.

9.32 In a random sample of 600 women, 300 indicate that they are in favor of American intervention in South Viet Nam. In a random sample of 400 men, 100 indicate that they are in favor of the same. Construct (a) 95 percent and (b) 99 percent confidence limits for the difference in proportions of all women and all men who are in favor of American intervention in South Viet Nam.

9.33 Twenty-five out of 250 TV picture tubes produced by Process I are defective, and 15 out of 180 TV picture tubes produced by Process II are defective. Construct (a) 95 percent and (b) 99 percent confidence intervals for the difference in proportions of defective tubes from the two processes. Comment on your results.

9.34 Of two groups of patients suffering from chronic headaches, A and B, consisting of 200 and 100 individuals, respectively, the first are given a new kind of headache pill and the second are given conventional pills; 15 of Group A find immediate relief and 12 of Group B find immediate relief. Construct a 99 percent confidence interval for the difference in proportions of immediate relief from the two kinds of medicine and comment on the result.

9.35 The variance of the breaking strength of 100 cables tested was found to be 32,000 pounds^2. Find the 95 percent confidence limits for variance of all cables of this make.

9.36 The standard deviation of the lifetime of a random sample of 200 portable radios of a certain make is 100 hours. Find the 99 percent confidence interval for the standard deviation of all such radios.

9.37 How large a sample should one take in order to be (a) 95 percent and (b) 99 percent sure that a population standard deviation will differ from a sample standard deviation by 0.1?

9.38 How should the probable error for the mean, for the proportion, and for the standard deviation be expressed mathematically?

9.39 Give the probable errors in problems 9.21, 9.25, and 9.36.

10

STATISTICAL TESTING

387–9
for test

10.1

Testing as a Decision Procedure

The purpose of the collection and analysis of data is often that of finding an objective criterion for deciding on a proper course of action. If, for example, there are two possible courses of action, A_1 and A_2, we would like to know which would be the better action in a given decision situation. We know, now, that such a decision depends upon the nature of a certain population or the probability distribution of a certain random variable. We also know, however, that the precise probability distribution of a random variable is often unattainable and, therefore, we are forced to make our choice between A_1 and A_2 on the basis of sample information. The procedure for selecting an appropriate course of action, in accordance with a decision criterion derived from the evaluation of sample results, is the classical analysis of *statistical testing* or *testing hypotheses.* We shall, in this section, cover some of the highlights of testing as a decision procedure by following through a simple testing problem.

Ex Suppose that the manager of a radio-parts store is considering whether or not to purchase a lot of radio tubes of a certain type. This lot contains 2000 units. The seller frankly admits that there is a high percentage of defectives in the lot. In addition, the seller is not at all sure whether the lot contains 1200 or 800 defective units; that is, $p = 0.60$ or $p = 0.40$. He quotes a price of \$1.00 per tube. At this price, the manager estimates that he could make \$1.00 profit per perfect tube and he could lose, obviously, \$1.00 per defective tube. Thus, if 60 percent of the lot consists of nondefective tubes and if he decides to buy it, he would make a total profit of

$$(2000)(0.6)(2) - (2000)(0.4)(1) = \$400.00$$

If, however, only 40 percent of the tubes were perfect and if he decides to buy the lot, he would suffer a total loss of

$$(2000)(0.4)(2) - (2000)(0.6)(1) = -\$400.00$$

Now, noting that the lot of radio tubes is a binomial population with $p = 0.6$ or $p = 0.4$ (where p is the probability of obtaining a nondefective tube at each random selection from the lot), and denoting the decision to buy as A_1 and not to buy as A_2, we may then summarize all pertinent information in Table 10.1 for later use.

For the above decision problem, what procedure can the manager adopt? A natural way to proceed would be to examine an unconditional random sample of, say, 25 tubes to see if the sample information obtained is consistent with the state of nature that $p = 0.6$ or with the state of nature that $p = 0.4$. These states of nature may be called *statistical hypotheses.* Furthermore, $p = 0.6$ may be named as the *null hypothesis*, denoted as H_0, and $p = 0.4$ as the *alternative hypothesis*, denoted as H_1. To focus attention on two possible population values is a convenient way of specifying what is required of a decision procedure, or of evaluating a proposed procedure. The designations of H_0 and H_1 are often arbitrary. There are, nevertheless, as we shall state in the following section, some important considerations on this point. For the time being, we shall simply write

TABLE 10.1 POSSIBLE GAINS AND LOSSES FROM THE PURCHASE OF 2000 RADIO TUBES

Possible State of Nature	Manager's Decision	
	A_1	A_2
$p = 0.60$	\$400.00	0
$p = 0.40$	−\$400.00	0

$$H_0: p = 0.60 \text{ nondefectives}$$
$$H_1: p = 0.40 \text{ nondefectives}$$

Having set up the hypotheses, the manager should next test the validity of H_0 against that of H_1. He does this by selecting first an appropriate *testing statistic* whose sampling distribution is known under the assumption that H_0 is true. In this case, the testing statistic is the number of successes (nondefective tubes) in the sample, and it has a binomial distribution, with $p = 0.6$ and $n = 25$.

With the appropriate testing statistic selected, we must establish a *decision criterion* or *rule* in accordance with which H_0 may be either accepted or rejected. Here a decision criterion involves a division of all possible sample results into two mutually exclusive categories: (1) those which lead to the rejection of H_0, and (2) those which lead to the acceptance of H_0. The first kind of result is said to be in the *region of rejection*, or *critical region;* the second result is said to be in the *region of acceptance*. Now, suppose the manager adopts this decision rule: Reject H_0 if and only if the number of nondefectives in the sample is equal to or less than 9; accept H_0 if and only if the number of nondefectives in the sample is greater than 9. That is,

$$\text{Reject } H_0\text{: } p = 0.60, \text{ if } X \leq 9$$
$$\text{Accept } H_0\text{: } p = 0.60, \text{ if } X > 9$$

A moment's reflection will reveal that, in following such a decision rule, the manager may make two types of errors due to random fluctuations in the testing statistic. On the one hand, the null hypothesis is in fact true, but because sample data appear to be inconsistent with it, it is rejected. The error of rejecting a true H_0 is generally referred to as an *error of the first kind*, or a *Type I error*. The probability of making a Type I error is called the *level of significance* and is denoted as α. Since, when a Type I error is committed we would take the wrong action, A_2, the level of significance can be defined as the probability of deciding on A_2 given that H_0 is true. Consequently, we state

$$\alpha = P(H_1|H_0) = P(A_2|H_0) = P(\mathrm{I}) \tag{10.1}$$

On the other hand, we may find that H_0 is in fact false; yet, on the basis of sample observations, we are led to accept it as true. The error of accepting a false null hypothesis is called the *error of the second kind*, or a *Type II error*. The probability of committing a Type II error is denoted as β. When we make errors of the second kind, we are led to take the wrong action, A_1, given that the alternative hypothesis is true. Thus,

$$\beta = P(H_0|H_1) = P(A_1|H_1) = P(\mathrm{II}) \tag{10.2}$$

Conclusions regarding α and β may be summarized as follows:

TABLE 10.2 DECISIONS VERSUS HYPOTHESES

State of Nature	Decision	
	Accept: A_1	*Reject:* A_2
H_0 is true	Correct decision	Type I error
H_0 is false	Type II error	Correct decision

Type I and Type II errors are used to evaluate various decision rules. For the decision rule formulated for our illustration, we have

$$\alpha = P(X \leq 9|p = 0.6) = \sum_{x=0}^{9} \binom{25}{x} (0.6)^x(0.4)^{25-x} = \sum_{x=16}^{25} \binom{25}{x} (0.4)^x(0.6)^{25-x}$$
$$= 1 - \sum_{x=0}^{15} \binom{25}{x} (0.4)^x(0.6)^{25-x} = 1 - 0.96561 = 0.03439$$

and

$$\beta = P(X > 9|p = 0.4) = 1 - P(X \leq 8|p = 0.4)$$

$$= 1 - \sum_{x=0}^{8} \binom{25}{x} (0.4)^x(0.6)^{25-x} = 1 - 0.27357 = 0.72643$$

The above value of β indicates that the probability of accepting the null hypothesis when the alternative hypothesis is actually true is almost 0.73. That is, the manager is running a great risk of losing \$400 by the present decision rule. Naturally, the manager would feel that such a risk is exceedingly large. To reduce his risk, he must set up another decision rule. As a matter of fact, the best decision rule, among all permissible decision rules, is one which can minimize both α and β in a given decision situation. In the present case, it may be argued that a more proper or "fairer" test would be to reject H_0 if $X \leq 13$ or 12, since each of these two numbers is close to halfway between 10 (the expected number of successes in 25 observations when $p = 0.4$) and 15 (the expected number of nondefectives in 25 observations when $p = 0.6$). Now, if the manager decides upon the following new decision rule,

Reject H_0, if and only if $X \leq 13$
Accept H_0, if and only if $X > 13$

then we would have

$$\alpha = P(X \leq 13|p = 0.6) = 1 - \sum_{x=0}^{11} \binom{25}{x} (0.4)^x(0.6)^{25-x} = 0.26772$$

and

$$\beta = P(X > 13|p = 0.4) = 1 - \sum_{x=0}^{12} \binom{25}{x} (0.4)^x(0.6)^{25-x} = 0.15377$$

We can, in fact, for various values of K consider this test: Reject H_0 if and only if $X \leq K$. For various values of K, we may compute α and β as in Table 10.3. From graphs, such as Figure 10.1 (suggested by the values of α and β in this table), we may also study how the sizes of testing errors change as K is changed.

TABLE 10.3 α AND β RISKS FOR THE TEST $X \leq K$ WITH $n = 25$

K	$\alpha = P(X \leq K\|p = 0.6)$	$\beta = P(X > K\|p = 0.4)$	$1 - \beta$
3	0.00	1.00	0.00
5	0.00	0.99	0.01
7	0.01	0.97	0.03
9	0.03	0.73	0.27
11	0.15	0.41	0.59
13	0.27	0.15	0.85
15	0.58	0.03	0.97
17	0.85	0.00	1.00
19	0.97	0.00	1.00

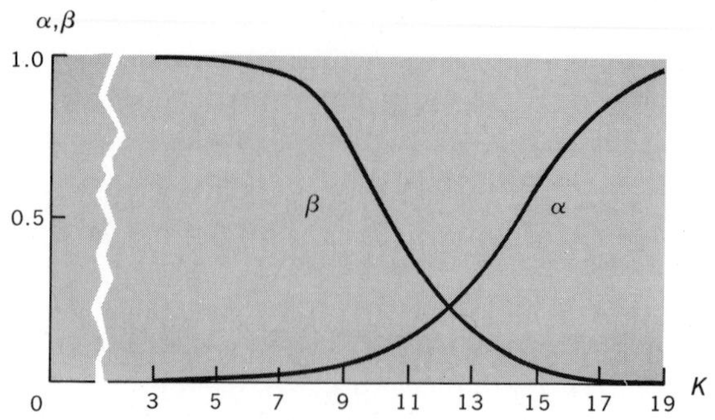

Figure 10.1 Values of α and β for testing $p = 0.6$ with $X \leq K$ and $n = 25$.

Let us pause for a moment to make a few observations about the relationship between Type I and Type II errors.

First, a Type I error is computed with the hypothetical value for H_0, and a Type II error is computed with the hypothetical value for H_1.

Second, from Figure 10.1, we may clearly see that, given the sample size, in an attempt to make β small we must increase α. Under the condition that n is fixed, we would prefer a test whereby both α and β could be made relatively small. In our illustrative case, the graph for both types of errors reveals that the decision rules of rejecting H_0 if and only if $X \leq 12$, or if and only if $X \leq 13$, are good ones, since in each case both α and β are quite small. Furthermore, had we selected as the testing statistic a sample proportion whose distribution was continuous, then the best decision rule would have been to reject the null hypothesis $\pi = 0.6$ if and only if $p \leq 0.5$, where p was the sample proportion. Can you explain why by looking at Figure 10.1? As we shall show later, if we are willing to let n vary, then it is possible to specify both α and β at any desired level.

Third, the probability that sample results will fall in the critical region under the null hypothesis when the alternative hypothesis is true is called the *power of the test*, or the *power function of the test.* Alternatively, the power of the test is the probability that a particular alternative hypothesis will be accepted when it is the true parameter in the case. Thus, if we are testing the pair of hypotheses

$$H_0: \theta = \theta_0$$
$$H_1: \theta = \theta_1$$

with testing statistic S and critical region CR, then

$$\text{Power function} = P(S \text{ falls in } CR|\theta = \theta_1) = P(A_2|\theta = \theta_1) = 1 - \beta \quad \textbf{(10.3)}$$

Thus, the power function of a test is the likelihood function for the critical region of the test and is related to a Type II error. In particular, when H_0 is true then the power of the test is α. Now, since the power of a test is the probability of accepting a hypothesis which is true, we would evidently like to have this probability large or β small. Suppose we have two tests or rules, R_1 and R_2, for deciding between two hypotheses. If R_1 and R_2 have the same size of α, R_1 has β_1 and R_2 has β_2, and if $\beta_1 < \beta_2$, then we say that R_1 is a more powerful test (or decision rule) than R_2. In general the best test is one which minimizes β for a given level of α. It can be shown that, in selecting a sample with $n = 25$ to test $p = 0.6$ against $p = 0.4$, the most powerful test at $\alpha = 0.27$ is: Reject H_0 if and only if $X \leq 13$. It is fortunate that this test, which seems so reasonable intuitively, is indeed the most powerful decision rule at the given α level and the given data.

From the preceding comments, it may be deduced that the "goodness" of a testing procedure is judged by the level of significance and the power function of the test. In general, given α, we would like the power of test to be large or the value of β to be small. In many economic and business decision situations, the best decision rule may be considered the one for which α and β risks are so balanced as to yield maximum gains or minimum losses. Thus, if we wish to evaluate the two decision criteria employed by the manager of the radio-parts store, we must first recall that if $p = 0.6$ and that if he decides to buy the lot, he would make a profit of \$400. If $p = 0.4$ and if he decides to buy, however, he may suffer a loss of \$400. Of course, if he decides not to buy, he would gain nothing and lose nothing. Here, to avoid extra implications, we are ignoring the opportunity lost due to failure to purchase a profitable lot of tubes.

The manager's first rule, to reject H_0 if and only if $X \leq 9$, can be evaluated as follows:

(1) If H_0: $p = 0.6$ is true and if it is rejected, the manager would select A_2 (do not buy), and he would gain nothing. The probability of making this wrong decision is α, which is equal to 0.03439. This means that he expects to gain nothing by this rule about 3 percent of the time.

(2) If H_0: $p = 0.6$ is true and if it is accepted, the manager would stand to gain \$400. The probability of making this right decision is $1 - \alpha = 1 - 0.03439 = 0.96561$. That is, he expects to gain \$400 by this rule more than 96 percent of the time.

From these observations, we see that if the manager adopts the first decision rule, his expected gain each time in the long run, denoted as $E(G;R_1)$, is

$$E(G;R_1) = (1 - \alpha)(\text{gain}) + \alpha(\text{loss}) = (0.96561)(400) + (0.03439)(0) = \$386.24$$

(3) If H_1: $p = 0.4$ is true and he decides upon A_2, he expects to lose nothing. The probability that he would make this correct decision (reject H_0 or accept H_1) is the power of the test, which is equal to $1 - \beta = 1 - 0.72643 = 0.27357$; that is, he expects to lose nothing about 27 percent of the time.

(4) If H_1: $p = 0.4$ is true and if the manager decides to take A_1, he would face a loss of \$400. The probability that he would make this wrong decision (accept H_0 when H_1 is true) is β, which is equal to 0.72543. Thus, he expects to lose \$400 about 73 percent of the time.

From (3) and (4), it is clear that if the manager adopts the first decision rule, his expected loss each time, $E(L;R_1)$, in the long run is

$$\begin{aligned} E(L;R_1) &= \beta(\text{loss}) + (1 - \beta)(\text{gain}) \\ &= (0.72643)(-400) + (0.27357)(0) = -\$290.57 \end{aligned}$$

Now, since the manager does not have any information about the population except possibly that $p = 0.6$ or that $p = 0.4$, he may reasonably assume that both parameter values are equally likely. Under this assumption, the average expected gain for the first decision rule may be considered as the mean of $E(G;R_1)$ and $E(L;R_1)$; that is,

$$\text{Average Expected Gain for } R_1 = \frac{386.24 + (-290.57)}{2} = \$47.83$$

The second decision rule is to reject H_0: $p = 0.6$ if and only if $X \leq 13$. Following the same evaluation procedure as before, we have

$$E(G;R_2) = (1 - \alpha)(\text{gain}) + \alpha(\text{loss}) = (0.73228)(400) + (0.26772)(0) = \$292.91$$

$$\begin{aligned} E(L;R_2) &= \beta(\text{loss}) + (1 - \beta)(\text{gain}) \\ &= (0.15377)(-400) + (0.84623)(0) = -\$61.51 \end{aligned}$$

$$\text{Average Expected Gain for } R_2 = \frac{292.91 + (-61.54)}{2} = \$115.70$$

The above calculations indicate that R_2 is better than R_1; but is R_2 the best decision rule in this problem situation?

10.2 A General Procedure of Testing

We should now study more rigorously and in a more orderly fashion some of the problems of statistical testing introduced in the previous section.

Testing hypotheses or the construction of decision criteria involves very complicated methods. For the sake of simplicity at an elementary level of statistical analysis, we shall concentrate on those testing problems in which there are only two courses of action open to the decision maker. Such problems are sometimes called *two-action* testing problems; for example, buying or not buying a given commodity, changing or not changing a process of production, accepting or rejecting a lot of merchandise, continuing or discontinuing production, using or not using a more expensive channel of advertising, and so forth. Starting from this premise, we shall try to develop a general procedure of testing. In principle, all statistical tests are similar to those we are about to discuss. Thus, when the reader fully understands the basic ideas underlying this general procedure, he should encounter no further difficulties with respect to the tests introduced in this chapter and subsequent chapters.

FORMULATING HYPOTHESES

A *statistical hypothesis*, in formal terms, is an assumption made about the distribution of a random variable. In applied situations, a hypothesis is usually given in a form that specifies one or more of the parameters associated with the population under consideration.

The first step in statistical testing is the formulation of hypotheses. As indicated in our illustrative problem in the previous section, for two-action problems we need to focus on two possible parameter values: one is stated as the null hypothesis and the other as the alternative hypothesis. In evaluating this proposed decision procedure, we associate A_1 with the acceptance of H_0, and A_2 with rejection of H_0 (or acceptance of H_1).

Hypotheses formation may not be an easy task in a given problem situation. In any event, hypotheses must be in agreement with existing knowledge or theory about the problem. This requirement that hypotheses be "predesignated" is logical. If we were to use sample information in formulating hypotheses, we would be using sample data to *formulate* as well as to *test* the hypotheses. Obviously, with such a procedure, we could hardly expect the sample data to be inconsistent with the hypotheses; thus, it is grossly illogical to employ sample data to test hypotheses which have been suggested by the same data.

On the other hand, it is important to note that the requirement that hypotheses be stated before observations are made does not mean that we should never use sample data to suggest possible hypotheses. As a matter of fact, after the specified hypotheses have been tested by the sample data collected later, we should examine the results thoroughly to discover if there is any possible hypothesis that had not been anticipated. Such a hypothesis, of course cannot be tested rigorously by the data that suggest it. It can, however, be regarded as a new hypothesis to be tested in the future by sample data collected independently. We should realize that although we must designate hypotheses for a given situation before observations are made, we should not ignore the possibility that useful new hypotheses or theories may be suggested by data employed to test the predesignated hypotheses.

In our later work, we shall use the simple, convenient, and most frequently adopted way of testing a null hypothesis against an alternative hypothesis. Even though the designations of H_0 and H_1 are often arbitrary, the most useful way in which the null hypothesis can be formulated is one that will enable sample data to discredit it and to cause its rejection; as a general rule, we always find it easier to discredit a hypothesis than to verify it. Often, a hypothesis can be rejected with a high degree of significance; that is, with α (the probability of rejecting a true H_0) being very small, even almost zero. When we accept a hypothesis, however, we may never know how reliable our action is.

To illustrate, if we desire to discover whether a new production process is more or less efficient than the old, we state as the null hypothesis that the two processes are equally efficient. If we would like to discover whether Drug A has a greater germ-killing power than Drug B, we state as the null hypothesis that they have the same germ-killing power. If we are concerned as to whether the mean income of professors during the past ten years has changed, we state as the null hypothesis that it has remained unchanged. Thus, a hypothesis is null in the sense that it "nullifies" the possible differences. Because of this, the form of the null hypothesis is typically precise, such as $\theta = \theta_0$. Occasionally, however, we may find it appropriate to state the null hypothesis in a less precise form, such as $\theta \leq \theta_0$ or $\theta \geq \theta_0$.

The practice of testing the null hypothesis against a single alternative suggests that the rejection of the former means the acceptance of the latter. Thus, the alternative hypothesis is in effect the statistical hypothesis that we suspect to be true and that we wish to establish; whereas the null hypothesis is what we suspect to be false and what we wish to discredit.

Usually, the alternative hypothesis is stated in a nonspecific form, such as $\theta < \theta_0$, $\theta > \theta_0$, or $\theta \neq \theta_0$. The first two types of designations for H_1 are called *single-tail tests.* More precisely, when H_1 is given as $\theta < \theta_0$, we have a *left-tail test.* When H_1 is stated as $\theta > \theta_0$, we have a *right-tail test.* The last form of an alternative hypothesis, $\theta \neq \theta_0$, indicates what is called a *two-sided* or *two-tail test.* Which of these designations we should use in a given case depends on the nature of the decision problem at hand. This point will become clear to the reader after we have presented examples of a few statistical tests.

The various ways of formulating hypotheses lead us to distinguish between what are called simple and composite hypotheses. A *simple hypothesis* is a complete specification of a probability distribution. For example, the hypotheses, H_0: $p = 0.6$ and H_1: $p = 0.4$ (which we had before) are both simple. A hypothesis that is not simple is a *composite hypothesis;* thus, it is convenient to think of a composite hypothesis as made up of a number of simple hypotheses. For example, the hypothesis, say, $\mu \geq 1000$ hours is composite, since it could be any value ranging from 1000 to ∞ hours. As we shall see, if we know a simple hypothesis to be true, then the probability distribution of the testing statistic is completely defined. If we know a composite hypothesis to be true, however, we still do not know the exact distribution of the testing statistic. Also, when we say that a composite hypothesis is true we mean that one of the simple hypotheses which make it up is true.

In concluding the discussion of the first step of the general procedure of testing, we note that even though the common terminology calls H_0 the hypothesis being tested and H_1 the alternative hypothesis, in reality we are testing to decide between two hypotheses, H_0 and H_1. That is, in a statistical test concerning a population, we try to find a rule which will enable us to choose between two courses of action.

SPECIFYING THE LEVEL OF SIGNIFICANCE

The second step of our general procedure is to select a level of significance: the specification of the value of α with respect to the sample size given. The α value is commonly set at 0.05 or 0.01. If $\alpha = 0.05$ and H_0 is rejected, for example, we say that the sample results are significantly different from H_0 at the α level of 0.05. That is, if we use this procedure to make decisions, in the long run the sample results could be expected to cause rejection of true null hypotheses 5 percent of the time.

When the null hypothesis is rejected at $\alpha = 0.05$, the test result is said to be "significant." When the null hypothesis is rejected at $\alpha = 0.01$, the test result is said to be "highly significant." These two phrases are used without explicit definitions, but in statistical terms, the word "significant" does have a precise meaning. The concept "level of significance" is employed to indicate merely that a statistical test evaluates the "significance" of a difference between the hypothetical value of a parameter and the sample result.

But which of the two values of α risk should be used in a given test? The classical position on this point is ambiguous. It evades the responsibility by saying that the determination of α level is in the hands of policy decision makers, not in the hands of statisticians. The statistician, given the α risk that the decision maker is willing to take, should only provide the decision maker with useful information for his policy formulation. Then, the student may ask; How does a policy maker decide on the size of risk to be taken, and whether or not to reject the null hypothesis? The answer here depends upon the consequences of Type I and Type II errors. We shall develop this observation by way of some examples.

Suppose that it has been established without doubt by medical scientists that if the average nicotine content of cigarettes is 25 milligrams or more, lung cancer is certain to develop in a smoker. If, however, the average nicotine content of cigarettes is less than 25 milligrams, the user is relatively safe. Suppose that you are a habitual smoker and are willing to take your chances; you would certainly try to find a brand of cigarettes whose average nicotine content is less than 25 milligrams. In this case, you should formulate your hypotheses as follows:

$$H_0: \mu_0 = 25$$
$$H_1: \mu_1 < 25$$

For this test, a Type I error is made if you conclude that the mean is less than 25 when it is in fact equal to 25. Clearly, this is a serious mistake to make, since you would select a brand of cigarettes that would endanger your life. You should try to avoid such a mistake as much as you can by employing a small α risk of 0.01, or of smaller values such as, say, 0.001 or even 0.0001. Here, if a Type II error is committed, you make the wrong conclusion that the mean is equal to 25; actually it is less than 25. You would then not use this brand of cigarettes even though it is considered safe. Type II error in this case is not serious, since the only trouble you have would be to test another brand.

Thus, in a situation where the consequence of committing a Type I error is serious, we should specify α at 0.01, or at an even lower risk than this if the occasion demands it. In such a case, it is clearly not necessary for us to be deeply concerned with β risk or the power of test.

Now, suppose a cannery specifies that 10 ounces of green asparagus tips be put in each can and adjusts the automatic measuring device accordingly. If the average weight is below this figure, the cannery may run into trouble with government inspectors and be subject to heavy fines. But if the average weight is higher than this figure, the cannery will make less profits. From time to time the cannery should test the following hypotheses to see whether the measuring device is appropriately adjusted:

$$H_0: \mu_0 = 10$$
$$H_1: \mu_1 \neq 10$$

In this test, a Type I error seems to be less important than a Type II error.

If the null hypothesis is rejected when it is true, all that is wasted is the time spent in checking the measuring device. If, however, H_0 is accepted when H_1 is actually true, the cannery is confronted with the risk of large fines and possible loss of goodwill. Consequently it is advisable to set α at 0.05 or at an even higher level, such as 0.10. Note that, with fixed n, the larger α is the smaller β is, a desirable condition in the present case.

To generalize, in a situation where the consequence of committing a Type II error is more serious than that of committing a Type I error, we should fix α at a relatively high level, thereby increasing the power of the test or reducing β, for every possible simple hypothesis in the composite alternative.

Situations may arise, of course, where errors of the first and the second kinds are equally serious. For instance, in the preceding example, we may find that the time lost in examining the measuring device is costly because stoppage of production is entailed. If this is the case, then it would be desirable to have both α and β small. Probabilities of committing both types of error can be set at any desired level, provided the sample size is permitted to vary.

We shall subsequently discuss statistical testing by, first following the procedure of specifying the level of significance alone, then moving on to test problems for which both α and β are evaluated, and finally taking up the method for finding the appropriate sample size so that prescribed α and β risks for a given test can be met. We might point out now, however, that the simple testing procedure of specifying α risk alone makes good sense because of the way in which we formulate hypotheses; it is quite adequate in many situations.

SELECTING THE TESTING STATISTIC

Having formulated hypotheses and determined the level of significance, we now turn to the third step of our general testing procedure: the selection of a testing statistic whose sampling distribution is known under the assumption that the null hypothesis is true. The common practice is to choose as the testing statistic that measure which is used to estimate the parameter in the terms of which H_0 and H_1 are stated. For example, if H_0 and H_1 are stated in terms of population mean, then a sample mean would be selected as the testing statistic.

When we have selected, say, $\hat{\theta}$ to test hypotheses concerning θ, we may use $\hat{\theta}$ to test H_0 against H_1 directly, or we may use its equivalent, $Z = (\hat{\theta} - \theta_0)/SE(\hat{\theta})$, where θ_0 is the assumed parameter in H_0, to do the job. The employment of Z as the testing statistic is especially convenient if the sampling distribution of $\hat{\theta}$ is normal, since then Z has the standard normal distribution.

ESTABLISHING DECISION CRITERIA

At this point—as a fourth step—a decision criterion, or rule, must be established for the purpose of rejecting or accepting H_0. This involves the division of the sampling distribution of the testing statistic into two parts: the *critical* or *rejection region*, and the *acceptance* or *nonrejection region*. This division depends upon the designation of the alternative hypothesis, the level of α, and the sampling distribution of the testing statistic—all of which have already been predetermined.

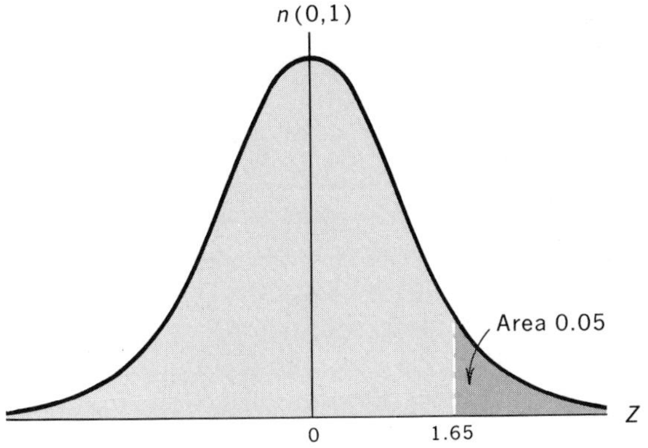

Figure 10.2 A one-sided critical region for testing $H_0: \theta = \theta_0$ against $H_1: \theta > \theta_0$ at $\alpha = 0.95$.

Throughout our discussion in this chapter, we shall assume that the parent population is normal or the sample is large enough for the central limit theorem to apply; namely, we shall assume that the sampling distribution of the testing statistic is normal. Then the task becomes one of discovering a *critical value* of the testing statistic which would separate the critical region from the region of acceptance at a given level of significance and for a particular alternative hypothesis. We shall now show how such a critical value can be found for each of the three most frequently used forms of alternative hypotheses.

First, if we test H_0: $\theta = \theta_0$ against H_1: $\theta > \theta_0$, then only large values of $\hat{\theta}$, the testing statistic, would support the validity of H_1. Consequently, given the level of significance, we try to determine the critical value, K, so that $P(\hat{\theta} > K) = \alpha$, where

$$K = \theta_0 + Z_{1-\alpha} SE(\hat{\theta}) \tag{10.4}$$

That is, if we are concerned with a right-tail test, the decision rule is: Reject H_0 if and only if $\hat{\theta} > K$; do not reject H_0 if $\hat{\theta} \leq K$. Since Z is a standard normal variate in (10.4), this decision rule can also be stated as: Reject H_0 if and only if

$$\frac{\hat{\theta} - \theta_0}{SE(\hat{\theta})} > Z_{1-\alpha} \tag{10.5}$$

The critical region for a right-tail test is shown in Figure 10.2.

Second, if we test H_0: $\theta = \theta_0$ against H_1: $\theta < \theta_0$, then H_1 is supported by small values of $\hat{\theta}$ only. Therefore, the critical value, K, should be chosen so that $P(\hat{\theta} < K) = \alpha$, where

$$K = \theta_0 - Z_\alpha SE(\hat{\theta}) \tag{10.6}$$

Namely, if we are concerned with a left-tail test, the decision criterion is to reject H_0 if and only if $\hat{\theta} < K$. If Z is used as the testing statistic, then this rule becomes: Reject H_0 if and only if

$$\frac{\hat{\theta} - \theta_0}{SE(\hat{\theta})} < Z_\alpha \tag{10.7}$$

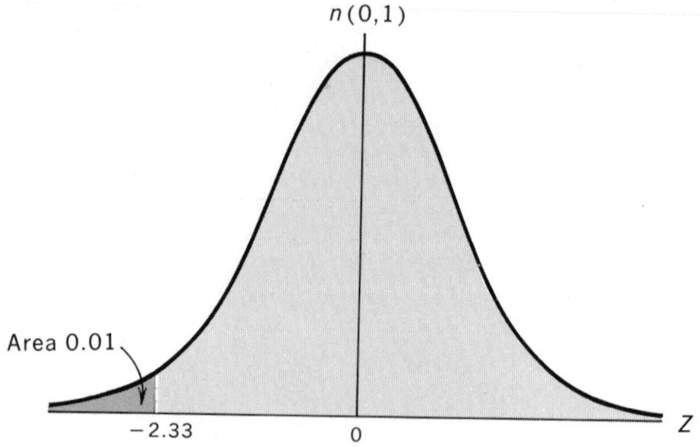

Figure 10.3 A one-sided critical region for testing $H_0: \theta = \theta_0$ against $H_1: \theta < \theta_0$ at $\alpha = 0.01$.

The critical region for a left-tail test is shown by Figure 10.3.

Finally, if we are testing H_0: $\theta = \theta_0$ against H_1: $\theta \neq \theta_0$, we have a two-tail test. In this case, both small and large values of the testing statistic tend to support H_1. There is reason to expect then that the critical region of this test, as shown by Figure 10.4, is divided equally between both ends of the density function of the testing statistic. We have two critical values, K_1 and K_2, instead of only one value to be determined. It is easy to show that, given the level of significance,

$$K_1 = \theta_0 - Z_{\alpha/2} SE(\hat{\theta}), \text{ and} \tag{10.8a}$$

$$K_2 = \theta_0 + Z_{1-\alpha/2} SE(\hat{\theta}) \tag{10.8b}$$

That is, now we have this decision rule: Reject H_0 if and only if $\hat{\theta} < K_1$ or $\hat{\theta} > K_2$.

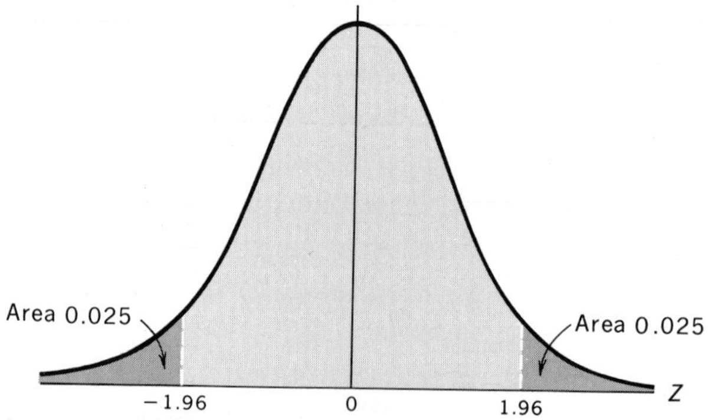

Figure 10.4 A two-sided critical region for testing $H_0: \theta = \theta_0$ against $H_1: \theta \neq \theta_0$ at $\alpha = 0.05$.

TABLE 10.4 RULES FOR TESTING H_0: $\theta = \theta_0$ AGAINST H_1 IN TERMS OF $\hat{\theta}$

H_1	Reject H_0 if and only if	$\alpha = 0.05$	$\alpha = 0.01$
$\theta > \theta_0$	$\hat{\theta} > K$	$K = \theta_0 + 1.65 SE(\hat{\theta})$	$K = \theta_0 + 2.33 SE(\hat{\theta})$
$\theta < \theta_0$	$\hat{\theta} < K$	$K = \theta_0 - 1.65 SE(\hat{\theta})$	$K = \theta_0 - 2.33 SE(\hat{\theta})$
$\theta \neq \theta_0$	$\hat{\theta} < K_1$ or	$K_1 = \theta_0 - 1.96 SE(\hat{\theta})$ or	$K_1 = \theta_0 - 2.58 SE(\hat{\theta})$
	$\hat{\theta} > K_2$	$K_2 = \theta_0 + 1.96 SE(\hat{\theta})$	$K_2 = \theta_0 + 2.58 SE(\hat{\theta})$

In terms of Z values, we state: Reject H_0 if and only if

$$\frac{\hat{\theta} - \theta_0}{SE(\hat{\theta})} < Z_{\alpha/2}, \text{ or} \tag{10.9a}$$

$$\frac{\hat{\theta} - \theta_0}{SE(\hat{\theta})} > Z_{1-\alpha/2} \tag{10.9b}$$

For convenience, the preceding decision criteria, under the assumption that the sampling distribution of the testing statistic is normal, are summarized in Tables 10.4 and 10.5 for $\alpha = 0.05$ and 0.01. Table 10.4 contains critical values in terms of the testing statistic directly, and Table 10.5 gives the corresponding critical values in terms of Z.

DOING COMPUTATIONS

Having taken the first four steps, we have completely designed a statistical test. We may now proceed to a fifth step—performance of various computations, from a random sample of size n, necessary for the test. These calculations include the testing statistic, the standard error of the testing statistic, and the critical value of the testing statistic.

TABLE 10.5 RULES FOR TESTING H_0: $\theta = \theta_0$ AGAINST H_1 IN TERMS OF Z

H_1	Reject H_0 if and only if	
	$\alpha = 0.05$	$\alpha = 0.01$
$\theta > \theta_0$	$\dfrac{\hat{\theta} - \theta_0}{SE(\hat{\theta})} > 1.65$	$\dfrac{\hat{\theta} - \theta_0}{SE(\hat{\theta})} > 2.33$
$\theta < \theta_0$	$\dfrac{\hat{\theta} - \theta_0}{SE(\hat{\theta})} < -1.65$	$\dfrac{\hat{\theta} - \theta_0}{SE(\hat{\theta})} < -2.33$
$\theta \neq \theta_0$	$\dfrac{\hat{\theta} - \theta_0}{SE(\hat{\theta})} < -1.96$ or	$\dfrac{\hat{\theta} - \theta_0}{SE(\hat{\theta})} < -2.58$ or
	$\dfrac{\hat{\theta} - \theta_0}{SE(\hat{\theta})} > 1.96$	$\dfrac{\hat{\theta} - \theta_0}{SE(\hat{\theta})} > 2.58$

MAKING DECISIONS

Finally, as a sixth step, we may draw statistical conclusions, and management may make decisions. On the basis of previous computations, the statistician would decide to reject H_0 if the testing statistic fell in the critical region and to accept H_0 if the testing statistic fell in the acceptance region. Then, the policy maker would make a managerial decision—that is, he would select an appropriate course of action in view of the statistical conclusion.

Statistical decision is made on the basis of the value of the testing statistic within the framework of established decision criteria. If the testing statistic falls in the critical region, we consider the difference between the sample statistic and the hypothetical parameter significant. In other words, we think the sample result is so rare that it cannot be explained by chance variations alone. We then decide to reject H_0 and state: "The null hypothesis is false"; or "The sample observations are not consistent with the null hypothesis." (The student is reminded that the rejection of H_0 is in effect the acceptance of H_1.) On the other hand, if the testing statistic falls in the region of nonrejection, we reason that the difference between the sample result and the hypothetical parameter can be explained by chance variations and therefore is not significant statistically. Consequently, we decide not to reject H_0 and state: "The sample result is not inconsistent with the null hypothesis."

Careful readers may have already noted that the rejection statement is much stronger than the acceptance statement. When we reject a hypothesis, we believe that there is a definite reason for us to do so, whereas when we accept a hypothesis, we simply behave as if the sample data are not sufficient evidence to doubt the truth of the hypothesis. This difference in attitudes arises essentially from the fact that, in logic, it is always easier to prove something false than to prove it true. In testing, Type I error is easily asserted, whereas Type II error is always difficult to evaluate because of the way we design our alternative hypotheses. Hence, we have adopted the rule of rejecting all hypotheses if we can; but if we are forced to accept them, we do not want to commit ourselves. Justification for this practice is the fact that the risk of rejecting a true hypothesis is completely under control, limited to the value of α. Furthermore, if the sample is sufficiently large, we can be fairly sure that our decision of rejection is right.

We should also note that if the difference between the observed value of the testing statistic and the hypothetical parameter is very great, we can reject H_0 with an extremely high level of significance. This point is indicative of an important characteristic of statistical testing: A test can never be considered positive proof of the truth of a hypothesis. The central core of testing is to give statistical evidence an opportunity to disprove a hypothesis. However, we must not blind ourselves to the fact that there is still the probability, though very small, that the rejection may be the wrong decision. What we should remember is that a hypothesis may be rejected with a degree of confidence that the acceptance of a hypothesis can never have.

The managerial decision which follows the statistical decision is beyond the authority of the statistician. It is mentioned here merely to stress that the basic purpose of testing is to assist management in reaching rational economic and business decisions. Generally, statistical decision is the basis on which a decision maker selects an appropriate course of action. His rule is to associate his decision with testing before the testing is conducted; he decides in advance that one course of action will be taken if H_0 is accepted and another course of action if H_0 is rejected. The important thing to realize is that "statistical significance" does not necessarily imply "practical importance." Statistically, we may find that there is a "significant difference"; but the decision maker may feel that this difference is not of any importance from a practical point of view. As a result, he may simply refrain from taking any action.

SUMMARY

The general procedure of statistical testing just outlined consists of the following six steps: (1) formulating the null and the alternative hypotheses, (2) specifying the level of significance, (3) selecting the testing statistic, (4) establishing decision criteria, (5) doing computations, and (6) making decisions. We shall apply this procedure, in the next three sections, to a number of statistical tests concerning μ, π, $\Delta\mu$, and $\Delta\pi$. Then we shall move on to discuss the evaluation of β given the sample size, and the determination of sample size given both α and β.

10.3

Testing the Hypothesis $\mu = \mu_0$

The population mean is a very important decision parameter. Often we wish to know if a population mean has increased, or decreased, or remained unchanged. Or, we may be interested in determining whether a population mean is significantly greater than, or less than, some assumed value. Has the real per capita income in Burma increased since its independence? Can a new assembly procedure really reduce the mean time of assembling a mechanism? Has there been no change in the life span of American males during the past ten years? Each of these questions can be answered by testing $\mu = \mu_0$ against some appropriate alternative hypotheses.

To test whether a specified μ_0 is reasonable in view of sample results, the testing statistic used is the sample mean or, more conveniently for our general testing procedure, the critical ratio $Z = (\bar{x} - \mu_0)/\sigma_{\bar{x}}$.

AN UPPER-TAIL TEST

A manufacturing process used by a factory during the past few years yields an average output of 100 units per hour with a standard deviation of 8 units. A new machine for making this type of product has just been put on the market. While it is very expensive compared to the one now in use, if the mean output of the new machine is more than 150 units per hour, its adoption would be quite profitable. To decide whether the new machine should be purchased to replace the old, the management of the factory buys 35 of the new machines as an experiment. The final decision is to be made according to the following procedure:

1. Formulating Hypotheses: Let us assume that the management is inclined to believe that the average output per hour of the new machines is greater than 150 and that it is also anxious to avoid the error of buying when it should not; then the hypotheses should be stated as

$$H_0\colon \mu = 150 \text{ units}$$
$$H_1\colon \mu > 150 \text{ units}$$

2. Specifying the Level of Significance: In this case, the rejection of H_0 if it is true would lead to serious consequences, since then the wrong action (buying the expensive machines) would be taken. In order to guard against the purchase of new machines that might not actually be as productive as expected, the management decides to set α at 0.01, a low probability of rejecting H_0 when it is true.

3. Selecting the Testing Statistic: To test the hypotheses stated in (1), the statistician decides to use as the testing statistic the critical ratio $Z = (\bar{x} - 150)/\sigma_{\bar{x}}$, which is a standard normal variable.

4. Establishing the Decision Criterion: At $\alpha = 0.01$, and for a right-tail test, the decision rule is: Reject H_0 if and only if $Z > 2.33$; otherwise, accept H_0.

5. Doing Computations:

a. Assume that the 35 new machines constitute a random sample, and that it has been found that $\bar{x} = 160$ units per hour.

b. Assuming that the standard deviation of outputs for the new machines is identical with that for the old (that is, $\sigma = 8$ units), then, with $n = 35$, we have

$$\sigma_{\bar{x}} = \frac{8}{\sqrt{35}} = 1.35$$

c. The critical ratio is

$$Z = \frac{\bar{x} - \mu_0}{\sigma_{\bar{x}}} = \frac{160 - 150}{1.35} = 7.41$$

6. Making Decisions: Since Z falls in the critical region, the difference between the sample mean and the hypothetical mean is significant. H_0 is not true. The rejection of H_0 is the acceptance of H_1. According to management's predesignated criterion, the new machine should be purchased to replace the old. (Note: When $Z = 7.41$, we know that if H_0 were true, the probability of obtaining a difference as large as 10 between $\bar{x}$ and μ_0 is practically zero; that is, the significance of this test is actually that $\alpha = 0$. What does this observation mean or imply?)

A LOWER-TAIL TEST

The average distance for stopping a certain make of automobile when it is travelling 30 miles an hour is 65 feet. The engineering department of the company has designed a new brake system thought to be more effective than the type currently in use. To test this invention, the new brake system is installed in 64 cars, and tests show that the average distance for stopping a car at a speed of 30 miles an hour is 63.5 feet, with a standard deviation, s, of 4 feet. Does the difference of 1.5 feet prove that the new brake system is more effective than the old? Now, if the management wishes to avoid the mistake of not adopting the new brake system if it really should, the question can be answered by the following test:

1. **Hypotheses:** H_0: $\mu = 65$ feet
   H_1: $\mu < 65$ feet

2. **Level of Significance:** $\alpha = 0.05$

3. **Testing Statistic:** $Z = (\bar{x} - \mu_0)/\sigma_{\bar{x}}$

4. **Decision Rule:** Reject H_0 if and only if $Z < -1.65$.

5. **Computations:** The sample mean, with $n = 64$, was given before as 63.5 feet. The standard deviation of the population is unknown in this case; but the sample is large enough for us to estimate it by s, which was given as 4 feet. Thus,

$$\hat{\sigma}_{\bar{x}} = \frac{s}{\sqrt{n}} = \frac{4}{\sqrt{64}} = 0.5$$

and

$$Z = \frac{63.5 - 65}{0.5} = -3.00$$

6. **Decision:** Since $Z = -3.00 < -1.65$, H_0 is rejected. The new brake system is more effective than the old. The management should seriously consider its adoption.

A TWO-TAIL TEST

A floor-wax factory has nationwide selling outlets. In the past, its salesmen earned a mean commission of \$600 per month, with a standard deviation of \$80. Recently, new brands of floor wax produced by other factories have entered the market.

This factor tends to reduce the volume of sales per salesman and, thus, to reduce the salesmen's income. Meanwhile, the list price of the firm's output has increased along with general price inflation; this factor tends to increase the salesmen's commission per sale. The management of the firm is anxious to discover the net effect of these two factors on its salesmen's commissions. To achieve this, the following decision procedure is adopted:

1. **Hypotheses:** $H_0: \mu = \$600$
$H_1: \mu \neq \$600$

2. **Level of Significance:** $\alpha = 0.05$

3. **Testing Statistic:** $Z = (\bar{x} - \mu_0)/\sigma_{\bar{x}}$, which, as before, is a standard normal variable.

4. **Decision Rule:** With $\alpha = 0.05$, for a two-tail test the decision rule becomes: Reject H_0 if and only if $|Z| > 1.96$. That is, reject H_0 if and only if $Z < -1.96$ or $Z > 1.96$.

5. **Computations:** Suppose that an unconditional random sample of $n = 100$ is drawn from all the salesmen's commission accounts, and it is found that $\bar{x} = \$585$. Also, since $\sigma = \$80$, we have

$$\sigma_{\bar{x}} = \frac{80}{\sqrt{100}} = 8$$

and

$$Z = \frac{585 - 600}{8} = -1.88$$

6. **Decisions:** Now, $Z = -1.88$, which falls in the acceptance region; therefore, H_0 is not to be rejected. This result indicates that the difference of \$15 may be attributed to random fluctuations in the sample statistic and that it is not significant at an α risk of 0.05. Since sample results lead us to conclude that the salesmen's monthly mean income has not changed, the management may no longer be concerned with this problem.

It is perhaps more appropriate for this problem to state the alternative hypothesis as $\mu < \$600$, since the management would have no trouble if the mean commission were equal to or greater than \$600. Had the alternative hypothesis been so designated, we would reject H_0, for the same level of significance of 0.05, if and only if $Z < -1.65$. Recall that the critical value for the present test is -1.88, which is less than -1.65. Thus, the same difference that is insignificant for a two-tail test may become significant for a single-tail test. Further, had we decided to conduct a left-tail test, the managerial decision would have been entirely different for the same sample results. Instead of choosing to do nothing (because H_0 was accepted), the management might (because of the acceptance of H_1) act in a number of ways: step up its advertising campaign in order to increase its sales; increase the percentage of commission paid; improve the quality of its output in order to improve its competitive strength; or take more than one of these actions, for the purpose of keeping its sales force intact.

These observations are indicative of the importance of formulating hypotheses and specifying level of significance, thereby establishing the decision rule, before the collection and observation of data. Unless the required procedure is followed, we may be led to conduct a statistical test within the framework of an improper decision procedure, which is both suggested and tested by the same sample results.

10.4 Testing the Hypothesis $\pi = \pi_0$

We have seen that situations often arise in which sampling is used to determine an attribute rather than the mean value of a population. In such a case, observations are qualitative in nature. Sampling attributes, as we already know, reveals information as to whether workers are males or females, whether products are defective or nondefective, and so forth. A population, of course, may contain more than two mutually exclusive categories. However, it is always possible to group the elementary units in a population into two mutually exclusive qualitative classes. For instance, voters may be Republican, Democratic, Liberal, or Independent, but all of those who are eligible to vote can be classified as being in favor or against a given candidate.

In any event, when we analyze qualitative data, we are interested in verifying an assumption made about the population proportion of successes, π. Here, we would like to test the hypothesis $\pi = \pi_0$ with the sample proportion p as the testing statistic. Also, in such a test, we compute the standard error of p by using π_0 in the null hypothesis. This is a logical practice because in our test the sampling distribution of the testing statistic is determined under the assumption that the null hypothesis is true.

A TWO-TAIL TEST

The manufacturer of a patent medicine claims that its product is more than 70 percent effective in giving immediate relief from skin itches, from any cause. The American Medical Association, hoping to verify this claim, appoints a committee of seven skin specialists to evaluate the effectiveness of this medicine. Some members of the committee are confident that the manufacturer's claim could not be possibly true (due to their past experience with similar drugs), but others suspect that the claim might be valid, so it is finally agreed that both alternatives should be considered. Also, after some thought, it is decided to set α at 0.01 and to try the medicine on 200 patients who have skin disorders. In other words, the testing procedure is designed as follows:

1. **Hypotheses:** $H_0: \pi = 0.70$
$H_1: \pi \neq 0.70$

2. **Level of Significance:** $\alpha = 0.01$

3. **Testing Statistic:** $Z = (p - \pi_0)/\sigma_p$, which is a standard normal variate if H_0 is true.

4. **Decision Rule:** Reject H_0 if and only if $Z < -2.58$ or $Z > 2.58$.

5. **Computations:** Suppose that among the 200 patients treated by this medicine, 125 experienced immediate relief; then $p = 0.625$. Also, for this test,

$$\sigma_p = \sqrt{\frac{\pi_0(1 - \pi_0)}{n}} = \sqrt{\frac{0.7(0.3)}{200}} = 0.0324$$

and

$$Z = \frac{0.625 - 0.700}{0.0324} = -2.31$$

6. **Decisions:** Since $Z = -2.31 > -2.58$, H_0 cannot be rejected. A difference of -0.075 between the sample proportion and the hypothetical value of π is not sufficient to disprove the manufacturer's claim. If the new drug does not have any undesirable side effects, it should be recommended for use.

A LEFT-TAIL TEST

Suppose that one of Philadelphia's diaper services has advertised in a newspaper as follows: "There are more than 10 diaper services in Philadelphia, but 30 percent of the families who use diaper services are our customers.—The Smile Diaper Service." As may happen in reality, we can imagine that other diaper services will complain to the Better Business Bureau, challenging the validity of this claim.

Before the BBB can act on the complaint, it must decide about the truth of the claim made by The Smile Diaper Service. To aid in arriving at this decision, the Bureau selects a random sample of 150 families, without replacement, from the thousands currently using diaper services. It finds that 39, or 26 percent, of the 150 families are customers of The Smile Diaper Service. Now, the question is: Is the difference of 4 percent between the sample result and the claimed share of the market statistically significant? To answer this question, the BBB adopts the following decision procedure:

1. **Hypotheses:** $H_0: \pi = 0.30$
$H_1: \pi < 0.30$

Note that the alternative hypothesis here suggests a left-tail test. It is appropriate because the BBB is primarily interested in verifying Smile's claim against the alternative that its market share is not as large as claimed.

2. Level of Significance: In deciding the value of α, the BBB is confronted with two conflicting forces. On the one hand, Smile would like the Bureau to adopt a decision criterion that would minimize the probability of its incorrectly rejecting the null hypothesis. This course demands a small α. On the other hand, the Bureau itself and other diaper services are more concerned with an unwarranted acceptance of the null hypothesis. Thus, they would like to have a small β or large α. After thoughtful consideration, the Bureau finally decides on the 5 percent level of significance as a balanced criterion, namely, $\alpha = 0.05$.

3. Testing Statistic: $Z = (p - \pi_0)/\sigma_p$

4. Decision Rule: With $\alpha = 0.05$ for a left-tail test, reject H_0 if and only if $Z < -1.65$. (We are assuming here that the BBB wants to reach a decision one way or another. Otherwise, the decision rule may be stated as; Reject H_0 if and only if $Z < -1.65$, but reserve judgment or conduct another test if $Z \geq -1.65$.)

5. Computations: $p = {}^{39}\!/_{150} = 0.26$

$$\sigma_p = \sqrt{\frac{0.3(0.7)}{150}} = 0.0374$$

$$Z = \frac{0.26 - 0.30}{0.0374} = -1.07$$

6. Decision: The value of Z falls in the acceptance region. If $\pi = 0.3$, it is not unlikely for $p = 0.26$ to occur. The claim made by The Smile Diaper Service is not unfounded in terms of sample observations. There are, therefore, no grounds for censure.

10.5

Testing the Hypothesis $\theta_1 = \theta_2$

One often meets such decision problems as those suggested by the following questions: Is the difference between the mean income of lawyers and that of physicians significant? Has a population of smokers a greater susceptibility to lung cancer than a population of nonsmokers? Is there any real difference between voting habits of men and women? Does one TV program have more appeal to the public than another? Is the new method of therapy for a certain class of patients more effective than the old? Has one make of aluminum channels a greater variability in stiffness than another make? Has the rate of economic growth in Red China been considerably reduced because of Mao Tse-tung's "cultural revolution?" Has teachers' real income in the United States changed since 1945?

These questions are concerned with the comparison of two populations. One action is called for if they can be thought of as the same population; another action is appropriate if there is a difference, or if some change has taken place from one population to another. Our interest here may focus on the difference between population means, between population proportions, or between population variances. We shall discuss statistical tests about the first two types of difference in this section, and take up the last kind in the next chapter.

COMPARISON OF TWO POPULATION MEANS

To test whether there is a significant difference between two population means, μ_1 and μ_2, we formulate the null hypothesis that there is no difference between them; that is, H_0: $\mu_1 = \mu_2$. The testing statistic used here is the difference between two sample means computed from two random samples independently drawn. Under the assumption that H_0 is true, the critical ratio $Z = (\Delta\bar{x} - \Delta\mu)/\sigma_{\Delta\bar{x}} = (\bar{x}_1 - \bar{x}_2)/\sigma_{\Delta\bar{x}}$ has the standard normal distribution. Note that if the null hypothesis $\mu_1 = \mu_2$ is true, then $\Delta\mu = \mu_1 - \mu_2 = 0$.

To ascertain whether a new fertilizer is more effective than the old for wheat production, a tract of land was divided into 100 squares of equal areas, all of the same quality. The new fertilizer was applied to 50 squares and the old fertilizer was applied to the other 50 squares. The mean number of bushels of wheat harvested per square of land using the new fertilizer was 25.5 bushels with a variance of 22. The corresponding mean and variance for the squares using the old fertilizer were 24.6 and 19 respectively. Is the new fertilizer more efficient than the old at $\alpha = 0.01$?

1. Hypotheses: H_0: $\mu_1 = \mu_2$
H_1: $\mu_1 < \mu_2$

2. Level of Significance: $\alpha = 0.01$

3. Testing Statistic: $Z = (x_1 - x_2)/\sigma_{\Delta\bar{x}}$

4. Decision Rule: Reject H_0 if and only if $Z < -2.33$.

5. Computations: Since both n_1 and n_2 are quite large, sample variances can be used as estimates of population variances to compute the standard error of $\Delta\bar{x}$. That is,

$$\hat{\sigma}_{\Delta\bar{x}} = \sqrt{\frac{s_1{}^2}{n_1} + \frac{s_2{}^2}{n_2}} = \sqrt{\frac{22}{50} + \frac{19}{50}} = 0.9$$

With this value of $\sigma_{\Delta\bar{x}}$, we have

$$Z = \frac{\bar{x}_1 - \bar{x}_2}{\hat{\sigma}_{\Delta\bar{x}}} = \frac{25.5 - 24.6}{0.9} = 1$$

6. **Decisions:** Z falls in the acceptance region and H_0 cannot be rejected. The new fertilizer is not more efficient than the old at the 1 percent level of significance.

An automobile manufacturing company has recorded observations of two different makes of batteries, both of which are being used on its cars: 40 observations of Make A indicated a mean life of 32 months; 45 records of Make B showed a mean life of 30 months. Also, past experience indicates that the standard deviation for both makes of batteries is the same, 4 months. Is there a real difference in mean life between the two makes of batteries at $\alpha = 0.05$?

The appropriate hypotheses in this case are H_0: $\mu_A = \mu_B$ and H_1: $\mu_A \neq \mu_B$. Also,

$$Z = \frac{\bar{x}_A - \bar{x}_B}{\sigma_{\Delta\bar{x}}} = \frac{32 - 30}{\sqrt{\dfrac{4^2}{40} + \dfrac{4^2}{45}}} = 2.30$$

which is greater than 1.96 and, therefore, H_0 is rejected. The difference of 2 months in mean life cannot be explained by chance variations alone.

For this problem, it may be interesting to recall that the 95 percent confidence interval discussed in Section 9.7 is $0.3 < \Delta\mu < 3.7$, which does not include $\Delta\mu = 0$. This indicates that a confidence interval with $1 - \alpha$ corresponds to a two-tail test at α. In general, for example, the inequality

$$\hat{\theta} - 1.96SE(\hat{\theta}) < \theta < \hat{\theta} + SE(\hat{\theta})$$

states that before a sample is selected and $\hat{\theta}$ determined, the probability is 0.95 that the interval computed from $\hat{\theta}$ will include the unknown but constant θ. When we test the null hypothesis $\theta = \theta_0$ against the alternative $\theta \neq \theta_0$ and assume that the variance σ^2 for H_0 is the same as that for H_1, we can actually adopt the decision rule of rejecting H_0 if θ_0 does not lie in the confidence interval. Namely, H_0 is rejected if

$$|\theta_0 - \hat{\theta}| > 1.96SE(\hat{\theta})$$

This is the same as saying, however, that the testing statistic $\hat{\theta}$ falls farther away from θ_0 than a certain quantity; this is actually a two-tail test we have been conducting. In the confidence coefficient $1 - \alpha$, α is the level of significance of a test. To show this relationship for a test against a single-tail alternative, we have, of course, to construct a one-sided confidence interval as introduced in the next chapter.

COMPARISON OF TWO POPULATION PROPORTIONS

To compare two population proportions, we conduct a test for H_0: $\pi_1 = \pi_2$ against some appropriate alternative. The testing statistic employed is usually $(p_1 - p_2)/\sigma_{\Delta p}$, where the sample proportions are derived from two independent random samples. Under the assumption that $\pi_1 = \pi_2 = \pi$,

$$\sigma_{\Delta p} = \sqrt{\frac{\pi(1 - \pi)}{n_1} + \frac{\pi(1 - \pi)}{n_2}} = \sqrt{\pi(1 - \pi)\left(\frac{n_1 + n_2}{n_1 n_2}\right)}$$

But the population proportions are usually unknown to us; we take the weighted mean of the two sample proportions as the estimate. Call this estimate $\bar{p}$; we then have

$$\bar{p} = \frac{n_1 p_1 + n_2 p_2}{n_1 + n_2}$$

Substituting $\bar{p}$ into the previous expression, we have as an estimate of $\sigma_{\Delta p}$ the following:

$$\hat{\sigma}_{\Delta p} = \sqrt{\bar{p}(1 - \bar{p})\left(\frac{n_1 + n_2}{n_1 n_2}\right)} \qquad \textbf{(10.10)}$$

In the spring of 1954, the Federal Reserve System and the Research Center of the University of Michigan jointly sampled 3000 spending units in a survey of consumer finances. A similar study was made in the spring of 1955, when 3120 spending units were interviewed. These two surveys, among other things, revealed that in 1954, 7.7 percent of those interviewed intended to purchase television sets, whereas in 1955, only 5.9 percent of those questioned planned to do so. Had there been a significant change in consumers' intentions to buy television sets between the spring of 1954 and that of 1955? To answer this question, we conduct the following test:

1. **Hypotheses:** H_0: $\pi_1 = \pi_2$
   H_1: $\pi_1 \neq \pi_2$

2. **Level of Significance:** $\alpha = 0.05$

3. **Testing Statistic:** $Z = (p_1 - p_2)/\sigma_{\Delta p}$, which has the standard normal distribution if H_0 is true.

4. **Decision Rule:** Reject H_0 if and only if $Z < -1.96$ or $Z > 1.96$; otherwise accept H_0.

5. **Computations:** Sample data give

1954	1955
$n_1 = 3000$	$n_2 = 3120$
$p_1 = 0.077$	$p_2 = 0.059$

We have also

$$\bar{p} = \frac{(3000)(0.077) + (3120)(0.059)}{3000 + 3120} = 0.0678$$

By (10.10),

$$\hat{\sigma}_{\Delta p} = \sqrt{0.0678(1 - 0.0678)\left(\frac{3000 + 3120}{(3000)(3120)}\right)} = 0.0064$$

Therefore,

$$Z = \frac{0.077 - 0.059}{0.0064} = 2.81$$

6. **Decisions:** Z falls in the rejection region; there had been a significant change in consumers' spending intentions between the spring of 1954 and that of 1955 insofar as the purchase of TV sets was concerned.

A sample poll of 200 college students and 250 high school students showed that 59 percent and 55 percent, respectively, were against the draft system as it applied in 1967. Was this difference of 4 percent in sample proportions sufficient evidence to support the contention that a greater percentage of college students was dissatisfied with the draft system at a level of significance of 0.01?

For this problem,

$$\bar{p} = \frac{(200)(0.59) + (250)(0.55)}{200 + 250} = 0.568,$$

$$\hat{\sigma}_{\Delta p} = \sqrt{0.568(1 - 568)\left(\frac{200 + 250}{(200)(250)}\right)} = 0.0469, \text{ and}$$

$$Z = \frac{0.59 - 0.55}{0.0469} = 0.85$$

Since the value of the critical ratio is less than 2.33, the null hypothesis, $\pi_1 = \pi_2$, cannot be rejected in favor of the alternative, $\pi_1 > \pi_2$. In other words, the difference of 4 percent is not sufficient evidence to support the contention that the college students were more dissatisfied with the draft system than the high school students in 1967, at $\alpha = 0.01$ with $n_1 = 200$ and $n_2 = 250$. However, there might be a real difference and, if so, this difference could perhaps be revealed statistically by larger samples.

10.6 Evaluation of β

Up to now we have been concerned with statistical testing by specifying α alone and have said very little about β. Whenever the consequences of both Type I and Type II errors are serious, we can no longer ignore β. The evaluation of β is often made in two different ways. One is to specify both α and β at desired levels, and then to try to find a sample size so that they can be satisfied. The other is to calculate β risks for various possible simple hypothetical values in the composite alternative, given the level of significance and the sample size. These procedures will lead us to consider what are called the *power* and the *operating characteristic functions of the test,* and these are the topics for discussion in this section.

Suppose that the purchasing agent of a department store is interested in buying a certain make of small cardboard containers which, according to the manufacturer's specifications, have a mean crushing strength of 30 pounds and a standard deviation of 3 pounds. These specifications are exactly what the purchasing agent desires. To decide whether or not to buy, he has two alternative approaches, depending upon his attitude toward α risks. First, he does not want to forego the opportunity to buy a useful product. According to this attitude, the appropriate designations for the null and the alternative hypotheses should be

$$H_0\colon \mu \geq 30 \text{ pounds}$$
$$H_1\colon \mu < 30 \text{ pounds}$$

Second, the agent may wish to avoid the mistake of buying when he should not; that is, the mean crushing strength may be less than 30 pounds, and he is fearful of buying a useless product. With this attitude, the set of hypotheses to be evaluated becomes

$$H_0\colon \mu \leq 30 \text{ pounds}$$
$$H_1\colon \mu > 30 \text{ pounds}$$

We shall now evaluate these two approaches by assuming that $\alpha = 0.05$ and $n = 36$ in both cases.

TESTING THE HYPOTHESIS $\mu \geq 30$

The first approach is to test the null hypothesis, $\mu \geq 30$, against the alternative, $\mu < 30$. For this, the α and β errors are as follows:

Action	State of Nature	
	$\mu_0 \geq 30$	$\mu_1 < 30$
A_1: Buy	Correct action	$\beta = P(A_1 \mid \mu < 30)$
A_2: Do not buy	$\alpha = P(A_2 \mid \mu \geq 30)$	Correct action

The logical testing statistic in each case is obviously the sample mean, with $n = 36$.

Establishing the decision rule: Having formulated the hypotheses, specified the significance level, and chosen the testing statistic, just as in the general procedure of testing, we must now decide upon the critical value of $\bar{x}$ that would enable us to accept or to reject the null hypothesis. For a left-tail test, at $\alpha = 0.05$, we would reject H_0, according to (10.6), if and only if

$$\bar{x} < K = \mu_0 - 1.65\sigma_{\bar{x}} = 30 - 1.65\frac{3}{\sqrt{36}} = 29.175 \text{ pounds}$$

That is, the decision rule so established means

If $\bar{x} \geq 29.175$ pounds, take A_1: buy.
If $\bar{x} < 29.175$ pounds, take A_2: do not buy.

Computing β risks: We now are ready to evaluate the Type II error. Recall that β is the probability of accepting H_0 when H_1 is in fact true. In general, when we are testing against an alternative $\theta < \theta_0$, we would accept H_0 if $\hat{\theta} \geq K$; and if $\hat{\theta}$ is normally distributed, then

$$\beta = P(A_1|H_1) = P(\hat{\theta} \geq K) = 1 - N\left(\frac{K - \theta_1}{SE(\hat{\theta})}\right) \tag{10.11}$$

where θ_1 is any particular simple alternative hypothesis in the composite H_1.

If, for example, we assume in our illustrative case, that $\mu_1 = 30$ pounds, then by (10.11)

$$\beta = 1 - N\left(\frac{K - \mu_1}{\sigma_{\bar{x}}}\right) = 1 - N\left(\frac{29.175 - 30}{0.5}\right) = 1 - N(-1.65) = 0.95$$

which indicates that the probability of taking A_1 when H_1 is true is as high as 0.95. It is, in other words, impossible to distinguish between H_0 and H_1 with $\mu_1 = 30$. This is a natural result since we have assumed here that $\mu_0 = \mu_1 = 30$.

Now, suppose we let $\mu_1 = 28.175$, then

$$\beta = 1 - N\left(\frac{29.175 - 28.175}{0.5}\right) = 1 - N(2) = 0.0228$$

which reveals that our ability to distinguish between $\mu_0 = 30$ and $\mu_1 = 28.175$ is very good. Why?

In similar fashion, β risks for any other simple hypothetical value in H_1 can be computed. A few such values are given in the first two columns of Table 10.6. β risks for $\mu_1 = 30$, 29, and 28 are graphed in Figure 10.5 so that the reader can get some feel for the problem of Type II errors in terms of areas.

The power function: For our previous discussion we have arbitrarily stated that $\alpha = 0.05$. However, the null hypothesis under investigation, $\mu \geq 30$, is composite, and there is actually no uniquely defined α risk for it. The probability of rejecting H_0 when it should not be rejected depends upon which simple hypothesis in H_0 we use; if any one of the simple hypotheses is true, it should not be rejected, and the probability of rejection varies from one simple hypothesis to another. If, when $\mu_0 = 30$, $\alpha = 0.05$, then when $\mu_0 > 30$, $\alpha > 0.05$; that is, if we allow μ_0 to vary, though the testing statistic and critical region remain the same as stated in the general procedure, Type I error should be interpreted as the maximum value of the power of the test when H_0 is true. Namely, the variation of the

TABLE 10.6

H_1	β	$1 - \beta$
30.0	0.95	0.05
29.5	0.74	0.26
29.0	0.48	0.52
28.5	0.09	0.91
28.0	0.01	0.99

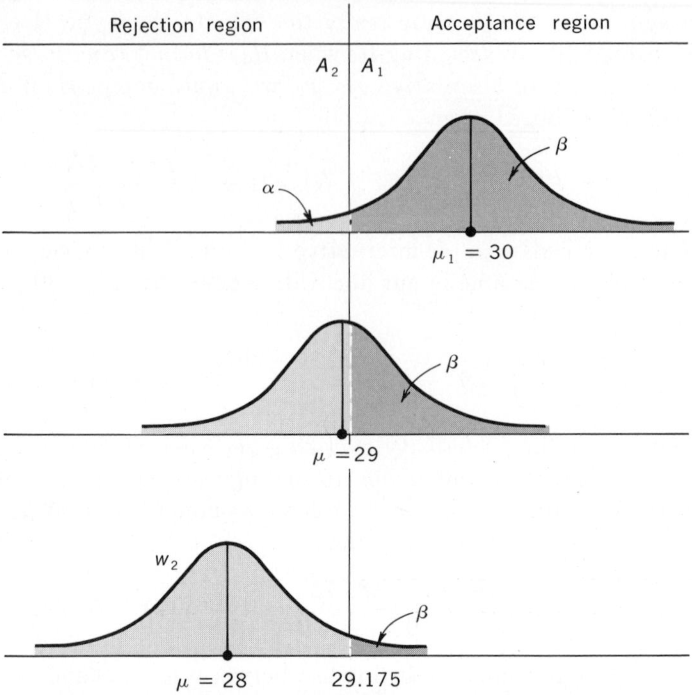

Figure 10.5 Type II error for testing $\mu \geq 30$ against $\mu < 30$.

probability of rejection, as a function of which the simple hypothesis is assumed to be true, is embodied in the power function.

The power function of a statistical test, as has already been noted, is the probability of making the correct decision of selecting A_2 when the true state of nature is H_1. Moreover, we know that

$$\underline{\text{Power function} = P(A_2|H_1) = 1 - \beta}$$

But β risk can be uniquely defined only if the alternative is a simple hypothesis. When, as in our illustrative example, the alternative hypothesis is composite, β risk would vary from one simple hypothesis to another in H_1. As a consequence, the power of a test would also vary with the variations in β. The power curve for our present test is given by Figure 10.6, which is suggested by the few values of $1 - \beta$ in Table 10.6.

Figure 10.6 shows that the farther away the simple hypothesis in H_1 is from H_0, the easier it is to distinguish H_1 from H_0, and vice versa. That is, the farther H_1 is from H_0, the greater is the power (probability) to detect that H_1 is the true state of nature when it is in fact the case. We must keep in mind that this particular power curve is obtained by fixing μ_0 at 30, α at 0.05, n at 36, and $\bar{x} < K$ at 29.175, but allowing μ_1 to vary.

In the ideal situation, when H_0 is true, the probability of rejecting it should be 0; and when H_0 is false, the probability of rejecting it should be 1. In other words,

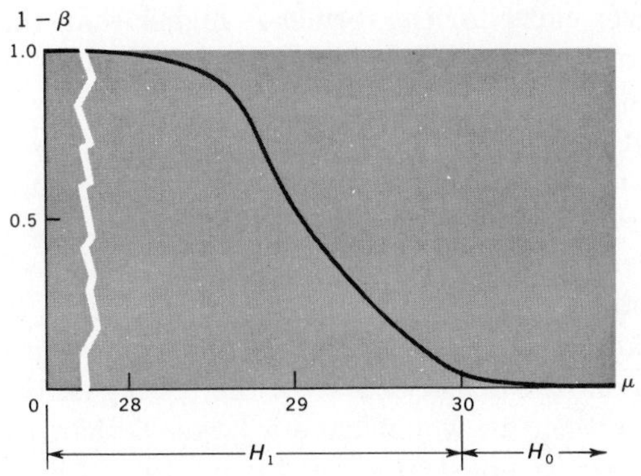

Figure 10.6 Power function for testing $H_0: \mu \geq 30$ against $H_1: \mu < 30$ at $\alpha = 0.05$ and with $n = 25$.

when we test $\theta \geq \theta_0$ against $\theta < \theta_0$, the ideal power function should take the form shown in Figure 10.7.

Typically, however, a power curve for a fixed n looks similar to the one in Figure 10.6. It is low over H_0 and high over H_1. This pattern indicates that when the simple hypothesis in H_1 deviates farther from H_0, the power of our decision rule increases or the probability of making the right decision increases.

Given α and n, one decision rule is said to be better or more powerful than

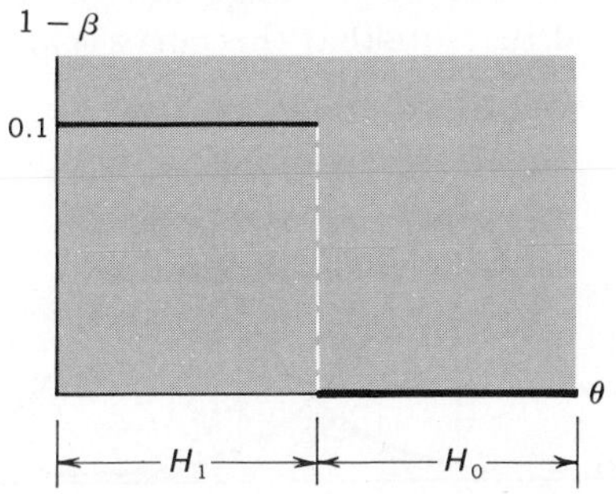

Figure 10.7 Ideal power curve for testing $\theta \geq \theta_0$ against $\theta < \theta_0$.

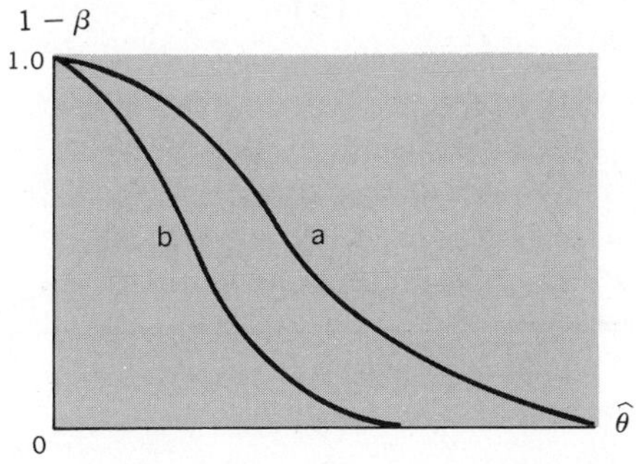

Figure 10.8 Typical power curves.

another if the power curve for the former is higher than that for the latter. For example, in Figure 10.8, the test which has power curve (a) is better than the test which has power curve (b). In this connection, we say that a decision rule is the *uniformly most powerful test* if it yields the highest power curve possible.

It turns out, other things being equal, that the power function of a test depends upon the selection of the critical region (decision rule). If the critical region is at either tail end (if the test is one-sided) of the distribution of the testing statistic, then the test would be uniformly most powerful, since then β is the smallest possible, given α and n. If, however, the critical region consists of any other portion of the area under the density curve of the testing statistic, the test would be less powerful, since then we would have a larger β than the minimum possible. These statements are demonstrated by Figure 10.9, which shows that while α values in (a) and (b) are identical, the critical region in (b) is unwisely chosen and the effect on β is disastrous.

The operating characteristic function: In practice, instead of a curve showing the power function of a test, a curve called the *operating characteristic function* (usually abbreviated to "OC function") is shown. The OC function of a test shows the probability of accepting H_0 as a function of the hypothesized parameter used to compute β. Since the probability of acceptance and that of rejection of H_0 must add up to unity, the OC function is clearly the reverse of the power function. Thus, we see that the OC function $=$ 1 $-$ power function $= \beta$.

The curve for the operating characteristic function, called the OC curve, for $n = 36$, $\alpha = 0.05$, and $K = 29.175$, as suggested by the few values of β in Table 10.6, is shown in Figure 10.10. Note that this curve is low over H_1, indicating a low

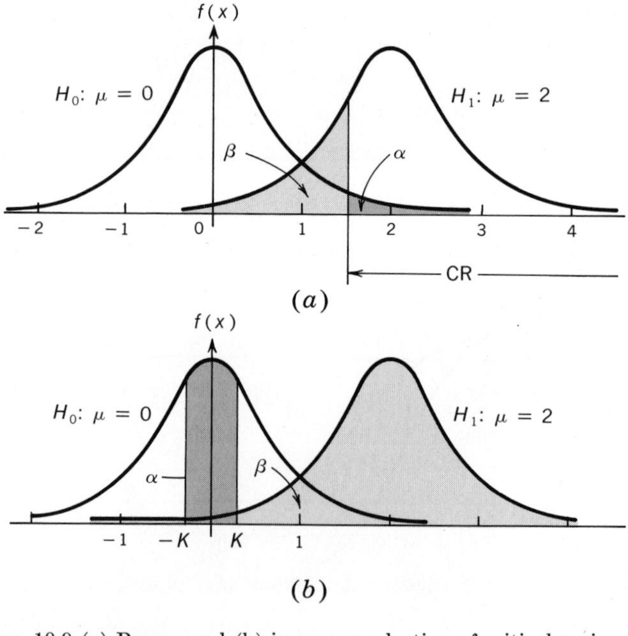

Figure 10.9 (a) Proper and (b) improper selection of critical region, CR.

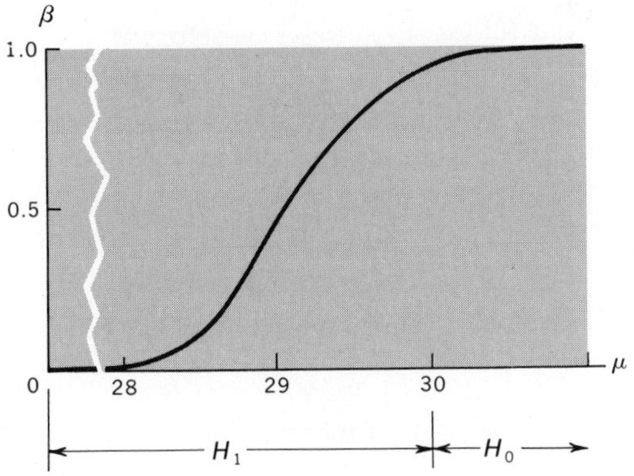

Figure 10.10 The OC curve for testing $\mu \geq 30$ against $\mu < 30$ with $n = 25$, $\alpha = 0.05$, and $K = 29.175$.

probability of accepting H_0 when H_1 is true, and high over H_0, indicating a high probability of accepting H_0 when H_1 is false.

In general, an OC curve reveals how the test or decision rule operates, regardless of whether that operation is correct or not. The ideal OC curve, corresponding to the ideal power curve, for a left-tail test would be one as portrayed by Figure 10.11.

While OC and power functions are asymmetric names, their difference in reality is arbitrary since H_0 and H_1 are arbitrarily designated. The OC and power functions for H_0: $\theta \geq \theta_0$ and H_1: $\theta < \theta_0$ would become the power and OC functions for the same pair of hypotheses with the null and the alternative hypotheses interchanged.

Typically, OC curves are similar to those in Figure 10.12. The height of an OC curve, as has already been mentioned, is the probability of accepting H_0. As such, when the curve is high near θ_0 and decreases rapidly as we move away from θ_0, as represented by curve (a) in Figure 10.12, the decision rule is said to be a good one because then the ability of the decision rule to distinguish θ_0 from θ_1 is good. On the other hand, a decision rule is said to be poor if the OC curve decreases slowly and remains high, as shown by curve (b) in Figure 10.12, for values far away from θ_0, and our ability to distinguish H_1 from H_0 is very weak.

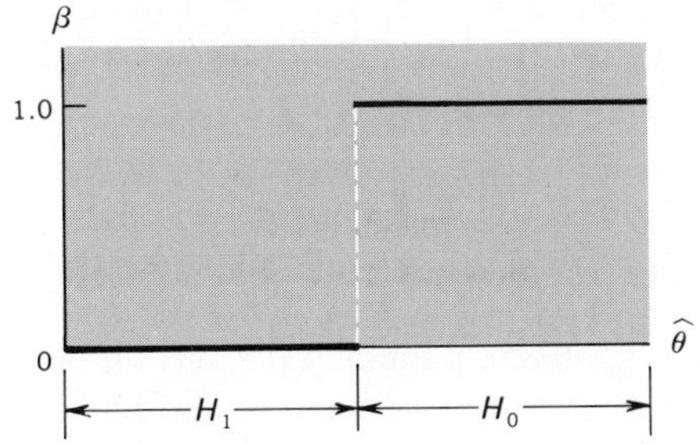

Figure 10.11 Ideal OC curve for testing $H_0{:}\,\theta \geq \theta_0$ against $H_1{:}\,\theta < \theta_0$.

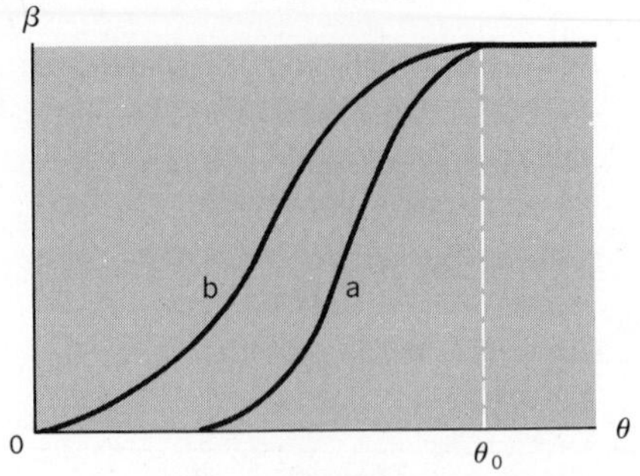

Figure 10.12 Typical OC curves.

To summarize: The decision rule established by the first approach is as follows:

(1) Take a random sample of size $n = 36$.
(2) Set α at 0.05.
(3) If $\bar{x} \geq 29.175$, take A_1: buy;
 $\bar{x} < 29.175$, take A_2: do not buy.
(4) β risks for various simple alternative hypotheses are shown by 10.10; or the power of this test is shown by Figure 10.6.

TESTING THE HYPOTHESIS $\mu \leq 30$

According to this approach, the purchasing agent is anxious to avoid the mistake of buying when he should not. In view of the hypotheses stated now, the α and β risks are as shown below:

	State of Nature	
Action	$\mu_0 \leq 30$	$\mu_1 > 30$
A_1: Do not buy	Correct action	$\beta = P(A_1 \mid \mu > 30)$
A_2: Buy	$\alpha = P(A_2 \mid \mu \leq 30)$	Correct action

In this case, α risk is the error of buying when the purchase should not be made, and β risk is the error of not buying when the purchase should be made.

Now, assuming $n = 36$, $\alpha = 0.05$, and $\sigma = 3$ (as before) for a right-tail test, as suggested by (10.4) or by information in Table 10.4, the null hypothesis $\mu \leq 30$ will be rejected if and only if

$$\bar{x} > K = \mu_0 + 1.65\sigma_{\bar{x}} = 30 + 1.65(0.5) = 30.825$$

From this critical value, the decision rule now becomes

If $\bar{x} \leq 30.827$ pounds, take A_1: do not buy.
If $\bar{x} > 30.825$ pounds, take A_2: buy.

To calculate β for this test, we note that, in general, when we conduct a right-tail test the null hypothesis is accepted if and only if $\hat{\theta} < K$. Therefore,

$$\beta = P(A_1|H_1) = P(\hat{\theta} < K) = N\left(\frac{K - \theta_1}{SE(\theta)}\right) \tag{10.12}$$

Now, applying this formula to our present case by assuming, say, $\mu_1 = 30.500$, we have

$$\beta = N\left(\frac{K - \mu_1}{\sigma_{\bar{x}}}\right) = N\left(\frac{30.825 - 30.500}{0.5}\right) = N(0.65) = 0.7422$$

A few more β risks, calculated in similar fashion for this decision rule, are entered in Table 10.7. Power values corresponding to these β risks are given in the same table.

TABLE 10.7

H_1	β	$1 - \beta$
30.0	0.95	0.05
30.5	0.74	0.26
31.0	0.48	0.52
31.5	0.09	0.91
32.0	0.01	0.99

The power curve and OC curve for this test are shown in Figures 10.13 and 10.14, respectively. It is interesting to note that these are the OC and power curves, respectively, for the previous test. Intuitively, we can also see that the second approach is more conservative than the first, since the critical value of $\bar{x}$ in the second is larger (30.825) than that in the first (29.175). That is, it is easier to take the act of buying from the first approach than from the second approach. For example, if from a sample of 36, we obtain $\bar{x} = 29.75$ pounds, by the first decision rule we would decide to buy, while by the second decision rule we would decide not to buy.

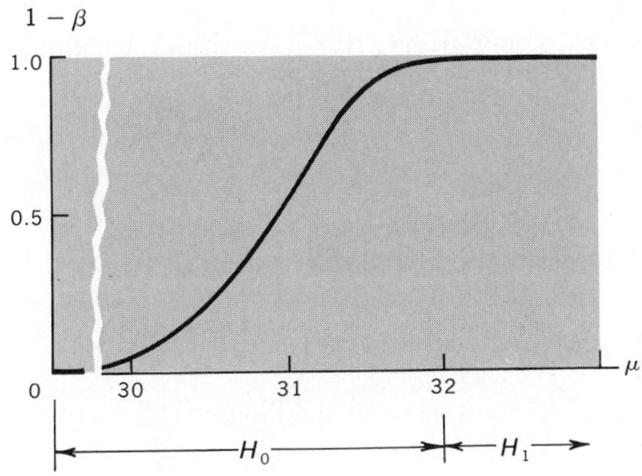

Figure. 10.13 Power curve for testing $\mu \leq 32$ against $\mu > 32$ with $n = 36$ and $\alpha = 0.05$.

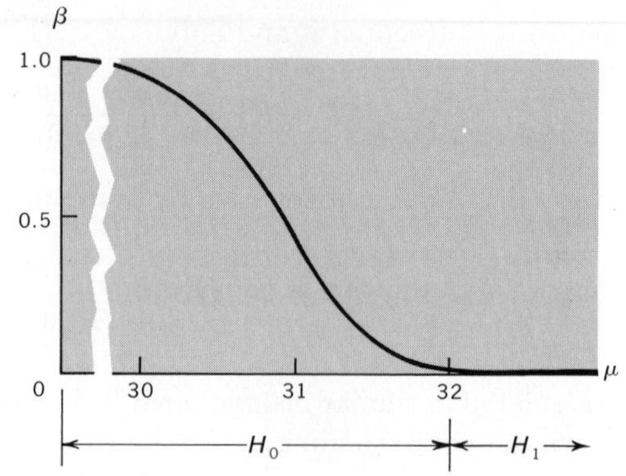

Figure 10.14 OC curve for testing $\mu \leq 32$ against $\mu > 32$ with $n = 36$ and $\alpha = 0.05$.

CALCULATION OF β RISKS FOR TWO-TAIL TESTS

The mean monthly family income in a certain city was known to be \$400, with a standard deviation of \$50, during the late 1950s. The government of that city now wishes to ascertain whether there has been any significant change in this average, and plans to do so by taking a random sample of 100 families and testing the following hypotheses at $\alpha = 0.05$:

$$H_0: \mu = 400$$
$$H_1: \mu \neq 400$$

Under the conditions specified, the decision rule (as suggested by Table 10.4) becomes: The null hypothesis will be rejected if and only if

$$\bar{x} < K_1 = \mu_0 - 1.96\sigma_{\bar{x}} = 400 - 1.96\left(\frac{50}{\sqrt{100}}\right) = \$390.20, \text{ or}$$

$$\bar{x} > K_2 = \mu_0 + 1.96\sigma_{\bar{x}} = 400 + 1.96(5) = \$409.80$$

In general, for a two-sided alternative hypothesis, we would accept $H_0: \theta = \theta_0$ if and only if $K_1 < \hat{\theta} < K_2$ and, therefore,

$$\beta = P(H_0|H_1) = P(K_1 < \hat{\theta} < K_2) = N\left(\frac{K_2 - \theta_1}{SE(\theta)}\right) - N\left(\frac{K_1 - \theta_1}{SE(\theta)}\right) \quad \textbf{(10.13)}$$

Now, in the present case, if the alternative value is assumed to be $\mu_1 = 395$, then

$$\beta = N\left(\frac{K_2 - \mu_1}{\sigma_{\bar{x}}}\right) - N\left(\frac{K_1 - \mu_1}{\sigma_{\bar{x}}}\right) = N\left(\frac{409.8 - 395}{5}\right) - N\left(\frac{390.2 - 395}{5}\right)$$
$$= N(2.96) - N(-0.96) = 0.9985 - 0.1685 = 0.83$$

If, $\mu_1 = 415$, then

$$\beta = N\left(\frac{409.8 - 415}{5}\right) - N\left(\frac{390.2 - 415}{5}\right) = N(-1.04) - N(5) = 0.1492$$

TABLE 10.8

H_1	β	$1 - \beta$
375.0	0.001	0.999
382.5	0.062	0.938
387.5	0.340	0.660
392.5	0.677	0.323
397.5	0.926	0.074
400.0	0.950	0.050
402.5	0.926	0.374
407.5	0.677	0.323
412.5	0.340	0.660
417.5	0.062	0.938
425.0	0.001	0.999

β risks for a few other values of μ_1 are calculated in the same manner and entered in Table 10.8. Figure 10.15 gives areas of β for $\mu_1 = 400$, 395, 415, and 375. The power and OC curves for this test are shown by Figures 10.16 and 10.17, respectively. It is interesting to note that both β and $1 - \beta$ are symmetrical about μ_0 in the case of two-sided alternatives.

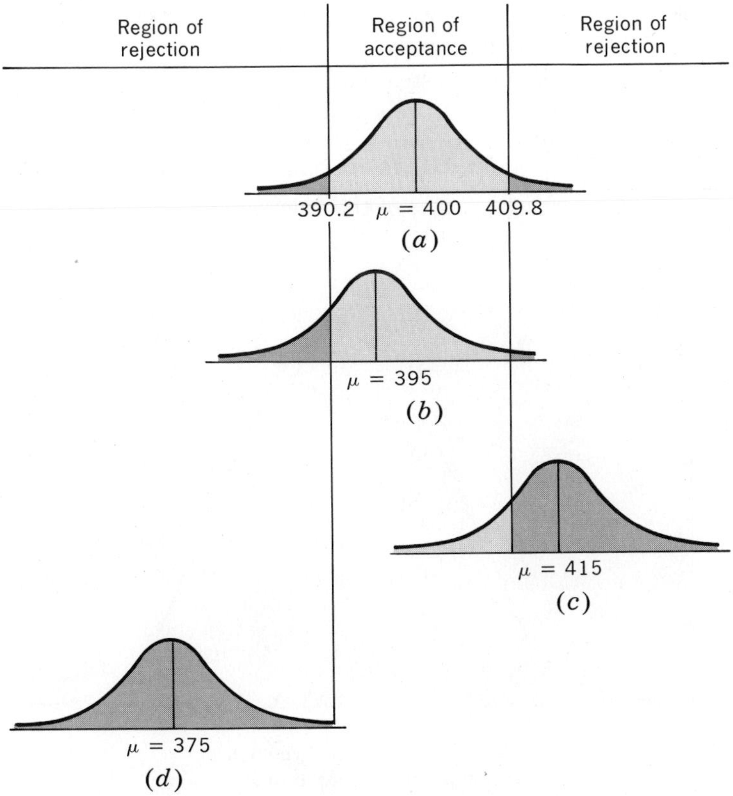

Figure 10.15 The probability of accepting the hypothesis that $\mu = \$400$ under the assumption of various alternative values of μ.

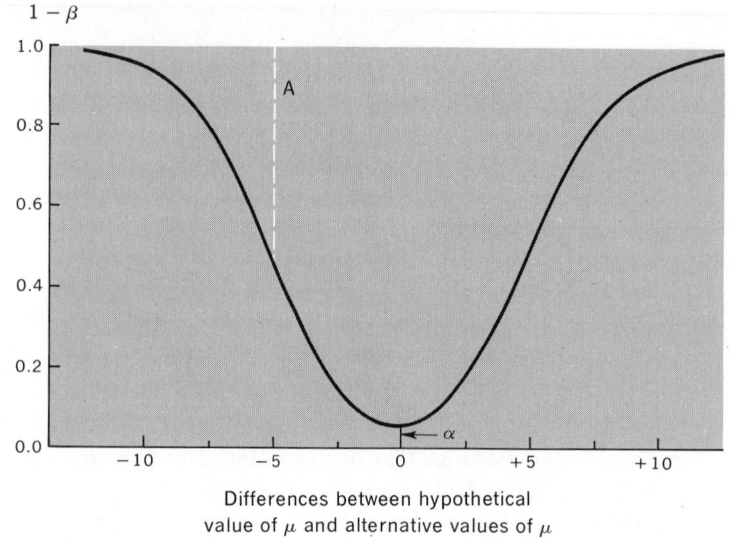

Figure 10.16 Power curve for $H_0{:}\mu = \$400$ and $H_0{:}\mu \neq \$400$.

In general, as we have already noted, single-tail tests are uniformly most powerful tests. There is no uniformly most powerful test with the two-sided alternative, but a two-tail test is a reasonable one on many counts. First of all, from the OC curve (Figure 10.17), we see that this test is unbiased in the sense that the largest probability of accepting H_0 is at H_0. This property is not possessed by single-tail tests. Furthermore, as Figure 10.18 shows, the two-tail test has some chance of detecting excesses of the true value of the parameter above θ_0,

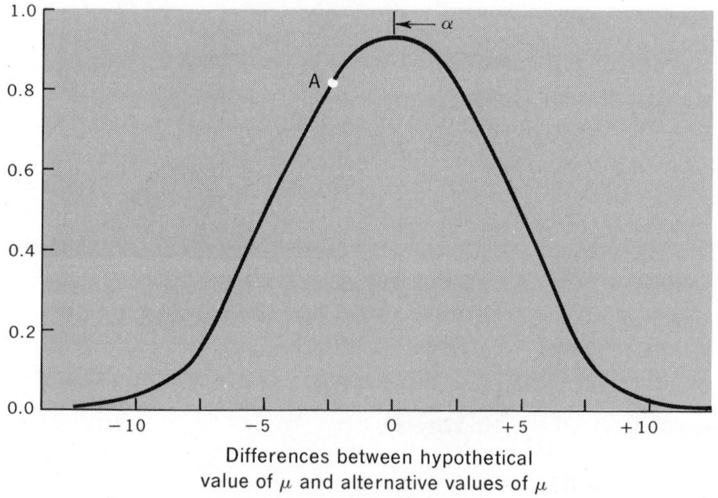

Figure 10.17 Operation characteristic curve for $H_0{:}\mu = \$400$ and $H_0{:}\mu \neq \$400$.

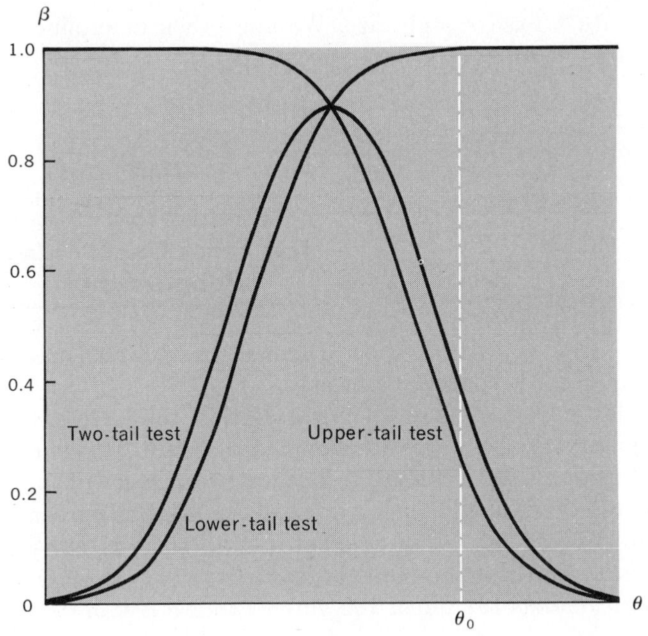

Figure 10.18 OC curves of three tests about μ with $H_0\colon \theta = \theta_0$ at $\alpha = 0.01$.

though not as good a chance as the right-tail test. However, unlike the right-tail test, the two-tail test has a fairly good chance of detecting shortages of the true parameter below θ_0. Similar comparisons can be made between a left-tail test and a two-tail test. Thus, in a given situation, the test selected should be the one whose OC curve comes closest to giving the proper decisions for various values of θ that might prevail.

RELATIONSHIP BETWEEN α AND β FOR FIXED n

From our discussion of β so far, we may note the following conclusions:

(1) For a given sample size, the probability of committing a Type II error is automatically determined, once the level of significance is selected.

(2) For a given sample size, if it is desired to reduce the error of the first kind, the error of the second kind is correspondingly increased.

(3) A Type I error is completely under control in a given test. Furthermore, the common practice of choosing $\alpha = 0.05$ or 0.01 indicates that the error of rejecting a true null hypothesis is seldom made. A Type II error is frequently made if the true value of a parameter is close to θ_0. However, in such a situation, the mistake of accepting a false null hypothesis is not serious. (Why?) When the true parameter deviates greatly from θ_0, it is a very serious matter to commit a Type II error. Fourtunately, however, in such a situation, the probability of accepting a false null hypothesis is very slight.

(4) If it is desired to fix α and β at specific low levels, a statistical test cannot

be conducted with a fixed sample size. We must then determine the particular size of sample that could satisfy these specifications. The method of determining n, given α and β, is discussed in Section 10.8.

10.7

Evaluation of β for Tests Concerning Proportions

Decision procedures concerning proportions are similar to those concerning means.

During the last days preceding a political election, the campaign manager for a certain candidate wants to decide whether it is necessary to expend some extra effort. Toward this end, he decides that if at least 51 percent of the population of voters are for his candidate, then no extra effort need be contemplated; but if the percentage is 51 or less, extra effort should be made in order to ensure the election of his candidate. Furthermore, he decides to verify the population proportion by taking a random sample of 100 voters at an α risk of 0.05. For this procedure, the α and β risks are then as follows:

	State of Nature	
Action	$\pi_0 \leq 0.51$	$\pi_1 > 0.51$
A_1: Extra effort	Correct decision	$\beta = P(A_1 \mid \pi_1)$
A_2: No extra effort	$\alpha = P(A_2 \mid \pi_0)$	Correct decision

According to this decision framework, H_0: $\pi \leq 0.51$ will be rejected if and only if

$$p > K = \pi_0 + 1.65\sigma_p = 0.51 + 1.65\sqrt{\frac{(0.51)(0.49)}{100}} = 0.51 + 1.65(0.05) = 0.5925$$

The operating characteristic function is now a function of π, since we must here identify the values of π and the simple admissible hypotheses. Assuming $\pi_1 = 0.5$, then

$$\text{OC function} = \beta = P(H_0|H_1) = P(p < K) = N\left(\frac{K + \frac{1}{2n} - \pi_1}{\sigma_p}\right)$$

$$= N\left(\frac{0.5925 + 0.005 - 0.5}{0.05}\right) = N(1.95) = 0.9744$$

Note that a continuity correction factor of $\frac{1}{2}n$ has been introduced in the above computation. Also note that the OC curve, identified as $p > 0.5925$ in Figure 10.19, gives a high probability of acceptance over H_0 and a low probability of acceptance over H_1, as should be the case.

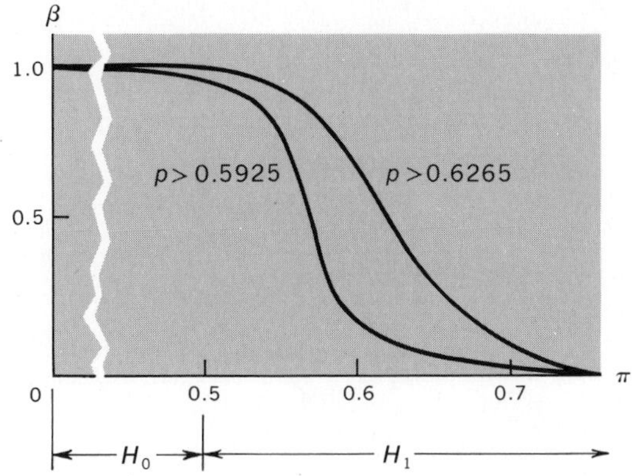

Figure 10.19 OC curves for the political campaign example at $\alpha = 0.05$, $\alpha = 0.01$, and $n = 100$.

The campaign manager's decision rule, therefore, is as follows: (1) Take a random sample of 100 voters; (2) take A_1 (extra effort) if $p \leq 0.5925$, and take A_2 (no extra effort) if $p > 0.5925$; (3) the level of significance is 0.05 and the OC function is given by Figure 10.19; (4) if there are, say, 55 out of 100 in favor of the candidate, A_1 should be the choice, but if there are, say 60 out of 100 in favor of the candidate, A_2 should be selected.

Now, if for some reason, the campaign manager wishes to conduct this test with $\alpha = 0.01$, then H_0 would be rejected if and only if

$$p > K = 0.51 + 2.33(0.05) = 0.6265$$

Again, suppose $\pi_1 = 0.5$, then for this critical value,

$$\text{OC function} = \beta = N\left(\frac{0.6265 + 0.005 - 0.5}{0.05}\right) = N(2.63) = 0.9957$$

Thus, as observed before, given the sample size, in an attempt to reduce α, β is correspondingly increased. So that the whole OC curve for $\alpha = 0.01$ and $n = 100$, the one labeled as $p > 0.6265$ in Figure 10.19, lies above that of the previous decision rule.

In the preceding example, if the campaign manager is more anxious not to make the extra effort because of the heavy costs it would entail, then the appropriate test becomes

$$H_0\colon \pi \geq 0.51$$
$$H_1\colon \pi < 0.51$$

The α and β risks are now as shown below:

	State of Nature	
Action	$\pi_0 \geq 0.51$	$\pi_1 < 0.51$
A_1: No extra effort	—	$\beta = P(A_1 \mid \pi_1)$
A_2: Extra effort	$\alpha = P(A_2 \mid \pi_0)$	—

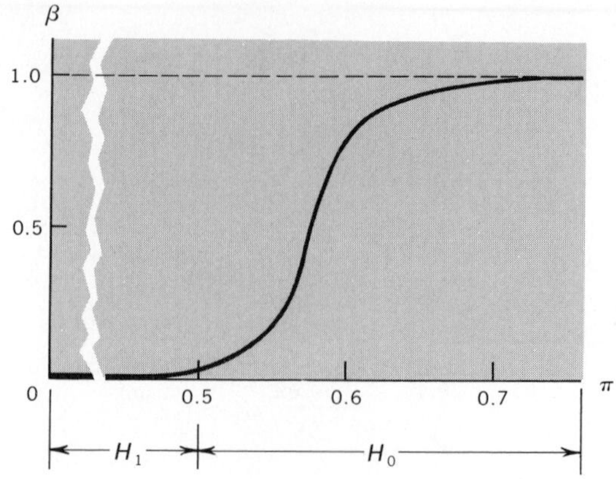

Figure 10.20 OC curve for the political campaign example with $H_0: \pi \geq 0.51$ at $\alpha = 0.05$ and $n = 100$.

According to this framework, the campaign manager will decide not to make the extra effort if there is evidence that more than 51 percent of the voters are in favor of his candidate. Let $\alpha = 0.05$ and $n = 100$ as before; H_0 will be rejected now if and only if

$$p < K = 0.51 - 1.65(0.05) = 0.4275$$

The decision rule here is: If $p < 0.4275$, take A_2 (make extra effort), and if $p \geq 0.4275$, take A_1 (make no extra effort). For this rule, $\alpha = 0.05$ and the OC function is

$$\text{OC function} = \beta = P(A_1 | \pi < 0.51) = P(p \geq K)$$
$$= 1 - N\left(\frac{0.4275 + 0.005 - \pi_1}{0.05}\right)$$

whose graph is shown by Figure 10.20. Note that this OC curve is the power curve for testing $H_0: \pi \leq 0.51$ against $H_1: \pi > 0.51$ for the same level of significance of 0.05 and the same size of sample of 100.

10.8

Tests with Fixed α and β

Our evaluation of β has been made so far by assuming that both α and n are arbitrarily fixed. Now, given the level of significance, how would β be affected if the sample size were increased?

EFFECTS ON β OF CHANGING n

If the sample size is increased, the size of the β risk will be reduced. The reason for this is obvious: the larger the sample, the smaller the standard error of the testing statistic. And, the smaller the standard error, the more compressed the sampling distribution of the testing statistic becomes. If, for example, the student looks at Figure 10.15 reflectively, he will realize that the movement of a given distance along the horizontal axis will remove a greater area of each of the density curves from the region of acceptance, or the smaller the area measuring β will become. This is to say that the larger the sample size, the greater the ability of our decision rule to distinguish the null hypothesis from the alternative hypothesis. Let us illustrate this point by way of an example.

Iron bars produced by a certain process are known to have a mean breaking strength of 300 pounds and a standard deviation of 24 pounds. To verify a belief that the mean breaking strength can be considerably increased by a newly developed manufacturing process, it seems appropriate to test the following pair of hypotheses:

$$H_0: \mu = 300 \text{ pounds}$$
$$H_1: \mu > 300 \text{ pounds}$$

Now, suppose we arbitrarily set α at 0.01 for this test, and let n vary from 10 to 100 to 500 for the purpose of observing the effects upon β of changing n. If $n = 10$, then H_0 will be rejected if

$$\bar{x} > K = 300 + 2.33 \frac{24}{\sqrt{10}} = 300 + 2.33(7.59) = 317.68$$

If $n = 100$, then H_0 will be rejected if

$$\bar{x} > K = 300 + 2.33 \frac{24}{\sqrt{100}} = 300 + 2.33(2.4) = 305.59$$

If $n = 500$, then H_0 will be rejected if

$$\bar{x} > K = 300 + 2.33 \frac{24}{\sqrt{500}} = 300 + 2.33(1.07) = 302.49$$

The OC function or the β risks for the various possible simple alternative hypotheses in the composite H_1 can be computed with these critical values, a few of which are given in Table 10.9.

The OC curves of this test at $\alpha = 0.01$, with $n = 10$, 100, and 500 respectively,

TABLE 10.9

	OC Function: β		
H_1	$n = 10$	$n = 100$	$n = 500$
290	1.000	1.000	1.000
295	0.999	1.000	1.000
300	0.990	0.990	0.990
305	0.953	0.749	0.010
310	0.844	0.159	0.000
315	0.639	0.004	0.000
320	0.378	0.000	0.000

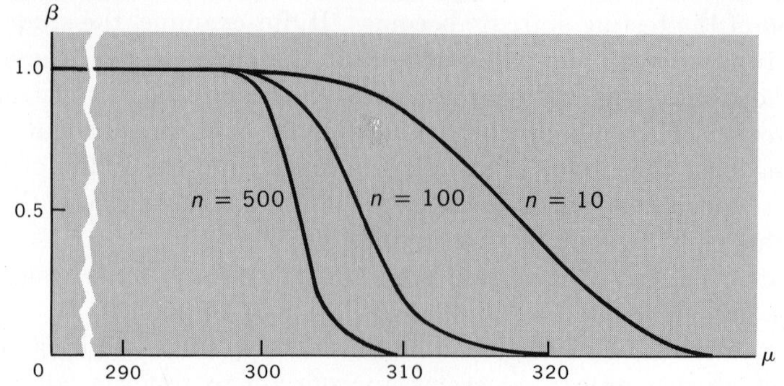

Figure 10.21 Effects of sample size on OC functions.

are given in Figure 10.21. See how the OC curve approaches its ideal form as n increases. These curves reveal that the largest sample gives the most discriminating test—the most nearly ideal test.

DETERMINING n FOR FIXED α AND β

Having noted the effects upon β of changing n, with α fixed, we can see why an appropriate sample size can be found so that prescribed α and β risks can be met.

Suppose that in the salesmen's commission illustration evaluated in Section 10.3 the management of the floor-wax factory decides that remedial actions are required if its salesmen's monthly mean commission is \$560, and that no remedial action will be taken if the mean is greater than this figure. Furthermore, suppose that the management is willing to assume an α risk of 0.05 and a β risk of 0.10. Then the decision framework can be summarized as follows:

Action	State of Nature	
	H_0: $\mu_0 = 600$	H_1: $\mu_1 = 560$
A_1: No remedial action	Correct decision	$\beta = 0.10$
A_2: Remedial action	$\alpha = 0.05$	Correct decision

To determine the required sample size, we first note that H_0 will be rejected if and only if

$$(1)\quad \bar{x} < K = \mu_0 - Z_{1-\alpha}\sigma_{\bar{x}} = 600 - 1.65\frac{80}{\sqrt{n}} = 600 - \frac{132}{\sqrt{n}}$$

Next, recall that β is the probability of accepting the null hypothesis when the alternative hypothesis is true. In this case, where β is specified to be 0.10, H_0 will be accepted only if

$$(2)\quad \bar{x} > K = \mu_1 + Z_{\beta}\sigma_{\bar{x}} = 560 + 1.28\frac{80}{\sqrt{n}} = 560 + \frac{102.4}{\sqrt{n}}$$

Note that $Z_\beta = 1.28$ here, since 10 percent of the values of $\bar{x}$ are located at the tail end of the density curve for $\bar{x}$ about 1.28 standard errors away from $E(\bar{x}) = 560$.

Finally, the values of n and K that will satisfy both of these two equations will furnish us with a basis for our decision rule. We can solve for n by equating both expressions for K as follows:

$$600 - \frac{132}{\sqrt{n}} = 560 + \frac{102.4}{\sqrt{n}}$$

$$40 = \frac{132 + 102.4}{\sqrt{n}}$$

$$\sqrt{n} = \frac{234.4}{40} = 5.86$$

$$n = 34.34 \text{ or } 35$$

The critical value for $\bar{x}$ can now be found by substituting n into either of the two equations for K. For example,

$$K = 560 + \frac{102.4}{\sqrt{35}} = 560 + \frac{102.4}{5.91608} = 577.31$$

With n and K so determined, the decision rule may now be stated: (1) Take a random sample of size $n = 35$; (2) compute the sample mean; (3) if $\bar{x} > \$577.31$, take A_1: no remedial actions are required (the probability that this action may be the wrong one is $\beta = 0.10$); (4) if $\bar{x} \leq \$577.31$, take A_2: start remedial actions (the probability that this is the wrong decision is $\alpha = 0.05$).

To verify that n and K, determined before, can satisfy the desired α and β risks, we note that

$$\alpha = P(\text{Reject } H_0 \text{ given that } H_0 \text{ is true}) = P(\bar{x} \leq 577.31|\mu = 600)$$

$$= N\left(\frac{577.31 - 600}{80/\sqrt{35}}\right) = N(-1.65) = 0.05$$

and that

$$\beta = P(\text{Accept } H_0 \text{ given that } H_1 \text{ is true}) = P(\bar{x} > 577.31|\mu = 560)$$

$$= 1 - N\left(\frac{577.31 - 560}{80/\sqrt{35}}\right) = 1 - N(1.28) = 1 - 0.8980 \doteq 0.10$$

In deciding whether a lot of 100,000 rifles should be accepted, the U. S. Army adopts the following decision rule: If only 1 percent of the lot is defective, accept. If 3 percent of the rifles in the lot are defective, reject. If the defective rifles comprise more than 1 percent or less than 3 percent of the lot, take another sample before making a final decision. Now, suppose we focus our attention on acceptance or rejection of the lot only; then the pertinent hypotheses become

$$H_0: \pi = 0.01$$
$$H_1: \pi = 0.03$$

Furthermore, if we assume that the Army is willing to take a risk of 0.01 to reject a useful lot, and to take a risk of 0.05 to accept a lot in which 3 percent of the items are defective, then the α and β errors are as shown below:

Action	State of Nature	
	$\pi_0 = 0.01$	$\pi_1 = 0.03$
A_1: Accept	Correct decision	$\beta = 0.05$
A_2: Reject	$\alpha = 0.01$	Correct decision

With $\alpha = 0.01$, H_0 will be rejected if and only if

$$p > K = \pi_0 + Z_{1-\alpha}\sigma_p = 0.01 + 2.33\sqrt{\frac{(0.01)(0.99)}{n}} = 0.01 + \frac{0.2318}{\sqrt{n}}$$

With $\beta = 0.05$, H_0 will be accepted if and only if

$$p \leq K = \pi_1 - Z_\beta\sigma_p = 0.03 - 1.65\sqrt{\frac{(0.01)(0.99)}{n}} = 0.03 - \frac{0.1642}{\sqrt{n}}$$

Solving these two equations simultaneously, we find

$$n = 392$$

and

$$K = 0.022$$

Thus, the stipulation that $\alpha = 0.01$ and $\beta = 0.05$ can be met by taking a random sample of 392 (size n). With this rule, A_1 (accept the lot) should be taken if $p \leq 0.022$, and A_2 (reject the lot) should be taken if $p > 0.022$.

A process adjusted to fill seasoned tomatoes into cans is said to be in control if the net weight in each can is within the range of 16 ounces $\pm$ 1 ounce. According to past experience, the standard deviation is quite stable and is equal to 0.5 ounces. When the process is in control, it is left alone; when the process is out of control, it is stopped so that adjustment can be made. To check whether the process is in control, from time to time, how large a sample should be taken if the management is willing to assume an α risk of 0.01 and a β risk of 0.05?

First, let us first summarize the decision framework here as follows:

Action	State of Nature	
	H_0: $\mu_0 = 16$	H_1: $\mu_1 = 15$ or $\mu_2 = 17$
A_1: Leave the process alone	Correct	$\beta = 0.05$
A_2: Stop the process	$\alpha = 0.01$	Correct

The α risk in this case is the probability of stopping the process when $\mu = 16$. This has been set at 0.01. To commit this mistake is to reject H_0; with a two-sided alternative, H_0 will be rejected if and only if

$$\bar{x} < K_1 = 16 - 2.58\frac{0.5}{\sqrt{n}} = 16 - \frac{1.29}{\sqrt{n}}$$

or

$$\bar{x} > K_2 = 16 + 2.58\frac{0.5}{\sqrt{n}} = 16 + \frac{1.29}{\sqrt{n}}$$

These are the relationships n, K_1, and K_2 must satisfy in order to meet the required probability, 0.01, for stopping the process.

When we direct our attention to β, we note that β now is the probability of leaving the process alone when $\mu = 15$ or when $\mu = 17$. Since β is set at 0.05, the management is willing to assume the risk of leaving the process alone 2.5 percent of the time if $\mu = 15$, and it is willing to assume the same risk (at the level of 0.025) if $\mu = 17$. But this error of the second kind is committed only if H_0 is accepted, and the acceptance of H_0 requires that

$$\bar{x} > K_1 = 15 + 1.96\frac{0.5}{\sqrt{n}} = 15 + \frac{0.98}{\sqrt{n}}$$

or that

$$\bar{x} < K_2 = 17 - 1.96\frac{0.5}{\sqrt{n}} = 17 - \frac{0.98}{\sqrt{n}}$$

We now have two equations for each of the two critical values. The required sample size can be obtained by equating either the pair of expressions for K_1 or that for K_2. Thus, for example, by equating the two statements for K_1, we have

$$16 - \frac{1.29}{\sqrt{n}} = 15 + \frac{0.98}{\sqrt{n}} \qquad \sqrt{n} = 2.27 \qquad n \doteq 5$$

The critical values may be found by substituting n into either pair of the equations for K_1 and K_2.

$$K_1 = 15 + \frac{0.98}{\sqrt{5}} \qquad K_2 = 17 - \frac{0.98}{\sqrt{5}}$$

$$= 15 + 0.43 \qquad = 17 - 0.43$$

$$= 15.43 \text{ ounces} \qquad = 16.57$$

The decision rule for this example may now be summarized as follows: (1) Take a random sample of 5 cans from the production line, say, every hour, and determine the average weight of the contents; (2) if this average is between 15.43 and 16.57 ounces, leave the process alone; (3) if this average is either less than 15.43 or greater than 16.57 ounces, stop the process for readjustment.

To conclude our discussion on the determination of the sample size and the critical value (or values) so that given levels of α and β can be satisfied, we may make the following generalizations:

First, when we are concerned with a right-tail test we focus attention on two states of the world, θ_0 and θ_1, but with $\theta_0 < \theta_1$, the general expressions for determining n and K, are given as

$$K = \theta_0 - Z_\alpha SE(\hat{\theta})$$
$$K = \theta_1 + Z_\beta SE(\hat{\theta})$$

Second, when we are concerned with a left-tail test we also focus attention on two states of the world, but with $\theta_0 > \theta_1$, the equations yielding n and K then become

$$K = \theta_0 + Z_{1-\alpha} SE(\hat{\theta})$$
$$K = \theta_1 - Z_\beta SE(\hat{\theta})$$

Finally, when we conduct a two-tail test, we focus our attention on three states of the world, with θ_0 for the null hypothesis and θ_1 and θ_2 for the alternative hypothesis, with $\theta_1 < \theta_0 < \theta_2$. The general procedure is to let the two alternative values be equidistant but in opposite directions from θ_0. With this procedure, the specified value of α is distributed equally at both tail ends of the testing statistic, with $E(\hat{\theta}) = \theta_0$. The specified value of β is also equally divided; half is located at the right-tail end for the distribution of the testing statistic with $E(\hat{\theta}) = \theta_1$, and half is located at the left-tail end of the distribution of the testing statistic with $E(\hat{\theta}) = \theta_2$. For convenience, we may call the first and second parts of β, β_1 and β_2, respectively. Now, when α and β risks are specified, then n, K_1, and K_2 can be determined from the following equations:

$$K_1 = \theta_0 - Z_{\alpha/2} SE(\hat{\theta})$$
$$K_2 = \theta_0 + Z_{1-\alpha/2} SE(\hat{\theta})$$
$$K_1 = \theta_1 + Z_{\beta_1} SE(\hat{\theta})$$
$$K_2 = \theta_2 - Z_{\beta_2} SE(\hat{\theta})$$

In this set and in the previous two sets of equations, Z is the value of the standard normal deviate for an area, in one or both tails, equal to α or to β. Furthermore, all these expressions are suggested by those which we used to determine critical regions as summarized in Tables 10.4 and 10.5.

Glossary of Formulas

(10.1) $\alpha = P(H_1|H_0) = P(A_2|H_0) = P(I)$ α is the level of significance of a statistical test. It is the probability of committing a Type I error. Type I error is the rejection of the null hypothesis (or acceptance of the alternative hypothesis) when H_0 is in fact true. When a Type I error is made, the decision procedure will lead to taking the wrong action, A_2.

(10.2) $\beta = P(H_0|H_1) = P(A_1|H_1) = P(II)$ The probability of committing a Type II error is called β. Type II error refers to the acceptance of a false null hypothesis. To make an error of the second kind will lead to the wrong action, A_1.

(10.3) Power function $= 1 - \beta$ The power function of a test is the likelihood function for the critical region of the test and is related to Type II error, which is the probability of rejecting H_0 when H_1 is true. Thus, the power of a test is the probability of making the correct decision.

(**10.4**) $K = \theta_0 + Z_{1-\alpha}SE(\hat{\theta})$ and (**10.5**) $\dfrac{\hat{\theta} - \theta_0}{SE(\hat{\theta})} > Z_{1-\alpha}$

In testing against a right-tail alternative, H_0 is rejected if and only if the testing statistic has a value greater than K as defined by (10.4), or if and only if $Z > Z_{1-\alpha}$ as defined by (10.5). In both expressions, $Z_{1-\alpha}$ refers to the area at the upper tail of the density curve for the testing statistic which is equal to α. The testing statistic is normally distributed under the assumption that H_0 is true and Z is a standard normal variable.

(**10.6**) $K = \theta_0 - Z_\alpha SE(\hat{\theta})$ and (**10.7**) $\dfrac{\hat{\theta} - \theta_0}{SE(\hat{\theta})} < Z_\alpha$

The critical region for a left-tail test is defined by a critical value of the testing statistic which is less than K as defined by (10.6), or less than Z_α as given by (10.7). Z_α here refers to the area at the left-tail end of the density curve for $\hat{\theta}$, which is equal to α. Again, Z is assumed to be a standard normal variable.

(**10.8a**) $K_1 = \theta_0 - Z_{\alpha/2}SE(\hat{\theta})$ and (**10.8b**) $K_2 = \theta_0 + Z_{1-\alpha/2}SE(\hat{\theta})$

For a two-tail test, the critical region is defined for $\hat{\theta} < K_1$ and for $\hat{\theta} > K_2$. Common practice is to distribute α equally at both tail ends of the distribution of $\hat{\theta}$.

(**10.9a**) $\dfrac{\hat{\theta} - \theta_0}{SE(\hat{\theta})} < Z_{\alpha/2}$ and (**10.9b**) $\dfrac{\hat{\theta} - \theta_0}{SE(\hat{\theta})} > Z_{1-\alpha/2}$

These two expressions define the critical region for a two-tail test in terms of the standard normal distribution. Now, H_0 is rejected if and only if $Z < Z_{\alpha/2}$ or $Z > Z_{1-\alpha/2}$.

(**10.10**) $\hat{\sigma}_{\Delta p} = \sqrt{\bar{p}(1 - \bar{p})\dfrac{n_1 + n_2}{n_1 n_2}}$ This is an estimate for the standard error of a difference between two sample proportions. This formula is derived by assuming $\pi_1 = \pi_2 = \pi$ which, furthermore, is estimated by a weighted average between the two sample proportions defined as $\bar{p} = (n_1p_1 + n_2p_2)/(n_1 + n_2)$.

(**10.11**) $\beta = 1 - N\left(\dfrac{K - \theta_1}{SE(\hat{\theta})}\right)$ This expression is employed to calculate β when we are conducting a left-tail test; here θ_1 may be any one of the possible simple hypotheses in the composite H_1.

(**10.12**) $\beta = N\left(\dfrac{K - \theta_1}{SE(\hat{\theta})}\right)$ The formula for computing β when the alternative hypothesis is of the right-tail variety.

(**10.13**) $\beta = N\left(\dfrac{K_2 - \theta_1}{SE(\hat{\theta})}\right) - N\left(\dfrac{K_1 - \theta_1}{SE(\hat{\theta})}\right)$ With a two-sided test, β is calculated by this expression.

Problems

10.1 Differentiate the following pairs of concepts: (a) Type I and Type II errors; (b) null and alternative hypotheses; (c) α and β risks; (d) critical region and region of acceptance; and (e) confidence level and level of significance.

10.2 How does a single-tail test differ from a two-tail test?

10.3 Why is it possible, given the sample size, to adjust Type I and Type II errors properly by specifying the level of significance alone? How is it done?

10.4 What do we mean by "the uniformly most powerful test"? Why is there no uniformly most powerful test for a two-sided test?

10.5 What is the justification for the common practice of specifying α alone without explicit evaluation of β in statistical tests?

10.6 Sketch the ideal power and OC curves for an upper-tail test.

10.7 How is the power function related to the OC function? Are they different functions? Explain.

10.8 Why is a single-tail test biased, while a two-tail test is unbiased?

10.9 Given the level of significance, how is the power of a test affected by increasing sample size? Explain.

10.10 What generalizations, if any, can you make about the different ways of formulating null and alternative hypotheses, as developed in this chapter, in the following three cases? (1) Tests with α and n given without evaluating β. (2) Tests with α and β given with evaluation of β. (3) Tests with α and β specified with varying n.

10.11 We suspect that the IQs of the students at a certain college tend to run above the national average of 100. It is also known, according to past experience, that IQ is normally distributed with $\sigma = 10$.
a. If we wish to verify this by statistical testing, how should the null and the alternative hypotheses be formulated?
b. If a random sample of size $n = 25$ is drawn from the student body of the college and a mean IQ of 104 is observed, what conclusion can we make with $\alpha = 0.01$?

10.12 Medical reports have linked cigarette smoking with lung cancer and other ill effects. Suppose it is known to the cigarette producers that before these reports the adult smokers in this country had smoked 10 cigarettes per day on the average, with a standard deviation of 1.5 cigarettes. The producers are anxious to evaluate the effects of the medical reports upon the cigarette consumption of adult smokers. To do this, suppose that they have drawn a simple random sample of 144 adult smokers throughout the country and found that the sample mean was 8.5 cigarettes per day. How should the null and the alternative hypotheses be designated in this case? What conclusion can you draw with $\alpha = 0.01$ and in accordance with your hypotheses?

10.13 The Transit Authority of New York City uses tens of thousands of light bulbs every year. The brand that has been used up to now has a mean life of 1000 hours with a standard deviation of 90 hours. A new brand is offered to the Transit Authority at a price much lower than what it has been paying. The Authority decides that the new brand should be bought now, unless it has a mean life of less than 1000 hours at a level of significance of 0.05. Subsequently, 100 bulbs of the new brand are tested, yielding an average of 990 hours. Assuming that the standard deviation for the new brand is the same as that for the old, what should be the decision of the Transit Authority?

10.14 Forty-nine American soldiers observed at random yield a mean weight of 160 pounds, with a standard deviation (s) of 11 pounds. Are these observations consistent with the assumption that the mean weight of all American soldiers is 170 pounds?

10.15 In order for a manufactured part to fit other parts, a certain critical dimension is designed to be 5 inches. The variance of the manufacturing process is known to be 0.0064 inches^2. If a given day's sample of 49 yields $\bar{x} = 5.05$, what is the 95 percent confidence interval for μ? In view of this interval, should the null hypothesis, $\mu_0 = 5$, be rejected in favor of the alternative hypothesis, $\mu_1 \neq 5$?

10.16 Certain steel pipes are supposed to have a mean outside diameter of 10 inches with a standard deviation of 0.018 inches. We wish to test H_0: $\mu = 10$ against H_1: $\mu \neq 10$. If $\bar{x} = 9.94$ for a given day's sample of 36 pipes, what is the corresponding 99 percent confidence interval for μ? Does this interval indicate the rejection or acceptance of the null hypothesis at $\alpha = 0.01$?

10.17 A manufacturer claimed that at least 90 percent of the machine parts that it supplied to a factory conformed to specifications. An examination of 200 such parts revealed that 160 parts were not faulty. Determine if the manufacturer's claim is legitimate at the 1 percent level of significance.

10.18 A dairy is contemplating a change of its milk containers from glass to paper. The change will not be made, however, unless at least 70 percent of its customers so prefer. When a survey is made of 200 of its customers, 120 of them are in favor of the change. How should the dairy decide if it is willing to assume an α risk of 0.05?

10.19 A pair of dice is tossed 100 times and "sevens" turn up 23 times. Test the hypothesis that the dice are fair at $\alpha = 0.05$.

10.20 A man who plans to open a restaurant near a residential district of a certain city tells the bank, from which he wishes to borrow the necessary capital, that at least 50 percent of the residents in the district will patronize his restaurant from time to time after it opens. Suppose you are the loan officer in the bank and wish to verify this claim at $\alpha = 0.05$. Furthermore, suppose that in a random sample of 50 residents in the district, 44 percent indicate their intention to patronize the proposed restaurant.
a. What conclusions would you draw from the above result?
b. Suppose the sample had been 200 instead of 50, and the sample proportion were still 0.44, would you conclude differently?
c. Discuss the limitations of statistical testing in the light of your answers to (a) and (b).

10.21 A sample of 80 steel wires produced by Factory A yields a mean breaking strength of 1230 pounds, with a standard deviation of 120 pounds. A sample of 100 steel wires produced by Factory B yields a mean breaking strength of 1190 pounds, with a standard deviation of 90 pounds. Is there a real difference in mean strengths of the two makes of steel wires at $\alpha = 0.05$?

10.22 An examination was given to 50 seniors at College A and to 60 seniors at College B. At A, the mean grade was 75 with a standard deviation of 9. At B, the mean grade was 79 with a standard deviation of 7. Is there a significant difference between the performances of the students at A and those at B, given that $\alpha = 0.05$? $\alpha = 0.01$?

10.23 For data in problem 10.21, is the mean breaking strength of Make A steel wire significantly greater than that of Make B steel wire at the 5 percent level of significance?

10.24 For data in problem 10.22, is the performance of students at College A significantly worse than that of students at College B, with $\alpha = 0.05$? with $\alpha = 0.01$?

10.25 The packaging machinery in a dry-cereal firm is known to pour dry cereal into economy-size boxes with a standard deviation of 0.6 ounces. Constant checks of the net weights in the boxes are made in order to maintain the adjustment of the machinery that controls the net contents. Two samples taken on two different dates yield the following information:

First Sample	Second Sample
$n_1 = 30$	$n_2 = 35$
$\bar{x}_1 = 18.7$ ounces	$\bar{x}_2 = 21.9$ ounces

Use $\alpha = 0.05$ throughout in answering the following questions:
a. Test the hypothesis that, on the first date, the machine was adjusted to fill 20 ounces.
b. Test the hypothesis that, on the second date, the machine was adjusted to fill 20 ounces.
c. Test the hypothesis that there was no change in the adjustment of the machine between the two dates.
d. What is the relationship of your answer to (c) and the answers to (a) and (b)?

10.26 A random sample of 100 men was selected from New York State in 1967 and 60 were found to be in favor of a "more modern" divorce law. A random sample of 100 women selected from the same state at the same time revealed that 40 were in favor of such a new law. Is the proportion of men favoring a new divorce law greater than that of women in the said state, given that $\alpha = 0.05$? given that $\alpha = 0.01$?

10.27 To determine whether a certain type of fertilizer is effective, 100 plants out of 500 are left unfertilized. Of the 100, 52 are found to have satisfactory growth, and out of 400 fertilized plants, 275 are found satisfactory. What conclusion can you draw at $\alpha = 0.05$? 0.01?

10.28 Two groups of certain type of patients, A and B, each consisting of 200 people, are used to test the effectiveness of a new serum. Both groups are treated identically except that Group A is given the serum while Group B is not. It is found that 140 and 120 of Groups A and B, respectively, recover from the disease. Is this observed result sufficient evidence for the conclusion that the new serum helps to cure the disease if we are willing to assume an α risk of 0.01?

10.29 During the spring of 1968, students in one of this author's classes were instructed to conduct a survey in New York City of the voters' reactions toward various policies of the national Democrat administration, and the following results were obtained:

Satisfied with the Administration's:	Men ($n_1 = 500$)	Women ($n_2 = 400$)
Vietnam policy	210	145
Civil Rights policy	397	232
Space policy	311	227
Economic policy	278	189
Defense policy	286	192

a. Assuming that the samples are random, do you think there is a significant difference, at $\alpha = 0.05$, between the reactions of men and women toward each of the policies mentioned above?
b. What conclusions can you make insofar as all New Yorkers' reactions toward these policies are concerned?

10.30 The resistance of steel wires of a certain make is known to have a standard deviation of 0.02 ohms. A factory decides to buy this make of wire if the mean resistance per unit length is 0.4 ohms or more, and decides not to buy it if the mean resistance is less than 0.4 ohms. Furthermore, the factory management desires a decision rule such that $\alpha = 0.05$ and $n = 100$.
a. State the appropriate hypotheses.
b. Show the α and β errors schematically.
c. Establish the decision rule.
d. Construct the power and OC curves for this test.

10.31 A company manufactures cables whose breaking strengths have a mean of 100 pounds and a standard deviation of 6 pounds. A newly developed manufacturing process is believed to be both more efficient and capable of increasing the mean strength. The company wishes to adopt the new process if it is indeed more efficient, and decides to check—at the 1 percent level of significance—25 cables produced by the process.
a. State the appropriate hypotheses.
b. Show the α and β errors schematically.
c. Establish the decision rule.
d. Construct the power and OC curves for this test.

10.32 Rework the preceding problem on the basis of testing 100 cables. What conclusions can you draw regarding the power of test when sample sizes are increased?

10.33 In a manufactured part that is to fit other parts, a certain critical dimension is designed to be 4.6 inches. The variability in manufacturing is indicated by a variance of 0.25 in^2. The process is considered to be in control and is continued if it produces this part with a mean of 4.6 inches for the critical dimension. Otherwise, the process is considered out of control and it is stopped. To check whether the process is in control, the plant statistician is satisfied with $\alpha = 0.05$ and $n = 25$, taken every other hour.
a. Formulate the appropriate hypotheses.
b. Show the α and β errors schematically.
c. Establish the decision rule.
d. Construct the power and OC curves.

10.34 A manufacturer who produces flashlight batteries considers that his process is in control when the mean life of his product is 35 hours, and out of control when the mean life is not 35 hours. The process is known to have a variance of 36 hr^2. Establish a decision rule for checking this process with $\alpha = 0.05$ and $n = 36$, and construct the power and OC curves for it.

10.35 A department store has set up this criterion for accepting or rejecting shipments of purchases: A shipment is accepted if 5 percent or less of the items are defective; otherwise, the shipment is rejected. The criteria also specify that α should be 0.05 and n should be 100.
a. Formulate the appropriate hypotheses.
b. Show the α and β risks schematically.
c. Establish the decision rule.
d. Construct the power and OC curves.

10.36 Rework the preceding problem by using $\alpha = 0.01$. Compare your results with those in the last problem in connection with the relationship between α and β when n is fixed.

10.37 The U.S. Navy decides to accept a shipment of 200,000 raincoats if the percentage of defectives in the lot is less than 10. Furthermore, the Navy is anxious to accept the lot and is willing, by observing a random sample of 1500 selected from the lot, to assume a risk of 0.05 of rejecting the shipment which meets its specification. How should the hypotheses be stated in this case? What is the decision rule for this test? What are the values of the OC function for $\pi_1 = 0.12$? 0.11? 0.10? 0.09? 0.08? 0.07? 0.05?

10.38 To test the hypothesis that a coin is fair, the following decision rule is adopted: (1) Accept the hypothesis if the number of heads in 100 tosses is between 40 and 60 inclusive; (2) reject the hypothesis otherwise.

a. What is the alternative hypothesis?

b. What is the value of α for this test?

c. What are the critical values for this decision rule in terms of a normal distribution?

d. How should β risks be calculated in this case?

e. Calculate β for $\pi_1 = 0.1, 0.2, 0.3, 0.4, 0.5, 0.6, 0.7, 0.8$, and 0.9.

10.39 Rework the last problem under the assumptions that $\alpha = 0.10$ and that critical values are to be determined.

10.40 In planning whether or not to have a branch in a certain city, the management of a department store has set up the following criteria: Build the branch if the mean monthly family income in that city is \$500; do not build it if the mean monthly income is \$450; $\alpha = 0.05$ and $\beta = 0.10$. Establish the decision rule (that is, find the critical value and sample size) for this case (assuming $\sigma = \$90.00$).

10.41 The variability in manufacturing a certain part for wall clocks is indicated by a standard deviation of 0.02 inches, which, according to past experience, remains stable. This part is specified to have an outside diameter of 2.06 ± 0.03 inches. When this specification is not met the production process is considered out of control, and it must be stopped for adjustment. If $\alpha = 0.01$ and $\beta = 0.10$, how large a sample is required for the purpose of checking whether or not the process is in control? What are the critical values for this decision rule?

10.42 A manufacturer who produces bolts considers his process in control if the average diameter is 12 cm. and if the individual bolts do not deviate from this mean diameter by more than 1 cm. The standard deviation is known to be 0.25 cm. Given that $\alpha = 0.01$ and $\beta = 0.05$, what are the required sample size and critical values for checking whether or not the manufacturer's process is in control?

10.43 Suppose, with reference to a binomial population, we wish to test $H_0: \pi = 0.4$ against $H_1: \pi = 0.6$. Furthermore, suppose we would like to have $\alpha = 0.01$ and $\beta = 0.10$; how large should the sample be? What is the critical value?

10.44 Suppose that the President of the United States would introduce legislation for socialized medicine if sample results indicated that 70 percent of the voters desired it, and would not introduce such legislation if sample results indicated that 50 percent of the voters desired it. Furthermore, suppose that he would like to assume an α risk of 0.01 and a β risk of 0.05. How large a sample is required? What is the critical value for this decision rule?

11

SAMPLING AND SAMPLING DESIGNS

11.1

Sampling and Reasons for Sampling

In this chapter we shall introduce the elements of sampling and alternative sampling designs. These topics are retrospective in the sense that they should be considered once the decision problem is defined in statistical terms and before data are collected. We have waited until now to discuss them because the student can more readily understand and appreciate some of the concepts connected with these topics when he knows something about probability and decision procedures.

Sampling is simply the process of learning about the population on the basis of a sample drawn from it. As such, its study includes a wide territory, as suggested by the following series of questions:

1. What are the reasons for sampling?
2. What is the theoretical basis for sampling?
3. How should samples be designed?
4. What are the methods of collecting sample data?
5. How can information desired be squeezed out from sample data?
6. How can sample information be utilized to draw conclusions about the parent population?
7. How can the reliability of generalizations made about the population from the sample be evaluated or measured?

From this list, it can be seen that sampling is practically identical with the statistical procedure as defined in Chapter 1; moreover, most of these topics have already been taken up previously. Organization of data and computations of sample statistics were studied in Chapters 2, 3, and 4. Problems of making inferences from samples about their parent populations and inductive conclusions are evaluated in terms of probability and theoretical probability distributions, which were covered by Chapters 5 through 10. What is left for analysis in this chapter is therefore the search for answers to the first three questions posed in the foregoing list.

Let us start with the question, What are the reasons for sampling?

With the great strides made in the theory of sampling during the past few decades, it is now possible to measure properties of mass data with calculated accuracy on the basis of samples. Consequently, today nearly all statistical surveys, whether they are for decision making in business, for policy formulation in government, or for the development of social and economic theories, are samplings.

Besides the fact that reliable results can be obtained from sound sampling procedures, there are other important reasons for their wide adoption. In the first place, populations under investigation may be infinite, and in such cases sampling is the only possible procedure. Furthermore, even in the case of a finite population, very often it may be found that sampling is the only practical procedure. This is so because a finite population may consist of tens of thousands or even millions of elements and its complete enumeration is practically impossible. Consider the case of consumers' preferences for the styling of automobiles or the case of the outcome of a presidential election. The population of the first includes millions of automobile buyers and that of the second contains tens of millions of the eligible voters. If a census were to be taken of either population, the costs of locating, visiting, and interviewing would be prohibitive.

Also, the measurement of a population often requires the destruction of the elements in it. For example, if the producer wants to find out whether the tensile strength of a lot of steel wires meets the specified standard, pressure is put on the steel wire until it breaks. A census then would mean complete destruction of all the steel wires, and there would be no products left after the completion of examination.

What should not be forgotten, either, is that for many types of data the population is not accessible. In practice, we have to deal with whatever part of the data is available. In time series analysis, for instance, studies are inevitably made of samples because only since the recent past have reliable data been available. It may be possible to extend the record of earlier dates but this can be done only by introducing more serious errors due to the unreliability of the data and without much reduction in sampling error. Obviously, we can extend the record into the future only by waiting patiently and watchfully.

Finally, even when it is financially, practically, and physically possible to observe the entire population, sampling may still be the most efficient procedure. Results obtained by the study of samples may be just as or even more accurate than the findings from a complete count of the universe. As will be explained presently, any statistical survey—sampling or census—always contains some error. Statistical errors are of two kinds: nonsampling and sampling. Nonsampling error is usually large for a census but it can be reduced greatly with well-exercised sampling. Moreover, nonsampling error cannot be estimated whereas the measurement of sampling error is possible. For these reasons, not only may the total error be expected to be smaller in a sample survey but sample results can also be used with a greater degree of confidence because of our knowledge of the probable size of the error. This point is vividly illustrated by the effective use of samples by the Bureau of Census to check the accuracy of the census.

In concluding these introductory remarks, we may point out that all the reasons for sampling may be summarized by a general principle: In dealing with numerical data there always comes a point beyond which the increase in information from additional observations is not worth the increase in costs.

11.2 Theoretical Basis of Sampling

In sampling we find a powerful instrument with which to predict and generalize the behavior of mass phenomena. It is indeed a great scientific achievement "toward an intellectual mastery of the world around us," commented Roy Jastram in *Elements of Statistical Inference*, "to generalize logically and precisely about thousands of values which we have not seen, simply upon the evidence afforded by, say, fifty or a hundred of those values." The theory of sampling that makes possible this kind of inference has as its foundation the permanent characteristics of mass data, which can be summarized simply and precisely by the phrase "diversity in unity."

On the one hand, the elementary units of any population are affected by a multiplicity of forces. This vast complex of forces, moreover, though related, acts upon the individual elements with a considerable degree of independence. These causes explain variations from unit to unit in the population. We thus find that brothers, although similar in most respects, may differ from each other, with varying degrees, in personality, temperament, habits, intelligence, physical features, and so forth. Oranges from the same tree may differ in size, in color, in weight, and in sweetness. Differences in diameters, in weight, in tensile strength, and in length always exist among steel bars coming off the same production process. No golf player has ever driven a ball exactly the same distance twice. In the stock market, on any day, one finds that some stock prices increase, others decrease, and still others remain relatively unchanged. And so on—the list can be extended indefinitely.

Although diversity is a universal quality of mass data, there is no statistical population whose elements would vary from each other without limit. Thus, rice varies, to a limited extent, in length, weight, color, protein content, and so forth, but it can always be identified as rice. Adolescents in a human population may differ from each other in height, but no ordinary individual is as short as, say, three feet or as tall as, say, eight feet. Economic goods can never have prices as low as zero and, practically, they can never have prices that are beyond the reach of even millionaires.

The facts that any population has characteristic properties and that variations in its elements are definitely limited make it possible for us to select a relatively small, unbiased random sample that can portray fairly well the traits of the population.

Another interesting and important property of mass data is regularity or uniformity. The related but independent forces, which produce variability in a population, are often so balanced and concentrated that they tend to generate equal values above and below some central value around which most of the values tend to cluster. In any school, both "A" and "F" students are in the minority, but most are "C" students. Balances of savings accounts in any bank may vary from a few dollars to over ten thousand. However, only a small number of the balances would be less than $100 or more than $10,000, and the majority would cluster within a small range somewhere between these two values. Most business firms carrying the same line of merchandise usually stock their inventories at some intermediate level; few would adopt a high or a low inventory policy. The concentration of inventory size at some intermediate level is the result of trial and error. Those who keep inventories too low would often be out of stock and those keeping inventories too high would lose interest on tied-up capital or suffer physical depreciation and obsolescence of merchandise. Thus, paradoxically, the individual items in a population tend to vary from each other and at the same time to conform to some mass standards. We have therefore both diversity and uniformity in mass data.

Because of statistical regularity, if a large random sample is selected, characteristics of this sample will differ very little from those in the population. Because of diversity, if a number of random samples are taken, although quite similar in many respects (because of uniformity), the samples can never have strict regularity.

Uniformity or regularity refers to the tendency of the measurable characteristics to cluster around some "center of gravity." The measure for such a central tendency, as you may recall, is an average from which individual observations diverge in some definite pattern.

Averages are more stable than individual values. Moreover, averages become more stable when more observations are included in the sample. Thus if a coin is tossed ten times, it may come up all heads or no heads at all. If a coin is tossed twenty or thirty times, however, such extreme occurrences are less likely. When the number of trials is increased infinitely, we would eventually approach the completely stable limit of half heads and half tails. Again, the number of highway traffic accidents per month in any state or group of states selected at random may be considerably larger this year than it was last. It is also conceivable that there are still other states with high monthly rates of highway accidents and that the number of states with large accident increases in this year's count is not entirely offset by decreases. However, the relative variation in all states will decrease when more states are considered; it may therefore be stated that the total number of traffic accidents in all states fluctuates much less than the number of traffic accidents in a single state or in a small group of states.

These examples illustrate the most important property of large numbers. That is, the behavior of large numbers tends to be uniform and stable. This is the famous *law of large numbers* in brief. To state this law more explicitly in connection with the theory of sampling, we may say that although samples vary, averages vary less in large samples than in small ones, all taken from the same population. There are at least two implications of this law that are of fundamental significance to the dependability of samples and the adequacy of sampling methods:

(1) If a larger and larger random sample is drawn from the same population, the characteristics of each enlarged sample—that is, averages, variations, skewness, and so on—will tend to differ less and less from the characteristics of the population; that is, the larger the sample the more representative it is of the population.

(2) If a large number of large random samples is taken from the same population, not only will each sample differ from the population very little but the samples themselves will differ very little from each other; that is, the averages of large random samples will tend to cluster around the population average within a small range.

Stability of large numbers, therefore, makes it possible to determine the size of sample in order that its characteristics may, within prescribed limits of precision, be attributed to the whole population.

In practice, in making inferences we usually take only one sample. Whether the sample is large or small, we are almost certain that its characteristics are not exactly those of the population. How can we be sure, then, about the degree of dependability of our conclusions? The answer to such a question is "Randomness." Objective measurement of sampling errors requires that the sample be random. Nearly any method of sampling has some pattern of variability. But the nature of the underlying frequency distribution may be known only if the sampling is random. This is so because this knowledge can be obtained only by applying the mathematical laws of probability, which, in turn, can be applied only to random samples.

Let us now summarize this section: (1) A random sample may be expected to represent its parent population fairly well because every population has relatively fixed traits and definitely limited variations. (2) Random samples selected from the same population will have much in common but they can never be expected to have strict uniformity because of the universal property of diversity in unity of mass data. (3) A large sample will differ less than a small sample from the parent population, and when samples become larger the variability among them will also decrease because of stability in large numbers. (4) Sampling variability can be measured if and only if the sampling is random because the laws of probability are applicable to random samples and thus make it possible to know the nature of the underlying frequency distribution in terms of which the sample results must be interpreted and evaluated.

11.3

Basic Concepts of Sampling

A number of technical terms associated with the theory of sampling must be properly understood. Some of them have already been introduced but now we should have a closer look at them.

THE FRAME OR THE SAMPLED POPULATION

The definition of the population in connection with a given problem situation, as pointed out in Chapter 1, is in a sense the identification of elementary units in the population. Elementary units must be precisely defined and identified before we can observe them. Once this is done we would then have some idea whether or not the definition of the population is operationally feasible; for example, whether or not the population is accessible.

In many problem situations, access to the population presents no problem. If, for instance, we wish to evaluate the quality of a production process, the elementary units would be the outputs being turned out continuously by the process. Again, if we would like to determine the feelings of employees of a firm toward a contemplated managerial policy affecting their well-being, the elementary units would be all the workers in the firm. Occasions may arise, however, when the population may be easily defined but its elementary units may be quite difficult to locate. For example, a population may be defined as all makes of television sets in use, but we may not know exactly where all the television sets are located.

Whenever access to the population presents difficulties, the first thing we must do is prepare some sort of list from which to identify the elementary units. Such a list may be called the *frame*, or the *working population*, or the *sampled population*. The last two names suggest that the frame is the operationally feasible population or the population that can actually be sampled. In contrast, the population originally defined or the population we intend to sample is referred to as the *target population*.

The target population may or may not be the sampled population. The success or failure of a statistical survey depends upon the working population that is available. Unless a reasonably adequate working population can be found, the proposed study should be abandoned. If the frame is different from the target population, but the former can be judged to contain adequate information, the study may be continued. It must be remembered, however, that statistical inferences are concerned with the sampled population. That is, statistical procedures can be applied to form conclusions about the sampled population, but these conclusions cannot be applied equally well to the target population unless the two populations happen to be the same.

For a survey of human populations, there are available a large number of such ready-made frames as tax assessors' lists, city directories, lists of registered voters, telephone directories, automobile registration records, customer lists. However, except for the first two sources mentioned, they are usually imperfect frames for general public surveys. Their imperfection may be due either to duplication or to incompleteness. A frame that may contain duplicates of certain elementary units can often be improved by thorough checks, although such a process may prove tedious and costly. A frame that does not include all elementary units of the target population presents more serious problems. In the first place, incompleteness of the frame can hardly be detected in the course of the study. Another still more serious matter is that incompleteness may represent absence of a group or groups of elementary units possessing some special properties not possessed by those listed in the frame. Finally, it is often impossible to correct an incomplete frame by adding the missing elementary units, which may be inaccessible. Thus knowing that the frame is incomplete does not provide a basis for its improvement but merely affords a basis from which to judge whether the incompleteness will invalidate the study and to act accordingly.

Unawareness or deliberate nonrecognition of the imperfection of the frame often leads to useless sample studies that may even have harmful results. A classical illustration of such a biased study is the prediction made of the 1936 presidential election poll conducted by the *Literary Digest*. From an extremely large sample of more than two million ballots, the *Digest* predicted that Roosevelt would receive only 40.9 percent of the votes in the election and that Landon would win by a landslide. The actual returns showed, however, Roosevelt received 60.7 percent of the votes and won overwhelmingly. Thus the *Digest's* poll made an error of more than 19 percentage points. This error was due to the use of a sample taken from an imperfect frame. The target population in this case was clearly all the eligible voters in 1936. However, the *Digest's* mailing list of names, which was taken from the telephone directories and automobile registrations, formed a universe whose members were predominately upper- and middle-class people. The *Digest's* findings might have been quite revealing for the working population from which the sample was drawn, but it was certainly not the target population of all eligible voters. In other words, the use of telephone directories and lists of automobile owners as the principal sources of the *Digest's* poll excluded most of the "ill-clad, ill-fed, and ill-housed" lower-income classes who were overwhelmingly in favor of Roosevelt because they were the main beneficiaries of the New Deal policies.

SAMPLES

A sample has been previously defined rather loosely as a part of a population. A more precise definition must now be developed in order to throw some light on the process of sampling. Items included in a population are called *sampling units,* which can be classified into two kinds: elementary and primary sampling units. *Elementary sampling units,* as we have already learned, are all the items contained in the population whose characteristics are to be measured or counted, such as workers of the steel industry. *Primary sampling units* may be the elementary units themselves or the groups or clusters of the elementary units, such as steel plants where the elementary units, the workers, are clustered. A *sample* is a collection of primary sampling units selected as a representative microcosm from which inferences about the population may be made. So defined, the concept of "sample" has three implications. First, a sample may contain primary sampling units, such as firms of an industry, even though its purpose is to observe the properties of the elementary units, such as the workers in the industry. Second, a sample is not a "chunk," a slice of the population selected because of convenience and ready availability, but a microcosm, a small universe or subset, which is expected to be representative of the parent population. Third and last, the ultimate purpose of sampling is not merely to secure sample statistics but to make inferences concerning the population.

In the theory of sampling, it is important to distinguish random samples from judgment samples. A *random sample* is selected by the probability method, according to which neither the investigator nor the sampling units can decide which items will be included in the sample. The selection is achieved by the operation of chance alone. Thus, in a random sample the probabilities of selection are known and it is thus appropriately called a probability sample. Moreover, for a probability sample, the sampling error can be measured and controlled by the theory of probability, and biases of selection, nonrespondence, and estimation are eliminated or at least contained within known limits. In contrast, a *judgment sample,* as its name implies, is taken according to personal judgment. Elements included in a judgment sample are results of the investigator's expert judgment as to their "representativeness." Consequently, the probability of each individual element's being drawn in the sample is unknown and the reliability of its results is not amenable to probability analysis but must depend on personal judgment.

The number of items included in a sample may be of any size, varying from one to all the elements in the population. The actual size to be taken depends mainly, as the reader knows now, on the variability of the population and the degree of precision required. What may be mentioned here is that a sample containing only one sampling unit may often yield useful information but measures of statistical inferences cannot be applied to its results. A sample that contains all the units in the population is called an *exhaustive sampling,* which, of course, is only another name for *census.*

STATISTICAL ERRORS

An *error* in a statistic means the difference between the value of a statistic and that of the corresponding parameter. Various causes combine to produce deviations of statistics from parameters, and errors, in accordance with the different causes, are classified into nonsampling and sampling errors.

The term *nonsampling error* derives from the fact that this type of error can occur in any survey, whether it be a complete enumeration or sampling. Nonsampling errors include biases and mistakes. A *bias* may be said to exist when the value of a sample statistic shows a persistent tendency to deviate in one direction from the value of the parameter. That is, if this type of error is present, the sample results tend to be always above or always below the parameter and, therefore, cannot be canceled out by averaging. Factors that cause biases include careless definition of the population, imperfection of the frame, indefiniteness of the questionnaire, a vague conception regarding the information desired, exaggerated or irrelevant responses to prestige or loaded questions, inaccurate methods of interviewing, and so forth. *Mistakes* of a survey arise when responses are entered in wrong columns or cells by interviewers, when wrong or inadequate responses are given by respondents, and when wrong computations and entries are made when the data are processed. Clearly, the main source of mistakes is the carelessness of all types of statistical workers for the survey.

Sampling error results from the chance selection of sampling units. This type of error occurs, then, because only a partial observation of the universe is made. If a census is taken, sampling error could be expected to disappear. By *sampling error* we mean precisely the difference between the sample result and that of the census when both results are obtained by using the same procedures. The most important property of sampling errors is that they follow chance variations and tend to cancel each other out when averaged. It is because of this attribute that, even though a given sample statistic may differ from the parameter and even though the results obtained from different samples of the same kind taken from the same population may differ from each other, the average of the sampling distribution of a statistic, that is, the *expected value* of that statistic, can be expected to equal the corresponding parameter.

If the expected value of a statistic is equal to the corresponding parameter being estimated, the statistic is said to be an *unbiased estimator*. The arithmetic mean of a simple random sample, for instance, is an unbiased estimate of the mean of the parent population. So is the proportion of a simple random sample. However, if the sample is not truly random or if it is biased, the sample mean or proportion would be a biased estimate. For example, the *Literary Digest's* estimate of votes for Roosevelt, cited earlier, was biased downward. That is, if the *Digest* were to repeat the sampling operation consistently by using telephone directories and automobile registrations as the frame, the mean of the successive estimates of the proportion of votes for Roosevelt would have been lower than the true proportion of the population. From this illustration, we see why we have defined *bias* previously as the difference between the expected value of the estimate and the value of the population parameter being estimated.

The sum of nonsampling and sampling errors is the total error of a statistical survey. One major concern in sampling is to make the total error as small as possible. The previous discussion of nonsampling error reveals that it could be reduced appreciably by defining precisely the population and the traits to be measured, by carefully preparing the frame, by pretesting the questionnaire, by following up nonrespondents, by thoroughly editing returns, by using tests of reasonableness, by painstakingly checking all the steps in processing the data and, most important of all, by thoroughly training interviewers. Note that the reduction of nonsampling errors requires small samples because most of the precautionary measures mentioned above are costly. For instance, the training of qualified interviewers is usually time-consuming and expensive and can be made available to only a relatively small group of people. However, a small sample tends to have a larger sampling error. Hence those responsible for sampling surveys have an important decision to make when they allocate their limited human and financial resources in such a way as to provide the highest possible reliability for the statistical results.

PRECISION VERSUS ACCURACY

Sampling errors or sampling variations are measured by what is called *precision.* Precision, which is also referred to as *reliability,* of sample results is the degree to which successive statistics computed from successive random samples of the same size differ from each other. Thus precision is but the size of the standard error of the statistic. The smaller the standard error, the greater the precision or reliability of the estimate.

Accuracy of a sample estimate is the difference between the sample statistic and the true parameter. Accuracy of an estimate is not directly measurable because the true parameter is an unknown constant. Accuracy does, however, depend upon both the degree of precision and the absence of bias. Accuracy is the goal of a survey, but precision is what can be measured. The control and calculation of precision of sample results by laws of mathematical probability are the basic function of sampling because they help to attain accuracy.

Biases, it may be noted, do not necessarily affect precision if the sources of biases are constant and thereby tend to generate reproducibility of estimates made from successive samples. That is, a biased procedure may actually have a lower standard error than an unbiased one. Consequently, the chances of getting a more accurate estimate may be greater with a biased estimate than with an unbiased one. What should be realized is that when a biased procedure is employed, we must know clearly the nature of the bias and interpret the results accordingly in decision-making procedures. That a biased sample may actually yield a more accurate estimate is one of the reasons for the use of sample designs other than random sampling.

EFFICIENCY OF SAMPLING DESIGNS

A sampling design is said to be *efficient* if the desired results are obtained at the lowest costs possible. One sampling design is said to be more efficient than another if the former yields the same precision at lower costs, or greater precision at the same costs, or, obviously, greater precision at lower costs, than the latter.

In simple random sampling, estimates of measurable precision are obtainable because the standard error is computable. Precision of a simple random sample can be increased to any level by increasing the sample size. What should be noted, however, is that precision of a random sample does not increase directly and proportionately with the sample size. Instead, as implied by the formula for standard errors, it increases with the increase in the square root of the sample size. Thus, if it is desired to double the precision, it is necessary to quadruple the sample size. As an illustration, if there are 100 items in a sample and it is desired to double its precision (or to reduce the standard error by half), a sample of 400 items is required. For the square root of 100 is 10, which should be multiplied by 2, obtaining 20, in order to increase its reliability by 100 percent or to reduce the standard error by 50 percent, and the 20 must then be taken from under the radical by squaring it so that the required number of items, 400, can be obtained. Following the same reasoning, the student should be able to find, for instance, that if a sample is to be three times as reliable as one of 100 items, 900 items are needed for that sample.

We are often confronted with situations where increasing the size of sample becomes so expensive with simple random sampling that we have to use other alternative sampling designs to increase precision without increasing the size of the sample. It is always important, in selecting among alternative sampling designs, to balance precision against cost.

11.4 Alternative Sampling Designs

Sampling designs can be conveniently grouped as random sampling and nonrandom sampling. *Random sampling* is also referred to as *probability sampling,* since if the sampling process is random, the laws of probability can be applied; thus the pattern of sampling distribution needed to interpret and evaluate a sample is provided. The term *random sample* is not used to describe the data in the sample but the process employed to select the sample. Thus, randomness is a property of the sampling procedure instead of an individual sample. As will be shown, randomness can enter a process of sampling in a number of ways and therefore random samples may be of many kinds.

Nonrandom sampling is a process of sample selection without the use of randomization. A nonrandom sample, in other words, is selected on a basis other than probability considerations, such as expert judgment, convenience, or some other criteria. The most important aspect of nonrandom sampling to remember is that it is subject to sampling variability, but there is no way of knowing the pattern of variability in the process.

We shall now turn to describe various sampling procedures by following this outline:

- I. Random sampling
  - 1. Unrestricted or simple random sampling
  - 2. Restricted random sampling
    - a. Stratified sampling
    - b. Cluster sampling
    - c. Systematic sampling
  - 3. Double, multiple, and sequential sampling
- II. Nonrandom sampling
  - 1. Judgment sampling
  - 2. Quota sampling
  - 3. Convenience sampling

11.5

Unrestricted Random Sampling

As has already been pointed out several times, *unrestricted* or *simple random sampling* is a procedure in selecting a sample by which every item in the population has an equal and independent chance of being included in the sample. Furthermore, if the sample of size n is taken, any possible sample of n items has the same probability of being drawn as any other combination of n items.

A simple random sample is drawn by chance selection, which must be differentiated from selection in a "haphazard" or "hit-and-miss" manner. Unless this distinction is observed, random methods may be misapplied and a sample may be assumed to be random that is not random at all. Consequently, serious errors would be made in sampling. Chance selection that assures randomness of sampling procedure can be maintained only by deliberate and orderly methods. This is to say that in order to ensure chance selection, the process of sampling must be carefully controlled.

Theoretically, in the selection of a simple random sample every individual item drawn must be measured, recorded, and returned to the population before another selection is made. Thus, each item can be chosen more than once in the same sample. However, in practice, this procedure is rarely observed. There are two basic reasons for this. In the first place, if the population is large compared with the sample, the error made by not returning the individual item is of no practical significance. Second, in many types of sampling survey, individual items selected are completely destroyed by investigation and it is therefore impossible to return them to the population. However, if the universe is very small, the procedure of returning selected items must be maintained. For instance, if it is desired to draw a random sample of four from a deck of fifty-two cards, we would select one card, record its value, return the card, shuffle the deck well, select another card, record its value, return it to the deck and repeat the process in the same manner until four cards are chosen. It is also interesting to note at this point that "each item has an equal chance of being selected" is not the same as "each measurement has an equal chance of being selected." For instance, since there are more number cards than face cards in a deck, the chance that number cards (small measurements) will be included in the sample is evidently greater than the chance that face cards (large measurements) will be included.

The previous example of drawing a sample from a deck of cards suggests a method of random selection that may be called the "goldfish bowl" procedure. According to this procedure, all the elements in the universe are numbered, and the numbers are written on chips, balls, cards, or other articles that are physically homogeneous. These articles are put into a bowl and mixed thoroughly. Then the specified numbers of items to be included in the sample is chosen either with or without replacement. With replacement, the probability of each element's being drawn is $1/N$; without replacement, the probability of selecting the subsequent items increases as N decreases by the number of items previously drawn. Without replacement, for instance, the probability of each item's being chosen at the first drawing is $1/N$, at the second drawing, $1/(N - 1)$, at the third drawing, $1/(N - 2)$, and so on.

The "goldfish bowl" procedure can be shortened and money and time saved in drawing a sample by using a table of *random numbers*. The construction of such a table and the process of drawing a simple random sample with the aid of it were discussed in Section 7.6.

Simple random sampling is certainly an efficient procedure if the population is not large and if it is relatively easy and inexpensive to find the sampling units. It could also be a practical procedure for large populations whose elements are concentrated within a small area. For instance, the study of all expense vouchers of a large corporation, the investigation of the attitudes of all the employees toward a pension plan in a big company, the survey of students' study habits at a given university, and so on, can easily be made by a simple random sample.

One drawback of applying simple random sampling to large populations is that the population must be numbered. Theoretically we could obtain simple random samples by numbering all the eligible voters, all the households in America, all automobile owners, all retail stores, all refugees from other countries, and so forth. Such a procedure, however, would be extremely costly and might even be physically impossible. As a matter of fact, many economic and business data cannot be effectively utilized by simple random sampling. Consequently, instead of using simple random sampling, we often employ restricted random sample designs.

Greater efficiency and the desire to have a probability sample when it is impossible to use simple random selection are the basic reasons for the preference for restricted random sampling designs, even though there are other advantages to be gained by their application.

11.6 Stratified Sampling

Stratified random sampling is one of the restricted random methods which, by using available information concerning the data, attempts to design a more efficient sample than that obtained by the simple random procedure. The process of stratification requires that the population be divided into homogeneous groups or classes called *strata*. Then a sample may be taken from each group by simple random methods, and the resulting sample is called a *stratified sample*.

A stratified sample may be either *proportional* or *disproportionate*. In a proportional stratified sampling plan, the number of items drawn from each group is proportional to the size of the group. For instance, if the population is divided into four groups, their respective sizes being 10, 20, 30, and 40 percent of the population, and a sample of 500 is to be drawn, the desired proportional sample may be obtained in the following manner:

From stratum one	500(0.10) =	50 items
From stratum two	500(0.20) =	100 items
From stratum three	500(0.30) =	150 items
From stratum four	500(0.40) =	200 items
Size of the entire sample	=	500 items

It is evident from the above example that proportional stratification yields a sample that represents the universe with respect to the proportion in each stratum in the population. This procedure is satisfactory if there is no great difference in dispersion from stratum to stratum. But it is certainly not the most efficient procedure, especially when variations differ substantially in the various classes, for, as has been repeatedly pointed out, the most important consideration for sample size is not the size but the variation of the population. This indicates that in order to obtain maximum efficiency in stratification, we should assign greater representation to a stratum with a large dispersion and smaller representation to one with small variation. For instance, in conducting a poll for a presidential election, we may take the 50 states as our strata. If the eligible voters of a given state are 5 percent of all the eligible voters, but according to our knowledge the result in that state is already a foregone conclusion, we may then take, say, merely 2 percent or even 1 percent of our sample from that state. If, however, the outcome of another state is highly doubtful, we may decide to give it a much greater representation in the sample than its relative size. A sample thus obtained is a *disproportionate stratified sample.* Disproportionate stratified sampling also includes procedures of taking an equal number of items from each stratum irrespective of its size and of giving only a small representation to one or more strata whose members are too expensive to investigate but some representation of them is nevertheless valuable. The term *optimum allocation* is often used for stratified methods that take both variation and size of each stratum into consideration in determining its representation in the sample.

Efficiency in stratification can be further increased, when the nature of the data permits, by classifying the strata into substrata, which in turn may be subdivided into smaller groups. For instance, knowing that voting preferences are often influenced by differences in sex, religion, race, educational standards, and economic status, in conducting a presidential election poll we may subdivide the various states according to these aspects. This refinement in stratification of grouping the population with respect to several relevant characteristics is referred to as *cross-stratification.*

Previous discussions may have indicated that stratification is most effective in dealing with heterogeneous or highly skewed populations, such as income data or retail sales. In such situations, we can stratify the population in such a manner that (1) within each stratum there is as much uniformity as possible and (2) among various strata the differences are as great as possible. Consequently, we are able to obtain a sample with a smaller sampling error or a smaller sample with the same precision as compared with simple random sampling. In other words, in handling populations with extreme values, stratification is a more efficient design than simple random sampling. This is true because in an unrestricted random sample extreme values, say high incomes, may be excluded while they will certainly be represented in a stratified sample. It is hardly necessary to point out now that stratification may not have any advantage over simple random sampling in studying moderately skewed or symmetrical distributions, such as testing the durability of a shipment of automobile tires. Hence, before deciding on stratification, we must have some knowledge of the traits of the population. Such knowledge may be based upon past data, preliminary observation from pilot studies, expert judgment or simply intuition or good guesses.

Subjective judgment used in dividing the universe into strata does not mean that stratified samples are not probability samples. By employing random selection from each stratum, the probability of each item's being drawn is ensured and this probability is known, even though in disproportionate sampling, the probabilities that individual items will be selected are not equal.

As probability samples, the weighted arithmetic average of the strata sample means is an unbiased estimate of the universe mean. In the case of proportional sampling, where each item has an equal probability of being chosen, the estimate of the universe mean is merely the arithmetic average of the entire sample because it is weighted in just these proportions. However, the mean of the disproportionate sample must be derived with proper weights or serious biases may result. The proper weights are the relative proportions of the population in the various strata.

Finally, precision of estimates in stratified sampling can be measured provided the sample is sufficiently large. Appropriate formulas for evaluating sampling errors in stratification can be found in more advanced books on sampling.

11.7 Cluster Sampling

Cluster sampling refers to the procedure of dividing the population into groups and drawing a sample of groups to represent the population. When the clusters, which are primary units, are drawn, we can either include in the sample all the elementary units in the selected groups or take a sample of smaller primary units or elementary units from sampled clusters. When all the elementary units in the sampled groups are observed, we have what is known as the *single-stage sampling*. When a sample of elementary units is drawn from the clusters, we have a type of design called *two-stage sampling* or *subsampling*. At both stages a random sample is selected. When cluster sampling involves more than two stages in selecting the final sample, it is called *multistage sampling*. For instance, we may take the colleges—primary units—as the first stage, then draw departments as the second stage, and choose students—the elementary units—as the third and last stage. When the clusters—the primary units—are geographical areas, cluster sampling becomes the widely used *area sampling*. In any case, our objective is to study the elementary units even though initially primary units are taken and random methods are used in selection at each stage.

Principles that dictate maximum efficiency in cluster sampling are the opposite of those used in stratification. In cluster sampling, it is efficient to have (1) differences among the elementary units in the same class as large as possible and (2) differences among the primary units, clusters, as small as possible. Evidently these requirements are very difficult to meet, for in most populations similar elementary units tend to cluster together. For instance, rich people tend to live in the same neighborhood in a city, while the poor families are concentrated in another area. Again, better students tend to be in the departments of natural sciences and engineering, whereas less able students usually take liberal arts and social sciences. (This last illustration is a sad but nevertheless a true situation in this country.)

This difficulty causes cluster sampling to have a greater standard error than stratified sampling has, even though the former also yields unbiased estimates. The fact that cluster sampling provides less precision for a sample of equal size does not mean that it cannot be more efficient than other random designs. If the cost per elementary unit is much lower in cluster sampling many more elementary units can be included in the sample. (And it is usually the case that cost per interview in cluster sampling is lower than for other types of random designs. Lower cost is most evident in area sampling, when interviewers' time and traveling are cut down.) Consequently, it is possible to have a larger cluster sample that can yield the same precision with lower cost than a simple or stratified random sample. Alternatively, it is possible, with the same cost, to have a cluster sample large enough to have greater reliability than either simple or stratified sampling. In other words, despite the drawback in achieving heterogeneity within each cluster and uniformity among different clusters, cluster sampling can still be a more efficient design. What the student should understand is that the purpose of cluster sampling is not to obtain the most reliable sample in relation to its size, as in the case of stratification, but to have the greatest precision in estimate in terms of cost per elementary unit.

11.8 Systematic Sampling

Another frequently used random plan is *systematic sampling*. It is a procedure of selecting the kth item from the population with a random start. Here, the number k is merely the sampling ratio, that is, n/N. For instance, if 100 items are to be included in the sample from a population of 1000 elements, k, the count number, is 10. To have a random start when the sampling ratio is 10, we select a random number from 1 through 10 by means of the table of random numbers or by other random methods. If the random number 3 is selected, we begin with this and include every tenth item from it, namely, the thirteenth, the twenty-third, the thirty-third, and so on, in the sample. Simplicity in design is a main advantage of systematic sampling. Also, from the method itself, it can be seen that greater efficiency in systematic sampling can be achieved if (1) the items that are close to each other have greater uniformity than (2) the items that are far away from each other. These requirements are quite similar to those for stratification. Indeed, if populations are sufficiently large, systematic sampling can often be expected to yield results that are similar to those yielded by proportional stratified sampling. Because of its simplicity in design, systematic sampling may be preferred over stratification if the population can be easily put into some orderly arrangement. When a list of corporate earnings is available, for example, we can rearrange the earnings according to their magnitude. The application of systematic sampling in such a case actually ensures proportionate representation of all earning-size classes without stratification.

A systematic sample tends to be more representative than a simple random sample of the same size if, as was pointed out in the preceding paragraph, elements in the universe tend to, or can be arranged to, resemble other nearby elements more than they resemble others farther away. For instance, if we are to estimate the average family income in Los Angeles, we may use systematic sampling and select every hundredth household in our sample. The resulting sample would be more representative than a simple random sample. People of the same economic status tend to live in the same neighborhood; hence systematic sampling would give more appropriate representation to every income group, even though simple random sampling may select a relatively large number of households with similar incomes from a certain area.

Systematic sampling becomes a less representative design than simple random sampling if we are dealing with populations having hidden *periodicities*. For example, if the sales of every seventh day of the calendar year are included, the sample will contain, say, all Mondays or all Fridays. If there is a definite repetitive weekly pattern in sales (which is usually the case), our sample is not representative at all of sales for the whole year and consequently the sample results may be seriously biased. Similarly, biases will result if we are to test the quality of a product coming off an assembly line by taking every fiftieth item for observation when the machinery happens to have a defect that produces imperfections in every fiftieth piece.

Up to now we have discussed various random procedures as independent designs. In practice, we often combine two or more of these methods into a single design. For instance, in sampling for an estimate of the labor force, the Bureau of Census uses a plan that includes all types of random sampling and other devices. In this survey, 68 areas were first decided upon. Then, approximately 2000 primary units—clusters that are whole counties or contiguous counties—were set up and grouped into 55 primary strata. After that, one primary unit was selected from each stratum. Next, these selected primary units were divided into segments to determine the elementary units (households) to be included in the final sample. At each stage, probability methods were used for selection so that proper formulas could be derived to measure the reliability of the sample results.

11.9

Double, Multiple, and Sequential Sampling

All sampling designs, except multistage clustering sampling, presented previously may be referred to as *single sampling*, since each design is used to obtain a single sample from which an estimate is made or a hypothesis is tested. The application of sampling for testing the quality of incoming material from a vendor or of a particular lot of products coming off a manufacturing process or of clerical work during the past two or three decades has led to popular use of double, multiple, and sequential sampling designs.

The choice between accepting or rejecting a lot can often be made on the basis of less than the full sample size. For instance, in a given situation we may adopt the following decision rule:

Take a random sample of 100 from the lot. If there are more than 5 items defective in the sample, the lot is rejected. If 5 or less items are defective in the sample, the lot is accepted.

In following such a decision rule, however, we may find that there are 5 defectives in the first 10 items examined. It is then clearly unnecessary to examine the remaining 90 items. This possibility has led to *double sampling*—a process by which a small sample is observed first and a decision may be made to accept, to reject, or to take a second sample. For example, in double sampling, we may formulate this decision rule:

Select and examine a random sample of 40 items. If 2 or less defective items are found, accept the lot; if 4 or more defectives are found, reject the lot; if 3 defectives are found, take a second sample of 60 items (total now is 100). If the total number of defectives in the combined sample is 4 or less, accept the lot; if more than 4, reject the lot.

If a decision can be made between accepting or rejecting a lot on the basis of a small first sample, costs would be considerably reduced, especially if the sampling is destructive. Thus double sampling is better than single sampling. From the same reasoning, however, it is easy to see that double sampling is more expensive than triple sampling, which, in turn, is more costly than quadruple sampling. This philosophy has suggested what is called *multiple sampling*—a procedure that utilizes a series of small samples such that the cumulative number of defects is compared against an "accept" and a "reject" criterion after each sample is drawn until a decision can finally be made. A multiple sampling plan is illustrated in Table 11.1. Observe that the choice between acceptance and rejection must be made on the last sample.

A logical extension of multiple sampling is to make an observation of one item at a time, deciding after each observation whether to accept or reject the lot or to continue sampling. Such a technique, called *sequential sampling*, is especially

TABLE 11.1 EXAMPLE OF MULTIPLE SAMPLING PLAN

		Combined Sample		
Sample	Sample Size	*Cumulative Size*	*Acceptance Number*	*Rejection Number*
First	30	30	*	3
Second	30	60	1	4
Third	30	90	3	6
Fourth	30	120	5	7
Fifth	30	150	6	8
Sixth	30	180	7	9
Seventh	30	210	8	9

Note: The asterisk for the first sample indicates that a sample of only 30 items cannot present sufficient evidence for acceptance

appropriate for inspecting products that are very expensive and whose investigation results in their destruction.

Acceptance sampling, whether single, double, or multiple, involves the same risks of making wrong decisions as the testing of hypotheses involves. That is, we may either reject a lot that should have been accepted (Type I error) or accept a lot that should have been rejected (Type II error). It may be noted that single sampling acceptance plans can be easily constructed on the basis of a table of binomial probabilities for various sample sizes. Construction of double, multiple, and sequential sampling plans is more difficult and complicated. In any event, risks of incorrect decisions in acceptance sampling can be specified at reasonably low levels of probability and the necessary sample size can be determined from these requirements.

There are available numerous acceptance sampling plans based on several criteria: the *acceptance quality level* (AQL)—the worst quality the buyer is prepared to accept is a lot with a stipulated high probability; the *lot tolerance percentage defective* (LTPD)—the quality above which there is only a small probability that a lot will be accepted; and so on. Widely used catalogues of tabulated acceptance sampling plans include Dodge and Romig, *Sampling Inspection Tables: Single and Double Sampling;* Statistical Research Group, Columbia University, *Sampling Inspection;* and *Military Standards* 105*A*: *Sampling Procedures and Tables for Inspection by Attributes* (MIL-STD-105A).

11.10 Judgment Sampling

Judgment sampling, the type of nonrandom sampling we shall consider, is also called *purposive sampling.* It is the method of selecting the items in a sample on the basis of the judgment of the sampler, who thinks that the results will be representative. In other words, the items in the population that are thought to be typical or representative, according to the knowledge of the investigator, are drawn into a judgment sample. In judgment sampling, then, the probability that an individual item will be chosen is unknown. Consequently, confidence intervals cannot be constructed and judgment sampling leads to point estimates only. Furthermore, only the expert sampler can interpret the precision of results obtained from this plan; these results cannot be objectively evaluated.

Even though the principles of sampling theory are not applicable to judgment sampling, they are often used in solving many types of economic and business problems. The use of judgment sampling is justified under a variety of circumstances. When only a small number of sampling units is in the universe, simple random selection may miss the more important elements, whereas judgment selection would certainly include them in the sample. A study of the steel industry would be questionable if the United States Steel Corporation were not included in the sample. Similarly, without General Motors in the sample, a survey of the automobile industry would be equally unsatisfactory.

Judgment sampling is sometimes used because we must keep the size of the sample small: it may be difficult to locate some of the sampling units or necessary to keep sampling costs down. In such situations, probability sampling becomes impractical because wide confidence intervals are attracted to small random samples. Thus the sample estimates, though unbiased and measurable, are of little value in decision making.

Again, in solving everyday business problems and making public policy decisions executives and public officials are often pressed for time and cannot wait for probability sampling designs. Judgment sampling is then the only practical method, since estimates can be made available quickly that will enable businessmen and governmental officials to arrive at solutions to their urgent problems that are better than decisions made without any statistical data.

And again, when we want to study some unknown traits of a population, some of whose characteristics are known, we may then stratify the population according to these known properties and select sampling units from each stratum on the basis of judgment. This method is used to obtain a more representative sample. Its justification lies in the assumption that since a sample is more representative with respect to the known traits of the population, it may also be more representative with respect to the unknown characteristics of that population.

Finally, judgment sampling may be used to conduct pilot studies. It has been mentioned that, in order to have effective stratification, knowledge of the population traits is essential. Pilot studies are one way of getting this desired information. Judgment samples, like pilot studies, are then merely exploratory surveys made before other sampling designs, such as stratified sampling, are adopted.

In any case, the reliability of sample results in judgment sampling depends on the quality of the sampler's expert knowledge or judgment. If it is good and is carefully and skillfully applied, judgment samples may be expected to be representative and to yield valuable results. On the other hand, when a sample is obtained with poor judgment, serious biases will be present.

11.11 Quota Sampling

Under the procedure known as quota sampling each interviewer is given a quota—a certain number of persons—to be interviewed. The interviewers are often instructed to distribute their interviews among the individuals within their quotas in accordance with some specified characteristics, such as so many in each of several income groups, so many in each race, so many with certain political or religious affiliations, and so on. One important reason for employing quota sampling is to obtain some of the merits of stratification at lower costs. More often than not, however, in the selection of sampling units an interviewer's judgment becomes the overriding factor. Interviewers often tend to interview better-educated or neatly dressed persons, or to question anyone who happens to be at home, or to ignore those sampling units who are not readily accessible. As a consequence, a quota sample actually depends on the interviewer's judgment and convenience. Because results always contain unknown biases, the sample cannot be treated as a random variable.

Quota sampling is widely used in public opinion studies. It occasionally produces satisfactory results if the interviewers are carefully trained and if they follow their instructions closely. More often, however, the satisfactory results obtained by quota sampling are due to the "success by gratuity of the universe," which means a lack of correlation between the selection of elements and the characteristics under study. In other words, quota sampling is sometimes successful because of the chance circumstance that certain traits of individuals induce interviewers to include those individuals in the sample but have no relationship whatsoever to the traits being investigated.

11.12 Convenience Sampling

The method of convenience sampling is also called the *chunk*. A chunk is a fraction of the population taken for investigation because of its convenient availability. Thus, a chunk is selected neither by probability nor by judgment but by convenience. A sample obtained from readily available lists, such as telephone directories or automobile registrations, is a convenience sample and not a random sample, even if the sample is drawn at random from the lists. Failure to recognize this difference often leads samplers astray.

A chunk—which is merely a convenient slice of the universe—can hardly be representative of the population. Its results are usually greatly biased and unsatisfactory. Formerly, the chunk was frequently used in public opinion surveys when interviewers stopped near the railroad station or the bus depot or in front of office buildings to interview people. Today, accountants still use convenience sampling to analyze or audit accounts. Their practice is to take for analysis a small slice of, say, accounts payable, in some convenient manner, such as all accounts under the letter "L," to verify the total population, say, of all accounts payable of the firm.

Finally, it may be mentioned that convenience sampling is also useful in making pilot studies. Questions may be tested and preliminary information may be obtained by the chunk before the final sampling design is decided upon.

Problems

11.1 What are the reasons for sampling?

11.2 Explain the relationship of each of the following concepts to the theory of sampling: (a) "diversity in unity"; (b) law of large numbers; and (c) randomization.

11.3 Differentiate the following pairs of terms:
a. sampling error and nonsampling error
b. elementary and primary sampling units
c. target and sampled population
d. probability sample and judgment sample
e. precision and accuracy
f. expected value and population parameter

11.4 How can stratification increase efficiency?

11.5 Under what conditions can cluster sampling be more efficient than other types of random sampling designs?

11.6 What is systematic sampling? What are its advantages and drawbacks?

11.7 What is the philosophy that leads to double, multiple, and sequential sampling?

11.8 In what sense do we say that acceptance sampling is similar to testing?

11.9 Define judgment sampling, quota sampling, and convenience sampling. Under what conditions can each of these designs be used to advantage?

11.10 Do you agree with the following statements? Explain.
a. Sample studies can never be as accurate as complete counts of populations.
b. Purposive sampling is sometimes better than random sampling.
c. A random sample of workers in a firm may be obtained by taking every tenth name from the payroll list of the firm.
d. A random sample of the households in a state may be obtained by taking every thousandth name from the state's automobile registration directory.

11.11 Determine the population, the frame, and the sampling design most appropriate for the following situations:
a. An accountant, auditing the books of a department store, wishes to confirm the accounts receivables as shown on the company books.
b. The Federal Reserve Board wishes to select a sample of commercial banks to analyze the effects of a tight money policy upon the discount and loan policies of the commercial banks.
c. A manufacturer wishes to control the quality of his products by taking samples from time to time for examination.
d. The Bureau of Labor Statistics wishes to construct a monthly consumer price index on the basis of a sample taken each month.
e. A market research director plans to prepare a questionnaire to explore the market potential for a new product. To help the design of the questionnaire a sample of the present users and nonusers of a close substitute for the new product is to be interviewed.
f. A labor union is to select a sample among its members to determine the attitude of the members toward a recently introduced social activities program.
g. The state government of West Virginia wants to determine the level of unemployment in the state on the basis of a sample of 2500 households.

12

CHI-SQUARE, F, AND STUDENT'S DISTRIBUTIONS

12.1

Exact Sampling Theory

Up to now, we have been concerned with making inferences from large samples. The concept "large sample," as the reader is aware by now, is a term relative to the sampling distribution of the statistic under consideration. For instance, when we make inferences about μ, a sample is considered large if its size is greater than 30, or if the parent population is normal and σ is known. In the case of proportions, a sample is large if $n(\pi) > 5$. For variances, however, a sample is said to be large if $n > 100$ and if the parent population is normal. In general, large sampling theory refers to a class of situations in which the probability distribution of a sample statistic is normal or approximately normal, either because the population under investigation is normal or the sample size is sufficiently large for the central limit theorem to be operative.

Sometimes, however, we may be called upon to make inferences when the underlying assumptions for large sampling theory are not available. Situations may arise in which either the population standard deviation is unknown because we are confronted with a new problem or a new theory, or the sample size is small due to such physical limitations as those related to medical research or to such practical limitations as the high costs involved in sample observations.

Although with small samples we do not have as much information as we would like, we are not destitute. The proper course is to draw conclusions or to make decisions from the sample by taking into account the scanty nature of the evidence. As a matter of fact, procedures of statistical inference with small samples are the same as those presented in the preceding two chapters, except that now we cannot apply the central limit theorem and assume that sampling distributions are normal. The study of statistical inferences with small samples is called the *small sampling theory* or, more appropriately, *exact sampling theory* since results obtained for small samples hold for large samples as well. The basic difference between large and small sampling is mainly between sampling distributions: for large samples, sampling distributions are assumed to be normal; for small samples, sampling distributions may take a number of forms. In addition to the binomial distribution (which has been discussed in detail before), there are three important probability distributions which the distribution of a statistic with small n may assume. They are chi-square, F, and Student's distributions. All three models are related to the normal probability model and are defined in terms of the number of degrees of freedom. In the discussions that follow, we shall first introduce the concept of "degrees of freedom," then present the properties of these three distributions, and finally take up their applications.

12.2 Degrees of Freedom

We have seen that the sample variance computed as

$$s^2 = \frac{1}{n} \sum (x - \bar{x})^2$$

is a biased estimate of σ^2, since

$$E(s^2) = \frac{n-1}{n} \sigma^2 = \sigma^2 - \frac{\sigma^2}{n}$$

The bias here is measured by $-\sigma^2/n$, which is quite pronounced when n is small. For example, if $n = 4$, then

$$E(s^2) = \frac{4-1}{4} \sigma^2 = \frac{3}{4} \sigma^2$$

Thus, with $n = 4$, s^2 underestimates, on the average, σ^2 by $\frac{1}{4}$ or 25 percent. To correct this bias, we may compute sample variance as

$$\hat{s}^2 = \frac{1}{n-1} \sum (x - \bar{x})^2$$

since then

$$E(\hat{s}^2) = E\left(\frac{n}{n-1}s^2\right) = \left(\frac{n}{n-1}\right)\left(\frac{n-1}{n}\sigma^2\right) = \sigma^2$$

We may say that $ns^2/(n - 1)$ is an unbiased estimate of σ^2.

To compute $\hat{s}^2$, we divide the sum of squared deviations by what is called $n - 1$ degrees of freedom. The *number* of *degrees of freedom* (df) is a property of a sum of squares and is determined by the number of independent linear comparisons which can be made among n observations. A rigorous discussion of this mathematical concept is beyond the scope of this book. Nevertheless the essence of the concept of degrees of freedom can be brought out by way of a simple illustration.

From the definition of $\hat{s}^2$, we see that it is a measure based on squared deviations from the sample mean. We may recall also that $\Sigma(x - \bar{x}) = 0$. Bearing these two facts in mind, if we are asked to guess about the deviations from $\bar{x}$ of a sample of, say, $n = 5$, we are "free" to assign any value to the first four deviations, such as

$$\begin{aligned}(x_1 - \bar{x}) &= -5\\ (x_2 - \bar{x}) &= 3\\ (x_3 - \bar{x}) &= 6, \text{ and}\\ (x_4 - \bar{x}) &= -1\end{aligned}$$

Once these values are freely assigned, we are no longer "free" to assign any value to the fifth deviation, since this last deviation must be such that the sum of all five deviations is zero. Thus, the value of the fifth deviation cannot be any value other than

$$(x_5 - \bar{x}) = 0 - ((-5) + 3 + 6 + (-1)) = -3$$

The foregoing statement leads us to conclude that only $n - 1$ independent linear comparisons can be made among n things. That is, given the values of the $n - 1$ deviations from $\bar{x}$, the nth deviation is completely determined. So we say that the sum of the squares, $\Sigma(x - \bar{x})^2$, has $n - 1$ degrees of freedom associated with it. However, the quantity $\Sigma(x - \mu)^2$ has n degrees of freedom, since it contains one more possible linear comparison, that of $\bar{x}$ with μ.

The last observation also leads us to another intuitive meaning of the number of degrees of freedom of a statistic. It may be considered as the number of n independent observations in the sample minus the number of m parameters (required to compute the statistic) which must be estimated by sample observations. Denote the number of degrees of freedom by δ; then for any statistic, $\delta = n - m$. We might mention that in other texts, the number of degrees of freedom is often denoted as ν (nu).

In the case of $\hat{s}^2$, the number of independent observations is n, but to compute it we need μ, which is estimated by $\bar{x}$. Thus, $m = 1$, and $\hat{s}^2$ has $\delta = n - 1$ as before. Similarly, each of the following two statistics has $n - 1$ degrees of freedom:

$$\hat{s}^2 = \frac{\Sigma(x - \bar{x})^2}{n - 1}, \text{ and}$$

$$\hat{\sigma}_{\bar{x}}^2 = \frac{s^2}{n - 1}$$

The former is an unbiased estimate of σ^2 and the latter is an unbiased estimate of the variance of the distribution of the mean.

Now, suppose that two independent samples with n_1 and n_2 are drawn from two normal populations which are assumed to possess identical variance; that is, $\sigma_1^2 = \sigma_2^2 = \sigma^2$, then $\hat{s}_1^2 = n_1s_1^2/(n_1 - 1)$ and $\hat{s}_2^2 = n_2s_2^2/(n_2 - 1)$ are, respectively, unbiased estimates of σ_1^2 and σ_2^2. Since it has been assumed that the two populations have the same variance, a better estimate of this common variance can be obtained by pooling the two sample variances. This is done by computing what is called *a pooled variance,* which is a weighted mean of the two sample variances; namely,

$$\hat{\sigma}^2 = \frac{n_1s_1^2 + n_2s_2^2}{n_1 + n_2 - 2}$$

which is an unbiased estimate of the common population variance with $\delta = n_1 + n_2 - 2$. Two degrees of freedom are lost in this case because $m = 2$.

It is interesting to note that this pooled variance can also be employed to estimate the standard error of a difference between two sample mean. Since

$$\hat{\sigma}_{\Delta\bar{x}} = \sqrt{\frac{\hat{\sigma}_1^2}{n_1} + \frac{\hat{\sigma}_2^2}{n_2}} = \sqrt{\frac{n_1s_1^2 + n_2s_2^2}{n_1 + n_2 - 2}\left(\frac{1}{n_1} + \frac{1}{n_2}\right)} = \hat{\sigma}\sqrt{\frac{1}{n_1} + \frac{1}{n_2}} = \hat{\sigma}\sqrt{\frac{n_1 + n_2}{n_1n_2}} \tag{12.1}$$

which also has $n_1 + n_2 - 2$ degrees of freedom.

12.3 Chi-Square Distributions

If $X_1, X_2, \cdots, X_n$ are independent standard normal variables, then the statistic

$$\chi^2 = X_1^2 + X_2^2 + \cdots + X_n^2$$

is said to have a χ^2 (chi-square) distribution with $\delta = n$ df. There are infinitely many χ^2 distributions, one corresponding to each positive integer n. The χ^2 distribution corresponding to the number δ will be denoted as χ_δ^2 and it is defined by the following density function:

$$f(\chi^2) = \frac{(\chi^2)^{[\delta/2-1]}e^{-[\chi^2/2]}}{2^{\delta/2}\Gamma(\delta/2)}, \text{ for } \chi^2 \geq 0$$

$$= 0, \text{ otherwise}$$

In this expression, δ, as before, is the number of df. For the quantity $\Gamma(\delta/2)$, read "gamma function of $\delta/2$," which refers to a value that is dependent on δ and that is $(\delta/2 - 1)!$ This density function makes that for the normal variable look simple. Fortunately, a full understanding of it is required only in theoretical statistics and for plotting curves. We need have little concern with it here.

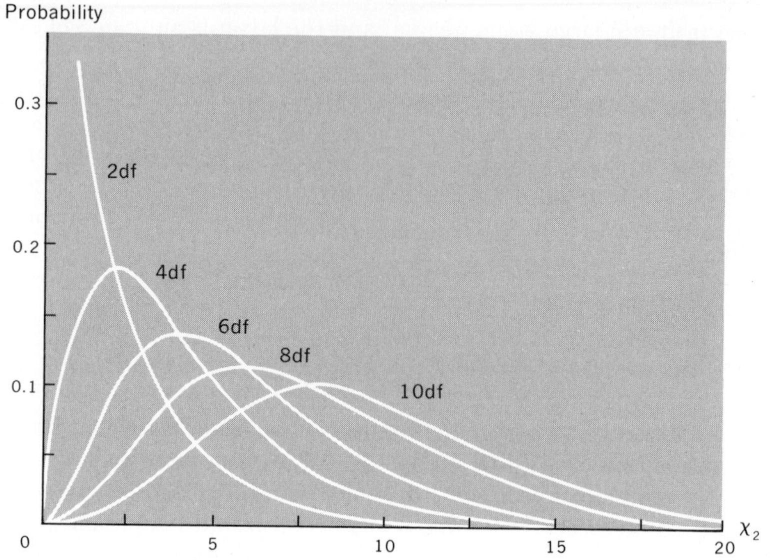

Figure 12.1 χ^2 distribution curves for selected degrees of freedom.

We shall now state the important properties of chi-square distributions, without proof, for our later use. Some of these properties can be deduced by studying Figure 12.1 reflectively, which gives the density curves of chi-square distributions for a number of selected δ.

(1) Chi-square, though denoted by Greek letter χ^2, is a statistic instead of a parameter. As a matter of fact, there is no corresponding parameter for χ^2.

(2) If X has a standard normal distribution, then X^2 has the distribution of $\chi_1{}^2$; that is, chi-square distribution with 1 df.

(3) If X_1 is $\chi_{\delta_1}{}^2$ and X_2 is $\chi_{\delta_2}{}^2$ and if X_1 and X_2 are independent, then $X_1 + X_2$ is $\chi^2_{\delta_1+\delta_2}$. This is commonly called the *additive property* of chi-square distributions.

(4) If X is a standard normal variable and $x_1, x_2, \cdots, x_n$ are n observations of a random sample, then $\sum_1^n x_i{}^2$ has the distribution $\chi_n{}^2$.

(5) If X is any normal variable and if $x_1, x_2, \cdots, x_n$ are n observations of a random sample, then $\sum_1^n \left(\frac{x_i - \mu}{\sigma}\right)^2$ has the distribution $\chi_n{}^2$.

It may be noted that properties (4) and (5) are equivalent statements. In (4) a random sample is drawn from $n(0,1)$; in (5) a random sample is drawn from $n(\mu,\sigma^2)$, and $(x_i - \mu)/\sigma$ transforms (5) into (4).

(6) A χ^2 distribution ranges from 0 to infinity, since it is the sum of squared values.

(7) A χ^2 distribution is completely defined by the number of degrees of freedom. If X is $\chi_\delta{}^2$, then its mean and variance are $\mu = \delta$ and $\sigma^2 = 2\delta$, respectively.

(8) Chi-square distributions are positively skewed. As δ gets large, however, χ_δ^2 approaches the normal distribution with $\mu = \delta$ and $\sigma = \sqrt{2\delta}$. In practice, probabilities for chi-square distributions, when $\delta > 30$, can be computed by employing normal approximations. For example, if X is χ_{50}^2, then X is approximately normal with $\mu = 50$ and $\sigma = 10$. Supposing that we desire to find $P(X < 40)$, we may find that

$$P(X < 40) \doteq N\left(\frac{40 - 50}{10}\right) = N(-1) = 0.1587$$

A more accurate method of approximating chi-square probabilities by the normal model is to use the fact that

$$Y = \sqrt{2X} - \sqrt{2\delta - 1} \tag{12.2}$$

is approximately normal. By employing this expression, we have, for the previous illustration

$$\begin{aligned} P(X < 40) &= P(2X < 80) = P(\sqrt{2X} < \sqrt{80}) = P(\sqrt{2X} < 8.94427) \\ &= P(\sqrt{2X} - \sqrt{99} < 8.94427 - \sqrt{99}) = P(Y < 8.94427 - 9.94987) \\ &= P(Y < -1.01) \doteq N(-1.01) = 0.1562 \end{aligned}$$

which is smaller, and a better estimate, than the value obtained before. It can be checked from chi-square probability tables. Table A-VIII in the Appendix gives various percentile points for chi-square distributions for $1 \leq \delta \geq 30$. For instance, if X is χ_{15}^2, then from Table A-VIII, we find that

$$\begin{aligned} P(X > 14.339) &= P(14.339 < \chi_{15}^2 < \infty) = 0.50 \\ P(X > 8.547) &= P(8.547 < \chi_{15}^2 < \infty) = 0.90 \\ P(X > 24.996) &= P(24.996 < \chi_{15}^2 < \infty) = 0.05 \end{aligned}$$

and so forth. Thus, this table is in the cumulative form and gives upper-tail probabilities for chi-square distributions.

12.4 F Distributions

If $X_1, X_2, \cdots, X_m$ and $Y_1, Y_2, \cdots, Y_n$ are all independent standard normal variables, then the statistic

$$F = \frac{n}{m} \cdot \frac{X_1^2 + X_2^2 + \cdots + X_m^2}{Y_1^2 + Y_2^2 + \cdots + Y_n^2}$$

is said to have the F distribution with (m,n) degrees of freedom. It turns out that an F distribution is the ratio of two chi-square distributions with δ_1 as the number of df for the numerator and with δ_2 as the number of df for the denominator. It

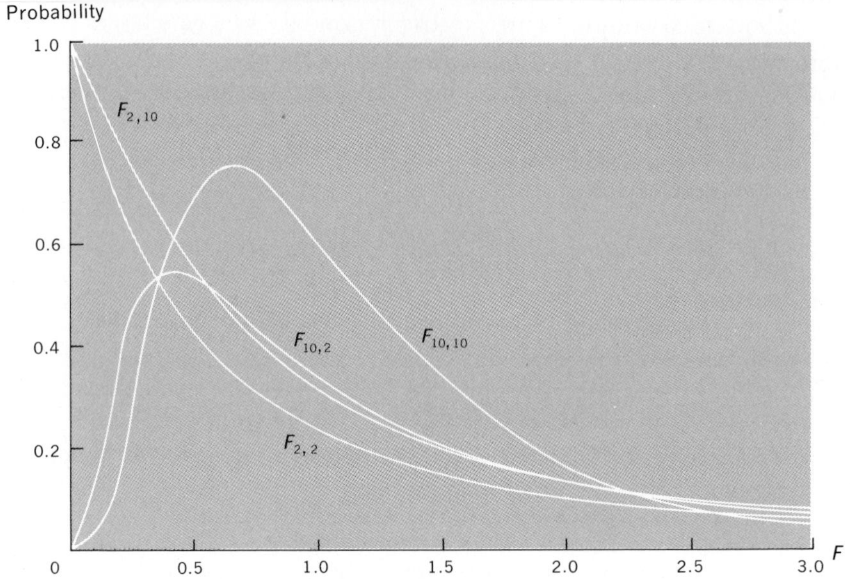

Figure 12.2 F distribution curves.

may be denoted as $F_{\delta_1\delta_2}$. $F_{\delta_1\delta_2}$ is defined by the density

$$f(F) = \frac{\Gamma((\delta_1 + \delta_2)/2)}{\Gamma(\delta_1/2)\Gamma(\delta_2/2)} (\delta_1/\delta_2)^{\delta_1/2} F^{(\delta_1/2)-1} \left(1 - \frac{\delta_1 F}{\delta_2}\right)^{-(\delta_1+\delta_2)/2}, \quad 0 \leq F \leq \infty$$

The only important thing for us to note about this formidable expression is that an F distribution has two parameters, δ_1 and δ_2, the numbers of df. Density curves for $F_{2,10}$, $F_{2,2}$, $F_{10,2}$, and $F_{10,10}$ are shown in Figure 12.2.

Let us now observe some of the properties of F distributions. First of all, since an F distribution is a ratio of two squared quantities, its value ranges from 0 to infinity. Second, there is an F distribution for each pair of positive integers, δ_1 and δ_2. Third, an F distribution is positively skewed, but its skewness decreases with increases in the numbers of df. Fourth, the mean of $F_{\delta_1\delta_2}$ is $\mu = (\delta_2/(\delta_2 - 2))$ when $\delta_2 > 2$, and the variance of $F_{\delta_1\delta_2}$ is $\sigma^2 = 2\delta_2{}^2(\delta_1 + \delta_2 - 2)/\delta_1(\delta_2 - 2)^2(\delta_2 - 4)$ when $\delta_2 > 4$. An F distribution has no mean when $\delta_2 \leq 2$, and it has no variance when $\delta_2 \leq 4$. Fifth and last, if X is $F_{\delta_1\delta_2}$, then $Y = 1/X$ is $F_{\delta_2\delta_1}$. This is the famous *reciprocal property* of F distributions.

Percentage points for the right tail of the F distribution at 10, 5, and 1 percent levels are given in Table A-IX in the Appendix. When X is, say $F_{10,7}$, then

$$P(X > 3.64) = P(F_{10,7} > 3.64) = 0.05$$

To say that $P(F_{10,7} > 3.64) = 0.05$ is the same as to say that

$$P(0 < X < 3.64) = P(0 < F_{10,7} < 3.64) = 1 - 0.05 = 0.95$$

See, for example, Figure 12.3.

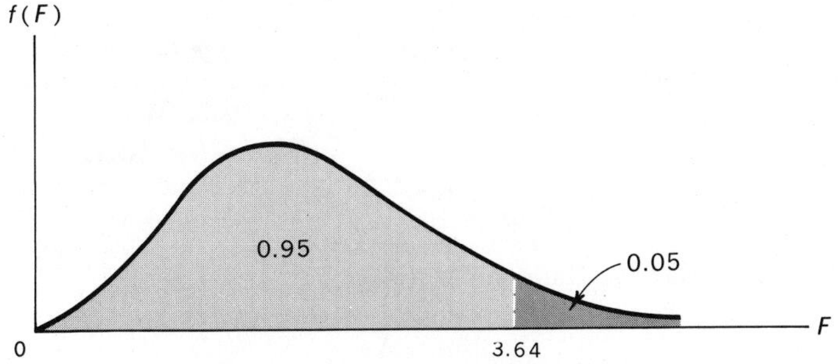

Figure 12.3 Upper-tail probability of $F_{10,7}$.

Similarly,

$$P(F_{10,7} > 6.62) = 0.01, \quad \text{and} \quad P(0 < F_{10,7} < 6.62) = 0.99$$

Table A-IX can also be employed to find probabilities for the left-tail points of F distributions by following the reciprocal property of F distributions. Suppose X is $F_{15,10}$, we may let Y be $F_{10,15}$. From Table A-IX, we see that

$$P(Y > 2.54) = P(F_{10,15} > 2.54) = 0.05$$

Then, by the reciprocal property,

$$P\left(\frac{1}{Y} < \frac{1}{2.54}\right) = 0.05$$

and since $1/Y$ has the same distribution as X, we have

$$P\left(X < \frac{1}{2.54}\right) = P\left(F_{15,10} < \frac{1}{2.54}\right) = P(F_{15,10} < 0.41) = 0.05$$

By the same procedure, of course, the 1 percent left-tail point can be found (see Figure 12.4). In general, to find the left-tail critical value for $F_{\delta_1\delta_2}$ at a given percentage point, we first find that for $F_{\delta_2\delta_1}$ at the percentage point, and then take the reciprocal value.

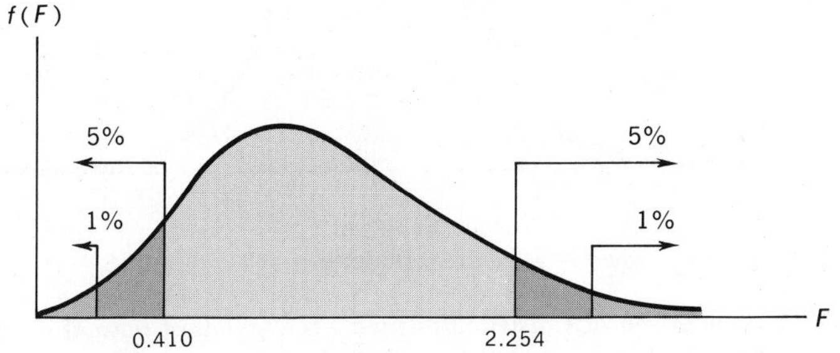

Figure 12.4 Illustration of the reciprocal property of F distributions with $F_{15,10}$.

12.5

Student's Distributions

If $X_0, X_1, X_2, \cdots, X_n$ are independent standard normal variables, then the statistic

$$t = \frac{X_0}{\sqrt{\frac{1}{n}(X_1^2 + X_2^2 + \cdots + X_n^2)}}$$

is said to have a Student's, or t, distribution with n df. It is interesting to note that

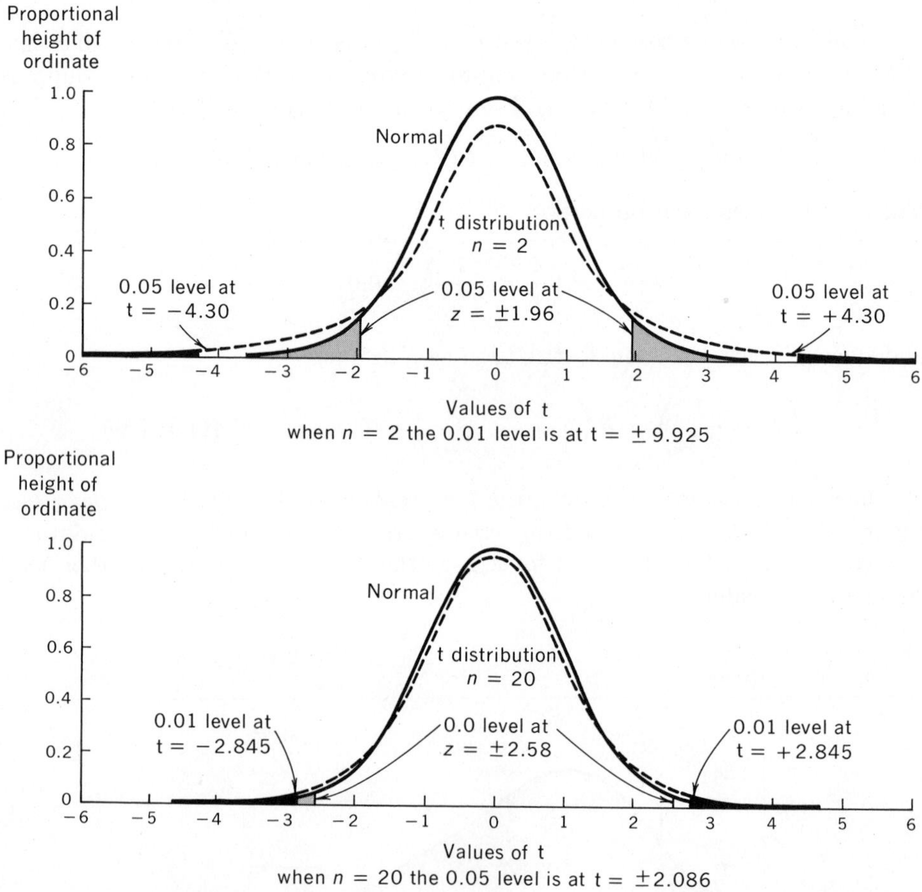

Figure 12.5 Comparison of t distributions with the standard normal distribution for $n = 2$ and $n = 20$.

t is a ratio of a normal distribution to the square root of a chi-square distribution. There is a t distribution corresponding to each positive integer. The density for a t distribution with δ, the number of df, is given by

$$f(t) = \left(\frac{1}{\sqrt{\delta\pi}}\right)\frac{\Gamma(\delta+1)/2}{\Gamma(\delta/2)}\left(1+\frac{t^2}{\delta}\right)^{-(\delta+1)/2}, \quad -\infty < t < \infty$$

Density curves for two t distributions, one with $\delta = 2$ and one with $\delta = 20$, are plotted against the standard normal distribution in Figure 12.5.

There are a number of important properties of t distributions worth noting. First, a t distribution ranges from $-\infty$ to ∞ just as does a normal distribution. Second, like the standard normal distribution, a t distribution is symmetrical with mean, mode, and median equal to zero, except that t_1 (t distribution with 1 df) has no mean. Third, the variance of t_δ, with $\delta > 2$, is $\delta/(\delta - 2)$; when $\delta \leq 2$, t_δ possesses no variance. Fourth, a t distribution has a greater dispersion than the standard normal distribution. However, when δ gets large, t_δ approaches the standard normal distribution as a limit. These observations are evident from the standard deviation of t_δ, which is $\sqrt{\delta/(\delta-2)}$. This quantity is always greater than 1, but it comes closer to 1 as δ becomes larger. In practice, we may treat t_δ as $n(0,1)$ when $\delta > 30$.

Table A-X in the Appendix gives critical values at both tail ends of t distributions for a number of selected percentage points. For example, if X is t_2, then

$$P(-4.303 < X < 4.303) = P(-4.303 < t_2 < 4.303) = 0.95$$

From this we see that

$$\begin{aligned} P(t_2 < -4.303) &= 0.025, \text{ and} \\ P(t_2 > 4.303) &= 0.025 \end{aligned}$$

See Figure 12.5 for $\delta = 2$.

Similarly, if X is t_{20}, we have

$$\begin{aligned} P(-2.845 < t_{20} < 2.845) &= 0.99 \\ P(t_{20} < -2.845) &= 0.005 \\ P(t_{20} > 2.845) &= 0.005 \\ P(t_{20} > 2.528) &= 0.01 \\ P(t_{20} < -1.725) &= 0.05 \end{aligned}$$

and so on. As a matter of fact, the column headings of Table A-X are levels of significance for two-tail tests. Entries in the body are the critical values of t_δ for such tests. If we would like to have a single-tail test with, say, $\alpha = 0.05$, the critical value is found in the column with $2\alpha = 0.10$ for the appropriate number of df. For example, for $P(t_{20} < -1.725) = 0.05$, the number -1.725 is found in the column headed by 0.10 corresponding to $\delta = 20$.

12.6 Statistical Inference about Population Variance

The unbiased point estimate of variance with a sample of size n was defined as

$$s^2 = \frac{1}{n-1} \sum (x - \bar{x})^2$$

If we divide both sides of this expression by the population variance σ^2, then we would have

$$\frac{s^2}{\sigma^2} = \frac{1}{n-1} \sum \left(\frac{x - \bar{x}}{\sigma}\right)^2$$

which can be written as

$$\frac{(n-1)s^2}{\sigma^2} = \sum \left(\frac{x - \bar{x}}{\sigma}\right)^2$$

The right side of this result, from property (5) of chi-square distributions, would have a chi-square distribution with $n - 1$ df, provided that the population is normally distributed.

Thus, we can formally state: If s^2 is the variance of a random sample with size n taken from a normal population, the ratio $(n - 1)s^2/\sigma^2$ is χ^2_{n-1}. Or, we can write

$$\chi^2 = \frac{(n-1)s^2}{\sigma^2} \tag{12.3}$$

with $\delta = n - 1$. This ratio is an appropriate testing statistic for hypotheses about population variances.

Suppose that the Transit Authority of New York City plans to purchase light bulbs for its subway system. The Authority wishes to have bulbs that possess not only long life but also a high degree of uniformity. It decides, on the basis of past experience, that the variance should not exceed 250 hours. Now, a test of 20 bulbs of a certain make yields a mean life of 1000 hours, which is considered to be satisfactory, but with a variance of 300 hours². Does this result indicate that the population variance exceeds 250?

Following the general procedure of testing developed in Chapter 10, we have now the following test:

1. **Hypotheses:** $H_0: \sigma^2 \leq 250$
   $H_1: \sigma^2 > 250$

2. **Level of Significance:** Arbitrarily set at 0.05.

3. **Testing Statistic:** $\chi^2 = \dfrac{(n-1)s^2}{\sigma^2}$, which has a chi-square distribution with $\delta = n - 1 = 19$

4. **Decision Rule:** Since $P(\chi_{19}^2 > 30.14) = 0.05$, H_0 will be rejected if and only if the observed chi-square value is greater or equal to 30.14, and H_0 will be accepted if and only if the observed chi-square value is less than 30.14.

5. **Computations:**

$$\chi^2 = \frac{(20-1)300}{250} = 22.80$$

6. **Decisions:** Since the observed χ^2 falls in the acceptance region, H_0 cannot be rejected. This sample result, in other words, is not sufficient evidence for us to conclude that $\sigma^2 > 250$. The Authority may, therefore, consider this make of light bulb satisfactory and decide on its purchase.

Tests on variances are typically of the upper-tail variety, since we are usually concerned with the fact that the variance may be too large. When we wish to conduct a two-sided test with chi-square distributions, we divide the region of rejection into two equal parts. For instance, if $\delta = 19$ and $\alpha = 0.05$ as in the previous test but the alternative is stated as $\sigma^2 \neq 250$, we would reject H_0 if $\chi^2 < 8.91$ or if $\chi^2 > 32.85$. In other words, we find the χ^2 value which has 97.5 percent of the area to the right of it (8.91) and the value has 2.5 percent of the area to the right of it (32.85). Note that $P(8.91 < \chi_{19}^2 < 32.85) = 0.95$.

To have an interval estimate of σ^2, as before, we try to construct a confidence interval such that it will contain σ^2 at a specified confidence probability, $1 - \alpha$. Such an interval may be made by using the information that if equation (12.3) is solved for σ^2, we would have

$$\sigma^2 = \frac{(n-1)s^2}{\chi^2}$$

Now, inserting the values of $(n - 1)$, s^2, and χ^2, we can estimate the lower and upper confidence limits for σ^2. Specifically, the confidence interval for σ^2 is

$$P\left[\frac{(n-1)s^2}{\chi^2_{(n-1,\alpha/2)}} < \sigma^2 < \frac{(n-1)s^2}{\chi^2_{(n-1,1-\alpha/2)}}\right] = 1 - \alpha \qquad \textbf{(12.4)}$$

That is, for σ^2, the lower (L) and upper (U) confidence limits are

$$L = \frac{(n-1)s^2}{\chi^2_{(n-1,\alpha/2)}}, \quad \text{and} \quad U = \frac{(n-1)s^2}{\chi^2_{(n-1,1-\alpha/2)}}$$

These chi-square values depend upon the number of degrees of freedom and the confidence level. These are indicated by the subscripts to χ^2 in the foregoing expressions. What we should note once more is that the χ^2 table gives the probability that a given χ^2 value is exceeded. Thus, if we desire to have a 95 percent confidence interval, we should choose the value of $\chi^2_{\delta,0.025}$ for the lower confidence limit and the value of $\chi^2_{\delta,0.975}$ for the upper confidence limit. Similarly, if $1 - \alpha = 0.99$, then values of $\chi^2_{\delta,0.005}$ and $\chi^2_{\delta,0.995}$ should be used for the lower and upper confidence limits, respectively. It is interesting to observe that the larger chi-square value is used for L and the smaller chi-square value is used for U in constructing a confidence interval for σ^2 because, as can be seen from equation (12.3), χ^2 and σ^2 are inversely related.

As an example, let us now construct a 95 percent confidence interval for σ^2, using the illustrative material from the preceding test. Recalling that, in this instance, $n = 20$ and $s^2 = 300$, we have, therefore,

$$L = \frac{(20 - 1)(300)}{\chi^2_{19,0.025}} = \frac{(19)(300)}{32.85} = 173.5, \text{ and}$$

$$U = \frac{(20 - 1)(300)}{\chi^2_{19,0.995}} = \frac{(19)(300)}{8.91} = 639.7$$

Or,

$$173.5 < \sigma^2 < 639.7$$

The confidence interval for the standard deviation can be derived directly from that for the variance. This is done by taking the square roots of the limits for σ^2. Thus, the corresponding 95 percent confidence interval for our illustrative problem is

$$\sqrt{173.5} < \sigma < \sqrt{639.7}; \quad \text{that is } 13.2 < \sigma < 25.3$$

Sometimes, especially in the case of estimating σ^2, we are concerned with only an upper confidence limit, not with the lower. In such a situation, we state that the variance is equal to or less than some specific value. Namely, we are interested in constructing a confidence interval of the form $0 < \sigma^2 < U$ with a confidence probability of $1 - \alpha$. Here, U can be found by the expression

$$U = \frac{(n - 1)s^2}{\chi^2_{(n-1,1-\alpha)}} \tag{12.5}$$

where the $\chi_\delta{}^2$ is that value which has $1 - \alpha$ of the area to the right of it. Here, $0 < \sigma^2 < U$ is called a *one-sided confidence interval*, with $1 - \alpha$, and corresponds to an upper-tail test at α.

The 95 percent one-sided confidence interval for the light bulb illustration has

$$U = \frac{(20 - 1)(300)}{\chi^2_{19,0.95}} = \frac{(19)(300)}{10.12} = 563$$

With this value of U, we see that $0 < \sigma^2 < 563$, which contains the hypothetical value of 250 in the null hypothesis for the previous test. This explains once again why H_0 was not rejected.

There is naturally a corresponding confidence interval of the form $L < \sigma^2 < \infty$, which is simply called the other one-sided confidence interval but which is of little practical significance. Why?

In concluding this section, we point out that chi-square is an amazingly versatile statistic. In addition to its use for inference with variances, it can also be used to test any hypothesis concerning categorical data, of which the binomial form is but a special case. With categorical data, observations consist mainly of frequencies occurring in the various classes or compartments of a table of simple or multiple classifications. Owing to the differences in classification and in reasons for testing, chi-square tests can be identified as those most useful to determine frequencies of one-way classification within two or more categories, the goodness-of-fit of a specific probability or density function to sample data grouped into several classes, the independence of cross classifications (or contingency tables), and homogeneity. These chi-square tests constitute a part of what is called nonparametric statistics, a topic to be surveyed in Chapter 14.

12.7

F Test of the Equality between Variances

We turn now to consider the problem of comparing two population variances. If we have two populations, X_1 and X_2, with variances σ_1^2 and σ_2^2 respectively, we would naturally wish to ascertain whether these variances are equal or one is greater than the other. One convenient way to do this is to compute the difference between two sample variances. Another obvious procedure is to form a ratio of sample variances.

If the two samples are independent, then sample variances would be independent. Furthermore, if populations are normal (or approximately so) and the samples are large, $n_1 \geq 100$ and $n_2 \geq 100$, we could then construct a large sample test for $\sigma_1^2 - \sigma_2^2$ on the basis of $s_1^2 - s_2^2$, which is an asymptotically normally distributed statistic.

Usually, when samples are small, though other assumptions mentioned above are fulfilled, we would use the ratio of the sample variances as the basis for comparison of population variances. This procedure has two advantages: First, the ratio is independent of scale, provided that both populations are given in the same units. Second, the probability distribution of this ratio, under the null hypothesis that $\sigma_1^2 = \sigma_2^2$, is independent of population parameters. Let us see how this ratio is distributed.

When we have available two sample variances, s_1^2 and s_2^2, derived from two independent samples with n_1 and n_2 drawn from two normal populations with σ_1^2 and σ_2^2, then $n_1s_1^2/(n_1 - 1)$ would be an unbiased estimator of σ_1^2 and it would have a χ^2 distribution with $\delta_1 = n_1 - 1$. Similarly, $n_2s_2^2/(n_2 - 1)$ would be an unbiased estimator of σ_2^2 and be distributed as χ^2 with $\delta_2 = n_2 - 1$. Clearly, the ratio of these two estimators would be distributed as $F_{\delta_1\delta_2}$. That is,

$$F_{\delta_1\delta_2} = \frac{n_1s_1^2/(n_1 - 1)}{n_2s_2^2/(n_2 - 1)} \tag{12.6}$$

To use (12.6) as the testing statistic for comparing two population variances, we state the null hypothesis either as $\sigma_1^2/\sigma_2^2 = 1$ or as $\sigma_1^2 = \sigma_2^2$. If the alternative is one-sided, it usually takes the form of an upper-tail test; namely, $\sigma_1^2 > \sigma_2^2$. Under this alternative, H_0 would be rejected if F is too large for the specified level of α. If the alternative is two-sided, that is, $\sigma_1^2 \neq \sigma_2^2$, H_0 would be rejected if F is either too small or too large. We may note also, with F tests, that the larger sample variance is used as the numerator due to the designation of H_0.

Suppose that the purchasing agent for a truck company is considering buying either Make A or Make B tires, which appear to have the same mean durability, but he is not quite sure about their variability. To determine whether their variances are the same, he draws one sample from each make of tire and obtains the following results:

$$n_A = 16 \qquad n_B = 21$$
$$s_A^2 = 9{,}965 \text{ miles}^2 \qquad s_B^2 = 5{,}261 \text{ miles}^2$$

There is evidently a difference in sample variances, but is it significant or is it just due to chance variations?

1. **Hypotheses:** $H_0: \sigma_A^2 = \sigma_B^2$
   $H_1: \sigma_A^2 > \sigma_B^2$

2. **Level of Significance:** $\alpha = 0.10$, arbitrarily set.

3. **Testing Statistic:** $\dfrac{n_A s_A^2/(n_A - 1)}{n_B s_B^2/(n_B - 1)}$ which is distributed as $F_{15,20}$

4. **Decision Rule:** Reject H_0 if and only if $F > 1.84$, since $P(F_{15,20} > 1.84) = 0.10$.

5. **Computations:** With data given before,

$$F_{15,20} = \frac{16(9965)/15}{21(5261)/20} = 1.92$$

6. **Decisions:** The critical value of F falls in the rejection region. H_0 is false. Make B tires should be bought; if the make does not differ from A in other aspects.

Typically, the comparison between two variances is made with an upper-tail test rather than with a two-tail test, since it is obviously more meaningful to ascertain if one variance is significantly greater than another, $\sigma_1^2 > \sigma_2^2$, rather than to determine if two variances are significantly different from each other, $\sigma_1^2 \neq \sigma_2^2$.

If, for some reason, a two-tail test is desired, it can be made with the F test easily. For example, if we wish to test $\sigma_A^2 = \sigma_B^2$ against $\sigma_A^2 \neq \sigma_B^2$ with data in the previous illustration, with $\alpha = 0.10$ as before, then H_0 will be rejected if the computed F value is greater than 2.20 or if it is less than 1/2.33. These critical values are found by noting that $P(F_{15,20} > 2.20) = 0.05$, and that $P(F_{15,20} < 1/2.33) = 0.05$. The smaller rejection value is obtained by utilizing the reciprocal property of F distributions. Since we know that $P(F_{20,15} > 2.33) = 0.05$, we know then that $P(F_{15,20} < 2.33) = 0.05$. Note that the computed F value for this case is 1.92, which falls in the acceptance region $1/2.33 < F_{15,20} < 2.20$. Therefore, H_0 cannot be rejected at $\alpha = 0.10$ for the two-sided alternative. This illustrates once again that the same sample data may be significant for a single-tail test but insignificant for a two-tail test.

12.8 Inferences with Student's Distributions

We shall in this section discuss the application of t distributions in making statistical inferences about expected values and about the difference between two expected values.

INFERENCES WITH μ WHEN σ IS UNKNOWN AND n IS SMALL

When we test hypotheses about μ, because of large sample size we often assume that the unknown population standard deviation may be replaced by the sample standard deviation; namely, instead of $(\bar{x} - \mu)/(\sigma/\sqrt{n})$, we use $(\bar{x} - \mu)/(\hat{\sigma}/\sqrt{n})$ as the testing statistic. The latter expression is assumed to be distributed approximately as a standard normal variable. The denominator in this ratio, $\hat{\sigma}/\sqrt{n} = s/\sqrt{n-1}$, however, is a random variable and it can hardly be considered as a constant when samples are small. But noting that $(\bar{x} - \mu)$ is a normal variable with zero mean, and that $s/\sqrt{n-1}$ is the square root of a chi-square variable with $\delta = n - 1$, by definition, then the ratio

$$t = \frac{\bar{x} - \mu}{s/\sqrt{n-1}} \tag{12.7}$$

is distributed as t_{n-1}.

To illustrate the t test concerning μ, let us assume that a medical research center announces that it has succeeded in developing a treatment for high blood pressure that can reduce a patient's blood pressure by more than 20 points. Learning about this, a doctor tries the new treatment on ten of his patients who are suffering from high blood pressure. As a result, he finds that their blood pressure has been reduced by 21.5 points on the average, with a standard deviation of 1.5 points. If the doctor is interested in finding out whether the new treatment is as effective as claimed by the research organization, by using the following t test he could determine this on the basis of the information he has obtained from his trials.

1. **Hypotheses:** H_0: $\mu \leq 20$
H_1: $\mu > 20$

2. **Level of Significance:** $\alpha = 0.01$

3. **Testing Statistic:** $t = (\bar{x} - \mu)/(s/\sqrt{n-1})$ which is distributed as t_9.

4. **Decision Rule:** Since $P(t_9 < 2.821) = 0.01$, he would accept H_0 if $t < 2.821$. Note: the critical value 2.821 is found from the column headed by 0.02 and the row headed by 9df in the t table. The column under 0.02 is used for our upper-tail test at $\alpha = 0.01$ because, as stated before, the t table is constructed with two-sided critical values.

5. **Computations:**

$$n = 10, \quad \bar{x} = 21.5, \quad s = 1.5, \quad \hat{\sigma}_{\bar{x}} = \frac{1.5}{\sqrt{10-1}} = 0.5$$

and

$$t = \frac{\bar{x} - \mu}{\hat{\sigma}_{\bar{x}}} = \frac{21.5 - 20.0}{0.5} = 3$$

6. **Decisions:** The difference, 1.5, between the sample mean and the hypothesized population mean is significant. H_0 is rejected in favor of H_1. If there are no undesirable side effects from the use of the new treatment, the doctor may decide to use it exclusively henceforth.

The confidence interval for μ with coefficient $1 - \alpha$ in terms of t distributions is given as

$$P\left(-t_{n-1;\alpha} < \frac{\bar{x} - \mu}{\hat{\sigma}_{\bar{x}}} < t_{n-1;\alpha}\right) = 1 - \alpha$$

which can be rearranged to yield

$$P(\bar{x} - t_{n-1;\alpha}\hat{\sigma}_{\bar{x}} < \mu < \bar{x} + t_{n-1;\alpha}\hat{\sigma}_{\bar{x}}) = 1 - \alpha$$

From this, we see that the desired confidence limits for μ can be written as follows:

$$\bar{x} \pm t_{n-1;\alpha}\hat{\sigma}_{\bar{x}} \tag{12.8}$$

As an illustration, supposing we wish to have an interval estimate of the mean tensile strength of malleable iron castings produced by a certain factory, we may first draw a random sample of, say, 17 castings for observation. Suppose, furthermore, we find that, for this sample,

$$\bar{x} = 57{,}000 \text{ psi}, \quad \text{and} \quad s = 1600 \text{ psi}$$

With these data, a 95 percent confidence interval may be constructed in accordance with (12.8), as below:

$$\bar{x} \pm t_{16;0.05}\hat{\sigma}_{\bar{x}} = 57{,}000 \pm 2.12\left(\frac{1600}{\sqrt{17-1}}\right)$$

or

$$56{,}152 < \mu < 57{,}848$$

Confidence intervals constructed with t distributions can be interpreted in the same way as those with the normal distribution. It is important, however, to note two points: First, the confidence interval constructed with a small sample is wider or less precise than that with a large sample. This is not only because a small sample has a larger standard error but also because the t multiple is larger than the k multiple (for the normal case). Second, if σ is known, confidence intervals for many different random samples of the same size would have the same width, even though they might have different mid-points, which are the different sample means. The reason here is that the standard error is a constant when the population standard deviation is known. When σ is unknown, the standard error is a random variable itself which varies in value from sample to sample. As a result, confidence intervals for different sample means, computed from samples of the same size, will have not only different mid-points but also unequal widths. This last observation, however, should not lead to the wrong inference that confidence intervals constructed with estimated standard errors from small samples have a different meaning. They are still probability statements and should be interpreted in the same way as those constructed under the assumption of normality.

INFERENCES CONCERNING TWO POPULATION MEANS WITH SMALL SAMPLES

In making inferences about two population means when samples are small and population variances are not known, it is again appropriate to employ the t distribution. To test the hypothesis that two population means are equal, we use the following ratio as the testing statistic:

$$t = \frac{\bar{x}_1 - \bar{x}_2}{\hat{\sigma}_{\Delta\bar{x}}} \tag{12.9}$$

where the denominator is the estimated standard error of a difference between two sample means made under the assumption that the two population variances are equal as given by equation (12.1). Note that, in (12.9), the numerator is a normal variable and the denominator is the square root of a chi-square distribution with $n_1 + n_2 - 2$ degrees of freedom. Therefore, this ratio is distributed as $t_{n_1+n_2-2}$.

Suppose we are interested in discovering whether or not there is any real difference between the average annual earnings of carpenters and house painters in a certain city. We draw two independent samples which yield the following information:

Carpenters	House Painters
$n_1 = 12$ men	$n_2 = 15$ men
$\bar{x}_1 = \$6000$	$\bar{x}_2 = \$5400$
$s_1 = \$750$	$s_2 = \$600$

Based on these data, we may conduct the following t test:

1. **Hypotheses:** H_0: $\mu_1 = \mu_2$
   H_1: $\mu_1 \neq \mu_2$

2. **Level of Significance:** $\alpha = 0.05$

3. **Testing Statistic:** $t = (\bar{x}_1 - \bar{x}_2)/\hat{\sigma}_{\Delta\bar{x}}$ which is distributed as $t_{n_1+n_2-2} = t_{25}$.

4. **Decision Rule:** With $\alpha = 0.05$ for a two-sided test, the acceptance region is defined as $-2.06 \leq t_{25} \leq 2.06$. Reject H_0 if $t < -2.06$ or if $t > 2.06$.

5. **Computations:** Estimator of pooled variance:

$$\hat{\sigma} = \sqrt{\frac{n_1 s_1^2 + n_2 s_2^2}{n_1 + n_2 - 2}} = \sqrt{\frac{(12)(750)^2 + (15)(600)^2}{12 + 15 - 2}} = 697$$

Estimate of standard error of the difference between two means:

$$\hat{\sigma}_{\Delta\bar{x}} = \hat{\sigma}\sqrt{\frac{n_1 + n_2}{n_1 n_2}} = 697\sqrt{\frac{12 + 15}{(12)(15)}} = 269.74$$

Critical ratio:

$$t = \frac{\bar{x}_1 - \bar{x}_2}{\sigma_{\Delta\bar{x}}} = \frac{6000 - 5400}{269.74} = 2.224$$

6. **Decisions:** The critical ratio is significant. The null hypothesis that carpenters and house painters have the same annual earnings is rejected at the 5 percent level of significance in favor of the alternative hypothesis that their annual average earnings are different.

As another example, suppose an agricultural experiment showed that 7 test plots planted with one variety of corn yielded, on the average, 95.5 bushels per acre with a standard deviation of 6 bushels per acre; whereas 6 plots planted with another variety of corn yielded 86.7 bushels per acre with a standard deviation of 5.5 bushels per acre. In order to ascertain whether the mean yield of the first variety is significantly greater than that of the second, the research scientist conducted the following test:

1. **Hypotheses:** H_0: $\mu_1 \leq \mu_2$
   H_1: $\mu_1 > \mu_2$

2. **Level of Significance:** $\alpha = 0.05$

3. **Testing Statistic:** $(\bar{x}_1 - \bar{x}_2)/\sigma_{\Delta\bar{x}}$, which is distributed as t_{7+6-2}.

4. **Decision Rule:** Reject H_0 if $t > 2.201$, since $P(t_{11} > 2.201) = 0.05$.

5. **Computations:** Sample information gives

$$\begin{aligned} n_1 &= 7 & n_2 &= 6 \\ \bar{x}_2 &= 95.5 \text{ bu.} & \bar{x}_2 &= 86.7 \text{ bu.} \\ s_1 &= 6.0 \text{ bu.} & s_2 &= 5.5 \text{ bu.} \end{aligned}$$

We have then

$$t_{11} = \frac{95.5 - 86.7}{\sqrt{\dfrac{7(6)^2 + 3(5.5)^2}{7 + 6 - 2}}\sqrt{\dfrac{7 + 6}{(7)(6)}}} = 2.521$$

6. **Decisions:** Reject H_0. The first variety of corn has a greater mean yield.

In our first illustration of t tests concerning two population means, we concluded that the average annual earnings between carpenters and house painters are different. If we desire to know the range within which such a difference would fall at a given level of probability, we then need to construct a confidence interval for $\Delta\mu$. In general, with a confidence coefficient of $1 - \alpha$, the confidence interval for $\Delta\mu$ is given as

$$\Delta\bar{x} \pm t_{\delta;\alpha}\hat{\sigma}_{\Delta\bar{x}} \tag{12.10}$$

where $\delta = n_1 + n_2 - 2$.

Now, if we wish to construct a 95 percent confidence interval for the carpenter and house painter data, we have

$$\Delta\bar{x} \pm t_{25;0.05}\sigma_{\Delta\bar{x}} = (6000 - 5400) \pm 2.201(269.74),$$

or,

$$\$6.57 < \Delta\mu < \$1193.43$$

Thus, the chances are 95 out of 100 that the true difference in annual average earnings between carpenters and house painters ranges from 6.57 to 1193.43 dollars.

INFERENCES CONCERNING TWO POPULATION MEANS WITH DEPENDENT SAMPLES

So far, comparison between two expected values has been made under the assumption of independent samples. Two samples are said to be independent of each other if the elements in one are not related to those in the other in any significant or meaningful manner. Occasions may arise where we are called upon to compare two population means on the basis of sample means computed from two dependent samples. Two samples are *dependent* if they are paired in the sense that each observation in one is associated with some particular observation in the other. Pairing may result when individuals on whom observations are made are matched by age, by sex, by political affiliation, or by some other criterion, or when the same individual element has a performance score or characteristic in each sample. In other words, if two samples are dependent they must contain the same elementary units but different sets of paired observations are made on these units.

To conduct a test with two dependent samples, we proceed as before by formulating the null hypotheses that the two population means are the same. We must now derive an appropriate testing statistic. To do this, we first take the difference between two matched observations for each element, d_i, obtaining a new variable D. We then reason that if the two samples are not different from each other, then D must average zero; that is, $\bar{d} = \Sigma d/n = 0$, where n is the number of pairs in the two samples. Next, we need the standard deviation for D for the estimation of the standard error of $\bar{d}$. The standard deviation for D can be computed in the usual manner as $s_d = \sqrt{\Sigma d^2/n - (\Sigma d/n)^2}$. With s_d, the unbiased estimate of the standard error of $\bar{d}$ becomes $s_{\bar{d}} = s_d/\sqrt{n - 1}$. Finally, it can be shown that the ratio of $\bar{d}$ to its standard error is distributed as the Student's distribution, with $\delta = n - 1$. Namely,

$$t_{n-1} = \frac{\bar{d}}{s_{\bar{d}}} \tag{12.11}$$

which is used as the testing statistic in this case.

Suppose a method engineer believes that he has perfected a training program which can considerably shorten workers' assembly time for a certain mechanism. To check this belief, he plans to select ten workers at random and to make time and motion studies on them before and after they have gone through his training program Then

1. **Hypotheses:** H_0: $\mu_1 = \mu_2$
H_1: $\mu_1 > \mu_2$

2. **Level of Significance:** $\alpha = 0.01$

3. **Testing Statistic:** $\bar{d}/s_{\bar{d}}$ which is distributed as t_{n-1}.

4. **Decision Rule:** Since $n = 10$ has already been determined, we have $P(t_9 > 2.821) = 0.01$. H_0 will be accepted if and only if $t \geq 2.821$.

5. **Computations:** Suppose that results of the time and motion studies before and after the training program are as shown in the first three columns of the following table, we may then subtract the second observation from the first for each worker to get ds and d^2s which are needed to effect this test.

Worker	First Study	Second Study	d	d^2
1	7 min.	8 min.	−1	1
2	8 "	8 "	0	0
3	10 "	7 "	3	9
4	11 "	6 "	5	25
5	18 "	10 "	8	64
6	16 "	9 "	7	49
7	12 "	9 "	3	9
8	12 "	8 "	4	16
9	6 "	7 "	−1	1
10	12 "	10 "	2	4
Total	112	82	+30	178

$$\bar{d} = \frac{\Sigma d}{n} = \frac{30}{10} = 3 \text{ minutes,}$$

$$s_d = \sqrt{\frac{\Sigma d^2}{n} - \left(\frac{\Sigma d}{n}\right)^2} = \sqrt{\frac{178}{10} - \left(\frac{30}{10}\right)^2} = 2.96648 \text{ minutes,}$$

$$s_{\bar{d}} = \frac{s_d}{\sqrt{n-1}} = \frac{2.96648}{\sqrt{10-1}} = 0.9888 \text{ minutes, and}$$

$$t_9 = \frac{\bar{d}}{s_{\bar{d}}} = \frac{3}{0.9888} = 3.034$$

6. **Decisions:** Since the computed t value falls in the critical region, H_0 is rejected. The training program is evidently of some help in reducing the average assembly time.

We can also compute a confidence interval for the true difference between two population means with two dependent samples in the familiar way

$$\bar{d} \pm t_{n-1;\alpha} s_{\bar{d}}. \tag{12.12}$$

12.9

Relationships among Sampling Distributions

We have now introduced all the major sampling distributions: the binomial, normal, chi-square, F, and t distributions. It is worth noting that these are not empirical but theoretical distributions; none is empirical in the sense that an experimenter has actually taken a large number of samples and discovered that the sample values occur according to the relative frequencies given by the function rule. Each is a theoretical distribution in the sense that a sampling distribution is deduced mathematically, or by a priori reasoning. We have argued that, if simple random sampling of independent observations is made from certain kinds of populations, various sample statistics must possess distributions given by the several function rules. As theoretical distributions, the assumptions, from which sampling distributions are derived, are of great importance since they are the justifications of our methods in making inferences.

In addition to the requirement of simple random sampling, the most basic assumption made in deriving sampling distributions is that the population must be a normal variable. All the three sampling distributions introduced in this chapter rest on the premise of normality. The normal distribution may thus be thought of as the parent distribution of these others. We have seen that the chi-square distribution is a sum of squares of independent standard normal variables. We have also shown that the distribution of sample variance can be solved explicitly on the elementary level only for a normal population. Furthermore, the distribution of s^2 depends on the chi-square distribution which, in turn, is based on the assumption of a normal distribution of single observations. The F distribution, as a ratio of two independent chi-square variables, clearly also rests on the assumption of normal populations.

The close relationships among the normal, chi-square, F and t distributions can be established in a number of ways. We shall give a few of these which can be readily understood. First of all, the t ratio, for instance, may be considered as that of a standard normal variable Z to the square root of a variable which is independent of Z and which is distributed as $\chi_\delta{}^2/\delta$. That is,

$$t = \frac{Z}{\sqrt{\chi_\delta{}^2/\delta}}$$

To show that this definition of t is valid, we see that

$$t = \frac{\bar{x} - \mu}{\sqrt{\hat{s}^2/n}} = \frac{(\bar{x} - \mu)/\sigma}{\sqrt{\hat{s}^2/n\sigma^2}} = \frac{(\bar{x} - \mu)/(\sigma/\sqrt{n})}{\sqrt{\hat{s}^2/\sigma^2}}$$

The numerator of this t ratio is clearly $n(0,1)$ and is the normal sampling distribution of means. Moreover, we know that

$$\frac{(n-1)\hat{s}^2}{\sigma^2} = \chi^2{}_{n-1} = \chi_\delta{}^2$$

and

$$\frac{\hat{s}^2}{\sigma^2} = \frac{\chi^2{}_{n-1}}{n-1} = \chi_\delta{}^2/\delta$$

It follows, therefore, that

$$t = \frac{Z}{\sqrt{\chi_\delta{}^2/\delta}}$$

The relationship between t and F distributions can be easily established by squaring both sides of the previous expression to obtain

$$t^2 = \frac{Z^2}{\chi_\delta{}^2/\delta}$$

By definitions, the numerator of t^2 is χ_1^2 and the denominator of t^2 is $\chi_\delta^2 = \chi^2_{n-1}$. Hence, t^2 qualifies as $F_{1,n-1}$. In general,

$$t_\delta^2 = F_{1,\delta_2}, \quad \delta = \delta_1$$

That is, the square of t with δ is an F variable with 1 and δ_2 degrees of freedom.

Finally, we know that F is related to chi-square distributions directly. If $\chi_{\delta_1}{}^2$ and $\chi_{\delta_2}{}^2$ are two independent chi-square variables, then, we can state that

$$F = \frac{\chi_{\delta_1}{}^2/\delta_1}{\chi_{\delta_2}{}^2/\delta_2}$$

which clearly satisfies the definition that

$$F = \frac{s_1^2}{s_2^2} = \frac{s_1^2/\sigma^2}{s_2^2/\sigma^2} = \frac{\chi_{\delta_1}{}^2/\delta_1}{\chi_{\delta_2}{}^2/\delta_2}$$

We note that division of both the numerator and denominator by σ^2 in the above derivation to form chi-square distributions automatically specifies the null hypothesis $\sigma_1^2 = \sigma_2^2$ in comparing two population variances.

Glossary of Formulas

(12.1) $\hat{\sigma}_{\Delta\bar{x}} = \hat{\sigma}\sqrt{\dfrac{n_1 + n_2}{n_1 n_2}}$ This is the unbiased estimator of the standard error of $\Delta\bar{x}$. Here $\hat{\sigma}^2$ is a pooled estimate of two population variances which are assumed to be identical, and it is a weighted mean of the two sample variances; namely, $\hat{\sigma}^2 = (n_1 s_1^2 + n_2 s_2^2)/(n_1 + n_2 - 2)$. Both $\hat{\sigma}_{\Delta\bar{x}}$ and $\hat{\sigma}^2$ are said to have $n_1 + n_2 - 2$ degrees of freedom since two population estimators have been employed to compute them.

(12.2) $Y = \sqrt{2X} - \sqrt{2\delta - 1}$ This expression states that if X is distributed as χ^2 with $\delta > 30$, then Y is approximately distributed as a normal variable. By utilizing this fact we can approximate chi-square probabilities by normal probabilities when the number of degrees of freedom is large.

(12.3) $\chi^2 = \dfrac{(n - 1)s^2}{\sigma^2}$ The ratio on the right side of this equation is distributed as χ^2_{n-1}. This statistic is used to test hypotheses concerning population variances. We note the value of σ^2 is that stated in H_0.

(12.4) $P\left[\dfrac{(n - 1)s^2}{\chi^2_{(n-1,\alpha/2)}} < \sigma^2 < \dfrac{(n - 1)s^2}{\chi^2_{(n-1;1-\alpha/2)}}\right] = 1 - \alpha$ This is the general statement of the confidence interval for population variance. Note that for this interval, $L = (n - 1)s^2/\chi^2_{(n-1;\alpha/2)}$ and $U = (n - 1)s^2/\chi^2_{(n-1;1-\alpha/2)}$. Thus, a larger chi-square value is used for the lower, and a smaller chi-square value is used for the upper, confidence limits for σ^2 because χ^2 and σ^2 are inversely related.

(12.5) $U = \dfrac{(n - 1)s_2}{\chi^2_{(n-1;1-\alpha)}}$ This is the upper confidence limit for σ^2 for a one-sided confidence interval, $0 < \sigma^2 < U$. This confidence interval corresponds to an upper-tail test for σ^2.

(12.6) $F_{\delta_1\delta_2} = \dfrac{n_1 s_1^2/(n_1 - 1)}{n_2 s_2^2/(n_2 - 1)}$ The ratio between two independent and unbiased estimators of two population variances is distributed as F with $\delta_1 = n_1 - 1$ and $\delta_2 = n_2 - 2$. This is used to test hypotheses concerning two population variances. Under the assumption that the two population variances are equal, we would expect this ratio to be close to 1. When F deviates greatly from 1, the assumption of equality should be rejected.

(12.7) $t = \dfrac{\bar{x} - \mu}{s/\sqrt{n-1}}$ When $n < 31$ and σ is unknown, we use this ratio as the testing statistic for hypotheses concerning means. This is one of the most important equations in statistical inferences because it employs only sample information. Note that the numerator of this t ratio is normally distributed with zero mean, and that the square of the denominator is an unbiased estimate of the variance of the numerator. Also, the denominator, estimated standard error of $\bar{x}$, is the square root of a chi-square variable with $\delta = n - 1$. Thus, if the numerator is normally distributed, then this ratio is distributed as t_{n-1}.

(12.8) $\bar{x} \pm t_{n-1;\alpha}\hat{\sigma}_{\bar{x}}$ This gives the confidence interval for μ in terms of t distributions. Note that the t multiple is determined by the number of degrees of freedom and the confidence coefficient $1 - \alpha$. Also note that $\hat{\sigma}_{\bar{x}} = s/\sqrt{n-1}$.

(12.9) $t = \dfrac{\bar{x}_1 - \bar{x}_2}{\hat{\sigma}_{\Delta\bar{x}}}$ This statistic, used to test hypotheses concerning two means is distributed as t with $\delta = n_1 + n_2 - 2$. Here, $\hat{\sigma}_{\Delta\bar{x}}$ is defined as (12.1).

(12.10) $\Delta\bar{x} \pm t_{\delta;\alpha}\hat{\sigma}_{\Delta\bar{x}}$ Confidence interval for the true difference between two population means. In this case, $\delta = n_1 + n_2 - 2$.

(12.11) $t_{n-1} = \dfrac{\bar{d}}{s_{\bar{d}}}$ This t ratio is used to test the equality between two population means when samples are dependent. Note, $\bar{d} = \Sigma d/n$, where n is the number of pairs in the two dependent samples, and $s_{\bar{d}}$ is the standard error of $\bar{d}$ and it is equal to

$$s_d/\sqrt{n-1} = \sqrt{\Sigma d^2/n - (\Sigma d/n)^2}/\sqrt{n-1}.$$

(12.12) $\bar{d} \pm t_{n-;\alpha}s_{\bar{d}}$ Interval estimate of the difference between two population means on the basis of two dependent sample means.

Problems

12.1 How does large sampling theory differ from exact sampling theory?

12.2 What do we mean by the number of degrees of freedom?

12.3 What are the important properties of chi-square, F, and t distributions, respectively?

12.4 Given that z_p is the percentile score of the standard normal distribution, show that

$$\chi_p{}^2 = \tfrac{1}{2}(z_p + \sqrt{2\delta})^2$$

12.5 If X is $\chi_{40}{}^2$, find $P(X < 30)$.

12.6 If X is $\chi_{50}{}^2$, find $P(X < 69.5)$.

12.7 For $\delta = 50$, find $\chi^2_{.95}$.

12.8 For $\delta = 90$, find $\chi^2_{.95}$, $\chi^2_{.99}$, $\chi^2_{.975}$, and $\chi^2_{.025}$. Compare your answers with the exact chi-square values.

12.9 In the past, scores on tests of elementary statistics have been found to be approximately normal with a variance of 100. A class of 31 students now taught by an outstanding statistician whose method of teaching seems to favor the bright students and to penalize the poor students is expected to have a wider dispersion of test scores. Suppose that this class of 31 students can be considered a random sample and suppose that the variance of scores of these students in the final examination is found to be 150, can we conclude that this instructor's method of teaching results in a wider dispersion at the 5 percent level of significance? At $\alpha = 0.01$?

12.10 Construct a 95 percent two-sided and a 95 percent one-sided confidence interval for σ^2 with data in the preceding problem.

12.11 Sixteen steel wires tested for tensile strength yielded a variance of 140 pounds². Is this result consistent with the belief that the population variance is greater than 125 pounds² at $\alpha = 0.10$? $\alpha = 0.05$?

12.12 Construct a 95 percent two-sided and a 95 percent one-sided confidence interval for σ^2 with data in the preceding problem.

12.13 Diameters of a certain make of tube are assumed to be normally distributed. Test the hypothesis $\sigma^2 = 0.2$ against $\sigma^2 > 0.2$, at $\alpha = 0.10$, given the 10 observations 5.5, 5.4, 5.4, 5.6, 5.8, 5.7, 5.4, 5.5, 5.4, and 5.6 inches. Would you reach the same conclusion at $\alpha = 0.05$? $\alpha = 0.01$?

12.14 A random sample of 250 units of canned tomatoes is drawn and the variance for weights of the cans is found to be 5 ounces². What is the 99 percent confidence interval estimate for σ^2?

12.15 The variance of the lifetimes of a sample of 200 electric light bulbs is 90,000 hours². Find (a) 95 percent and (b) 99 percent confidence limits for σ^2.

12.16 Find confidence intervals for σ with data in the preceding two problems for $1 - \alpha = 0.99$.

12.17 Two methods of performing a certain operation are compared. The results are assumed to be normal variables, with possible differences in expectations and variances. Suppose data obtained yield the following information:

$$\bar{x}_1 = 72.50 \qquad s_1^2 = 8.6 \qquad n_1 = 13$$
$$\bar{x}_2 = 66.05 \qquad s_2^2 = 6.1 \qquad n_2 = 7$$

Test the hypothesis that the two variances are equal at $\alpha = 0.10$.

12.18 With data in the preceding problem, can we accept H_0 if the alternative hypothesis is $\sigma_1^2 \neq \sigma_2^2$ at $\alpha = 0.10$?

12.19 Suppose there are two sources of raw materials under consideration. Both sources seem to have similar characteristics, but we are not sure about their respective uniformity. A sample of ten lots from Source A yields a variance of 250, and a sample of eleven lots from Source B yields a variance of 195. Is it likely that the variance of Source A is significantly greater than the variance of Source B at $\alpha = 0.10$?

12.20 Would you reach the same conclusion for a two-tail test with data in the preceding problem?

12.21 A random sample consisting of ten colleges in New York State yields the following minimum salaries for instructors: 5000, 5400, 6000, 5200, 4700, 7200, 5400, 9000, 8000, and 7500. A random sample of eight colleges in California gives the following information on minimum salaries: 7400, 6000, 6500, 6200, 7500, 7800, 8000, and 8200. Test the hypothesis $\sigma_n^2 = \sigma_c^2$ against $\sigma_n^2 \neq \sigma_c^2$ at $\alpha = 0.02$.

12.22 The producer of a certain make of flashlight dry-cell batteries claims its output has a mean life of more than 750 minutes. Suppose a sample of 15 of these batteries has been tested and yielded the following values: 730, 759, 725, 740, 754, 745, 750, 753, 730, 780, 725, 790, 719, 775, 700. From these data, can you conclude that the producer's claim is valid at a 1 percent level of significance?

12.23 A sample of ten measurements of the diameter of a sphere gives a mean of 4.08 inches and a standard deviation of 0.05 inches. Is this information consistent with the fact that the manufacturing process is adjusted to produce diameters with $\mu = 4$ inches, given $\alpha = 0.05$? (*Hint:* you should conduct a two-sided test here.)

12.24 A meat cannery has just installed a new filling machine. A random sample of 20 filled cans yields a mean weight of the contents of 16.05 ounces with a standard deviation of 1.5 ounces.

a. Make an interval estimate of the true mean with $1 - \alpha = 0.95$.

b. If the net contents of each can is supposed to be 16 ounces is the machine properly adjusted?

12.25 Construct a 95 percent confidence interval for μ with the data in Problem 12.23. In view of this interval, what can you say about the conclusion reached for Problem 12.23?

12.26 Two working designs are under consideration for adoption in a plant. A time and motion study shows that 12 workers using Design A have a mean assembly time of 300 seconds with a standard deviation of 12 seconds, and that 15 workers using Design B have a mean assembly time of 335 seconds with a standard deviation of 15 seconds. Is the difference in mean assembly time between the two working designs significant at a 1 percent level of significance?

12.27 Construct a 99 percent confidence interval for the true difference between mean assembly times of the working designs in the preceding problem.

12.28 Test the hypothesis that $\mu_1 = \mu_2$ against that $\mu_1 > \mu_2$ with $\alpha = 0.05$ for data in Problem 12.17 and construct a 95 percent confidence interval for $\Delta\mu$ with the same data.

12.29 Test the hypothesis that $\mu_1 = \mu_2$ against that $\mu_1 \neq \mu_2$ with data in Problem 12.21 at $\alpha = 0.02$. Construct a 98 percent confidence interval for $\Delta\mu$ with the same data.

12.30 To compare the efficiency of standard and electric typewriters, eight typists are chosen at random and thoroughly oriented to the operation of both kinds of typewriters. Then they are asked to type on each kind of typewriter for ten minutes and their speeds, measured as average number of words typed per minute, are observed. The results of this experiment are shown in table in first column. Can you conclude that the two types of typewriters are different in efficiency?

	Average Speed per Minute	
Typist	*Electric*	*Standard*
1	75	79
2	89	62
3	79	54
4	85	67
5	102	81
6	115	78
7	97	66
8	69	73

12.31 Suppose that a big corporation would decide to buy the standard typewriters for use unless the average speed of the electric typewriters is 30 words per minute greater than that of the standard typewriters with a probability of 0.99. In view of results obtained in the preceding problem, how would the corporation decide?

12.32 Ten students are selected at random and their final grades in physics and economics show these results:

Student	1	2	3	4	5
Physics	66	72	50	81	62
Economics	75	70	65	88	59
Student	6	7	8	9	10
Physics	73	55	90	77	85
Economics	85	60	97	82	90

Can you conclude that the mean grades in these two subjects are different? Construct a 95 percent confidence interval for the true mean difference for this illustration.

13

ANALYSIS OF VARIANCE

13.1

Introduction

We have now learned how to make inferences about expectations, proportions, and variances of one or two populations from samples selected from these populations. We have also discovered that the chi-square tests provide us with a method of comparing more than two sample proportions. Now we naturally wish to know if there exists a general procedure by which more than two sample means can be evaluated. Obviously, such a procedure is important in statistical analysis, because situations often arise in which the researcher may be interested in conducting an experiment for evaluating several designs, methods, products, and so on, rather than in designing an experiment for comparing a new theory or process with the standard one. For example, one may wish to compare the efficiency of three work designs for assembling a mechanism, or to evaluate the performance of four different makes of auto tires, or to study the relative productivity of k different kinds of fertilizers.

Clearly, it would be grossly inefficient to compare several samples by pairing them two at a time because, if there are, say, five samples, we would have $\binom{5}{2} =$

10 different tests to perform. Furthermore, even if we have made ten such tests, we may still not be able to generalize about all the five populations, from which the samples are drawn, at the same time. Worse still, it is easy to show that such a tedious and time-consuming procedure actually has a high probability of leading to wrong conclusions pairwise. If we were to have ten t tests on the ten possible pairs of the five sample means, the probability of our arriving at a correct decision of no significant difference would be 0.95, given a 5 percent level of significance. Thus, the probability for our arriving at correct conclusions for all the ten t tests would be 0.95^{10}; namely, the probability of obtaining at least one wrong conclusion would be as large as $1 - (0.95)^{10} = 0.401$. That is, more than 40 percent of the time we would make a Type I error. It is easy to see that the larger the number of samples to be evaluated, the greater the probability is of making wrong decisions. In short, the method of comparing only pairs of samples is at once uneconomical and inefficient for the ultimate purpose of comparing several populations together.

The efficient procedure that has been developed to compare several sample means is known as the *analysis of variance,* a name derived from the fact that this analysis is based on the comparison of variances estimated from various sources. This statement will become clear as the reader proceeds to learn about this new tool.

The analysis of variance originated in agrarian research and its language is thus loaded with such agricultural terms as *blocks* (referring to land) and *treatments* (referring to populations, or samples, which are differentiated in terms of varieties of seed, fertilizer, or cultivation methods). Today, procedures of this analysis find useful application in nearly every type of experimental design, in natural science as well as in social sciences. The reason for this widespread use is that the analysis of variance is amazingly versatile: it can be readily adopted to furnish, with broad limits, a proper evaluation of data obtained from a large body of experiments which involve several continuous random variables. It can give us answers as to whether different sample data classified in terms of a single variable are meaningful. It can also provide us with meaningful comparisons of sample data which are classified according to two or more variables. In each case, the answer would involve significance, or lack of it, for each variable classification as well as possible joint effects of the variables—treatment combinations.

In this chapter, no attempt will be made to have a full coverage of this vast field of statistical techniques. Our goal is to introduce the reader to some of the simple models that are frequently encountered in economic and business research. The models introduced here, in order of our presentation, are

(1) Completely randomized, one-variable classification model
(2) Randomized-blocks, one-variable classification model
(3) Completely randomized, two-variable classification without replication model
(4) Completely randomized, two-variable classification with replication model
(5) Randomized-blocks, two-variable classification with replication model

13.2

Completely Randomized, One-Variable Classification Model

THE MODEL

Suppose that we have c populations, $A_1, A_2, \cdots, A_c$, each normally distributed with mean μ_i and all possessing a common variance σ^2. These populations are often called *treatments*—representing c kinds of soybeans, c ways of teaching economics, or c makes of machines for producing a certain type of output. These c populations are assumed to possess a common variance, since there is reason to believe that different treatments used for the same purpose may differ in central tendencies but not in dispersion. Furthermore, we conceive that these populations together constitute a *grand* population with mean μ, called the *grand population mean*, defined as

$$\mu = \frac{1}{c}\sum_i \mu_i$$

Under the preceding two assumptions, we wish to test the null hypothesis that all the treatment means are equal; that is,

$$\mu_1 = \mu_2 = \cdots = \mu_c$$

If this H_0 is true, then we would expect $\mu_i = \mu$. However, if H_0 is false, then we would expect μ_i to deviate from μ by a quantity, say, α_i; that is,

$$\alpha_i = \mu_i - \mu, \quad i = 1, 2, \cdots, c$$

where α_i, for obvious reasons, are called *treatment effects*. By the property of the arithmetic mean, we must have

$$\Sigma\alpha_i = \Sigma(\mu_i - \mu) = 0$$

For the previous arguments we can also see that

$$\mu_1 = \mu_2 = \cdots = \mu_c$$

and

$$\alpha_i = 0$$

are equivalent statements, since when we say that the treatment means are equal, we imply that the effects due to treatments are nil.

Now, we assume that x_{ij}, the jth observation in the ith sample, are normally distributed about μ_i and we state that $x_{ij} = \mu_i + e_{ij}$. Here, e_{ij} are deviations of x_{ij} from μ_i due to chance fluctuations in random sampling. The e_{ij} are called the *error term*, or *residuals*, and are assumed to be independent and distributed normally with zero mean and a variance identical with the common variance, σ^2, for the treatment populations.

Finally, noting that $\mu_i = \mu + \alpha_i$, we have as the basic model for the completely randomized, one-variable classification experiment the following linear form:

$$x_{ij} = \mu_i + e_{ij} = \mu + \alpha_i + e_{ij} \tag{13.1}$$

THE ANALYSIS

The central core of the analysis of variance, irrespective of the type of model, lies in the partitioning of the total sum of squares into meaningful and distinct portions. To see how this is done for our present model, let us first arrange the data obtained from c samples, each with r observations, into a general form as shown by Table 13.1.

From the data in Table 13.1, two summary statistics can be computed.

(1) $\bar{x}_{i.}$ = the sample mean of the ith column

$$= \frac{1}{n_i} \sum_j x_{ij} = \frac{1}{n_i}(x_{i1} + x_{i2} + \cdots + x_{ir})$$

which is an unbiased estimate of μ_i. Note that the dot (.) in the notation $\bar{x}_{i.}$ indicates that the column mean is obtained by summing on the index j.

(2) $\bar{\bar{x}}$ = the grand sample mean

$$= \frac{1}{N} \sum_i \sum_j x_{ij} = \frac{1}{N}\left(\sum_j x_{1j} + \sum_j x_{2j} + \cdots + \sum_j x_{cj}\right)$$

which is an unbiased estimate of μ. Here, $N = n_1 + n_2 + \cdots + n_c$.

The variability of the set of i experimentally different samples can be partitioned into two distinct parts in terms of the sum of squared deviations about the grand sample mean. Each observation, x_{ij}, in sample i is expected to deviate from

TABLE 13.1 SAMPLE DATA OF THE LINEAR MODEL

	Sample			
Observation	*1*	*2*	⋯	c
1	x_{11}	x_{21}	⋯	x_{c1}
2	x_{12}	x_{22}	⋯	x_{c2}
·	·	·	⋯	·
·	·	·	⋯	·
·	·	·	⋯	·
r	x_{1r}	x_{2r}	⋯	x_{rc}
Total	$\sum_j x_{1j}$	$\sum_j x_{2j}$	⋯	$\sum_j x_{cj}$
n_i	n_1	n_2	⋯	n_c

the grand sample by the amount $(x_{ij} - \bar{\bar{x}})$ which can be thought of as comprised of two parts,

$$(x_{ij} - \bar{\bar{x}}) = (x_{ij} - \bar{x}_{i.}) + (\bar{x}_{i.} - \bar{\bar{x}})$$

The first component on the right of this expression is the deviation of x_{ij} from the mean of sample i. The second term is the deviation of the sample mean from the grand sample mean. Clearly, $(x_{ij} - \bar{\bar{x}})$ is an estimate of $(x_{ij} - \mu)$ and the two terms on the right are estimates of e_{ij} and α_i respectively.

Now, if we square the deviation from $\bar{\bar{x}}$ for each observation in the sample and total these squared deviations over all individuals j in all sample groups i, we would then have

$$\sum_i \sum_j (x_{ij} - \bar{\bar{x}})^2 = \sum_i \sum_j ((x_{ij} - \bar{x}_{i.}) + (\bar{x}_{i.} - \bar{\bar{x}}))^2$$
$$= \sum_i \sum_j (x_{ij} - \bar{x}_{i.})^2 + \sum_i \sum_j (\bar{x}_{i.} - \bar{\bar{x}})^2 + 2 \sum_i \sum_j (x_{ij} - \bar{x}_{i.})(\bar{x}_{i.} - \bar{\bar{x}})$$

The last term in this expression is 0, since the value represented by the term $(x_{i.} - \bar{\bar{x}})$ is the same for all j in group i, and the sum of $(x_{ij} - \bar{x}_{i.})$ must be 0 when taken over all j in any group i. Thus, this expression can be written as

$$\sum_i \sum_j (x_{ij} - \bar{\bar{x}})^2 \qquad \text{Total variation: SST}$$
$$= \sum_i \sum_j (x_{ij} - \bar{x}_{i.})^2 \qquad \text{Variation within (samples) or for error: SSE}$$
$$+ \sum_i \sum_j (\bar{x}_{i.} - \bar{\bar{x}})^2 \qquad \text{Variation between (samples) or between (columns): SSC}$$

This identity is often called the "partition of the sum of squares."

The logic of the partition of the sum of squares into two distinct parts can be put into intuitive terms quite easily. Observations in any sample will always show variability. Differences among individual observations here can come from two sources. Differences among individuals belonging to different treatment samples may result either from different treatments or from chance variation, or from both. The sum of squares between samples, SSC, reflects the contribution of both different treatments and chance to intersample variability. Individual observations in the same treatment sample, however, can differ from each other only because of chance variation, since each individual within the group receives exactly the same treatment. The sum of squares within samples, SSE, measures those intersample differences due to chance only. Therefore, in any group of i samples, it is possible to isolate the two kinds of variability: the sum of square between groups which reflects variation due to both treatments and chance, and the sum of square within groups which reflects chance variation alone.

The sums of squares are computed by expending the binomials which define them, as follows:

$$\text{SST} = \Sigma\Sigma(x_{ij} - \bar{\bar{x}})^2 = \Sigma\Sigma(x_{ij}^2 - 2x_{ij}\bar{\bar{x}} + \bar{\bar{x}}^2)$$
$$= \Sigma\Sigma x_{ij}^2 - 2\bar{\bar{x}}\Sigma\Sigma x_{ij} + N\bar{\bar{x}} = \Sigma\Sigma x_{ij}^2 - N\bar{\bar{x}}^2 \qquad \textbf{(13.2)}$$

$$\begin{aligned}\text{SSC} &= \Sigma\Sigma(\bar{x}_{i.} - \bar{\bar{x}})^2 = \Sigma n_i(\bar{x}_{i.} - \bar{\bar{x}})^2 = \Sigma n_i(\bar{x}_{i.}{}^2 - 2\bar{x}_{i.}\bar{\bar{x}} + \bar{\bar{x}}^2) \\ &= \Sigma n_i \bar{x}_{i.}{}^2 - 2\bar{\bar{x}}\Sigma n_i \bar{x}_{i.} + \bar{\bar{x}}^2 \Sigma n_i = \Sigma n_i \bar{x}_{i.}{}^2 - 2N\bar{\bar{x}}^2 + N\bar{\bar{x}}^2 = \Sigma n_i \bar{x}_{i.}{}^2 - N\bar{\bar{x}}^2\end{aligned} \tag{13.3}$$

$$\begin{aligned}\text{SSE} &= \Sigma\Sigma(x_{ij} - \bar{x}_{i.})^2 = \Sigma\Sigma(x_{ij}{}^2 - 2x_{ij}\bar{x}_{i.} + \bar{x}_{i.}{}^2) \\ &= \Sigma(\Sigma x_{ij}{}^2 - 2\bar{x}_{i.}\Sigma x_{ij} + n_i \bar{x}_{i.}{}^2) = \Sigma(\Sigma x_{ij}{}^2 - 2n_i \bar{x}_{i.}{}^2 + n_i \bar{x}_{i.}{}^2) \\ &= \Sigma\Sigma x_{ij}{}^2 - \Sigma n_i \bar{x}_i{}^2\end{aligned} \tag{13.4}$$

As a check, we note that SST = SSC + SSE. Also, SSE can be obtained by subtraction; that is, SSE = SST − SSC, if we wish.

The preceding formulas for the sums of squares are applicable for both equal and unequal sample sizes. If the null hypothesis that treatment effects are nil is true, then each of these three sums of squares, divided by an appropriate number of degrees of freedom, is an unbiased estimate of the common population variance σ^2. The total number of degrees of freedom in this model is $N - 1$, since SST is computed with a total of N observations by using the sample grand mean as an estimator of μ: one degree of freedom is thus lost. To estimate σ^2 on the basis of SSC—variation between column means—the number of degrees of freedom is $c - 1$, since there are c column means to compare, and again the grand sample mean is used as an estimator of μ for the calculation of SSC. The number of degrees of freedom for estimating σ^2 based on SSE is $N - c$, since SSE is obtained from all the N observations with the c group means as estimates of the population mean. It is interesting to note that, just as in the case of sums of squares, the numbers of degrees of freedom are also additive. Namely, we have here $N - 1 = (c - 1) + (N - c)$.

Estimates for the common population variance are often called mean squares in the analysis of variance. For the present model, as shall be seen immediately, we are interested in comparing two mean squares: the mean square between column means, MSC = SSC/$(c - 1)$, and the mean square for error, MSE = SSE/$(N - c)$. Note that both MSC and MSE are distributed as chi-squares with $c - 1$ and $N - c$ df, respectively.

TABLE 13.2 ANALYSIS OF VARIANCE TABLE: LINEAR MODEL

Source of Variation	SS	δ	MS
Between (columns)	SSC	$c - 1$	MSC = SSC/$(c - 1)$
Within; error	SSE	$N - c$	MSE = SSE/$(N - c)$
Total	SST	$N - 1$	—

It is customary to summarize calculations for sums of squares, together with their numbers of df and mean squares, in a table called "the analysis of variance table." For the present model, see Table 13.2.

THE TEST

For the present model, the pair of hypotheses to be tested is

H_0: All treatment means are equal, or $\alpha_i = 0$
H_1: Not all treatment means are equal, or not all α_i are zero

The testing statistic to be used here is the F ratio of MSC to MSE. The rationale here is quite simple. Under the assumption that e_{ij} are independent and distributed normally with 0 mean and a variance identical with σ^2, MSE must be an unbiased estimate of σ^2 whether or not H_0 is true, since MSE measures chance variation alone. On the other hand, MSC is an unbiased estimate of σ^2 if and only if H_0 is true. When the treatment effects are not nil, MSC actually estimates the quantity $\sigma^2 + a$, where a is a positive bias and measures the degree of differences in treatments. Under these circumstances, we would expect $F_{c-1,N-c} = \text{MSC}/\text{MSE} = 1$, or very close to 1, if H_0 is true, and would expect this ratio to be significantly greater than 1 if H_0 is false.

Thus, once again, the F test is of the upper-tail variety. When the computed value of F is, or is close to, 1, the null hypothesis, $\alpha_i = 0$, is accepted. When it is significantly greater than 1, we conclude that treatment effects are not nil.

FIRST EXAMPLE: EQUAL SAMPLE SIZE

Suppose that a manufacturing company has purchased three new machines of different makes and wishes to determine whether one of them is faster than the others in producing a certain output. Five hourly production figures are observed at random from each machine, and the results are presented in Table 13.3. What conclusions can be drawn at $\alpha = 0.05$? At $\alpha = 0.01$?

Following our general procedure of testing, the foregoing questions can be answered by the following test:

1. **Hypotheses:** H_0: $\mu_1 = \mu_2 = \mu_3$
   H_1: Not all three μ_i are equal

2. **Level of Significance:** $\alpha = 0.05, 0.01$

3. **Testing Statistic:** $F_{2,12} = \dfrac{\text{MSC}}{\text{MSE}}$. *Note:* for this F, $\delta_1 = c - 1 = 3 - 1 = 2$, and $\delta_2 = N - c = 15 - 3 = 12$.

4. **Decision Rule:** Since $P(F_{2,12} > 3.88) = 0.05$ and $P(F_{2,12} > 6.93) = 0.01$, H_0 will be rejected at the 5 percent level if the observed F is greater than 3.88, and it will be rejected at the 1 percent level of significance if the observed F is greater than 6.70.

5. **Computations:** Suppose that sample data are as recorded in the following table; then we may compute, quite conveniently, the column group means and the grand sample mean first. These summary statistics may be placed at the bottom of the table that contains the original data.

Next we should compute the sums of squares. In doing this, we see that there are only three quantities, $\Sigma\Sigma x_{ij}^2$, $\Sigma n_i \bar{x}_{i.}^2$, and $N\bar{\bar{x}}$, in the equations for the three sums in this model; therefore it is advisable to compute them first before the sums of squares are obtained.

TABLE 13.3 HOURLY OUTPUT VOLUMES OF THREE MACHINES FOR FIVE HOURS

Observations	A_1	A_2	A_3	Total
1	25	31	24	
2	30	39	30	
3	36	38	28	
4	38	42	25	
5	31	35	28	
Total	160	185	135	480
n_i	5	5	5	
$\bar{x}_{i.}$	32	37	27	$\bar{\bar{x}} = 32$

$$\Sigma\Sigma x_{ij}^2 = 25^2 + 30^2 + \cdots + 28^2 = 15{,}810$$
$$\Sigma n_i\bar{x}_{i.}^2 = n_1(\bar{x}_{1.}^2) + n_2(\bar{x}_{2.}^2) + n_3(\bar{x}_{3.}^2) = 5(32^2) + 5(37^2) + 5(27^2) = 15{,}610$$
$$N\bar{\bar{x}}^2 = 15(32^2) = 15{,}360$$

Finally, the three sums of squares are as follows:

$$\begin{aligned} \text{SST} &= \text{total variation} \\ &= \Sigma\Sigma x_{ij}^2 - N\bar{\bar{x}}^2 = 15{,}810 - 15{,}360 = 450 \\ \text{SSC} &= \text{variation between columns (machines)} \\ &= \Sigma n_i\bar{x}_{i.}^2 - N\bar{\bar{x}}^2 = 15{,}610 - 15{,}360 = 250 \\ \text{SSE} &= \text{variation within columns (error)} \\ &= \Sigma\Sigma x_{ij}^2 - \Sigma n_i\bar{x}_{i.} = 15{,}810 - 15{,}610 = 200 \end{aligned}$$

We may now present these sums of squares, along with their degrees of freedom and mean squares, in the following summary table:

TABLE 13.4 SUMMARY TABLE FOR MACHINE DATA

Source of Variation	SS	δ	MS
Machines (between columns)	250	$3 - 1 = 2$	125.00
Error (within columns)	200	$15 - 3 = 12$	16.67
Total	450	$15 - 1 = 14$	—

$$F_{2,12} = \frac{\text{MSC}}{\text{MSE}} = \frac{125.00}{16.67} = 7.50$$

6. **Decisions:** Since the computed value is greater than both 3.88 and 6.70, H_0 is rejected at both the 5 percent and 1 percent levels of significance. The three new machines are significantly different in their mean speeds; or, the machine effects are significant.

SECOND EXAMPLE: UNEQUAL SAMPLE SIZE

During the past four semesters, four different textbooks have been used in this author's class in mathematical economics, and the final grades for the students in these four classes are recorded in Table 13.5. In view of these data, can we conclude that students' scores are significantly affected by textbooks?

TABLE 13.5 FINAL EXAMINATION SCORES

	Textbook				
	(1)	(2)	(3)	(4)	
	60	80	97	67	
	80	81	84	84	
	69	73	93	90	
	65	69	79	78	
		75	92	61	
		72			
Total	274	450	445	380	$\Sigma\Sigma x_{ij} = 1549$
n_i	4	6	5	5	$N = 20$
$\bar{x}_{i.}$	68.5	75.0	89.0	76.0	$\bar{\bar{x}} = 77.45$

Computations for this example are as follows:

$$\begin{aligned}
\Sigma\Sigma x_{ij}^2 &= 60^2 + 80^2 + \cdots + 61^2 = 122{,}115 \\
\Sigma n_i \bar{x}_{i.}^2 &= 4(68.5^2) + 6(75.0^2) + 5(89.0^2) + 5(76.0^2) = 121{,}004 \\
N(\bar{\bar{x}}^2) &= 20(77.45^2) = 119{,}970 \\
\text{SST} &= 122{,}115 - 119{,}970 = 2145 \\
\text{SSC} &= 121{,}004 - 119{,}970 = 1034 \\
\text{SSE} &= 122{,}115 - 121{,}004 = 1111
\end{aligned}$$

TABLE 13.6 SUMMARY TABLE FOR TEXTBOOK DATA

Source of Variation	SS	δ	MS
Textbooks (between columns)	1034	3	344.67
Error (within columns)	1111	16	69.44
Total	2145	19	—

$$F_{3,16} = \frac{344.67}{69.44} = 4.96$$

Since $P(F_{3,16} > 3.24) = 0.05$ and $P(F_{3,16} > 5.29) = 0.01$, there is a significant difference between textbooks at $\alpha = 0.05$; the null hypothesis that there is no textbook effect cannot be rejected at $\alpha = 0.01$. These conclusions are only valid if the four classes constitute four independent random samples.

13.3

The Least Significant Difference

The analysis of variance proper shows only whether or not there exists any significant difference among treatment means. It does not reveal where the difference is. When the F test indicates the existence of a significant difference, we naturally wish to know which of the means causes the difference. When sample sizes are equal, as in our first illustration in the last section, a simple procedure, called the *least significant difference*, can aid us in locating this source.

When $n_1 = n_2 = \cdots = n_c$ and when the F test leads to the rejection of $\alpha_i = 0$, the least significant difference, denoted as LSD, is defined as the smallest difference which could exist between two significantly different sample means. To derive a formula for LSD, we recall that a test for the difference between two sample means is made by the statistic

$$t = \frac{\bar{x}_1 - \bar{x}_2}{\hat{\sigma}_{\Delta\bar{x}}}$$

If this ratio exceeds, say, $t_{\delta;\,0.05}$, we say the difference between the two sample means is significant at $\alpha = 0.05$. Thus, LSD can be considered as the difference between two sample means for which $t = t_{\delta;\,\alpha}$, where α is the level of significance. Namely,

$$\text{LSD} = t_{\delta;\,\alpha}\hat{\sigma}_{\Delta\bar{x}}$$

Furthermore, if $n_1 = n_2$ and $\sigma_1^2 = \sigma_2^2$ is assumed,

$$\hat{\sigma}_{\Delta\bar{x}} = \sqrt{\frac{\hat{\sigma}_1^2}{n_1} + \frac{\hat{\sigma}_2^2}{n_2}} = \sqrt{\frac{2\hat{\sigma}^2}{n}}$$

where, as it can be shown, $\hat{\sigma}^2 = \text{MSE}$, the error mean square. We have, therefore,

$$\text{LSD} = t_{\delta;\,\alpha}\sqrt{\frac{2}{n}\,\text{MSE}}$$

where $\delta = N - c$. But when t has δ_2, then $t_{\delta_2}{}^2 = F_{1,\delta_2}$. The formula for the least significant difference becomes, finally,

$$\text{LSD} = \sqrt{\frac{2}{n}\,(\text{MSE})F_{1,N-c;\,\alpha}} \tag{13.5}$$

Any two treatment sample means in the analysis of variance are said to differ from each other significantly at a given level of α if their absolute difference is greater than LSD as defined by (13.5).

To apply (13.5) to our first illustration, we note $n_1 = n_2 = n_3 = 5$, $\text{MSE} = 16.67$, and $F_{1,12;\,0.01} = 9.33$. Thus

$$\text{LSD} = \sqrt{\tfrac{2}{5}(16.67)(9.33)} = 7.89$$

From calculations in Table 13.3, we find that

$\lvert\bar{x}_{1.} - \bar{x}_{2.}\rvert = \lvert 32 - 37\rvert = 5 < 7.89$	Not significant
$\lvert\bar{x}_{1.} - \bar{x}_{3.}\rvert = \lvert 32 - 27\rvert = 5 < 7.89$	Not significant
$\lvert\bar{x}_{2.} - \bar{x}_{3.}\rvert = \lvert 37 - 27\rvert = 10 > 7.89$	Significant

Thus, among the three pairs of column means, only one pair $(\bar{x}_{2.}, \bar{x}_{3.})$ differs significantly. Other things being equal, the management should decide to use the second make of machines exclusively.

It is important to note that the LSD *tests* are also called *pairwise tests* and they should be employed only when the F test has led to the rejection of H_0. When LSD is erroneously employed in a case where F value is insignificant, it is quite possible to find that some pair or pairs of sample means differ by an amount greater than the value of LSD. However, such differences are still explained by chance variations in random sampling from the same population or populations with identical means.

In a case where samples are not of equal size, there is no simple generalized procedure for determining which pair of samples cause the significant difference in the F test. We must evaluate all possible pairs of sample means individually by t tests.

13.4 Randomized-Blocks, One-Variable Classification Model

Randomized-blocks experiment is a term that stems from agricultural research in which several "variables" or "treatments" are applied to different blocks of land for repetition, or replication, of the experiment. The main objective here is to establish significant differences among treatment effects, such as yields of different types of soybeans, or the quality of different makes of fertilizers. But differences in crop yield may be attributed not only to kinds of soybeans but also to differences in quality of the blocks of land. To isolate the "block effects," randomization—achieved by assigning treatments at random to plots of each block of land—is employed.

Despite its agricultural origin, the randomized-blocks design is widely used in many types of experiment. For instance, to determine the differences in productivity of c makes of machines (treatments), we may isolate the possible effects due to differences in efficiency among operators (blocks) by assigning the machines at random to randomly selected operators. The basic idea here is to compare all treatment effects within a block of experimental material by eliminating the environmental effects.

THE MODEL

Suppose we have c treatments, A_i, and r blocks, B_j, then x_{ij} may be thought of as samples, each with size 1, drawn from populations with μ_{ij}. There are, therefore, rc populations, one for each treatment-block combination. We assume that all the rc populations are normally distributed with a common variance σ^2. We next define a grand population mean as

$$\mu = \frac{\mu_{11} + \mu_{12} + \cdots + \mu_{rc}}{rc} = \frac{1}{rc} \sum_i \sum_j \mu_{ij}$$

TABLE 13.7 SAMPLE DATA: RANDOMIZED-BLOCKS, ONE-VARIABLE CLASSIFICATION MODEL

	A_1	A_2	A_3	$\cdots$	A_c	$t_{.j}$	$\bar{x}_{.j}$
B_1	x_{11}	x_{21}	x_{31}	$\cdots$	x_{c1}	$t_{.1}$	$\bar{x}_{.1}$
B_2	x_{12}	x_{22}	x_{32}	$\cdots$	x_{c2}	$t_{.2}$	$\bar{x}_{.2}$
.	.	.	.	$\cdots$	.	.	.
.	.	.	.	$\cdots$	.	.	.
.	.	.	.	$\cdots$	.	.	.
B_r	x_{1r}	x_{2r}	x_{3r}	$\cdots$	x_{cr}	$t_{.r}$	$\bar{x}_{.r}$
$t_{i.}$	$t_{1.}$	$t_{2.}$	$t_{3.}$	$\cdots$	$t_{c.}$	T = grand total	
$\bar{x}_{i.}$	$\bar{x}_{1.}$	$\bar{x}_{2.}$	$\bar{x}_{3.}$	$\cdots$	$\bar{x}_{c.}$	$\bar{\bar{x}} = \frac{1}{rc}T$	

The deviation of μ_{ij} from μ can then be reasonably explained as due to both treatment and block effects. Now, if we denote the ith treatment effect as α_i and the jth block effect as β_j, then we may assume that

$$\mu_{ij} = \mu + \alpha_i + \beta_j$$

Our samples, x_{ij}, are normally distributed about μ_{ij}. The deviation of each single sample observation from its corresponding population mean is the error in random sampling and is denoted as e_{ij}. Namely, the single observation made from the ijth population can be represented by

$$x_{ij} = \mu_{ij} + e_{ij} = \mu + \alpha_i + \beta_j + e_{ij} \tag{13.6}$$

In the above basic model, we have assumed a number of things. First, the treatment and block effects are assumed to be additive; that is, there is no joint effect between α_i and β_j except the sum of their simple effects. Second, as before, e_{ij} are assumed to be independent and distributed as $n(0,\sigma^2)$. Finally, for convenience of solution for this model, we shall also assume that

$$\Sigma\alpha_i = \Sigma\beta_j = 0$$

THE ANALYSIS

The analysis of this model begins, again, with partitioning of the total sum of squares into nonoverlapping and meaningful components. We shall begin with the aid of a tabular presentation of the sample data, Table 13.7.

In the above table,

A_i = Treatments, $i = 1, 2, \cdots, c$

B_j = Blocks, $j = 1, 2, \cdots, r$

$t_{i.}$ = Column sums; for example $t_{1.} = x_{11} + x_{12} + \cdots + x_{1r}$

$\bar{x}_{i.}$ = Column means $= \frac{1}{r}t_{i.}$; for example $\bar{x}_{2.} = \frac{1}{r}t_{2.}$

$t_{.j}$ = Row sums; for example $t_{.r} = x_{1r} + x_{2r} + \cdots + x_{cr}$

$\bar{x}_{.j}$ = Row means $= \frac{1}{c}t_{.j}$; for example $\bar{x}_{.1} = \frac{1}{c}t_{.j}$

The total variation, represented by the total sum of squares, SST, as before, measures the squared deviations of all observations from the grand sample mean. It is now composed of three independent parts as indicated by the following identity:

$\Sigma\Sigma(x_{ij} - \bar{\bar{x}})^2$	Total variation: SST
$= \Sigma\Sigma(\bar{x}_{i.} - \bar{\bar{x}})^2$	Variation between column means: SSC
$+ \Sigma\Sigma(\bar{x}_{.j} - \bar{\bar{x}})^2$	Variation between row means: SSR
$+ \Sigma\Sigma(x_{ij} - \bar{x}_{i.} - \bar{x}_{.j} + \bar{\bar{x}})^2$	Variation within or for error: SSE

In the above identity, SSC measures both chance variations and variations in treatment effects, if any, since the quantity $(\bar{x}_i - \bar{\bar{x}})$ is an estimate of $(\mu_{i.} - \mu) = \alpha_i$.

Similarly, the variation between row means, SSR, measures both error variations and block effects, β_j, since $(\bar{x}_{.j} - \bar{\bar{x}})$ reflects dispersion of row means from the grand sample mean. It can be thought of as, therefore, an estimate of $\mu_{.j} - \mu = \beta_j$.

Finally, the sum of squares for error is based on deviations from a cell mean for individuals treated in exactly the same way, and thus the only possible contribution to this sum of squares should be error variation. Alternatively, we may say that the quantity $(x_{ij} - \bar{x}_{i.} - \bar{x}_{.j} + \bar{\bar{x}})$ in effect measures the error term e_{ij}.

The foregoing sums of squares are computed as follows:

$$\text{C} = \text{correction factor} = \frac{1}{rc} T^2 \tag{13.7}$$

$$\text{SST} = \sum\sum x_{ij}^2 - \text{C} \tag{13.8}$$

$$\text{SSC} = \frac{1}{r} \sum t_{i.}^2 - \text{C} \tag{13.9}$$

$$\text{SSR} = \frac{1}{c} \sum t_{.j}^2 - \text{C} \tag{13.10}$$

$$\text{SSE} = \text{SST} - (\text{SSC} + \text{SSR}) \tag{13.11}$$

TABLE 13.8 ANALYSIS OF VARIANCE TABLE, RANDOMIZED-BLOCKS ONE VARIABLE-CLASSIFICATION MODEL

Source of Variation	SS	δ	MS
Column treatments	SSC	$c - 1$	MSC = SSC/$(c - 1)$
Row treatments (blocks)	SSR	$r - 1$	MSR = SSR/$(r - 1)$
Error	SSE	$(r - 1)(c - 1)$	MSE = SSE/$(r - 1)(c - 1)$
Total	SST	$rc - 1$	—

It can be shown again that under the assumption that treatment effects are nil, each of the three sums of squares which comprise SST divided by an appropriate number of df is an unbiased estimate of the common population variance. It is easy to see that for SSC, $\delta = c - 1$; for SSR, $\delta = r - 1$; and for SSE, $\delta = (r - 1)(c - 1)$. These three numbers of df, as usual, add up to the total number of df for SST, which is $rc - 1$. The analysis of variance table for this model is given as Table 13.8.

THE TEST

For this model, there seem to be two sets of hypotheses to be evaluated; these are

(1) Treatment effects test: H_0: $\alpha_i = 0$
H_1: Not all α_i are zero

(2) Block effects test: H_0: $\beta_j = 0$
H_1: Not all β_j are zero

In reality, however, we are interested in testing only the first set of hypotheses. We are not concerned with the problem of whether or not block effects are nil since the block means, under the assumptions of this model, are merely indicative of the differences in blocks of experimental material. This second test, however, is a meaningful one, as we shall see, in the completely randomized, two-variable classification model discussed in the next section.

Now, each of the treatment sample means, $\bar{x}_{i.}$, is an unbiased estimate of $\mu + \alpha_i$. Thus we say that if the treatment means are equal, then $\alpha_i = 0$. Again we note that MSE is an unbiased estimate of σ^2 whether or not $\alpha_i = 0$, and that MSC is an unbiased estimate of σ^2 only if $\alpha_i = 0$. When $\alpha_i = 0$ is false, MSC measures both differences in treatment effects and chance variation and thus tends to be much larger than MSE. Under these conditions, we would again accept H_0 if and only if $F_{c-1,(r-1)(c-1)} = \dfrac{\text{MSC}}{\text{MSE}} = 1$ or is close to 1. H_0 is rejected when the computed value of F is significantly greater than 1.

THE EXAMPLE

A motion study is to be conducted for determining the best work design for assembling wall clocks; five designs are under investigation. Four assemblers are selected at random from all assemblers in the plant and are thoroughly taught to work with all five designs. Then each worker follows each design for a day, and the number of clocks assembled is recorded. For this problem, the present model is appropriate. Here, the treatments are the effects of different work designs, and the blocks are the assemblers selected at random. The test for this experiment is made as follows:

1. Hypotheses: H_0: $\alpha_i = 0$
H_1: Not all α_i are zero

TABLE 13.9 SAMPLE DATA FOR WORK DESIGNS EXPERIMENT

	Work Design						
Assembler	*1*	*2*	*3*	*4*	*5*	$t_{.j}$	$\bar{x}_{.j}$
1	10	13	9	14	11	57	11.4
2	5	10	5	10	6	36	7.2
3	6	12	5	10	6	39	7.8
4	4	8	4	11	5	32	6.4
$t_{i.}$	25	43	23	45	28	164	$= T$
$\bar{x}_{i.}$	6.25	10.75	5.75	11.25	7.00	8.2	$= \bar{\bar{x}}$

2. Level of Significance: $\alpha = 0.05$; $\alpha = 0.01$

3. Testing Statistic: $F_{4,12} = \text{MSC}/\text{MSE}$. For this experiment, $c = 5$ and $r = 4$.

4. Decision Rule: Since $P(F_{4,12} > 3.26) = 0.05$ and $P(F_{4,12} > 5.41) = 0.01$, H_0 will be rejected at $\alpha = 0.05$ if the computed value of F is greater than 3.26, and it will be rejected at $\alpha = 0.01$ if the computed F value is greater than 5.41.

5. Computations: Suppose that the output figures (number of clocks assembled) for this experiment are as shown in Table 13.9, then we have

$$\text{C} = \frac{1}{rc} T^2 = \frac{1}{(4)(5)} 164^2 = 1344.8$$

$$\text{SST} = \sum\sum x_{ij}^2 - \text{C} = (10^2 + 5^2 + \cdots + 5^2) - 1344.8 = 191.2$$

$$\text{SSC} = \frac{1}{r} \sum t_{i.}^2 - \text{C}$$

$$= \tfrac{1}{4}(25^2 + 43^2 + 23^2 + 45^2 + 28^2) - 1344.8 = 108.2$$

$$\text{SSR} = \frac{1}{c} \sum t_{.j}^2 - \text{C}$$

$$= \tfrac{1}{5}(57^2 + 36^2 + 38^2 + 32^2) - 1344.8 = 73.2$$

$$\text{SSE} = \text{SST} - (\text{SSC} + \text{SSR}) = 191.2 - (108.2 + 73.2) = 9.8$$

The sums of squares, together with their df and mean squares, are summarized in Table 13.10.

$$F_{4,12} = \frac{\text{MSC}}{\text{MSE}} = \frac{27.05}{0.817} = 33.11$$

For this problem, $n = r$, and if $\alpha = 0.05$,

$$\text{LSD} = \sqrt{\frac{2}{r} (\text{MSE}) F_{1,12;0.05}} = \sqrt{\frac{2}{4} (0.817)(4.75)} = 1.39$$

TABLE 13.10 SUMMARY TABLE FOR WORK DESIGNS DATA

Source of Variation	SS	δ	MS
Between assemblers	73.2	$4 - 1 = 3$	24.40
Between designs	108.2	$5 - 1 = 4$	27.05
Error	9.8	$(4 - 1)(5 - 1) = 12$	0.817
Total	191.2	$20 - 1 = 19$	—

and for $\alpha = 0.01$,

$$\text{LSD} = \sqrt{2/4(0.817)(9.33)} = 1.95$$

Let us present $|\bar{x}_{i.} - \bar{x}_{j.}|$ in a tabular form, and indicate significant difference at $\alpha = 0.05$ by * and significant difference at $\alpha = 0.01$ by ** as follows:

	6.25	10.75	5.75	11.25	7.00
6.25	—				
10.75	4.50**	—			
5.75	0.50	5.00**	—		
11.25	5.00**	0.50	5.50**	—	
7.00	0.75	3.75**	1.25	4.25**	—

6. **Conclusions:** The null hypothesis is rejected at both 5 percent and 1 percent levels of significance. Among the ten pairs of treatments means, six pairs are found to be significantly different at $\alpha = 0.01$. Design 2 and Design 4 are by far the most efficient assembling methods.

13.5 Completely Randomized, Two-Variable Classification without Replication Model

This model refers to a completely randomized design in which sample data are classified in terms of two independent random variables and in which there is only one observation in each cell. The framework of this model is exactly the same as that for the model just discussed, except for the interpretation of data.

In the randomized-blocks, one-variable classification model, we are interested in investigating only treatment effects—the independent variable, A_i. The blocks are considered merely experimental material. In our present two-variable classification model, both treatments and blocks are independent variables and are evaluated simultaneously. Consequently, while A_i are treatments in the one-variable classification case, the treatments in the two variable-classification designs are treatment-block combinations, A_iB_j. Furthermore, in the two-variable classification without replication model, we select samples of size 1 from each treatment combination, and sample data may be presented as follows:

Treatment combination	A_1B_1	A_1B_2	$\cdots$	A_2B_1	$\cdots$	A_cB_r
Observation	x_{11}	x_{12}	$\cdots$	x_{21}	$\cdots$	x_{cr}

Expressed by a two-way table, this would become

	A_1	A_2	$\cdots$	A_c
B_1	x_{11}	x_{21}	$\cdots$	x_{c1}
B_2	x_{12}	x_{22}	$\cdots$	x_{c2}
$\cdots$	$\cdots$	$\cdots$	$\cdots$	$\cdots$
B_r	x_{1r}	x_{2r}		x_{cr}

The second presentation is identical with the scheme for the randomized-blocks, one-variable classification model. This same scheme, however, has two completely different interpretations. The first is as a one-variable classification with randomized-blocks in which A_i constitute the independent variables and B_j are blocks to which A_i are assigned at random. The second is as a completely randomized, two-variable classification model in which both A_i and B_j are independent variables, and x_{ij} are each a sample of size 1 drawn from a population corresponding to A_iB_j.

For example, in our analysis of the motion study earlier, we took the assemblers as blocks. They are assumed to be homogeneous; that is, selected from the same population. The differences in efficiency among the five work designs were then compared under this assumption. Now, if we wished to analyze the same data by our current model, we would investigate the possible differences in productivity among the assemblers as well as those in efficiency among the work designs. In other words, sample data are classified in accordance with two independent variables—work designs and assemblers. We no longer assume that the assemblers are homogeneous units. Now each row is not merely a block that has no effect on the number of clocks assembled, except for the possible differences in work-design efficiency. The observation, say, $x_{11} = 10$ should now be interpreted as one level of output of the first assembler and the first work design that has been drawn from a population of output levels corresponding to A_1B_1 whose mean is μ_{11}. Clearly, with this interpretation, x_{11} is now thought to be affected by both α_1 and β_1.

The analyses of variance table and testing procedures for this model are exactly the same as for the previous one, except that, in the present case, both tests, $\alpha_i = 0$ and $\beta_j = 0$, are now meaningful and should be conducted simultaneously. These two tests were stated before, but we shall repeat them here for convenience:

(1) Test for variable A_i (column treatments): $H_0\colon \alpha_i = 0$
H_1: Not all α_i are zero

$$\text{Testing statistic: } F_{c-1,(r-1)(c-1)} = \frac{\text{MSC}}{\text{MSE}}$$

(2) Test for variable B_j (row treatments): $H_0\colon \beta_j = 0$
H_1: Not all β_j are zero

$$\text{Testing statistic: } F_{r-1,(r-1)(c-1)} = \frac{\text{MSR}}{\text{MSE}}$$

As an illustration, let us use the data in Table 13.10 to test whether there is a significant difference between work designs and between assemblers.

(1) For work designs, A_i: $F_{4,12} = \dfrac{\text{MSC}}{\text{MSE}} = \dfrac{27.05}{0.817} = 33.11$

$$P(F_{4,12} > 5.41) = 0.01$$

(2) For assemblers, B_j: $F_{3,12} = \dfrac{\text{MSR}}{\text{MSE}} = \dfrac{24.40}{0.817} = 29.87$

$$P(F_{3,12} > 5.95) = 0.01$$

Thus both tests are significant at $\alpha = 0.01$. The work designs are different in efficiency and the assemblers are different in productivity.

The pairwise tests for A_i conducted earlier indicated that six out of the ten pairs of column treatment means were significantly different at $\alpha = 0.01$. For B_j now, we have

$$\text{LSD} = \sqrt{\frac{2}{c}\,\text{MSE}(F_{1,12;0.05})} = \sqrt{\frac{2}{5}\,(0.817)(4.75)} = 1.25$$

$$\text{LSD} = \sqrt{\frac{2}{c}\,\text{MSE}(F_{1,12;0.01})} = \sqrt{\frac{2}{5}\,(0.817)(9.33)} = 1.75$$

From the row treatment sample means in Table 13.9, we note that

	11.4	7.2	7.8	6.4
11.4				
7.2	3.2**	—		
7.8	3.6**	0.6	—	
7.4	5.0**	0.8	1.4*	—

Thus, three out of the six pairs of row means are significantly different at the 1 percent level of significance and one pair is significantly different at 5 percent level of significance. The first assembler is by far the most productive worker.

13.6

Completely Randomized, Two-Variable Classification with Replication Model

THE MODEL

This model, as was the preceding one, is concerned with two independent classes of treatments, or "factors." We have, however, two new points to consider.

First, in this model, instead of a single observation, we have two or more observations in each cell; namely, each cell contains a sample with $n_{ij} \geq 2$ selected from the treatment combination population A_iB_j.

Second, in the previous model, the column and row effects, α_i and β_j, were assumed to be independent or additive; but now we assume that they are dependent and "interact" upon each other. To explain this new "interaction" property, let us consider an example. Suppose an experiment is conducted to determine the durability of auto tires, and the variables of interest are three different methods of production and four different makes of rubber. In such a case, we are interested not only in the separate effects of each of these two factors on the durability of auto tires, but also in their possible joint effects. For instance, one type of rubber may work better than another in a particular manufacturing process. This differential effect is called *interaction.*

To generalize, let the two factors be A_i, $i = 1, 2, \cdots, c$, and B_j, $j = 1$ $2, \cdots, r$. Let the sample size be $n_{ij} = k$. Sample data for this model may then take the following form:

	A_1	A_2	$\cdots$	A_c
B_1	x_{111}	x_{211}	$\cdots$	x_{c11}
	x_{112}	x_{212}	$\cdots$	x_{c12}
	x_{11k}	x_{21k}	$\cdots$	x_{c1k}
	$\cdots$	$\cdots$	$\cdots$	$\cdots$
B_r	x_{1r1}	x_{2r1}	$\cdots$	x_{cr1}
	x_{1r2}	x_{2r2}	$\cdots$	x_{cr2}
	$\cdots$	$\cdots$	$\cdots$	$\cdots$
	x_{1rk}	x_{2rk}	$\cdots$	x_{crk}

Each cell above contains k observations—a sample drawn from a treatment combination population with mean μ_{ij}. There are all together rc populations which are normally distributed with a common variance σ^2. As before, the mean of these population means is the grand population mean, μ. Now, if there are no joint effects between A_i and B_j, then each of these cell population means can be represented by the expression

$$\mu_{ij} = \mu + \alpha_i + \beta_j.$$

If, however, A_i and B_j do interact, then we would have

$$\mu_{ij} = \mu + \alpha_i + \beta_j + I_{ij},$$

where I_{ij} stands for the interaction between A_i and B_j.

Now, if we assume that x_{ijk} (the kth observation in the sample drawn from the treatment combination population A_iB_j) is normally distributed about μ_{ij}, then we can state

$$x_{ijk} = \mu_{ij} + e_{ijk} = \mu + \alpha_i + \beta_j + I_{ij} + e_{ijk} \quad (13.12)$$
$$i = 1, 2, \cdots, c$$
$$j = 1, 2, \cdots, r$$
$$k = 1, 2, \cdots, n$$

where the error term, e_{ijk}, is considered as a normal variable with zero mean and variance σ^2. Again, for convenience in solution without loss of generality, we assume that

$$\Sigma\alpha_i = \Sigma\beta_j = \Sigma I_{ij} = 0$$

THE ANALYSIS

To assist our understanding of the partitioning of the total sum of squares for this model and the computations, we shall first give the general presentation of sample data in Table 13.11.

In Table 13.11, we have assumed that there are c of A_i, r of B_j, and have taken a sample (in each cell) of n observations from the treatment combination population A_iB_j. Also, in this table,

$t_{ij.}$ = cell sums. For example, $t_{11.} = x_{111} + x_{112} + \cdots + x_{11n}$

$t_{i..}$ = sums of cell sums added by columns. For example, $t_{2..} = t_{21.} + t_{22.} + \cdots + t_{2r.}$

$t_{.j.}$ = sums of cell sums added by rows. For example, $t_{.1.} = t_{11.} + t_{21.} + \cdots + t_{c1}$

$$T = \sum_i t_{i..} = \sum_j t_{.j.} = \sum_i \sum_j \sum_k x_{ijk} = \text{grand sum}$$

$\bar{x}_{ij.}$ = cell means. There is a cell mean corresponding to each cell A_iB_j. $\bar{x}_{ij.}$ are unbiased estimates of μ_{ij} respectively. For example,

$$\bar{x}_{11.} = \frac{1}{n}\sum_k^n x_{11k} = \frac{1}{n} t_{11.} = \hat{\mu}_{11}$$

$\bar{x}_{i..}$ = group column means. They are means of cell means obtained by

TABLE 13.11 SAMPLE DATA FOR COMPLETELY RANDOMIZED TWO-VARIABLE CLASSIFICATION WITH REPLICATION MODEL

	A_1	A_2	$\cdots$	A_c	$t_{.j.}$
B_1	x_{111}	x_{211}	$\cdots$	x_{c11}	
	x_{112}	x_{212}	$\cdots$	x_{c12}	
	$\cdot$	$\cdot$	$\cdots$	$\cdot$	
	$\cdot$	$\cdot$	$\cdots$	$\cdot$	
	$\cdot$	$\cdot$	$\cdots$	$\cdot$	
	$t_{11.}$	$t_{21.}$	$\cdots$	$t_{c1.}$	$t_{.1.}$
B_2	x_{121}	x_{221}	$\cdots$	x_{c21}	
	x_{122}	x_{222}	$\cdots$	x_{c22}	
	$\cdot$	$\cdot$	$\cdots$	$\cdot$	
	$\cdot$	$\cdot$	$\cdots$	$\cdot$	
	$\cdot$	$\cdot$	$\cdots$	$\cdot$	
	$t_{12.}$	$t_{22.}$	$\cdots$	$t_{c2.}$	$t_{.2.}$
	$\cdots$	$\cdots$	$\cdots$	$\cdots$	
B_r	x_{1r1}	x_{2r1}	$\cdots$	x_{cr1}	
	x_{1r2}	x_{2r2}	$\cdots$	x_{cr2}	
	$\cdot$	$\cdot$	$\cdots$	$\cdot$	
	$\cdot$	$\cdot$	$\cdots$	$\cdot$	
	$\cdot$	$\cdot$	$\cdots$	$\cdot$	
	$t_{1r.}$	$t_{2r.}$	$\cdots$	$t_{cr.}$	$t_{.r.}$
$t_{i..}$	$t_{1..}$	$t_{2..}$	$\cdots$	$t_{3..}$	T

columns. The $\bar{x}_{i..}$ are averages of the observations for A_i. For example, the average of all observations for A_2 is

$$\bar{x}_{2..} = \frac{1}{r}(\bar{x}_{21.} + \bar{x}_{22.} + x_{..} + \bar{x}_{2r.}) = \frac{1}{r}\sum_j \bar{x}_{2j.} = \frac{1}{r}t_{2..}$$

$\bar{x}_{.j.}$ = means of cells means obtained by rows. The $\bar{x}_{.j.}$ are averages of the observations for B_j. For example, the average of all observations for B_r is

$$\bar{x}_{.r.} = \frac{1}{c}(\bar{x}_{1r.} + \bar{x}_{2r.} + \cdots + \bar{x}_{cr.}) = \frac{1}{c}\sum_i \bar{x}_{ir.} = \frac{1}{c}t_{.r.}$$

$\bar{\bar{x}} = \frac{1}{rc}T$ = the grand sample mean, the unbiased estimate of grand mean μ.

With the preceding notations and definitions, we again define the total variation as the over-all dispersion of all observations from the grand mean; namely,

$$\text{SST} = \sum_i \sum_j \sum_k (x_{ijk} - \bar{\bar{x}})^2$$

This total sum of squares can be decomposed into four separate and independent sources of variations as follows:

$\Sigma\Sigma\Sigma(x_{ijk} - \bar{\bar{x}})^2$	Total variation: SST
$= \Sigma\Sigma\Sigma(x_{ijk} - \bar{x}_{ij.})^2$	Error variation: SSE
$+ \Sigma\Sigma\Sigma(\bar{x}_{i..} - \bar{\bar{x}})^2$	Variation between columns: SSC
$+ \Sigma\Sigma\Sigma(\bar{x}_{.j.} - \bar{\bar{x}})^2$	Variation between rows: SSR
$+ \Sigma\Sigma\Sigma(\bar{x}_{ij.} - \bar{x}_{i..} - \bar{x}_{.j.} + \bar{\bar{x}})^2$	Variation in interaction: SSI

The sum of squares between columns, SSC, shows the deviations of $\bar{x}_{i..}$ from the grand sample mean. It reflects, therefore, fluctuations in the column treatment effects, α_i, and in the error term, e_{ijk}.

The sum of squares between rows, similarly, measures the dispersion of $\bar{x}_{.j.}$ around the grand sample mean. Therefore, SSR reflects both variation in row treatment effects, β_j, and deviation in e_{ijk}.

The sum of squares for error, SSE, is based on the deviation cells means for individual observations treated exactly in the same manner. It thus measures random variations in sampling. For example, $x_{11k} - \bar{x}_{11.}$ reflects the deviation of the first observation in the first sample cell for A_1B_1 from its mean $\bar{x}_{11.}$. It measures therefore the error term e_{11k}.

The sum of squares for interaction, SSI, measures the dispersion in the interaction, or joint effects, between the treatment combinations A_iB_j as well as error. This is based on the fact that $(\bar{x}_{ij.} - \bar{x}_{i..} - \bar{x}_{.j.} + \bar{\bar{x}})$ estimates the quantity $I_{ij} = \mu_{ij} - (\alpha_i + \beta_j + \mu)$ which is the joint effect between A_i and B_j.

The above sums of squares are computed by expanding their respective definitional expressions as below:

$$\text{C} = \text{correction factor} = \frac{1}{rcn} T^2 \tag{13.13}$$

$$\text{SST} = \sum\sum\sum (x_{ijk} - \bar{\bar{x}})^2 = \sum\sum\sum x^2_{ijk} - \text{C} \tag{13.14}$$

$$\text{SSC} = \sum\sum\sum (\bar{x}_{i..} - \bar{\bar{x}})^2 = \frac{1}{rn} \sum t^2_{i..} - \text{C} \tag{13.15}$$

$$\text{SSR} = \sum\sum\sum (\bar{x}_{.j.} - \bar{\bar{x}})^2 = \frac{1}{cn} \sum t^2_{.j.} - \text{C} \tag{13.16}$$

$$\text{SSE} = \sum\sum\sum (x_{ijk} - \bar{x}_{ij.})^2 = \sum\sum\sum x^2_{ijk} - \frac{1}{n} \sum\sum t^2_{ij.} \tag{13.17}$$

$$\text{SSI} = \text{SST} - (\text{SSC} + \text{SSR} + \text{SSE}) \tag{13.18}$$

These sums of squares, together with their respective degrees of freedom and mean squares, are given in Table 13.12.

TABLE 13.12

Source of Variation	SS	δ	MS
Between columns	SSC	$c-1$	MSC = SSC/$c-1$
Between rows	SSR	$r-1$	MSR = SSR/$r-1$
Interaction	SSI	$(r-1)(c-1)$	MSI = SSI/$(r-1)(c-1)$
Error	SSE	$rc(n-1)$	MSE = SSE/$rc(n-1)$
Total	SST	$rcn-1$	—

THE TEST

We are interested in performing three tests in this model; they are

(1) The column treatment effects are nil; that is, $\alpha_i = 0$. The testing statistic here is $F_{c-1,rc(n-1)} = \text{MSC/MSE}$.

(2) The block treatment effects are nil; that is, $\beta_j = 0$. For this, the testing statistic is $F_{r-1,rc(n-1)} = \text{MSR/MSE}$.

(3) The joint effects are nil; that is, $I_{ij} = 0$. We test this null hypothesis by $F_{(r-1)(c-1),rc(n-1)} = \text{MSI/MSE}$.

The decision procedures are the same as in previous cases. When the computed F value is 1 or close to 1, H_0 is accepted. When F is significantly larger than 1, H_0 is rejected.

THE EXAMPLE

Suppose that a firm has available four different sources of raw materials, A_1, A_2, A_3, and A_4, and three machines of different makes, B_1, B_2, and B_3, to produce a new product. It is known that all three makes of machines are equally productive in terms of speed—number of outputs produced per hour—but it is not known whether they perform equally well in terms of the number of defectives produced among hourly outputs. In addition, it is unknown to the firm whether there are differences in quality of the raw materials from the four sources. Finally, it is suspected that the raw material from one source may have some particular effect on a particular machine, or vice versa. Thus, it is desired to establish whether A_i are different, whether B_j are different, and if there exist any joint effect, $A \times B$. To answer these questions, each make of machine is operated under identical conditions with each source of material for two hours, and the number of defective units for each hour is recorded. Results of this experiment are recorded in Table 13.13. In view of these data, what conclusions can be drawn? Which make of machine and which source of raw material should the firm purchase?

The three sets of hypotheses to be evaluated for this example are:

TABLE 13.13

		A_1	A_2	A_3	A_4	Total
B_1		9	6	8	5	
		5	6	4	7	
		14	12	12	12	50
B_2		4	5	2	1	
		2	3	6	5	
		6	8	8	6	28
B_3		10	8	7	9	
		6	9	8	5	
		16	17	15	14	62
	Total	36	37	35	32	140

(1) H_0: Mean numbers of defectives per hour for raw materials from all the four sources are the same; that is, $\alpha_i = 0$.

H_1: Not all α_i are zero.

(2) H_0: Mean numbers of defectives per hour for all the three makes of machines are equal; that is, $\beta_j = 0$.

H_1: Not all β_j are zero.

(3) H_0: Interaction between raw materials and machines is nil; that is, $I_{ij} = 0$.

H_1: Interaction is not nil; that is, $I_{ij} \neq 0$.

For data in Table 13.13, we have

$$\mathrm{C} = \frac{1}{rcn} T^2 = \frac{1}{(3)(4)(2)} (140^2) = 817$$

$$\mathrm{SST} = \sum\sum\sum x^2_{ijk} - \mathrm{C} = 9^2 + 5^2 + \cdots + 5^2 - 817 = 135$$

$$\mathrm{SSC} = \frac{1}{rn} \sum t^2_{i..} - \mathrm{C} = \frac{1}{(3)(2)} (36^2 + 37^2 + 35^2 + 32^2) - 817 = 2$$

$$\mathrm{SSR} = \frac{1}{cn} \sum t^2_{.j.} - \mathrm{C} = \frac{1}{(4)(2)} (50^2 + 28^2 + 62^2) - 817 = 74$$

$$\mathrm{SSE} = \sum\sum\sum x^2_{ijk} - \frac{1}{n} \sum t^2_{ij.} = 952 - \frac{1}{2} (14^2 + 12^2 + \cdots + 6^2 + \cdots + 14^2) = 54$$

$$\mathrm{SSI} = \mathrm{SST} - (\mathrm{SSC} + \mathrm{SSR} + \mathrm{SSE}) = 135 - (2 + 74 + 54) = 5$$

With data in the preceding analysis of variance table, we have

(1) For $\alpha_i = 0$,

$$F_{3,12} = \frac{\mathrm{MSC}}{\mathrm{MSE}} = \frac{0.67}{3.86} = 0.17, \text{ which leads to the acceptance of } \alpha_i = 0.$$

TABLE 13.14 SUMMARY TABLE FOR RAW MATERIALS AND MACHINE DATA

Source of Variation	SS	δ	MS
Materials (columns)	2	$4 - 1 = 3$	0.67
Machines (rows)	74	$3 - 1 = 2$	37.00
Interaction $A \times B$	5	$(3 - 1)(4 - 1) = 6$	0.83
Error	54	$(3)(4)(1) = 12$	3.86
Total	135	$(3)(4)(2) - 1 = 23$	—

(2) For $\beta_j = 0$,

$F_{2,12} = \dfrac{\text{SSR}}{\text{SSE}} = \dfrac{37.00}{3.86} = 9.59$, which leads to the rejection of $\beta_j = 0$ at $\alpha = 0.05$ since $F_{2,12;\,0.05} = 3.89$ as well as at $\alpha = 0.01$, since $F_{2,12;\,0.01} = 6.93$.

(3) For $I_{ij} = 0$,

$F_{6,12} = \dfrac{\text{SSI}}{\text{SSE}} = \dfrac{0.83}{3.86} = 0.22$, which leads to the acceptance of $I_{ij} = 0$.

Thus we see there are neither significant column (raw materials) treatment effects nor joint effects between column and row treatments. There exist, however, highly significant differences between machines of different makes. These conclusions indicate that the firm should purchase the second make of machines (B_2) (why?), and that it can select any of the four sources of raw materials because there is no difference.

13.7 Randomized-Blocks, Two-Variable Classification with Replication Model

THE MODEL

The last model can easily be adopted for a randomized-blocks design. All that one has to do is to randomize all the treatment combinations over a block of experimental material, and to repeat this basic experiment as often as required for achieving the desired accuracy.

To bring out the salient features of the randomized-blocks, two-variable classification with replication model more vividly, let us consider an example. Suppose that there are c varieties of soybeans, A_i, r types of fertilizers, B_j, and k blocks of land, C_k, which may be different in productivity. We wish to determine whether significant differences exist among A_i and among B_j. We do not wish our experimental results to be affected, however, by the possible differences that might exist among C_k. We achieve this by subdividing each block of land into rc plots and assigning them to the rc treatment combination populations, A_iB_j, by a random process.

By the preceding scheme, the sample size drawn from each treatment combination would be $n = k$—the number of blocks of land—and there would be $(r)(c)(n)$ cells, each representing a normal population with mean μ_{ijk} and a common variance σ^2. These μ_{ijk} are assumed to be distributed around a grand mean μ. Now, x_{ijk} are sample observations from populations with μ_{ijk}. The x_{ijk}, for example, are yields of soybeans, and we would have $x_{ijk} = \mu_{ijk} + e_{ijk}$. Again, if we assume that the effects between A_i, B_j, and C_k are additive, our basic model may be expressed as follows:

$$x_{ijk} = \mu + \alpha_i + \beta_j + \gamma_k + I_{ij} + e_{ijk} \tag{13.19}$$

where

$\alpha_i = \mu_{i..} - \mu$, effects between factor A_i
$\beta_j = \mu_{.j.} - \mu$, effects between factor B_j
$\gamma_k = \mu_{..k} - \mu$, effects between block C_k
$I_{ij} = \mu_{ij.} - (\alpha_i + \beta_j + \mu)$, interaction between A_i and B_j

We also assume, for convenience of solution, that

$$\Sigma\alpha_i = \Sigma\beta_j = \Sigma\gamma_k = \Sigma I_{ij} = 0$$

THE ANALYSIS

The partitioning of the total sum of squares for this model is similar to that for the previous model, and the same notations are employed. We have now, however, five instead of four components in the total sum of squares.

The total sum of squares, as always in the analysis of variance, measures the variation of all observations from the grand sample mean. It is now decomposed in the following way:

$$\sum_i^c \sum_j^r \sum_k^n (x_{ijk} - \bar{\bar{x}})^2 \qquad \text{Total variation: SST}$$

$$= \sum\sum\sum (\bar{x}_{i..} - \bar{\bar{x}})^2 \qquad \text{Variation between } A_i\text{: SSC}$$

$$+ \sum\sum\sum (\bar{x}_{.j.} - \bar{\bar{x}})^2 \qquad \text{Variation between } B_j\text{: SSR}$$

$$+ \sum\sum\sum (\bar{x}_{ij.} - \bar{x}_{i..} - \bar{x}_{.j.} + \bar{\bar{x}})^2 \qquad \text{Joint variation between } A_i \text{ and } B_j\text{: SSI}$$

$$+ \sum\sum\sum (\bar{x}_{..k} - \bar{\bar{x}})^2 \qquad \text{Variation between blocks: SSB}$$

$$+ \sum\sum\sum (x_{ijk} - \bar{x}_{..k} - \bar{x}_{.j.} + \bar{\bar{x}})^2 \qquad \text{Variation for error: SSE}$$

Interpretations of these sums of squares may have become familiar to us by now. We shall move on to introduce the formulas for their calculations as follows:

$$\text{C} = \text{correction factor} = \frac{1}{rcn} T^2 \qquad \textbf{(13.13')}$$

$$\text{SST} = \sum\sum\sum (x_{ijk} - \bar{\bar{x}})^2 - \text{C} = \sum\sum\sum x^2_{ijk} - \text{C} \qquad \textbf{(13.14')}$$

$$\text{SSC} = \sum\sum\sum (\bar{x}_{i..} - \bar{\bar{x}})^2 = \frac{1}{rn}\sum t^2_{i..} - \text{C} \qquad \textbf{(13.15')}$$

$$\text{SSR} = \sum\sum\sum (\bar{x}_{.j.} - \bar{\bar{x}})^2 = \frac{1}{rn}\sum t^2_{.j.} - \text{C} \qquad \textbf{(13.16')}$$

$$\text{SSB} = \sum\sum\sum (\bar{x}_{..k} - \bar{\bar{x}})^2 = \frac{1}{rc}\sum t^2_{..k} - \text{C} \qquad \textbf{(13.20)}$$

$$\text{SSE} = \sum\sum\sum (x_{ijk} - \bar{x}_{..k} - \bar{x}_{.j.} + \bar{\bar{x}})^2$$

$$= \sum\sum\sum x^2_{ijk} + \text{C} - \left(\frac{1}{rc}\sum t^2_{..k} + \frac{1}{n}\sum_i\sum_j t^2_{ij.}\right) \qquad \textbf{(13.21)}$$

$$\text{SSI} = \text{SST} - (\text{SSC} + \text{SSR} + \text{SSB} + \text{SSE}) \qquad \textbf{(13.22)}$$

It is interesting to note that C, SST, SSC, and SSR for this model are defined and computed in exactly the same way as they were for the previous one. The formulas for SSE and SSI are different, however, due to the introduction of SSB. We may also note that in the preceding identity between the total sum of squares and the sums of squares for the five sources of variations, by referring to Table 13.11,

$$\bar{x}_{..k} = \text{the mean of the } k\text{th observation in all the cell samples}$$

$$= \frac{1}{rc} t_{..k}.$$

For example,

$$\bar{x}_{..1} = \frac{1}{rc} t_{..1} = \frac{1}{rc}(x_{111} + x_{211} + \cdots + x_{cr1})$$

The sums of squares, together with their degrees of freedom and mean squares, are summarized in Table 13.15 on page 426.

THE TEST

In our current randomized-blocks, two-variable classification with replication model, as for the previous one, we are concerned with three tests; they are

TABLE 13.15 ANALYSIS OF VARIANCE TABLE: RANDOMIZED-BLOCKS, TWO-VARIABLE CLASSIFICATION WITH REPLICATION MODEL

Source of Variation	SS	δ	MS
Blocks	SSB	$n-1$	MSB = SSB/$n-1$
Columns	SSC	$c-1$	MSC = SSC/$c-1$
Rows	SSR	$r-1$	MSR = SSR/$r-1$
Interaction	SSI	$(r-1)(c-1)$	MSI = SSI/$(r-1)(c-1)$
Error	SSE	$(rc-1)(n-1)$	MSE = SSE/$(rc-1)(n-1)$
Total	SST	$rcn-1$	—

(1) On column treatment effects: H_0: $\alpha_i = 0$
H_1: Not all α_i are zero

(2) On row treatment effects: H_0: $\beta_j = 0$
H_1: Not all β_j are zero

(3) On interaction: H_0: $I_{ij} = 0$
H_1: Not all I_{ij} are zero

The respective testing statistics for these three tests are

$$(1)\quad F_{c-1,(rc-1)(n-1)} = \frac{\text{MSC}}{\text{MSE}}$$

$$(2)\quad F_{r-1,(rc-1)(n-1)} = \frac{\text{MSR}}{\text{MSE}}$$

$$(3)\quad F_{(r-1)(c-1),(rc-1)(n-1)} = \frac{\text{MSI}}{\text{MSE}}$$

Again, in each case, H_0 is rejected if and only if the computed value for F is significantly greater than 1.

In general, for randomized-blocks designs, we are not interested in any possible block effects, because they merely represent experimental material over which the treatment combinations are randomized.

THE EXAMPLE

A commercial bank which has a large volume of daily filing wishes to test two methods of filing and to establish whether or not there is any difference in efficiency between male and female file clerks. To effect this experiment, two experienced file clerks are selected at random, one male and one female. Each clerk is to use each method for one hour per day for a period of one week. (The days here constitute the blocks in our randomized-blocks design.) Results recorded as to number of documents filed per hour for each method during the whole week are shown in Table 13.16. In view of these data, can we conclude that there is a significant difference in efficiency between male and female operators? between

TABLE 13.16

Day	Method	Operator *Male*	*Female*
Monday	1	110	125
	2	123	140
Tuesday	1	108	120
	2	121	127
Wednesday	1	113	119
	2	126	137
Thursday	1	120	123
	2	126	126
Friday	1	105	117
	2	118	123

the two methods of filing? Can we also decide that there is significant interaction between filing methods and operators?

Since the variance is not affected if a constant is subtracted from all values in a group, we shall, for the sake of speed, deduct a constant of 100 from each of the figures in Table 13.16 before doing our computations.

TABLE 13.17

Day	Method	Male	Female	Total
Monday	1	10	25	35
	2	23	40	63
		33	65	98
Tuesday	1	8	20	28
	2	21	27	48
		29	47	76
Wednesday	1	13	19	32
	2	26	37	63
		39	56	95
Thursday	1	20	23	43
	2	26	26	52
		46	49	95
Friday	1	5	17	22
	2	18	23	41
		23	40	63
Total	1	56	104	160
	2	114	153	267
$t.$		170	257	427

For the above data, the sums of squares are

$$C = \frac{1}{rcn} T^2 = \frac{1}{(5)(2)(2)} (427^2) = 9116.45$$

$$\text{SST} = \sum\sum\sum x^2_{ijk} - C = (10^2 + 25^2 + \cdots + 23^2) - 9116.45$$

$$= 10{,}511 - 9116.45 = 1394.55$$

$$\text{SSC} = \frac{1}{rn} \sum t^2_{i..} - C = \frac{1}{(5)(2)} (170^2 + 257^2) - 9116.45 = 378.45$$

$$\text{SSR} = \frac{1}{cn} \sum t^2_{.j.} - C$$

$$= \frac{1}{(2)(2)} (98^2 + 76^2 + 95^2 + 95^2 + 63^2) - 9116.45 = 233.30$$

$$\text{SSB} = \frac{1}{rc} \sum t^2_{..k} - C$$

$$= \frac{1}{(5)(2)} (160^2 + 267^2) - 9116.45 = 572.45$$

$$\text{SSE} = \sum\sum\sum x^2_{ijk} + C - \left(\frac{1}{rc} \sum t^2_{..k} + \frac{1}{n} \sum_i \sum_j t^2_{ij.}\right)$$

$$= 10{,}511 + 9116.45 - \left[\frac{1}{(5)(2)} (160^2 + 267^2) + \frac{1}{2} (33^2 + 65^2 + 29^2 + \cdots + 40^2)\right]$$

$$= 105.05$$

$$\text{SSI} = \text{SST} - (\text{SSC} + \text{SSR} + \text{SSB} + \text{SSE})$$

$$= 1394.55 - (378.45 + 233.30 + 572.45 + 105.05) = 105.30$$

These sums of squares, together with their df and mean squares, are summarized in Table 13.18.

TABLE 13.18 SUMMARY TABLE FOR OPERATORS AND METHODS DATA

Source of Variation	SS	δ	MS
Days (blocks)	572.45	$2 - 1 = 1$	—
Operators (columns)	378.45	$2 - 1 = 1$	378.450
Methods (rows)	233.30	$5 - 1 = 4$	58.325
Interaction	105.30	$(5 - 1)(2 - 1) = 4$	26.325
Error	105.05	$((5)(2) - 1)(2 - 1) = 9$	10.505
Total	1394.55	$(5)(2)(2) = 19$	—

For operator effects, we have

$$F_{1,9} = \frac{\text{MSC}}{\text{MSE}} = \frac{378.450}{10.505} = 36.03$$

but

$$P(F_{1,9} > 5.12) = 0.05, \quad \text{and} \quad P(F_{1,9} > 10.56) = 0.01$$

Therefore, $\alpha_i = 0$ is rejected at both 5 percent and 1 percent levels of significance. Obviously, female file clerks seem to be more efficient than male clerks.

For method effects, we have

$$F_{4,9} = \frac{\text{SSR}}{\text{SSE}} = \frac{58.325}{10.505} = 5.52$$

but

$$P(F_{4,9} > 3.63) = 0.05, \quad \text{and} \quad P(F_{4,9} > 6.42) = 0.01$$

Hence, $\beta_j = 0$ is rejected at the 5 percent level of significance but it cannot be rejected at the 1 percent level of significance. It seems that the second method of filing is more efficient than the first. Why?

For interaction between operators and methods, we have

$$F_{4,9} = \frac{\text{MSI}}{\text{MSE}} = \frac{26.325}{10.505} = 2.51$$

which can be explained by chance variations alone. Thus, $I_{ij} = 0$ is accepted. This conclusion indicates that there seem to be no particular effects of a given filing method upon male or female operators in any significant or unique way, or vice versa.

13.8 Concluding Remarks

The aim of this chapter has been to introduce the reader to the basic principles of the analysis of variance and to impress upon him the power of this tool. Because a number of aspects of this amazingly versatile technique have been left untouched, it seems appropriate to make a few general remarks here about those things which have been omitted.

First, by now the reader may have gained the impression that the analysis of variance, especially for the two-way classification models, is mainly concerned with the generation of a series of F tests on the same set of data, while the partitioning of the total sum of squares into meaningful components is only a means to this end. This, however, would be a superficial viewpoint. The essence of the analysis of variance is that it permits the decomposition of all potential information in the data into exclusive and distinct parts, each reflecting only some specific aspect of the experiment. In the linear model, for example, the mean square between groups reflects variations in both systematic experimental manipulations and chance phenomena, but the mean square within groups measures only the unsystematic random variations which are attributable to any chance experiment. These two statistics are two independent ways of summarizing data—the information provided by one is not redundant of the other. Therefore, they furnish the researcher with a basis on which to decide whether or not significant treatment effects exist.

Similarly, in the two-way classification models, the mean squares for both treatment factors are independent of each other and are not redundant to the mean square for error or for interaction. The analysis of variance enables us to separate and to identify the factors contributing to variations with particular summary measures. This makes it possible for experimenters to retain balance among comparisons. The retention of balance is a device for finding the statistics that reflect particular and meaningful aspects of data. Thus, the process of partitioning, not the F ratios, is the heart of the analysis of variance.

In short, the analysis of variance sorts the information yielded by an experiment into neat, exclusive, and meaningful portions and helps us to judge the experimental treatments easily—however large the number of treatment factors is. This is exactly the reason that the multiple t tests are useless in comparing more than two populations. In other words, multiple t tests do not provide the nonredundant feature—the various differences between pairs of means do overlap in the information they give. As a result, one cannot assess the evidence for the over-all existence of treatment effects from a complete set of such differences derived from multiple t tests.

Next we note that, in this chapter, we have introduced models for the analysis of variance up to two experimental variables. In experiments with three or more factors, the basic ideas of partition, means of squares, and F tests are the same, but the models become much more involved. For example, for a three-factor model, in addition to mean squares representing the joint effects of particular pairs of the classification variables, there is also a mean square which measures the simultaneous interaction among all the three factors. Hence, experiments with many factors are very difficult, in terms of time and labor, to perform.

In addition to the practical difficulty mentioned above, there is also a theoretical limitation for complicated models. That is, in the analysis of variance, we have assumed that the F tests are independent. In reality, however, the F ratios are dependent chiefly because all F ratios in the same model share the same denominator—the mean square for error. It can be shown that if the assumption of independence holds, and if there are only three F tests, then we should expect about 3(0.01) or 0.03 of these tests to show significance at $\alpha = 0.01$ by chance alone. Moreover, the probability is $1 - (0.99)^2$ or about 0.03 that at least one of these tests will show spurious significance. However, when (as is generally true) these F tests are not independent, we have no accepted standards to calculate the number expected by chance; we know only that the probability is between 0.01 and 0.03 for at least one spuriously significant result. For really complicated models, where many F tests are made with the same data, the probability that at least one spuriously significant result will occur may be very large; what is worse, this probability and that for a Type I error cannot be determined in a simple and routine manner.

Due to practical difficulty and theoretical limitation, we must not yield to the temptation to include too many treatment factors in our analysis when sample data seem easy or cheap to obtain, or when we desire to have a model that is impressive. In business and economic research, as in any type of scientific investigation, we should be interested in obtaining meaningful results by simple designs if situations permit, rather than in the elegance or complexity of experiments that may furnish only results which have no clear interpretation.

The models introduced in this chapter are called *fixed effects* models, since in these models the different levels of each factor were chosen in advance of the experiment and the experiment treatments actually administered were thought of as exhausting all treatments of interests. For these models, therefore, the only conclusions that can be drawn are concerned with the particular levels and treatments or treatment combinations. By far, the most frequently confronted random experiments are of this kind.

In addition to fixed effects models, there are two other types of models: random effects and mixed models. A *random effects* model is one in which inferences are made about an entire set of distinct treatments on the basis of a random sample selected from such a treatments population. Thus, in the fixed effects models, treatments actually observed are considered as populations. Such treatments, however, constitute only a random sample in the random effects models. From a computational point of view, fixed effects and random effects models are the same. They differ from each other mainly in the inferences that are drawn.

A *mixed* model, as the name indicates, is a mixture of both fixed and random effects models. In such a model, one or more treatments have fixed effects and the remaining factors are sampled. Also, each observation now results in a score which is a sum of both fixed and random effects. Obviously, in mixed models, there must be at least two factors under investigation. For the simple two-factor design of mixed models, computational procedure requires no change from the fixed effects models, but there is an important difference in the F ratios used to test the different effects.

Finally, the analysis of variance has been developed under a set of rigid assumptions: (1) treatments or treatment combinations are normally distributed with common variance, (2) the treatment and enviromental effects are additive, and (3) the experimental errors are independent and distributed as $n(0,\sigma^2)$. Whenever any of these assumptions is not met, the F tests cannot be employed to yield valid inferences. It is indeed fortunate that many economic and business experiments do conform to these premises. It is not uncommon, however, to encounter experimental work where departures from these assumptions exist. In such a situation, the analysis of variance may still be applied by way of transformation. *Transformation* refers to a process of transforming the original data into some other form, such as square roots, reciprocals, inverse sines, or logarithms, before the analysis is made. The assumptions of normality, additivity, and homogeneity of variances are often violated together; hence a single transformation which corrects one of these will usually also result in the improvement of data in other respects.

Procedures of analysis of variance with transformed data are identical with those for original data; but the results obtained must be converted back into the original units before intelligent and appropriate interpretations can be given. While transformation techniques are quite simple and straightforward, the selection of an appropriate form of transformation for a particular set of data requires considerable theoretical knowledge in the field to which the analysis of variance is applied. It is perhaps worthwhile to repeat once again that competence in applied statistics depends not only upon an understanding of statistical methods but also upon a thorough orientation in the subject matter to which statistical methods are applied.

Glossary of Formulas

(13.1) $x_{ij} = \mu_i + e_{ij} = \mu + \alpha_i + e_{ij}$ This expression represents the completely randomized, one-variable model of the analysis of variance. For this model, we have c samples selected from the c treatment populations which are normally distributed with μ_i and common variance σ^2. Sample observations, x_{ij}, are assumed to be distributed normally about μ_i for fixed effects models. Residuals, e_{ij}, are chance variations in the experiment and are assumed to be independent and distributed normally with zero mean and variance σ^2. Here, α_i are treatment effects which can be considered as deviations of treatment population means from the grand population mean; that is, $\alpha_i = \mu_i - \mu$. For convenience of solution, we assume that $\Sigma\alpha_i = 0$. For this model, we desire to test H_0: $\alpha_i = 0$ against H_1: not all α_i are zero.

(13.2) SST $= \Sigma\Sigma x_{ij}^2 - N\bar{\bar{x}}^2$ **(13.3)** SSC $= \Sigma n_i\bar{x}_{i.}^2 - N\bar{\bar{x}}^2$ **(13.4)** SSE $= \Sigma\Sigma x_{ij}^2 - \Sigma n_i\bar{x}_{i.}^2$

These are the computational equations for the sums of squares for model (13.1). The heart of the analysis of variance is the partitioning of the total sum of squares into distinct and meaningful parts. For this model, the total sum of squares, defined as SST $= \Sigma\Sigma(x_{ij} - \bar{\bar{x}})^2$, is partitioned into the sum of the square between, SSC $= \Sigma\Sigma(\bar{x}_{i.} - \bar{\bar{x}})^2$, and the sum of square within, SSE $= \Sigma\Sigma(x_{ij} - \bar{x}_{i.})^2$. While SSE measures chance variation alone, SSC measures both chance variation and differences in treatments. Under the assumption that $\alpha_i = 0$ is true, each of these three sums of squares, divided by an appropriate number of df is an unbiased estimate of σ^2. Also, $\delta = N - 1$ for SST, $\delta = c - 1$ for SSC, and $\delta = N - c$ for SSE. It is interesting to note that just as SST = SSC + SSE, so $N - 1 = (c - 1) + (N - c)$. Finally, the null hypothesis that the treatment effects are nil is tested by the statistic $F_{c-1,N-c}$ = MSC/MSE.

(13.5) LSD $= \sqrt{(2/n)(\text{MSE})F_{1,N-c;\alpha}}$ This formula defines what is called the least significant difference in pairwise tests. It can be employed only when sample sizes are equal and when the F ratio leads to the rejection of $\alpha_i = 0$. Two sample means are said to be significantly different at α if their absolute difference is greater than LSD. Pairwise tests can also be made with this formula for randomized-blocks, one-variable classification, and completely randomized, two-variable classification without replication models.

(13.6) $x_{ij} = \mu_{ij} + e_{ij} = \mu + \alpha_i + \beta_j + e_{ij}$ This expression defines both the randomized-blocks, one-variable classification model and the completely randomized, two-variable classification without replication model. In both cases, we assume that μ, the grand population mean, is located at a position such that $\Sigma\alpha_i = \Sigma\beta_j = 0$. Furthermore, we also assume that (1) the rc populations are normal with μ_{ij} and common variance σ^2, (2) the column treatment effects, $\alpha_i = \mu_{i.} - \mu$, and the block treatment effects, $\beta_j = \mu_{.j} - \mu$, are additive, and (3) e_{ij} are independent and distributed normally with mean 0 and variance σ^2. These two models differ mainly in interpretations. For the former, we are interested in only the column effects and we treat blocks merely as experimental material to which column treatments have been assigned at random. Thus, we wish to test $\alpha_i = 0$ by F = MSC/MSE. In the latter, we consider both A_i and B_j as independent variables and evaluate them simultaneously. Now, we wish to test $\alpha_i = 0$ as well as $\beta_j = 0$. For $\beta_j = 0$, we employ the testing statistic, F = MSR/MSE.

(13.7) $C = \frac{1}{rc} T^2$ **(13.8)** $SST = \sum\sum x_{ij}^2 - C$ **(13.9)** $SSC = \frac{1}{r} \sum t_{i.}^2 - C$

(13.10) $SSR = \frac{1}{c} \sum t_{.j}^2 - C$ **(13.11)** $SSE = SST - (SSC + SSR)$

These are the computational equations for the sums of squares of model (13.6). Here, r refers to the number of blocks, c refers to the number of columns, T stands for the grand total, and $t_{i.}$ and $t_{.j}$ are the column and row sums, respectively. To estimate σ^2 on the bases of these sums of squares, we note that for SSC, $\delta = c - 1$, for SSR, $\delta = r - 1$, and for SSE, $\delta = (r - 1)(c - 1)$. These numbers of df add up to $rc - 1$, δ for SST.

(13.12) $x_{ijk} = \mu_{ij} + e_{ijk} = \mu + \alpha_i + \beta_j + I_{ij} + e_{ijk}$ This is the basic model for a completely randomized, two-variable classification with replication experiment. In the completely randomized, two-variable classification without replication model, it has been assumed that column and row effects are additive and independent. Now, with two or more observations in each sample for the present model, we also assume that there is joint effect, or interaction, between A_i and B_j in addition to the additivity of α_i and β_j. Under the assumptions of additivity and independence, we have $\mu_{ij} = \mu + \alpha_i + \beta_j$. If we further assume that joint effect exists, then the interaction term must be something in addition to α_i and β_j and it may be thought of as $I_{ij} = \mu_{ij} - (\mu + \alpha_i + \beta_j)$. In this model, we assume that the rc cells are random samples, each with $n \geq 2$, selected from the corresponding rc treatment combination populations which are normal with means μ_{ij} and variance σ^2. We also assume that e_{ijk} are independent and distributed normally, with mean zero and variance σ^2. Finally, we assume, without loss of generality, that $\Sigma\alpha_i = \Sigma\beta_j = \Sigma I_{ij} = 0$ for convenience in solution. For this model, we have three sets of tests to perform: (1) $\alpha_i = 0$, (2) $\beta_j = 0$, and (3) $I_{ij} = 0$.

(13.13) $C = \frac{1}{rcn} T^2$ **(13.14)** $SST = \sum\sum\sum x^2_{ijk} - C$ **(13.15)** $SSC = \frac{1}{rn} \sum t^2_{i..} - C$

(13.16) $SSR = \frac{1}{cn} \sum t^2_{.j.} - C$ **(13.17)** $SSE = \sum\sum\sum x^2_{ijk} - \frac{1}{n} \sum\sum t^2_{ij.}$

(13.18) $SSI = SST - (SSC + SSR + SSE)$

The sum of squares for model (13.12) is partitioned into four distinct and independent portions. SSC, with $\delta = c - 1$, measures variability in column treatment effects and error. SSR, with $\delta = r - 1$, measures variability in row treatment effects and error. SSI, with $\delta = (r - 1)(c - 1)$, measures variability in interaction effects and error. SSE, with $\delta = rc(n - 1)$, measures pure chance variation of the experiment. The sum of these four df corresponds to the total number of df, $rcn - 1$, for SST.

(13.19) $x_{ijk} = \mu + \alpha_i + \beta_j + \gamma_k + I_{ij} + e_{ijk}$ This expression defines the randomized-blocks, two-variable classification with replication model. In this model, all the treatment combinations, A_iB_j, are randomized over blocks of experimental material, C_k. We are here interested in the effects of factor A_i, α_i, the effects of factor B_j, β_j, and their interaction effects, I_{ij}; but not in the effects of C_k, γ_k. For this model, there are rcn normal populations with μ_{ijk} and common variance σ^2. The μ_{ijk} are assumed to be distributed about the grand mean μ. It is also assumed here that α_i, β_j, and γ_k are additive, and that $\Sigma\alpha_i = \Sigma\beta_j = \Sigma\gamma_k = \Sigma I_{ij} = 0$.

(13.20) $\text{SSB} = \frac{1}{rc}\sum t^2_{..k} - \text{C}$

(13.21) $\text{SSE} = \sum\sum\sum x^2_{ijk} + \text{C} - \left(\frac{1}{rc}\sum t^2_{..k} + \frac{1}{n}\sum_i\sum_j t^2_{ij.}\right)$

(13.22) SSI = SST − (SSC + SSR + SSB + SSE)

The sum of squares for model (13.20) is partitioned into five exclusive parts. The computational formulas for C, the correction factor, SST, SSC, and SSR are the same as (13.13) to (13.16), exclusively. In this model, the sum of squares for blocks, C_k, measures both block effects and error. SSI, as usual, is computed by subtraction because of the identity SST = SSC + SSR + SSB + SSE + SSI. Also, for SSB, $\delta = n - 1$; for SSE, $\delta = (rc - 1)(n - 1)$; for SSI, $\delta = (r - 1)(c - 1)$. The total number of df for the model is $(rcn - 1) = (c - 1) + (r - 1) + (n - 1) + (r - 1)(c - 1) + (rc - 1)(n - 1)$.

Problems

13.1 Why should multiple t tests not be employed to compare more than two means?

13.2 How is the F distribution employed to test the equality of means?

13.3 How are fixed effects, random effects, and mixed models differentiated?

13.4 What are the difficulties involved in the analysis of variance of complicated models?

13.5 What are the basic and common assumptions made for the analysis of variance?

13.6 Under what conditions can the least significant difference formula be applied to conduct pairwise tests?

13.7 Variable or variables used for classification are considered as the independent variables in the analysis of variance, but what constitutes the dependent variable?

13.8 Use the linear model to prove that the degrees of freedom are additive in the analysis of variance.

13.9 Use the linear model to show that $E(e_{ij}) = 0$ and that $E(\text{est. } \alpha_i) = \alpha_i$.

13.10 Use the linear model to show that the mean square within groups (that is, the error mean square) is an unbiased estimate of σ^2 whether or not $\alpha_i = 0$ is true.

13.11 Use the linear model to show that the mean square between groups, MSC, is an unbiased estimate of σ^2 if and only if $\alpha_i = 0$ is true.

13.12 An agricultural experimental station wishes to determine the yields of five varieties of corn. Each type of corn is planted on four plots of land with equal fertility, and the yields, in units of 100 bushels, are as follows:

CORN

A	B	C	D	E
4	7	5	11	4
6	13	4	10	8
4	10	4	9	6
10	12	9	14	10

In view of these results, can we conclude that the mean yields of the five varieties of corn are the same at the 5 percent level of significance?

13.13 Apply the pairwise test to the results of Problem 13.12.

13.14 Miss Smith, supervisor of a secretarial pool, has four typists under her supervision. She is concerned with the length of their coffee break, so she times them. Her observations, recorded in minutes, for each girl are as follows. Can the differences in average time that the four typists spent on coffee breaks be explained by chance variation?

TYPIST

A	B	C	D
27	20	24	20
35	22	22	18
18	30	25	26
24	27	25	19
28	22	20	26
32	24	21	24
16	28	34	26
18	21	18	159
25	23	32	
223	18	23	
	30	22	
	32	266	
	322		

13.15 We wish to test whether there are any differences in the durability of three makes of computers. Samples of size $n_i = 5$ are selected from each make, and the frequency of repair during the first year of purchase is observed. The results are as follows:

MAKE

A_1	A_2	A_3
10	6	5
11	7	3
9	8	4
9	6	2
8	5	1

In view of the above data, what conclusion can you draw? If a significant difference exists between A_i, which treatment mean seems to have produced the difference?

13.16 Four machines, A, B, C, and D, are used to produce a certain kind of cotton piece goods. Samples of size $n_i = 4$, with each unit as 100 square yards, are selected from outputs of the machines at random, and the number of flaws in each 100 square yards are counted with the following results:

A	B	C	D
20	13	8	9
24	17	10	5
19	15	12	7
21	14	9	4

In view of the above data, can the null hypothesis $\alpha_i = 8$ be rejected at the 5 percent level of significance? At the 1 percent level of significance? Which pairs of means are significantly different at $\alpha = 0.05$? at $\alpha = 0.01$?

13.17 The same examination is given to five classes in elementary statistics taught by five different instructors, A_i, and it is desired to test whether the average grades of the five classes are equal. Random samples of size $n_i = 5$ are drawn from each class, and the grades are shown below. What conclusion can you reach?

A_1	A_2	A_3	A_4	A_5
85	66	59	61	55
78	75	50	71	60
90	72	66	69	73
79	68	70	80	61
92	81	62	75	81

13.18 A random sample is selected from each of three makes of ropes, and their breaking strengths (in pounds) are measured, with the following results:

SOURCE

I	II	III
95	121	82
87	130	73
90	99	76
88	119	70
101	125	81
	127	71
	120	

Test whether there are any differences in the breaking strengths of the ropes.

13.19 How does the completely randomized model differ from the randomized-blocks model?

13.20 Explain the rationale behind the testing procedure of the randomized-blocks, one-variable classification model.

13.21 A motion study is conducted to determine the best method of assembling a simple mechanism; three methods are under evaluation. Five assemblers are selected at random from all the assemblers in the factory. After they are given training to familiarize them with the particular methods chosen for the experiment, a test is conducted; and number of assemblies completed in one hour is recorded, with the following results:

Assembler	Method I	II	III	$t_{.j}$	$\bar{x}_{.j}$
1	2	4	5	11	2.75
2	3	4	6	13	3.35
3	5	7	8	20	5.00
4	7	9	11	27	6.75
5	6	5	9	20	5.00
$t_{i.}$	23	29	39	91 (= T)	
$\bar{x}_{i.}$	4.6	5.8	7.8		

a. Make an analysis of variance of the above data.
b. Perform the pairwise test.
c. What is the population about which this test is generalized?
d. What is the importance of the random selection of the assemblers?

13.22 Suppose that we are interested in establishing the yield-producing ability of four types of soybeans, A_1, A_2, A_3, and A_4. We have three blocks of land which may be different in fertility. Each block of land is divided into four plots, and the A_i are assigned to the plots in each block by a random procedure. The following results are obtained:

	A_1	A_2	A_3	A_4
B_1	5	9	11	10
B_2	4	7	8	10
B_3	3	5	8	9

a. Test whether A_i are significantly different.
b. Perform the pairwise test for A_i.

13.23 A chemist is interested in determining the effects of storage temperature on the preservation of apples. The variable under consideration is the number of apples that are rotten after a month of storage. He decides to use five lots of apples as the blocks of experimental material. He selects 120 apples from each lot, divides them into four portions of equal size, and assigns the treatments at random to the portions. The treatments are arbitrarily set as follows:

A_1 = storage with temperature at 50°,
A_2 = storage with temperature at 55°,
A_3 = storage with temperature at 60°, and
A_4 = storage with temperature at 70°.

This experiment yields the following results in terms of the number of rotten apples:

Lots	Temperature Treatment A_1	A_2	A_3	A_4	Total
1	8	5	7	10	30
2	14	10	3	5	32
3	12	8	6	5	31
4	9	8	5	7	29
5	12	9	4	8	33
Total	55	40	25	35	155

a. Test whether or not there is any significant difference due to storage temperatures.
b. Perform the pairwise test for A_i.

13.24 Four different drugs, A_i, have been developed for a certain disease. These drugs are used in three different hospitals, and the results given below show the number of cases of recovery from the disease per 100 people who have taken the drugs. The randomized-blocks design has been employed to eliminate the effect of the different hospitals.

	A_1	A_2	A_3	A_4
B_1	10	11	12	10
B_2	19	9	18	7
B_3	11	8	23	5

What conclusion can you draw?

13.25 What are the similarities and differences between the randomized-blocks, one-variable classification model and the completely randomized, two-variable classification model?

13.26 Make an analysis of variance for data in Problem 13.21 according to the completely randomized, two-variable classification model. Also, perform the pairwise test for B_j.

13.27 Suppose that there are four types of

corn, A_i, and three types of fertilizer, B_j, and we wish to test the difference between the corn and between the fertilizers. Using the completely randomized, two-variable classification model, we obtain the following yields:

	A_1	A_2	A_3	A_4
B_1	12	11	13	15
B_2	6	8	5	9
B_3	4	3	2	4

a. Does there exist a significant difference between the types of corn?
b. Does there exist a significant difference between fertilizers?
c. Perform the pairwise test for A_i and for B_j.

13.28 Are the block effects for data in Problem 13.23 nil, assuming completely randomized design?

13.29 Suppose that the data in Problem 13.24 are designed for the completely randomized, two-variable classification model; then,
a. Test whether there is a significant difference in recoveries due to different drugs.
b. Test whether there is a significant difference in recoveries due to different hospitals.
c. If the answer to (b) is yes, what factors could have contributed to these "block effects"?

13.30 Explain, in your own words, the rationale behind the testing procedure of the completely randomized, two-variable classification with replication model.

13.31 There are two machines of different makes and three operators of different skills for the production of a certain product. We are interested in testing whether there is a difference in the machines A_i, in the operators, B_j, and if there is any interaction $A \times B$. Each combination of A_iB_j produces the items for a number of hours, and $n_{ij} = 2$ hourly output figures are selected at random, and the following results are obtained:

	A_1	A_2
B_1	9	16
	10	15
B_2	6	12
	8	10
B_3	2	11
	3	10

a. Test whether there is a significant difference between machines.
b. Test whether there is a significant difference between operators.
c. Test whether there is a joint effect, or interaction.

13.32 An experiment is conducted on the life of cutting tools where the variables of interest are the type of metal, A_i, $i = 1, 2, 3$, and speed of lathe, B_j, $j = 1, 2, 3$. Suppose that the following data, given in hundreds of hours of effective use of the tools, are obtained:

	Type of Metal		
Speed	*I*	*II*	*III*
High	6	9	10
	4	8	9
	7	6	5
Medium	11	13	14
	10	10	12
	9	11	15
Low	9	8	10
	7	10	9
	6	9	12

a. Test whether there is a significant difference between metals, A_i.
b. Test whether there is a significant difference between speeds B_j.
c. Test whether there is an interaction effect.

13.33 Suppose that there are four kinds of corn, A_i, and two types of fertilizer, B_j. Assume that the four kinds of corn are planted on land of equal fertility and samples of $n = 2$ are selected from each combination A_iB_j, with the following results:

	A_1	A_2	A_3	A_4
B_1	110	100	90	90
	125	95	90	80
B_2	120	97	85	80
	115	92	86	81

a. Test whether there is a significant difference in A_i.
b. Test whether there is a significant difference in B_j.

c. Test whether there is a significant interaction effect.

13.34 How does the randomized-blocks, two-variable classification with replication model differ from the completely randomized, two-variable classification with replication model?

13.35 Data below represent operating time (in seconds) of three operators, A_i, working on two machines of different make, B_j, on three lots of raw materials, C_k. The C_k constitute the blocks in the randomized-blocks design.

		Operator			
Lot	Machine	A_1	A_2	A_3	$t_{.j.}$
C_1	B_1	6	9	10	
	B_2	4	5	8	
		10	14	18	42
C_2	B_1	5	8	12	
	B_2	3	5	7	
		8	13	19	40
C_3	B_1	8	10	15	
		4	5	8	
		12	15	23	50
Total	B_1	19	27	37	83
	B_2	11	15	23	49
$t_{i..}$		30	42	60	132

a. Test that $\alpha_i = 0$.
b. Test that $\beta_j = 0$.
c. Test that $I_{ij} = 0$.

13.36 Suppose that there are three types of wheat, A_i, and two types of fertilizer. It is desired to test whether or not there is a significant difference in the wheat and in the fertilizer. There are four blocks of land which differ in fertility. Each block of land divided into (3)(2) $= 6$ plots, and the treatment combinations, A_iB_j, are used to obtain the data. Suppose that the following data are obtained. In view of these results, can we conclude that (a) there is a significant difference in the wheat? (b) there is a significant difference in the fertilizer? (c) there is a significant interaction effect?

		Wheat		
Block	Fertilizer	A_1	A_2	A_3
C_1	B_1	7	6	4
	B_2	5	7	7
C_2	B_1	9	8	6
	B_2	10	9	11
C_3	B_1	20	12	19
	B_2	19	15	21
C_4	B_1	29	19	22
	B_2	33	25	20

14

NONPARAMETRIC STATISTICS

14.1

Introduction

Our discussion of statistical inference has been presented, so far, under the assumption that samples have originated from populations having certain distributions which are known. Each theoretical distribution function depends upon one or more population parameters. Thus, we call the classical decision procedure the *parametric* case. Efficient use of the parametric methods of estimation and testing usually requires normality of population and stability of variance.

In many situations, however, it is impossible to specify the form of population distribution; we can only reasonably assume that it is continuous. The process of drawing conclusions directly from sample observations without formulating assumptions with respect to the mathematical form of the population distribution is called the *nonparametric* or *distribution-free theory*. Because many nonparametric statistics have been designed and have been widely accepted during the recent past, it seems fitting that we should have a survey of nonparametric theory in a text on modern statistical analysis.

Actually, the student is not completely ignorant of nonparametric methods by now. The chi-square tests mentioned in Chapter 12 are nonparametric techniques. Again, the Chebyshev inequality, although not strictly nonparametric since it involves μ and σ^2, is distribution-free since it holds for any probability distribution. In the pages that follow we shall present a number of nonparametric statistics that have been found to be useful and efficient.

14.2

Tests for Randomness

The classical theory of statistical inference rests on randomness. For classical procedures to apply, the sample must be a random sample: sample observations must be independent of each other, and each sample observation must be identically distributed with the population. This assumption may easily be violated either because the population changes with respect to one or more characteristics while the sample is being drawn, or because dependence has somehow crept into sample observations. In other words, the population may not be in statistical control or the sample may not be random. When this happens, the classical theory fails. Clearly, methods which can evaluate the randomness of a sample are of supreme importance for the justification of employing classical inference procedures.

In this section, we shall introduce two simple nonparametric tests for randomness based on the concept of "runs": (1) runs above and below the median, and (2) runs up and down.

RUNS ABOVE AND BELOW THE MEDIAN

Suppose that we have a sample of, say, 26 observations gathered as listed in the order below:

97, 89, 25, 81, 11, 83, 16, 96, 44, 32, 98, 19, 68,
33, 25, 54, 74, 82, 17, 49, 33, 22, 62, 20, 92, 80

A moment's study of this set of data reveals that the value of the median is $(54 + 49)/2 = 51.5$. We proceed to observe now whether each observation is above or below this median value. If above the median, we mark it as an a, and if below the median, we mark it as a b. For the above sample, we have the following sequence of as and bs:

aa–b–a–b–a–b–a–bb–a–b–a–bb–aaa–bbbb–a–b–aa

Each sequence of as or bs, uninterrupted by the other letter, is a *run* and we have, therefore, 17 runs for our data.

Now, there is reason to believe that if the sample is random the number of runs above and below the median should also be a random variable. Furthermore, if the population is continuous, n is even, and $n \geq 25$, the sampling distribution of this number of runs, R, is asymptotically normal with

$$E(R) = \frac{n + 2}{2} \tag{14.1}$$

and

$$V(R) = \frac{n(n-2)}{4(n-1)} \tag{14.2}$$

The null hypothesis to be tested here is usually stated as: "The population is in statistical control," or "The sample is a random sample." The alternative is often of the left-tail variety. This is especially appropriate if there are too few runs, since such a situation usually indicates a shifting population mean. We may occasionally wish to conduct a right-tail test when there are too many runs, revealing a systematic swing above and below the median from one observation to another, thus making us suspect that the data might have been manipulated to yield too perfect a result. When we have no specific ideas, except that H_1 should be the opposite of H_0, we may then proceed to make a two-sided test. In any event, when the normal theory can be applied to the distribution of R, the testing statistic is

$$Z = \frac{R - E(R)}{\sqrt{V(R)}} \tag{14.3}$$

Suppose that we wish to have a two-sided test for our illustrative data at $\alpha = 0.01$; we would have

$$E(R) = \frac{26+2}{2} = 14$$

$$V(R) = \frac{26(26-2)}{4(26-1)} = 6.24$$

$$Z = \frac{17-14}{\sqrt{6.24}} = 1.21$$

Since the observed value of Z falls between -2.58 and $+2.58$, H_0 cannot be rejected at the 1 percent level of significance. We conclude therefore that the sample is a random sample.

In closing our discussion of the test of randomness with runs above and below the median, let us note three things. First, when the sample is small, exact probabilities can be determined with the binomial theorem with $p = \frac{1}{2}$. Second, the condition that n be even can be violated to yield satisfactory result when n is large enough for normal approximation. Third, when there are sample values which are tied with (equal to) the median value, either they may be disregarded so that the number of runs and sample size are reduced, or they may be assigned *a*s or *b*s by a random process such as the toss of a coin.

RUNS UP AND DOWN

A test for runs up and down is similar to the preceding test. Given a sample of n observations recorded in the order obtained, we place either a plus sign (+) or a minus sign (−) between each pair of successive observations, depending on whether the latter observation is greater or less than the preceding observation.

Now, each string of +s or −s, uninterrupted by the other sign, is counted as a run. For example, using the same sample data, we have

$$-- \; + \; - \; + \; - \; + \; -- \; + \; - \; + \; -- \; +++ \; - \; + \; -- \; + \; - \; + \; -$$

a total of 19 runs. Be sure to note these signs are determined between two successive observations, and there are $(n - 1)$ such signs for a sample of size n. Again, ties may be broken by a chance process so as to make more probable the rejection of randomness.

Under the assumptions that the sample is random and that the population is continuous, the number of runs up and down, R, is asymptotically normal with

$$E(R) = \frac{1}{3}(2n - 1) \tag{14.4}$$

and

$$V(R) = \frac{1}{90}(16n - 29) \tag{14.5}$$

The testing statistic to be used becomes

$$Z = \frac{R - E(R)}{\sqrt{V(R)}} \tag{14.6}$$

Normal approximation in this case requires that $n \geq 20$.

Supposing we wish to test randomness against a shifting population mean, namely, against a left-tail alternative, with our illustrative data at $\alpha = 0.01$, we have then

$$R = 19$$

$$E(R) = \tfrac{1}{3}(2(26) - 1) = 17$$

$$V(R) = \tfrac{1}{90}(16(26) - 29) = 4.3$$

$$Z = \frac{19 - 17}{\sqrt{4.3}} = 1.4$$

Note that $1.40 > -2.33$, and randomness is accepted at $\alpha = 0.01$.

In general, a test based on runs above and below the median and a test based on runs up and down are expected to yield the same result.

14.3

Tests for Independence

Tests for independence are the same as the tests for randomness, since independence in sample observations is the same as the randomness of the sample. We have here distinguished these two types of tests merely for the purpose of convenience in presentation. The tests for randomness discussed before were mainly concerned with cross-section data. Now, in this section, we shall introduce procedures of testing randomness (independence) for times series samples.

Business and economic data ordered in time sequence can hardly be considered random samples. An observation on price, production, national income, and the like, in a month or year is usually dependent to some extent upon the value of that variable in the previous time period. This mutual dependence of successive observations in time raises a number of problems in statistical analysis. We would like to have some device by which to measure the degree of this dependence. One such measure is called *serial correlation*. Another simple but much more powerful method is the *Von Neumann ratio*. We shall now introduce these two concepts in the order mentioned.

SERIAL CORRELATION

The term serial correlation refers to the lag relationship in time series samples. Consider the trend equation

$$y_t = a + bx$$

for example. We say that the dependent variable Y is serially correlated if the value of Y in period 1, y_1, is correlated with the value of Y in period $t - 1$, y_{t-1}, or if y_t is correlated with y_{t-2}, y_{t-3}, and so on. Sometimes the terms *autocorrelation* and *lag correlation* are used to describe this dependent relationship. When a distinction is made, we take *autocorrelation* to mean the lag relationship in the population, *serial correlation* to mean the lag relationship in a sample, and *lag correlation* to mean the lag relationship between two different time series.

Given a time series with values $y_1, y_2, \cdots, y_n$, serial correlation between the successive terms is defined as

$$r_1 = \frac{\text{cov}\,(y_i, y_{i+1})}{\sqrt{V(y_i)V(y_{i+1})}} = \frac{\sum y_i y_{i+1} - \dfrac{(\Sigma y_i)^2}{n}}{\sum y_i^2 - \dfrac{(\Sigma y_i)^2}{n}} \tag{14.7}$$

Here, r_1 is called *serial correlation of order 1* because the time interval between y_i and y_{i+1} is one time unit. In general, serial correlation between y_i and y_{i+k} is called *serial correlation of order k*.

The statistic r_1 can be used to test the null hypothesis that a given time series is a random sample. When randomness is rejected, we conclude that the sample is not random and that there exists in the population significant trend and/or cyclical elements. Clearly, $-1 < r_1 < 1$. When r_1 is close to 0, we would tend to accept H_0. When r_1 is close to -1 or $+1$, we would tend to reject H_0.

Testing for randomness with r_1 is facilitated by a table of distribution of r_1 developed by R. I. Anderson, presented in the Appendix as Table A-XII. This table gives critical values of r_1 at both the 5 percent and the 1 percent levels of significance. If the value of r_1 exceeds the corresponding value, which depends upon n, H_0 is rejected and we conclude that there exists serial correlation in the population.

TABLE 14.1

Year	y_i	y_i^2	$y_i y_{i+1}$
1960	5	25	35
1961	7	49	42
1962	6	36	60
1963	10	100	100
1964	10	100	90
1965	9	81	99
1966	11	121	132
1967	12	144	60
Total	70	656	618

Let us consider an illustration. Suppose that the inventories of a certain store during the past ten years (given in thousands of dollars) are as shown in Table 14.1. Quantities required to compute r_1 can be easily obtained. These quantities are also shown in the same table.

Note that in the above calculations we have assumed that $y_{n+1} = y_1$. This is called the *circular definition*. This definition would affect the value of r_1 significantly when n is small and data are not detrended as in our present example. In any event, for our illustrative data, we have

$$r_1 = \frac{618 - \dfrac{70^2}{8}}{656 - \dfrac{70^2}{8}} = +0.126$$

Since for $n = 8$, $P(r_1 > 0.531) = 0.01$, we accept H_0. This conclusion, however, is quite doubtful due to the effects of the circular definition on r_1 for our data, which are not detrended and rather short.

THE VON NEUMANN RATIO

Another simple but more powerful method for testing independence in time series is based on the *mean-square-successive difference* and is called the Von Neumann ratio, defined as

$$K = \frac{m^2}{s_y^2} \tag{14.8}$$

where m^2 stands for the mean-square-successive difference and is obtained by

$$m^2 = \frac{\Sigma(y_{i+1} - y_i)^2}{n - 1}$$

and s_y^2 is the variance of Y,

$$s_y^2 = \frac{1}{n} \sum (y_i - \bar{y})^2$$

As implied by the definition of m^2, the ratio K is closely related to the variance of the first difference. When these differences are small, a small K will result and positive serial correlation in the population is indicated. When these differences are large, a large K will result and negative serial correlation is revealed. Thus, very large and very small values of K would lead us to the rejection of randomness or independence.

To effect the test of independence by the Von Neumann ratio, a table worked out by B. I. Hart, presented as Table A-XIII, may be used. This table gives the 5 percent and 1 percent points of the distribution of K. It gives two critical values, k and k', for each level of significance and each sample size. For instance, if $n = 10$ and $\alpha = 0.05$, $k = 1.1803$ and $k' = 3.2642$. If $K < k$, K is considered as significant and we reject H_0 and conclude that there exists positive serial correlation. If $K > k'$, we would reject H_0 and conclude that negative serial correlation exists. H_0 is accepted when $k < K < k'$.

Using the inventory data in Table 14.1 again as an illustration, we have the following results:

TABLE 14.2

Year	y_i	$(y_i - \bar{y})^2$	$(y_{i+1} - y_i)^2$
1	5	14.0625	4
2	7	3.0625	1
3	6	7.5625	16
4	10	1.5625	0
5	10	1.5625	1
6	9	0.0625	4
7	11	5.0625	1
8	12	10.5625	—
	70	43.5000	27

$$\bar{y} = 70/8 = 8.75$$

$$s_y^2 = \frac{43.5}{8} = 5.44$$

$$m^2 = \frac{27}{(8-1)} = 3.86$$

$$K = \frac{3.86}{5.44} = 0.71$$

We see from Table A-XIII, $0.71 < 0.7575 = k$ for $n = 8$; H_0 is rejected at $\alpha = 0.01$. Serial correlation exists in the population—there is a positive trend and/or a cyclical component in the population. (*Note:* For the same data, r_1 failed to reject randomness before. Clearly, the Von Neumann ratio is much more sensitive in detecting the existence of serial correlation.)

14.4

Population Median

For a continuous random variable X, the median is that number M such that $P(X > M) = P(X < M) = \frac{1}{2}$.

Suppose we wish to test the null hypothesis that the population median is equal to a fixed number; that is, H_0: $M = M_0$. If this is true, we should expect approximately half of the observations in a random sample of size n to lie on either side of M_0, namely, $P(X - M_0 > 0) = \frac{1}{2}$. We see then that the number of observations in a sample exceeding M_0 has a binomial distribution with $p = \frac{1}{2}$. Given the usual two-sided alternative in evaluating population medians, we would reject H_0 either because there are too few or too many observations in the sample in excess of M_0.

The mechanics of the test for the median are as follows:

(1) Given M_0 and n, arbitrarily select the critical values and then find the value of α by the binomial theorem.

(2) Construct this pattern of signs: "+" for an observation that exceeds M_0 and "−" for an observation that is short of M_0. When a tie occurs (when $x_i = M_0$), write down a sign selected by a random process.

(3) Count the number of plus signs. If it is less than the smaller, or greater than the larger, critical value determined in (1), reject H_0 at the level of α calculated in terms of these critical values.

Suppose that we have drawn a random sample of size $n = 10$ with the following results, listed according to the order in which they were obtained:

18.7, 25.0, 21.5, 16.5, 20.3, 19.2, 17.1, 23.5, 24.3, 15.7

Can we conclude from these data that

$$H_0\colon M = 21$$
$$H_1\colon M \neq 21$$

if we decide to reject H_1 when there are 1 or less, or 9 or more, plus signs?

With the above decision rule, the level of significance becomes

$$\begin{aligned}\alpha &= P(0) + P(1) + P(9) + P(10)\\ &= \binom{10}{0}(0.5)^{10} + \binom{10}{1}(0.5)^{10} + \binom{10}{9}(0.5)^{10} + \binom{10}{10}(0.50)^{10}\\ &= 0.02148\end{aligned}$$

From the original data, we have the following signs:

$$- \; + \; + \; - \; - \; - \; - \; + \; + \; -$$

There are four plus signs (four sample observations having values above 21). This number is between 1 and 9; therefore, we accept the hypothesis that the population median is 21 at $\alpha \doteq 0.02$.

An interval estimate for M can be obtained immediately after the sample observations have been arranged in an array. Again, no assumptions about the population are required except that it be continuous. Suppose that we have a sample of 10 items; we may arrange them in such order as $x_1 < x_2 < \cdots < x_{10}$ and say that M is between two of these observations. For example, we may state that M is between x_1 and x_{10}, or between x_2 and x_9, and so on. As we get closer to the center of the sample array, our interval becomes more precise but we become less confident that the interval will indeed include M. It is desirable therefore that we should attach a probability statement to our interval estimate of M.

A table of confidence intervals for M has been worked out by Banerjee and Nair and is presented in the Appendix as Table A-XIV. This table gives the largest values of k (depending on n), such that we are more than 95 percent confident or more than 99 percent confident that the population median is between x_k and x_{n+1-k}. The values of α in the table give us probabilities that our statements about confidence intervals are false. For example, for $n = 10$ and $\alpha \leq 0.05$, $k = 2$. Thus, if we construct a confidence interval with x_2 and x_9 as the lower and upper confidence limits, we would be about 98 percent confident. Namely, for $n = 10$, $P(x_2 < M < x_9) = 1 - 0.021$. This observation is clearly consistent with our decision to accept $M = 21$ with $n = 10$ and $\alpha = 0.02$ before. Can you explain why?

14.5 The Sign Test for Matched Pairs

Very often we have two or more sets of data which are not random samples but are matched in terms of one or more factors. We have encountered this type of data before. It first appeared in our t test for two dependent samples. It occurred again when we introduced the randomized-blocks experiment in the analysis of variance, where c treatment populations were compared when there were available r blocks of factors to which the treatments were assigned at random. When there are only two treatments and the observations are paired, a simple procedure called the *sign test* can be employed to test whether two population means are equal. This test is developed below.

Let A and B represent two treatments with n matched pairs of observations, (x_i, y_i). We note that the sign of $d_i = x_i - y_i$. If we are concerned with data which are not measurable but only comparable, the d_i is taken to be positive if A is better than B and negative if A is worse than B. Whenever a tie exists (that is, $d_i = 0$), the pair of tied observations is disregarded.

Under the assumption that the two populations have equal means, we should have approximately the same numbers of $+$s and $-$s. Furthermore, if we let S be the number of times the less frequent sign occurs in the sequence of d_i, then S has the binomial distribution with $p = \frac{1}{2}$.

It is important to note that, for the sign test to be applicable, it is not necessary

TABLE 14.3

Pair Number	Method 1	Method 2	Sign of d_i
1	17	10	+
2	28	25	+
3	19	10	+
4	20	21	−
5	11	13	−
6	27	20	+
7	21	16	+
8	20	20	0
9	29	24	+
10	30	27	+
11	31	25	+
12	28	30	−
13	40	27	+
14	29	25	+
15	33	21	+
16	35	32	+

to assume that differences are identically distributed as for the t test. The utility of the sign test is that it is permissible for each pair of observations to be collected under different conditions and thus to have different distributions.

The critical value for a two-sided alternative at $\alpha = 0.05$ can be conveniently found by the expression

$$K = \frac{(n-1)}{2} - (0.98)\sqrt{n+1} \tag{14.9}$$

H_0 is rejected if $S \leq K$ for the sign test.

Suppose that a production manager wishes to conduct an experiment to compare two methods of assembling a certain mechanism. He first pairs the workers according to age, and assigns the methods (treatments) to the workers at random. A sample of 16 pairs of workers in the experiment yields the mean number of units assembled per hour recorded in Table 14.3.

The above data indicate that the less frequent sign is minus and $S = 3$. Given that $n = 15$, is this result significant at $\alpha = 0.05$? (Note that $n = 15$ instead of 16 here because one pair yields $d = 0$ and is disregarded.)

To answer this question, we find the rejection number to be

$$K = \frac{15-1}{2} - (0.98)\sqrt{16} \doteq 4$$

Since $S \leq K \doteq 4$, the null hypothesis of equal means is rejected. Method 1 is superior to Method 2.

We can just as easily calculate the probability of obtaining as extreme a result as the observed S by the binomial theorem. Thus, for our example, we have

$$P(S \leq 3) = \sum_{S=0}^{3} b(S;15,0.5) = 0.0176$$

which is less than 0.05, and therefore H_0 cannot be accepted.

Furthermore, in our present illustration $np = 15(0.5) = 7.5 > 5$, a satisfactory result can also be obtained by the normal approximation. Hence

$$P(S \leq 3) = N\left(\frac{S + 0.5 - np}{\sqrt{npq}}\right) = N\left(\frac{3 + 0.5 - 7.5}{\sqrt{(15)(.5)(.5)}}\right) = N(-2.07) = 0.0192$$

which is again less than 0.05 and agrees with the exact probability quite well.

We note that the sign test is applicable only when n is not too small. For $\alpha = 0.05$, n should be at least 6, since if $n = 5$, even in the extreme case where all differences are either positive or negative, the probability is as large as $(0.5)^5 = 0.0625 > 0.05$; hence, with $n = 5$, we can never reject the null hypothesis of equal means at $\alpha = 0.05$. Needless to say, when $\alpha = 0.01$, n should preferably be larger than 6.

14.6 The Chi-Square Test on Frequencies

As mentioned in Chapter 12, the chi-square distribution can be employed to test any hypothesis concerning categorical data. We shall in this and the next few sections introduce some of the more important nonparametric tests via the chi-square distribution.

Generally, with a chi-square test, we first formulate the null hypothesis under which expected, or theoretical, frequencies are determined. Next, we investigate sample data for the purpose of establishing observed frequencies. Then we compare these two sets of frequencies by striking differences between them. Finally, on the basis of these differences, we may specify a decision criterion to judge whether the observed frequencies, on the average, diverge significantly from the expected frequencies. Namely, we wish to determine, from an appropriate criterion, whether these differences are due to chance variations in random sampling. If so, the null hypothesis is accepted; otherwise, it is rejected.

For this group of problems, it can be shown that the ratio of the sum of squared differences between observed frequencies, denoted as o_i, and expected frequencies, denoted as e_i, to the expected frequencies is distributed approximately as a chi-square. That is,

$$\chi^2 = \sum \frac{(o_i - e_i)^2}{e_i} \tag{14.10}$$

is approximately distributed as chi-square, with degrees of freedom equal to the number of classes or categories compared, less the number of restrictions imposed on the comparison. More precisely, for (14.10), $\delta = k - 1$ if expected frequencies can be computed without having to estimate population parameters from sample statistics. Here, k is number of classes; thus if we know $k - 1$ of the expected frequencies, the remaining frequency is uniquely determined because we must satisfy the condition that $\Sigma o_i = \Sigma e_i = n$, where n is the sample size or the total frequency. However, we would have in this case $\delta = k - 1 - m$ if the expected frequencies can be computed only by estimating m population parameters by sample statistics.

Two observations may be made here in connection with the use of (14.10) for various types of chi-square tests. First, whenever $\delta = 1$, especially when the total number of frequencies is rather small, say, less than 50, it is advisable to introduce a continuity correction factor of $\frac{1}{2}$ in computing the chi-square value. That is, when $\delta = 1$, we modify equation (14.10) as follows:

$$\chi^2 = \sum \frac{(|o_i - e_i| - \frac{1}{2})^2}{e_i} \tag{14.11}$$

Second, there must be at least 5 items in each theoretical frequency class. This requirement is to avoid inflated chi-square values due to the division of the squared differences by a small size of expected frequency. In a situation where there are one or more classes with theoretical frequencies less than 5, they may be combined into a single category before calculating the difference, $o_i - e_i$. With this practice, it is important to remember that the number of degrees of freedom is determined with the number of classes after the regrouping. For example, if we start out with ten classes and three of these have small theoretical frequencies, we may pool these three classes into one and we would have then $k = 8$ classes to compare.

In this section we discuss the application of chi-square to test hypotheses about frequencies. For this, two examples will be given: one is concerned with one-way classification with two classes; the other, with one-way classification with more than two classes.

A TEST ON FREQUENCIES OF TWO CLASSES

For this test, we shall use the same example as the one in Chapter 11, where the Better Business Bureau had to make a decision about the claim by The Smile Diaper Service that it shared 30 percent of the diaper market in Philadelphia. The sample data showed that among the 200 families surveyed, 52 of them were customers of The Smile Diaper Service. The BBB set out to verify whether it was true that every 3 out of 10 users of diaper service in Philadelphia used The Smile Diaper Service, and made the test under the assumption that the sample proportion was normally distributed. To verify the same claim by a chi-square test, we have

1. **Hypotheses:** H_0: Number of customers are distributed in the ratio 3:7
   H_1: The ratio is not 3:7

2. **Level of Significance:** $\alpha = 0.05$

3. **Testing Statistic:** $\chi^2 = \sum \frac{(|o_i - e_i| - \frac{1}{2})^2}{e_i}$ which is distributed as χ_1^2.

(*Note:* For this test, $k = 2$; therefore, $\delta = 2 - 1 = 1$. Because $\delta = 1$, the continuity correction factor is employed.)

4. **Decision Rule:** $P(\chi_1^2 > 3.84) = 0.05$; reject H_0 if and only if the computed chi-square value is greater than 3.84. Otherwise, accept H_0.

5. **Computations:**

Smile's Customers	o_i	e_i	$\lvert o_i - e_i\rvert - \frac{1}{2}$	$(\lvert o_i - e_i\rvert - \frac{1}{2})^2/e_i$
Yes	52	60	7.5	0.94
No	148	140	7.5	0.40
Total	200	200	—	1.34

$$\chi_1^2 = 1.34$$

6. **Decision:** The computed chi-square value is 1.34, which is less than 3.84. The null hypothesis cannot be rejected at $\alpha = 0.05$. The sample data are not sufficient evidence for doubting Smile's claim. It is interesting to note that this conclusion is identical with the one reached for the same problem when we used the normal probability law as the decision criterion. This should not be a surprising result, since we know that a chi-square statistic is the square of the normal statistic. The two critical regions correspond to one another, since for $\alpha = 0.05$ and $\delta = 1$,

$$\begin{aligned} P(Z < -1.645) &= P(Z > 1.645) = P(Z < -1.96) + P(Z > 1.96) \\ &= P(\chi_1^2 > 3.84) = 0.05 \end{aligned}$$

It may be observed also that for the present example, the hypotheses can be stated in the binomial notation or multinomial notation. In binomial notation, we have H_0: $p = 0.3$ against H_0: $p \neq 0.3$. In multinomial notation, we have H_0: $p_1 = 0.3$ and $p_2 = 0.7$ against H_1: $p_1 \neq 0.3$ and $p_2 \neq 0.7$.

A TEST ON FREQUENCIES OF MORE THAN TWO CLASSES

While the chi-square test can be considered as a generalization of the normal test when there are only two classes as "evealed by the preceding example, we are forced to use the chi-square test when there are more than two categories, as illustrated by the following example.

Five coins are tossed 100 times for the purpose of determining if all the five coins are fair. For convenience, we conduct this test by observing the number of samples which contains 0, 1, 2, 3, 4, or 5 heads. The appropriate chi-square test with six categories is as follows:

1. **Hypotheses:** H_0: All coins are fair; that is, the probability of obtaining heads in any random toss of any of the five coins is $\frac{1}{2}$
   H_1: Not all coins are fair

2. **Level of Significance:** $\alpha = 0.01$—arbitrarily assigned

3. Testing Statistic: $\sum \frac{(o_i - e_i)^2}{e_i}$ which is approximately distributed as χ_3^2.

Here $\delta = 4 - 1 = 3$, as shown by the table in step (5), since the first class is combined with the second and the fifth class is combined with the last due to small theoretical frequencies in the first and the last classes. Actually we have, after the combinations, $k = 4$.

4. Decision Rule: The critical region is separated from the acceptance region by 11.34 because $P(\chi_3^2 > 11.34) = 0.01$.

5. Computations: First, the five coins are tossed 100 times and this total number of tosses is distributed according to the number of heads as recorded in the first two columns in the following table; the second column, of course, contains the observed frequencies. Next, if all the five coins are fair, then in any toss of the five coins the number of heads should be distributed as a binomial distribution. Such a theoretical distribution is given in column three in the same table. With this theoretical distribution, the expected frequencies are automatically determined as shown in the fourth column. In the fifth column we have the computation for the chi-square value under consideration, and it is 26.88.

No. of Heads	o_i		$b(x;0.5,5)$	$e_i = 100\,[b(x)]$		$(o_i - e_i)^2/e_i$
0	1	11	0.0312	3.12	18.74	3.46
1	10		0.1562	15.62		
2	20		0.3125	31.25		4.05
3	36		0.3125	31.25		0.72
4	23	33	0.1562	15.62	18.74	18.65
5	10		0.0312	3.12		
Total	100		1.0000*	100.00*		26.88

** These totals are not exactly 1 and 100, respectively, because of rounding.*

6. Decisions: The observed value of χ_3^2 is greater than 11.34 and falls in the critical region. We therefore reject H_0 that all the five coins are fair. This implies that at least some of these coins are biased.

14.7 Testing Goodness-of-Fit

In many problem situations we can deduce a priori that prevailing conditions make it seem reasonable that the population follows a specific probability model; nonetheless, it is beneficial to have a statistical testing procedure to substantiate the adequacy of such a model. Statistical inferences discussed in the preceding three chapters were designed under the assumption that the population is normal or that samples are large enough for the central limit theorem to be operative. The assumption of normality is not a crucial one insofar as inferences about means are concerned, but it can seriously affect the drawing of conclusions about variances. While experience may tell us that a given random variable is normally or nearly normally distributed, we may still doubt the normality assumption and seek an appropriate procedure which can prove or disprove the assumption. The statistical procedure employed for this purpose is called *test of goodness-of-fit.* It begins, as does our general testing procedure, with formulating the null hypothesis that a given population has a specific probability or density function, such as the uniform, the binomial, the Poisson, or the normal model. Then a random sample is taken from the population and observed, which furnishes us with observed frequencies (empirical distribution) in terms of chi-square tests. Then the theoretical distribution, as specified in the null hypothesis, is fitted to the empirical distribution. The theoretical probability values, when multiplied by the total number of frequencies in the sample, become expected frequencies. After this, the chi-square test of goodness-of-fit follows exactly the same procedure as that on frequencies, except that theoretical frequencies are now usually computed with sample statistics as estimators for the corresponding population parameters.

TESTING THE FIT OF A UNIFORM DISTRIBUTION

The simplest, and a useful, test of goodness-of-fit is that with a discrete uniform distribution, such as the toss of a coin or the throw of a die. Let us consider an example.

It is known that the number of sales made by a salesman usually depends upon the number of calls he makes. Thus, salesmen of any firm are always encouraged to make as many calls as possible. If salesmen are equally ambitious, energetic, resourceful, and so on, all should be expected to make the same number of calls during the same period of time. The sales manager of a certain company is anxious to test the validity of this hypothesis. He selects at random six salesmen's records for a given week and finds that a total number of 210 calls was made, distributed as follows:

Salesman	Number of Calls
A	35
B	20
C	47
D	32
E	51
F	25
Total	210

The test:

1. **Hypotheses:** H_0: The number of calls is uniformly distributed; that is, $f(x;n) = 1/n,\ x = 1,2, \cdots, n$
   H_1: The number of calls is not uniformly distributed

2. Level of Significance: $\alpha = 0.01$

3. Testing Statistic: $\sum \frac{(o_i - e_i)^2}{e_i}$ which is distributed approximately as χ_5^2. Here, $\delta = 6 - 1$, since expected frequencies can be computed without estimating parameters by sample statistics.

4. Decision Rule: H_0 shall be rejected if and only if the computed chi-square value is greater than 15.09, since $P(\chi_5^2 > 15.09) = 0.01$.

5. Computations:

Salesman	o_i	$f(x;5)$	$e_i = 210\,[f(x)]$	$(o_i - e_i)^2/e_i$
A	35	1/6	35	0.00
B	20	1/6	35	6.43
C	47	1/6	35	4.11
D	32	1/6	35	0.26
E	51	1/6	35	7.31
F	25	1/6	35	2.86
Total	210	6/6	210	20.97

6. Decisions: Reject H_0, since the computed chi-square value falls in the rejection region. This conclusion means that the fit of a uniform distribution to the calls made by salesmen is bad. This inference also implies that salesmen may be different in ambition, energy, resourcefulness, and so on.

TESTING THE FIT OF A BINOMIAL DISTRIBUTION

For this case, we shall try to fit a binomial distribution to the same data of tossing five coins 100 times discussed in Section 14.6. To judge whether such a fit is good or bad, we have the following chi-square test:

1. Hypotheses: H_0: The number of heads is distributed as a binomial distribution. $b(x;p,n)$
H_1: The number of heads is not distributed as a binomial

2. Level of Significance: $\alpha = 0.01$

3. Testing Statistic: $\sum \frac{(o_i - e_i)^2}{e_i}$ which is seen to be distributed as χ_3^2. Here, $\delta = k - 1 - m = 5 - 1 - 1 = 3$. The number of classes is five since the first two of the six original classes are combined into a single category. Furthermore, $m = 1$ because we need the sample estimate of p, the probability of heads in any single toss of each of the five coins, to compute expected frequencies.

4. **Decision Rule:** H_0 will be rejected if and only if the computed chi-square value is greater than 11.34 since $P(\chi_3^2 > 11.34) = 0.01$.

5. **Computations:** For this problem the mean number of heads is $np = 5p$. For the actual or observed frequency distribution, the mean number of heads is

$$5\hat{p} = \frac{\Sigma fx}{\Sigma f} = \frac{0(1) + 1(10) + 2(20) + 3(36) + 4(23) + 5(10)}{100} = 3$$

Or, $\hat{p} = 0.6$, which may be considered as the best estimate of the binomial parameter under investigation. With this estimate, the fitted binomial distribution is given as $b(x;0.6,5) = \binom{6}{x}(0.6)^x(0.4)^{5-x}$. With this probability distribution, the expected frequencies are then determined and the chi-square value is computed.

Number of Heads	o_i		$b(x;0.6,5)$	$e_i = 100\,[b(x)]$		$(o_i - e_i)^2/e_i$
0	1	11	0.0102	1.02	8.70	0.61
1	10		0.0768	7.68		
2	20		0.2304	23.04		0.30
3	36		0.3456	34.56		0.06
4	23		0.2592	25.92		0.23
5	10		0.0778	7.78		0.63
Total	100		—	—		1.83

6. **Decision:** Since the computed chi-square value is less than 11.34, H_0 is accepted. This conclusion means that the fit of the data is good. (*Note:* The fit here is made with p estimated as 0.6. This conclusion is, therefore, consistent with our previous decision that not all of the five coins are fair in view of the same sample data.)

TESTING THE FIT OF A POISSON DISTRIBUTION

The number of clients who visited the office of a young lawyer each day during his first 100 days of practice is distributed as follows:

Number of clients	0	1	2	3	4	5	6
Number of days	40	36	16	7	2	1	0

Determine the goodness-of-fit of these data by a Poisson distribution.

First, to compute the expected frequencies we must have the Poisson probabilities which, in turn, require the value of the mean number of visitors per day. For data above, we have

$$\hat{\mu} = \frac{T}{N} = \frac{0(40) + 1(36) + 2(16) + 3(7) + 4(2) + 5(1) + 6(0)}{100} = 1 \text{ per day}$$

With this estimate, the Poisson distribution to be fitted to our sample data becomes

$$p(x;1) = e^{-1}\frac{1^x}{x!}$$

Next, we have the necessary computations:

Number of Clients	o_i	$p(x;1)$	$e_i = 100\,[p(x)]$	$(o_i - e_i)^2/e_i$
0	40	0.36788	36.788	0.28
1	36	0.36788	36.788	0.17
2	16	0.18394	18.394	0.21
3 or more	10	0.08030	8.030	0.48
Total	100	—	—	1.14

For this problem, $k = 4$ and $m = 1$; therefore, $\delta = 2$. We know that $P(\chi_2^2 > 5.99) = 0.05$ and $P(\chi_2^2 > 9.21) = 0.01$, so the fit here is good at $\alpha = 0.05$ as well as at $\alpha = 0.01$.

TESTING THE FIT OF A NORMAL DISTRIBUTION

Measurements of resistance for 360 units of a certain type of electrical product result in a frequency distribution as shown in the first two columns of the following table. For this sample, we find that $\bar{x} = 3.65$ ohms, and $s = 0.1334$ ohms. Suppose that it is desired to test the hypothesis that the population from which the sample is drawn is normal; then we may proceed by fitting a normal distribution to the sample data by using $\bar{x}$ and s as estimators of the corresponding population parameters. Using the normal probability table, we can easily compute expected frequencies under this null hypothesis. For example, let the variable under consideration be X, then

$$\begin{aligned} P(3.3 < X < 3.4) &= N\left(\frac{3.4 - 3.65}{0.1334}\right) - N\left(\frac{3.3 - 3.65}{0.1334}\right) \\ &= N(-1.87) - N(-2.62) = 0.0307 - 0.0044 \\ &= 0.0263 \end{aligned}$$

and $e_i = 360(0.0263) = 9.468$, and so forth.

Resistance (Ohms)	o_i	$f(3.65;0.1334)$	e_i	$(o_i - e_i)^2/e_i$
3.3–3.4	10	0.0263	9.468	0.03
3.4–3.5	37	0.1028	37.008	0.00
3.5–3.6	73	0.2222	79.992	0.61
3.6–3.7	114	0.2886	103.896	0.98
3.7–3.8	81	0.2222	79.992	0.01
3.8–3.9	29	0.1028	37.008	1.73
3.9–4.0	14	0.0263	9.468	2.17
Total	360	—	—	5.53

Thus the computed chi-square value is 5.53, with $\delta = k - 1 - m = 7 - 1 - 2 = 4$. This fit is good at both the 5 percent and 1 percent levels of significance, since $P(\chi_4^2 > 9.49) = 0.05$ and $P(\chi_4^2 > 11.14) = 0.01$.

In concluding this section, it is to be noted that chi-square tests on goodness-of-fit are typically single-tail tests. Either the computed chi-square value is too large to be attributable to chance, the fit is considered too bad, and the hypothesis is rejected; or, the computed chi-square value is not too large, the fit is considered good, and the hypothesis is accepted. However, this procedure should not obscure the fact that in effect there are two admissible alternative hypotheses: It is possible that the fit may be too good; that is, the differences between observed and expected frequencies may be smaller than those that could result from random sampling. Therefore, it has been suggested that chi-square tests on goodness-of-fit should be two-sided. Although we do not recommend this procedure, the student should always bear in mind that, in repeated sampling, too good a fit is just as likely as too bad a fit. When the computed chi-square value is too close to zero, we should suspect the possibility that the two sets of frequencies have been manipulated in order to force them to agree and, therefore, the design of our experiment should be thoroughly checked.

14.8 Tests of Independence: Contingency Table Tests

Tests on frequencies and on goodness-of-fit are both concerned with multinomial populations. In both cases, populations as well as samples are classified in accordance with a single attribute. We shall now see that the same technique of chi-square tests can be applied to data other than multinomial distributions.

When the population and the sample are classified according to two or more attributes, we may use *tests of independence* to determine whether the principles or criteria employed for cross classification are meaningful or effective. For instance, a random sample of n retail stores may be classified by size of capitalization and by type of ownership. The proportion of each of the two classes in the population is unknown. Our interest would be to establish whether there is any dependency relationship between a store's capitalization and its type of ownership. Clearly, in a case such as this, we would want to test the hypothesis that the size of capitalization is independent of the type of ownership against the hypothesis that they are related. In this test of independence, we try to verify the significance of a capitalization difference in type of ownership. If the difference is significant, we conclude that the type of ownership is independent of the size of capitalization. Otherwise, we say that these two criteria of classification are related or dependent.

Tests of independence are also called *contingency table tests*. So far, we have been concerned with only *one-way classification tables* since, in each case, the observed frequencies have occupied a single row or a single column. Also because the observed frequencies are distributed in k classes (columns or rows), one-way classification tables are called *1 × k* (read *1* by k) or *k × 1 tables.* Extending these ideas, we can arrive at 2 × 2, or *two-way classification tables* and, generally, $R \times C$ tables in which the observed frequencies occupy R rows and C columns. Such tables are often called *contingency tables.* Corresponding to each observed frequency in an $R \times C$ table, there is an expected frequency computed under the specified null hypothesis. Frequencies, observed or expected, which occupy the cells of a contingency table are called *cell frequencies.* The total frequency in each row or each column is called the *marginal frequency.*

To evaluate differences between observed and expected frequencies contained in contingency tables, we employ the same statistic as that for tests discussed in the previous section,

$$\chi_\delta^2 = \sum \frac{(o_i - e_i)^2}{e_i}$$

The sum is taken over all cells in a contingency table and, in general,

$$\delta = (R - 1)(C - 1)$$

In testing independence with a 2 × 2 contingency table, we would have $\delta = (2 - 1)(2 - 1) = 1$. As recommended before, the continuity correction factor of ½ should be used to compute the chi-square value. This is especially important with small samples so that some or all expected frequencies are less than 5. For large samples, corrected and uncorrected chi-square values may be practically the same and, as a consequence, the continuity correction factor may be ignored. Let us consider an example.

There is reason to believe that high-income families usually send their children to private colleges and low-income families often send their children to city or state colleges. To verify this, 1600 families are selected at random from California and the following results are obtained:

OBSERVED DATA

Income	College		Total
	Private	*Public*	
Low	506	494	1000
High	438	162	600
Total	944	656	1600

Here, we have two classifications: income, and type of college. At a glance, we see that, relatively, a greater number of high-income families send their children to private colleges than do low-income families. But is such a difference in proportions significant or is it merely the result of chance variations in random sampling? The chi-square test of independence can be used to answer questions such as this by comparing the set of observed frequencies with that of expected frequencies.

Under the assumption that family income and type of college are independent, we would then expect the proportion of all families which send their children to private colleges to be $944/1600$. The expected number of low-income families which send their children to private colleges is then

$$(944/1600)(1000) = 590$$

Now, since the expected frequencies must agree with the marginal frequencies of the observed frequencies, and since there are four cells in a 2×2 contingency table, we can automatically fill in the remaining three expected numbers once any one of the four expected numbers is determined. It is for this reason that there is only 1 degree of freedom associated with a 2×2 contingency table. The expected frequencies for our example are as follows:

EXPECTED DATA

	College		
Income	*Private*	*Public*	Total
Low	590	410	1000
High	354	246	600
Total	944	656	1600

Following our general testing procedure, the chi-square test of independence for data on hand becomes

1. **Hypotheses:** H_0: Income and type of college are independent
H_1: They are related or dependent

2. **Level of Significance:** $\alpha = 0.01$

3. **Testing Statistic:** $\sum \frac{(o_i - e_i)^2}{e_i}$, which is approximately distributed as χ_1^2

4. **Decision Rule:** The critical region of this test is $6.63 < \chi_1^2 < \infty$. Thus, accept H_0 if and only if the computed chi-square value is less than 6.63.

5. **Computations:** From observed and expected data given before, we have

$$\chi_1^2 = \sum \frac{(o_i - e_i)^2}{e_i} = \frac{(506 - 590)^2}{590} + \frac{(494 - 410)^2}{410}$$

$$+ \frac{(438 - 354)^2}{354} + \frac{(162 - 246)^2}{246} = 77.78$$

6. Decisions: Reject H_0. Family income and type of college are not independent. A greater proportion of high-income families send their children to private colleges.

As a second example, let us consider this: Are men different from women with respect to their concept of a successful life defined in terms of material well-being, M, spiritual well-being, S, and fame, F? Suppose that a random sample of 430 people has been selected across the United States and the following results have been obtained:

OBSERVED DATA

	Successful Life			
Sex	M	S	F	Total
Men	80	95	60	235
Women	60	70	65	195
Total	140	165	125	430

In this case, we are interested in testing the hypothesis that a person's concept of a successful life is independent of his sex against the hypothesis that one's sex is related to his idea of a successful life. The null hypothesis here says that men and women are of the same opinion insofar as the meaning of a successful life is concerned. Under this assumption, we would expect, for example, $140/430$ of the 430 people interviewed to define a successful life as material well-being. The expected numbers of men and of women who consider this a criterion for a successful life, respectively, are

$$(140/430)(235) = 76.51$$
$$(140/430)(195) = 63.49$$

Expected frequencies for other cells can be obtained in the same manner; they are recorded below:

EXPECTED DATA

	Successful Life			
Sex	M	S	F	Total
Men	76.51	90.17	68.32	235
Women	63.49	74 83	56.68	195
Total	140.00	165.00	125.00	430

There are six cells for a 2×3 contingency table. The requirement that cell frequencies must agree with marginal frequencies, however, gives us but 2 degrees of freedom. We are free to place one number in one of the three columns and to place another number in any one of the four cells of the remaining two columns.

After these two numbers are given, all numbers in the empty cells are uniquely determined from the indicated marginal frequencies. This conclusion is in agreement with the fact that $\delta = (R - 1)(C - 1) = (2 - 1)(3 - 1) = 2$.

For data given before, we find that

$$\chi_2^2 = \frac{(80 - 76.51)^2}{76.51} + \frac{(95 - 90.17)^2}{90.17} + \frac{(60 - 68.32)^2}{68.32} + \frac{(60 - 63.49)^2}{63.49} + \frac{(70 - 74.83)^2}{74.83} + \frac{(65 - 56.68)^2}{56.68} = 3.16$$

Since $P(\chi_2^2 > 5.99) = 0.05$ and $P(\chi_2^2 > 9.21) = 0.01$, we accept the null hypothesis that the concept of a successful life is independent of a person's sex at both $\alpha = 0.05$ and 0.01. Men and women interpret a successful life in the same way with respect to the three principles.

EFFICIENT FORMULAS FOR COMPUTING χ^2

There are available efficient formulas to compute chi-square values from 2×2 and 2×3 contingency tables which involve only observed frequencies.

Denote cell frequencies as A, B, C, and D, marginal frequencies as m_1, m_2, m_3, and m_4, and total number of observations as n, for a 2×2 contingency table as given below; then it can be shown that

OBSERVED DATA

A	B	m_3
C	D	m_4
m_1	m_2	n

$$\chi_1^2 = \frac{n(\text{AD} - \text{BC})^2}{m_1 m_2 m_3 m_4} \tag{14.12}$$

For instance, applying the above formula to our first example for test of independence, we have

$$\chi_1^2 = \frac{1600((506)(162) - (492)(438))^2}{(944)(652)(1000)(600)} = 77.78$$

which is as obtained before.

When the sample is small, say, less than 50, and when some or all cell frequencies are less than 5, it is desirable to compute χ_1^2 with a continuity correction factor. In this case, the efficient formula becomes

$$\chi_1^2 = \frac{n(\text{AD} - \text{BC} - \frac{1}{2}n)^2}{m_1 m_2 m_3 m_4} \tag{14.13}$$

For a 2×3 contingency table, given the following notations,

OBSERVED DATA

A	B	C	m_4
D	E	F	m_5
m_1	m_2	m_3	n

it can be shown that

$$\chi_2^2 = \frac{n}{m_4}\left(\frac{A^2}{m_1} + \frac{B^2}{m_2} + \frac{C^2}{m_3}\right) + \frac{n}{m_5}\left(\frac{D^2}{m_1} + \frac{E^2}{m_2} + \frac{F^2}{m_3}\right) - n \tag{14.14}$$

Applying (14.14) to data in our second example of test on independence, we have

$$\chi_2^2 = \frac{430}{235}\left(\frac{80^2}{140} + \frac{95^2}{165} + \frac{60^2}{125}\right) + \frac{430}{195}\left(\frac{60^2}{140} + \frac{70^2}{165} + \frac{65^2}{125}\right) - 430 = 3.16$$

as before.

In general, we note that

$$\chi_\delta^2 = \sum \frac{(o_i - e_i)^2}{e_i}$$

$$= \sum \frac{(o_i^2 - 2o_ie_i + e_i^2)}{e_i} = \sum \frac{o_i^2}{e_i} - 2\sum \frac{o_ie_i}{e_i} + \sum \frac{e_i^2}{e_i} = \sum \frac{o_i^2}{e_i} - n \tag{14.15}$$

This formula is valid for contingency tables of any dimension. Its use, however, requires expected frequencies. Hence it is recommended for only 3 × 3 or greater dimension tables. To see the generality of this expression, we may apply it to the data in our 2 × 3 contingency table test and find that

$$\chi_2^2 = \sum \frac{o^2}{e} - n = \frac{80^2}{76.51} + \frac{95^2}{90.17} + \frac{60^2}{68.32} + \frac{60^2}{63.49} + \frac{70^2}{74.83} + \frac{65^2}{56.68} - 430$$

$$= 3.16$$

as before.

14.9 Tests of Homogeneity

Tests of homogeneity are designed to determine whether two or more independent random samples are drawn from the same population or from different populations.

The chi-square test of homogeneity is an extension of the chi-square test of independence. In both cases we are concerned with cross-classified data. As we shall immediately see, too, the same testing statistic used for tests of independence is used for tests of homogeneity. These two types of test are, however, different in a number of ways. First, they are associated with different kinds of problems. Tests of independence are concerned with the problem of whether one attribute is independent of another, while tests of homogeneity are concerned with whether different samples come from the same population. Second, the former involve a single sample taken from one population; but the latter involve two or more independent samples, one from each of the possible populations in question. This second point also implies that, in the case of independence, all marginal frequencies are chance quantities and that, in the case of homogeneity, row totals are sample sizes which are chosen numbers.

To illustrate this type of test, let us suppose that three samples are taken: one consists of 115 professional people, one consists of 110 businessmen, and one consists of 125 farmers. Each individual chosen is asked to select, say, one of the three categories which best represents his feeling toward a certain national policy. Suppose these three categories are (1) in favor of the policy, F; (2) against the policy, A; (3) indifferent toward the policy, I. Now, assume that the results of the interviews are distributed as follows:

OBSERVED DATA

	Reaction			
Occupation	F	A	I	Total
Professionals	80	21	14	115
Businessmen	72	15	23	110
Farmers	69	31	25	125
Total	221	67	62	350

From the way the problem is posed, an appropriate null hypothesis to be tested seems to be: The three samples come from the same population; that is, the three classifications are homogeneous insofar as the opinion of three different groups of people about the national policy under consideration is concerned. This also means there exists no difference in opinion among the three classes of people on the issue. From the alternative expression of the null hypothesis for this problem, we can see why it is called a test of homogeneity. (When we say things are homogeneous we mean they have something in common or they are the same or they are equal.)

We note that if the null hypothesis stated before is true, then the best estimates for proportions specifying "in favor of the policy," "against the policy," and "indifferent toward the policy," respectively, should be $221/350$, $67/350$, and $62/350$. Thus, of the 115 professional people, expected frequencies for the three categories become

$$\begin{aligned} (221/350)(115) &= 72.61 \text{ in favor} \\ (67/350)(115) &= 22.01 \text{ against} \\ (62/350)(115) &= 20.37 \text{ indifferent} \end{aligned}$$

Expected frequencies for the other two groups of people can be computed in the same way. Expected data for the whole problem are as follows:

EXPECTED DATA

	Reaction			
Occupation	F	A	I	Total
Professionals	72.61	22.01	20.37	115
Businessmen	69.46	21.06	19.49	110
Farmers	78.93	23.93	22.14	125
Total	221.00	67.00	62.00	350

Again following our general procedure of testing, we have this test of homogeneity for the problem:

1. **Hypotheses:** H_0: The three samples are drawn from the same population
H_1: They are drawn from different populations

2. **Level of Significance:** $\alpha = 0.05$, arbitrarily selected

3. **Testing Statistic:** $\sum \frac{(o_i - e_i)^2}{e_i}$ which is distributed approximately as χ_4^2. Here, $\delta = (3 - 1)(3 - 1) = 4$.

4. **Decision Rule:** The rejection region is defined for this test as $9.49 \leq \chi_4^2 < \infty$, since $P(\chi_4^2 > 9.49) = 0.05$.

5. **Computations:** From observed and expected data given before, we have

$$\chi_4^2 = \frac{(80 - 72.61)^2}{72.61} + \frac{(21 - 22.01)^2}{22.01} + \frac{(14 - 20.37)^2}{20.37} + \frac{(72 - 69.46)^2}{69.46}$$
$$+ \frac{(15 - 21.06)^2}{21.06} + \frac{(23 - 19.49)^2}{19.49} + \frac{(69 - 78.93)^2}{78.93} + \frac{(31 - 23.93)^2}{23.93}$$
$$+ \frac{(25 - 22.14)^2}{22.14}$$
$$= 0.75 + 0.05 + 1.99 + 0.09 + 1.74 + 0.63 + 1.25 + 2.09 + 0.37 = 8.96$$

This computation can also be made with (14.15), the more efficient formula, where we need not obtain the differences between observed and expected frequencies. Using this formula here, we consider n as the total number of observations in all the samples. Thus,

$$\chi_4^2 = \sum \frac{o_i^2}{e_i} - n$$

$$= \frac{80^2}{72.61} + \frac{21^2}{22.01} + \frac{14^2}{20.37} + \frac{72^2}{69.46} + \frac{15^2}{21.06} + \frac{23^2}{19.49} + \frac{69^2}{78.93} + \frac{31^2}{23.93} + \frac{25^2}{22.14} - 350$$

$$= 8.96$$

6. **Decisions:** The computed chi-square value falls in the region of acceptance. The three samples are drawn from the same population. The views of professional people, businessmen, and farmers are homogeneous insofar as the national policy under discussion is concerned.

14.10

The Median Test for Two Samples

The median test is based on signs (see Section 14.5). However, this test is applicable to independent samples as well as to matched pairs. Furthermore, unlike the sign test, the median test does not require that the two samples be of the same size.

The median test is designed to test the null hypothesis that the two samples are drawn from populations with identical distributions, that is, $H_0: f(x) = g(x)$, against the alternative that the populations have different distributions, $H_1: f(x) = g(x + c)$. The procedure for this test may be summarized as follows:

(1) Combine observations in the two samples into a single series and arrange them in an array of descending order. Then find the grand median, denoted as $\bar{m}$, for the combined data.

(2) Count the number of observations in the first sample which are greater than $\bar{m}$ and those which are smaller than $\bar{m}$. Denote these numbers as a_1 and b_1, respectively. Do the same for the second sample and denote the resulting numbers as a_2 and b_2, respectively.

(3) The a_i and b_i are cell frequencies in a 2×2 contingency table. Let n_1 and n_2 be the respective sample sizes, m_1 and m_2 be the respective marginal frequencies for the rows, and n be the grand sample size; sample data now may be presented as follows:

	Sample 1	Sample 2	Total
Above $\bar{m}$	a_1	a_2	m_1
Below $\bar{m}$	b_1	b_2	m_2
Total	n_1	n_2	n

TABLE 14.4

Machine I	Machine II	Array of I and II	
69	98	109	
109	56	105	
63	62	98	73
90	75	93	71
73	68	91	71
76	80	90	70
85	70	88	69
78	87	87	68
77	59	85	66
91	66	81	63
61	70	80	63
88	63	78	62
105	81	77	61
93		76	59
71		75	56

Clearly, from this table, we see that the statistic

$$\chi_1^2 = \frac{(a_1b_2 - a_2b_1)^2 n}{n_1 n_2 m_1 m_2} \tag{14.16}$$

can be used to test the set of hypotheses mentioned before. It is interesting to note that this chi-square with $\delta = 1$ is sensitive only to the location of the median and is insensitive to the differences in the pattern of distributions. Also, when $n_1 \leq n_2 < 10$, a correction factor should be introduced to compute the chi-square value as discussed before.

Two different makes of machines are adjusted to produce a given product. Defective units produced daily by both machines are observed at random and recorded as in the first two columns of Table 14.4. Do these results indicate that the distributions of defective items of the two machines are identical at $\alpha = 0.05$?

From Table 14.4, we have

$$n_1 = 15 \qquad n_2 = 13 \qquad n = 28$$
$$\bar{m} = (76 + 75) = 75.5$$
$$a_1 = 10 \qquad a_2 = 5$$

These data are presented by the contingency table below:

	I	II	Total
Above 75.5	10	4	14
Below 75.5	5	9	14
Total	15	13	28

$$\chi_1^2 = \frac{(90 - 20)^2(28)}{(15)(13)(14)(14)} = 3.590$$

Since $P(\chi_1^2 > 3.841) = 0.05$, the observed chi-square value falls in the region of acceptance. In accepting that $f(x) = g(y)$, we conclude that the two samples come from populations with identical distributions of defective units produced daily.

14.11 The Median Test for More Than Two Independent Samples

The median test for two samples can be easily extended to compare several populations. As before, we assume that the population distributions are continuous so that the probability of a tie between two observations is zero.

The null hypothesis to be tested now is: The c populations are identical with respect to their distribution functions. The alternative is simply the opposite of H_0.

The testing procedure here is the same as in the two-sample case. The c independent random samples, with $n_1, n_2, \cdots, n_c$, are combined into a single distribution and the grand median is found. Then each sample observation in each sample is compared with $\bar{m}$. Denote those observations in the ith sample which are above $\bar{m}$ as a_i; the sample data may then be presented as a $2 \times c$ contingency table as below:

	Sample 1	Sample 2	$\cdots$	Sample c	Total
Above $\bar{m}$	a_1	a_2	$\cdots$	a_c	m_1
Below $\bar{m}$	b_1	b_2	$\cdots$	b_c	m_2
Total	n_1	n_2	$\cdots$	n_c	n

It can be shown that if $n \geq 20$ and $n_i \geq 5$ a satisfactory approximation to the exact significance level can be found by the expression

$$\chi^2_{c-1} = \frac{(n-1)}{m_1 m_2} \sum_{i=1}^{c} \frac{(na_i - n_i m_1)^2}{nn_i} \tag{14.17}$$

which has an approximate chi-square distribution with $\delta = c - 1$.

In using this test, we may note that ties often occur with observed data because of either rounding or lack of precision in measurements. A safe way to deal with ties here is to assign ties either as a or as b in a way that makes a_i as close as possible to b_i in the ith sample. This procedure has at least the merit of being conservative. Why?

Let us consider an example. A random sample of ten is drawn from the senior students who have taken the graduate record examination, based on a total of

TABLE 14.5

Sample 1	Sample 2	Sample 3
10	11	9
13	13	10
17	15	11
15	18	14
15	15	10
13	14	10
12	16	8
10	20	9
10	13	8
10	17	10

20 points, from each of three colleges. Observations on the grades of those students sampled are given in Table 14.5. In view of these data, can we conclude that the distributions of grades are identical for all three colleges at $\alpha = 0.05$?

For the above data, the grand sample median is found to be $\bar{m} = 12.5$. Comparing each observation in the three samples with $\bar{m}$, we obtain the following results:

	1	2	3	Total
Above 12.5	5	9	1	15
Below 12.5	5	1	9	15
Total	10	10	10	30

$$\chi^2_{3-1} = \frac{(30-1)}{(15)(15)}\left\{\frac{[(30)(5)-(10)(15)]^2}{(30)(10)} + \frac{[(30)(9)-(10)(15)]^2}{(30)(10)} + \frac{[(30)(1)-(10)(15)]^2}{(30)(10)}\right\} = 12.48$$

Thus we conclude that the population distributions are not identical, since $P(\chi_2^2 > 5.99) = 0.05$. The departure from identity may be inferred from this significant result: effects of experimental treatments are associated with the different colleges.

14.12

The Median Test for More Than Two Matched Samples

When there are c experimental treatments to be compared and these groups are matched in terms of one or more factors or levels, the median test for several independent samples introduced in the previous section can be adjusted to test the identity of population distributions with c matched groups. The procedure is outlined as follows:

First, sample data are arranged into c columns representing treatments and r rows representing blocks of experimental materials. Next, the median of each row is found. Third, each observation in each row is given a sign of + or −, depending upon whether or not its value is greater than or less than the median for that row. Finally, the number of plus signs occurring in the ith column, a_i, is determined. After this is done, sample data are then reduced into a $2 \times c$ contingency table as follows:

	Sample				
Sign	I	II	$\cdots$	c	Total
+	a_1	a_2	$\cdots$	a_c	$m_1 = ra$
−	b_1	b_2	$\cdots$	b_c	$m_2 = n - ra$
	n_1	n_2	$\cdots$	n_c	n

where r is the number of rows in the original data table and

$$a = c/2 \qquad \text{for } c \text{ even}$$
$$a = (c+1)/2 \qquad \text{for } c \text{ odd}$$

Now if we assume that the distributions of the row populations are identical, then we can test the hypothesis that the distributions of populations represented by the columns are identical. The testing statistic in this case is

$$\chi^2_{c-1} = \left[\frac{c-1}{ra(c-a)}\right] \frac{\sum^{c} (ca_i - ra)^2}{c} \tag{14.18}$$

which is approximately distributed as chi-square with $\delta = c - 1$.

Suppose that we wish to determine the relative yields of four varieties of corn. Moreover, let us suppose that there are six blocks of land available for this experiment. Each block of land is divided into four plots of equal size. Plots of each block of land are then assigned to the four types of corn at random. Suppose that the following data, in Table 14.6, are obtained:

TABLE 14.6

r: Block	I	II	III	IV	Row Median
1	4(−)	9(+)	10(+)	5(−)	7.0
2	3(−)	4(−)	9(+)	8(+)	6.0
3	4(−)	6(−)	11(+)	8(+)	7.0
4	5(−)	3(−)	10(+)	10(+)	7.5
5	7(+)	5(−)	8(+)	6(−)	6.5
6	6(−)	8(+)	9(+)	7(−)	7.5
a_i	1	2	6	3	—

Be sure to note that ratings of observations in each row are compared with the median of that row. From these ratings, can we conclude that the distributions underlying the four different types of corn are identical at $\alpha = 0.05$?

To answer this question, we note that in terms of the ratings above we have the following contingency table:

Sign	I	II	III	IV	
+	1	2	6	3	12 = ra
−	5	4	0	3	12 = $n - ra$
	6	6	6	6	24

$$\chi_3^2 = \frac{3}{12(2)}\left[\frac{((4)(1) - 6(2))^2}{4} + \frac{((4)(2) - (6)(2))^2}{4} + \frac{((4)(6) - (6)(2))^2}{4} + \frac{((4)(3) - (6)(2))^2}{4}\right] = 7.00$$

For $\delta = 3$, this value is less than the chi-square value of 7.81 required for the 5 percent level of significance. H_0 cannot be rejected.

We should note that what we have done in this section is similar to the test treatment effects against interaction in the analysis of variance. If we are interested in not only the fact that the treatment populations are identically distributed but also in the fact that the treatment means are equal, we should be prepared to assume that either the joint effects are nil or the data should be analyzed by the random effects or mixed models. It may be pointed out, too, that the procedure just introduced can also be extended to a two-variable classification with replication analysis where both the treatment and interaction effects are tested. However, such an extension requires many more assumptions and is quite time-consuming and laborious.

14.13

The Wald-Wolfowitz Test for Two Independent Samples

The Wald-Wolfowitz test is designed to test the identity between two population distributions with two independent samples on the basis of runs. Again assuming that the underlying variable is continuous, the test begins with grouping observations in the two samples, with n_1 and n_2, in a single array. For convenience, if we identify an observation from the first sample in the array as X and identify that from the second sample as Y, then we would have runs or clusterings of Xs and Ys. For example, consider these two samples,

First sample: 3, 2, 3, 4, 7, 6, 9, 7, 3, and 5
Second sample: 1, 1, 2, 3, 4, 4, 5, 2, 6, and 2

Now, combining these observations and arranging them in an array, together with their designations of Xs and Ys, we would have twelve runs in terms of Xs and Ys:

$$\begin{array}{cccccccccccccccccccc} Y & Y- & X- & Y & Y & Y- & X & X & X- & Y- & X- & Y & Y- & X- & Y- & X- & Y- & X & X & X \\ 1 & 1 & 2 & 2 & 2 & 2 & 3 & 3 & 3 & 3 & 4 & 4 & 4 & 5 & 5 & 6 & 6 & 7 & 7 & 9 \end{array}$$

Let us note, first of all, that in any ordering of observations from two samples, there must be at least two runs and, at most, $n_1 + n_2$ runs in the array. Next, if the two samples are drawn independently from two populations with identical distributions, the sample observations should be very well mixed when they are arranged according to magnitudes. We should expect, in other words, to have a large number of runs. If, however, the samples are selected from populations with different distributions, we would expect to have a small number of runs. These arguments provide us the foundation of testing based on fairly simple probability calculations.

Let R be the number of runs among the $n_1 + n_2$ observations; the sampling distribution of R is given in Table A-XV for $n_1 \leq 10$ and $n_2 \leq 10$ and $n_1 = n_2 = 11$ to 100. For example, for our illustration, $n_1 = n_2 = 10$ and $R = 12$ correspond to the 75.8th percentile of the distribution of R. This value is between 0.005 and 0.995; therefore, we accept the null hypothesis that the two parent populations are identically distributed at $\alpha = 0.01$.

The foregoing decision assumed a two-sided alternative. If we wish to reject H_0 only when the number of runs is too small, then we would have a left-tail test and, for $n = 10$, we would reject H_0 when $R \leq 5$, since $P(R \leq 5) = 0.01$, or when $R \leq 6$, since $P(R \leq 6) = 0.05$. See Table A-XV for these values.

R is asymptotically normally distributed for $n_1 \geq n_2 \geq 20$, with

$$E(R) = \frac{2n_1n_2}{n_1 + n_2} + 1 \tag{14.19}$$

and

$$V(R) = \frac{2n_1n_2(2n_1n_2 - n_1 - n_2)}{(n_1 + n_2)^2(n_1 + n_2 - 1)} \tag{14.20}$$

Thus, when samples are large enough, the Wald-Wolfowitz test may be conducted by the testing statistic

$$Z = \frac{R - E(R)}{\sigma_R} \tag{14.21}$$

in the usual manner.

It may be noted that this test is generally less powerful than the t test for equality between means and other nonparametric tests for identity between population distributions. The utility of the procedure is that it may be the only one available in some situations. The Wald-Wolfowitz test can also be applied to a single sample, and any criterion of producing an ordering of Xs and Ys may be employed. For example, we may wish to find whether there is a time-related trend in a time series. In this instance, we may first locate the median for the set of data ordered according to time, and then denote above-median and below-median values as Xs and Ys respectively. We may treat the occurrence of too many runs as the evidence of randomness, and the appearance of too few runs as the existence of a time-related trend, or as dependence among successive terms of a series. This procedure makes the "runs" test useful in a variety of circumstances. For employing this test for a single sample, the total observations are broken down into two equal parts as n_1 and n_2, or such that $n_1 = n_2 + 1$ or $n_1 = n_2 - 1$. Then Table A-XV may be used or normal approximations be made.

14.14

The Wilcoxon Test for Two Matched Samples

The problem of treating matched pairs was taken up previously under the sign test, which takes into consideration merely the sign of a difference between each pair of values. The sign test overlooks the fact that a pair of scores has not only a difference but also a rank among all such differences. The Wilcoxon test takes both features, signs and ranks of differences, into consideration; it is sometimes called the *rank-sum* test.

The rank-sum test is perhaps the most powerful design among all nonparametric tests for the identity between two population distributions with matched pairs. The mechanics of this test are just as simple as those of the sign test. When observations are given in n pairs, we first observe their differences, d_i, and then rank these differences according to their absolute magnitudes. Next, we attach the signs of these ranks for the corresponding differences. Finally, we find the testing statistic, T, defined as the sum of the ranks with the less frequent sign.

When the null hypothesis, $f(x) = g(y)$, is true, then each of the 2^n possible sets of signed ranks (obtained by arbitrarily assigning + and − signs 1 through n) is equally likely to occur. On this basis, the exact distribution of T can be found.*

To test $f(x) = g(y)$ against the CDF, $F(x) > G(y)$, the approximate significance levels for the smaller rank total may be obtained by the following expressions:

Denoting the 5 percent level of the lower rank total as $T_{0.05}$, we have

$$T_{0.05} = \frac{n^2 - 7n + 10}{5} \tag{14.22}$$

We reject H_0 at $\alpha = 0.05$ if $T \leq T_{0.05}$.

* See, for example, Frank Wilcoxon, "Individual Comparison by Ranking Method," *Biometrics Bulletin,* I (1945).

TABLE 14.7

Pair	Procedure 1	Procedure 2	d_i	Rank of $\|d_i\|$
1	109	61	48	10
2	100	78	22	9
3	77	57	20	8
4	69	64	5	5
5	59	60	−1	−1
6	69	75	−6	−6
7	87	85	2	2
8	88	84	4	4
9	88	85	3	3
10	91	83	8	7

Denoting the 1 percent level for the lower rank total as $T_{0.01}$, we have

$$T_{0.01} = \frac{11n^2}{60} + 5 - 2n \tag{14.23}$$

We reject H_0 at $\alpha = 0.01$ if $T \leq T_{0.01}$.

Now, an example is in order. Suppose that two auditing procedures are to be compared with ten pairs of auditors, matched by sex, and the number of accounts audited per hour are as given in columns two and three of Table 14.7; is there a significant difference in the two methods of auditing at $\alpha = 0.05$? at $\alpha = 0.01$?

The less frequent sign in this case is minus, and the sum of these minus ranks is

$$T = 1 + 6 = 7$$

$$T_{0.05} = \frac{10^2 - 7(10) + 10}{5} = 8$$

$$T_{0.01} = \frac{11(10)^2}{60} + 5 - 2(10) = 3.33$$

Thus, H_0 is rejected at $\alpha = 0.05$, but it cannot be rejected at $\alpha = 0.01$.

When $n \geq 10$, the sampling distribution of T is approximately normal with

$$E(T) = \frac{n(n+1)}{4} \tag{14.24}$$

and

$$V(T) = \frac{n(n+1)(2n+1)}{24} \tag{14.25}$$

Hence, for large samples, we use

$$Z = \frac{T - E(T)}{\sigma_T} \tag{14.26}$$

as the testing statistic in the rank-sum test. For example, with data in Table 14.7, we have

$$E(T) = \frac{(10)(10 + 1)}{4} = 27.5$$

$$V(T) = \frac{10(10 + 1)((2)(10) + 1)}{24} = 96$$

$$Z = \frac{7 - 27.5}{\sqrt{96}} = -2.09$$

Thus, $P(T \leq 7) = N(-2.09) = 0.0183$. This result indicates again that H_0 can be rejected at $\alpha = 0.05$ but not at $\alpha = 0.01$, as before. It is interesting to note that, according to Wilcoxon's Table, the exact probability, $P(T \leq 7)$, for $n = 10$ is 0.0186. The normal value is an excellent approximation for this exact probability.

When ties occur in this test they are disregarded and n is adjusted accordingly.

14.15 The Kruskal–Wallis *H* Test

W. H. Kruskal and W. A. Wallis have extended the Wilcoxon test for two samples to cover c samples by incorporating the Wilcoxon test with the analysis of variance of the one-variable classification model on the basis of ranked data. This procedure, called the H test, has been designed to test that the c independent samples are drawn from continuous populations with identical distributions.

The H test begins by arranging observations of all the samples in order of their magnitudes so that they may be ranked. Then the rank sum for each sample, T_i, is found. After this, the total of all group rank sums T is obtained. Namely, $T = \Sigma T_i$. It can be seen that

$$T = \frac{n(n + 1)}{2}$$

where $n = n_1 + n_2 + \cdots + n_c$. It is known that if $n_i \geq 8$, the statistic

$$H = \frac{12}{n(n + 1)} \sum \frac{T_i^2}{n_i} - 3(n + 1) \tag{14.27}$$

is approximately distributed as chi-square with $\delta = c - 1$.

When ties occur in the H test, we assign an average of the joining ranks to each of the tied observations. For example, if two observations are tied and if the ranks for them should be 5 and 6, then we assign to both observations a rank of $(5 + 6)/2 = 5.5$. Again if we encounter a tie of three observations, the ranks being, say, 10, 11 and 12, then we assign a rank of 11 to each of three observations. And so on.

An adjusted value of H, denoted as H', should be computed when there are ties. We define

$$H' = \frac{H}{C} \tag{14.28}$$

where C, the correction factor, is defined as

$$C = 1 - \frac{\sum^{k} (t_i^3 - t_i)}{n^3 - n}$$

In the definition for C, k is the number of tied sets and t_i is the number of tied observations in set i.

Suppose that a random sample has been taken from each of three sources of high tension steel wires, and that their tensile strengths in hundreds of pounds have been measured. We wish to test whether these observations indicate that the parent populations are identically distributed. The original data together with the ranks (in parentheses) are given in Table 14.8.

For the data below, we have

$$n = 33, \quad T = \frac{33(34)}{2} = 561$$

$$H = \frac{12}{33(34)}\left[\frac{(256.5)^2}{10} + \frac{(175.5)^2}{11} + \frac{(139)^2}{12}\right] - 3(34) = 15.53$$

We see from Table 14.8 that there are three sets of two tied observations, one

TABLE 14.8

	Sample 1	Sample 2	Sample 3
	29 (17.5)	36 (26.5)	24 (11.5)
	36 (26.5)	17 (2)	18 (3)
	37 (29.5)	19 (4)	20 (5)
	36 (26.5)	21 (7)	24 (11.5)
	36 (26.5)	26 (24)	25 (13)
	35 (24)	29 (17.5)	28 (16)
	39 (32)	27 (15)	31 (20)
	38 (31)	21 (7)	34 (23)
	40 (33)	32 (21)	30 (19)
	23 (10)	33 (22)	22 (9)
		37 (29.5)	21 (7)
			16 (1)
T_i	256.5	175.5	139.0
n_i	10	11	12

set of three tied observations, and one set of four tied observations. We should therefore compute H' before a decision is made.

$$C = 1 - \left[\frac{(3(2)^3 - 2) + ((3)^3 - 3) + ((4)^3 - 4)}{(33)^3 - 33}\right] = 0.996$$

and

$$H' = \frac{15.53}{0.996} = 15.59$$

The values of H and H' are practically the same in this example because the samples are rather large and there are not too many ties.

Now, since $P(\chi_2^2 > 9.21) = 0.01$, we conclude that the result is highly significant: the population distributions are not all identical.

The H test is superior to the median test in most situations and compares quite favorably with the F test in the analysis of variance. However, when samples are large, it is rather painful to rank the observations.

14.16

Parametric versus Nonparametric Statistics

We have in this final chapter introduced several of the most frequently used nonparametric methods. While the nonparametric theory arose as early as the middle of the nineteenth century, the well-known nonparametric method of testing independence and goodness-of-fit by the chi-square test was formulated by Karl Pearson in 1900. The true beginning of this theory, however, was Spearman's rank-correlation coefficient test for covariability popularized by Harold Hotelling and Margaret R. Pabst in 1936. Even then there was little interest in nonparametric statistics until 1945, when Wilcoxon proposed a test for the two-sample case which was distinguished by both its simplicity and its excellent results, even in cases where the population distribution is known to be normal.

There are many reasons for the wide adoption and the continuous growth of the nonparametric theory since 1945. Both theoretical and practical necessities are concerned with nonparametric problems. Theoretically, it is always desirable to discover procedures of statistical inference which have the minimum number of restrictions imposed by the underlying assumptions. Practically, it is important for scientific investigators to have at their disposal simple and widely adoptable techniques.

Nonparametric statistics, being primarily concerned with order relations, provides us with two important motives for its study and development. In the first place, data may occur naturally in the form of ranks. Such being the case, observations obtained give us relative magnitudes of the underlying ordinal property, while arithmetic differences between quantities provide no meaningful significance (direct ordinal analysis would then suffice for reaching a judgment about the independence of underlying variables). Moreover, when the order relations in data are considered alone, the reasoning behind each nonparametric method requires relatively simple probability theory so that the discrete sampling distributions can be easily found.

The basic importance of nonparametric theory from a theoretical viewpoint is perhaps that it requires only the simple assumption that the population distribution under investigation be continuous but does not require the specific form of the distribution, as implied by the classical theory. Thus, when data are gathered from populations whose distributions are unknown or cannot be assumed, we have little choice but to employ the nonparametric tests.

The foregoing remarks, however, are not meant to mislead the reader to believe that nonparametric methods are somehow superior to the classical decision procedures. As a matter of fact, in most situations where both the parametric and nonparametric techniques apply, the former is distinctly more desirable than the latter on at least two counts.

First, in general, the nonparametric tests have less than 100 percent power efficiency as compared to the parametric tests. The concept "power efficiency" may be explained by a simple example. Suppose that we have two alternative testing procedures for a given H_0, A and B. If A requires $n = 25$ but B requires $n = 50$ in order to reach a power of, say, 0.95 for a given true alternative at $\alpha = 0.05$, then we say that the power efficiency of B is only 50 percent of that of A. It can be shown, for normal populations, that the sign test, for example, requires 4 to 50 percent more observations than the t test, and the rank-sum test requires about 5 percent more scores than the t test in order to be equally powerful. Of course, when the assumption of normality is not met, the reverse may even be true.

Second, the null hypothesis tested by a nonparametric method is seldom equivalent to that tested by a parametric procedure. In general, with nonparametric tests, hypotheses are less precise and yield less information in conclusions. For example, when the usual assumptions for the linear model in the analysis of variance are met, we test the hypothesis that the treatment means are equal by the F ratio. In accepting this H_0, we have not only accepted the statement of equal means, but we have also implied that the treatment populations are identically distributed. When this H_0 is rejected, we know precisely that the populations are not all identical with respect to their central tendencies. The order method of comparing several samples, on the other hand, is concerned with the null hypothesis that the population distributions are the same. When this H_0 is rejected, we have a rather vague conclusion that the population distributions are not all identical. But are they different in terms of central tendency or in terms of variability? This question is left unanswered by the nonparametric tests. Thus, the departure of identity makes the nonparametric procedures highly insensitive. To be sure, this insensitivity can be reduced or even eliminated if we are willing to make some mild assumptions about population distributions. Such assumptions, however, may be just as offensive as those which the order statistics have been designed to avoid.

Glossary of Formulas

(**14.1**) $E(R) = \frac{n+2}{2}$ (**14.2**) $V(R) = \frac{n(n-2)}{4(n-1)}$ (**14.3**) $Z = \frac{R - E(R)}{\sqrt{V(R)}}$

With $n \geq 25$, the number of runs above and below the sample median is asymptotically normal with the expectation and variance as defined here. Thus, when the sample is large, the test of randomness can be made by the normal critical ratio, Z.

(**14.4**) $E(R) = \frac{1}{3}(2n - 1)$ (**14.5**) $V(R) = \frac{1}{90}(16n - 29)$ (**14.6**) $Z = \frac{R - E(R)}{\sqrt{V(R)}}$

When $n > 20$, the sampling distribution of runs up and down, in terms of $+$ and $-$ signs between each pair of successive observations in a sample, is asymptotically normally distributed with expectation variance as defined here.

(**14.7**) $r_1 = \frac{\text{cov}\,(y_i, y_{i+1})}{\sqrt{V(y_i)V(y_{i+1})}} = \frac{\sum y_i y_{i+1} - \frac{(\Sigma y_i)^2}{n}}{\sum y_i^2 - \frac{(\Sigma y_i)^2}{n}}$ This is the definition for serial correlation of order 1. This measure can be used to test the randomness of time series samples. However, due to the assumption that $y_{n+1} = y_1$ (circular definition) in the computation of r_1, r_1 may be quite insensitive in detecting the existence of serial correlation in the population when the sample is rather small and when the data are not detrended.

(**14.8**) $K = \frac{m^2}{s_y{}^2}$ This is the Von Neumann ratio—a ratio of mean-square-successive difference to the variance. Here, $m^2 = \Sigma(y_{i+1} - y_i)^2/(n-1)$ and $s_y{}^2 = (1/n)\Sigma(y_i - \bar{y})^2$. This ratio is used to test independence of time series samples. When a positive trend exists in the population, K will be very small. When a negative trend exists in the population, K will be very large. This statistic is more sensitive than r_1 in detecting the existence or absence of serial correlation.

(**14.9**) $K = \frac{n-1}{2} - (0.98)\sqrt{n+1}$ To test the equality between two population means with matched pairs by the sign test, the critical value for a two-sided alternative at $\alpha = 0.05$ can be obtained by this expression. H_0 is rejected when $S \leq K$, where S is the number of times the less frequent sign occurs. It is important to note that S has the binomial distribution with $p = 0.5$.

(**14.10**) $\chi^2 = \sum \frac{(o_i - e_i)^2}{e_i}$ and (**14.11**) $\chi^2 = \sum \frac{(|o_i - e_i| - \frac{1}{2})^2}{e_i}$

The ratio of the sum of squared differences between observed and expected frequencies to expected frequencies is used as the testing statistic for a number of chi-square tests. The number of degrees of freedom for χ^2 here varies from one case to another. In general, with tests on frequencies, $\delta = k - 1$, where k is the number of classes of the multinomial distribution. With tests of goodness-of-fit, $\delta = k - 1 - m$, where k is defined as previously and m is the number of parameters which are estimated by sample statistics for computing expected frequencies. With tests of independence, or contingency table tests and tests of homogeneity, $\delta = (R - 1)(C - 1)$, where R stands for the number of rows and C stands for the number of columns in the contingency table. To compute the chi-square value when $n \leq 50$ and when $\delta = 1$, a continuity correction factor of $\frac{1}{2}$ should be used.

(**14.12**) $\chi_1{}^2 = \frac{n(\text{AD} - \text{BC})^2}{m_1 m_2 m_2 m_4}$ and (**14.13**) $\chi_1{}^2 = \frac{n(\text{AD} - \text{BC} - \frac{1}{2}n)^2}{m_1 m_2 m_3 m_4}$

These are efficiency formulas for computing chi-square from a 2 $\times$ 2 contingency table, without (14.12) or with (14.13) a continuity correction. Here, n = sample size, A, B, C, and D are cell frequencies, and m_i are marginal frequencies for observed data.

$$(\mathbf{14.14})\ \chi_2^2 = \frac{n}{m_4}\left(\frac{A^2}{m_1} + \frac{B^2}{m_2} + \frac{C^2}{m_3}\right) + \frac{n}{m_5}\left(\frac{D^2}{m_1} + \frac{E^2}{m_2} + \frac{F^2}{m_3}\right) - n$$

This is the efficient formula to compute the chi-square value from observed frequencies directly for a 2 by 3 contingency table. In this expression, m_4 and m_5 are marginal frequencies for the two rows and m_1, m_2, and m_3 are the marginal frequencies for the three columns. A, · · · F are the cell frequencies and n, as before, is the sample size.

$$(\mathbf{14.15})\ \chi_\delta^2 = \sum \frac{o_i^2}{e_i} - n$$

This is the generalized chi-square expression for contingency tables of any dimension. The utility of this formula lies in the fact that we need not to compute the differences between observed and theoretical frequencies. In this formula n may stand for the sample size for contingency table tests or for the sum of sizes of the independent samples for tests of homogeneity.

$$(\mathbf{14.16})\ \chi_1^2 = \frac{(a_1b_2 - a_2b_1)^2 n}{n_1n_2m_1m_2}$$

This chi-square is used as the testing statistic for the identity between two population distributions with independent samples or matched pairs by the median test. Here a_i refers to the number of observations in the ith sample which are greater than the grand sample median, and $b_i = n_i - a_i$. Also, $n = n_1 + n_2$, and m_i refers to the row marginal frequencies in the 2 $\times$ 2 contingency table.

$$(\mathbf{14.17})\ \chi^2_{c-1} = \frac{(n-1)}{m_1m_2}\sum_{i=1}^{c}\frac{(na_i - n_im_1)^2}{nn_i}$$

The median test with c independent samples employs this statistic to test the identity among population distributions. Other symbols here represent quantities as in (14.16).

$$(\mathbf{14.18})\ \chi^2_{c-1} = \left[\frac{c-1}{ra(c-a)}\right]\frac{\sum^{c}(ca_i - ra)^2}{c}$$

The median test with c matched groups employs this statistic to test the identity among the c population distributions. Here r stands for the number of rows—experimental material; $a = c/2$ or $= (c+1)/2$, depending upon whether c is even or odd; a_i is the number of observations in sample i which exceed the grand sample median in value.

$$(\mathbf{14.19})\ E(R) = \frac{2n_1n_2}{n_1 + n_2} + 1 \quad (\mathbf{14.20})\ V(R) = \frac{2n_1n_2(2n_1n_2 - n_1 - n_2)}{(n_1+n_2)^2(n_1+n_2-1)} \quad (\mathbf{14.21})\ Z = \frac{R - E(R)}{\sigma_R}$$

R here is defined as the number of runs in the $n_1 + n_2$ observations in two independent random samples for the rank-sum test for the identity between population distributions. When $n_1 \geq n_2 \geq 20$, R is approximately normally distributed with expectation and variance as defined. Z is the usual normal critical ratio.

$$(\mathbf{14.22})\ T_{0.05} = \frac{n^2 - 7n + 10}{5} \quad \text{and} \quad (\mathbf{14.23})\ T_{0.01} = \frac{11n^2}{60} + 5 - 2n$$

To test $f(x) = g(y)$ against $F(x) > G(y)$ in the Wilcoxon test with matched samples, these expressions can be employed to obtain the approximate significance levels for the smaller rank T—the sum of the ranks with the less frequent sign. H_0 is rejected when $T \leq T_\alpha$.

$$(\mathbf{14.24})\ E(T) = \frac{n(n+1)}{4} \quad (\mathbf{14.25})\ V(T) = \frac{n(n+1)(2n+1)}{24} \quad (\mathbf{14.26})\ Z = \frac{T - E(T)}{\sigma_T}$$

The distribution of T, as defined in (14.23), is approximately normal with mean and variance as defined. For T, satisfactory result can be obtained with normal approximation even if the sample size is as small as ten pairs.

$$(\mathbf{14.27})\ H = \frac{12}{n(n+1)} \sum \frac{T_i^2}{n_i} - 3(n-1) \quad \text{and} \quad (\mathbf{14.28})\ H' = \frac{H}{\mathrm{C}}$$

This H is used to test the identity among several population distributions with independent samples. Ranks for observations in all samples are assigned in this case by considering them as a single series. Here, T_i is the sum of ranks in the ith sample, and n is the grand sample size, $n = \Sigma n_i$. When ties occur, tied observations are assigned mean ranks. When samples are rather small and many sets of ties appear, H' should be employed as the testing statistic. In the definition of H', the correction factor C is defined as follows:

$$\mathrm{C} = 1 - \frac{\sum^{k} (t_i^3 - T_i)}{n^3 - n}$$

where k is the number of tied sets and t_i is the number of tied observations in the ith set of ties. Remember that both H and H' are approximately distributed as chi-square with $\delta = c - 1$.

Problems

14.1 How are parametric and nonparametric statistics differentiated?

14.2 Are "distribution-free" and "nonparametric" identical concepts? If so, why? If not, why not?

14.3 What is the most important reason from a theoretical point of view for the use and development of nonparametric theory?

14.4 In what situations are the nonparametric methods necessary in the practice of statistical operations?

14.5 Why do we say that in most situations where both classical and order theories apply, the former is generally superior to the latter?

14.6 "Runs" in the different nonparametric tests are obtained in slightly different ways. Can you give this term a general definition of your own?

14.7 When n or n_i are large, most sampling distributions discussed in this chapter are asymptotically normally distributed. How then do nonparametric statistics differ from classical decision procedures?

14.8 Prove that, for a 2×2 contingency table, equation (14.12) holds.

14.9 Suppose that we are testing randomness against the alternative of a shifting level, using a sample of size 30. Furthermore, suppose that there are 10 runs above and below the sample median. Would this be cause for rejecting randomness at $\alpha = 0.01$?

14.10 A sample with 30 observations is listed in the order in which they were obtained:

80	34	10	20	40	64	50	47	44	25
86	66	49	53	65	77	28	28	55	74
81	90	67	53	38	71	72	73	75	78

In view of these data, can we conclude that the underlying population is in statistical control with a left-tail alternative at $\alpha = 0.01$ by the runs-up-and-down test?

14.11 If the runs-above-and-below-the-median test is applied to sample data in the preceding problem, what conclusion can you reach?

14.12 Take a sample of 28 two-digit numbers by any system you design from the Table of Random Digits, and test randomness at $\alpha = 0.01$ (a) by the runs-above-and-below-the-median test; (b) by the runs-up-and-down test.

14.13 A time series with ten terms yields the following sums: $\Sigma y_i = 50$, $\Sigma y_i y_{i+1} = 125$, and $\Sigma y_i^2 = 425$.
a. Determine the serial correlation coefficient of order 1, r_1.
b. Test if serial correlation exists in the population at $\alpha = 0.01$.

14.14 Suppose that a sample of time series consists of 30 observations from which we obtain $s_y^2 = 8$ and $m^2 = 5.508$. Do these results indicate independence among the successive terms in the sample by the Von Neumann ratio test?

14.15 Use data in Table 17.1 in Chapter 17 to test if the sample is a random sample by r_1 at $\alpha = 0.05$.

14.16 Use data in Table 17.1 again to test independence by the Von Neumann ratio.

14.17 Fifteen pairs of plots are planted with two varieties of soybeans and the observed yields are recorded in Table 14P-1. Does this information indicate equality between population means against a two-sided alternative at $\alpha = 0.05$ in terms of the sign test?

TABLE 14P-1

Pair	Variety I	Variety II
1	135	134
2	129	137
3	130	151
4	146	142
5	127	138
6	128	142
7	125	140
8	151	122
9	151	121
10	128	138
11	134	122
12	132	119
13	121	130
14	136	139
15	121	128

14.18 Suppose that normality holds for data in Table 14P-1, what conclusion can you reach by the t test? Comment.

14.19 In a paired feeding experiment, the gains in weight (in pounds) of hogs fed on two different diets, A and B, are as shown in Table 14P-2. Can either diet be considered as superior at $\alpha = 0.05$?

TABLE 14P-2

Pair	A	B
1	17	9
2	20	22
3	21	11
4	24	23
5	13	17
6	21	24
7	29	24
8	24	18
9	23	16
10	25	16

14.20 A motion study is designed to test the efficiency of two procedures, I and II, of assembling a certain mechanism. The workers are paired on the basis of intelligence quotients and number of years of education. Members of each pair are assigned by lot to one of the two procedures. Time in seconds for completing the assemblage is recorded in Table 14P-3.

TABLE 14P-3

Pair	I	II
1	44	41
2	40	44
3	40	29
4	31	27
5	44	30
6	41	36
7	26	25
8	26	30
9	32	34
10	38	23
11	37	29
12	29	25
13	29	29
14	20	20

a. Specify H_0 and H_1 for this test.
b. Apply the sign test to (a), indicating appropriate critical region.
c. Assuming normality holds, apply t test to (a).
d. Compare results obtained in (b) and (c).

14.21 Test $H_0: M = 70$ against $H_1: M \neq 70$ with the sample data in Problem 14.10. Suppose we wish to reject H_0 if there are 9 or less, or if there are 20 or more, plus signs, what is the level of significance for this test? Can H_0 be rejected at this α? What is the corresponding confidence interval for M?

14.22 Consider data for Variety I in Table 14P-1 as a random sample and test $M = 133$ against $M \neq 133$. What is the value of α if the decision rule is: Reject H_0 if there are 3 or less or if there are 9 or more observations above 133? Can H_0 be accepted at this α? Construct a greater than 95 percent confidence interval for M with this sample.

14.23 Out of 5000 portable radios examined, 200 had defective parts. Is this result in accord with the advertised standard that defectives represent 3 percent or less of total output in terms of a normal test? in terms of a chi-square test?

14.24 A department store employs 300 men and 400 women. In a given year 3100 days of absence were recorded for men and 4600 days of absence for women. Is the difference in rate of absenteeism between men and women significant at $\alpha = 0.05$ in terms of a chi-square test?

14.25 Suppose that a die is tossed 120 times with the following results:

Spots showing	1	2	3	4	5	6
Observed number	10	19	30	29	21	11

Is this result consistent with the hypothesis that the die is fair at $\alpha = 0.05$?

14.26 Suppose that 100 tosses of 10 coins in each toss yielded the following result:

No. of heads	0	1	2	3	4	5	6	7	8	9	10
Observed data	0	1	2	8	25	28	22	4	6	3	1

Would you conclude that the coins are fair?

14.27 Suppose that it is hypothesized that there are three times as many automobile accidents on Saturday and Sunday as on any other day of the week, what is the probability distribution of auto accidents throughout a week? From the record, 100 accidents are chosen independently of one another, and the distribution of accidents according to the days of the week is found to be

Mon.	Tue.	Wed.	Thu.	Fri.	Sat.	Sun.
5	9	10	8	11	30	27

Do the observed data contradict or confirm the hypothesis if $\alpha = 0.01$?

14.28 In Mendel's experiments with peas he observed 315 round and yellow, 108 round and green, 101 wrinkled and yellow, and 32 wrinkled and green. According to his theory of heredity, the numbers should be in the proportion 9:3:3:1. In view of his observed data, can we accept his theory if $\alpha = 0.01$?

14.29 A survey of 200 families with three children gave the following results:

Male births	0	1	2	3
Families	40	58	62	40

Are these data consistent with the hypothesis that male and female births are equally likely? (*Hint:* If this hypothesis is true, then the probabilities of 0, 1, 2, 3 male births are respectively $\frac{1}{8}$, $\frac{3}{8}$, $\frac{3}{8}$, $\frac{1}{8}$.)

14.30 Five coins are tossed 1000 times and, at each toss, the number of tails are counted. Results of this experiment are given below:

No. of tails	0	1	2	3	4	5
No. of tosses	38	144	342	287	164	25

Fit a binomial distribution to the above data and test the goodness-of-fit at $\alpha = 0.05$.

14.31 Fit a binomial distribution to data in problem 14.26 and test if the fit is good.

14.32 The number of automobile accidents on each day in a certain city is observed during a 50-day period and the following information is obtained:

No. of accidents	0	1	2	3	4	5
No. of days	25	15	8	0	1	1

Fit a Poisson distribution to these data and test if the fit is good.

14.33 In the accounting department of a bank 100 accounts are selected at random and examined for errors. Suppose the following result has been obtained:

No. of errors	0	1	2	3	4	5	6
No. of accounts	36	40	19	2	0	2	1

Does this information verify that the errors are distributed according to the Poisson probability law?

14.34 It is often stated that instructors tend to grade their students according to the "normal curve." Grades of 162 students in freshman mathematics classes during a certain year are found to be distributed as in Table 14P-4. Assuming that these grades constitute a random sample, can we conclude that this sample has been drawn from a population of grades which are normally distributed at $\alpha = 0.01$?

TABLE 14P-4

Class	Frequency
30 and under 40	5
40 and under 50	10
50 and under 60	28
60 and under 70	52
70 and under 80	31
80 and under 90	26
90 and under 100	10
Total	162

14.35 Four hundred independent random samples, each with 200 units, are drawn from a population and sample variances are computed. This experiment gives the following information. From these data can we conclude that the population is normally distributed?

Class	100–105	105–110	110–115	115–120	120–125	125–130	130–135	Total
f	21	56	61	125	70	47	20	400

(*Hint:* If the population is normally distributed, we would then expect the distribution of sample variance to be approximately normal with n as large as 200.)

14.36 A random sample consisting of 100 men and 100 women gives the following results in connection with their attitudes toward the New York City income tax introduced in 1966:

Sex	For	Against	Total
Men	25	75	100
Women	35	65	100
Total	60	140	200

In view of this information, can you conclude that New Yorkers' attitude toward the said tax is independent of their sex at $\alpha = 0.05$?

14.37 A market research firm desires to determine whether the inclusion of a coin in a questionnaire would increase the number of responses. Three hundred questionnaires, half with coins and half without, are sent to 300 persons selected at random; the following results are obtained:

	Responded	Not Responded	Total
Coin included	97	53	150
Coin not included	80	70	150
Total	177	123	300

Is there significant evidence to indicate that the inclusion of a coin is related to the number of responses to the questionnaire?

14.38 A sample of 100 small machine parts was taken from products produced by three different operators to determine whether they tended to produce the same distribution of output quality as defined by two quality classifications; the results follow:

	Operator		
Quality	*1*	*2*	*3*
Good	20	27	32
Average	8	6	7

How should you state the hypotheses? What is your conclusion?

14.39 A firm selling four products wishes to determine how sales of these products are distributed among four general classes of customers. A random sample of sales provides the following information:

	Product			
Customer Group	*1*	*2*	*3*	*4*
Professionals	85	23	56	36
Businessmen	153	44	128	75
Factory workers	128	26	101	45
Farmers	34	7	15	44

What conclusions can you draw?

14.40 For the purpose of comparing two manufacturing processes, 100 units are selected from outputs produced by each process and the following results are obtained:

Process	Nondefectives	Defectives
A	94	6
B	90	10

What are your conclusions? (*Hint:* This should be a test of homogeneity. Also, it is advisable to determine the chi-square value with the continuity correction factor in this case.)

14.41 Three random samples of students at a certain university are taken. The first contains 100 graduate students, the second 100 upperclassmen, and the third 100 lowerclassmen. Each is asked whether he considers the instruction he is receiving excellent, good, or average, and the following results are gathered:

Classification	Instruction		
	Excellent	*Good*	*Average*
Graduate	77	12	11
Upperclassmen	73	7	20
Lowerclassmen	85	10	5

What conclusions can you draw?

14.42 On a particular proposal of national importance, a random sample of 180 Democrats and a random sample of 150 Republicans yield the following information:

	In Favor	Opposed	Undecided
Democrats	89	65	26
Republicans	62	54	34

In view of the above data, can we conclude that there is no difference between the two parties insofar as this proposal is concerned at $\alpha = 0.05$? $\alpha = 0.01$?

14.43 A random sample of 10 units is selected from each of two suppliers of a certain electrical product. Results of resistance measured in ohms are given as follows:

A: 2.7 2.8 2.9 3.4 4.0 4.5 4.7 4.8 5.2 5.3
B: 3.5 3.6 3.7 3.8 3.9 4.0 4.2 4.3 4.4 4.9

a. Employ the median test to test $f(x) = g(y)$ against $f(x) = g(y + c)$ at $\alpha = 0.01$.
b. Assuming normality holds, employ the F test to test $\sigma_a^2 = \sigma_b^2$ against $\sigma_a^2 > \sigma_b^2$ at $\alpha = 0.01$.
c. Comment on results in (a) and (b).

14.44 Wire cable is being manufactured by two different processes. A sample is drawn at random from outputs produced by each process. Breaking strengths observed in pounds yield the following data:

Process I: 87 90 64 75 79 86 78
Process II: 69 70 66 68 76 80 67 72 74

a. Use the median test to test the identity between the two population distributions at $\alpha = 0.05$.
b. Use the t test, assuming that normality holds, to test $\mu_1 = \mu_2$ against $\mu_1 \neq \mu_2$ at $\alpha = 0.05$.
c. Comment on results obtained in (a) and (b).

14.45 Apply the median test to data in Table 14P-3 at $\alpha = 0.05$. Compare conclusion here with that obtained before by the sign test.

14.46 Three methods of making concrete are available. Eight sample blocks are made by each method. The compressive strength in pounds per square inch is measured and the following data result:

Method I: 147 140 149 146 152 143 150 155
Method II: 146 133 148 131 140 125 127 130
Method III: 158 150 160 165 158 171 140 165

Use the median test for independent samples to test the identity among the population distributions at $\alpha = 0.01$.

14.47 Measurements of height (in inches) of five adult males of each of four different nationalities are given below:

I: 61 65 62 67 64
II: 63 67 66 68 70
III: 60 69 68 71 72
IV: 64 68 70 74 71

Use the median test for independent samples to determine if the underlying populations are identical at $\alpha = 0.05$.

14.48 Suppose that in an experiment evaluating four different methods of teaching statistics, 10 matched groups of 4 students each had been used. The students were matched in terms of their intelligence quotients and freshman mathematics grades. At the end of the course, final grades were rated with respective row medians, and the over-all table of frequencies of plus and minus categories within columns is shown below:

	I	II	III	IV	
+	9	7	2	2	$20 = ra$
−	1	3	8	8	$20 = n - ra$

Are the above results consistent with the hypothesis that the population distributions (populations being grades of the four methods of teaching) are identical at $\alpha = 0.01$?

14.49 Apply the median test for matched groups to data in Problem 13.22. Compare your result with that obtained before.

14.50 Apply the Wald-Wolfowitz test to test the identity between population distributions with data in Problem 14.44 at $\alpha = 0.05$.

14.51 Suppose that a manufacturing process is turning out steel pipes whose outside diameters are to be measured. Of the first 50 measured, there is a total of 20 runs above and below the median of this group. Use the Wald-Wolfowitz test to test the hypothesis that the machinery is turning out pipes whose outside diameters vary randomly at 0.05.

14.52 Weekly family income sampled in two different cities yielded the following information:

City A: 155, 169, 148, 200, 165, 100, 75
City B: 205, 145, 195, 198, 210, 127, 167

a. Apply the Wald-Wolfowitz test to test the identity between the underlying population distributions at $\alpha = 0.05$.
b. Apply the t test to test the equality between the two population means, assuming normality holds, at $\alpha = 0.05$.
c. Comment on results in (a) and (b).

14.53 Suppose that data in Problem 14.10 constitute a time series sample, employ Wald-Wolfowitz test to test that there exists no serial correlation in the population.

14.54 Employ the Wilcoxon T test to test $f(x) = g(y)$ against $F(x) > G(x)$ with data in Problem 14.20.

14.55 The following data are results of planting corn with two types of fertilizer on 8 pairs of plots:

Pair	1	2	3	4	5	6	7	8
A	164	165	158	149	170	155	154	162
B	150	143	152	160	151	145	148	134

a. What are the populations whose distributions are generalized if the T test is applied?
b. What conclusion can you draw at $\alpha = 0.05$?

14.56 In a paired experiment, speed (in number of items packed per hour) by two different systems are given in Table 14P-5. Can we reject the null hypothesis $f(x) = g(y)$ at $\alpha = 0.05$? at $\alpha = 0.01$?

TABLE 14P-5

Pair	System I	System II
1	10	7
2	11	6
3	9	12
4	8	9
5	15	13
6	12	7
7	11	13
8	13	9
9	15	8
10	10	11
11	13	9
12	12	8
13	14	13
14	13	10
15	14	8
16	12	7
17	14	7
18	12	12

(*Note:* When $d_i = 0$, disregard. When $d_i = d_{i+1}$, and so on, assign a mean rank to the tied differences.)

14.57 Data in Table 14P-6 represent differences in heights between husband and wife of three different nationalities. Do they come from identical continuous populations at $\alpha = 0.01$ in accordance with the H test?

14.58 To determine whether the distributions of three makes of gasoline are identically distributed in terms of number of miles travelled per gallon, the following data are obtained:

I:	7.7	6.8	8.5	9.0	11.0	8.9	10.1	9.5
II:	8.7	9.2	10.5	9.8	11.4	12.6	10.9	9.9
III:	10.7	9.8	12.5	11.4	10.8	13.5	14.1	12.7

What conclusion can be drawn in accordance with the H test?

TABLE 14P-6

	French	Japanese	American
	2.1 (13.5)	3.4 (26)	5.1 (33)
	0.8 (2)	1.5 (6.5)	2.7 (22)
	2.5 (18)	1.7 (9)	4.2 (32)
	0.7 (1)	1.9 (11.5)	2.7 (22)
	2.4 (16)	2.5 (18)	4.1 (31)
	2.6 (20)	3.1 (25)	2.9 (24)
	1.1 (4)	1.5 (6.5)	2.7 (22)
	1.3 (5)	1.9 (11.5)	2.3 (15)
	1.8 (10)	1.6 (8)	2.5 (18)
	0.9 (3)	3.6 (28)	3.5 (27)
			2.1 (13.5)
			3.7 (29.5)
			3.7 (29.5)
n_i	10	10	13
T_i	92.5	150	318

15

INDEX NUMBERS

15.1

Measurement of Aggregates

One of the most important problems in studying economics and business is how to measure the quantity of some heterogeneous aggregate. The aggregate may be one of physical quantity, such as stocks in the balance sheet or flows in the income statement. The aggregate may also be a list of prices, such as prices paid for the purchases of various types of inputs by a firm or as prices received by a department store from its sales. In each case, the problem of measurement is to derive a single figure that is descriptive of the volume of, or change in, a given aggregate through time or from place to place. The statistical device for such measurement is called index numbers. *Index numbers,* in effect, relate a variable or variables in a given period to the same variable or variables in another period, called the *base period.* An index, the simplified name for index numbers, which is computed from a single variable, is called a *simple index,* whereas an index which is constructed from a group of variables is considered a *composite index.* As will be shown immediately, a simple index can be easily constructed and its meaning is readily apparent, but serious problems arise when we try to combine many variables in some meaningful way to present a composite index.

15.2 An Illustration of a Simple Index

Even though the main concern of this chapter is to present the techniques and problems of constructing composite indices, we shall first give an illustration of a simple index in order to bring out the basic nature and function of index numbers. Let us consider the hypothetical production data in Table 15.1. Index A is computed by selecting 1964 as the base period. The index numbers are then found by dividing each year's production by the production in 1964 and multiplying by 100. Thus, the index numbers are in the form of percentages. The index number for each year is interpreted in terms of the base period. For instance, the index number for 1968 means that production in 1968 was 133 percent of production in 1964. Index B is computed by using the average production of 1958 and 1959 (that is, 10.5 thousand pounds) as the base. This base is divided into each year's production and the result multiplied by 100 to form Index B.

TABLE 15.1 ILLUSTRATION OF SIMPLE PRODUCTION INDEX

Year	Production (thousand pounds)	Index A (1964 = 100)	Index B (1958–1959 = 100)
1958	8.5	63	81
1959	12.5	93	119
1960	9.4	70	90
1961	10.7	79	102
1962	13.6	101	130
1963	15.3	113	146
1964	13.5	100	129
1965	12.8	95	122
1966	14.7	109	140
1967	16.7	124	159
1968	18.0	133	171

Clearly, the main function of a simple index number is to transform the absolute quantities of a variable into relative numbers so that comparisons of the changes in the variable through time can be readily made.

15.3 Types of Composite Indices

There are many theoretical and practical considerations about the construction of composite indices. Before we take up these problems, however, we shall concentrate on the technical aspects of constructing composite indices.

Composite indices are of many types and may be constructed in many ways. We shall be unable to cover the whole variety of indices in a single chapter. Instead, we shall confine our discussion to some of the more frequently used indices in economics and business. These indices, in the order of their discussion here, are classified as follows:

1. Simple aggregative indices
   a. Simple aggregative price index
   b. Simple aggregative quantity index

2. Simple average of relatives indices
   a. Price index of simple average of relatives
   b. Quantity index of simple average of relatives

3. Weighted aggregative indices
   a. Weighted aggregative price index
   b. Weighted aggregative quantity index

4. Weighted average of relatives indices
   a. Weighted price index of average of relatives
   b. Weighted quantity index of average of relatives

5. Special indices
   a. Value index
   b. Productivity index

15.4 Symbols and Data for Constructing Indices

In order to avoid repetition, we shall define here all the symbols to be used in our discussion of indices:

p = price of a single commodity
q = quantity of a single commodity
p_0 = base-period price of a commodity
q_0 = base-period quantity of a commodity
p_n = given-period price of a commodity. Clearly, n refers to periods 1, 2, 3, etc.
q_n = given-period quantity of a commodity
P = a price index
Q = a quantity index
P_b = a price index derived by using base-period quantity as weights

P_n = a price index derived by using given base-period quantity as weights
Q_b = a quantity index derived by using base-period price as weights
Q_n = a quantity index derived by using given base-period price as weights
V = a value index
E = a productivity index

To simplify our numerical illustrations, we shall use the hypothetical data of Table 15.2. We suppose that a manufacturing firm, established in 1966, produces three types of products, A, B, and C. Price is defined as the annual average selling price, and production refers to annual total production. The firm wishes to measure the changes in its selling prices and in the physical volume of production in the aggregate from year to year.

TABLE 15.2 ILLUSTRATIVE PRICE AND PRODUCTION DATA, 1966–1968

		Price per unit			Quantity		
Item	Unit	*1966*	*1967*	*1968*	*1966*	*1967*	*1968*
A	ounce	$ 1.00	$ 1.25	$ 1.50	10,000	12,500	13,000
B	ton	10.00	11.75	13.50	1,000	1,100	1,250
C	pound	4.00	5.00	4.50	500	500	400

15.5 Simple Aggregative Indices

The computation of a price index by the simple aggregative method is a very simple matter. We first add the various prices for each time unit to obtain Σp_0 and Σp_n. Next, the total of each given period is divided by the base-period total. Finally, the results are multiplied by 100 in order to transform the proportions into percentages. These computations are done in Table 15.3. The general formula is

$$P = \frac{\Sigma p_n}{\Sigma p_0}(100) \tag{15.1}$$

TABLE 15.3 COMPUTATION OF SIMPLE AGGREGATIVE PRICE INDEX

Item	1966 p_0	1967 p_1	1968 p_2
A	$ 1.00	$ 1.25	$ 1.50
B	10.00	11.75	13.50
C	4.00	5.00	4.50
Total	$15.00	$18.00	$19.50
Index number	100.00	120.00	130.00

From formula (15.1) it can be seen that the simple aggregative price index attempts, in our example, to ascertain the total sales receipts in each year under the assumption of selling one unit of each commodity and to express this total as a percentage of the base-year revenue. As such, the simple aggregative assigns equal importance to the absolute change of each price. This is where the main defect of this method rests, since it permits a commodity with a high price to dominate the index. As our illustrative data stand now, the price of B exerts much more influence than the price of C, which, in turn, dominates the price of A in the index numbers. These influences, as can be readily seen, would be reversed if all the prices were quoted in the same units, say, ounces. Thus, the unit by which each price happens to be quoted creeps into the simple aggregate of prices as a concealed weight that is often of no economic significance. This illogical concealed weight limits the practical usefulness of the simple aggregative price index.

The formula for the simple aggregative quantity index is

$$Q = \frac{\Sigma q_n}{\Sigma q_0} (100) \tag{15.2}$$

This formula, apparently, cannot be used for an aggregate in which the items are expressed in different units, since it would be meaningless when we add tons, pounds, and ounces together. If we used this formula to form an index for a group of, say, consumers' goods quoted in the same units, we would be comparing the cost in a given year with the cost in the base year of purchasing the goods actually bought in each year as if the price of each item each year were $1.00 per unit. Evidently, this is a highly unrealistic assumption. Consequently, this formula is seldom employed to measure changes in quantity.

15.6

Simple Average of Relatives Indices

As the name of this type of index implies, it consists of averaging the price or quantity relatives. To compute a simple average of relatives price index, as illustrated by Table 15.4, we follow these steps: (1) We obtain the price relative by dividing the price of each item in a given period, p_n, by its base-period price, p_0. The result is expressed as a percentage. (2) We obtain the sums of the relatives of the years and divide each by the number of items in the aggregate. The simple average of relatives is, in effect, an arithmetic mean of relatives. This method is summarized by the formula below:

$$P = \frac{\sum \left(\frac{p_n}{p_0}\right) 100}{n} \tag{15.3}$$

The procedure for computing a simple average of relatives quantity index is

TABLE 15.4 COMPUTATION OF SIMPLE AVERAGE OF RELATIVES PRICE INDEX (1966 = 100)

Item	1966 $\left(\frac{p_0}{p_0}\right)100$	1967 $\left(\frac{p_1}{p_0}\right)100$	1968 $\left(\frac{p_2}{p_0}\right)100$
A	100.0	125.0	150.0
B	100.0	117.5	135.0
C	100.0	125.0	112.5
Total	300.0	367.5	397.5
Index number	100.0	122.5	132.5

the same as that for the corresponding price index. This is illustrated by Table 15.5. The analogous quantity index equation is

$$Q = \frac{\sum \left(\frac{q_n}{q_0}\right) 100}{n} \tag{15.4}$$

From the results obtained in Tables 15.4 and 15.5 by the method of simple averages of relatives, it may be said that the prices have increased, on the average, by 32.5 percent and that the quantities have increased, on the average, by 12 percent over the three-year period.

What can be said about the simple average of relatives indices as compared with the simple aggregative indices?

First, it may be noted that the simple average of price relatives has avoided the difficulty encountered in the simple aggregative price index. That is, the former is no longer influenced by the units in which prices are quoted or by the absolute level of individual prices. Relatives are pure numbers and are therefore divorced from the original units. Consequently, index numbers computed by the relative method would be the same regardless of the way in which the prices are quoted. This simple average of price relatives is said to meet what is called the *units test*. Also, it may be noted that the simple average of relatives method can now be used to compute a quantity index for an aggregate of items that are not quoted in the same units.

TABLE 15.5 COMPUTATION OF SIMPLE AVERAGE OF RELATIVES QUANTITY INDEX (1966 = 100)

Item	1966 $\left(\frac{q_0}{q_0}\right)100$	1967 $\left(\frac{q_1}{q_0}\right)100$	1968 $\left(\frac{q_2}{q_0}\right)100$
A	100	125	130
B	100	110	125
C	100	100	80
Total	300	335	335
Index number	100	112	112

Despite these merits, the simple average of relatives is still a very unsatisfactory method, since it presents a new difficulty and a new defect. The new difficulty is concerned with the selection of an appropriate average. The use of the arithmetic average is considered as questionable sometimes because it has an upward bias. We shall comment on this point in section 15.14. Meanwhile, it suffices to say here that this difficulty is not a very serious one. The new defect is that the relatives are assumed to have equal importance. This is again a kind of concealed weighting system that is highly objectionable, since economically some relatives are more important than others. It is interesting to note that, in our example, the quantity relative for C exerts an influence in the 1968 quantity index way out of line with its practical importance. This is so because from the quantities and units of A, B, and C, we can see that C is the least important item in the group; yet we assign to the quantity relative the same weight as those for A and B. As a consequence, the rather large absolute increase in quantities of A and B have been unduly offset by the rather small absolute (but large relative) decrease of the quantity of C.

In concluding this section, it should be noted that the main objection to both simple aggregative and simple average of relatives indices is to the concealed weights that are usually of no economic importance. Thus, improvements of an index rest with the introduction of appropriate weighting systems for its construction.

15.7 Weighted Aggregative Indices

In a later section we shall give a rather detailed discussion on weighting to suit the purpose of a given index. Here, we may merely note that the most frequently used weights for aggregative price indices are base-year quantities of commodities and that the usual weights for aggregative quantity indices are either base-year or given-year prices of commodities.

The weighted aggregative price index with base-year quantity as weights is called the *Laspeyres index* and is given by the formula

$$P_b = \frac{\Sigma p_n q_0}{\Sigma p_0 q_0}(100) \qquad \textbf{(15.5)}$$

The application of this formula, as shown by Table 15.6, involves the following three simple steps:

(1) Multiply the price of each item in each year by the base-year quantity of that item to obtain p_0q_0 for the base year and p_nq_0 for each given year.

(2) Obtain the sums of the products computed in step (1).

(3) Divide the total of each given year by the total of the base and multiply the results by 100.

In Table 15.6, the value of 121.6 for the Laspeyres index of 1967 may be interpreted in this way: "The list of products sold in 1966 would yield 21.6 percent more at 1967 selling prices than it actually did yield in 1966." In other words, according to this index, selling prices on the average rose 21.6 percent from 1966 to 1967. If the Laspeyres formula is used to compute a consumers' price index, the result would measure the difference between the theoretical cost in a given year and the actual cost in the base year of maintaining the standard of living of the base year. In general, the Laspeyres index attempts to answer the question, "What is the change in aggregate value of the base period's list of goods when valued at given-period prices?"

The weighted aggregative price index with given-period quantity as weights is sometimes referred to as the *Paasche index*. It is defined by the formula

$$P_n = \frac{\Sigma p_n q_n}{\Sigma p_0 q_n}(100) \tag{15.6}$$

Computation of the Paasche indices is illustrated by Table 15.7. The index value of 140.9 for 1968 should now be interpreted thus: "The list of products sold in 1968 yielded 40.9 percent more than the same list of products would have yielded at 1966 prices." The Paasche formula, when used to calculate a consumers' price index, compares the actual cost in the given period with the theoretical cost in the base period of keeping the standard of living in the given period. In general, this formula answers the question, "What would be the value of the given-period list of goods when valued at base-period prices?"

A weighted aggregative quantity index is the counterpart of the analogous aggregative price index. The weights to be used are prices. The Laspeyres and Paasche aggregative quantity indices are given by the following formulas:

$$Q_b = \frac{\Sigma p_0 q_n}{\Sigma p_0 q_0}(100) \tag{15.7}$$

$$Q_n = \frac{\Sigma p_n q_n}{\Sigma p_n q_0}(100) \tag{15.8}$$

Note that all the sums of the products for these two formulas have already been obtained in Tables 15.6 and 15.7. Thus, by referring to these tables, the aggregative quantity index for 1968, by using base-year prices as weights, we have

$$Q_b = \frac{\Sigma p_0 q_2}{\Sigma p_0 q_0}(100) = \frac{27{,}100}{22{,}000}(100) = 123.2$$

The quantity index, by using given-period price as weights for 1968, is

$$Q_n = \frac{\Sigma p_2 q_2}{\Sigma p_2 q_0}(100) = \frac{38{,}175}{30{,}750}(100) = 124.1$$

The first result means that, at 1966 prices, the volumes of output increased 23.3 percent between 1966 and 1968. The second result suggests that, at 1968 prices, the volume of output rose 24.1 percent between 1966 and 1968. Thus, generally

TABLE 15.6 CONSTRUCTION OF AN AGGREGATIVE PRICE INDEX, USING BASE-YEAR QUANTITY WEIGHTS

	Quantity	Price			Theoretical Value of 1958 Quantities at Given-Year Prices		
	1966	1966	1967	1968	1966	1967	1968
Item	q_0	p_0	p_1	p_2	p_0q_0	p_1q_0	p_2q_0
A	10,000	1.00	1.25	1.50	10,000	12,500	15,000
B	1,000	10.00	11.75	13.50	10,000	11,750	13,500
C	500	4.00	5.00	4.50	2,000	2,500	2,250
Total					22,000	26,750	30,750
Index number					100.0	121.6	139.8

TABLE 15.7 CONSTRUCTION OF AN AGGREGATIVE PRICE INDEX, USING GIVEN-YEAR QUANTITY WEIGHTS

	Quantity			Price			Actual Value in Given Year			Theoretical Value of Given-Year Quantities at 1958 Prices		
	1966	1967	1968	1966	1967	1968	1966	1967	1968	1966	1967	1968
Item	q_0	q_1	q_2	p_0	p_1	p_2	p_0q_0	p_1q_1	p_2q_2	p_0q_0	p_0q_1	p_0q_2
A	10,000	12,500	13,000	1.00	1.25	1.50	10,000	15,625	19,500	10,000	12,500	13,000
B	1,000	1,100	1,250	10.00	11.75	13.50	10,000	12,925	16,875	10,000	11,000	12,500
C	500	500	400	4.00	5.00	4.50	2,000	2,500	1,800	2,000	2,000	1,600
Total							22,000	31,050	38,175	22,000	25,500	27,100
Index number										100.0	121.8	140.9

speaking, a weighted aggregative quantity index answers the question, "If we buy (or sell) varying quantities of the same items in each of the two periods, but at the same price, how much would be spent (or received) in the given period relative to the base period?"

An interesting observation may now be made. The Paasche and Laspeyres formulas employed to compute either the price or the quantity index usually yield different results. This is, of course, due to the difference in the weights introduced. The difference, however, is not fortuitous. It does not make any particular sense to ask which formula is accurate or better. Each of them is meaningful in the sense that it has a simple and precise physical interpretation. In a way, we can actually consider the difference as a meaningful range of values. If, for instance, the price index computed by one method is 110 and by another method is 130, we may then say that the price level has changed from 100 to between 110 and 130. This statement gives us useful, though not completely precise, information: the price level has not decreased or increased more than 30 percent. It should be observed that this range becomes narrower, and therefore more precise, if the difference between base-year quantity (or price) and given-year quantity (or price) is smaller. During a short interval of time, say, from two to five years, neither prices nor quantities may change by large amounts and, as a result, it does not make much difference in theory whether the Laspeyres or the Paasche index is constructed. (In practice, however, there is a great deal of difference between these two formulas. For example, the use of the Paasche formula to construct a consumer price index requires an annual survey of expenditure patterns to determine current weights.) It should also be pointed out that the problem of approximation exists in all types of measurement. If the ranges are wider in the case of economic and business aggregates, it is not a difference of kind but is inherent in the nature of the problem.

15.8

Weighted Average of Relatives Indices

Weighted average of relatives indices can be computed in a way similar to the calculation of simple average of relatives, except that proper weights are introduced. Weights used here are dollar values of items in the aggregate. Thus, if base-year values are used as weights, the weights are p_0q_0. If given-year values are employed, the weights are p_nq_n. If theoretical values are used as weights, the weights are p_nq_0 or p_0q_n. Values rather than quantities or prices are used in order to produce weighted relatives that are all in the same units. In the construction of weighted average of price relatives, for instance, if quantities are employed as weights, the product of relatives times the quantities quoted in different units would yield weighted price relatives in original units, and these could not be added together.

As is true of the weighted averages, the weighted average of relatives is computed by multiplying each relative by its weight and dividing the sum of the products by the sum of the weights.

The Laspeyres averages of relatives indices are

$$P_b = \frac{\sum (p_0 q_0) \frac{p_n}{p_0} (100)}{\Sigma p_0 q_0} \tag{15.9}$$

$$Q_b = \frac{\sum (p_0 q_0) \frac{q_n}{q_0} (100)}{\Sigma p_0 q_0} \tag{15.10}$$

The Paasche average of relatives indices are

$$P_n = \frac{\sum (p_0 q_n) \frac{p_n}{p_0} (100)}{\Sigma p_0 q_n} \tag{15.11}$$

$$Q_n = \frac{\sum (p_n q_0) \frac{q_n}{q_0} (100)}{\Sigma p_n q_0} \tag{15.12}$$

Note that (15.10) is equivalent to Laspeyres aggregative quantity index (15.7), (15.11) to (15.6), and (15.12) to (15.8).

Since the procedures of constructing the four indices are similar, only one example is furnished here. A price index is constructed in Table 15.8 by using formula (15.9). Note that the index numbers obtained by this formula are identical with those obtained by formula (15.7). This should not be a surprising result. It can be readily seen that the weighted aggregative formulas and their analogous weighted average of relatives formulas are algebraically equivalent. For instance,

$$\frac{\sum (p_0 q_0) \frac{p_n}{p_0} (100)}{\Sigma p_0 q_0} = \frac{\Sigma p_n q_0}{\Sigma p_0 q_0} (100)$$

and so forth.

TABLE 15.8 COMPUTATION OF AVERAGE OF RELATIVES PRICE INDEX WITH BASE-YEAR VALUE AS WEIGHTS

	Weights	Price Relatives		Weighted Price Relatives	
	1966	1967	1968	1967	1968
Item	$p_0 q_0$	$\frac{p_1}{p_0}(100)$	$\frac{p_2}{p_0}(100)$	$p_0 q_0 \frac{p_1}{p_0}(100)$	$p_0 q_0 \frac{p_2}{p_0}(100)$
A	10,000	125.0	150.0	1,250,000	1,500,000
B	10,000	117.5	135.0	1,175,000	1,350,000
C	2,000	125.0	112.5	250,000	225,000
Total	22,000			2,675,000	3,075,000
Index number	100.0			121.6	139.8

However, several distinct advantages of weighted average of relatives indices over weighted aggregative indices must be observed:

(1) The price or quantity relatives for each single item in the aggregate are, in effect, themselves a simple index that often yields valuable information for analysis.

(2) When a new commodity is introduced to replace one formerly used, the relative for the new item may be spliced to the relative for the old one, using the former value weights.

(3) When an index is computed by selecting one item from each of the many subgroups of items, the values of each subgroup may be used as weights. Then only the method of weighted average of relatives is appropriate.

(4) When different index numbers are constructed by the average of relatives method, all of which have the same base, they can be combined to form a new index.

(5) When index numbers are to be constructed with data that are not analogous as to prices and quantities—ratios of profits to net worth, to median income, to cost of production; or ratios of defective items to output, to name a few—they can be made comparable by being expressed as percentages of some base.

15.9 Value Indices: Consistency between Price and Quantity Indices

The value of a single commodity is the product of its price and its quantity, that is, $v = pq$. Analogously, the value of an aggregate of commodities is the sum of individual values of the commodities, that is, $\Sigma v = \Sigma pq$. The change in the value of an aggregate of values is measured by value index, which is defined as

$$V = \frac{\Sigma p_n q_n}{\Sigma p_0 q_0}(100) \tag{15.13}$$

In this formula, both given-period price and quantity are variables in the numerator. It is not necessary to introduce any special weights; they are inherent in the value figures.

The value index is not in wide use, although, because of the unsatisfactory nature of price and quantity indices, it has been occasionally suggested that they be replaced by the value index. This temptation, however, must be resisted, since the concepts of price level and quantity level answer questions that cannot be answered by the value level. Furthermore, an aggregate of values may be viewed as the product of a price level and a quantity level. The division of an aggregate of values into its price and quantity factors may be arbitrary, but this arbitrariness need not create confusion of thought as long as our concepts of the two factors are consistent. The test of consistency is that the product of the price and quantity indices must produce the value index. These depend on appropriate weighting for the two indices. Following the notation used previously, we have two actual dollar values: $\Sigma p_0 q_0$ and $\Sigma p_n q_n$; and two theoretical values: $\Sigma p_n q_0$ and $\Sigma p_0 q_n$. With these quantities, four possible indices can be constructed:

$$P_b = \frac{\Sigma p_n q_0}{\Sigma p_0 q_0}$$

$$P_n = \frac{\Sigma p_n q_n}{\Sigma p_0 q_n}$$

$$Q_b = \frac{\Sigma p_0 q_n}{\Sigma p_0 q_0}$$

$$Q_n = \frac{\Sigma p_n q_n}{\Sigma p_n q_0}$$

It can be readily seen that P_b and Q_n are consistent indices, since

$$P_b Q_n = \frac{\Sigma p_n q_0}{\Sigma p_0 q_0} \times \frac{\Sigma p_n q_n}{\Sigma p_n q_0} = \frac{\Sigma p_n q_n}{\Sigma p_0 q_0} = V$$

the value index measures the change in actual values between the base and the given periods. Similarly, P_n and Q_b are consistent measures. However, P_b and Q_b, or P_n and Q_n are not consistent. Thus a rule can be derived from these demonstrations: If the price index is constructed with base-period quantity weights, the quantity index must be constructed with given-period price weights, and vice versa, in order to make the price and quantity levels consistent.

15.10 Productivity Indices

Productivity means efficiency in production. It is measured in terms of the ratio of output to inputs. If this ratio rises, that is, more units of output are produced with the same units of inputs, productivity increases. The measurement of changes in output is a simple matter. If only one product is involved, changes in output are merely the changes in the number of units produced. If an aggregate of products is under consideration, changes in outputs can be readily measured by a production index. The measurement of changes in inputs, however, presents very complicated problems: inputs are of great variety—different kinds of labor, various types of raw materials, investments in machines and equipment, management skill, and so on. Possibly an index of some sort could be constructed to measure the changes in aggregate factors of production, but appropriate weighting for such an index is extremely difficult, even physically impossible in some cases. In practice, therefore, a productivity index is usually constructed on the basis of a single input that is judged as the most important factor. The input selected is often labor, since, on the average, the wage bill consists of about two thirds of the total costs of production in many types of operations. In addition, labor data are more readily available, and labor units—usually man-hours—can be defined and interpreted more precisely than can other kinds of input data.

Labor is defined as a ratio—man-hours per unit of output. Units of output per man-hour define a measure of productivity. A productivity index, however, can be constructed by using either ratio. The construction of a productivity index, using man-hours per unit of output and base-year quantity as weights, is illustrated in Table 15.9. The formula used is

$$E_b = \frac{\Sigma r_n q_0}{\Sigma r_0 q_0}(100) \tag{15.14}$$

where r_0 and r_n refer to man-hours per unit of output in the base and given periods respectively.

This index, by assuming that output remains constant, measures changes in man-hours per unit of output. Thus, the result of our example means that man-hours per unit of output have decreased by 14 percent between the two periods.

It is interesting to note that the ratio of input to output is the reciprocal of the ratio of output to input: $\frac{\text{input}}{\text{output}} = \frac{1}{\text{output/input}}$. It follows, therefore, that the result of taking the reciprocal of the previously computed index—that is, $1/0.86 = 1.16$ or 116 percent—is an index that measures the changes in output per man-hour.

What should also be noted is that the index of labor per unit of output, or of output per unit of labor, must not be considered as merely a labor productivity index: labor efficiency also depends upon the quality of investment and management. In other words, changes in labor required to produce a given quantity of output are often the joint effects of all factors of production.

TABLE 15.9 CONSTRUCTION OF A PRODUCTIVITY INDEX, USING BASE-YEAR QUANTITY WEIGHTS

	Units Produced	Man-hours per Unit		Man-hours Required to Produce Base-Year Quantity	
		1966	1968	1966	1968
Items	q_0	r_0	r_n	$r_0 q_0$	$r_n q_0$
A	10,000	$\frac{1}{2}$	$\frac{2}{5}$	5,000	4,000
B	1,000	5	$\frac{9}{2}$	5,000	4,500
C	500	$\frac{3}{2}$	$\frac{3}{2}$	750	750
Total				10,750	9,250
Index number				100	86

15.11

Special Topics

In this section we shall introduce three special topics: (1) changing the base of an index, (2) splicing two overlapping indices, and (3) chain index procedures.

Sometimes we wish to change the base of an index either to make the base more recent or to make two indices with different bases comparable. To change the base of an index is a very simple matter. For instance, if an index was constructed with 1945 as the base and we wish to change the base to 1960, we merely divide each index number in the series by the index value of 1960; a new index with 1960 as the base would then result.

The weights of an index may become out of date and we may wish to use new weights to construct the index. Thus two indices result. However, we wish to make these two indices continuous. This calls for *splicing* the overlapping indices. The procedure of splicing is illustrated by Table 15.10, which contains two indices, the first with 1963 as the base and the second with 1967 as base.

As shown by the last two columns of Table 15.10, two overlapping indices can be spliced in two ways: either by making them continuous with the old index or by making them continuous with the new. If we wish to derive an index that is continuous with the old, we splice at the year that is the base of the new index—1967 in our illustration. In this year the ratio of the old to the new index is 115 to 100 and therefore this relationship prevails for years that follow the base year of the new index as

$$1.15:1.00 = x:1.10$$
$$x = 1.265$$

That is, to change the new index to the base of the old, we must multiply the future values of the new index by 1.15.

To make the old index continuous with the new is the same as to change the original base of the old index to the base of the new index. Thus, in our example, the base of the first index must be changed from 1963 to 1967. This is done by dividing each value of the first index by the 1967 value, which is 1.15.

TABLE 15.10 PROCEDURE OF SPLICING TWO INDICES

Year	First Index	Second Index	Spliced Indices *1963 = 100*	Spliced Indices *1967 = 100*
1963	100		100	87
1964	95		95	83
1965	101		101	88
1966	110		110	96
1967	115	100	115	100
1968		110	126.5	110

New commodities are introduced almost continuously into markets; it is therefore desirable in some respects to revise the list of items and the system of weights from time to time. For this purpose, we employ what is called the *linking* procedure, in which an index is constructed with the immediately preceding period as a base. Such an index is called the *link index*. The link index numbers, if we wish, may be chained back to a common base period by a process of multiplication. For instance, if we consider 1963 as the base period of an index, then 1964 is period 1, 1965 is period 2, and so on. Naturally, the index for 1963 is 100. The index for 1964 is computed in the usual manner, by either the Laspeyres or the Paasche formula, usually the former in the relative form. Assume the link index for 1964 is 108. The link index for 1965 is then computed by using 1965 as the base. Assume the resulting value is 110. Now, if we wish to chain the 1965 link relative back to 1963 as the base, we observe that the 1965 index is 110 percent of 1964, which, in turn, is 108 percent of 1963. Therefore, the 1965 index is 110 percent of 108 percent of 1963, the base period. That is, the index for 1965 with 1963 as the base is 118.8 (= 1.10 × 108).

From the previous illustration, a general procedure of chaining emerges. A chain index number, $I_{n:o}$, that is, the index for period n with period o as base, is obtained by calculating first the link index for period n with the preceding period, $n - 1$, as base, $I_{n:n-1}$, which is then multiplied by the chain index of the previous period, $I_{n-1:o}$. The general expression may be written as

$$I_{n:o} = (I_{n:n-1})(I_{n-1:o}) \tag{15.15}$$

Although the link-chain procedure is useful in that it permits changes in the composition of the index from period to period, its utility should not be exaggerated. As a matter of fact, strict comparability is confined to chain index numbers that immediately follow the fixed base. When outmoded commodities are continuously replaced by new ones, the meaning of the chain index becomes increasingly doubtful through time—we may not be able then to describe just what the index measures.

15.12 Important Current Indices

The majority of published indices are price indices. The Wholesale Price Index and the Consumer Price Index constructed by the United States Bureau of Labor Statistics are by far the most important. Important quantity indices currently available are the Federal Reserve Board's Index of Industrial Production and the United Nation's World Index of Industrial Production. We shall present a brief description of these four indices in this section.

THE BLS WHOLESALE PRICE INDEX

The BLS Wholesale Price Index was first calculated in 1920 but estimates made have carried it back to 1890. It has the main purpose of showing the general movements of prices at primary market levels. As such, the price data used to construct this index are drawn from those of sales in large lots in the primary markets—that is, prices prevailing at the first important commercial transaction for each commodity. Most of the prices included in the index are selling prices of representative manufacturers or producers as well as the prices quoted on organized exchanges or markets.

This index is based on a large sample of about 2000 commodity prices selected from 15 major groups and 88 subgroups of commodities classified by product. The purpose of such an extremely large sample is to provide sufficient data to compute price indices for commodity subgroups, such as indices of prices of wholesale processed food, textile products and apparel prices, metal and metal product prices.

The Wholesale Price Index is constructed by a modified version of the Laspeyres formula, with 1957–1959 as the base period for prices. Weights employed are the total transactions as reported in the Census of Manufactures. These weights are reviewed and revised each time complete census data become available. In between censuses, the weighting pattern may also be modified if it is considered desirable. The frequent changes in weighting patterns are accomplished by the link-chain procedure described in the last section.

The Wholesale Price Index, its components, and the individual price series are published monthly and can be found in the *Survey of Current Business*. Information furnished by these indices are invaluable to economists, managers and government policy makers. These data enable the economist to study the general fluctuations in price level, to evaluate the imbalance between aggregate demand and supply, to analyze the price structure of the economy and the changes in relationships among individual commodities, and so on. Businessmen benefit from this information in determining production costs, in planning investment programs, in formulating production schedules and sales policies, in evaluating inventories, and in purchasing raw materials. Government planners find these data indispensable formulating policies for economic stability, as well as long-range policies for economic growth and other economic programs.

THE BLS CONSUMER PRICE INDEX

As its full title—"Index of Change in Prices of Goods and Services Purchased by City Wage-Earner and Clerical-Worker Families to Maintain Their Level of Living"—indicates, the index measures the average change in the price of a fixed "market basket" of commodities and services purchased by families of urban wage earners and salaried clerks.

The index was initiated during World War I under the demand for wage increases to meet rising costs of living, especially in the shipbuilding centers. It has been published regularly since 1921 with a few major revisions. The last new sample survey was carried out in 1950. A fully revised index, based on new expenditures and family income surveys for 1960–1961, was constructed in 1964. As now constituted, this index is constructed from a collection of approximately 300 carefully selected retail prices that include sales and excise taxes. These prices are reported regularly from 46 cities. Separate indices are published monthly for the 20 large cities as is a single "all-item" index for the whole nation.

This index is also computed by a modified Laspeyres formula with 1957–1959 as the reference base period. In addition, as is true of the Wholesale Price Index, the link-chain procedure is used to offer freedom of substitution at intervals so that the indices may coincide with realities of consumption and markets. The actual weight base period, however, is the 1952 spending pattern—projected from 1950 data. The estimated 1952 spending pattern is based on an average family size of 3.3 persons and an average income per family of $4160 after taxes.

Of the thousands of statistics published by United States government agencies, the Consumer Price Index is probably the most important single statistic. As will be shown in the next section, it is used for a variety of purposes by private organizations as well as public agencies. However, it should be remembered that this index is calculated specifically to measure the average change in prices of goods consumed by families of urban wage earners and clerical workers. Severe limitations are encountered, therefore, in applying this index to the very rich or the very poor, or to groups whose living and spending patterns differ from those of typical urban worker families.

THE FRB INDEX OF INDUSTRIAL PRODUCTION

The Index of Industrial Production has been published by the Board of Governors of the Federal Reserve System since 1927. Its main purpose is to measure changes in the physical volume of manufacturing and mineral production and in the output of the gas and electricity industry.

This index was last revised in 1959. In its present form, it is a weighted average of relatives index of more than 200 different series, each of which represents the output of a particular product or industry or the man-hours worked in that industry. The current base for the index is 1957–1959, but the index has been shown both on a 1957 and on a 1947–1949 base. Data used are adjusted for the number of working days in the month, and the quantity relatives are weighted in proportion to the value added in production by each industry in 1957. The total index and all its components are available with or without adjustment for seasonal variations in the *Federal Reserve Bulletin*.

The Federal Reserve Index has become one of the most quoted business indices now being published. Two reasons are mainly responsible for its wide use: It is by far the best index available of the activity in a highly important sector of the economy, for it is comprehensive, well constructed, and up to date; moreover it has symptomatic value. Manufacturing, mining, and utilities contribute about one third to the total national income and account for about the same proportion of all nonagricultural employment. Furthermore, many other sectors of the economy are engaged in supplying materials to, or using the products of, manufacturing. There is also the fact that manufacturing and mining are more cyclically sensitive than is any other type of business activity. All these factors make the Federal Reserve Index a widely used barometer of general business conditions, despite the fact that it is not an accurate measure of total business activity.

THE UN WORLD INDEX OF INDUSTRIAL PRODUCTION

The World Index of Industrial Production, published by the Statistical Office of the United Nations, includes production in mining and manufacturing of all countries except the Iron Curtain nations. The World Index is supplemented by separate regional indices and is calculated as an arithmetic average of national indices. It now has 1953 as its base and is weighted by value added in 1948 in the manufacturing and mining industries. The value of this index lies in the fact that it is the only international production index available. Further information about this index can be obtained in issues of the *Statistical Yearbook of the United Nations*.

15.13 Some Applications of Price Indices

Early interest in indices came mainly from economists engaged in studying inflation and deflation, the structure of individual prices, the behavior and interrelationships of important economic variables, and the like. Through the years, however, indices, especially price indices, have come to play an increasingly important role in many decision-making problems for private enterprises, labor unions, and governments. This section will present some of the important applications of price indices.

PURCHASING POWER

It is the basic traditional function of a price index to measure the purchasing power, or value, of money. Since the value of money is the power of money over goods and services in exchange, we may deduce that the higher the level of prices, the lower the purchasing power of money—that is, the less will a given income buy in terms of goods and services. More precisely, if the price index for a defined group of goods and services doubles, a given income will buy half as much of that group of commodities and services. This is the same as saying that the purchasing power of money is the reciprocal of the price index. The general expression may be given thus:

$$\text{purchasing power of money} = \frac{1}{\text{price index}} \tag{15.16}$$

For example, if the value of the Consumer Price Index in a given year is 124.5, then the purchasing power of the consumer dollar becomes $1/1.245 = 0.80$, or 80 cents. This means that the purchasing power of the consumer dollar in that year is 80 cents as compared with the base period. Or a given amount of consumer income in that year can purchase only 80 percent of the goods and services that it could have purchased in the base period. The purchasing power of the consumer dollar has decreased by 20 percent from the base period to that year.

REAL WAGES

Real wages are simply the purchasing power of money wages. Thus a real wage index can be constructed to show the changes in the purchasing power of the money wage income of the workers. This is done by constructing a money wage index with the same base as the Consumer Price Index and dividing the former by the latter. That is,

$$\text{index of real wages} = \frac{\text{index of money wages}}{\text{Consumer Price Index}}(100) \tag{15.17}$$

In formula 15.17, the numerator may be either a city index or the national index. As an illustration, suppose that the index of money wages for a given type of worker in the United States, constructed with 1957–1959 as the base, stood at 150 in January, 1968, and that for the same month, the Consumer Price Index had the value of 123.8; then the real wage index for these workers in that month would be

$$\frac{150}{123.8}(100) = 121.2$$

this means that these workers' real wages had increased by 21.2 percent, which is much less than what the money wage index indicated.

ESCALATOR CLAUSES

The most widely publicized use of the Consumer Price Index is for the automatic adjustment of wages under "escalator clauses" in collective bargaining agreements. An escalator clause provides for an automatic amount of wage change, say, 1 cent an hour, with a given change in the percentage points of the Consumer Price Index, say, 0.8 point. Usually, no ceiling is placed on the extent that wages can increase, but the amount that wages can decline is limited so that they cannot fall very far below the wage rate at the time the escalator clause went into effect. The reason for such a clause, obviously, is to protect the workers' standard of living from price changes.

The escalator clause was first introduced by General Motors and the United Auto Workers in 1948. Now it has become a widely used device in collective bargaining. Even in areas where collective bargaining does not exist, wages of workers and clerks in private and public organizations are often tied in directly or indirectly with the Consumer Price Index today.

TERMS OF EXCHANGE

Just as the real wages are ratios of the money wage index to the Consumer Price Index, the terms of exchange for a commodity or a group of commodities are ratios of the price index of one commodity or a group of commodities to the price index of another commodity or group of commodities. The terms of exchange may thus be considered as an index of relative price that measures the change in the price of one in terms of the price of another. Or, as a formula,

$$\begin{array}{l}\text{index of relative price} \\ \text{of A in terms of B}\end{array} = \frac{\text{price index of A}}{\text{price index of B}}(100) \tag{15.18}$$

If, for example, the price index for fuel, power, and lighting materials has increased from its base period by 20 percent but the Consumer Price Index has risen by 50 percent, then the prices of the first group of products have actually decreased by 20 percent in terms of the latter. That is,

$$\begin{array}{l}\text{index of relative prices of fuel,} \\ \text{power, and lighting materials}\end{array} = \frac{120}{150}(100) = 80$$

A special application of terms of exchange that is of interest to students of foreign trade is the concept of "terms of trade." By the "terms of trade" we mean the ratio at which commodities are traded internationally. When only two commodities are being traded, the terms of trade are the same as the ratio of their international prices. When many articles are entering trade, the terms of trade are defined as the ratio of the index of import prices to the index of export prices. As such, the terms of trade may be interpreted as the quantity of imports obtainable for one unit of exports. When the average price of a country's imports rises less rapidly than the average price it receives from exports, the nation's terms of trade are said to have become more favorable, since a greater amount of imports can now be obtained per unit of exports.

Another index ratio, more important from the viewpoint of the American economy, is the *parity ratio*, defined thus:

$$\text{parity ratio} = \frac{\text{index of prices received by farmers}}{\text{index of prices paid by farmers}}(100) \tag{15.19}$$

The parity ratio, since the late 1920s, has been the hitching post of American agricultural policy. Present laws stipulate that farmers who conform to certain requirements can sell most staple crops to the government at a price that is a given percentage of the parity.

Both indices are prepared by the United States Department of Agriculture with 1910–1914 as the base—"the golden age of farming." From formula 15.19 it is obvious that when the prices-received index is higher than the prices-paid index, agricultural prices are above parity. When the converse is true, agricultural prices are below parity. Suppose this parity ratio stood at 83, in a given year; it means that in that year, the purchasing power of agricultural commodities with respect to what farmers buy was 83 percent of what it was in the base period.

THE CONSTANT DOLLAR

The value of an aggregate of goods is the product of price and quantity. Thus the value of an aggregate of goods may change because of changes in prices or in quantities or in both. Often we are interested only in the changes in the physical volume of goods and services: the changes in the value of an aggregate of items at constant dollars. In other words, we are interested in ascertaining what the value of goods and services would be if their prices had not changed or the value of money had remained constant. To achieve this, we employ what is called the procedure of *statistical deflation*, by which the effects of changes in prices upon the value of a list of commodities are removed. This procedure is defined by the following expression:

$$\begin{array}{l}\text{physical volume in}\\ \text{constant dollars}\end{array} = \frac{\text{volume in current dollars}}{\text{appropriate price index}}(100) \tag{15.20}$$

Inasmuch as the denominator is a pure number, the quotient obtained in formula (15.20) is also in terms of dollars, which, however, by virtue of the division, have been transformed into constant dollars. As a concrete illustration, suppose that the value of sales of a department store in 1967 is \$5,585,000 and the index for selling prices of the same store in that year is 115, with 1960 = 100, then

$$\begin{array}{l}\text{Value of sales for 1967}\\ \text{in constant dollars}\end{array} = \frac{\$5{,}585{,}000}{115}(100) = \$4{,}856{,}522$$

which means that if the selling prices in 1967 had been the same as those in 1960, the sales volume in 1967 would have been \$4,856,522. Also, if the sales in 1960 were \$4,627,236, we would say that the physical volume of sales increased by 5 percent from 1960 to 1967.

The price index used in statistical deflation is called the *deflator*. The deflator does not have to be constructed with all the prices of the items whose value is to be deflated. For instance, in deriving the real Gross National Product—that is, the GNP in constant dollars—the Department of Commerce uses deflators that are constructed on the basis of samples of prices that enter into the GNP. The actual procedure used is to break down the GNP into its main expenditure components—consumption, investment, and government expenditures—and deflate each component with a price index of a sample of prices specially constructed for the component. The deflated components are then added up to form the total GNP in constant dollars.

15.14 Problems of Constructing Indices

Up to this point, we have been concerned with some of the techniques of constructing indices and their possible applications. Now we turn our attention to some of the important theoretical and practical problems encountered in constructing indices.

The first problem is sampling. The most important thing to be noted here is that random sampling is seldom used. This is so because to sample from a population of literally millions of commodities and services, the random procedure could be neither practical nor representative. Typically, indices are constructed from samples deliberately selected. The representativeness of an index depends, therefore, on the fact that all or most of the prices or commodities judged to be important in the population are included in its construction. Clearly, the index maker's judgment and knowledge of the data under investigation are of paramount importance. He must select the commodities or their prices to be included in the index. He must decide, in the case of the price index, how the prices should be defined and where and when the price quotations should be collected.

As to the number of items to be included in the sample, the decision is made with the primary objective of the index in mind. If our aim is to describe the movements of general business activity, for instance, we would want a broadly representative index, such as the Wholesale Price Index, which may behave rather sluggishly because it reflects the behavior of all types of business. If our purpose is to stress symptoms, we choose only a few series that we think to be particularly significant as indicators of the future course of business. The FRB Index of Industrial Production is an example of such an index, which is more sensitive and can therefore be used as a business barometer or a forecasting index.

The second problem is concerned with the selection of the base period. The base period of an index is used for the computation of relatives and as a basis of reference both in describing the behavior of individual index numbers and in comparing them. Thus it must be carefully selected so that misleading results and interpretations may not arise. There are two simple rules to follow in choosing the base. One is that the base value should be judged as typical or "normal." It should be neither too high nor too low relative to values in other periods. When the base value is too high, the whole index would appear chronically depressed because most of the index numbers will fall well below it. If the base value is too low, opposite distortions may be created. A base value may be considered typical if it coincides with the trend rather than conforms to cyclically high or low values. When it is difficult to select a single period as the base, the average value of a few periods—perhaps covering one complete cycle—may be used as the base.

The other rule for selecting the base period to keep in mind is that the base of an index should be relatively recent. This is desirable because we are usually more interested in comparing current fluctuations with some economic framework similar to the present. Moreover, a base period that is far in the past makes the recent index numbers less representative because the individual values contained in the index tend to disperse more and more with the passage of time. Finally, a more recent base period is needed in order to accommodate necessary changes in weights and lists of items in the index.

The third is the problem of appropriate weighting. Weighting the different variables according to their economic importance is always an important but troublesome task. Solution here relies heavily upon knowledge of economic theory. Fortunately, it may be noted, for an index to be useful in practice only approximate accuracy in weights is demanded. In addition to the weighting systems introduced previously in this chapter, many others can be used: "average quantities (or prices) of base and given years," "average quantities (or prices) of several years," "average quantities (or prices) of all years," and "hypothetical quantities (or prices)" are a few. Each of these weighting systems has its theoretical and practical merits as well as its drawbacks. Although the relative merits of these weighting systems are beyond the scope of our present discussion, it is important to note these points: (1) To change the weights also changes the meaning of the index. Thus, the type of weights we want to use depends upon the question we seek to answer. (2) Inasmuch as two types of weights can yield similar information, we always select the one that involves less computational effort and/or permits more precise interpretation and greater theoretical consistency.

The fourth problem refers to the selection of the average. This problem has two aspects: Which average should be used? How representative is the index number as an average?

Theoretically, all five types of averages previously introduced can be used to average the weighted aggregative quantities or weighted relatives. However, in practice, the median and the mode are never employed because these two measures are difficult to find and are not very meaningful when only a few variables are involved. Also, being position measures, they cannot be used for further computations. This leaves us with the three computed averages. From the point of view of mathematical logic, in averaging relatives both geometric and harmonic averages are superior to the arithmetic mean. Yet the arithmetic mean is the most frequently used average in constructing indices. This is so because no other indices have such a simple and apparent economic meaning as the weighted arithmetic indices. There are no other indices, either, that are so easy to compute.

The question whether an index as an average is representative or meaningful depends on the distribution pattern of the relatives. It has been argued that an average is meaningful only if there is a real central tendency. To the extent that the values move together through time, the use of index numbers to describe their mass behavior is legitimate. However, if the values are widely dispersed, the index number may lose its meaning. According to many studies made on this subject, it has been demonstrated that the frequency distribution of relatives computed from a base period in the recent past has a pronounced central tendency and the proportion of relatives under the modal class is rather large. However, the distribution of relatives with reference to a more remote base period becomes more dispersed and more negatively skewed, with a smaller proportion of relatives included in the modal class. These findings suggest that the index is more representative when its base is more recent. If we ignore the time element, we have also found a more marked central tendency in a group of items that are more homogeneous, such as agricultural products, consumers' durable goods, or utilities, than exists in a group of greater heterogeneity. This indicates that an index is more meaningful if it is constructed from a group of items that have a greater tendency toward homogeneity.

Ideally, an index, therefore, like any other average, should be accompanied by a measure of dispersion. Unfortunately, indices are seldom so accompanied. Still, in using and interpreting index numbers we must keep the degree and type of dispersion in mind by analyzing the components of the index. For instance, the Wholesale Price Index was fairly stable between 1952 and 1955. This was the combined effect of rising industrial prices and falling agricultural prices. After 1955 the decline in agricultural prices ceased, and the over-all Wholesale Price Index rose significantly to reflect the continued increases in industrial prices. Similarly, in studying the major components of the Consumer Price Index, we find that the most marked rises in consumer prices during the decade of the 1950s were in services rather than in commodities.

The fifth problem is that of changing products. In a dynamic economy old products are continuously being improved in quality or are replaced by new products. Since the significance of an index depends on the retention of the significance of the assortment of commodities comprising it, the comparison of either price levels or quantity levels from two separate and distant points of time becomes difficult if not completely meaningless. Because of this, measurements of price levels over long periods of time are confined to staple commodity prices, such as grains and metals, that remain much the same from age to age. It is also for this reason that the link-chain procedure has been designed to cover the transition between periods of different product mixes in measuring price and quantity levels. In this connection, it is interesting to note that, according to the July 21, 1961, *Report of the Subcommittee on Economic Statistics* to the Joint Economic Committee of Congress, with the acceleration of the introduction of new items, a high research priority to devising measures of quality change is widely regarded as necessary. The *Report* added:

> It was felt that the research into how quality changes can be measured is essential to a better understanding of changes in the present Consumer Price Index. It was pointed out that there is a widespread feeling that because of improvements in quality which are not taken into account by present methods of compiling the Consumer Price Index, the index may be overstated by some undefined amount of the rise in prices which has occurred in the last decade. But it also was made clear that there is no reliable quantitative evidence to support this claim, nor in fact is there empirical evidence to measure the extent of downward bias which also might be occurring in the case of deterioration of quality of some goods and services. There was quite general agreement that everything possible should be done to determine whether or not useful, objective measures of quality change could be developed.

The sixth and the last is the problem of the accuracy of formulas. Different formulas usually produce different results when applied to the same price or quantity data. There thus arises the question, Which formula is accurate? The accuracy of a formula, according to some theoretical statisticians, depends upon whether or not it meets certain mathematical tests. Of these tests, probably the most important are the time reversal test, the circular test, and the factor reversal test.

The *time reversal test* seeks to ascertain whether the index number for period 0 relative to period 1 is the reciprocal of the index number for period 1 relative to period 0. In other words, if the product of two index numbers computed by the same formula, with the functions of the base period and given period interchanged, is equal to 1, we say that the formula meets the time reversal test. For example, if the formula meets this test, when the index for 1960, with 1950 as the base, is

200, then the index for 1950 with 1960 as the base must equal 50. Surprisingly enough, this seemingly important and desirable property of an index is not met by most formulas. It can be verified that among formulas introduced in this chapter, only the simple aggregate index satisfies this requirement.

The *circular test* is an extension of the time reversal test. A formula is said to meet this test if, for example, the 1960 index, with 1955 as the base, is 200, and the 1955 index, with 1950 as the base, is again 200; then the 1960 index, with 1950 as the base, must be 400. In other words, we should be able to get a consistent index for 1960 relative to 1950 by multiplying the 1960 index relative to 1955 by the corresponding index for 1955 relative to 1950. Clearly, the desirability of this property is that it enables us to adjust the index values from period to period without referring each time to the original base. Among the index formulas so far discussed, only the simple aggregative index and an aggregative index with constant weights meet this test.

The *factor reversal test* holds that the product of a price index and the quantity index should equal the corresponding value index. We have already discussed this property in section 15.9 in some detail. What may be pointed out here is that none of the formulas so far discussed meets this test.

Attempts have been made, especially by Professor Irving Fisher, to design formulas that would meet most of these requirements. Of these, the most frequently noted is Fisher's "ideal index," which meets both the time reversal and the factor reversal tests but not the circular test. The ideal index is given by the formula

$$I_{(\text{Fisher})} = \sqrt{\frac{\Sigma p_n q_0}{\Sigma p_0 q_0} \times \frac{\Sigma p_n q_n}{\Sigma p_0 q_n}} = \sqrt{L \times P} \qquad \textbf{(15.21)}$$

This formula, it may be noted, is the geometric mean of the Laspeyres and the Paasche indices. The trouble with the ideal index is that it is laborious to calculate. More serious is the objection that it is difficult to give a precise interpretation of what it actually attempts to measure.

The fact that most of the commonly used indices, especially the weighted indices, do not meet these tests does not necessarily mean that their usefulness is impaired. Failure to meet these test does not prove that the index values do not have logically desirable properties. As has been mentioned before, both the Laspeyres and the Paasche formulas, inaccurate as they may be in some sense of mathematical logic, do yield precise answers to specific questions. It has also been suggested that index values should be considered as ranges that are only approximations but are all we are entitled to because of the nature of the problem inherent in measurements.

Glossary of Formulas

(15.1) $P = \frac{\Sigma p_n}{\Sigma p_0}(100)$ The simple aggregate price index is expressed as a relative by dividing the sum of prices in a given period by the sum of prices in the base period. The relative is multiplied by 100 so that the index is changed into a percentage—the usual form of an index number.

(**15.2**) $Q = \frac{\Sigma q_n}{\Sigma q_0}(100)$ The simple aggregate quantity index. This formula can be used only for an aggregate of physical quantities that are expressed in the same units.

(**15.3**) $P = \frac{\sum \left(\frac{p_n}{p_0}\right)(100)}{n}$ The price index computed as the arithmetic average of price relatives. This formula has the advantage over formula 15.1 in that it is not influenced by absolute prices. However, it has a defect equally undesirable since it assigns equal importance to each relative.

(**15.4**) $Q = \frac{\sum \left(\frac{q_n}{q_0}\right)(100)}{n}$ The simple average of relatives quantity index can be used for an aggregate of items quoted in different units. It is still a very unsatisfactory method, however, because it has the same undesirable concealed weights as has equation (15.3).

(**15.5**) $P_b = \frac{\Sigma p_n q_0}{\Sigma p_0 q_0}(100)$ The Laspeyres formula: an aggregative price index weighted by base-period quantities. It measures the change in the aggregate value of the base period's assortment of goods when valued at given-period prices.

(**15.6**) $P_n = \frac{\Sigma p_n q_n}{\Sigma p_0 q_n}(100)$ The Paasche formula: an aggregative price index weighted by given-period quantities. It measures changes in value of the given period's list of goods when valued at base-period prices. Although this formula yields satisfactory results, because of several limitations, it is not so frequently used as formula (15.5). Its drawbacks, compared with those of (15.5), are these: (1) Current quantities are often difficult to obtain promptly. (2) Computation labor is nearly doubled. (3) With changing weights, even though comparison between a given period and the base period is clear and meaningful, the comparisons between different periods are not entirely valid.

(**15.7**) $Q_b = \frac{\Sigma p_0 q_n}{\Sigma p_0 q_0}(100)$ and (**15.8**) $Q_n = \frac{\Sigma p_n q_n}{\Sigma p_n q_0}(100)$

These are analogous Laspeyres and Paasche formulas for aggregative quantity indices. Note that prices are weights in these formulas. Generally, a weighted aggregative quantity index seeks to answer the question: If we buy (or sell) varying quantities of the same items in each of the two periods, but at the same prices, how much would be spent (or received) in the given period relative to the base period?

(**15.9**) $P_b = \frac{\Sigma(p_0 q_0)\frac{p_n}{p_0}(100)}{\Sigma p_0 q_0}$ and (**15.10**) $Q_b = \frac{\Sigma(p_0 q_0)\frac{q_n}{q_0}(100)}{\Sigma(p_0 q_0)}$

The Laspeyres price or quantity indices can be computed as the weighted arithmetic average of price or quantity relatives, the weights being base-year values.

(**15.11**) $P_n = \frac{\Sigma(p_0 q_n)\frac{p_n}{p_0}(100)}{\Sigma p_0 q_n}$ and (**15.12**) $Q_n = \frac{\Sigma(p_n q_0)\frac{q_n}{q_0}(100)}{\Sigma p_n q_0}$

The Paasche price or quantity index can also be computed as weighted arithmetic average of price or quantity relatives, the weights being theoretical values—base-year price multiplied by given-year quantity for the price index, and the given-year price multiplied by the base-year quantity for the quantity index.

(**15.13**) $V = \frac{\Sigma p_n q_n}{\Sigma p_0 q_0}(100)$ The value index measures the change in actual values of an aggregate of goods and services. For this index, the weights are inherent in the value figures.

(**15.14**) $E_b = \frac{\Sigma r_n q_0}{\Sigma r_0 q_0}(100)$ A productivity index computed with labor inputs and weighted with base-period quantity. Here r may be considered either as units of output per man-hour or as man-hours per unit of output. Care must be taken to observe the definition of r in interpreting the index value.

(**15.15**) $I_{n:o} = (I_{n:n-1})(I_{n-1:o})$ The chain index number for any period is computed by computing the link index number for that period first and then multiply the result by the chain index for the period immediately preceding it.

(Formulas (15.16) to (15.20) are omitted here because they are self-evident.)

(**15.21**) $I_{(\text{Fisher})} = \sqrt{\frac{\Sigma p_n q_0}{\Sigma p_0 q_0} \times \frac{\Sigma p_n q_n}{\Sigma p_0 q_n}} = \sqrt{L \times P}$ The Fisher ideal index. It is an "ideal" in the sense that it meets both the time reversal and the factor reversal tests. However, its practical utility is limited by laborious computation and its vague meaning.

Problems

15.1 Give a critical evaluation of each of the following statements:
a. Neither the simple aggregative nor the simple average of relatives indices is weighted.
b. When quantities in an aggregate are expressed in uniform units, both the simple aggregative and the simple relatives methods furnish satisfactory indices.
c. The Laspeyres and the Paasche formulas not only provide meaningful results but also measure the same thing.
d. Since the value of the base is always 100, it does not make any difference which period is selected as the base on which to construct an index.
e. The fact that none of the commonly used index formulas meets all the essential mathematical tests does not mean that no logically desirable results can be obtained.
f. The Wholesale Price Index is a better measure than the Consumer Price Index if we wish to determine the general purchasing power of the American dollar.
g. Since the 1940s, the average salary of college professors has nearly tripled. Clearly the professors' standard of living must have increased by about 200 percent.
h. A productivity index computed with man-hours per unit of output is actually a measure of labor efficiency alone.

15.2 What does each of the following indices measure?
a. the Laspeyres price index
b. the Paasche price index
c. the Laspeyres quantity index
d. the Paasche quantity index
e. the Wholesale Price Index
f. the Consumer Price Index
g. the FRB Index of Industrial Production

15.3 If you wish to construct a price index by using a fixed quantity, q_a, for each price as weight, how would you state the formula? What would such a price index mean?

15.4 If you are employed to construct a price index for a department store that sells thousands of items, (a) how would you decide on which items to include? (b) how would you define the prices? (c) what weights would you use? (d) which formula would you select?

15.5 If a productivity index is computed with r to mean output per man-hour and the average of two years' quantities as weights, what would be the corresponding index value with r to mean man-hours per unit of output? Give a verbal interpretation of each index value.

15.6 A value index measures changes in actual values between two points of time. Can it be transformed into an index that measures the changes in physical volumes? If so, how? If not, why not?

15.7 What purposes can be served by studying the components of an index?

15.8 Why is it necessary to shift the base period of an index from time to time?

15.9 In your studies of economics or management, where do you think index numbers may be useful? Be specific.

15.10 A manufacturer purchases more than a thousand different types of raw materials for production. He is interested in measuring changes in costs and physical volumes of purchases. Assume that it has been decided that five types of raw materials, A, B, C, D, E, are the important ones and that prices paid and quantities bought are as follows:

		Unit Price		Quantity	
Item	Unit	*1965*	*1968*	*1965*	*1968*
A	pound	$ 0.50	$ 0.65	4,500	4,000
B	ton	50.00	52.00	50	65
C	gallon	1.50	2.00	1,250	1,500
D	yard	4.00	5.50	400	350
E	ounce	0.25	0.15	80,000	120,000

a. Construct, with 1965 as the base,
1) a simple aggregative price index;
2) a simple average of relatives price index;
3) a simple average of relatives quantity index;
4) a weighted aggregative price index with base-year quantities as weights;
5) a weighted aggregative price index with given-year quantities as weights;
6) a weighted aggregative quantity index with base-year prices as weights;
7) a weighted aggregative quantity index with given-year prices as weights;
8) a weighted average of relatives price index with base-year values as weights; and
9) a weighted average of relatives quantity index with base-year values as weights.
b. Construct, with 1968 as the base,
1) an aggregative price index weighted by base-year quantities; and
2) a weighted average of relatives quantity index with the Paasche formula.
c. Give an appropriate interpretation of each of the results you have obtained above.
d. Which index or indices constructed above would you actually recommend for the problem in hand?

15.11 Construct for the following data a spliced index continuous with Index A, and a spliced index continuous with Index B.

Year	Index A	Index B
1	100	
2	95	
3	110	
4	125	100
7		105
6		98

15.12 Construct a chain index from the following link relatives:

Year	Link Index
1	100
2	105
3	95
4	115
5	102

16

INTRODUCTION TO TIME SERIES ANALYSIS

16.1

Reasons for Time Series Analysis

Nations must plan for the future in order to survive and grow. Even installment purchases of a household involve plans for the future. Plans for sales, production, investment, marketing, and so on, are made daily by business firms in order to meet current competition and to maintain steady growth. The federal government must plan for future revenues and expenditures, not only for the sake of performing its routine functions efficiently but also for the purpose of influencing the aggregate business activity so that the economic progress of the nation may not be slowed by inflation or deflation.

A business or an economic action taken today is based on yesterday's plan and tomorrow's expectations. Plans for the future cannot be made without forecasting events and the relationships they will have. Not only can forecasting be made for a given line of activity independently; the forecast of one type of event can also be made on the basis of other forecasts. Thus, the projection of population growth for the next decade is an element in the forecast of the future demand for steel which, in turn, is the basis for plans for expanding plant capacities. Similarly, forecasts of national income have been used by the government to estimate its future tax revenue, and by various industries to predict their relative shares of the national market. Needless to say, an individual firm can base its forecast of sales on the forecast of sales for the whole industry.

Business executives and government officials seem to have recognized the importance of forecasting as the basis of rational decisions and actions concerning the future. Hardly a single day goes by without reading some news in the financial columns about this kind of forward planning. Yet forecasting remains more an art than a science. Those who remember the presidential prediction that prosperity was just around the corner in 1930 or the government economists' forecast of mass unemployment for the spring of 1946 may frown at every mention of the term. Whatever misgivings there may be, the need for the forecast, be it implicit or explicit, is there. The question is not: "Forecast or no forecast?" Instead it is: "What kind of forecast?" It may also be mentioned here that the value of a forecast is not merely its accuracy but the fact that making it requires a balanced consideration of factors influencing future developments, right or wrong.

Forecasting techniques vary from simple expert guesses to complex analyses of mass data. In any case, a forecast must be based upon what has happened in the past. The most promising method of knowing about the past so that inferences can be made for the future is the analysis of a time series. A *time series* may be defined as a collection of readings, belonging to different time periods, of some economic variable or composite of variables, such as production of steel, per capita income, gross national product, price of tobacco, or index of industrial production. A time series, then, portrays the variations of the variable quantity or price through time. Like all kinds of economic behavior, the movements of time series are generated by the systematic and stochastic (the opposite of systematic) logic of the economy. As such, if we find that a certain underlying and persistent trend of a series has continued for decades, we would be unwise to ignore the probability that it will continue. The belief that past behavior of a series may continue into the future forms a rational basis for statistical forecasting. Forecasting is also the main purpose of analyzing time series.

In this chapter we shall comment on the general nature of time series and their analyses. Techniques of measuring various types of economic change as embodied in time series will be presented in the next two chapters.

16.2

Types of Economic Change

The actual movements of a time series are caused by a variety of factors, some economic, some natural, and some institutional. Furthermore, some of the factors tend to affect only the long-run movement of a series, whereas others tend to produce its short-run fluctuations. Thus contained in a time series are different types of change. Looking now at the diagram in Figure 16.1, first of all we observe that even though the movements of the original series seem quite irregular, on the average they tend to move continuously upward through time. This general movement, persisting over a long

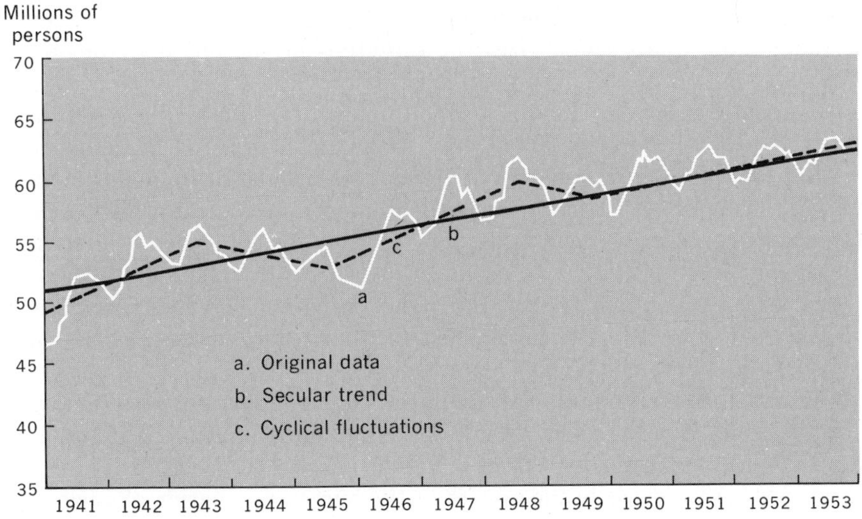

Figure 16.1 Employment in the United States, Bureau of Census estimates, 1941–1953, by month. (Source: *Federal Reserve Charts on Bank Credits, Money Rates and Business*, November 1953.)

period of time, is called a *secular trend*. It is represented by the diagonal line drawn through the irregular curve.

Next, if we study the irregular curve year by year, we see that in each the curve starts with a low and reaches a high about the middle of the year and then decreases again. This type of fluctuation, which completes the whole sequence of change within the span of a year and has about the same pattern year after year, is called a *seasonal variation*.

Furthermore, looking at the broken curve superimposed on the original irregular curve, we find pronounced fluctuations moving up and down every few years throughout the length of the chart. These are known as *business cycles* or *cyclical fluctuations*. They are so called because they comprise a series of repeated sequences, just as a wheel goes round and round.

Finally, the little saw-tooth irregularities on the original curve represent what are referred to as *irregular movements*.

We shall now comment on the nature of each of these four types of economic change.

16.3

Nature of Secular Trend

Secular trend, as a general and persistent movement, disregards the ups and downs of cyclical fluctuations. Short-lived phases of the business cycles pass, leaving a path that, when averaged, reflects the general drift of events over a long period of time. Like other representative measures, the secular trend summarizes the essentials of the life story of economic variables. It measures the average change or growth of time series per unit of time. As such, the secular trend is a sound index of vitality of the organism represented by the time series. Thus, the economic history of an industrialized nation can be told largely in terms of trends established for its major industries. Trends in the major industries are not only of importance to the owners, managers, and workers in these particular fields; they are also of significance as an indication of the economic life of the whole nation, which is mainly made up of these basic economic organs. For this reason, the trend measurements today are used primarily to study growth itself in contrast to their principal use for the study of business cycles before the Great Depression.

To establish a trend line that is significant for a time series the period must be sufficiently long. But the exact length of the period is undefined. It could be as short as 20 days, as in the growth of a population of yeast cells, or it could be as long as a few centuries, as in the growth of the human population of a nation. For most economic time series, the consensus is that the period must cover two or three complete cycles in order to produce any significant trend line.

The trend line for a given time series may be positive, such as the growth of gross national income of the United States, or negative, such as the ratio of agricultural employment to total labor force in any industrialized nation. It could also be linear or curvilinear. In any case, however, the trend is irreversible and smooth. As an irreversible movement, the trend line does not change direction so frequently as the cyclical fluctuation, although through its gradual and smooth course, the rate of change may vary, or even the direction may eventually change, as the time series passes from its expansion phase into long-term saturation or decline.

In selecting a method of trend fitting we must avoid the common fallacy of thinking that the facts will speak for themselves. They never do. Every method of observing them, every technique for analyzing them is an implicit economic theory. Thus, instead of merely looking at the chart of a time series and concluding that a certain type of line would fit the data well, we must start, a priori, with some hypothesis that seems to describe the behavior of the series adequately. Usually, we have different trend hypotheses for the growth of an industry, the growth of the economy, and the growth of the individual firm.

First, what theory can we formulate for the growth of new industries? Empirical observations and theoretical considerations have produced some typical models for the growth of new industries. At the initial stage of a new industry there are only a few pioneering firms that are mainly concerned with exploring the market potential, perfecting its production techniques, and so on; thus its growth is slow. Then the new industry goes through a period of accelerating growth, when it provides the greatest stimulus to net new investment. As it approaches maturity, there is less need to expand capacity, and investment is mainly for replacement and modernization. Consequently, its growth rate continues to decline until the industry reaches a stage of saturation.

This typical behavior of single industries can be described by a number of

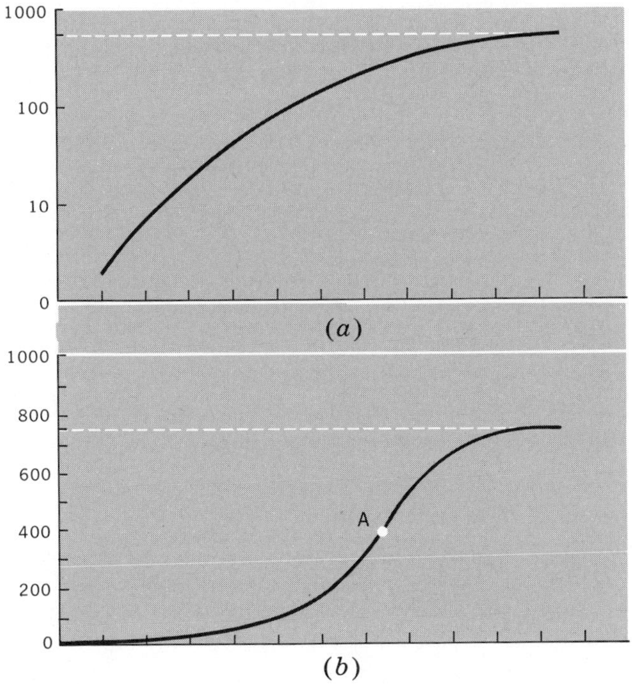

Figure 16.2 Gompertz curves drawn to (a) semilogarithmic and (b) arithmetic scales.

mathematical curves, all of which trace out an S-shaped pattern of movement. Of these, the best-known and most frequently used is the Gompertz curve, the equation of which is given in section 17.8.

The Gompertz curve, as shown by Figure 16.2, describes a trend in which the growth of increments of the logarithms declines at a constant rate. Therefore, the natural values of the trend would show a declining ratio of increase, even though by neither a constant amount nor a constant percentage. This curve has an upper and a lower asymptote. The lower asymptote of an upward trend is zero at negatively infinite time. The ordinate of the inflection point of the curve is about 37 percent of the upper asymptote. This curve is not symmetrical about its inflection point. Its first differences are skewed to the right.

The Gompertz curve fitted to per capita consumption of rayon and acetate fibers in the United States for the years 1922–1960, drawn on semilogarithmic paper, is shown in Figure 16.3. This figure shows that consumption increased at a rate decreased gradually. It may also be seen from this figure that the Gompertz trend is a monotonic function of time; it moves either up or down but not in both directions for the same function. When it moves down, its curvature is convex to the origin, and when graphed on semilog paper, it has a lower asymptote, usually not zero.

Another frequently used mathematical curve that may be fitted to a series to yield a declining percentage rate of growth is the "logistic" growth curve. A logistic curve fitted to the Index of Industrial Production of the Standard Statistics Company is shown in Figure 16.4. We shall comment on this chart a little later. Meanwhile, it may be noted that the Gompertz and logistic curves, according to detailed studies by Kuznets and others, give a very close approximation to the growth pattern of individual industries. These curves are often used to describe the "law of growth" of population and of production in certain industries, and are also used extensively in the field of biology. Two points must be kept in mind when we use and interpret the growth curves of individual industries. First, it may be necessary to give these curves, such as those in Figures 16.3 and 16.4, an upward tilt in order to reflect the fact that a steady rising national income could lead to some corresponding rise in the saturation level for a particular industry. Second, because of many factors, such as changes in technology, discoveries of new sources of supply or exhaustion of old sources of raw materials, or the sudden emergence of new and strong competition, the growth trend of a

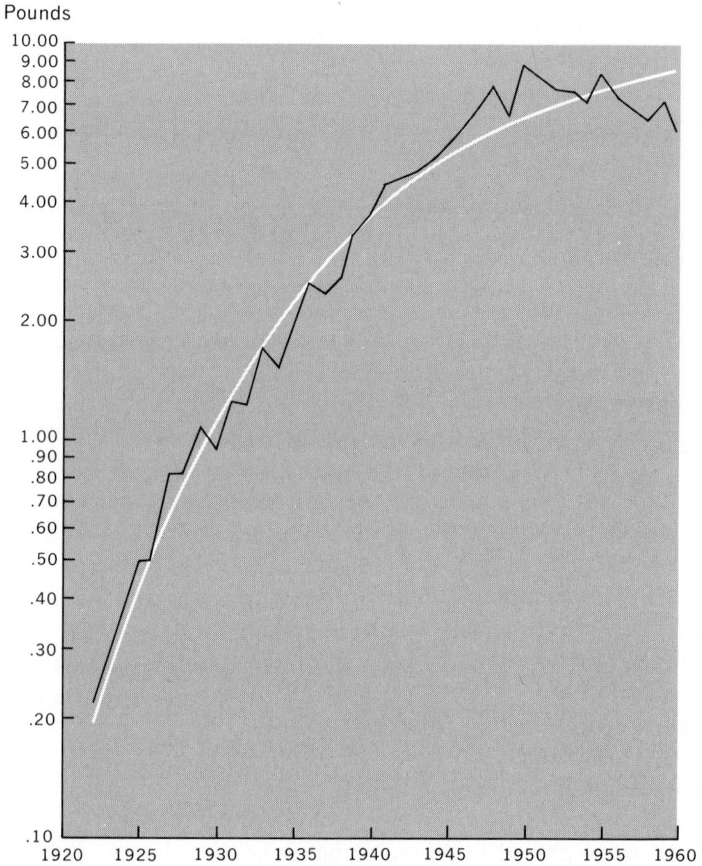

Figure 16.3 Gompertz trend fitted to per capital consumption of rayon and acetate fibers in the United States, 1922–1960. (Source: *Textile Organon*, March 1961.)

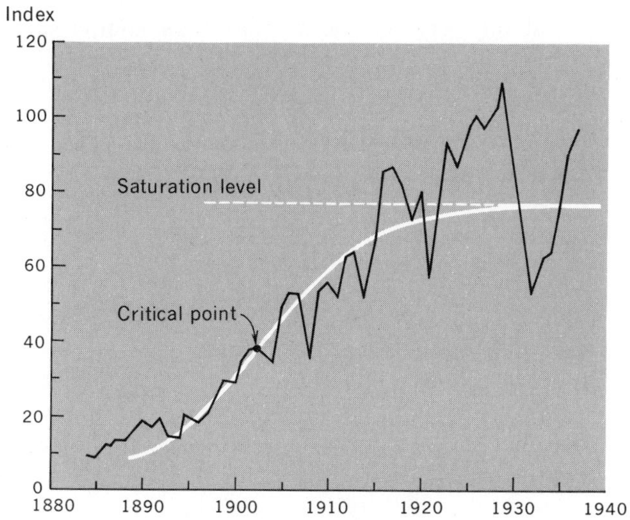

Figure 16.4 Logistic curve fitted to index of industrial production. (Source: H. T. Davis, *The Analysis of Economic Time Series*. Copyright, 1914, by the Cowles Commission for Research in Economics.)

particular industry may be subjected to shifts that are quite abrupt. Herein lies the potential danger of projecting the trend of an industry.

The economic growth of a whole nation is the result of the growth of all individual industries. For the economy as a whole, then, the trend should be measured most logically by the series of the Gross National Product. Other types of data, though less comprehensive, are also frequently used to reflect the nation's general economic activity. One method is to select one phenomenon so pervasive that it reveals the amount of activity in many different kinds of transactions. For instance, freight car loadings or the production of cardboard containers follows a pattern of change similar to the output of material products. Bank credit outside New York City is also a fairly accurate measure of the total volume of business of all kinds. A second method is to use a composite index of activities of the basic and vital industries. One of the most widely used indices for this purpose is the FRB Index of Industrial Production described in the last chapter.

On theoretical grounds, the growth of the aggregate economy is quite different from the typical pattern of individual industries. For instance, fitting a "growth curve" to total production leaves much to be desired. For example, looking at Figure 16.4, we find that it gives a very poor fit for the period before World War I. More serious is the objection that it flattens out too rapidly and at a level too low for the latter period.

The inevitable and eventual retardation of the growth of an individual industry is influenced by many factors. The more important of these are discoveries of new processes, new products, new materials, changes in consumer tastes, and shifts in population distribution. The very factors that tend to retard old and mature industries, however, also stimulate the establishment of new ones. The stagnation of the coal industry has been more than compensated for by the emergence of the oil and power industries. The decline of textile industries that use cotton and silk as raw materials was primarily caused by discoveries of the synthetic fabrics and the growing production of rayon, nylon, and other chemically produced materials.

Thus, in a dynamic economy characterized by unceasing experimentation and innovation, the general drift of total production may be expected to continue upward. It is possible, furthermore, that this general upward drift could progress along a straight line, either indicating a constant amount of increase per year or reflecting a constant rate of increase annually as revealed by Figure 16.5 and 16.7. Our theoretical considerations, however, do not rule out the possibility that the rate of change of the whole economy may vary from period to period. For instance, Siberling has developed a curve of American business experience over the period 1700–1940 that shows a growth rate of less than 3 percent prior to the middle of the last century, the increase then rising to about 3.8 percent in the last decade of the century, and a slightly lower rate of approximately 3.5 percent characterizing the growth of the first forty years of the present century.

It may also be noted that, among the factors counted as critical for the long-run growth of the aggregate economy, the most significant is the rate of population growth. Not only does population growth generate continuous increase in the demand for consumer goods and investment expenditures; it also enlarges the labor force of a nation. Other principal aggregate growth factors include natural resources, capital formation, technological development, type of economic system, the sociopolitical fabric of the society, and the religious and psychological attributes of the population.

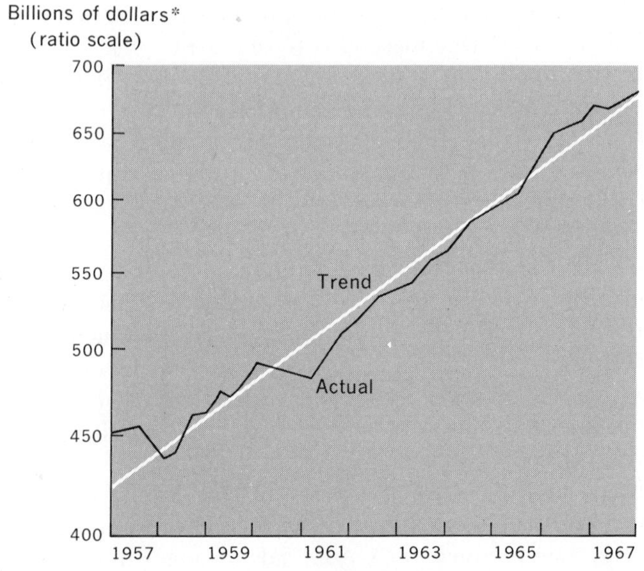

Figure 16.5 Gross National Product in 1958 prices, 1957–1967. (Source: Adapted from *The Economic Report of the President,* February 1968, p. 61.)

The growth of an individual enterprise may be reflected in its sales, its production, or its employment. Other things being equal, obviously the most important influence on the growth of individual concerns is the quality of management, which dictates the firm's long-term flexible adaptation to a changing economy. Keeping differences in the quality of management in mind, three possible generalizations can be made about the growth patterns of individual firms. In the first place, the most efficient firms may continue to grow when the industry is subject to retardation. Their continuous growth may be maintained by absorbing weak competitors or by controlling a progressively larger share of the market through shrewd marketing or other effective policies. This process may go on for such a long period as to give the individual concern a trend of indefinite constant growth, as described either by the arithmetic lines or by geometric straight lines.

At the other extreme are the least efficient firms. Such firms may grow when the industry is advancing rapidly, but they face decline when the industry reaches maturity. Trend for this type of behavior, among others, can be described by a second-degree parabola curve. This curve, together with other types of straight-line tends, will be discussed in the following chapter.

Then there is the middle possibility. Firms of average efficiency may experience a growth pattern similar to that of the industry of which they are a part.

16.4 Nature of Seasonal Variations

Seasonal variations follow a pattern of regular recurrence over time. The word "seasonal" has a broad meaning. A seasonal pattern may be a daily one (average temperatures for 24 hours), or a weekly one (store sales), or an annual one (employment). Nevertheless, seasonal patterns for many types of economic time series are related to the changing seasons of the year. Also, like the seasons, they tend to repeat themselves, even though the precise pattern may change through the passage of time. Figure 16.6 contains some illustrations of seasonal patterns of consumer goods output and industrial production.

Weather and social customs are the most important factors of seasonal variations. Variations in weather are clearly an important cause of seasonal patterns in agricultural production, construction work, logging and lumbering activities, and their related series. The relatively large seasonal variation experienced by the retail trade is the joint effect of customary buying seasons and weather-induced changes in consumer demand.

Seasonal variations generate problems for the individual firms as well as for the economy as a whole. They are expensive and wasteful because they necessitate surplus plant capacity and idle labor during seasonal slack periods. For the

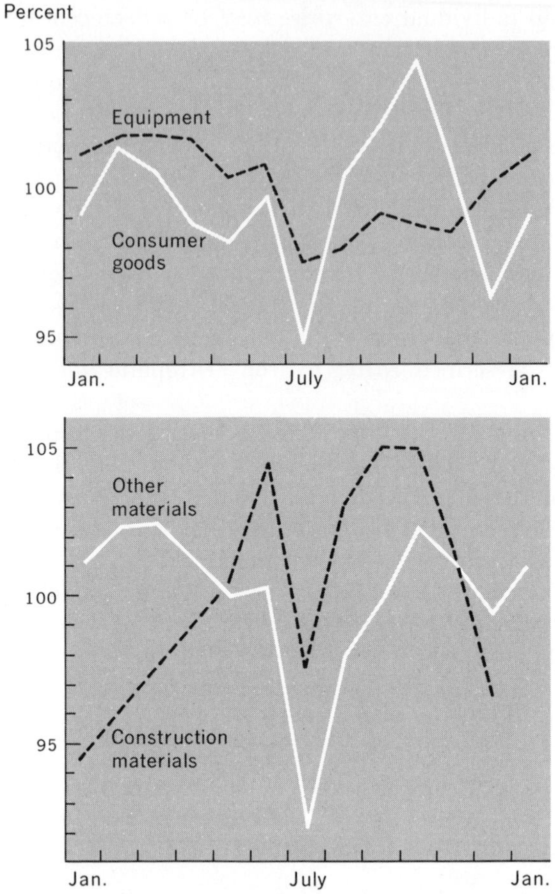

Figure 16.6 Seasonal patterns in industrial production, 1959. (Source: Board of Governors of the Federal Reserve System, *Industrial Production: 1959 Revision*. Washington, D.C., July 1960.)

individual firm there is perhaps no other phase of time series analyses so vital to the day-to-day planning of business operations as seasonal analysis. Management must allow for seasonal variations in its purchases of raw materials and its employment of workers in order to offset these variations or to take advantage of them. Excess capacity in individual industries adds up to an enormous waste of resources for the entire economy. Severe seasonal fluctuations in industrial production, as shown in Figure 16.6, indicate that for a great part of the year a considerable amount of plant capacity is not utilized. Whether or not a comprehensive solution for seasonal variation is possible, an understanding of its nature and an accurate method of measuring it are certainly a necessary first step toward a possible solution.

Our interest in seasonal analysis is not confined to its own significance only. If we are to succeed in studying business cycles, ability to analyze seasonal influences and to remove their effects on the raw data is essential. Unless seasonally adjusted data are used, the location of the turning points of cycles can never be precisely determined.

A technical problem exists for the students of business cycles because seasonal patterns tend to change through time. The change may be gradual because of technological advancements or because of changes in weather, custom, consumer tastes, government policy, or business practices. The change may be abrupt because of some powerful and rather prolonged irregular forces, such as World War II. With this changing characteristic, we may fail to measure the exact pattern of seasonal variations. Although methods are available to measure changing seasonals, these techniques are based on the rather unrealistic assumption that the seasonals are changing in some regular and systematic pattern. Fortunately, cyclical movements are usually so pronounced that they show up clearly when conventional methods, as presented in Chapter 18, are used to eliminate seasonal variations. Difficulties do arise from time to time, however, when we attempt to trace the movement of cycles that are brief in duration and mild in intensity.

16.5 Nature of Business Cycles

The term *business cycles* refers to cyclical fluctuations in the aggregate economic activity of a nation. As such, it should be represented by series in the GNP or in other indices that reflect general business activity. Cyclical fluctuations in production or sales of an individual industry or firm are usually referred to as "specific cycles." Methods used to measure business and specific cycles are, of course, the same.

Arthur F. Burns and Wesley C. Mitchell of the National Bureau of Economic Research have written one of the most frequently quoted definitions of business cycles:

> Business cycles are a type of fluctuation found in the aggregate economic activity of nations that organize their work mainly in business enterprises: a cycle consists of expansions occurring at about the same time in many economic activities, followed by similarly general recessions, contractions, and revivals which merge into the expansion phase of the next cycle; this sequence of changes is recurrent but not periodic; in duration business cycles vary from more than one year to ten or twelve years; they are not divisible into shorter cycles of similar character with amplitudes approximating their own.—Arthur F. Burns and Wesley C. Mitchell, *Measuring Business Cycles.*

Figure 16.7 furnishes an example of conventional trend elimination. The upper chart shows the American Telephone and Telegraph Index of Industrial Activity without eliminating the trend. The lower chart presents the index as percentage

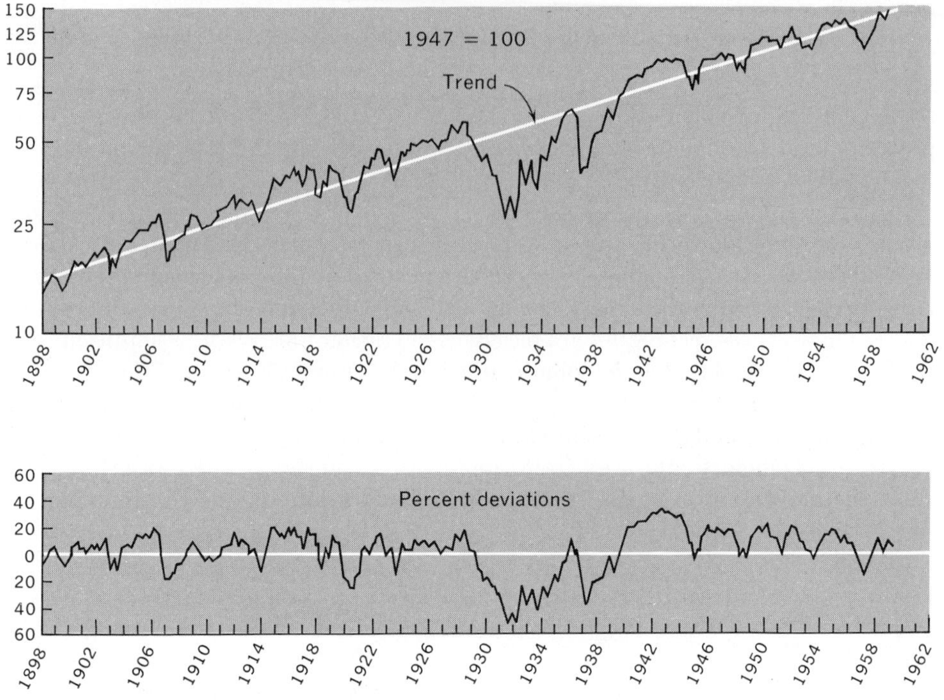

Figure 16.7 Index of industrial activity in the United States, 1899–1960, before and after correction of trend. (Source: American Telephone and Telegraph Company.)

deviations from the computed trend. These percentage deviations portray the general cyclical movements of the index uncorrected for seasonal and irregular fluctuations.

From the definition quoted above and with the aid of Figure 16.7, we may appreciate the complex nature of business cycles and the difficulty of studying them. There is in each cycle, following a period of decline, an upturn that gradually quickens its pace until it develops into a period of prosperity; then this reaches a limit; whereupon a downturn ensues that culminates in a depression. These four phases of a cycle are often termed *recovery*, *prosperity*, *recession*, and *depression*. The first two phases constitute a cumulative *upswing*, and the last two phases are the cumulative *downswing* of a cycle. The peak and trough of a cycle are often referred to as the *downturn* and *upturn*, respectively. A cycle is measured either from trough to trough or from peak to peak.

The most striking characteristic of cycles, general as well as specific, is that each cycle is a unique historical phenomenon. Each cycle differs from all the rest in duration, amplitude, and causes. This may account for the existence of almost as many business cycle theories as the number of recorded cycles. (There are probably more theories than recorded cycles.) Thus we shall not even attempt to mention here the main forces that lead to this type of fluctuation. The interested student may find the information in any standard textbook on business cycles.

16.6

Nature of Irregular Movements

Irregular movements are a type of fluctuation that is caused by random forces or specific and sporadic causes. A random movement, as its name implies, is completely unpredictable, though it does contribute to period-to-period variations. Random movements, fortunately, are relatively unimportant, so that in practice we may simply consider them merely as part of the total cyclical swing or as seasonal variations. Although it is impossible to identify or explain the random forces, their effects can be clearly seen. They appear as slight saw-tooth irregularities on cyclical movements in most monthly series.

Irregular movements caused by sporadic changes are more serious and their causes can usually be traced. For instance, strikes, earthquakes, floods, wars, and the like, produce sharp breaks in the underlying cyclical sweeps in most economic time series. Therefore, in measuring seasonals and cycles, these specific and sporadic causes must be recognized and due allowance made for them if possible.

16.7

Time Series Models

Analysis of time series, either to study one component in its own right or to eliminate one or more components from the original series, calls for the decomposition of a series. To decompose a series, we must assume that some type of relationship exists among the four components contained in it. Usually we proceed on the assumption that an economic time series is made up of several additive or multiplicative components. Trends, cycles, and seasonals are in some sense considered as rather stable functions of time, the irregular movements are not.

The *additive model* assumes that the value of original data is the sum of the four components. Thus, let

Y = value of the original time series
T = value of trend
C = value of cycle
I = value of irregular
S = value of seasonal

then the additive model can be expressed as

$$Y = T + S + C + I \tag{16.1}$$

The *multiplicative model* assumes that the value of original data is the product of the values of the four components. That is,

$$Y = TSCI \tag{16.2}$$

What are the differences between these two models?

First, in the additive model, all the components are treated as residuals and are expressed in the original units. In the multiplicative model, only one component, usually trend, is in units of original data; but seasonal and cyclical components are treated as relatives or percentages whose average value is 100.

Second, the word "additive" signifies lack of interaction effects among the components. It is assumed in the additive model, therefore, that the value of one component does not affect, and is not affected by, the values of other components. In contrast, the multiplicative model implies mutual dependence among the components in an algebraic sense; that is, it is assumed here that both seasonal and cyclical fluctuations are functions of trend.

Third, we can often distinguish the additive model from the multiplicative model by observing the effects of trend values upon seasonal fluctuations and cyclical variations. In the additive model, seasonal swing remains constant as trend increases. In the multiplicative model, the ratio of seasonal to trend remains constant; that is, seasonal swings increase in magnitude as trend increases. Also, generally speaking, the cyclical residuals tend to become larger numerically as the trend values become larger, while the cyclical relatives remain more or less constant.

Although the additive assumption is undoubtedly true in some cases, the multiplicative assumption characterizes the majority of economic time series. Consequently, the multiplicative model is not only considered the standard, or traditional, assumption for time series analysis; it is more often employed in practice than all other possible models combined, including the additive model. For this reason, we shall use only the multiplicative model in our subsequent discussion.

16.8 Methods of Analyzing Time Series

There are available two common approaches to analysis of time series. One is offered by Warren M. Persons; the other by the National Bureau of Economic Research. The Persons method is in common use and is also the method we shall employ in the next two chapters. Persons, in 1919, generalized and described a method of distinguishing trends from cycles that had long before been extensively used. The idea is a very simple one.

First of all, a simple type of linear function, frequently though not always, is selected to represent the average long-run underlying movement of a time series.

The main technique used to fit the trend line is that of least squares. Next, an effort is made to discover if the data contain seasonal variations. If so, they are eliminated by dividing the original data by a seasonal index constructed for the purpose. Of course, if annual data are used, seasonals do not show up and therefore this step is unnecessary. Finally, the deseasonalized data are expressed as deviations, usually in percentage form, from the trend line. These relative deviations represent the cyclical fluctuations in the series. From these, the turning points and the amplitude of cycles in the series may then be examined.

Even though the Persons method is related to specific time series, it can be applied equally well to indices that represent general business activities. That is, although this method is not directly applicable to the problem of dating business cycles, we could still use it to study changes in general business conditions by applying it to composite index numbers or to several composites, each representing a major field of business activity.

The National Bureau of Economic Research has engaged in the study of time series for about half a century under the leadership of Wesley C. Mitchell and Arthur F. Burns. It has developed a method of its own for analyzing time series in order to study the behavior of cyclical fluctuations only. This method, for various reasons, has been used by few investigators outside the Bureau. Nevertheless, the Bureau has made invaluable contributions to our understanding of the behavior of business cycles. Its findings are frequently quoted in writings on economic fluctuations and forecasting. They are also often used by firms in their planning and are a basis of many government policies. We recommend to the interested student that he consult *Measuring Business Cycles*, by Burns and Mitchell, from which we quoted in Section 16.5.

16.9 Editing Time Series Data

Before computing a trend equation or a seasonal index, we often find it necessary to make certain adjustments in raw data. The usual adjustments to be made are for price changes, population changes, comparability, and calendar variations.

Raw data must be edited for price changes if we have a value series and if we are interested only in the quantity changes. The sales of a firm or of an industry may appear to have increased continuously for a number of years and thus show a positive trend. Yet this growth in sales may be due not to an increase in the physical volume of goods sold but to rising prices. If we are to have a more reliable picture of the growth of sales through time, the effects of price changes must be eliminated. Since value is the product of price and quantity, the effect of price changes can be eliminated by dividing each item in the value series by an appropriate price index. This, of course, is the familiar deflation process discussed in the preceding chapter.

The comparisons of production, income, consumption, and other series among nations or among regions in the same country require that raw data be adjusted for population changes. For example, we may find that the national income of a country is rising but the per capita income of that nation is actually falling because of a faster growth in population. To avoid distortions in the figures and confusion in our comparative analysis, we should express the series on a per capita basis. This is done by dividing each value in the series by the appropriate total population figures.

Sometimes a series that covers a few decades may not be comparable with another, for a variety of reasons. First, figures for different periods of the series may be collected and reported by different agencies. Second, the definition of an item or the quality of a product may have changed through time. Changes in definitions have the same impact as changes in the limits of a population. The effects of changing quality are to produce noncomparable price quotations. Finally, a series may be reported in different manners at different time intervals. For instance, production figures may be reported as annual monthly averages for some years and, for others, as annual totals. Clearly, to edit data for comparability is a complicated task. It differs from series to series. What we must note here is that it is often difficult and sometimes impossible to get strictly comparable data and, therefore, comparability should not be taken for granted. Also, it must be remembered that comparability of data throughout the period under examination is necessary if meaningful results are to be obtained from analysis.

The use of the Gregorian calendar introduces an irregular yet consistent pattern of variations in monthly time series. Monthly data are skewed because of the fact that the months have varying numbers of days. For example, the month of February can be expected to be a special one in the statistical arrays. Again, national holidays and five-Sunday months produce the same sort of distortion. A series subjected to this kind of illogical and spurious fluctuation should be edited for calendar variations. Such an adjustment is made by a change that makes all months the same length. This is done in two simple steps. First, each monthly figure is divided by the number of days in that month (or by the number of working days in that month if desired) in order to obtain a daily average for each month. Next, each of the daily averages is multiplied by the average number of days in a month in order to obtain comparable monthly values. The average number of days in a month is the adjusting factor, which is 30.4167 for ordinary years and 30.5 for leap years.

16.10

Serial Correlation in Time Series

Before moving on to the traditional techniques of decomposing a time series, we must comment briefly on a difficult problem concerning the analysis of time series. The difficulty with which we are concerned is mainly the fact that a time series is not a random sample drawn from a population. It is, instead, a series of ordered observations over time.

As we shall see, in the case of estimation by association, if the population satisfies the regression population model, we know that the method of least squares gives us the best unbiased linear estimate of the dependent variable if the deviations of observations from the regression—the population mean—are independently distributed. Furthermore, if the deviations are both normally and independently distributed, confidence limits for the estimates of parameters can be constructed and significance tests can be conducted.

An observation on production, price, stocks, savings, inventory, or other economic variables in a given time unit is not independent of, but correlated with, the value of that same variable in a previous time unit. When the values of the adjacent observations in the same time series are correlated, we refer to this type of correlation as *serial correlation*. Sometimes, the term *autocorrelation* is used for the same concept. However, some statisticians use the two terms in two distinct ways. When this distinction is made, serial correlation is taken to mean the lag correlation in a sample and autocorrelation to mean the lag correlation in a population.

Serial correlation enters into economic time series in a number of ways. First, when time units are too short, random terms are automatically correlated. For instance, if a series is reported in time units of months or weeks, then the random terms have to absorb the effects of the months' being different in length, weather, and holidays—effects that are not random in the short period but that follow with the recurrence of a year. Second, the existence of the trend element in a series also produces serial correlation. The trend values appear in ordered sequence, and each value is, in a sense, determined by the value that precedes it. Finally, cyclical variations impose a regularity among successive observations of the variable over time and thus introduce the same effects into the series as does the trend.

In serially correlated series, successive observations do not give independent determinations of their respective population means. It must be realized, however, that autocorrelation does not refer to the fact that the population mean of the series is changing. It refers, instead, to the phenomenon that the deviations of the successive observations from the population mean are correlated.

With the existence of serial correlation, it is still possible to provide reasonably good point estimates of the parameters of the trend line. However, the phenomenon of mutual dependence of successive observations reduces greatly the number of degrees of freedom that are related to the number of independent observations. As a consequence, serial correlation invalidates the usual estimates of standard errors and thus prevents interval estimates and tests of significance.

The fact that residuals are correlated when the time intervals are too short may produce illusive appearances of cycles. Thus, an irregular force may raise the series above the mean for a number of months before the series is moved back to its former position. We can imagine a situation, therefore, where a time series may not contain any cyclical elements at all; but if it covers a long enough period, it would eventually form a cyclical pattern because of serial correlation in residuals.

The presence of serial correlation also creates difficulty in comparing two time series. Conformity or nonconformity of a number of points between the series may only be a single instance seen repeatedly. Such illusions created by serial correlation may lead us to unwarranted conclusions.

Methods do exist for estimating the degree of serial correlation in time series. If the degree of correlation is known, it is also possible to estimate the standard errors for serially correlated series. Some possible tests of the randomness of time series samples were introduced in Chapter 14.

In conclusion, we may say that the existence of serial correlation makes it impossible to use the methods previously described in this book to forecast future behavior of time series on the basis of its past and present behavior. Traditional methods can offer only descriptive measures for time series analysis. Such a conclusion may make us feel exasperated but not completely hopeless. In spite of the absence of a complete body of statistical theory for forecasting in time series, the venture can still be made. All that we have to remember is that the application of forecasting methods to be discussed in the subsequent chapters will depend mainly upon careful formulation of problems and the judgment of experts in the subject matter to which the series is related.

Glossary of Formulas

(**16.1**) $Y = T + S + C + I$ and (**16.2**) $Y = TSCI$ These two formulas represent the most frequently used time series models. Formula (16.1), the additive model, states that the magnitude of the four components in a series are additive. That is, the magnitude of one element would not be affected by the magnitude of any other. Formula (16.2) is the multiplicative model. It states that the actual value of a variable at any time is the product of the four components. It assumes that the four components are interrelated.

Problems

16.1 Why is forecasting important?

16.2 Define the following terms: (a) time series, (b) trend, (c) seasonal variations, and (d) irregular movements.

16.3 What hypothesis or hypotheses can we formulate for the secular trend of an industry? of the whole economy? of the individual firm? Give reasons for each.

16.4 Why do we wish to measure seasonal variations?

16.5 Distinguish the following pairs of concepts: (a) business cycles and specific cycles; (b) duration and amplitude of cycles.

16.6 What are the two types of irregular movements? How do they affect the movements of cycles?

16.7 Which measure of the time series components—trend, cycle, or seasonal—do we need in order to make a decision in each of the following problem situations?

a. An estimate of the demand for automobiles for 1975 is needed as a basis for planning for plant capacity.

b. A manufacturer must schedule his monthly production for the forthcoming year.

c. The federal government desires to estimate the corporate profits taxes for the next year in order to prepare its budget.

d. A speculator wishes to decide, on the basis of past prices and earnings, whether or not to buy a certain common stock for possible short-run capital gains.

e. An investor wishes to decide, on the basis of past prices and earnings, whether or not to purchase a certain common stock for continuous returns in the long run.

16.8 How do the additive and multiplicative models of time series differ from each other? Why is the multiplicative model the most commonly used assumption in time series analysis?

16.9 Since the Persons method is related to specific time series, it is completely incapable of measuring fluctuations in general business activity. True?

16.10 Why is it sometimes necessary to edit time series data for prices changes, population changes, comparability, or calendar variations? How is each of these adjustments made?

16.11 What is meant by autocorrelation? What are the sources of autocorrelation? What difficulties does autocorrelation create in the analysis of time series?

16.12 At the present state of knowledge, forecasting with time series depends almost exclusively on personal judgment. Discuss.

17

MEASURES OF LONG-RUN GROWTH

17.1

Introduction

This chapter will be devoted to the measurement of secular trend, the long-run growth, and the decline of time series.

Among the various possible patterns of secular trends in time series, the most important and basic is perhaps the straight line because the growth or decline of many series can be assumed to proceed gradually without abrupt changes in direction. A straight line on an arithmetic chart represents a constant amount of change per time unit. A straight line on a semilogarithmic chart indicates that the rate of change is the same for each time period.

When neither the arithmetic nor the geometric straight-line trend can describe the underlying drift of the series, a smooth curve on either an arithmetic or a semilogarithmic chart may have to be used. Nevertheless, it is interesting to note here that, very often, even if the trend of a series has curvilinearity, linear trend may still be used to advantage. This is done by fitting several straight lines, each covering a portion of the whole range of the series. The justification for this practice is the fact that changes in amounts or in rates of growth or decline are usually very slow; thus a straight line can describe adequately the average behavior within a short range.

The use of several straight lines to approximate a curvilinear trend is not merely for the sake of simplicity as well as ease in computation. This practice actually offers two real advantages in trend analysis. In the first place, for many purposes, it is desirable to have a measure of average (amount or rate) growth for a shorter period so that comparisons among the changes in different periods can be made. Second, a smooth curve of higher degrees is often of little value for extrapolation, since such a curve tends to be explosive at large values of the time units. For the purpose of forecasting, in other words, it may be much more desirable to project from a straight line that covers the most recent period of the series.

For these reasons, therefore, in the discussion that follows we shall deal primarily with linear trends even though methods of fitting trends by the second-degree parabola and moving averages will also be introduced.

A time series treats a variable quantity, Y, as a function of time, X. The time unit in economic time series is usually 1 year, 1 quarter, or 1 month. Now, if we use y_t to represent the trend values, the linear trend is given by the expression

$$y_t = a + bx \tag{17.1}$$

where, a is the trend value when $x = 0$, and b is the slope of the trend line, or the change in the value of y_t per unit of time.

In a time series chart, the origin of the X axis is at the point where $X = 0$, since a time series is plotted by using time as the X variable. In time series analysis, any unit of time can be chosen as the origin. If, for instance, we designate 1960 as the origin, it would have a value of "0"; 1959 would be "-1"; 1961 would be "$+1$"; and so on.

Obviously, to fit a straight line trend to a series, all we have to do is derive the values for the Y intercept and the slope. These values can be obtained by any of the following methods:

(1) Freehand method
(2) Selected points method
(3) Semiaverage method
(4) Least squares method

In the following sections we shall discuss these methods in turn.

17.2 Freehand Method

The freehand method is also called the *graphic method* in the sense that the trend line is determined by inspecting the graph of the series. According to this method, the trend values are determined by drawing freehand a straight line through the time series data that is judged by the investigator to represent adequately the underlying long-term movement in the series. As an aid in establishing the line, a transparent ruler or a piece of string may be laid on the surface of the chart and adjusted until the properties of the line are determined.

Once the freehand trend line is drawn, a trend equation for the line may be approximated. This is done by first reading off the trend values of the first and last period from the chart with reference to the freehand line. For this method, the first period is usually considered the origin. Thus, the trend value for the first period is the value of a for the equation. Then the difference between the trend values of the first and last period is obtained. This difference represents the total change in variable Y throughout the whole duration of the series. Therefore, when this difference is divided by the number of periods in the series, a result that represents the average change in Y per time unit—the value of b—is obtained.

In an expert's hand, the trend line for many series may be satisfactorily drawn, especially when the fluctuations around the general drift are so mild that the path of the trend is clearly defined. The trouble with this method is that it is too subjective. Even in the hands of researchers of long experience, different lines may be drawn for the same series. There is no formal statistical criterion, either, whereby the adequacy of such a line can be judged. Furthermore, although this method appears simple and direct, in actuality, as experienced statisticians can verify, it is very time-consuming to construct a freehand trend if a careful and conscientious job is done. For these reasons, the freehand method is not recommended for beginners.

17.3

Selected Points Method

To construct a trend line by the selected points method, we select two actual points in the series, p_1 and p_2, usually taken, respectively, from the beginning and the end of the series, which are deemed characteristic of the original data; a straight line is then drawn through these two points. In this method, p_1 may be conveniently considered as the origin of the X axis. The original value for p_1, then, is the Y intercept. To get the value of the slope of the straight line we divide the difference between the two selected points by the number of time units separating them, t; that is,

$$b = \frac{p_2 - p_1}{t} \tag{17.2}$$

As an illustration of this method, we shall use the data in Table 17.1 that has been plotted in Figure 17.1. From this figure, it can be observed that a straight line that goes through the two points of 1957 and 1965 seems to describe the general drift of the series quite well. Thus, we may decide to select 1957 as p_1 and

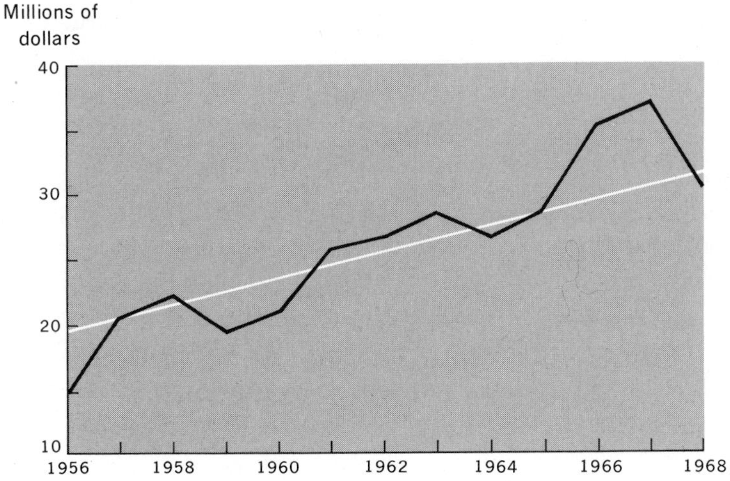

Figure 17.1 Trend line by selected points for new plant and equipment expenditures, 1956–1968. (Source: Table 17.1.)

1965 as p_2 and from Table 17.1, we find

$$p_1 = a = \$20.6 \text{ million}$$
$$p_2 = \$28.7 \text{ million}$$

Consequently,

$$b = \frac{28.7 - 20.6}{1965 - 1957} = \$1.0125 \text{ million}$$

TABLE 17.1 NEW PLANT AND EQUIPMENT EXPENDITURES OF ALPHA CORPORATION, 1956–1968 (MILLIONS OF DOLLARS) COMPUTATION OF STRAIGHT-LINE TREND BY SELECTED POINTS METHOD

Year	Expenditure (y)	x	Trend Value (y_t)
1956	14.8	−1	19.6
1957*	20.6	0	20.6
1958	22.1	1	21.6
1959	19.3	2	22.6
1960	20.6	3	23.6
1961	25.6	4	24.6
1962	26.5	5	25.7
1963	28.3	6	26.7
1964	26.8	7	27.7
1965*	28.7	8	28.7
1966	35.1	9	29.6
1967	37.0	10	30.7
1968	30.5	11	31.7

** Selected points.*

With this information, we can write the trend equation for the series as

$$y_t = 20.6 + 1.0125x$$

(Origin, 1957; time unit, 1 year; Y, annual total new plant and equipment expenditures of Alpha Corporation, 1956–1968.)

With the trend equation so determined, the trend values for all periods can now be obtained by simply substituting values of X into the equation. It should be remembered here that this method provides only approximations of trend values.

This method is a refinement of the freehand method. It is a useful device with which to get a quick approximation of the trend. However, it also suffers from subjectivity because the selection of representative points is left to the judgment of the investigator.

17.4

Semiaverage Method

To determine the trend values by the semiaverage method, first the series in question is divided into two equal segments; then the arithmetic mean for each part is computed. Finally, a straight line passing through these two averages is drawn to provide the trend for the series. Each average furnishes the trend value for the middle time period of the corresponding segment. When the series includes an odd number of periods, there are three ways of separating the series:

(1) Add half of the value of the middle period to the total value of each part.
(2) Add the total value of the middle period to the total value of each part.
(3) Drop the value of the middle period from the computations of the averages.

With the semiaverage method, the middle time unit is considered as the origin, and the values of the Y intercept and slope of the straight line are derived by applying the following equations:

$$a = \frac{\Sigma_1 + \Sigma_2}{t_1 + t_2} \tag{17.3}$$

$$b = \frac{\Sigma_2 - \Sigma_1}{t_1(n - t_2)} \tag{17.4}$$

where t_1 and t_2 refer, respectively, to the number of time units for the first and second segment in the series, and n is the total number of period in the series.

As an illustration of this method, we use the data in Table 17.2. This series contains 15 years and is divided into two parts with 7 years in each, the middle year being dropped. The arithmetic mean for the first half is 12.6 and that for the second, 18.0. The straight-line trend drawn through these two points is shown in Figure 17.2. Furthermore with these two average values, we have

$$a = \frac{88.2 + 125.8}{7 + 7} = \$15.3 \text{ million}$$

$$b = \frac{125.8 - 88.2}{7(15 - 7)} = \$0.67 \text{ million}$$

Thus, the trend equation becomes

$$y_t = 15.3 + 0.67x$$

(Origin, 1961; time unit, 1 year; Y, volume of department store sales in the South in the United States, 1954–1968.)

As before, the trend value for each year can now be determined by substituting the x value for that period in the trend equation. Trend values for all the years for the entire series are entered in the last column of Table 17.2.

Obviously, the semiaverage method is not a subjective one. The slope of the trend line now depends upon the difference between the averages that are computed from the original values, with each average as typical of the level of that segment of the data. This method, however, is not entirely free from drawbacks. The most serious defect here is inherent in the nature of the arithmetic mean, which can be unduly affected by extreme values in either half of the series. If it so happens that one part contains more depressions or fewer prosperities than the other, then the trend line is not a true representation of the secular movements of the series. This danger is more severe the shorter the time period represented by the average. Consequently, trend values obtained are not precise enough for the purpose either of forecasting the future trend or of eliminating trend from original

TABLE 17.2 VOLUME OF DEPARTMENT STORE SALES IN THE SOUTH IN THE UNITED STATES, 1954–1968 (MILLIONS OF DOLLARS) COMPUTATION OF TREND BY SEMIAVERAGE METHOD

Year	x	Export	Semiaverage	Trend Value
1954	−7	9.5		10.6
1955	−6	14.3		11.3
1956	−5	12.5		11.9
1957	−4	11.9	12.6	12.6
1958	−3	10.1		13.3
1959	−2	14.9		13.9
1960	−1	15.0		14.6
1961	0	15.1		15.3
1962	1	15.0		16.0
1963	2	15.4		16.6
1964	3	18.9		17.3
1965	4	20.7	18.0	18.0
1966	5	17.7		18.6
1967	6	17.6		19.3
1968	7	20.5		20.0

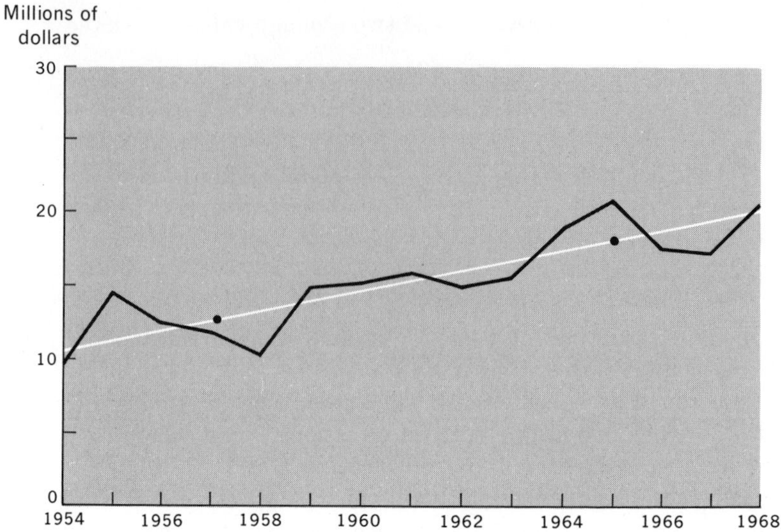

Figure 17.2 Trend line by semiaverage method for volume of department store sales in the South in the United States, 1954–1968. (Source: Table 17.2.)

data. We shall now turn to the fourth method of computing trend, one that promises to yield more satisfactory results.

17.5

Least Squares Method: General Comments

Although the three methods so far discussed have certain defects, each is capable, under certain circumstances, of providing a satisfactory trend for practical purposes. Their application is nevertheless rather limited. The main objection to all subjective methods is that the investigator determines what answer he seeks and then proceeds to determine it.

A convenient device for obtaining an objective fit of a straight line to a series of data is the method of least squares. An important theorem in mathematical statistics, known as the Markoff theorem, states that for a given condition the line fitted by the method of least squares is the line of the "best" fit in a well-defined sense. "Best" is employed here to mean that the estimates of the constants A and B are the best linear unbiased estimates of those constants.

The method of least squares accomplishes two objectives simultaneously. The first is that the sum of the vertical deviations of the observed values from the fitted straight line equals zero. This objective can, of course, be met by any straight line that passes through $\bar{x}$ and $\bar{y}$. Second, the sum of the squares of all these deviations is less than the sum of the squared vertical deviations from any other straight line. This property, however, is not shared by other straight lines. Furthermore, when a line is fitted to meet the second requirement, the first is automatically fulfilled. It is because of this second property that the name "least squares" is derived.

The method of least squares has the desirable characteristic of being simple to apply. The simplicity of application is due to the fact that the best-fitted line is automatically obtained when derived formulas for the least squares method are applied to observed data.

When the trend component can be expressed properly by a straight line of the form

$$y_t = a + bx + e$$

where a and b are constants, X is the time variable, and e is an independent random variable with 0 mean and constant variance, then the method of least squares will yield the best linear unbiased estimates of A and B. Thus, the most satisfactory measurement of trend is obtained by the least squares method when the deviations from the trend values are distributed normally. In practice, however, the es do not meet the criteria of being independent and random variables with constant variance. This is mainly because the raw data, to which we fit the trend line, have a cyclical component. Thus, e contains both cyclical and random movements.

Because of failure of time series data to meet the necessary assumptions, we cannot say that the least squares trend line is better than any other trend line. Moreover, in the absence of these conditions, very nearly the same result may be obtained by careful application of any of the previous methods introduced. Nevertheless, statisticians have become accustomed to using least squares; because the method does have the advantage of objectivity, it is frequently employed.

The method of least squares can be used not only to fit a straight line but also to fit nonlinear trends. We shall use it in the next two sections to calculate arithmetic and geometric straight line trends and, later, to compute a second-degree parabola line.

To derive the values of constants a and b in a linear equation by the least squares method we are required to solve the following two normal equations simultaneously:

$$\Sigma y = na + b\Sigma x$$
$$\Sigma xy = a\Sigma x + b\Sigma x^2$$

The solution of a and b from these equations in the case of time series analysis is greatly simplified by using the middle of the series as the origin. Since the time units in a series are usually of uniform duration and are consecutive numbers, when the middle period is taken as the origin, not only the mean of the time variable, $\bar{x}$, but also the sum of time units, Σx, will be zero. As a result, the two normal equations become

$$\Sigma y = na$$
$$\Sigma xy = b\Sigma x^2$$

from which we can then derive

$$a = \frac{\Sigma y}{n} \quad \text{and} \tag{17.5}$$

$$b = \frac{\Sigma xy}{\Sigma x^2} \tag{17.6}$$

It will be recognized that (17.5) is the equation for the arithmetic mean of Y. Thus, a, the value of Y at the origin, is the arithmetic mean of the Y variable. The value of b, of course, is the average amount of change in the trend values per unit of time.

The student is again reminded that in computing trend it is convenient to use the middle of the series as the origin. Now, we may note that if the series contains an odd number of periods, the origin is the middle of the middle period. Thus, if the series extends from 1956 to 1968, in years, the origin is taken at the end of June, 1962. If an even number of periods is involved, the origin is set between the two middle periods. Thus if the series covers a period from 1956 to 1967, again in years, the origin then falls between 1961 and 1962; that is, midnight, December 31, 1961. In the next section we shall present the general procedure for computing the arithmetic straight-line trend by the least squares method.

17.6

Least Squares Arithmetic Straight-Line Trend

According to our theoretical knowledge of an economic time series, if we think it appropriate to compute a straight-line trend, we first plot the data as a further check on the justification of using a straight line to describe the average long-run behavior of the series. If we are satisfied with our conclusion, we may proceed to compute the trend by the least squares method. This method is illustrated by the data in Table 17.3, which contains an odd number of years. The general procedure is as follows:

(1) Locate the middle of the time period (in our example it is 1962) and assign to it an X value of zero. Thus, the X value for 1961 becomes -1; for 1960, -2; and so on. Conversely, to every year more recent than the year of origin, consecutive positive values of X are assigned. (See column 2.)

(2) Square the xs and sum up the products to obtain Σx^2. (See column 3.)

(3) Sum up the original values of the Y variable to obtain Σy. (See column 4.)

(4) Multiply x by y for each year and sum up these products to obtain Σxy. (See column 5.)

(5) Now we have all the quantities to compute the values of a and b. For our

example,

$$a = \frac{\Sigma y}{n} = \frac{1{,}201.3}{13} = 92.4 \quad \text{millions}$$

$$b = \frac{\Sigma xy}{\Sigma x^2} = \frac{331.3}{182} = 1.82 \text{ millions}$$

(6) Write the equation of the trend line by substituting the results in step (5). The origin, time unit, and information about the Y variable must be attached to the trend equation. For our example, we have

$$y_t = 92.4 + 1.82x$$

(Origin, 1962; time unit, 1 year; Y, annual total coffee production, 1956–1968, in millions of pounds.)

(7) Compute the trend values by substituting the X values for each year in the trend equation. For instance, in our example, the trend value for 1956 is

$$y_{1956} = 92.4 + 1.82(-6) = 81.5 \text{ millions}$$

(8) Plot two trend values on the chart for the original data and draw a straight line through the two trend values and the original data as shown in Figure 17.3.

The result of our example seems to justify the application of the least squares method to construct a straight-line trend for coffee production data. Not only is the general drift of the data clearly linear in nature; there are no extreme deviations from the trend.

If we wish to compute a least squares arithmetic straight-line trend for a series that contains an even number of periods we follow the same general procedure described above except for the method of assigning values to the time units.

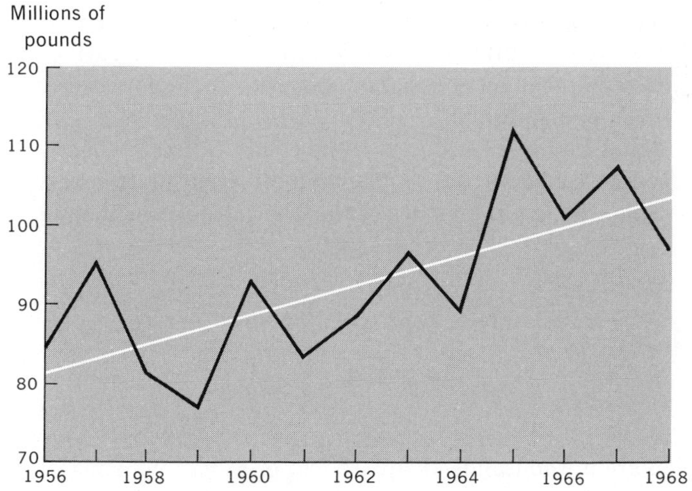

Figure 17.3 Least squares arithmetic straight-line trend for coffee production in South America, 1956–1968. (Source: Table 17.3.)

TABLE 17.3 COFFEE PRODUCTION IN SOUTH AMERICA, 1956–1968 (IN MILLIONS OF POUNDS) COMPUTATION OF LEAST SQUARES ARITHMETIC STRAIGHT-LINE TREND

Year (1)	x (2)	x^2 (3)	Production y (4)	xy (5)	Trend Value y_t (6)
1956	−6	36	82.3	−493.8	81.5
1957	−5	25	95.6	−478.0	83.3
1958	−4	16	81.3	−325.2	85.1
1959	−3	9	76.4	−229.2	87.0
1960	−2	4	92.8	−185.6	88.9
1961	−1	1	83.4	− 83.4	90.6
1962	0	0	88.2	00.0	92.4
1963	1	1	96.1	96.1	94.2
1964	2	4	89.1	178.2	96.0
1965	3	9	112.2	336.6	97.9
1966	4	16	100.4	401.6	99.7
1967	5	25	106.9	534.5	101.5
1968	6	36	96.6	579.6	103.3
Total	0	182	1201.3	331.3	

Consider the series in Table 17.4, which has 12 years. Now, if we decide that the origin falls at the middle of the series, it would be between 1951 and 1952. Thus, the middle of 1951 is six months less than the origin and the middle of 1952 is six months more than the origin. Furthermore, if we wish to keep the time unit at 1 year, then clearly the x value for 1951 is -0.5, that is, a half unit before the origin. The middle of 1950 is one unit and a half before the origin and thus has a value of -1.5, and so on. Similarly, consecutive positive values of 0.5, 1.5, 2.5, and so on should be assigned to every year more recent than the origin. An alternative method is to consider the time unit as six months. Then each year has two time units and for each year prior to the origin, consecutive negative values of -1, -3, -5, and so on, should be assigned. Likewise, for more recent years, the X values are 1, 3, 5, and so forth.

Table 17.4 and Figure 17.4 illustrate the computation of a least squares arithmetic straight-line trend for data that have an even number of years. For the data in this table we have the values of the Y intercept and the slope of the trend line as

$$a = \frac{38{,}737}{12} = 3228 \quad \text{and}$$

$$b = \frac{-28{,}703}{143} = -200.7$$

both of which are in thousands of square feet of radiation. Thus, the trend equation becomes

$$y_t = 3{,}228 - 200.7x$$

(Origin, 1951–1952; time unit, 1 year; Y, monthly average shipment of radiators and convectors, 1946–1957, in thousands of square feet of radiation.)

TABLE 17.4 SHIPMENTS OF RADIATORS AND CONVECTORS, 1946–1957 (MONTHLY AVERAGE IN THOUSANDS OF SQUARE FEET OF RADIATION) COMPUTATION OF LEAST SQUARES ARITHMETIC STRAIGHT-LINE TREND

Year (1)	x (2)	x^2 (3)	Shipment y (4)	xy (5)	Trend Value y_t (6)
1946	−5.5	30.25	3,179	−17.584	4,332
1947	−4.5	20.25	4,727	−21,272	4,131
1948	−3.5	12.25	5,028	−17,598	3,930
1949	−2.5	6.25	2,991	− 7,478	3,730
1950	−1.5	2.25	4,015	− 6,022	3,529
1951	−0.5	0.25	3,543	− 1.772	3,328
1952	0.5	0.25	3,075	1,538	3,128
1953	1.5	2.25	2,639	3,958	2,927
1954	2.5	6.25	2,412	6,030	2,726
1955	3.5	12.25	2,572	9,002	2,526
1956	4.5	20.25	2,464	11,088	2,325
1957	5.5	30.25	2,074	11,407	2,124
Total	0.0	143.00	38,737	−28,703	

Source: Business Statistics, 1958.

As has been previously pointed out, our interest in trend analysis is not confined to determining the growth pattern of a series in the past. We are also concerned with forecasting trend values. To forecast trend values, we use the same technique that we use to compute the trend values for periods covered by the series. For instance, if we are interested in forecasting the trend value of shipments

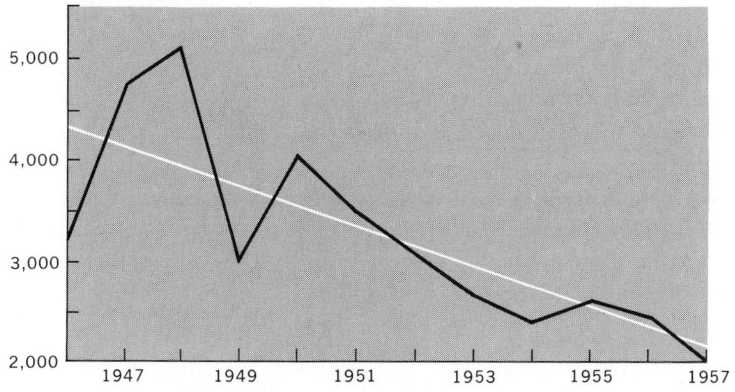

Figure 17.4 Least squares arithmetic straight-line trend for shipments of radiators and convectors in the United States, 1946–1957. (Source: Table 17.4.)

of radiators and convectors for 1960, we simply observe that 1960 is 8.5 years ahead of the origin for the trend equation we have just established, and thus

$$y_{1960} = 3{,}228 - 200.7(8.5) = 1522$$

We may then state that, under the assumption that the same trend factors that produced the trend equation for 1946 and 1958 will remain operative, the average monthly shipments of radiators and convectors will have a trend value of 1522 thousands of square feet of radiation in 1960.

17.7 Least Squares Geometric Straight-Line Trend

The arithmetic straight-line trend is appropriate when there is reason to believe that the time series is changing, on the average, by equal absolute amounts each time period. Sometimes, however, we may believe the growth influence to be such that it causes the absolute amounts of the Y variable to increase more rapidly in later time units than in earlier ones. When this is the case, if a trend is to be fitted to the original data plotted on the natural number-scale, a curvilinear instead of the simple linear description would be necessary. The same data when plotted on a semilogarithmic scale, however, may very often reveal linear average relationship. (A straight line on a semilogarithmic scale, as may be recalled, is referred to as a geometric straight line, indicating that the change in Y per unit of X is at a constant rate.) In other words, a straight line would seem to describe the trend of the series plotted on a semilogarithmic chart. If theoretical considerations permit, we may then express the growth pattern of the series by a geometric straight line.

To fit a geometric straight line, trend values are computed from the logarithms of the data instead of the original data. Thus the trend values obtained by this method will be logarithms of trend values and could be related to logarithms of the data. It is, however, more useful in practice to find the antilogs that will give trend values in natural numbers and that can be compared with the actual values of the data.

In fitting a straight line to the logarithms of the data by the least squares method, the procedure employed is identical with that for an arithmetic straight line presented in the previous section. The only difference is that logarithms of the original data instead of the natural numbers are used throughout. The normal equations are now written as follows:

$$\begin{aligned} &1. \quad \Sigma \log y = n\,(\log a) + \log b(\Sigma x) \\ &2. \quad \Sigma x\,(\log y) = \log a(\Sigma x) + \log b(\Sigma x^2) \end{aligned} \qquad (17.7)$$

Again, as before, if we set the origin at the middle of the series, the formulas for the Y intercept and slope become

$$\log a = \frac{\Sigma \log y}{n} \tag{17.8}$$

$$\log b = \frac{\Sigma x (\log y)}{\Sigma x^2} \tag{17.9}$$

With the values of a and b, the logarithmic straight line equation can now be written

$$\log y_t = \log a + x (\log b) \tag{17.10}$$

As an example of fitting a geometric straight-line trend, we use the data of monthly average production of crude petroleum in the United States from 1945 to 1958. The original data and the various computations for the trend are presented in Table 17.5. The general procedure for calculating this trend is outlined below:

(1) Plot the natural values of both xs and ys on semilogarithmic paper, with the logarithmic scale extending along the Y axis.

(2) Examine the data analytically on the basis of economic theory and observe them graphically to determine if a geometric straight line can reasonably describe the general drift of the series. If the answer is affirmative:

(3) Set the origin at the middle of the series and assign the X values to each year.

(4) Square the X values and obtain $\Sigma(x^2)$.

(5) Find the logarithms of all Y values and obtain $\Sigma \log y$.

(6) Multiply the logarithms of each Y value by the X value of that year and sum up these products to get $\Sigma(x \log y)$.

(7) Compute the Y intercept and the slope with the quantities obtained in the previous steps. For our illustrative series, they are

$$\log a = \frac{\Sigma \log y}{n} = \frac{31.5667}{14} = 2.25477$$

$$\log b = \frac{\Sigma x (\log y)}{\Sigma x^2} = \frac{3.34296}{227.5} = 0.014694$$

(It may be noted that a is the logarithmic value of trend at the origin and that b is the logarithm of the rate of change per unit of time stated as a ratio.)

(8) Write out the trend equation in logarithmic terms. For our example, it is

$$\log y_t = 2.25477 + 0.014694x$$

(Origin, 1951–1952; time unit, 1 year; Y, monthly average crude petroleum production, 1945–1958, in millions of barrels.)

TABLE 17.5 CRUDE PETROLEUM PRODUCTION IN THE UNITED STATES, 1945–1958 (MONTHLY AVERAGES IN MILLIONS OF BARRELS) COMPUTATION OF GEOMETRIC STRAIGHT-LINE TREND

Year (1)	x (2)	x^2 (3)	y (4)	log y (5)	x (log y) (6)	log y_t (7)	y_t (8)
1945	−6.5	42.25	142.8	2.15473	−14.005745	2.15926	144.3
1946	−5.5	30.25	144.5	2.15987	−11.879285	2.17395	149.3
1947	−4.5	20.25	154.7	2.18949	− 9.852706	2.18865	154.4
1948	−3.5	12.25	168.3	2.22608	− 7.791280	2.20334	159.7
1949	−2.5	6.25	153.5	2.18611	− 5.465275	2.21804	165.2
1950	−1.5	2.25	164.5	2.21617	− 3.324255	2.23273	170.9
1951	−0.5	0.25	187.3	2.27254	− 1.136270	2.24742	176.8
1952	0.5	0.25	190.8	2.28058	1.140290	2.26211	182.9
1953	1.5	2.25	196.4	2.29314	3.439710	2.27681	189.2
1954	2.5	6.25	192.9	2.28533	5.713325	2.29150	195.7
1955	3.5	12.25	207.0	2.31597	8.105895	2.30620	202.4
1956	4.5	20.25	218.1	2.33846	10.523070	2.32089	209.4
1957	5.5	30.25	218.1	2.33846	12.861530	2.33559	216.6
1958	6.5	42.25	204.1	2.30984	15.013960	2.35028	224.0
Total	0.0	227.50	2543.0	31.56677	+ 3.342965		

Source: Business Statistics, 1959.

(9) Get the trend values in logarithmic terms. This is done simply by substituting the X values in the trend equation. (See column 7 in Table 17.5.)

(10) Look up the antilogs of the logarithmic trend values. The results are, of course, trend values in natural numbers.

(11) Plot two or more natural number trend values to locate the trend line on the semilogarithmic chart.

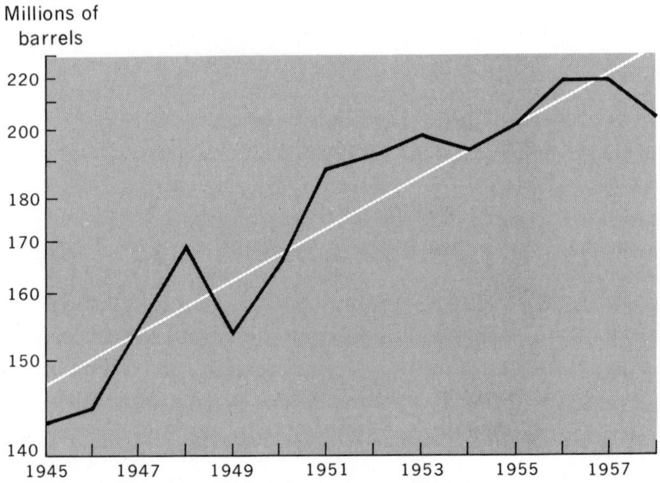

Figure 17.5 Geometric straight-line trend for United States crude petroleum production, 1945–1958. (Source: Table 17.5.)

In fitting the trend with this procedure, we satisfy the condition that the sum of the squares of the logarithmic deviations is a minimum. That is, the deviations to which this condition relates are the differences between logarithms of the original values and the logarithms of the trend values. The geometric straight line, therefore, should not be thought of as the same as the least squares arithmetic straight line for which the sum of the squares of natural number deviations is a minimum.

When we use this method, it should also be remembered that both the Y intercept and the slope are logarithmic values. It can be seen, from equation (17.8), that $\log a$ is the logarithm of the geometric mean of the series. The antilog of this value is the trend value in natural numbers at the origin. In our illustrative problem, the antilog of 2.25477 is 179.8, which is, therefore, the trend value between 1951 and 1952.

Log b is the logarithm of the ratio of each time unit to the preceding time unit. In our example the value of $\log b$ is 0.014694, which is the logarithm of 1.035. This value of 1.035 is the ratio of the trend value of each year to the preceding year. To facilitate the interpretation of this value, we usually express it as a percentage and then deduct 100 from it. The result is the constant rate of change in Y per unit of X. For the series under consideration, we have $103.5 - 100 = 3.5$ percent. Thus we state: The average monthly crude petroleum production increases, on the average, at a constant rate of 3.5 percent per year.

The property of b for a geometric straight line is of considerable significance. When it has been transformed into a pure percentage it defines the average annual rate of growth or decline of the series. Being an abstract measure, it permits comparison of trends, all described by straight lines on ratio paper, of series with different original units. Series, such as production of all kinds, population, or national income, become immediately comparable, and conclusions about the direction and magnitude of economic activities can be readily drawn. This measure, consequently, provides an effective device for the study of the socioeconomic changes in a nation.

It may also be noted that the original trend equation for the geometric straight line is of the exponential form

$$y_t = ab^x$$

which, by using logarithms, becomes

$$\log y_t = \log a + x (\log b)$$

Thus, the illustrative result

$$\log y_t = 2.25477 + 0.014694x$$

can also be written in the form

$$y_t = 179.8(1.035)^x$$

For further discussion of the properties of the exponential equation, see Chapter 20.

17.8

Curvilinear Trends

The straight-line trends belong to a family of simple polynomials whose general expression is

$$y_t = a + bx + cx^2 + dx^3 + ex^4 + \cdots \tag{17.11}$$

where

a = the Y intercept
b = the slope, or the rate of change in Y, at the origin
c = the rate of change of the slope at the origin—the degree to which the curve is bent
d = the rate of change in the degree to which the curve is bent at the origin
e = the rate of change in the rate of change in the degree $\cdots$; and so forth

The straight line is the special and simple case where all constants beyond b are zero. It is called a power curve of the first degree. A convenient way to remember the degree of power curves is to relate it to the exponent of x in the last term of the equation. A power curve has the same degree as the exponent of x in the last term. The straight line has only the first two terms on the right of the equality sign in expression (17.11). With three terms on the right, we have a second-degree curve, and so on. Some of the common types of nonlinear trends that might be used for particular types of data are shown in Figure 17.6.

Although the straight line is not always an appropriate representation of trend, it is nevertheless generally the most useful member of the family of trends. When-

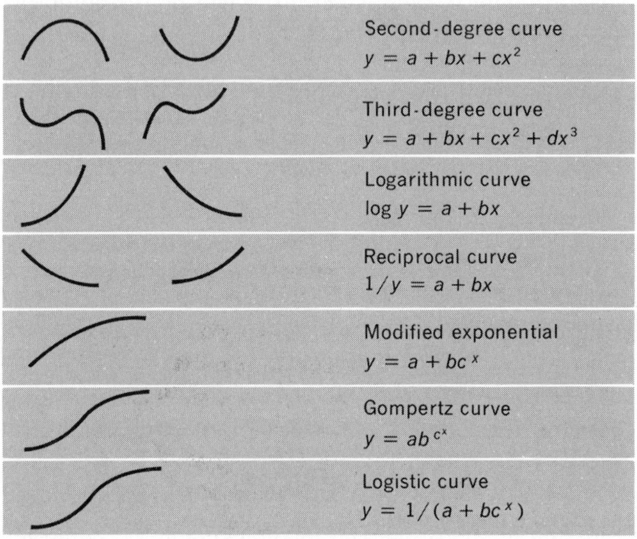

Figure 17.6. Illustration of some typical forms of trend equations.

ever a curve instead of a straight line is necessary to describe the general drift of a series, we must be careful not to fit a high-degree parabola to the data because, if we do, we are almost certain to mix trend with cycle. Moreover, observation of Figure 17.6 will show that none of polynomials of second or higher degrees can be projected very far without going off the page. We shall therefore give only an illustration of fitting a second-degree parabolic trend, and comment briefly on the nature of the third-degree polynomial. No more will be said about higher-degree polynomial trends, since they are not very useful in practice.

The general equation of the second-degree parabolic trend is

$$y_t = a + bx + cx^2 \tag{17.12}$$

where a is still the Y intercept, b is the slope of the curve at the origin, and c is the rate of change in the slope. It should be noted that, just as b is a constant in the first-degree curve, c is a constant in the second-degree curve. Here the c value determines whether the curve is concave or convex and the extent to which the curve departs from linearity. The second-degree curve, as shown in Figure 17.6, has curvilinearity without change of inflection and c determines the existence or nonexistence of curvilinearity under these conditions.

Since there are three constants in the second-degree power curve, if it is to be fitted by the method of least squares to a series, three normal equations, as given below, are required:

$$\begin{aligned} &1. & \Sigma y &= na + b\Sigma x + c\Sigma x^2 \\ &2. & \Sigma xy &= a\Sigma x + b\Sigma x^2 + c\Sigma x^3 \\ &3. & \Sigma x^2 y &= a\Sigma x^2 + b\Sigma x^3 + c\Sigma x^4 \end{aligned} \tag{17.13}$$

When the middle of the time series is taken as the origin, then all the sums of the odd powers of x become zero, and the three normal equations are reduced to

$$\begin{aligned} &1. & \Sigma y &= na + c\Sigma x^2 \\ &2. & \Sigma xy &= b\Sigma x^2 \\ &3. & \Sigma x^2 y &= a\Sigma x^2 + c\Sigma x^4 \end{aligned}$$

A little algebraic manipulation with these equations will show that

$$c = \frac{n\Sigma x^2 y - \Sigma x^2 \Sigma y}{n\Sigma x^4 - (\Sigma x^2)^2} \tag{17.14}$$

$$a = \frac{\Sigma y - c\Sigma x^2}{n} \tag{17.15}$$

$$b = \frac{\Sigma xy}{\Sigma x^2} \tag{17.16}$$

To obtain the constant values for a second-degree parabolic trend, therefore, we need five sums, which are Σx^2, Σx^4, Σy, Σxy, and $\Sigma x^2 y$. Table 17.6 illustrates

TABLE 17.6 SYNTHETIC RUBBER PRODUCTION IN THE UNITED STATES, 1944–1958 (MONTHLY AVERAGES IN THOUSANDS OF LONG TONS) COMPUTATION OF SECOND-DEGREE PARABOLIC TREND

Year (1)	x (2)	x^2 (3)	x^4 (4)	(y) (5)	xy (6)	x^2y (7)	(y_t) (8)
1944	−7	49	2,401	63.6	−445.2	3,116.4	63.1
1945	−6	36	1,296	68.4	−410.4	2,462.4	58.2
1946	−5	25	625	61.7	−308.5	1,542.5	54.4
1947	−4	16	256	42.4	−169.6	678.4	51.8
1948	−3	9	81	40.7	−122.1	366.3	50.3
1949	−2	4	16	32.8	− 65.6	131.2	50.0
1950	−1	1	1	39.7	− 39.7	39.7	51.0
1951	0	0	0	70.4	0.0	0.0	53.1
1952	1	1	1	66.5	66.5	66.5	56.3
1953	2	4	16	70.7	141.4	282.8	60.8
1954	3	9	81	51.9	155.7	467.1	66.4
1955	4	16	256	80.9	323.6	1,294.4	73.2
1956	5	25	625	90.0	450.0	2,250.0	81.2
1957	6	36	1,296	93.2	559.2	3,355.2	90.3
1958	7	49	2,401	87.9	615.3	4,307.1	100.6
Total	0	280	9,352	960.8	750.6	20,360.0	

Source: Business Statistics, 1959.

the procedure of obtaining these totals. For that series, we have

$$c = \frac{(15)(20{,}360) - (280)(960.8)}{(15)(9352) - (280)^2} = 0.58785$$

$$a = \frac{960.8 - (0.58785)(280)}{15} = 53.08$$

$$b = \frac{750.6}{280} = 2.6807$$

With these values, the trend equation becomes

$$y_t = 53.08 + 2.6807x + 0.58785x^2$$

(Origin, 1951; time unit, 1 year; Y, average monthly synthetic rubber production, 1944–1958, in thousands of long tons.)

The trend values are obtained by substituting the xs and x^2s into the trend equation. To draw curvilinear trends, it is advisable to plot all the trend values on the chart. The trend drawn through the original data for our example is shown in Figure 17.7.

It may be pointed out that the value of b in a second-degree power curve is identical with that in the straight-line trend for the same series if the second-degree curve is fitted from the middle of the series. In our present situation, furthermore, the positive value for b is responsible for the general upward drift, and the positive value of c makes the curve convex to the X axis. Conversely, negative values of b and c will make the curve have a downward drift concave to the base. It is the characteristic of a second-degree curve that it changes its direction only once.

It is also important to note that when the values of x^2 become larger at the extremes of the series, the influence of c upon the trend increases and thereby the departure from linearity becomes quite marked. This limits somewhat the usefulness of this type of trend for the purpose of forecasting trend values.

Now, a few words on the third-degree parabolic trend. This trend curve has this general equation

$$y_t = a + bx + cx^2 + dx^3 \tag{17.17}$$

which has four constants, and therefore to fit it four normal equations are required. The addition of the constant d reflects the fact that the concavity of the cubic curve changes from one point on the curve to another. Thus, it is quite possible for a curve of this type, by changing its concavity, to reverse the direction of the bend. It may actually assume an S shape. It is therefore a very flexible curve and provides a reasonably good description of some economic time series, especially if the period covered is relatively long. However, as has already been mentioned, the flexibility of higher-degree polynomials, such as this, is actually its weakness as well. This is so because when such a curve follows the data too closely it may fail to reveal the persisting and irreversible general drift of the series under the strong influences of cyclical fluctuations.

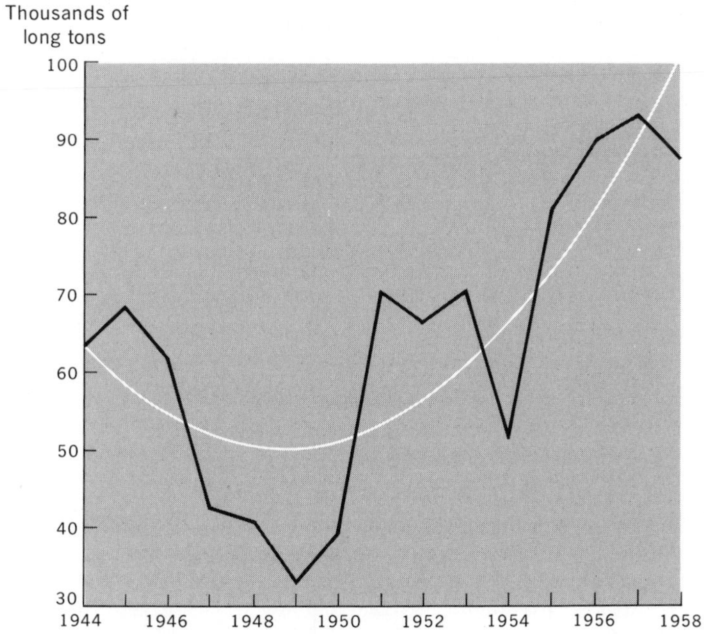

Figure 17.7 Second-degree parabolic trend for United States synthetic rubber production, 1944–1958. (Source: Table 17.6.)

17.9

Method of Moving Averages

A moving average may be considered as an artificially constructed time series in which each periodic figure is replaced by the mean of the value of that period and those of a number of the preceding and succeeding periods. Although the computation of moving averages is simple and straightforward, this measure is not without theoretical and technical interest. The properties and utility of moving averages can perhaps be brought out more vividly by a hypothetical illustration. Hypothetical data in Table 17.7 are assumed to have a uniform cyclical duration of 5 years and equal amplitude of 2 units. Three-year and five-year moving averages are fitted to the data. The procedure for calculating three-year moving averages is as follows:

(1) Compute the three-year moving totals. This is done by adding up the values of the first three years and centering at the second year. This is the first three-year moving total. Then the value of the first year is dropped and the value of the fourth year is added to form the second three-year moving total, which is centered at the third year. And so the computation moves through the end of the series. The three-year moving totals are entered in column 3 of Table 17.7.

TABLE 17.7 COMPUTATION OF THREE-YEAR AND FIVE-YEAR MOVING AVERAGES FOR HYPOTHETICAL DATA OF A FIVE-YEAR CYCLE WITH UNIFORM DURATION AND AMPLITUDE

Year (1)	Original Value (2)	Three-Year m.t. (3)	Three-Year m.a. (4)	Five-Year m.t. (5)	Five-Year m.a. (6)
1	1				
2	2	6	2.0		
3	3	7	2.3	9	1.8
4	2	6	2.0	10	2.0
5	1	5	1.7	11	2.2
6	2	6	2.0	12	2.4
7	3	9	3.0	13	2.6
8	4	10	3.3	14	2.8
9	3	9	3.0	15	3.0
10	2	8	2.7	16	3.2
11	3	9	3.0	17	3.4
12	4	12	4.0	18	3.6
13	5	13	4.3	19	3.8
14	4	12	4.0	20	4.0
15	3	11	3.7	21	4.2
16	4	12	4.0	22	4.4
17	5	15	5.0	23	4.6
18	6	16	5.3	24	4.8
19	5	15	5.0		
20	4				

(2) The three-year moving averages are obtained simply by dividing each of the three-year moving totals by 3. These values for our example are entered in column 4 of Table 17.7.

A similar procedure is used to compute the five-year moving averages. In a five-year moving average, the value of each year is replaced by the mean of the values of the five successive years of which two precede and two succeed the given year. Both five-year moving totals and moving averages are, of course, centered in the middle of the respective five-year periods, with the first five-year moving total and the moving average entered in the third year. Columns 5 and 6 in Table 17.7 illustrate the computation of five-year moving averages. It may also be pointed out in passing that in computing moving averages for an even number of periods, the procedure becomes more complicated. Such a procedure will be discussed in the next chapter.

The results of our calculations are plotted with original data in Figure 17.8. Both sets of moving averages may be considered as the statistical expression of secular movement of our hypothetical series.

By studying Table 17.7 and Figure 17.8, we can reach several conclusions about the characteristics of moving averages:

(1) If the data show a periodic fluctuation, moving average of equal length period will completely eliminate the periodic variations.

(2) If the series changes on the average by a constant amount per time unit and its fluctuations are periodic, a moving average of equal length will be linear.

(3) Even if the data show a periodic fluctuation, a moving average of unequal length period, no matter how small the difference is between the duration of periodicity of original series and the duration of the moving average, the moving average cannot obliterate completely the periodic variations in the original series.

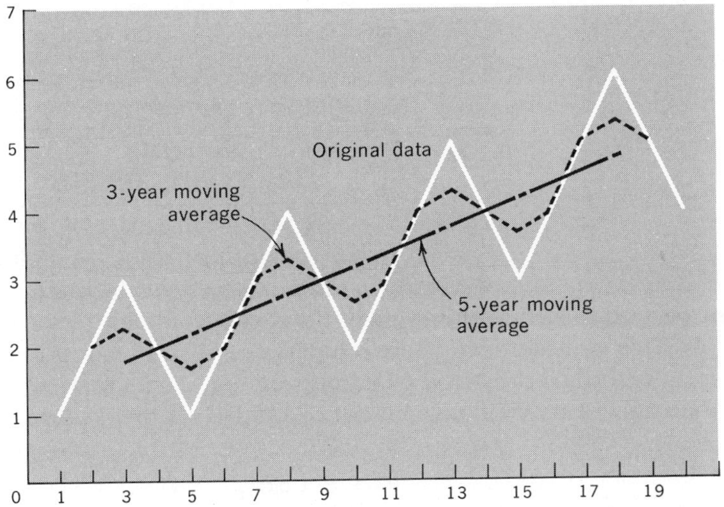

Figure 17.8 Three-year and five-year moving average trends fitted to hypothetical data of a five-year cycle with uniform duration and amplitude.

The averaging process then only tends to smooth out somewhat the short-run highs and lows.

(4) From the preceding paragraph it may be deduced that if the data fluctuates without periodicity, a moving average would only have the effect of eliminating the lesser variations to some extent and therefore smooth the general appearance of the time series.

With these properties in mind, we can see that the moving average may constitute a satisfactory trend for a series that is basically linear and that is regular in duration and amplitude. But these conditions are seldom met by any economic time series. In connection with the use of the moving average as a trend measure, there are two disadvantages. The first is that in computing moving averages we lose some years at each end of the series. For instance, we lose two years at each end with the five-year moving average. Had we computed a seven-year moving average, we would have lost three years at each end.

The second drawback, which is still more serious, is that the moving average is not represented by some mathematical function and, therefore, is not capable of objective projection into the future. Since one of the main objectives of trend analysis is that of forecasting, the moving average, lacking this desirable property, is no longer in wide use as a trend measure.

Nevertheless, the method of moving averages is a very useful technique in analyzing time series. First of all, in problems in which the trend of the time series is clearly not linear and in which we are concerned only with the general movements of the time series, be it a trend, or a cycle, or possibly both, it is customary to study the smoothing behavior of the series by applying a moving average. Second, as will be shown in the next chapter, the characteristic of a moving average is the basis of seasonal analysis. In that context, a twelve-month moving average is calculated to eliminate seasonal variations which have a periodicity of one year.

17.10 Shifting the Trend Origin

In computing trends, the middle of the time series is always used as the origin in order to short-cut the computation. But very often we need to change the origin of the trend equation to some other point in the series. This is either to facilitate comparisons of trend values among neighboring years or to convert a trend equation from an annual to a monthly basis. Shifting the origin is a very simple matter. Consider the trend equation for data in Table 17.3:

$$y_t = 92.4 + 1.82x$$

(Origin, 1962; time unit, 1 year; Y, coffee production in South America in millions of pounds, 1956–1968.)

The origin of 1962 is taken to mean that the precise origin is the center of that year: July 1. Now, if we wish to shift the trend equation to 1956, we note that this year precedes the stated origin of 1962 by 6 time units. Consequently, we must deduct 6 times the annual increment, that is, $b(-6)$, from the trend value of 1962 as below:

$$y_t = 92.4 + 1.82(-6) = 81.5$$

The value of 81.5 becomes the trend value at the new origin, 1956, and the trend equation may now be written as

$$y_t = 81.5 + 1.82x$$

(Origin, 1956; time unit, 1 year; Y, coffee production in South America in millions of pounds.)

Again, as another example, consider the trend equation for shipments of radiators and convectors in Table 17.4,

$$y_t = 3{,}228 - 200.7x$$

(Origin, 1951–1952; time unit, 1 year; Y, monthly average shipment in thousands of square feet of radiation.)

The origin of 1951–1952 means that the precise origin is the center of this period: January 1, 1952. Suppose we wish to express the trend equation with an origin at 1952; that is, at July 1, 1952. Since X unit is a whole year, the new origin desired is 0.5 unit succeeding the old origin. To achieve this we must then add one half of the annual increase, or $b(\frac{1}{2})$, to the trend value at the old origin. Then we have

$$y_t = 3228 - 200.7(\tfrac{1}{2}) = 3128$$

The trend equation reads:

$$y_t = 3128 - 200.7x$$

(Origin, 1952; time unit, 1 year; Y, monthly average shipment in thousands of square feet of radiation.)

It may be noted that the shifting of trend origin amounts to finding the new Y intercept, that is, the new value of a. The value of b remains unchanged, since the slope of the trend line is the same, irrespective of the origin.

From the preceding discussion, the procedure of shifting the origin may be generalized by the expression

$$y_t = a + b(x \pm k) \tag{17.18}$$

where k is the number of time units shifted. If the origin is shifted forward in time, k is positive; if the converse, k is negative. Thus, to change the origin of the trend equation for coffee production in South America, $y_t = 92.4 + 1.82x$, from 1962 to 1956, we have

$$\begin{aligned} y_t &= 92.4 + 1.82(x - 6) \\ &= 92.4 + 1.82x - 10.92 \\ &= 81.5 + 1.82x \end{aligned}$$

(Origin, 1956.)

Formula 17.18 can also be expanded to cover parabolic trend equations. As an example, we may consider the following second-degree trend equation for the average monthly synthetic rubber production:

$$y_t = 53.08 + 2.6807x + 0.58785x^2$$
(Origin, 1951.)

If we wish to shift the origin for this equation to 1944, we may follow a procedure suggested by formula 17.18:

$$\begin{aligned} y_t &= 53.08 + 2.6807(x - 7) + 0.58785(x - 7)^2 \\ &= 53.08 + 2.6807(x - 7) + 0.58785(x^2 - 14x + 49) \\ &= 63.11957 - 5.5492x + 0.58785x^2 \end{aligned}$$
(Origin, 1944.)

17.11

Reducing Annual Trend Values to Monthly Values

For simplicity and ease of computation, trends are usually fitted to annual rather than quarterly or monthly data. For the purpose of eliminating trend from a time series so that seasonal or cyclical patterns may be studied, monthly trend values are needed. Fortunately, monthly trend equations can be derived from annual trend equations without any loss of accuracy. The conversion of annual trend equations to monthly values depends upon the units in which the original data are expressed. There are two possible situations. One is that Y units are annual totals, such as the GNP, annual total production of steel, or total population. The other is that the Y units are monthly averages, such as the monthly average petroleum production presented in Table 17.5. The monthly averages of petroleum production, like other monthly average data, are annual productions for each year divided by 12. We shall discuss these two situations in turn.

The case where the time unit is one year and the annual value is a total of monthly values is illustrated by our previous example of pneumatic casing production. Since in such a situation, the a value is expressed as the sum of the production of each of the 12 months of the year, it must be divided by 12 in order to reduce the a value to terms of monthly Y values. The a value so computed is clearly the arithmetic mean of the monthly totals. The origin of the monthly series is $1/12$ of the a value computed from annual totals.

The value of b in an annual trend equation for annual totals measures the annual total change in Y. When b is divided by 12, it has been reduced to annual change in monthly magnitudes. When it is divided by 12 again the annual change in monthly magnitudes is further reduced to monthly change in monthly magnitudes. The necessity of dividing b twice by 12, that is, by 144 altogether, must be clearly understood. The division of annual change by 12 gives us only the change from some month in a given year to the corresponding month in the following year, or the annual change in monthly magnitudes. However, we are here seeking for an expression of the change in each and every month, that is, monthly change in monthly magnitudes. Thus, b must be divided again by 12.

In conclusion, to reduce an annual trend equation to a monthly trend equation when the original data are given as annual totals, a is divided by 12 and b is divided by 144. Once again, the annual trend equation for coffee production in South America may be cited as an illustration. Recall that it is

$$y_t = 92.4 + 1.82x$$

(Origin, 1962; time unit, 1 year; Y, coffee production in millions.)

which can be converted to a monthly basis as

$$y_t = \frac{92.4}{12} + \frac{1.82}{144}x = 7.7 + 0.013x$$

(Origin, July 1, 1962; time unit, 1 month; Y, monthly production in millions.)

Note that the origin for this monthly trend equation is the beginning of July. Usually, we would have the origin at the center of a period in trend analysis. Thus, we may rewrite the equation in the following form:

$$\begin{aligned} y_t &= 7.7 + 0.013(x + \tfrac{1}{2}) \\ &= 7.7065 + 0.013x \end{aligned}$$

(Origin July, 1962.)

Where data are given as monthly averages per year, the a value in the annual trend equation is already the arithmetic mean of the twelve-month total. In other words, it is already at the monthly level. The b value now represents the annual change in monthly magnitudes. As a result, to convert an annual trend equation when annual data are expressed as monthly averages, a would remain unchanged and b is divided by 12 only once. Consider the annual trend equation for production data of crude petroleum,

$$\log y_t = 2.25477 + 0.014694x$$

(Origin, 1951–1952; time unit, 1 year; Y, monthly average production in millions of barrels.)

To convert this equation into monthly terms, we have

$$\log y_t = 2.25477 + \frac{0.014694}{12}(x)$$

$$= 2.25477 + 0.0012245x$$

(Origin, January 1, 1952; time unit, 1 month; Y, monthly production in millions of barrels.)

We may, on occasion, need to convert an annual trend equation into quarterly values. The procedure involved is similar to what has just been discussed. This is left as an exercise for the student.

17.12 Choice of Period and Mathematical Expression for Trend

Before concluding this chapter, we must mention two more analytical considerations in measuring trend: the period to be studied must be carefully chosen, and the mathematical expression employed to describe the secular movement must be decided upon.

Generally speaking, the longer the period covered the more significant the trend. Secular movements cannot be expected to reveal themselves clearly within a short span of time. When the period is too short, the general drift of the series may be unduly influenced by the cyclical fluctuations. This would make it difficult to separate the various sources of variations in time series. As a minimum safety guard, it may be said that to compute trend the period must cover at least two or three complete cycles.

Furthermore, in order to produce accurate values of the Y intercept and the slope of the trend equation, the period must be so selected that the number of prosperous years must be about the same as the number of depressed years, and that the first and last years of the series must be on the opposite sides of the cycle, that is, if the series starts with an upswing it should end with a downswing. If the number of prosperous years is greater than the number of depressed years, the Y intercept and, therefore, the level of the trend would be too high. If the situation is reversed, then the level of trend would be too low. If the series starts and ends with a cyclical upswing, the value of slope would be too large. If both the first and last years of the series coincide with a cyclical downswing, then the value of the slope would be too small. Correct value of the slope, therefore, can be derived only for a series that starts with an upswing and ends with a downswing, or vice versa. It may also be noted that usually bias in the slope is more damaging than bias in the level, since a trend becomes progressively worse as X increases if b is inaccurate.

The selection of the mathematical expression for a trend is always a difficult matter, because there is really no positive, objective test that can be applied. In any case, as was suggested in the preceding chapter, in the selection of the mathematical expression to be employed in trend fitting for a given series, the very first thing to do is to establish a hypothesis or theory about the long-run average behavior of the series. Next, we should plot the data and subject them to economic analysis in accordance with the hypothesis we have formulated. If observation of the graph seems to support our hypothesis about the data, then a mathematical expression and the method of fitting it may be finally determined in conjunction with the following criteria:

(1) The mathematical expression, subjected to modification in view of the other criteria, must provide the best fit.

(2) The mathematical expression must be able to provide reasonable extrapolated values.

(3) The mathematical expression must be able to provide a trend that is free from cyclical and irregular influences.

The selection of a mathematical expression, therefore, cannot be defended by mathematical tests of goodness of fit alone. It is for this reason that more complex curves are avoided in applied work. Such curves may on occasion provide excellent fits, but they are neither free from cyclical influences nor capable of yielding reasonable extrapolated values.

The truth of the matter is that trend values are affected by a myriad of economic and social forces, legal framework, and natural phenomena; hence any attempt to express the aggregate impact of these by a simple formula is often fruitless and frustrating.

We may conclude our discussion by saying that trend fitting is at best an elusive process. It can contribute only an approximate concept of long-run forces. For this reason, it appears, more time should be spent in studying the underlying socioeconomic, technological, and legal factors that contribute to the general drift than in attempting to improve the fit of a curve. For this purpose, the freehand curve that seems to average out the cyclical-irregular movements in an approximate fashion may often prove sufficient. Although the freehand method is not recommended for beginners, it has considerable merit in the hands of experienced statisticians and is widely used in applied situations.

Glossary of Formulas

(17.1) $y_t = a + bx$ This formula represents the general linear trend equation. When a and b are assigned specific numerical values, the formula then characterizes a specific trend line. In it, y_t refers to computed trend values.

(17.2) $b = \dfrac{p_2 - p_1}{t}$ The slope of a linear trend line is determined by this formula when the line is constructed by selected points of the actual series. In this method, the first point, p_1, is conveniently set as the origin. The value for p_1, then, is the Y intercept. In this equation, t refers to the number of years separating the two selected points.

(17.3) $a = \dfrac{\Sigma_1 + \Sigma_2}{t_1 + t_2}$ and **(17.4)** $b = \dfrac{\Sigma_2 - \Sigma_1}{t_1(n - t_2)}$ When a linear trend is constructed by the semiaverage method, the values of the Y intercept and slope of the trend equation are estimated by these two formulas. In these, t_1 stands for the number of years from which the first average is computed, t_2 is the number of years from which the second average is computed, and n is the total number of years contained in the time series. It should also be remembered that by the semiaverage method the middle of the series is chosen as the origin.

(**17.5**) $a = \frac{\Sigma y}{n}$ and (**17.6**) $b = \frac{\Sigma xy}{\Sigma x^2}$ To fit an arithmetic straight-line trend by the least squares method and by taking the middle period as the origin, not only $\bar{x}$ but also Σx in the normal equations becomes zero. As a result the computations of the Y intercept and the slope, as indicated by these two formulas, are simplified.

(**17.7**) 1. $\Sigma \log y = n (\log a) + \log b(\Sigma x)$ 2. $\Sigma x (\log y) = \log a(\Sigma x) + \log b(\Sigma x^2)$
These are the pair of normal equations that must be solved simultaneously when the least squares method is used to fit a geometric or semilogarithmic straight-line trend.

(**17.8**) $\log a = \frac{\Sigma \log y}{n}$ and (**17.9**) $\log b = \frac{\Sigma x (\log y)}{\Sigma x^2}$ These equations are used to compute the values of the y intercept and the slope for a geometric straight-line trend by the least squares method. They are derived from the pair of normal equations in (17.7) after the terms containing Σx are eliminated by setting the middle of the series as the origin.

(**17.10**) $\log y_t = \log a + x (\log b)$ This is the geometric straight-line trend equation. It is interesting to note that this formula is the logarithmic expression of an exponential equation of the form $y_t = ab^x$, which is the general expression of a geometric progression. Thus, a geometric straight line represents a constant rate of change in Y per unit of X.

(**17.11**) $y_t = a + bx + cx^2 + dx^3 + ex^4 + \cdots$ This is the general expression of polynomials, or power curves. When all the constants, except a and b, are zero, we have the equation of the straight line. When c assumes a value other than zero, it becomes a second-degree parabola. When a nonzero d is added, a third-degree parabola emerges. And so on.

(**17.12**) $y_t = a + bx + cx^2$ As indicated just above, this represents a second-degree power curve. This trend equation is useful as a general method if all that can be determined is that the trend has curvilinearity without change in concavity. Furthermore, c determines the existence or absence of curvilinearity under these conditions.

(**17.13**) 1. $\Sigma y = na + b\Sigma x + c\Sigma x^2$
2. $\Sigma xy = a\Sigma x + b\Sigma x^2 + c\Sigma x^3$
3. $\Sigma x^2 y = a\Sigma x^2 + b\Sigma x^3 + c\Sigma x^4$
These three normal equations are derived for the purpose of fitting a second-degree parabolic trend. It is interesting to note that the first two of these would become the normal equations for the straight line when the last term is eliminated.

(**17.14**) $c = \frac{n\Sigma x^2 y - \Sigma x^2 \Sigma y}{n\Sigma x^4 - (\Sigma x^2)^2}$ (**17.15**) $a = \frac{\Sigma y - c\Sigma x^2}{n}$ (**17.16**) $b = \frac{\Sigma xy}{\Sigma x^2}$
In fitting a second-degree parabolic trend by the least squares method, the sums of the odd powers of x in the preceding three normal equations will vanish. As a consequence, the formulas for these three constants can be derived from simplified normal equations into the present forms. It should be remembered that a equals the trend value at the origin; b equals the slope of the curve at the origin when the origin is located at the middle of the series; and c equals the rate of change in the slope. As was mentioned in connection with (17.12), c determines the existence or nonexistence of curvilinearity. In other words, if $c = 0$, curvilinearity does not exist.

(**17.17**) $y_t = a + bx + cx^2 + dx^3$ This is, as the reader can readily recognize now, the trend equation of the third degree. It differs from the second-degree equation in that another constant, d, is added. The constant d reflects the fact that the concavity of the cubic curve changes once. Thus, d measures the rate of change in the degree to which the curve is bent at the origin.

(**17.18**) $y_t = a + b(x \pm k)$ This equation is used to shift the origin of trend equation. Here k is the number of time units shifted. It may be positive or negative, depending on whether the origin is shifted forward or backward in time. This expression can be extended to cover parabolic equations by substituting $(x \pm k)$ for x, expanding the results, and recombining the terms.

Problems

17.1 What does the trend of a series measure?

17.2 What objections do we have for each of the following methods of measuring trend: (a) the freehand method, (b) the selected points method, and (c) the semiaverage method?

17.3 The freehand method can be used only to approximate a straight-line trend. True?

17.4 What are the mathematical properties of a trend line fitted by the least squares method?

17.5 How does the semilogarithmic straight-line trend differ from the arithmetic straight-line trend?

17.6 What are the properties of moving averages? Under what conditions would a moving average provide a satisfactory trend?

17.7 What are the limitations of moving averages as a measure of trends?

17.8 In each of the following series, what mathematical expression would you choose to fit a trend? Give reasons.
a. the population growth of the United States since 1800
b. the growth of sales of television sets since World War II
c. the growth of production of General Motors since its founding
d. the GNP series of Canada since 1900
e. purchases of ordinary life insurance in the United States since 1940
f. consumption of cigarettes since the 1920s.

17.9 Give a critical evaluation of each of the following statements:
a. The only standard of judging the appropriateness of a given fitted trend is the mathematical test of "goodness-of-fit."
b. "Extrapolation" means inferring values on a curve beyond the available data.
c. The only consideration we must have in selecting the period for trend analysis is that the period must cover at least ten years.
d. In the last analysis, the freehand method has considerable merit.
e. If a higher-degree polynomial gives a good fit to data, it must always be used in order to yield reasonable extrapolated values.

17.10 Give a verbal statement for each of the following trend equations:
a. $y_t = 36{,}000 + 1200x$
(Origin, 1950; time unit, 1 year; Y, annual total production in hundreds of tons.)
b. $y_t = 1{,}200 - 125x$
(Origin, 1945–1946; time unit, 1 year; Y, monthly average production in thousands of pounds.)
c. $\log y_t = 1.73 + 0.02200x$
(Origin, 1942; time unit, 1 year; Y, monthly average consumption in millions of barrels.)
d. $y_t = 1500 + 36x$
(Origin, 1950; time unit, 1 quarter; Y, total quarterly sales in thousands of dollars.)

17.11 Use the trend equations in problem 17.10 and perform the following: Convert each trend equation into monthly values and change the origins of monthly trend equations so derived to June 15, 1950.

17.12 Use data in Table 17.1 to perform the following:
a. Plot the data on arithmetic paper.
b. Compute an arithmetic straight-line trend by the least squares method.
c. Draw the trend line on the chart of the original data.
d. Compare your result with the trend for the same data constructed by the selected method.

17.13 Drop 1954 from the data in Table 17.2 and then do the same as for the preceding problem, except that the results should be compared with those of the semiaverage method for the same data.

17.14 Construct a geometric straight-line trend and plot the result from the data below:

AVERAGE MONTHLY DOMESTIC DEMAND FOR GASOLINE, 1940–1958 (MILLIONS OF BARRELS)

Year	Demand	Year	Demand	Year	Demand
1940	49.1	1946	61.3	1952	96.4
1941	55.6	1947	66.3	1953	100.5
1942	49.1	1948	72.6	1954	102.6
1943	47.4	1949	76.1	1955	111.2
1944	52.7	1950	82.9	1956	114.4
1945	58.3	1951	90.8	1957	116.1
				1958	118.2

Source: Business Statistics, 1959.

a. What annual rate of growth is implied in your trend equation?
b. What is the extrapolated value of 1960 for this series? Examine a recent issue of *Business Statistics* and explain why the extrapolated value is so different from the actual value.

17.15 Bring data in the preceding problem up to date and compute a new geometric straight-line trend. Compare the result with that previously obtained.

17.16 Find a time series of your own from an appropriate source to which a second degree parabolic trend can be fitted.

18

MEASURES OF SHORT-RUN FLUCTUATIONS

18.1

Introduction

The measurement of long-run growth is but one of the problems connected with time series analysis. Such series are also subject to short-run fluctuations, seasonal and cyclical in nature, that are of great interest and importance to economic research and business policy formulation. We shall deal with the measurements of both types of variations in this chapter.

Seasonal variations in economic time series are true periodicities and are all-pervasive. An understanding of seasonal patterns, as will be indicated later, aids the making of managerial decisions in a variety of fields and the formulation of economic policies that attempt to reduce waste in employing a nation's resources. Furthermore, accurate measurements of seasonal changes are necessary if we wish to adjust statistically for them, thereby leaving only trends, cycles, and irregular movements in the series. Deseasonalized data are often easier to interpret, since, for example, one is less likely to confuse a seasonal downswing with deterioration of general business activity.

The measurement of cyclical fluctuations is regarded by many as the central task in analyzing time series. Interest in cycles may either be in a given series—with cyclical movements of steel production, of retail prices, of volumes of sales—or in the complex patterns of cycles in general business activity. Our subsequent cycle analysis will be exclusively concerned with individual time series. We must realize, however, that cyclical changes in individual series are closely related to business cycles. Various studies have shown that the majority of economic time series that reflect such movements do conform, with varying degrees of leads and lags, to cycles in the economy at large. Thus, although our effort will be directed at identifying the unique individual cyclical patterns, the results thus obtained may also contribute to an understanding of such variations in aggregate business activity.

18.2 The Problem of Editing Monthly Data

Seasonal variations are computed from quarterly and, more frequently and appropriately, monthly data. These variations, which recur every year, do not show up if data are lumped together by years or by longer time periods. The use of monthly data often requires the editing of data for calendar variations and for number of working days. Of these, the correction for calendar variation is of greater importance, for if uncorrected it may produce a type of quasi seasonal that tends to blur the true nature of seasonal variations. The method of correcting data for calendar variation was discussed in Chapter 16. However, this correction is not always necessary.

The most important factor determining whether or not monthly data should be edited for calendar variation is the purpose for which the seasonal index is constructed and employed. Our interest in a seasonal index may be either for the purpose of analyzing the total monthly values or for studying the daily rate of activity. If we merely wish to estimate monthly total production, employment, or sales, or to approximate total values for a quarter or half year, the correction is not necessary. On the other hand, if the seasonal index is to be used as a basis for scheduling incoming raw materials, outgoing shipments, employment for rush seasons, or production, we are then concerned with the rate of activity per day and the correction must be made.

In the following sections on computing seasonal indices, correction for calendar variation is not necessary and is therefore not made. However, it should be noted that the method of constructing seasonal indices is the same for edited and unedited data. That is, no additional modifications of the method are required when applied to edited data.

18.3 Types of Seasonals

Seasonal variations in economic series are of two types: those whose pattern remains stable over a number of years, and those whose pattern changes gradually over a period of time. The measurement of constant seasonals is a simple matter, but the analysis of changing seasonals involves rather complicated theoretical as well as practical considerations.

A *seasonal index* consists of twelve numbers, one for each month of a year, or for each of the months of a number of years, showing the relative amount of activity for a year, or a number of years, that has taken place typically in each month. Thus, a seasonal index may be specific or typical. A *specific seasonal index* refers to the seasonal changes during a particular year. A *typical seasonal index* is obtained by averaging a number of specific seasonals. It is thus a generalized expression of seasonal variations for a series.

By observing specific seasonals one can determine if the seasonal pattern of a series is stable or is changing gradually or abruptly. The representativeness of a typical index demands that the specific seasonals must be stable. With changing specific seasonals, the construction of a typical seasonal index is highly questionable and of little practical utility.

18.4

Calculation of the Typical Seasonal Index

There are available several methods of isolating seasonal variations from a series, each having merits and defects. Among these, the *method of ratio to moving average* is by far superior from a theoretical point of view as well as for practical considerations. We shall therefore first give a detailed treatment of this method before turning to evaluate alternative techniques.

The method of ratio to moving average starts with the multiplicative assumption of time series. That is, $Y = TSCI$. It first attempts to estimate TC by employing a twelve-month average. After this, TC is to be eliminated in order to estimate SI. This is done by dividing TC into the original data. That is,

$$SI = \frac{TSCI}{TC}$$

Finally, irregular movements are eliminated from SI by an averaging process.

We shall use the series of frozen vegetables (stocks, cold storage, end of month), 1952–1958, to illustrate the application of the method of ratio to moving average for computing seasonals. These data are entered in column 1 of Table 18.1. The details of this method are given in the next few sections.

18.5

Estimate of Trends and Cycles

The first step of the ratio to moving average method is to estimate TC. To do this we may note again that seasonal variations have a rigid periodicity of twelve months. Also, random and sporadic factors usually produce changes in data that last less than a year. Furthermore, as was shown in the previous chapter, a moving average will smooth out fluctuations that are uniform in duration and amplitude if the term of the moving average is equal to, or is an integral multiple of, the period of the fluctuation. There is then reason to believe that a 12-month moving average will eliminate nearly all the seasonal and most of the irregular influences in the series. It cannot, however, remove all the seasonal and irregular forces. This is so because seasonal changes are periodicities, but their amplitudes may vary somewhat from year to year. For instance, during a warm winter in Miami Beach the hotel business there should be expected to be better than during a rather cold winter. Furthermore, the periodicity of irregular movements is always unknown but it is very unlikely that it would be an integral multiple of 12 months. Remembering this limitation we may say that a 12-month moving average, which attempts to eliminate seasonal and irregular variations, represents the remaining components—trend and cycle—in the series. Thus the 12-month moving average may be called the approximate *TC* curve.

Despite limitation mentioned above, it is interesting to note that the 12-month moving average for the frozen-vegetable data seems to have ironed out almost completely the seasonal and irregular movements in the series. It can be seen in Figure 18.1 that not only does the 12-month moving average form a smooth curve; it also seems no longer to follow the seasonal peaks and troughs in the original data. Indeed, this smooth curve seems to reflect the trends and cycles of the series quite well.

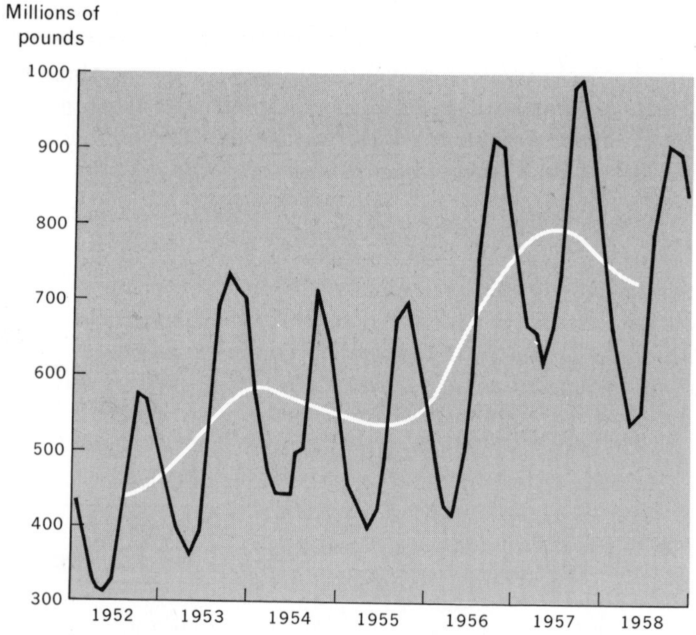

Figure 18.1 Stock of frozen vegetables in the United States, 1952–1958, and twelve-month moving average. (Source: Table 18.1.)

The calculation of a 12-month moving average starts, as does the calculation of any moving average, with adding up the first 12 monthly values in the series to form a 12-month moving total. The second moving total is obtained by dropping the value of the first month from, and adding the value of the thirteenth month to, the first moving total, and so on until all the moving totals have been obtained from the series. Then each moving total is divided by 12 to get the 12-month moving averages. The moving totals and averages so obtained fall in between two months. For instance, in our example, the first moving total and average fall between June and July of 1952, the second pair between July and August of the same year, and so on. However, data that are typical of a month should be centered at the middle of the month. Thus, to compute moving averages when an even number of time units is used, an additional step is required to center the average at the middle of each time unit. For a 12-month period, centered averages are obtained by adding the averages two at a time and dividing each sum by 2. The first centered 12-month moving average would fall on the seventh month of the series, the second on the eighth month, and so on. The procedure of computing a 12-month moving average is illustrated in columns 2 through 4 in Table 18.1.

This method of computing a 12-month moving average is rather tedious. It has been introduced merely for the purpose of revealing more clearly the complete process. A simpler procedure is to derive one average that is itself centered. This is done by calculating the sum of the values of 13 consecutive months, counting each of the 11 central months twice at intervals of approximately 12 months. The centered 12-month averages are then obtained by dividing these sums by 24. Thus let M_7 and $\bar{m}_7$ denote the 24-month moving total and the centered 12-month moving average centered opposite the seventh month of the series, respectively; and let M_8 and $\bar{m}_8$ denote the 24-month moving total and centered 12-month moving average centered opposite the eighth month of the series, respectively; and so on; then we have

$$\bar{m}_7 = \frac{M_7}{24} = \frac{y_1 + 2y_2 + 2y_3 + 2y_4 + 2y_5 + 2y_6 + 2y_7 + 2y_8 + 2y_9 + 2y_{10} + 2y_{11} + 2y_{12} + 2y_{13}}{24} \tag{18.1}$$

Also, it may be noted that

$$\bar{m}_8 = \frac{M_8}{24} = \frac{M_7 - y_1 - y_2 + y_{13} + y_{14}}{24} \tag{18.2}$$

$$\bar{m}_9 = \frac{M_9}{24} = \frac{M_8 - y_2 - y_3 + y_{14} + y_{15}}{24} \tag{18.3}$$

$$\bar{m}_{10} = \frac{M_{10}}{24} = \frac{M_9 - y_3 - y_4 + y_{15} + y_{16}}{24} \tag{18.4}$$

and so forth. These computations can be easily performed on a calculating machine or an adding machine with a subtotal device.

TABLE 18.1 FROZEN VEGETABLES, STOCKS, COLD STORAGE, END OF THE MONTH, 1952–1958 (MILLIONS OF POUNDS)
COMPUTATION OF RATIOS TO CENTERED 12-MONTH MOVING AVERAGES

Year and Month	Stock (1)	12-Month Moving Total (2)	12-Month Moving Average (3)	Centered 12-Month Moving Average (4)	Ratio to Moving Average (5)
1952					
Jan.	444	—	—	—	—
Feb.	399	—	—	—	—
Mar.	348	—	—	—	—
Apr.	314	—	—	—	—
May	302	—	—	—	—
June	337	5204	434	—	—
July	385	5255	438	436.0	88.3
Aug.	463	5306	442	440.0	105.2
Sept.	530	5378	448	445.0	119.1
Oct.	577	5448	454	451.0	127.9
Nov.	570	5507	459	456.5	124.9
Dec.	535	5554	463	461.0	116.1
1953					
Jan.	495	5637	470	466.5	106.1
Feb.	450	5748	479	474.5	94.8
Mar.	420	5906	492	485.5	86.6
Apr.	384	6066	506	499.0	77.0
May	361	6218	518	512.0	70.5
June	384	6388	532	525.0	73.1
July	468	6524	544	538.0	87.0
Aug.	574	6638	553	548.5	104.6
Sept.	688	6731	561	577.0	123.5
Oct.	737	6817	568	564.5	130.6
Nov.	722	6902	575	571.5	126.3
Dec.	705	6962	580	577.5	122.1
1954					
Jan.	631	6987	582	581.0	108.6
Feb.	564	7015	585	583.5	96.7
Mar.	513	7025	585	585.0	87.7
Apr.	470	6998	583	584.0	80.5
May	446	6965	580	581.5	76.7
June	444	6909	576	578.0	76.8
July	493	6855	571	573.5	86.0
Aug.	602	6795	566	568.5	105.9
Sept.	698	6740	562	564.0	123.8
Oct.	710	6697	558	560.0	126.8
Nov.	689	6647	554	556.0	123.9
Dec.	649	6622	552	553.0	117.4
1955					
Jan.	577	6612	551	551.5	104.6
Feb.	505	6615	551	551.0	91.7
Mar.	457	6590	544	547.5	83.5
Apr.	427	6573	548	546.0	78.2
May	396	6547	546	547.0	72.4
June	419	6522	544	545.0	76.9
July	483	6503	542	543.0	89.0
Aug.	605	6494	541	541.5	111.7
Sept.	673	6487	541	541.0	124.4
Oct.	693	6484	540	540.5	128.2

TABLE 18.1 FROZEN VEGETABLES, STOCKS, COLD STORAGE, END OF THE MONTH, 1952–1958 (Continued)

Year and Month	Stock (1)	12-Month Moving Total (2)	12-Month Moving Average (3)	Centered 12-Month Moving Average (4)	Ratio to Moving Average (5)
1955					
Nov.	663	6503	542	541.0	117.0
Dec.	624	6565	547	544.5	114.6
1956					
Jan.	558	6670	556	551.5	101.2
Feb.	496	6816	568	562.0	88.3
Mar.	450	7012	584	576.0	78.1
Apr.	424	7234	610	597.0	71.0
May	415	7476	623	616.5	67.3
June	481	7710	643	633.0	76.0
July	588	7939	662	652.5	90.1
Aug.	751	8165	680	671.0	111.9
Sept.	869	8380	689	689.0	126.1
Oct.	915	8612	718	708.0	129.2
Nov.	905	8822	735	726.5	124.6
Dec.	858	8998	750	742.5	115.6
1957					
Jan.	787	9137	761	755.5	104.2
Feb.	722	9260	772	766.5	94.2
Mar.	665	9376	781	776.5	85.6
Apr.	656	9454	788	784.5	83.6
May	625	9506	792	790.0	79.1
June	657	9530	794	793.0	82.8
July	727	9524	794	794.0	91.6
Aug.	874	9500	792	793.0	110.2
Sept.	985	9457	788	790.0	124.7
Oct.	993	9378	781	784.5	126.6
Nov.	957	9289	774	777.5	123.1
Dec.	882	9182	765	769.5	114.6
1958					
Jan.	781	9106	759	762.0	102.5
Feb.	698	9025	752	755.5	92.4
Mar.	622	8901	742	747.0	83.3
Apr.	577	8812	734	734.5	78.6
May	536	8754	730	732.0	73.2
June	550	8719	727	728.5	75.5
July	651	—	—	—	—
Aug.	793	—	—	—	—
Sept.	861	—	—	—	—
Oct.	904	—	—	—	—
Nov.	899	—	—	—	—
Dec.	847	—	—	—	—

Source: Various issues of Business Statistics.

It should be observed that in the short method the desired average is obtained from the sum of two 12-month moving totals, the first covering the months 1 to 12 and the second the months 2 to 13. It is here that one operation of division is saved. The short and the long methods, of course, yield identical results. Furthermore, both methods clearly indicate that a centered 12-month moving average is actually a weighted 13-month moving average, the weights being 1 for the first and last months and 2 for each of the central months. It is also important to note that the centering process does not in any way influence the degree of freedom from seasonal influence in the moving average because, in this average, each of the twelve months is equally represented.

18.6 Estimate of Seasonal and Irregular Movements

With the estimate of TC, the next step involves their elimination from original data so that SI can be isolated. The estimate of SI, as has already been mentioned, is made by dividing the original data by the centered 12-month moving averages, that is, $SI = TSCI/TC$. The resulting ratios are usually expressed in percentage form, as shown in column 5 of Table 18.1. From this step the name "method of ratio to moving average" is derived.

In computing the centered 12-month moving average, 6 months are lost at each end of the series. For each remaining month there is a ratio of original data to moving average. The twelve monthly ratios of each year are the specific seasonal indices for that year, which are theoretically free from trend and cyclical effects. They are therefore estimates of SI. These estimates, however, are slightly biased, mainly because the moving averages may have smoothed out some of the cyclical changes at peaks and troughs. Also, as was mentioned previously, the movement of the moving averages may still contain some irregular influences.

TABLE 18.2 SPECIFIC SEASONALS FOR END-OF-MONTH STOCK OF FROZEN VEGETABLES, 1952–1958

Month	Year						
	1952	*1953*	*1954*	*1955*	*1956*	*1957*	*1958*
Jan.		106.1	108.6	104.6	101.2	104.2	102.5
Feb.		94.8	96.7	91.7	88.3	94.2	92.4
Mar.		86.6	87.7	83.5	78.1	85.6	83.3
Apr.		77.0	80.5	78.2	71.0	83.6	78.6
May		70.5	76.7	72.4	67.3	79.1	73.2
June		73.1	76.8	79.9	76.0	82.8	75.5
July	88.3	87.0	86.0	89.0	90.1	91.6	
Aug.	105.2	104.6	105.9	111.7	111.9	110.2	
Sept.	119.1	123.5	123.8	124.4	126.1	124.7	
Oct.	127.9	130.6	126.8	128.2	129.2	126.6	
Nov.	124.9	126.3	123.9	117.0	124.6	123.1	
Dec.	116.1	122.1	117.4	114.6	115.6	114.6	

Source: Table 18.1.

Consequently, just as we consider the centered 12-month moving average as an approximate estimate of TC, so we consider the ratios to moving averages as an approximation of SI.

Before we attempt to isolate S from SI, or to construct a typical seasonal index, we must examine the specific seasonals to see whether such an attempt is justified, since the justification for such an attempt depends upon their stability. If the seasonals change through time, it would be of little practical value to have a generalized expression for them.

To examine the stability of specific seasonals, we first put them in a tabular form and then observe them together with their graphs. The specific seasonals for our illustrative problem are recorded in Table 18.2.

As graphic aid to the study of the data in Table 18.2, the ratios may be either drawn as a continuous series or represented by a *tier graph* as shown by Figure 18.2. In a tier graph, the specific seasonal indices are piled up one after the other. In the tier diagram, the horizontal scale is identical for all years. The vertical scale is a movable one. That is, while the vertical scales for all the years are the same size, the scales for adjacent years are separated by a constant vertical distance. The actual vertical scales need not be shown, since we are interested in studying only the general movements and timing of the specific seasonal patterns.

From Figure 18.2 it can be seen that all the specific seasonal indices have

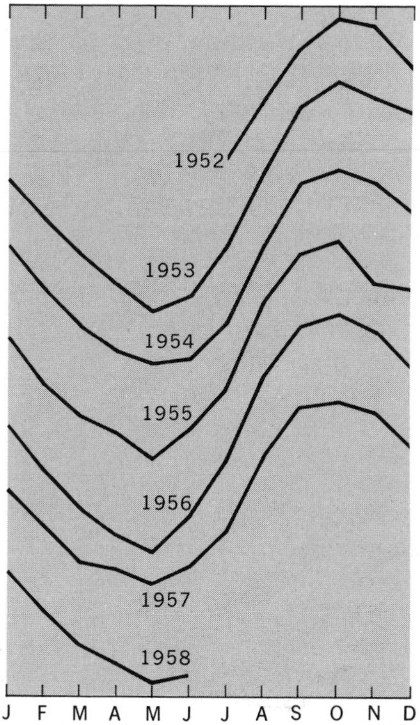

Figure 18.2 Tier chart showing specific seasonal patterns for end-of-month stock of frozen vegetables. (Source: Table 18.2.)

identical turning points: the month of May is their seasonal low and October is their seasonal high. Besides, although the amplitude of each seems to differ slightly from the others, they do give the general impression of a high degree of uniformity. These mild differences in amplitude should actually be expected since the specific seasonals contain irregular movements as well. Finally, if we read Table 18.2 horizontally month by month, we find that even though values of the same month for various years vary from each other, these variations seem to follow a random pattern without any noticeable trend. That is, all the January values, all the February values, and so on, do not seem to increase or decrease on the average through time. From these observations we may conclude that the seasonal pattern of frozen vegetable stock may be considered a stable one. As a result we decide that the construction of a typical seasonal index for the data is justified.

18.7 Eliminating the Irregular Movements

Under the assumption that the irregular component of SI is a normally distributed random variable with mean zero, then the irregular movements will be averaged out by monthly means of the SI values, leaving only the seasonal component. This assumption would be true if the irregular variations were caused by random forces alone, but we know otherwise. Irregular fluctuations may also be influenced by sporadic forces that often produce extreme values. The conventional arithmetic mean, being affected by the extreme observations all the time, cannot then be expected to eliminate from SI all the influences unrelated to seasonal fluctuation. One way out of this difficulty is to use the median of the monthly values as estimates of S. A still better alternative is the employment of a modified arithmetic mean that combines the advantages of the mean and the median.

To compute the modified monthly means we first array the monthly SI values as shown in Table 18.3. Next we disregard the largest and the smallest values of each array and include only the central values in the calculation of the monthly means. The number of values to be excluded from both ends depends upon the data on hand. In our illustration we note that some months the SI values are quite close to each other; other months have one extreme high and/or one low entry. Therefore, it is sufficient to disregard only one value from each end of the arrays. In other words, the modified means for our present illustration may be computed from the central four percentages for each month. These averages are the remaining effects in SI and thus they are estimates of S—the typical seasonal index, or simply the seasonal index. The sum of the twelve values in a seasonal index should add up to 1200, or an average 100.0. This total in our illustration is 1199.5, which is close enough for most practical and theoretical analyses of seasonal variations. Nevertheless, many people are trained to believe that everything should attain the highest accuracy. To accommodate this, we adjust the 12 modified means by multiplying each by a ratio of 1200 to the sum of the modified

TABLE 18.3 COMPUTATION OF SEASONAL INDEX FOR STOCK OF FROZEN VEGETABLES, 1952–1958

Rank	Jan.	Feb.	Mar.	Apr.	May	June	July	Aug.	Sept.	Oct.	Nov.	Dec.
1	101.2	88.3	78.1	71.0	67.3	73.1	86.0	104.6	111.9	126.6	117.0	114.6
2	102.5	91.7	83.3	77.0	70.5	75.5	87.0	105.2	123.5	126.8	123.1	114.6
3	104.2	92.4	83.5	78.2	72.4	76.0	88.3	105.9	123.8	127.9	123.9	115.6
4	104.6	94.2	85.6	78.6	73.2	76.8	89.0	110.2	124.4	128.2	124.6	116.1
5	106.1	94.8	86.6	80.5	76.7	76.9	90.1	111.7	124.7	129.2	124.9	117.4
6	108.6	96.7	87.7	83.6	79.1	82.8	91.6	111.9	126.1	130.6	126.3	122.1
Total of 4 central items	417.4	373.1	339.0	314.3	292.8	305.2	354.4	433.0	496.4	512.1	496.5	463.7
Modified means	104.4	93.3	84.8	78.6	73.2	76.3	88.6	108.2	124.1	128.0	124.1	115.9
Adjusted seasonal index	104.4	93.3	84.8	78.6	73.2	76.3	88.6	108.2	124.2	128.1	124.2	115.9

Source: Table 18.1.

means. This adjusting factor for our illustration is

$$\frac{1200}{1199.5} = 1.00041884$$

The adjusted seasonal index for our example appears in the last row of Table 18.3.

A graph of the seasonal index appears in Figure 18.3, which shows that the seasonal pattern of our illustrative series is quite pronounced, with a seasonal low associated with May that is slightly less than 27 percent below, and a seasonal high in October that is more than 28 percent above an average of 100. A moment's reflection will show that the seasonal pattern of frozen vegetable stock is closely associated with the rate of fresh vegetable production and the rate of vegetable consumption. The production of fresh vegetables normally reaches its seasonal peak in early summer and its seasonal trough in late fall. On the other hand, the rate of consumption of vegetables remains relatively constant throughout the year. Consequently, during the months of winter and spring, when harvests of fresh vegetables continue to decline but the level of vegetable consumption remains stable, frozen vegetable stocks are bound to shrink progressively. Throughout the summer and fall, when the production of vegetables is increasing without a corresponding rise in vegetable consumption, the stock of frozen vegetablse would evidently increase.

18.8 Uses of the Seasonal Index

A seasonal index may be used either analytically or synthetically. Analytically, a seasonal index is employed to adjust original data in order to yield deseasonalized data that permit the study of short-

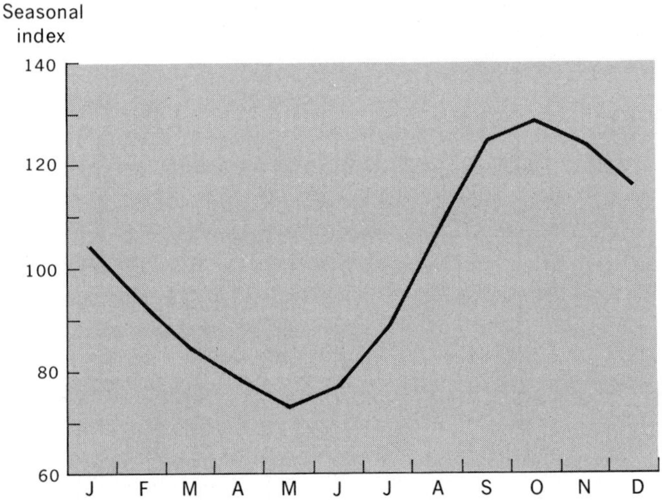

Figure 18.3 Seasonal index of frozen vegetable stock in the United States, 1952–1958.

run fluctuations of a series not associated with seasonal variations. Many economic time series, such as the Index of Industrial Production of the Federal Reserve Board, are reported in the deseasonalized form. The procedure of adjusting data for seasonal variations is a simple one. It involves merely the division of each of the original observations by the appropriate seasonal index for that month. That is, $TCI = TSCI/S$.

Synthetically, a seasonal index may be used for economic forecasting and managerial control. Let us now comment on the use of a seasonal index for synthesis in control and policy.

Management usually benefits from examining the seasonal patterns of its own business, patterns that directly influence its employment, production, purchase, sales, and inventory policies. As a simple example of the controlling function of a seasonal index, consider a manufacturer of a certain product, the sales of which follow violent seasonal variations with a peak in the middle of the fall and two troughs in the late winter and early spring. With such a seasonal variability in sales, there are many possible ways of scheduling production, two of which will be discussed.

The firm may maintain a constant monthly rate of production regardless of the month-to-month changes in sales. In other words, in each month it would produce one twelfth of the estimated total annual sales. Under this production schedule, plant capacity could be kept to a minimum compared with other plans of operating, since it would be fully utilized throughout the year. In addition, this plan would avoid layoffs of the labor force during slack seasons. This may be a great advantage if the kind of labor required is highly skilled and relatively scarce. This plan, however, will evidently raise the monthly inventory levels and intensify seasonal swings in the inventory. If costs of keeping inventories are high, this may be a serious drawback.

Another alternative is to schedule production so that it roughly corresponds to changes in sales. That is, during months of high sales, production would be speeded up, and vice versa. Such a schedule would reduce considerably both the inventory levels and their seasonal variations. The disadvantage of this plan is that a much greater plant capacity would be required in order to accommodate the peak monthly production; a consequence would be partial plant utilization in other months. Also, this schedule would necessitate some loss of personnel through layoffs during low production months.

The decision as to which of these plans should be adopted depends upon the cost estimates of the accountants. The schedule to be chosen is naturally the one that costs the least or saves the most.

Basically, the problem posed by seasonal variations is economic waste. One important source of waste attributable to seasonality is represented by idle capacities necessary to meet seasonal peak demands but not fully utilized throughout the year. According to Kuznets' *Seasonal Variations in Industry and Trade*, the true size of excess capacity caused by seasonality is estimated at about 8 to 9 percent in manufacturing and 10 to 12 percent in mineral production. The economic resources and effort devoted to the erection of extra capacity because of seasonal fluctuations might otherwise have been employed in efforts that would have enhanced the general welfare of the nation. Seasonal variations in production and employment also put heavy burdens and costs upon workers whose jobs vary with the intensity of business activity in the season. From the standpoint of practical utility, therefore, the most important synthetic use of seasonal indices is to study them with an eye to reducing or eliminating economic waste produced by seasonal fluctuations.

The attempts that have been made to reduce or eliminate the costs of seasonality are mainly initiated by individual firms. Solutions for seasonal variations in individual firms, however, usually involve the transfer of the cost to some other sector or group in the economy without real reduction of waste for the society as a whole. Although it may be difficult, even impossible in many situations, a real solution is one that can actually reduce seasonally induced costs to society.

Possible solutions for seasonality available to individual firms are numerous. By special price and advertising policies a producer confronted with a strong seasonal demand for his product may try to stabilize sales by encouraging off-season consumption. Off-season sales, such as air conditioners in the spring, heaters and furnaces in the summer, and furs in August, may be introduced. Preseason discounts on coal and oil and off-season rates at resort hotels are other examples of such efforts. Manufacturers may also offer special prices to retailers who will order and accept delivery before the start of the rush season for the outputs. Special quotations are frequently made by builders to those who will accommodate their plans to offset the construction seasonal.

Seasonal demand can also be smoothed to a great extent by managerial innovations. For instance, department store sales may be increased on weekdays and reduced on Saturdays by closing later during weekdays. The automobile companies have been able to modify the seasonal demand for cars by introducing new models in the fall instead of in the spring. The results have been two seasonal peaks for automobile sales instead of one and an increase in the stability of production and employment.

The most promising solution for seasonality is diversification. It benefits not only the firm but also society at large. Whenever diversification is possible, real costs of seasonal variations can be reduced or even eliminated. Diversification involves the development of production lines having complementary seasonal movements: while some expand seasonally, others contract. Consequently, labor and facilities can be transferred from one line to another as seasonal changes take place. However, diversification is possible only in those lines of production that have approximately the same labor and equipment requirements. The classical example of diversification is the ice and coal company, which uses the same manpower, equipment, and supervision the year round for the major portion of its operations. More recent instances are the combination of refrigerators and air conditioners with furnaces and stoves, the use of the same drivers for ice-cream wagons in the summer and for fuel-oil trucks in the winter, the featuring of the winter attraction of skiing by summer resorts, and vice versa—the promotion of air-conditioned "package" summer tours by winter resorts.

On occasion governments have initiated public policies aimed at reducing or eliminating seasonality or altering its incidence. For example, one of the main reasons for the establishment of the Federal Reserve System was to ease the strain of seasonal movements of funds between rural and urban areas in conformance with harvest seasons. Many states limit unemployment insurance to a fraction of a year, evidently in the expectation of assisting seasonally rather than cyclically unemployed workers. Because seasonal variations are highly wasteful and their solutions by individual firms can usually only shift the burden instead of eliminating it, the government may recognize the need to find real economic solutions to cope with seasonality. Meanwhile it is clear that accurate measurement of seasonal variations is necessary not only for the study of other types of fluctuations in time series, but also for the solution of the problem they present. We must recognize, in other words, the importance and magnitude of seasonality as a problem in its own right.

18.9 Alternative Methods of Measuring the Seasonal Index

Two other methods of computing the seasonal index merit some mention, especially because one of them brings out indirectly the advantages of the ratio to moving average method.

The simplest possible method of computing a seasonal index is to calculate the average value for each month (for January, for February, and so on) and express the averages as percentages so that the percentages of the average for all 12 months can add up to 1200, or average 100.

This is indeed a very simple but a very crude method. It assumes there is no trend component in the series; that is, $Y = CSI$. This, however, is an unjustified assumption. Most economic series have trends, and, therefore, the seasonal index computed by this method is actually an index of trends and seasonals. Furthermore, the effects of cycles on the original values may or may not be eliminated by the averaging process. This depends on the duration of the cycle and the term of the average, that is, on the number of months included in the average. Thus, this method is seldom of any value.

The second alternative is the method of *ratios to trend*. This method assumes that seasonal variation for a given month is a constant fraction of trend. Starting with the model $Y = TSCI$, some argue that trend can be eliminated by dividing each observation by its corresponding trend value. The ratios resulting from this operation compose SCI. Each of these ratios to trend is a one-based relative, that is, a pure number with a unity base. Next, an average is computed for the ratios for each month. This averaging process eliminates cycle and random influences from the ratios to trend. Thus, the 12 averages of ratios to trend contain only the seasonal component. These averages, therefore, constitute the seasonal index. As is true of the method of ratio to moving average, some slight corrections are necessary in order to adjust these ratios to average unity.

The ratio to trend method is certainly a more logical procedure than the crude measurement of seasonals mentioned above. It has an advantage over the moving average procedure too, for it has a ratio to trend value for each month for which data are available. Thus there is no loss of data as occurs in the case of moving averages. This is a distinct advantage, especially when the period covered by the time series is very short.

The main defect of the ratio to trend method is that if there are pronounced cyclical swings in the series, the trend—whether a straight line or a curve—can never follow the actual data as closely as a 12-month moving average does. As a result, the ratios to trend will tend to be more variable than the ratios to a 12-month moving average. Under these circumstances, it is likely that the averaging process will have a greater capacity to remove irregular movements from the ratios to moving averages than it has to isolate seasonals from the ratios to trend. This is another way of saying that a seasonal index computed by the ratio to moving average method may be less biased than one calculated by the ratio to trend method.

In conclusion, it may be said that because of its theoretical and practical advantages, the ratio to moving average method is probably the most widely used procedure. Its application is not confined to periodic fluctuations of 12-month duration. It can be used equally well for any periodic fluctuation, such as hourly variations of temperature within the day or daily variations in sales within the week.

18.10

Changing Seasonals

As mentioned before, the use of the conventional methods to measure the typical variations is appropriate and significant only if the variations are stable from year to year. But, unfortunately, seasonal patterns have an undesirable trait of changing through time. Changes in seasonals may show up either in the amplitude or—and this is more serious—in timing. These changes are sometimes quite abrupt and sometimes quite gradual. We are concerned here with the appropriate measurements of such changing seasonals.

Abrupt changes in seasonal patterns occur in a situation in which the seasonal variations of a series may remain stable for a number of years and then change suddenly into new patterns that are quite different from the old and that may remain stable for years before they change again. This type of change is due either to the impact of some powerful sporadic factors or to deliberate managerial policies. During World War II, which necessitated an all-out effort in many production lines, seasonals in these lines disappeared but re-established themselves afterward. Changes in the seasonal variations of automobile sales were observed when the automobile industry altered the design of cars to make them less dependent on weather and when it changed to the fall instead of early spring the dates of introducing new models.

A simple but satisfactory method of dealing with this kind of seasonal variation is to analyze the variations independently for the two periods separated by the abrupt change. The seasonal index for each separated period may be computed either by the ratio to moving average method or by the ratio to trend device. Evidently, when we use such indices in analysis, each index should be used for that period for which the index is constructed. In synthesis, we would ignore the index computed before the abrupt change and employ only the one established after the change.

Gradual changes in seasonals are fairly common. They are brought about by modifications of custom, development of technology, changes in business practices, and other forces that change slowly but steadily. New designs of automobiles have greatly reduced the amplitude of gasoline consumption. The emergence of air conditioning has increased the summer use of electricity. Changes in consumers' tastes, aided by intensive advertising, have gradually eliminated the pronounced seasonal fluctuations in ice-cream production. The policy of an elastic credit supply and the central banking reserves of the federal reserve banks have progressively reduced the seasonal swings in short-term interest rates almost to the vanishing point.

Gradual changes in seasonal patterns are usually secular in nature. As such, no average can be a typical representation of the seasonal pattern, since errors that run in one direction cannot be removed by averaging. When the change in a seasonal is gradual, instead of deriving a typical seasonal index illogically, we study the steady and progressive change in the specific seasonal indices month by month and obtain a measure of change in the index for each month. When the specific seasonal indexes—that is, ratios to moving averages—have been obtained, the ratios of each given month are plotted as an independent time series graph. We would have 12 such graphs, with the X axis scaled in years, which would characterize the changes in seasonal pattern and therefore enable us to measure such changes. We reason that if there is a changing seasonal pattern, then progressive change would be reflected in the movements of the ratios for each month in the form of a discernible general drift from year to year. That is, the movement in the values of each month for the years can be appropriately defined by a trend line. The trend line here serves, as usual, the purpose of an average. The trend value for each month in each year is the first approximation of the seasonal index for that month in that year. We therefore end our analysis of gradually changing seasonal patterns with as many specific seasonal indexes as there are years in the period under examination. These specific index values are trend values. The preliminary specific seasonal indices for each year, as before, must be adjusted to average 100. To adjust data for seasonal variations, each monthly original observation is divided by the adjusted seasonal index for that month in the same year. For forecasting, we use the projected trend values by extending the trend line for the desired month.

Whenever the values of the same month for the years tend to change gradually by a constant absolute amount, straight-line trends are appropriate. In many cases, however, values for each given month may not behave linearly. One frequently used device is to draw a freehand trend line through plotted points for a single month; the trend values are then read from the chart. This is actually used by the Federal Reserve System in dealing with changing seasonals for the construction of its industrial production index. Another widely used method is to compute a moving average of five or seven terms. This procedure is used by the National Bureau of Economic Research. Here, as elsewhere in statistical analysis, whatever trend line is to be determined depends upon the nature of the data and the judgment and experience of the investigator.

18.11

Measuring Cycles from Annual Data

Cyclical fluctuations do not repeat themselves periodically as do seasonal variations. Neither do they behave fortuitously as do the irregular movements. The cycles of a specific series usually embrace a certain broad pattern that shows some similarity but always contains some differences in duration and intensity. Because of this lack of uniformity in cycles, it is possible to isolate them but is is impossible to project them into the future. This, as will be seen the the next chapter, constitutes the most difficult part of forecasting with time series.

As to the problem of isolating cycles from annual data, it is a relatively simple matter. In annual data there are only two components: the trend and the cycle. Seasonal variations, being monthly changes that complete their pattern within a year, do not show up in annual data. Irregular movements, which are also short-run influences compared to a year, produce negligible influences upon annual data and their positive and negative effects tend to offset each other during the course of a year. With annual data, the trend value is considered as the *statistical normal*—the value which represents the normal annual growth if there had been no cyclical influences. Conversely, then, we may consider cyclical fluctuations as the deviations from the normal values—the trend.

Thus the problem of isolating cycles from annual data is the same as the elimination of trend. Trend can be eliminated by subtracting the trend value from the original data if we assume an additive model, $Y = T + I$. The results are cyclical residuals. Or, if we assume a multiplicative model, $Y = TC$, cycles are isolated by dividing original data by the trend values, obtaining cyclical relatives. In order to facilitate comparison of cyclical fluctuations among different time series, both cyclical residuals and relatives are expressed in terms of percentage deviations from the trend. The percentage deviations can be obtained by either of the following two formulas:

$$\text{Percentage deviations of cyclical residuals} = \frac{y - y_t}{y_t}(100) \qquad (18.5)$$

$$\text{Percentage deviations of cyclical relatives} = \left(\frac{y}{y_t} - 1\right)100 \qquad (18.6)$$

TABLE 18.4
COFFEE PRODUCTION IN SOUTH AMERICA, 1956–1968

Year	Production (y)	Trend (y_t)	Cyclical Residuals ($y - y_t$)	Cyclical Relatives [$y/y_t(100)$]	Percentage Deviations
1956	82.3	81.5	+ 0.8	101.0	+ 1.0
1957	95.6	83.3	+12.3	114.8	+14.8
1958	81.3	85.1	− 3.8	95.5	− 4.5
1959	76.4	87.0	−10.6	87.8	−12.2
1960	92.8	88.9	+ 3.9	104.4	+ 4.4
1961	83.4	90.6	− 7.2	92.1	− 7.9
1962	88.2	92.4	− 4.2	95.5	− 4.5
1963	96.1	94.2	+ 1.9	102.0	+ 2.0
1964	89.1	96.0	− 6.9	92.8	− 7.2
1965	112.2	97.9	+14.3	114.6	+14.6
1966	100.4	99.7	+ .7	100.7	+ 0.7
1967	106.9	101.5	+ 5.4	105.3	+ 5.3
1968	196.6	103.3	− 6.7	93.5	− 6.5

Source: Table 17.3.

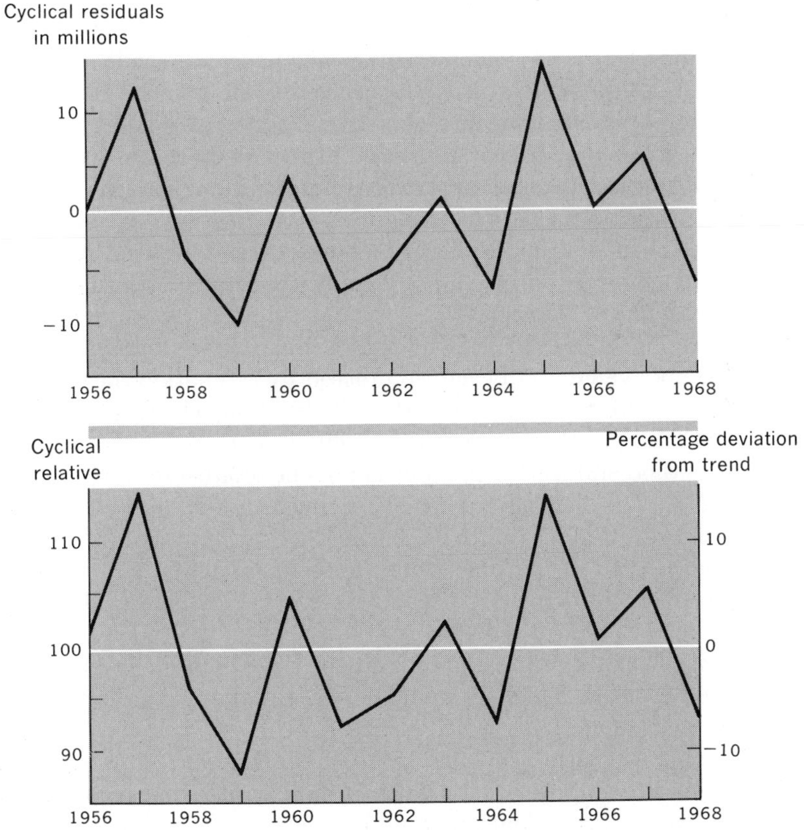

Figure 18.4 Coffee production in South America, 1956–1968.

It may be noted that values of percentage deviations computed by both formulas are the same.

As an illustration of computing cycles under the assumption of the additive and the multiplicative models, respectively, we shall use the coffee production data. As it may be recalled, from the previous chapter, the trend equation for these data, covering 1956–1968, is $y_t = 92.4 + 1.82x$, with origin at the middle of 1962. Computations of cyclical residuals, cyclical relatives, and percentage deviations are all shown in Table 18.4. The resulting cyclical residuals, the cyclical relatives, and the percentage deviations are portrayed in the lower panel of Figure 18.4.

18.12 Measuring Cycles from Monthly Data

A great deal of work in the analysis of cyclical fluctuations makes use of annual data, but this practice leaves much to be desired. Annual series obscure two of the most critical aspects of cycles: the location of the turning points and the measurement of the amplitude of fluctuations between the turning points. In order to bring out the relevant information on cycles, therefore, it is much more desirable to use quarterly or monthly data, especially the latter. However, as soon as data that record variations within the year are used, the distortions caused by seasonal and irregular movements enter the picture. Thus, to measure cycles from monthly data, we must attempt to remove seasonal variations in addition to the correction for trend. Alternatively, to measure cycles from monthly data, we are mainly concerned with the estimate of cyclical and irregular movements.

To estimate CI under the multiplicative assumption; that is, $Y = TSCI$, three methods are available. In each we proceed by estimating the trend and the seasonal pattern first for the purpose of their elimination. The three methods differ from each other only in the sequences of eliminating T and S.

The first method starts with the division of the original data by trend values to obtain CSI: $CSI = TSCI/T$. Then the estimate of CI is derived by dividing the seasonal index into data that have been adjusted for trend: $CI = CSI/S$.

The second method is just the reverse, for original data are first adjusted for seasonal and then corrected for trend. Here, (1) $TSCI/S = TCI$, and (2) $TCI/T = CI$.

The third method begins with an estimate of TS, which is, of course, the product of trend and seasonal. Then TS is eliminated from the original data by dividing the latter by the former. That is, $CI = TSCI/TS$.

All three methods yield the same results. The first one, however, is seldom employed because there is no practical use for CSI values. The second is convenient when we attempt to estimate CI with data that have already been deseasonalized. The third is often followed in measuring cycles from original data. It has two advantages over the first method, one of which is trifling: it is slightly less laborious to perform one multiplication and one division than to divide twice. The second advantage is more important: the division of original data by TS reveals more clearly that CI are deviations from statistical normal.

In general, cyclical relatives remain more or less constant through time, whereas cyclical residuals tend to become larger numerically with increases in trend values. The reason is that when the activity of a given line expands it is also susceptible to larger absolute cyclical swings. But larger absolute cyclical deviations from higher trend values may have the same relative magnitude of smaller cyclical deviations from lower trend values. This is one of the reasons why we often prefer to adjust for trend by division. Careful examination of the two charts in Figure 18.4 shows that this difference clearly exists in our illustrative data. For instance, the cyclical residual for 1965 is much farther away from the trend than that for 1957. Yet the cyclical relatives for the same two years are of approximately the same magnitude.

Another reason for the preference for relatives over residuals in adjustment for trend is that percentages are usually more convenient and satisfactory when the cycles of one series are compared with those of another. This is especially true when the series are expressed in different units.

Before we move on to the next section, it should be noted that the procedure just described is appropriate whether the trend is an arithmetic straight-line or a geometric straight-line or any other type of mathematical trend function. In each case, when the data have been adjusted for trend, the trend line becomes a straight line horizontal to the base, suggesting a zero or a 100 percent normal line. The percentage values on the Y axis express the cyclical influences that are the percentage of the normal obtained in each year.

In monthly data, statistical normal is defined as the product of trend and seasonal, TS. Table 18.5 illustrates the estimate of CI, by the third method, for factory cheese production in the United States, 1947–1955. It will be observed that, for the sake of condensation, only computations for 1947 and 1948 are shown in the table.

The original data of cheese production are entered in column 1. Column 2 contains the trend values for the series, which are derived by an arithmetic straight-line trend equation, $y_t = 93.29 + 0.165x$, with origin at January, 1947. The seasonal pattern of this series is stable, and a typical seasonal index is computed by the ratio to moving average method, the monthly index values being modified means of the four central items. These index values are recorded in column 3 in ratio form. Column 4 presents the values of statistical normal values for the series. These values are obtained by multiplying trend values by seasonal index. When we divide the TS values into original observations in column 1 and multiply the results by 100, we obtain the estimates of cyclical-irregular fluctuations in percentage form, as shown in column 5. The cyclical-irregular percentages are plotted in the lower chart in Figure 18.5. These percentages are often looked upon as deviations from normal—the values that could have been had there been no cyclical and irregular fluctuations at all.

Sometimes we may wish to eliminate the major portion of irregular move-

TABLE 18.5 CYCLICAL IRREGULAR MOVEMENTS FOR UNITED STATES FACTORY PRODUCTION OF CHEESE, 1947–1955

Year and Month	Production (millions of pounds) (y) (1)	Trend* (y_t) (T) (2)	Seasonal† Index (S) (as ratios) (3)	Statistical Normal (TS) (4)	Cyclical-Irregular Movements ($y \div TS$) (%) (5)
1947					
Jan.	74.3	93.3	0.780	72.774	102.1
Feb.	78.3	93.5	0.770	71.995	108.8
Mar.	100.2	93.6	0.968	90.605	110.6
Apr.	114.1	93.8	1.099	103.086	110.7
May	140.6	94.0	1.410	132.540	110.7
June	148.1	94.1	1.416	133.246	111.1
July	133.3	94.3	1.233	116.272	114.6
Aug.	104.0	94.4	1.096	103.462	100.5
Sept.	87.7	94.6	0.933	88.261	99.4
Oct.	80.9	94.8	0.839	79.537	101.7
Nov.	60.9	94.9	0.712	67.569	90.1
Dec.	60.4	95.1	0.744	70.754	85.4
1948					
Jan.	63.7	95.3	0.780	74.334	85.7
Feb.	65.2	95.4	0.770	73.458	88.8
Mar.	80.3	95.6	0.968	92.541	86.8
Apr.	94.0	95.8	1.099	105.284	89.3
May	129.8	95.9	1.410	135.210	96.0
June	131.9	96.1	1.416	136.078	96.9
July	114.7	96.3	1.233	118.738	96.6
Aug.	108.5	96.4	1.096	105.654	112.7
Sept.	89.4	96.6	0.933	90.128	99.2
Oct.	81.5	96.8	0.839	81.215	100.4
Nov.	67.3	96.9	0.712	68.993	97.5
Dec.	72.1	97.1	0.744	72.242	99.8
.	.	.	.	.	.
.	.	.	.	.	.
.	.	.	.	.	.

Source: Business Statistics.
* $y_t = 93.29 + 0.165x$ *(origin, January, 1947).*
† *Modified mean of 4 middle ratios of actual values to centered 12-month moving averages for 1947-1955.*

ments from the CI curve. This is often done by smoothing the curve with freehand drawing or by the calculation of a moving average. To estimate C by taking a moving average of CI, the terms used are often 3 months or 5 months. Odd terms make computation easier, and short durations avoid the tendency to smooth out too much of the cycles.

The isolation of cyclical fluctuations by a moving average is actually not a satisfactory method. A 3-month moving average may smooth out only a negligible part of the cycle, and the resulting curve may not be very smooth. A 5-month moving average may produce a smoother estimate of cycles, but the portion of cycles smoothed out may be considerable.

The fact that there is no satisfactory method of removing irregular movements, coupled with some theoretical considerations—on which we shall here comment briefly—has led to the practice of analyzing cyclical and irregular fluctuations as a single entity. Irregular movements caused by random forces, as was mentioned in Section 16.6, are relatively unimportant. Furthermore, random forces occur quite frequently and are usually from unknown sources. Thus it is neither possible nor necessary to identify them: it is best to treat them as part of the cyclical swing that we actually observe. It seems reasonable to consider the slight sawtooth effect of cyclical movement as a part of the cycles, but can we also ignore the short-term sporadic changes that rise and fall with considerable magnitude just as the cycles do? Part of the answer to this question is the basic differ-

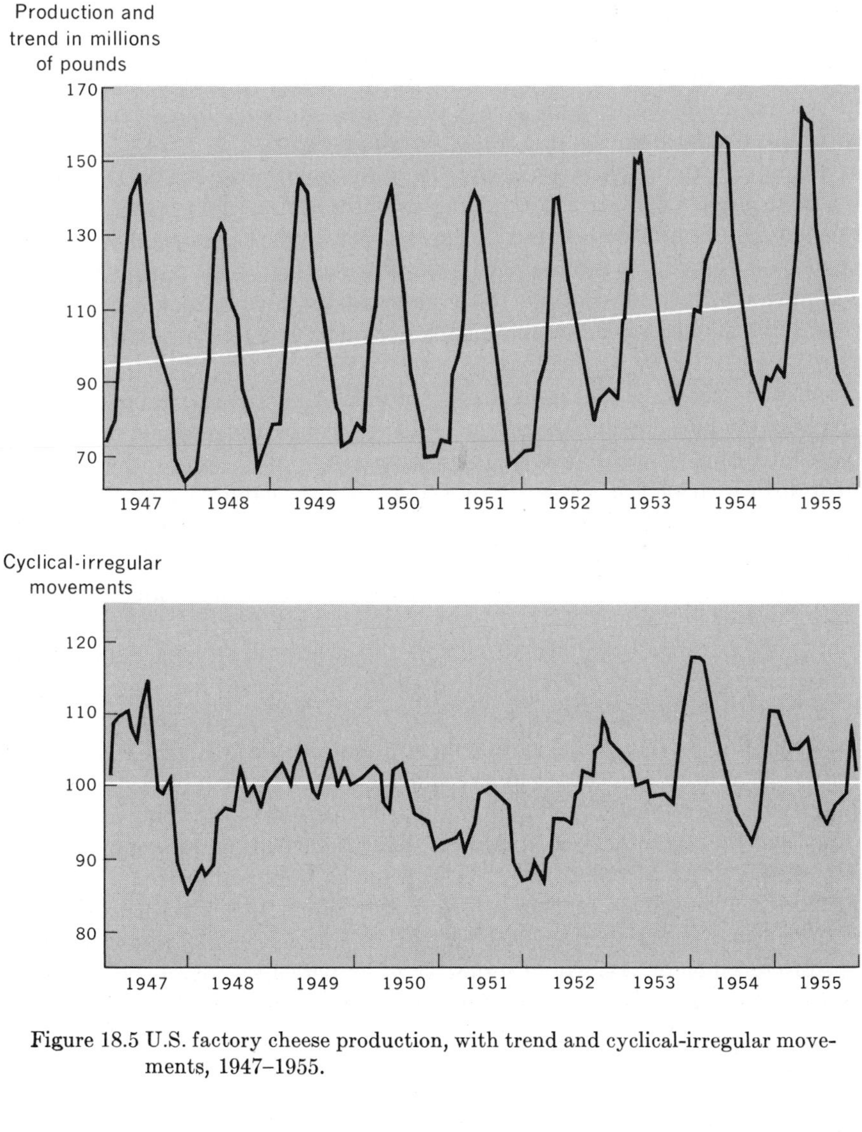

Figure 18.5 U.S. factory cheese production, with trend and cyclical-irregular movements, 1947–1955.

ence between these two types of variation. It is the main property of cycles that they are cumulative and self-generating. Thus, if the fluctuations in the short run are such that they rise abruptly and then immediately fall back toward the previous level, then they are not cumulative and should be considered as irregular variations. Sometimes sporadic forces are quite capable of generating fluctuations that are cumulative for a number of months. Prolonged strikes, speculative booms, and war scares are examples of such irregular forces. Under these circumstances, we may insist that, to be qualified as cycles, there must be a minimum number of months, say longer than 12 as used by the National Bureau of Economic Research, for the cumulative expansions and contractions. Here we come to the realization that cyclical causes are diversified, complicated, and dynamic. They can never be expected to produce smooth cycles as portrayed by the function of the sines and cosines of the time scale. Irregular movements are part of the aggregate of factors that create the ragged cyclical course we observe in reality. Our attention to irregular movements should warn us that not every rise and fall comprise a cycle. When we mark off the turning points of cycles, we must be careful that they reveal faithfully the underlying cyclical forces at work.

In examining the cyclical and irregular movements of our illustrative series as plotted in Figure 18.5, it is interesting to observe first the upper chart, which shows the original data with trend. It can be seen that the seasonal variations of cheese production are so pronounced that the original data hardly reveal any cyclical movements at all. But the lower chart for CI tells quite a different story. This indicates the necessity for eliminating seasonals from the original data before cycles can be profitably studied.

From the lower chart in Figure 18.5, it is clear that random influences on the CI curve are evident, but breaks produced by sporadic changes seem to be absent. The percentages of cyclical and irregular movements of cheese production define the first downturn of the series in July 1947, whose value is about 15 percent above normal. Then it follows a sharp downswing that reaches a subnormal activity about 14 percent in December 1947. In January of the following year, a recovery carries the series above the 100 percent trend line in August of the same year until a peak is reached in May 1949, which is only slightly more than 5 percent above the normal. The downswing from this peak was prolonged and gradual. The trough is reached in the last month of 1950. There are at least two interesting points of the 1948–1950 cycle to be noted: It is the longest cycle of the whole series, almost four years. And it gives a flat-top impression, with the top persisting for almost two years along the trend line. The next cycle, which starts in December 1950, and ends in March 1952, is marked by brief duration and subnormal activity. The upswing that commences in March 1952 seems quite pronounced, but it actually moves up to a peak of only about 9 percent above the trend. Thereafter a mild recession sets in. It reaches a low of only more than 2 percent of activity under the normal in October 1953. The peak in January 1954 marks the highest level of production for the whole series. It is 18 percent above normal activity. The downswing that follows is also quite severe. In intensity it is next to the first of the series. There does not seem to be any particular feature in the last whole cycle for the series. In general we may say that the cyclical pattern of cheese production seems to have an average duration of less than two years, ranging from about one to four. The amplitudes of the cycles are quite moderate. The first two thirds of the series, except the first eight months in 1947, reveal much lower levels of activity than the last one third.

It is interesting to note here the turning points and the duration of American business cycles. In approximately the same period, according to the National Bureau's dates, the general business cycles had peaks in November 1948, July 1953, and July 1957, and troughs in October 1949, August 1954, and April 1958. Thus, the cyclical fluctuations in American factory cheese production do not conform to the general cycles of American business at all. What is the explanation for this nonconformity? A possible one is that cheese is a dairy product that is more affected by weather and its own price than by general business conditions. Why would cheese production, then, be incorporated in a two-year cycle? The answer to that may be given by the "cobweb theorem" advanced by agricultural economists. This theorem claims that low price of a given kind of food (or agricultural production) in a given year would curtail its production and, therefore, that the supply of that kind of food will decrease in the following year. Unless the demand for it decreases correspondingly, its price will rise, which would encourage an increase in the supply in the next production period. High production will depress price again, and again the cycle will start all over.

18.13 Comments on the Estimate of *CI*

The irregular nature of cyclical movements defies any attempt to find an average cycle that could be used as a typical representation of their effects on a time series. The best—or the least unsatisfactory—approach is to measure these fluctuations indirectly by removing effects of trend and seasonal forces. The remaining variations are considered as cyclical variations, not forgetting that they also reflect irregular influences. This method is admittedly imperfect, not only because a cycle estimated by it cannot be projected into the future but also because the estimate is to some degree biased.

To estimate CI, we must first estimate T and S. The actual behavior of the cyclical and irregular movements depends on the trend line as well as on the seasonal index. When we determine the trend line we usually have two purposes in mind. One is extrapolation. Another is adjustment of the original data by subtraction or division. But these two functions of the trend are often contradictory. We may often find that a given trend equation may isolate cycles satisfactorily but is inadequate for forecasting, or vice versa. Furthermore, the same mathematical trend function usually yields different trend values when the initial and terminal years selected differ and when the period covered varies. Consequently, the cyclical relatives computed from deviations from trend are affected by the choice of trend function and of the period to which it is fitted.

The adjustment for seasonals is also a rather arbitrary matter. In the first place, the seasonal patterns of many series are subject to change without notice. Even if the seasonals remain relatively constant, the values of typical seasonal indices always differ with different methods of computation. This means the appearance of deseasonalized data is affected by the decision on whether to use changing or typical seasonal indices and on the method of computation.

The resulting CI is influenced, finally, by our assumption made for decomposition of the series. Have we used the multiplicative model or the additive model? When choosing one, we may often be unable to know if the other is closer to the relationship of the components of the series.

From these considerations, we may say that the estimate of CI not only contains the cyclical and irregular forces, but also includes errors in defining the trend, in determining the seasonal pattern, and in decomposing the original series. Since these errors can never be precisely determined, there must always be some uncertainty and speculation in the interpretation and use of the statistical results that we call "cycles."

In concluding our analysis of time series, we must recognize that secular trend, seasonal variations, and cyclical fluctuations are all real and significant in many economic and social time series, even though our measures for them are only approximations. Approximations as they may be, they can nevertheless be of great use in both economic research and business management. These approximate measures can enable us to forecast the future behavior of time series.

Glossary of Formulas

(18.1) $$\bar{m}_7 = \frac{M_7}{24} = \frac{y_1 + 2y_2 + 2y_3 + 2y_4 + 2y_5 + 2y_6 + 2y_7 + 2y_8 + 2y_9 + 2y_{10} + 2y_{11} + 2y_{12} + y_{13}}{24}$$

(18.2) $$\bar{m}_8 = \frac{M_8}{24} = \frac{M_7 - y_1 - y_2 + y_{13} + y_{14}}{24}$$

(18.3) $$\bar{m}_9 = \frac{M_9}{24} = \frac{M_8 - y_2 - y_3 + y_{14} + y_{15}}{24}$$

(18.4) $$\bar{m}_{10} = \frac{M_{10}}{24} = \frac{M_9 - y_3 - y_4 + y_{15} + y_{16}}{24}$$

.
.
.

This series of equations shows the direct computations of centered 12-month moving averages. $M_7, M_8, \cdots$, are the sums of 13 consecutive monthly values with each of the central 11 months counted twice and with M_7 as the first 24-month moving total centered opposite the seventh month of the series, $\bar{m}_7, \bar{m}_8, \cdots$, are the centered 12-month moving averages, with $\bar{m}_7$ as the first average, which is centered at M_7. It can be seen that each subsequent moving total is obtained by deducing the first two monthly values from the previous moving total and adding to it the value of the last month in, and that of the next month immediately following, the previous moving total.

(**18.5**) Percentage deviations of cyclical residuals $= \frac{y - y_t}{y_t}(100)$

(**18.6**) Percentage deviations of cyclical relatives $= \left(\frac{y}{y_t} - 1\right) 100$

In order to facilitate comparisons among the cyclical fluctuations of different series in annual units, the cyclical deviations from trend are transformed as percentage deviations from normal. Although cyclical residuals usually increase with the increase in trend values and cyclical relatives usually remain stable, the percentage deviations of both are the same, since $(y - y_t)/y_t = (y/y_t) - 1$.

Problems

18.1 Monthly data often need editing before a seasonal index can be computed. What is the most important type of correction? Under what conditions must such a correction be made? Under what conditions is it unnecessary to make the correction?

18.2 How does a specific seasonal differ from a typical seasonal index?

18.3 What do we mean by "changing seasonals"? What are the causes of changing seasonal patterns?

18.4 When is it appropriate to compute a typical seasonal index?

18.5 What does a 12-month moving average measure? Why?

18.6 What do the ratios of original data to 12-month moving averages estimate? Are such ratios biased estimates? Why or why not?

18.7 Why do we usually select the modified or positional mean in seasonal analysis?

18.8 How would you identify changes in the seasonal pattern of a series?

18.9 How should you treat seasonals that change abruptly? How should you treat those that change gradually?

18.10 Why is seasonality a problem not only for the individual firm but also for the whole economy?

18.11 Why is diversification the best solution for seasonal fluctuations? Under what conditions is this solution possible?

18.12 What are the merits and disadvantages of the ratio to trend method compared with the ratio to moving-average method of computing a seasonal index?

18.13 What do we mean by the term "deseasonalized data"? What elements do such data contain?

18.14 Many published economic time series are in the deseasonalized form, but only a small percentage of published series are adjusted for trend. Why is the correction for seasonal made so much more frequently than for secular trend?

18.15 Why can we assume that annual data contain only trend and cyclical forces?

18.16 Why do we often prefer to estimate cycles as relative deviations from the trend rather than as amounts of deviations from the trend?

18.17 Why is it more desirable to measure cycles from monthly data than from annual data?

18.18 What are the reasons for considering the estimate of CI as the estimate of C?

18.19 Why is the estimate of CI biased?

18.20 If you are asked to estimate cycles from a 12-month moving average, how would you go about it?

18.21 It is sometimes believed that a better estimate of cycles can be obtained by adjusting original data by specific seasonal indexes. What do you think is the reason behind it? Do you agree with the reason? Defend your answer.

18.22 Consult a copy of *Business Statistics* in your library and select a series of your own. The series should be about ten years in length. Then do the following:

a. Compute an appropriate trend line for the series, with the first month of the series as

the origin. Plot the original series and its trend in one diagram.
b. Compute a typical seasonal index for the series by the ratio to moving average method. Use the short cut to calculate your moving averages. Plot the results.
c. Estimate CI and plot the results.
d. Give a verbal description of your CI curve. (*Note:* If the series you select is one of changing seasonals, then estimate C by dividing the specific seasonal indices into data that have been adjusted for trend.)

18.23 Compute percentage deviations of cyclical residuals and percentage deviations of cyclical relatives for data in Table 17.1.

18.24 Do the same as for the preceding problem for data in Table 17.5.

18.25 Follow the instructions for Problem 18.23, using data in Table 17.6.

19

ASSOCIATION ANALYSIS: SIMPLE LINEAR MODELS

19.1

Association between Variables

So far we have been concerned with analysis of *univariate data;* that is, a single observation is made on each elementary unit of the sample, with a sample of values for a single variable. The sample statistics computed are then used to make inferences about the corresponding parameters of the parent *univariate population.* In this chapter and the one that follows, we shall study *bivariate* and *multivariate populations.* A *bivariate population* is one that contains two measurements on each elementary element. For example, we may observe the height and weight of each individual in a population of adult men. All the measured heights would form a variable, say, X; all the weights, another variable, say, Y. When each element of a population can yield three or more measurements, each referring to a specific characteristic, we have what is called *multivariate data.* For instance, lengths, diameters, and breaking strengths of steel bars produced by a certain process give us three variables with which to work. The main problem in analyzing bivariate or multivariate data is to discover and measure the association or covariation between the variables and to determine how the variables vary together. Thus, we may observe that taller men are usually heavier and shorter men are usually lighter. This is, of course, only true on the average, since there are exceptions, such as short and heavy men, or tall and light men. If, however, this average relationship can be established in some functional form mathematically, we will be in a position to estimate quite accurately, on the average, the weight of a man on the basis of his height. Such a procedure is called *estimation by association.*

The technique of estimation by association is far more sophisticated and useful than our homely example of heights and weights. This procedure is actually one of prediction, and prediction is the central function of sciences. The main task of any scientific study is to discover the general relationships between observed variables, and to state the nature of such relationships precisely in mathematical terms, so that the value of one variable can be predicted on the basis of that of another. Let us give a few examples to show the important place that decision by association holds in business and economic statistics.

A direct test of the quality of many kinds of product is destructive. However, if it can be found that a given property of a product is related to another characteristic of the product, the former can then be evaluated in terms of the latter. Thus, the breaking strength of metal parts manufactured from compressed metal powder can be determined by density as measured in weight; the shearing strength of joints on airplane wings can be tested by observing the diameter of welds; and so forth. These examples illustrate that decision by association often reduces costs in decision making.

We know that average consumption expenditures are associated with a variety of factors: disposable income, interest rate, population size, age composition, size and distribution of liquid assets, and so on. However, to relate all these variables to consumption is extremely cumbersome. Moreover, these variables may not be operating on consumption at the same time, or to the same significant degree. Under these circumstances, the most fruitful tactic is to relate consumption to the most important single factor. This is determined by studying the nature and degree of association of each factor to consumption. If we find, say, that disposable income exerts a clear-cut and most powerful influence upon consumption, we may then predict consumption expenditures on the basis of disposable income alone. This example shows that we are sometimes able to find a sufficiently good explanatory variable when restricting our investigation to the bivariate analysis.

To estimate the selling expenses of each branch office of a chain retailing organization on the basis of mean selling expenses may be quite unsuccessful because of an extremely large standard deviation. Since there is reason to believe that selling expenses are usually associated with the volume of sales, more accurate estimates of the former can often be made on the basis of the latter. Here, we have an illustration of the possibility of increasing precision in prediction by employing association analysis.

The employment policy of hiring personnel on probation is usually time-consuming and expensive. If it can be established that the performance of workers is associated with the results of certain tests, then time and money can be saved by adopting the policy of predicting an applicant's performance on the job on the basis of the related variable, test scores.

There are two related but distinct aspects of the study of association between variables. The first is called *regression analysis,* which attempts to establish the "nature of relationship" between variables; that is, in regression analysis, we study the functional relationship between the variables so that we may be able to predict the value of one on the basis of another or of others. Conventionally, the variable or variables which are the basis of prediction are called the *independent variable* or *variables,* and the variable that is to be predicted is referred to as the *dependent variable.*

The second aspect of association analysis is referred to as *correlation analysis,* which is concerned with determining the "degree of relationship" between variables. In the study of correlation the designation of dependent and independent variables is purely a personal choice and is of no practical significance.

It may also be noted now that association analysis may be distinguished as simple and multiple. *Simple association analysis* refers to a situation in which there is only one independent variable; that is, bivariate data. *Multiple association analysis* studies the relationship between three or more variables. Association analysis can also be differentiated as *linear* and *nonlinear,* according to the types of relationships that the variables have. In this chapter, we shall be concerned with simple linear association models. Multiple linear, and simple and multiple nonlinear, association models will be presented in the following chapter.

19.2

The Linear Bivariate Regression Model

In regression analysis, as in other types of statistical studies, we usually proceed by observing the sample data and using the results obtained as estimates of the corresponding population relationship. To make valid inferences, we must assume some population model. For a bivariate population, there are many possible models that can be constructed to describe the mutual variations of the two variables. The particular one in which we are interested at this time is called the *simple linear regression model,* which is constructed under the following set of assumptions:

(1) The value of the dependent variable, Y, is dependent in some degree upon the value of the independent variable, X. The dependent variable is assumed to be a random variable, but the values of X are assumed to be fixed quantities that are selected and controlled by the experimenter. The requirement that the independent variable assume fixed values, however, is not a critical one. Useful results can still be obtained by regression analysis in the case where both X and Y are random variables.

(2) The average relationship between Y and X can be adequately described by a linear equation whose geometrical presentation is a straight line, as in Figure 19.1. The height of the line tells the average value of Y at a fixed value of X. When $X = 0$, the average value of Y is equal to A. The value of A is called the Y intercept, since it is the point at which the straight line crosses the Y axis. The slope

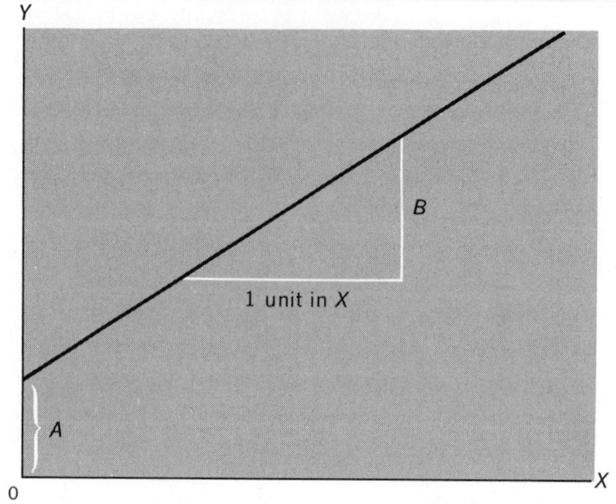

Figure 19.1 Properties of a straight line.

of the line is measured by B, which gives the average amount of change of Y per unit change in the value of X. The sign of B also indicates the type of relationship between Y and X.

(3) Associated with each value of X there is a subpopulation of Y. The distribution of each Y may be assumed to be normal or nonspecified in the sense that it is unknown. In any event, the distribution of each subpopulation Y is conditional to the value of X.

(4) The mean of each subpopulation Y is called the expected value of Y for a given X; $E(Y|X) = \mu_{yx}$. Furthermore, under the assumption of a linear relationship between X and Y, all values of $E(Y|X)$, or μ_{yx}, must fall on a straight line. That is,

$$E(Y|X) = \mu_{yx} = A + Bx \tag{19.1}$$

which is the *population regression equation* for our bivariate linear model. In this equation, A and B are called the *population regression coefficients*. Assumptions (3) and (4) are portrayed by Figure 19.2.

(5) An individual value in each subpopulation Y, as indicated by Figure 18.3, may be expressed as

$$y = E(Y|X) + e \tag{19.2}$$

where e is the deviation of a particular value of Y from μ_{yx} and is called the *error term* or the *stochastic disturbance term*. The errors are assumed to be independent random variables because Ys are random variables and independent. The expectation of these errors are zero; $E(e) = 0$. Moreover, if Ys are normal variables, the error term can also be assumed to be normal.

(6) Finally, we assume that variances of all subpopulations, called *variances of the regression*, are identical. Namely,

$$\sigma_1{}^2 = \sigma_2{}^2 = \cdots = \sigma_n{}^2 = \sigma^2 = \sigma_{yx}{}^2 = E(Y - E(Y|X))^2$$

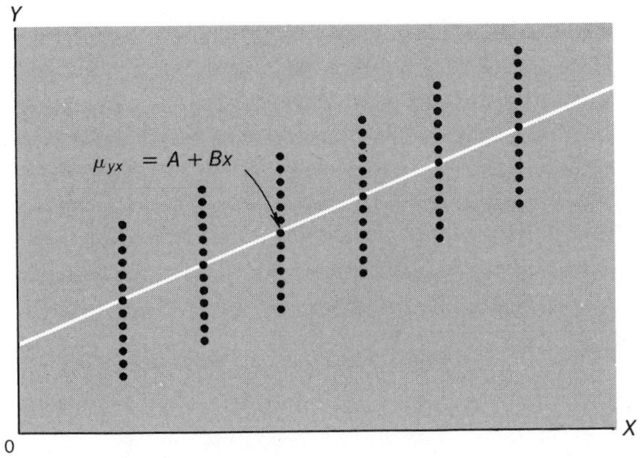

Figure 19.2 Distribution of Ys associated with Xs.

It is interesting to note that the common variance of the regression is also identical with the variance of the error term. Namely,

$$V(Y) = V(e) = \sigma_{yx}^2$$

since

$$\sigma_{yx}^2 = (y - E(Y|X))^2 = e^2 = E(e - E(e))^2 = V(e)$$

Due the fact that e shows the error of Y with respect to μ_{yx}, σ_{yx}^2 is often called the *residual variance*. The definitional equation for this value is

$$\sigma_{yx}^2 = \frac{\Sigma(y - \mu_{yx})^2}{N} \tag{19.3}$$

The square root of σ_{yx}^2 is called the *standard deviation of the regression* and is denoted as σ_{yx}.

To summarize: The simple linear regression model consists of the following

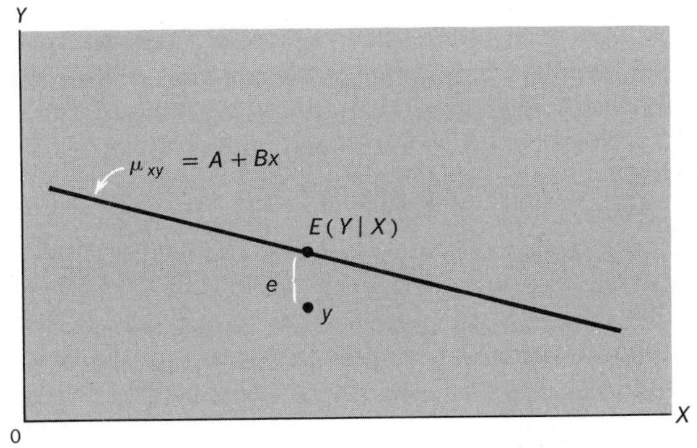

Figure 19.3 Deviation of an observation of Y from the regression.

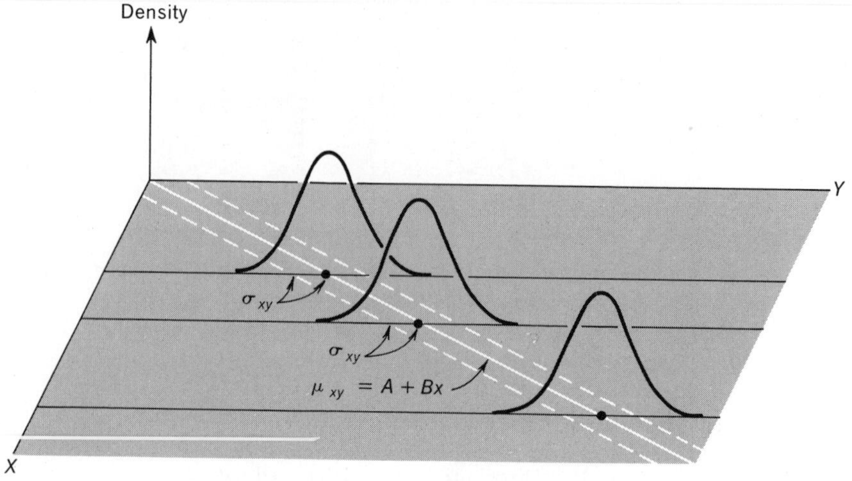

Figure 19.4 The linear regression model.

three equations:

$$\mu_{yx} = A + Bx$$

$$y = E(Y|X) + e$$

$$\sigma_{yx}{}^2 = \frac{\Sigma(y - \mu_{yx})^2}{N}$$

These equations describe and summarize regression behavior of bivariate population data. Furthermore, if the subpopulations are normally distributed, this model can be portrayed by Figure 19.4.

19.3 Estimating the Population Regression Equation

With the linear regression analysis, the main task is to estimate the population regression coefficients A and B in equation (19.1) on the basis of n pair sample observations: (x_1,y_1), (x_2,y_2), $\cdots$, (x_n,y_n). Let us denote the estimates of A and B by a and b respectively; then the *sample regression equation* becomes

$$y_c = a + bx \tag{19.4}$$

Since $\mu_{yx} = A + Bx$, y_c, or $(a + bx)$, is in effect an estimator of μ_{yx}. Here $(a + bx)$ is called a *linear estimator* because it represents a linear relationship.

From our discussion on the properties of estimators in Chapter 9, we can say that $(a + bx)$ is an unbiased estimator of μ_{yx} if

$$E(a + bx) = A + Bx$$

which implies that

$$E(a) = A, \text{ and}$$
$$E(b) = B$$

Furthermore, if there are many procedures available to estimate $(A + Bx)$, $(a + bx)$ is said to be the minimum variance, or the best, estimator of μ_{yx} if the scatter of $(a + bx)$ around $(A + Bx)$ is less than the dispersion of any other estimator obtained by any other procedure. If the distributions of subpopulations, Ys, are not specified, and if they possess equal variances corresponding to fixed X values, then according to the Gauss-Markov theorem, the best linear estimator of $(A + Bx)$ can be obtained by the method of least squares. On the other hand, if the subpopulation distributions have been known or assumed to be normal, then we may apply the maximum likelihood method to obtain an unbiased estimator of $(A + Bx)$ because we now know the form of the density functions of Ys.

Let us proceed to see how the best estimator of μ_{yx} can be obtained by the least squares method. First, define the estimator of e as

$$\hat{e} = y - y_c = y - a - bx$$

Namely, $\hat{e}$ is the residual of y from y_c. Then the procedure of estimating the population regression coefficients by the least squares method is to find a and b that will minimize the sum of squared deviations of $\hat{e}$. That is, a and b must be found such that

$$\Sigma\hat{e}^2 = \Sigma(y - a - bx)^2 = \text{minimum}$$

This requirement is met, according to the Gauss-Markov theorem, if a and b are determined by solving simultaneously the following set of normal equations:

$$1.\ \ \Sigma y = na + b\Sigma x$$
$$2.\ \ \Sigma xy = a\Sigma x + b\Sigma x^2$$

It turns out that the estimation of $(A + Bx)$ by the maximum likelihood method also requires the solution for a and b from the foregoing set of normal equations.

All quantities in the above normal equations can be computed from sample data. For the sake of simplicity, however, we usually solve for a and b first in terms of other quantities in the normal equations. Then values of a and b can be computed directly from sample data. To solve for b, we multiply the first equation by Σx and the second by n, and then subtract the first result from the second so that a can be eliminated. As a result, we have

$$b = \frac{n\Sigma xy - \Sigma x\Sigma y}{n\Sigma x^2 - (\Sigma x)^2} \tag{19.5}$$

Substituting the right side of the foregoing expression for b in either one of the two normal equations yields

$$a = \frac{\Sigma x^2 \Sigma y - \Sigma x \Sigma y}{n\Sigma x^2 - (\Sigma x)^2} = \frac{\Sigma y}{n} - b\frac{\Sigma x}{n} = \bar{y} - b\bar{x} \tag{19.6}$$

where $\bar{y}$ and $\bar{x}$ are the means of Y and X, respectively.

To illustrate, let us suppose that a manufacturer of spot welds of aluminum of high shearing strength wishes to predict shearing strength by the diameters of spot welds instead of by destroying the product for that purpose. A sample of ten welds, taken to establish the relationship between the two variables, yields the following result:

TABLE 19.1 DIAMETERS AND SHEARING STRENGTH OF SPOT WELDS OF ALUMINUM

Weld Diameter (in inches)	Shearing Strength (in thousands of pounds)
2.4	7.0
1.8	5.3
1.6	4.2
1.0	3.3
1.2	3.8
2.8	8.5
1.6	6.6
1.5	4.5
2.3	8.8

The first step in regression analysis is to construct a diagram of the sample data to see whether or not the dependent variable does depend to some extent upon the independent variable, and whether the average relationship between them can indeed be explained by a straight line. Such a diagram is called a *scatter diagram*, in which each point corresponds to a pair of values, (x_i, y_i). The scatter diagram for our illustrative data is shown in Figure 19.5.

Observing carefully the scatter diagram, we can obtain some distinct impressions. First, there is a clear positive relationship between diameters and shearing strength of the welds; that is, the former, on the average, seem to increase with the increase of the latter. Second, the points seem to scatter from each other to a great extent, but they also seem to have a linear relationship among them; that is, their average relationship can be adequately described by a straight line. These observations strengthen our belief that linear regression analysis can be fruitfully applied to the sample data on hand.

The next step is to fit a straight line through the points in the scatter diagram. This is done by computing a and b, the sample regression coefficient, with formulas (19.5) and (19.6). To compute these values, we need to accumulate Σx, Σy, Σxy, and Σx^2. For later calculations, we also need Σy^2. All these sums can be arranged in one table; for our example, they are accumulated in Table 19.2. The last two columns in this table are for use in checking the accuracy of our calculations as

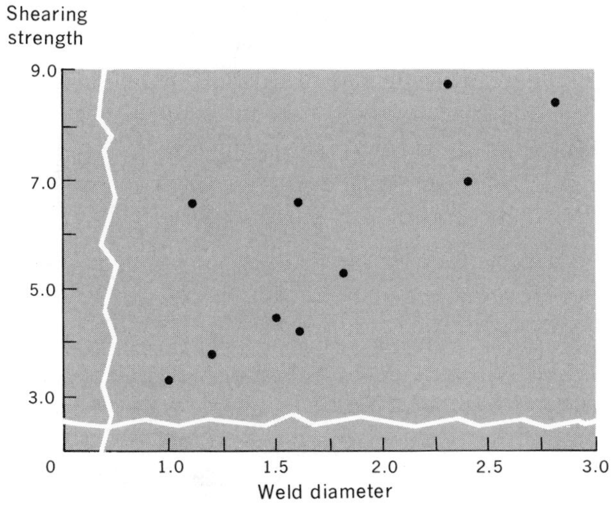

Figure 19.5 Scatter diagram for weld diameters and shearing strength.

carried out below the table. From the sums in Table 19.2, we have

$$b = \frac{(10)(109.53) - (17.3)(58.6)}{(10)(33.15) - (17.3)^2} = 2.531 \text{ thousand pounds per inch}$$

and

$$a = 5.86 - (2.531)(1.73) = 1.481 \text{ pounds}$$

Our estimate of the population regression equation is therefore

$$y_c = 1.481 + 2.531x$$

TABLE 19.2 SUMS REQUIRED FOR REGRESSION COMPUTATIONS

x	y	xy	x^2	y^2	$(x+y)$	$(x+y)^2$
2.4	7.0	16.80	5.76	49.00	9.4	88.36
1.8	5.3	9.54	3.24	28.09	7.1	50.41
1.6	4.2	6.72	2.56	17.64	5.8	33.64
1.0	3.3	3.30	1.00	10.89	4.3	18.49
1.2	3.8	4.56	1.44	14.44	5.0	25.00
1.1	6.6	7.26	1.21	43.56	7.7	59.29
2.8	8.5	23.80	7.84	72.25	11.3	127.69
1.6	6.6	10.56	2.56	43.56	8.2	67.24
1.5	4.5	6.75	2.25	20.25	6.0	36.00
2.3	8.8	20.24	5.29	77.44	11.1	123.21
17.3	58.6	109.53	33.15	377.12	75.9	629.33

$n = 10$ pairs of observations
$\Sigma x = 17.3$
$\Sigma y = 58.6$
$\Sigma xy = 109.53$
$\Sigma x^2 = 33.15$
$\Sigma y^2 = 377.12$

$\bar{x} = 1.73$
$\bar{y} = 5.86$

For checks, we note:
$17.3 + 58.6 = 75.9$
$33.15 + 2(109.53) + 377.12 = 629.33$

The estimate of A, as in this particular instance, has no practical significance. In general, the a value has only the mathematical significance that it, together with b, locates the position of the regression line. The b value, however, is always of importance. As in the present case, it is an estimate of the average amount that the shearing strength is increased by each additional inch in the diameter of welds. Specifically, the value 2.531 means that the shearing strength is estimated to increase by an average amount of 2.531 thousand pounds per inch increase in diameter.

19.4

Computing *a* and *b* with Coded Data

When values of X and Y are large, it becomes quite time-consuming to compute the Y intercept and the slope of the sample regression equation. There are two ways to save time and effort in computing a and b. The first is referred to as a *shift of the origin*. This is done by deducting a constant, say, c from each value of X, which results in a new variable, X'; and by deducting a constant, say, k from each value of Y, which results in a new variable, Y'. Then calculations are made with X' and Y'.

As is shown by Figure 19.6, the shift of the origin from (0,0) to (c,k) does not affect the slope but it does affect the Y intercept. Thus, while

$$b = b' = \frac{n\Sigma x'y' - \Sigma x'\Sigma y'}{n\Sigma x'^2 - (\Sigma x')^2} \tag{19.7}$$

without further adjustment, a should be computed as

$$a = (\bar{y}' + k) - b(\bar{x}' + c) = \bar{y}' - b\bar{x}' + (k - bc) = a' + (k - bc) \tag{19.8}$$

since $X' = X - c$, and $Y' = Y - k$.

The second short-cut is called *reducing the scale*. According to this procedure, both X and Y are divided by a constant, say, g, to obtain the new variables, X' and Y'. Again, calculations with X' and Y' will have no effect upon b, and again a is affected and must be converted back to the original scale by multiplying a' by g. That is,

$$a = a'g \tag{19.9}$$

As an extension of reducing the scale, the scale can also be *inflated* before calculations are made. In this case, both X and Y are multiplied by a constant to yield X' and Y'. This procedure is especially appropriate when we wish to get rid of the decimal points in the original data. For example, in our illustrative data, we may multiply both X and Y by 10 before a and b are computed. Here again, $b = b'$ but $a = a'/m$, where m is the common multiplier.

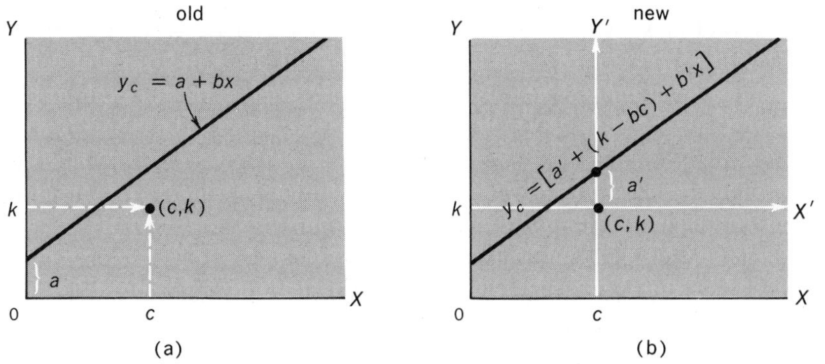

Figure 19.6 Effects of shifting origin from (0,0) — (*a*) to (*c*,*k*) — (*b*).

Needless to say, both methods of shifting the origin and reducing the scale can be combined in order to simplify computations of the estimators of population regression coefficients. For instance, we may let $X' = X/g - c$ and $Y' = Y/g - k$, or we may let $X' = mX - c$ and $Y' = mY - k$ for computational purpose. In each of these two possibilities, does $b = b'$ as before? How should a' be adjusted accordingly to get a in each case?

19.5 Estimating the Variance and the Standard Deviation of the Regression

The sample regression equation, $y_c = a + bx$, is called the *predictive equation* because its main function is to predict the mean value of Y, or the value of an individual observation of Y, associated with a given value of X. But how good is it as a predictive device? The answer to this question evidently rests upon the variability of the individual Y values from the computed values (averages) of Y associated with the X values. One way to visualize such dispersion is to draw a regression line through the points in the scatter diagram and to connect the former with the latter by vertical lines, as depicted in Figure 19.7. To draw the regression line, it is only necessary to plot two y_c values with, perhaps, a third as a check. The regression line, of course, represents estimated values of μ_{yx}. Examining Figure 19.7, we find that the line seems to fit the points fairly well, but the points still deviate somewhat from the line. The numerical measure of such deviations is the estimate of the population variance, or the estimate of the population standard error, of the regression. The $\sigma_{yx}{}^2$ is estimated by using the minimum variance, or maximum likelihood, estimators of a and b, as follows:

$$\hat{\sigma}_{yx}^2 = \frac{\Sigma(y - a - bx)^2}{n - 2} = \frac{\Sigma(y - y_c)^2}{n - 2} \tag{19.10}$$

This is an unbiased estimator of σ_{yx}^2. Here, $(n - 2)$ degrees of freedom is used as the denominator. The two degrees of freedom lost correspond to the number of regression coefficients. The positive square root of the estimator $\hat{\sigma}_{yx}^2$ is called the estimator of the population standard deviation of the regression, or the *sample standard deviation of regression*. It is also sometimes, rather misleadingly, named the *standard error of estimate.*

Quantities required to compute $\hat{\sigma}_{yx}^2$ are given in Table 19.3. From this table, and by (19.10), we have for our illustrative example

$$\hat{\sigma}_{yx}^2 = \frac{13.1468}{10 - 2} = 1.64$$

and

$$\hat{\sigma}_{yx} = \sqrt{1.64} = 1.28 \text{ thousand pounds}$$

Equation (19.10) is the theoretical formula for $\hat{\sigma}_{yx}^2$ and it involves a great deal of work, especially when samples are large. For practical work, this measure is usually obtained by the following expression:

$$\hat{\sigma}_{yx}^2 = \frac{\Sigma y^2 - a\Sigma y - b\Sigma xy}{n - 2} \tag{19.11}$$

This formula indicates that when a and b have been calculated, the only new quantity needed to compute $\hat{\sigma}_{yx}^2$ is Σy^2. Applying it to our example, we have

$$\hat{\sigma}_{yx}^2 = \frac{377.12 - (1.481)(58.6) - (2.531)(109.53)}{10 - 2} = 1.64$$

as before.

A precise interpretation of $\hat{\sigma}_{yx}^2$ or $\hat{\sigma}_{yx}$ will be given later in connection with our discussion on the "coefficient of determination." For the time being, we may consider the sample variance, or the sample standard deviation, of the regression as a measure of the "closeness" of the relationship between the two variables. The smaller either measure is, the closer are the Y values to the regression line, and thus the more accurate is the regression equation as a predictive device. We should also note here that larger samples will aid in locating the regression more accurately but will not reduce the standard deviation of regression very much.

An interesting indication of the utility of the regression analysis that may be seen here is in the comparison between the standard deviation of the regression and the standard deviation of Y computed with $\bar{y}$. In our illustration, $\hat{\sigma}_y$ of shearing strength does not take diameter into consideration. Let us compare Figures 19.7 and 19.8, both plotted with the data of our present example. Figure 19.7 shows the vertical deviations of ys from the regression line, and Figure 19.8 shows the vertical deviations of ys from their mean $\bar{y}$. It is clear from these two diagrams that $\Sigma(y - y_c)^2$ is much smaller than $\Sigma(y - \bar{y})^2$. To put this difference in numerical

TABLE 19.3 COMPUTATION OF THE ESTIMATE OF STANDARD DEVIATION OF REGRESSION

Observed values				
Diameter x	*Shearing Strength* y	Calculated Shearing Strength y_c	$y - y_c$	$(y - y_c)^2$
(1)	(2)	(3)	(4)	(5)
2.4	7.0	7.56	−0.56	0.3136
1.8	5.3	6.04	−0.74	0.5476
1.6	4.2	5.53	−1.33	1.7689
1.0	3.3	4.01	−0.71	0.5041
1.2	3.8	4.52	−0.72	0.5184
1.1	6.6	4.26	+2.34	5.4756
2.8	8.5	8.57	−0.07	0.0049
1.6	6.6	5.53	+1.07	1.1449
1.5	4.5	5.28	−0.78	0.6084
2.3	8.8	7.30	+1.50	2.2500
17.3	58.6	58.60	0	13.1468

terms, we find, as can be verified from the column for Y in Table 19.2, that

$$\hat{\sigma}_y = \sqrt{\frac{\Sigma(y - \bar{y})^2}{n - 1}} = \sqrt{\frac{33.724}{9}} = 1.94 \text{ thousand pounds}$$

which is 0.66 (= 1.94 − 1.28) thousand pounds greater than $\hat{\sigma}_{yx}$. This rather large reduction in variations suggests that the size of diameter does help considerably in estimating shearing strength. In other words, it suggests that the relationship between shearing strength and diameter is rather close, and that the average shearing strength may vary to a large extent with diameter. We shall have occasion to return to this point later in this chapter.

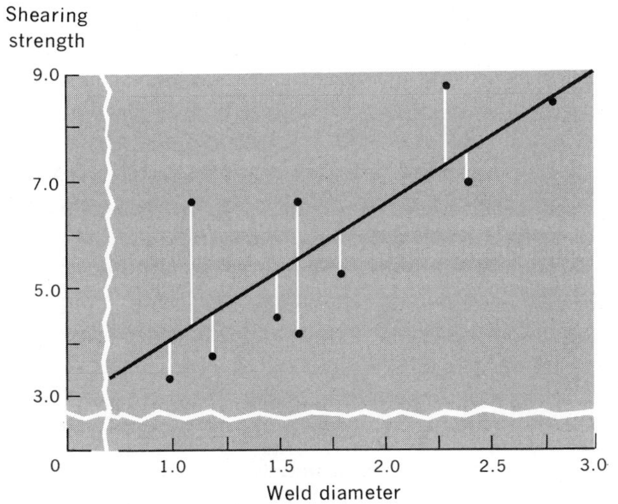

Figure 19.7 Deviations from the regression.

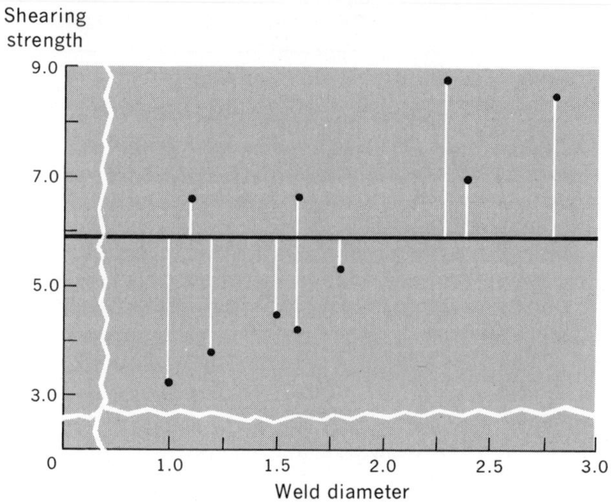

Figure 19.8 Deviations from the mean.

19.6 Inferences about Population Regression Coefficients

Having obtained the sample regression equation, which shows the existence of a certain functional relationship, and having concluded that the regression equation may be a useful one on the basis of the sample standard deviation of the regression, the reader may jump to the conclusion that now the sample regression equation can be used to predict Y values on the basis of X values. But, at this point, the reader should realize that even if the sample regression equation is identical with the true regression equation, prediction may still contain some error because the relationship between the two variables in bivariate population may not be perfect. The size of this error is measured by the population standard deviation of regression. This type of error is not due to sampling, and there is no way to evaluate it because the population regression coefficients and standard deviation of the regression are unknown. To predict with the sample regression equation, moreover, there is an additional error due to variability in sampling. For repeated sampling with the same sample size, the estimates of A and B would tend to vary from sample to sample. Sampling errors of a and b, as usual, are evaluated in terms of their repective sampling distributions.

SAMPLING DISTRIBUTION OF a

The distribution of a has as its expectation $E(a) = A$ and as its variance $V(a) = \sigma_a^2 = (\sigma_{yx}^2)(\Sigma x^2)/n\Sigma(x - \bar{x})^2$, where the sum, Σ, is taken over the sample. Since a is often of no practical importance, we shall not discuss inferences about A on the basis of a, except to note that tests and estimations can be made in a fashion similar to that for B, discussed below.

SAMPLING DISTRIBUTION OF b

The estimated slope, b, is a crucial value in making predictions. It may be positive or negative, suggesting a positive or negative relationship between the two variables. But, b, being a sample statistic, is subject to sampling variability. That is, because of chance variations, b may assume positive or negative values but the population slope may actually be zero. If this is true, then the sample regression equation is of no value whatsoever as a predictive device. Thus, we must have knowledge of the nature of the sampling error in b and its size before we can use the sample regression equation with confidence.

The sampling distribution of b, to state without proof, is normal if Y (or e) is normal or if $n \geq 30$. When $n < 30$, b is distributed as Student's t with $n - 2$ degrees of freedom. Furthermore, in both cases, the distribution of b has as its expectation, $E(b) = B$, and as its variance $V(b) = \sigma_b^2 = \dfrac{\sigma_{yx}^2}{\Sigma(x - \bar{x})^2}$. The unbiased estimator of $V(b)$ is

$$\hat{\sigma}_b^2 = \frac{\hat{\sigma}_{yx}^2}{\Sigma(x - \bar{x})^2} = \frac{\sigma_{yx}^2}{\Sigma x^2 - (\Sigma x)^2/n} \tag{19.12}$$

The square root of the unbiased estimate of $V(b)$ is the unbiased estimate of the standard error of b and is denoted as $\hat{\sigma}_b$.

When the distribution of Y is unspecified, the distribution of b cannot be specified either. In this case, therefore, testing and estimation of b, either with the normal or the t distribution, can be considered only very rough measures of significance and confidence.

TEST ABOUT, AND CONFIDENCE INTERVAL FOR, B

Conventionally, to test the significance of b we formulate the null hypothesis that the population slope is zero, that is, $B = 0$; and the alternative hypothesis that the population slope is not zero, that is, $B \neq 0$. When samples are large, we use as the testing statistic

$$Z = \frac{b - B}{\hat{\sigma}_b} \tag{19.13}$$

When samples are small, we use

$$t = \frac{b - B}{\hat{\sigma}_b} \tag{19.14}$$

as the testing statistic. Here, for t, $\delta = n - 2$.

Suppose that we wish to test $B = 0$ against $B \neq 0$ with our illustrative data at, say, $\alpha = 0.05$, with $\delta = 10 - 2 = 8$; then the critical values would be ± 3.306.

And

$$t = \frac{2.531 - 0}{\sqrt{\dfrac{1.64}{33.15 - (17.3)^2/10}}} = 3.55$$

which is greater than 2.306. Thus we conclude that the difference between the estimate slope and the hypothetical population slope cannot be explained by chance variation alone, and the null hypothesis is rejected at the 5 percent level of significance. This conclusion suggests that H_1 is true and that the sample regression equation should not be discarded. It can be used for predictive purposes.

The corresponding 95 percent confidence interval for B is

$$b \pm 2.306\hat{\sigma}_b = 2.531 \pm 2.306(0.713) = 0.089 \text{ to } 4.175$$

or

$$0.089 < B < 4.175$$

The positive value of the lower confidence limit here indicates that we have established, at the 95 percent level of confidence, that the true slope is positive; namely, $B \neq 0$.

19.7 Confidence Limits of Prediction

After we have decided that the sample regression equation should be retained, we may then use it to predict the mean of Y on X, (μ_{yx}), or a new observation of Y, (y), given the value of X. For this, we must be able to answer the question: How reliable is our prediction? The reply here again is made with our understanding of the distribution of y_c and y.

CONFIDENCE INTERVAL FOR μ_{yx}

The estimator of μ_{yx} is y_c. It can be shown that the sampling distribution of y_c is a t distribution, with $\delta = n - 2$. The expectation of y_c is

$$E(y_c) = E(a + bx) = E(a) + E(b)x = A + Bx = \mu_{yx}$$

Thus, we see that y_c is an unbiased estimate of μ_{yx}. In this derivation, we should remember that X is not a random variable and, consequently, $E(b)x = Bx$. The variance of y_c can be derived from the sample regression equation,

$$y_c = a + bx$$

where

$$a = \bar{y} - b\bar{x}$$

and, therefore,

$$y_c = \bar{y} + b(x - \bar{x})$$

From this new expression for the sample regression equation, we have as the variance of y_c, given x,

$$V(y_c) = \sigma_{y_c}{}^2 = V(\bar{y} + b(x - \bar{x})) = V(\bar{y}) + V(b(x - \bar{x})) = \frac{\sigma_{yx}{}^2}{n} + V(b)(x - x)^2$$

$$= \frac{\sigma_{yx}{}^2}{n} + \frac{\sigma_{yx}{}^2}{\Sigma(x - \bar{x})^2}(x - \bar{x})^2 = \sigma_{yx}{}^2\left[\frac{1}{n} + \frac{(x - \bar{x})^2}{\Sigma(x - \bar{x})^2}\right]$$

The square root of $V(y_c)$ is the standard error of y_c and is denoted as σ_{y_c}. The unbiased estimate of $V(y_c)$ is

$$\hat{\sigma}_{y_c}{}^2 = \hat{\sigma}_{yx}{}^2\left[\frac{1}{n} + \frac{(x - \bar{x})^2}{\Sigma(x - \bar{x})^2}\right] \tag{19.15}$$

and the positive square root of this quantity is the unbiased estimate of the standard error of y_c, denoted as $\hat{\sigma}_{y_c}$.

Examination of (19.15) shows that the variance or standard error for y_c depends upon the given value of X. If $x = \bar{x}$, it has its minimum value, since then $V(y_c) = \sigma_{yx}{}^2/n$. The farther x is from $\bar{x}$, the greater is the variance or the standard error. Consequently, predictions of μ_{yx} are more precise the closer the fixed values of X are toward $\bar{x}$. We become less and less sure about the relationship between the two variables the more extreme the X values become.

With the standard error of y_c estimated, a confidence interval for μ_{yx}, given the value of X, can be constructed in the usual way. For example, if we wish to construct a 95 percent confidence interval for our illustrative data, given that $x = 1.48$, we would have first,

$$\hat{\sigma}_{y_c} = 1.28\sqrt{\frac{1}{10} + \frac{(1.48 - 1.73)^2}{3.221}} = 0.442$$

Next, we note, for $x = 1.48$,

$$y_c = 1.481 + 2.531(1.48) = 5.23$$

Finally,

$$y_c \pm t_{n-2;\,\alpha/2}\hat{\sigma}_{y_c} = 5.23 \pm 2.306(0.442) = 4.21 \text{ to } 6.25$$

or

$$4.21 < \mu_{yx} < 6.25$$

Confidence intervals constructed for μ_{yx} at various fixed values of X are called a *confidence band for* μ_{yx}. A 95 percent confidence band is computed in Table 19.4 and the results are plotted in Figure 19.9. To speed up calculations, we may note that the values of X are selected so that they are symmetrical with respect to $\bar{x}$. Thus, the calculations of the second half are duplicates of those in the first half. Note how the general shape of the confidence ban follows previous comments on the standard error of y_c. It should be clear that this confidence band is for the true population regression line; not for the individual observations of Y, ys. The result here should be interpreted as follows: We are 95 percent certain that the population regression line will fall within the confidence band, as shown by Figure 19.9. It is interesting to observe also that our confidence band is rather wide, but

TABLE 19.4 COMPUTATION OF 95 PERCENT CONFIDENCE INTERVAL FOR TRUE REGRESSION LINE

x	$x - \bar{x}$	$(x - \bar{x})^2$	$\hat{\sigma}_{y_c}$	y_c	$y_c \pm 2.306(\hat{\sigma}_{y_c})$
0.98	−0.75	0.5625	0.672	3.96	2.41–5.51
1.23	−0.50	0.2500	0.540	4.59	3.45–5.84
1.48	−0.25	0.0625	0.442	5.23	4.21–6.25
1.73	0.00	0.0000	0.404	5.86	4.93–6.79
1.98	+0.25	0.0625	0.442	6.49	5.47–7.51
2.23	+0.50	0.2500	0.540	7.13	5.88–8.38
2.48	+0.75	0.5625	0.672	7.76	6.21–9.31

it is still not wide enough to contain a regression line of zero slope. This reinforces our confidence once again that the true regression is indeed to some extent positively sloped.

CONFIDENCE INTERVAL FOR AN INDIVIDUAL OBSERVATION y

Sometimes we are interested in predicting the value of an individual observation of Y instead of the average value of all Y values associated with a given value of X. For example, given a student whose IQ is 105, what is his predicted performance at college? Here, we are not asking for the average performance of all students whose IQ is 105. The problem of predicting about an individual value is dealt with by constructing a confidence interval for y.

For the construction of a confidence interval for y, we start with an evaluation of the distribution of the difference $y_c - y$. Given the value of X, the expected value of this difference is

$$E(y_c - y) = E(y_c) - E(y) = \mu_{yx} - \mu_{yx} = 0$$

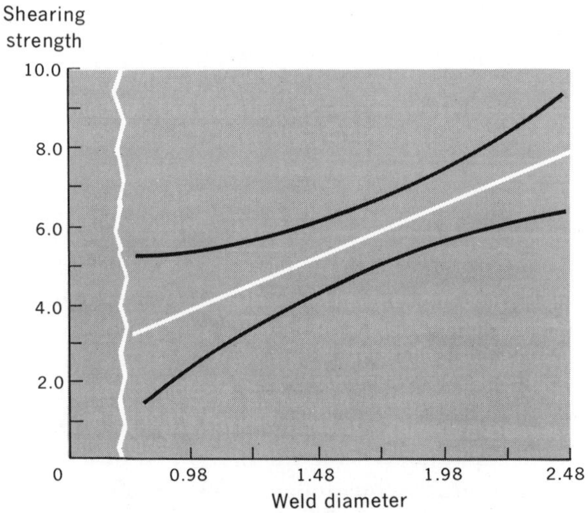

Figure 19.9 Confidence interval for the true regression line.

and the variance of this difference is

$$V(y_c - y) = \sigma^2_{y_c - y} = V(y_c) + V(y) = V(y) + V(b(x - \bar{x}))$$
$$= \sigma_{yx}^2 + \frac{\sigma_{yx}^2}{n} + \frac{\sigma_{yx}^2}{\Sigma(x - \bar{x})^2}(x - \bar{x})^2 = \sigma_{yx}^2\left[1 + \frac{1}{n} + \frac{(x - \bar{x})^2}{\Sigma(x - \bar{x})^2}\right]$$

Again, using $\hat{\sigma}_{yx}^2$ as an estimate of σ_{yz}^2, we have the estimate of $V(y_c - y)$ as follows:

$$\hat{\sigma}^2_{y_c - y} = \hat{\sigma}_{yx}^2\left[1 + \frac{1}{n} + \frac{(x - \bar{x})^2}{\Sigma(x - \bar{x})^2}\right] \tag{19.16}$$

Note that $V(y_c - y)$ is greater than $V(y_c)$ by a factor of 1. This implies that the prediction of y is less precise than that of y_c. Again the positive square root of (19.16) is an unbiased estimate of the standard error of $(y_c - y)$ and is denoted as $\hat{\sigma}_{y_c - y}$.

It can be shown that the ratio $\dfrac{y_c - y}{\hat{\sigma}_{y_c - y}}$ is distributed as t_{n-2}. From this fact, we can easily derive a confidence interval for y as follows:

$$P\left(-t_{n-2;\alpha/2} < \frac{y_c - y}{\hat{\sigma}_{y_c - y}} < t_{n-2;\alpha/2}\right) = 1 - \alpha$$

which, by rearranging terms, becomes

$$y_c - t_{n-2;\alpha/2}\hat{\sigma}_{y_c - y} < y < y_c + t_{n-2;\alpha/2}\hat{\sigma}_{y_c - y}$$

Again as before, given $x = 1.48$, we have as our point estimate of an individual or new observation y

$$y_c = 1.481 + 2.531(1.48) = 5.23$$

for which, the standard error is

$$\hat{\sigma}_{y_c - y} = 1.28\sqrt{1 + \frac{1}{10} + \frac{(1.48 - 1.73)^2}{3.221}} = 1.36$$

With these values, a 95 percent confidence interval for y with $x = 1.48$ becomes

$$5.23 - 2.306(1.36) < y < 5.23 + (2.306)(1.36)$$

or

$$2.09 < y < 8.37$$

A confidence band can also be computed for y at various values of X. The interpretation here is the same as that for the regression line, except that, for obvious reasons, the confidence band for ys is much wider.

To conclude this section, we should note that in the prediction of either μ_{yx} or y, we should not fix the value of X so that it is outside the range of Xs employed to determine the sample regression equation. This is due the fact that the relationship between the two variables below or above the range of the X values may not be the same as that which has been observed within the range. In general, the narrower the prediction limits, the more useful is the independent variable as the basis for predicting the values of the dependent variable. The factors that influence the width of prediction limits for both the average value of ys and individual new observations are the same and are indicated in equations for their respective variances. These factors are summarized below:

(1) The estimated standard deviation of the regression varies directly with the prediction intervals. The size of sample standard deviation of regression, however, does not depend upon the sample size very much; it is inherent in the particular bivariate population.

(2) The prediction interval varies inversely with n for two reasons: The larger n is, the larger is the quantity $\Sigma(x - \bar{x})^2$ and, therefore, the smaller is the standard error; the larger n is, the smaller is the critical value of t_{n-2} and, therefore, the narrower is the confidence interval.

(3) The greater the variability of the independent variable is, the narrower will be the interval, since the quantity $\Sigma(x - \bar{x})^2$ will be large.

(4) As has already been pointed out, given the first three factors, the prediction limits become wider as the fixed X value departs farther from $\bar{x}$. This is so because the quantity $(x - \bar{x})^2$ will increase as the difference between x and $\bar{x}$ increases, and therefore increases the standard error.

19.8 Population Correlation Coefficient

Up to now, we have confined our discussion to the type of relationship between two variables. In some problem situations, however, we may be interested in discovering the degree of relationship between the two variables. For example, the federal government, in deciding on educational grants to a state or city government, may consider per capita income as a measure of the latter's ability to finance its own schools. Thus, the federal government may decide that per capita income is an appropriate criterion for the decision because there is a high degree of relationship between per capita income and expenditures per student in each locality. Otherwise, another decision criterion should be chosen.

The measure for the degree of relationship between two variables is called the *correlation coefficient*, universally denoted as ρ, the Greek letter *rho*. The assumptions that constitute a bivariate linear correlation population model are briefly noted below:

(1) Both X and Y are random variables. As such, the two variables do not have to be designated as dependent and independent. Any designation will yield the same result.

(2) The bivariate population is normal. A bivariate normal population is one in which both X and Y are normally distributed. The expectation and variance of X are $E(X) = \mu_x$ and $V(X) = \sigma_x^2$, respectively, and the expectation and variance of Y are $E(Y) = \mu_y$ and $V(Y) = \sigma_y^2$, respectively.

(3) The relationship between X and Y is linear. This assumption of linearity implies that all the means of Ys associated with each and every X value, μ_{yx}, fall on the regression line of Y on X: $\mu_{yx} = A + Bx$. Also, all the means of Xs associated with each and every Y value, μ_{xy}, fall on the regression line of X on Y: $\mu_{xy} = A' + B'y$. Furthermore, the population regression coefficients in the two equations are the same if and only if the relationship between X and Y is perfect. Otherwise, with Y dependent, intercepts and slope will differ from the regression equation with X dependent.

Under the foregoing assumptions, the population correlation coefficient is defined in terms of the covariance between X and Y as

$$\rho = \frac{\text{cov}\,(X, Y)}{\sigma_x \sigma_y} = \frac{E(X - \mu_x)(Y - \mu_y)}{\sqrt{E(X - \mu_x)^2}\,\sqrt{E(Y - \mu_y)^2}} \tag{19.17}$$

A number of things may be observed from the above definition. First of all, it is a function with five parameters: μ_x, σ_x, μ_y, σ_y, and ρ. The last, as mentioned before, is the correlation coefficient for the bivariate normal population. Secondly, ρ is symmetric with respect to X and Y, with the origin of ρ at μ_x and μ_y. Finally, ρ is independent of units of measurement since it is the covariance between two standard normal variables. In other words, ρ is a pure number due to the fact it is defined as the ratio of the covariance between X and Y to the product of their respective standard deviations. As such, when $\text{cov}(X,Y) = 0$, ρ would be 0, indicating that there is no relationship between the two variables. When there is perfect covariability between X and Y, and X and Y vary in the same direction, $\rho = 1$. Similarly, when there is perfect covariability but X and Y vary in opposite directions, $\rho = -1$. When some degree of covariability exists between X and Y, we would have $-1 < \rho < 0$ or $0 < \rho < 1$.

It is convenient to think of the bivariate normal distribution as a three-dimensional surface instead of a two-dimensional frequency curve. Figure 19.10 is a generalized portrait of a joint normal distribution. This figure reveals that our particular bivariate normal surface has some degree of covariability, since the mound-shaped surface is not parallel to either the X or the Y axis. If we slice this population surface at a given X value, we have a frequency curve giving the frequency of Y at that X value. That is, we have the distribution of the subpopulation Y that corresponds to the selected X value. Similarly, if we slice the population surface at a selected Y value, a subpopulation of X associated with this particular Y emerges. Now, if we take any cross section of the surface parallel to the XY plane, this cross section will be an ellipse. Furthermore, this cross section will approach a circle when $\rho = 0$, a straight line when $\rho = \pm 1$, and some in-between shape when imperfect correlation exists. These cases are illustrated by the shapes in Figure 19.11.

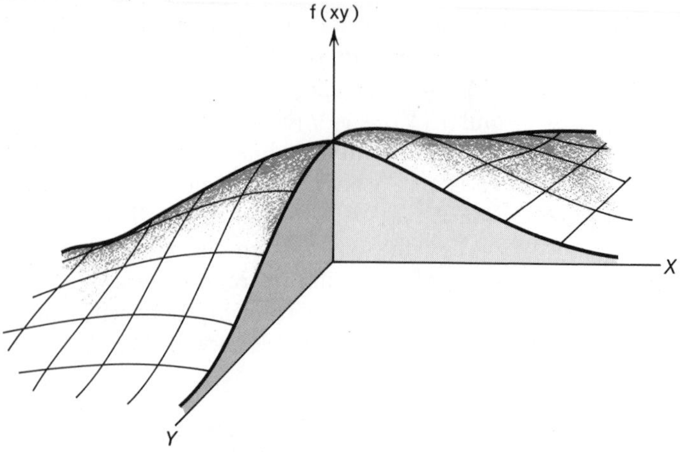

Figure 19.10 Bivariate normal distribution.

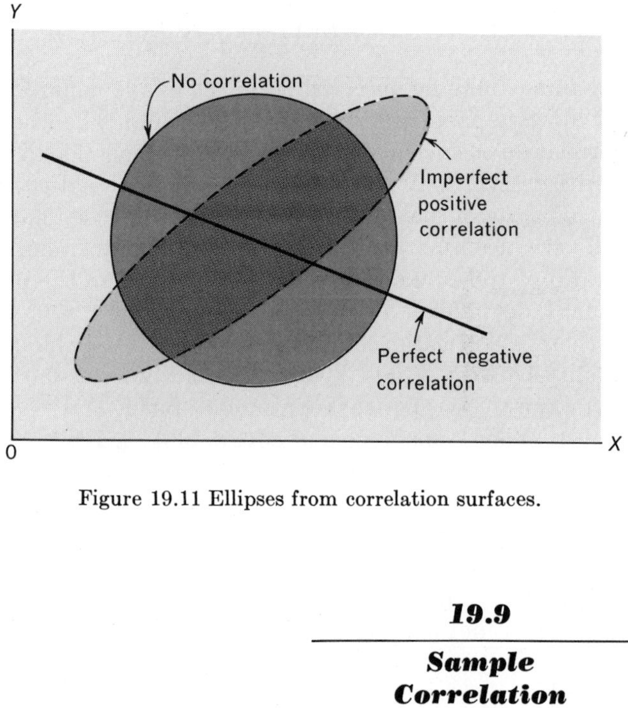

Figure 19.11 Ellipses from correlation surfaces.

19.9 Sample Correlation Coefficient

When a sample of n pairs of observations is drawn, each X value is a random observation from the X population, and each Y value is a random observation from the Y population, but the two are not necessarily independent; and when the assumption of bivariate normal population is met, the maximum likelihood estimator of ρ, usually denoted as r, is obtained by the following expression:

$$\hat{\rho} = r = \frac{\Sigma(x - \bar{x})(y - \bar{y})}{\sqrt{\Sigma(x - \bar{x})^2}\sqrt{\Sigma(y - \bar{y})^2}} = \frac{\Sigma xy - n\bar{x}\bar{y}}{\sqrt{(\Sigma x^2 - n\bar{x}^2)(\Sigma y^2 - n\bar{y})^2}}$$

$$= \frac{n\Sigma xy - \Sigma x \Sigma y}{\sqrt{[n\Sigma x^2 - (\Sigma x)^2][n\Sigma y^2 - (\Sigma y)^2]}} \tag{19.18}$$

which may vary from -1 through 0 to $+1$.

We note that all the quantities in (19.18) are the same for regression analysis. Referring to the sums in Table 19.2, we have, for our illustrative data,

$$r = \frac{(10)(109.53) - (17.3)(58.6)}{\sqrt{[10(33.15) - (17.3)^2][10(377.12) - (58.6)^2]}} = +0.78$$

Since r is a pure number, shifting the origin and changing the scale of unit do not affect its value. Thus, we may let $X' = X - c$ and $Y' = Y - k$, or let $X' = X/g$ and $Y' = Y/g$, or let $X' = mX$ and $Y' = mY$, and can compute r with (19.18) with X' and Y' in each of these cases without additional adjustment. Needless to say, two or more of these coding methods can be employed to compute r. A special procedure of coding is of particular interest here: If we define $X' = X - \bar{x}$ and $Y' = Y - \bar{y}$, then both Σx and Σy become zero, and (19.18) is reduced to

$$r = \frac{n\Sigma x'y'}{\sqrt{(n\Sigma x'^2)(n\Sigma y'^2)}} = \frac{\Sigma x'y'}{\sqrt{(\Sigma x'^2)(\Sigma y'^2)}} \tag{19.19}$$

which is sometimes called the *product-movement* formula for the sample correlation coefficient.

19.10 Inferences about ρ

The estimator r, like other statistics, is subject to sampling variations. When we obtain a positive or negative r, we may not be at all sure, because of sampling error, that the corresponding value of ρ is also positive or negative. This situation calls for a test of significance.

For the purpose of testing, we note that the sampling distribution of r varies with the value of ρ and the sample size. Under the assumption that the sample is drawn from a normal bivariate population, r tends to be distributed normally as the sample size increases. This tendency is much stronger for values of ρ that are close to 0 than for values of ρ that are close to -1 or $+1$. When $\rho = 0$ and n is large, in other words, the distribution of r closely resembles a normal distribution. Figure 19.12 illustrates this point.

The property that, when $\rho = 0$ and n is large, r is approximately normal,

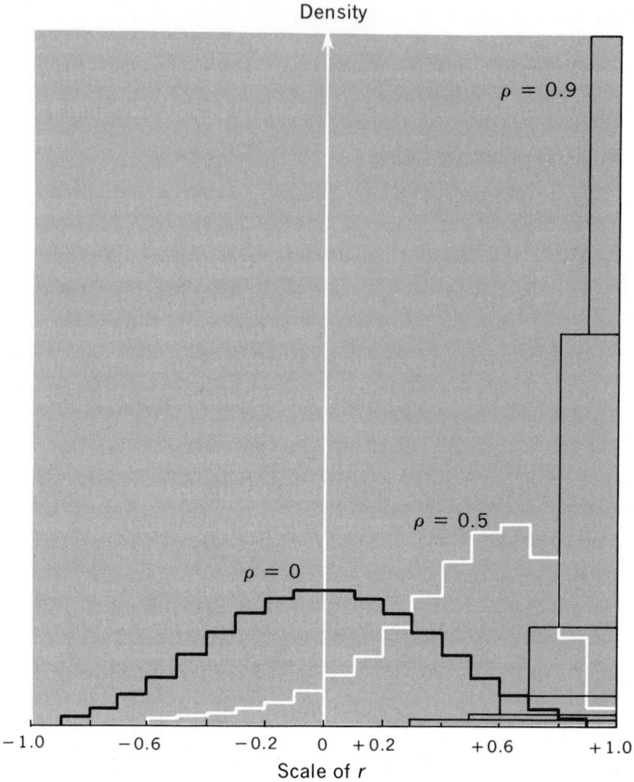

Figure 19.12 Sampling distributions of coefficient of correlation samples of size 10; $\rho = 0$, $\rho = 0.5$, and $\rho = 0.9$.

enables us to conduct the simplest and perhaps the most useful test on the null hypothesis that there is no correlation in the bivariate population, $\rho = 0$, against the alternative that $\rho \neq 0$. When $n \leq 30$, the statistic used to test this H_0 is

$$t = r\sqrt{\frac{n - 2}{1 - r^2}} \tag{19.20}$$

which is distributed as t_{n-2} if the null hypothesis is true.

Applying such a test to our illustrative data at $\alpha = 0.05$, we would accept H_0: $\rho = 0$ if $-2.306 < t_8 < 2.306$. For the value of r obtained in the last section, however,

$$t = 0.78\sqrt{\frac{10 - 2}{1 - (0.78)^2}} = 3.48$$

Consequently, the null hypothesis that there is no correlation in the population is rejected at the specified level of significance. This means that a value of r as large as 0.78 with $\delta = 8$ cannot be the result of chance variations. It is significant.

With the t test just introduced, we cannot state explicitly the extent of the relationship between the two variables when the null hypothesis is rejected. If we desire to test a hypothesis that ρ has some value other than 0, or if we desire to construct a confidence interval for ρ, we may employ what is called the z transformation. That is, we may make a transformation of the asymmetric distribution of r into an approximately normal distribution as follows:

$$z_r = \frac{1}{2} \ln \frac{1+r}{1-r} = \frac{1}{2} 2.3026 \log \frac{1+r}{1-r} \tag{19.21}$$

where ln is a natural logarithm and is approximately equal to 2.3026 log. Here, log stands for the common logarithm of base 10. It can be shown that z_r is approximately normally distributed with $E(z_r) = z_\rho$, and the estimated standard error is

$$\hat{\sigma}_z = \frac{1}{\sqrt{n-3}} \tag{19.22}$$

To test a hypothesis about ρ by r, we now have the testing statistic

$$Z = \frac{z_r - z_\rho}{\hat{\sigma}_z} \tag{19.23}$$

which is approximately $n(0,1)$.

To avoid the computations with logarithms, we employ a table of values for z_r corresponding to various values of r as given by Table A-XI.

Now, suppose that we wish to test $H_0: \rho = 0.85$ against $H_1: \rho \neq 0.85$ for our illustrative example; we refer to Table A-XI and find

$$z_\rho = 1.25615 \quad \text{for} \quad \rho = 0.85$$
$$z_r = 1.04537 \quad \text{for} \quad r = 0.78$$

Hence,

$$Z = \frac{1.04534 - 1.25615}{1/\sqrt{10-3}} = -0.55$$

Since the critical value is greater than -1.96, H_0 is accepted at $\alpha = 0.05$.

Confidence limits may be computed for z_ρ as follows:

$$P(z_r - Z_{\alpha/2}\hat{\sigma}_z < z_\rho < z_r + Z_{\alpha/2}\hat{\sigma}_z) = 1 - \alpha \tag{19.24}$$

The conversion of z_r into r in this expression can be made by looking at the entries in Table A-XI for the value of z_ρ and then checking the stub and column headings to find the corresponding value of r. Due to rounding and/or limited entries in the table, this conversion may be only an approximation.

As an illustration of applying (19.24), we shall now make a 95 percent confidence interval for the problem on hand:

$$1.04537 - 1.96(0.392) < z_\rho < 1.04537 + 1.96(0.392)$$

or

$$0.28097 < z_\rho < 1.80977$$

From Table A-XI, we find

z_ρ	r
0.28	0.485
1.81	0.955

Hence, the 95 percent confidence interval becomes

$$0.485 < \rho < 0.955$$

which includes $\rho = 0.85$.

Also, the significance of a difference between two sample correlation coefficients obtained from two independent samples and their difference can be tested by the statistic

$$Z = \frac{z_{r1} - z_{r2}}{\sqrt{\hat{\sigma}_{z_1}{}^2 + \hat{\sigma}_{z_2}{}^2}} \tag{19.25}$$

where

$$\hat{\sigma}_{z_1} = \frac{1}{\sqrt{n_1 - 3}} \quad \text{and} \quad \hat{\sigma}_{z_2} = \frac{1}{\sqrt{n_2 - 3}}$$

19.11

Population Coefficient of Determination

We have defined the correlation coefficient of bivariate normal population as the degree of convariability between X and Y. This definition led us to compute r as the maximum likelihood estimator of ρ. Supposing that we are now concerned with a bivariate population whose distribution is unspecified, we can then define the population correlation coefficient as a measure of closeness-of-fit of the regression line. This new definition is based on the concept of coefficient of determination, which we are about to introduce with the aid of Figure 19.13. This figure attempts to show the relationship between an individual value of Y, y, and the population regression line of Y on X, as well the relationship between y and the mean of Y, μ_y. From this figure, we see that the following relationship holds:

$$\begin{array}{lcccc} & \text{Total error} & = & \text{Unexplained error} & + \text{Explained error} \\ (1) & y - \mu_y & = & (y - \mu_{yx}) & + \quad (\mu_{yx} - \mu_y) \end{array}$$

In the above expression, the deviation of y from μ_y is the *total error*. The second term on the right side of (1), $\mu_{yx} - \mu_y$, is called the *explained error* because its pattern is definite in the sense that it measures the amount of error which has been removed after the regression line has been fitted. The term $(y - \mu_{yx})$ is called the *unexplained error* because it is a random quantity and is unpredictable; it measures that part of the total error which still remains or is unexplained after the regression line has been fitted.

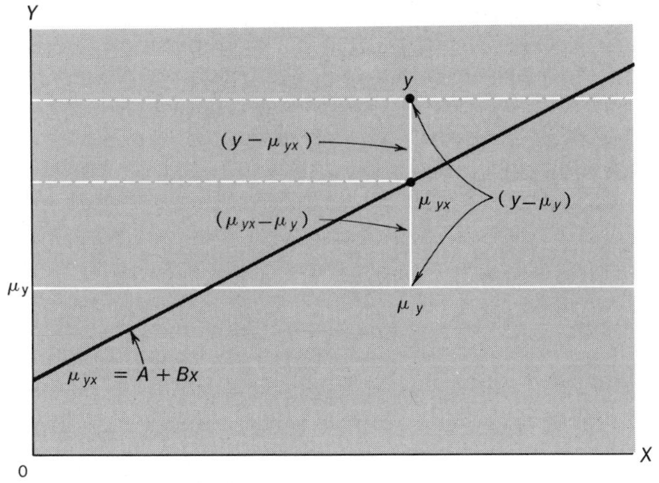

Figure 19.13 Partition of $(y - \mu_y)$.

If we decompose each and every point in the scatter diagram as indicated by (1) summing all the points over the population, and squaring both sides of the equation of sums, we would then obtain

$$(2) \quad \Sigma(y - \mu_y)^2 = \Sigma(y - \mu_{yx})^2 + \Sigma(\mu_{yx} - \mu_y)^2 - 2\Sigma(y - \mu_{yx})(\mu_{yx} - \mu_y)$$

in which the last sum is zero, since

$$\begin{aligned}\Sigma(y - \mu_{yx})(\mu_{yx} - \mu_y) &= \Sigma(y - A - Bx)(A + Bx - \mu_y)\\ &= A\Sigma(y - A - Bx) + A\Sigma x(y - A - Bx) - \mu_y\Sigma(y - A - Bx) = 0\end{aligned}$$

Thus, (2) becomes

$$\Sigma(y - \mu_y)^2 = \Sigma(y - \mu_{yx})^2 + \Sigma(\mu_{yx} - \mu_y)^2 \tag{19.26}$$

which says that the *total sum of squares* $\Sigma(y - \mu_y)^2$ is partitioned into two independent parts: the *unexplained sum of squares* $\Sigma(y - \mu_{yx})^2$, and the *explained sum of squares* $(\mu_{yx} - \mu_y)^2$.

Our interest in decomposing the total sum of squares rests on the fact that the ratio of the explained sum of squares to the total sum of squares gives us a parameter of a bivariate population called the *coefficient of determination*. Denoting this parameter as ρ^2, we have

$$\rho^2 = \frac{\Sigma(\mu_{yx} - \mu_y)^2}{\Sigma(y - \mu_y)^2} \tag{19.27}$$

There are a number of properties of the population coefficient of determination which are of interest to us. First of all, the value of ρ^2 ranges from 0 to 1. From Figure 19.14, we see that its maximum value is at the point where $\mu_{yx} = y$, since at this point $\mu_{yx} - \mu_y$ is at its maximum. Thus,

$$\rho^2 = \frac{\Sigma(\mu_{yx} - \mu_y)^2}{\Sigma(y - \mu_y)^2} = \frac{\Sigma(y - \mu_y)^2}{\Sigma(y - \mu_y)^2} = 1$$

The value of $\mu_{yx} - \mu_y$ is at its minimum when $\mu_{yx} = \mu_y$, and

$$\rho^2 = \frac{\Sigma(\mu_{yx} - \mu_y)^2}{\Sigma(y - \mu_y)^2} = \frac{\Sigma(\mu_y - \mu_y)^2}{\Sigma(y - \mu_y)^2} = 0$$

Therefore, $0 < \rho^2 < 1$.

Second, we see from equations (19.26) and (19.27) that if the unexplained error is zero, the total error is equal to the explained error and, therefore, $\rho^2 = 1$. Thus, ρ^2 is a measure that shows the relative reduction of the total error when a regression line is fitted. For example, if $\rho^2 = 0.8$, it means that 80 percent of the variations in Y are "explained" by the variations in X. When $\rho^2 = 1$, then we know that $y = \mu_{yx} = e = 0$, indicating that all points fall on the regression line. When $\rho^2 = 0$, then the explained error equals zero and $\mu_{yx} = \mu_y$. Graphically, this means that the regression line is parallel to and coincides with, the line for μ_y. See Figures 19.14(a) and (b). It is important to note, in this connection, that the coefficient of determination is a measure of the improvement of reducing the total error. It is not a measure of the covariability between X and Y.

Third, when ρ^2 is close to 1, the individual points will be scattered closely and give the appearance of a straight line. When ρ^2 is close to 0, the points will be widely dispersed and will not resemble a straight line. Thus, ρ^2 can also be considered as a measure of *linearity* of the points. It is important to observe here, however, that the consideration of linearity requires that the points must have a scatter or a distribution and that $B \neq 0$. In the case in which Y is a constant variable (that is, each individual Y value is equal to, say, c), for each value of X, $y = c$ and $\bar{y} = c$. As such, as shown by Figure 19.15, all the points would fall on the regression line, which is identical with the horizontal line for μ_y. However, now the ys do not have a distribution and $B = 0 = \Sigma(y - \mu_y)^2$. As a result, the formula defining the coefficient of determination becomes meaningless. There is, in other words, no regression problem at all.

Finally, as we have pointed out before, the population standard deviation of regression is a measure of the scatter of ys around μ_{yx}. However, if the subpopulations, Ys, do not possess any specific form of distribution, an explicit evaluation of σ_{yx} cannot be made. In this case, ρ^2 is a much more obvious measure of the scatter

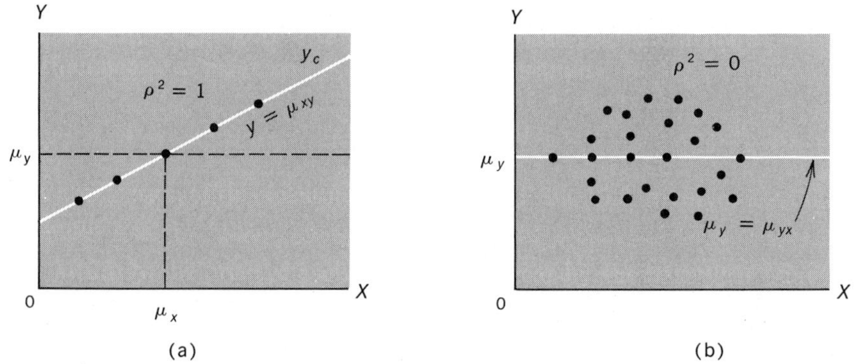

Figure 19.14 ρ^2 as a measure of reducing total error.

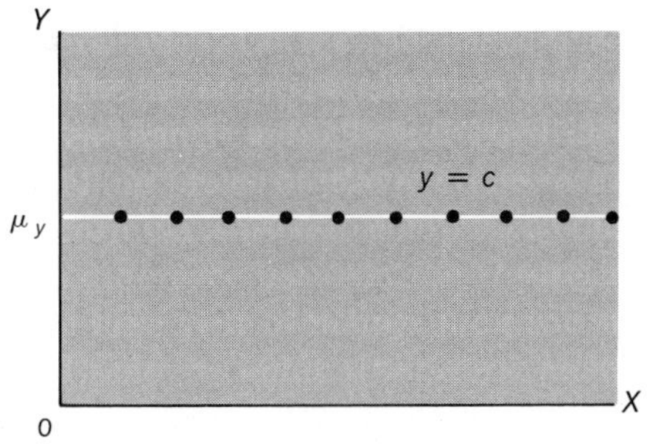

Figure 19.15 Graphic illustration of the absence of regression problem.

or of the goodness-of-fit. It is easy to show, fortunately, that σ_{yx} and ρ^2 are closely related. For bringing out this relationship, let us divide both sides of the equation for the sums of squares by $\Sigma(y - \mu_y)^2$ in order to obtain

$$1 = \frac{\Sigma(y - \mu_{yx})^2}{\Sigma(y - \mu_y)^2} + \frac{\Sigma(\mu_{yx} - \mu_y)^2}{\Sigma(y - \mu_y)^2}$$

The second term on the right side of this expression is our definition for ρ^2, thus

$$\frac{\Sigma(y - \mu_{yx})^2}{\Sigma(y - \mu_y)^2} = 1 - \rho^2$$

Now, recall that

$$\sigma_{yx}^2 = \frac{1}{N} \sum (y - \mu_{yx})^2$$

and

$$\sigma_y^2 = \frac{1}{N} \sum (y - \mu_y)^2$$

We have, therefore,

$$\sigma_{yx}^2 = (1 - \rho^2)\sigma_y^2 \tag{19.28}$$

This result states that, after the regression line has been fitted, the variance of Y has been reduced by $100\rho^2$ percent, and there remains a residual of $(1 - \rho^2)$ percent of σ_y^2 which is not explained. When $\rho^2 = 1$, all of σ_y^2 has been removed, or all the variations in Y have been explained by the variations in X. When $\rho^2 = 0$, there has been no improvement at all by fitting the regression line, and $\sigma_{yx}^2 = \sigma_y^2$.

In concluding this section, let us note that the square root of the population coefficient of determination is the population correlation coefficient; namely

$$\sqrt{\rho^2} = \pm\rho \tag{19.29}$$

where the sign of the correlation coefficient is the same as that for B, the population regression slope. When we have a bivariate population, ρ or ρ^2 can be interpreted as a measure of covariability as well as of closeness-of-fit. When Xs are fixed values, however, cov(X,Y) is undefined, and ρ or ρ^2 can only be interpreted as a measure of closeness-of-fit.

19.12

Sample Coefficient of Determination

Given the minimum variance estimator of the population regression line, $y_c = a + bx$ (as suggested by Figure 19.16), we have the following valid relationship:

$$\underset{\text{Total error}}{\Sigma(y - \bar{y})^2} = \underset{\text{Unexplained error}}{\Sigma(y - y_c)^2} + \underset{\text{Explained error}}{\Sigma(y_c - \bar{y})^2} \tag{19.30}$$

From this, we define the *sample coefficient of determination* as

$$r^2 = \hat{\rho}^2 = \frac{\Sigma(y_c - \bar{y})^2}{\Sigma(y - \bar{y})^2} \tag{19.31}$$

The sample coefficient of determination, as an estimator for ρ^2, measures the closeness-of-fit of the sample regression line to sample observations. It can be shown that r^2 is a consistent and unbiased estimator of ρ^2 as n becomes large. We may also note that the square root of r^2 is the sample correlation coefficient,

$$\sqrt{r^2} = \pm r \tag{19.32}$$

where the sign is the same as that for b.

For data below, X stands for the age of a tree in years and Y stands for the height of the tree in feet:

X:	1	2	3	4	5
Y:	2	2	4	3	5

For these data, the sample regression equation is

$$y_c = 1.10 + 0.7x$$

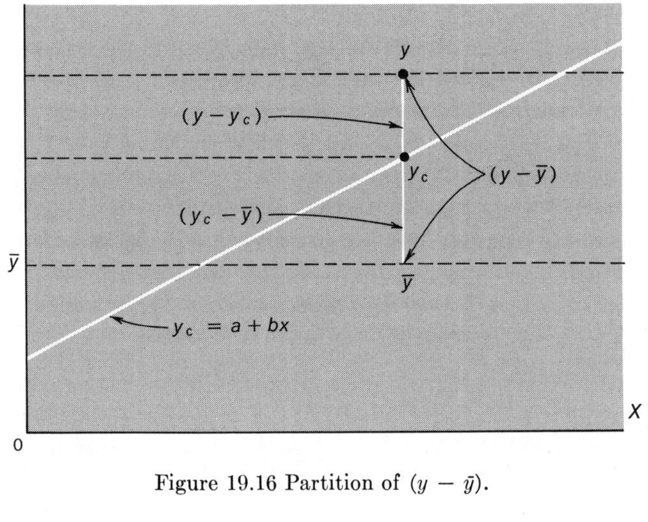

Figure 19.16 Partition of $(y - \bar{y})$.

TABLE 19.5

$(y - \bar{y})$	$(y - \bar{y})^2$	$(y - y_c)$	$(y - y_c)^2$	$(y_c - \bar{y})$	$(y_c - \bar{y})^2$
−1.2	1.44	+0.2	0.04	−1.4	1.96
−1.2	1.44	−0.5	0.25	−0.7	0.49
+0.8	0.64	+0.8	0.64	0.0	0.00
−0.2	0.04	−0.9	0.81	+0.7	0.49
+1.8	3.24	+0.4	0.16	+1.4	1.96
0.0	6.80	0.0	1.90	0.0	4.90

Now, for the purpose of checking the relationship between the sums of squares for error and computing the sample coefficient of determination, we accumulate the necessary sums in Table 19.5. From quantities therein, we find that

$$\begin{aligned} \Sigma(y - \bar{y})^2 &= \Sigma(y - y_c)^2 + \Sigma(y_c - \bar{y})^2 \\ 6.8 &= 1.9 + 4.9 \end{aligned}$$

and that

$$r^2 = \frac{4.90}{6.80} = 0.72$$

The value of r^2 indicates that 72 percent of the changes in the height of the trees are associated with the changes in the ages of the trees. There remains $100 - 72 = 28$ percent variability in the height of the trees that is unexplained by the regression. Also, since b is positive in this case, we have

$$r = \sqrt{0.72} = +0.85$$

Here, X values are fixed; the value of r^2 or r is thought of as a measure of closeness-of-fit.

Using the data in Table 19.3 concerning spot welds of aluminum, we have

$$r^2 = \frac{20.67}{33.72} = 0.61$$

In this case, where both X and Y are random variables and we have a bivariate distribution, r^2 may be thought of as a measure of closeness-of-fit as well as of the degree of covariability.

19.13 The Analysis of Variance for Regression Analysis

From our discussion in the preceding section, the reader may have gained the correct impression that the linear regression model must be related to the linear model of the analysis of variance. Indeed, it is not difficult to show that these two analytical methods can be applied to the same set of data with similar underlying logic. Nevertheless, they are different in terms of objectives of investigation. When we are interested in determining the improvement made by the regression line, we obviously would employ the regression analysis. When we are concerned with the variations among the subpopulation means, μ_ys, we use the analysis of variance. Granting this difference in objectives, we note that both methods are based on the sums of squares which measure variability from different sources. For instance, in the linear regression model, we have

$$\Sigma(y - \bar{y})^2 = \Sigma(y - y_c)^2 + \Sigma(y_c - \bar{y})^2$$

which corresponds to the partition of the total sum of squares for the fixed effects, completely randomized, one-variable classification model in the analysis of variance:

$$\Sigma\Sigma(x_{ij} - \bar{\bar{x}})^2 = \Sigma\Sigma(x_{ij} - \bar{x}_{i.})^2 + \Sigma\Sigma(\bar{x}_{i.} - \bar{\bar{x}})^2$$

Thus, we see that the "within" sum of squares corresponds to "unexplained error," and the "between" sum of squares corresponds to "explained error," in regression analysis.

To apply the analysis of variance terminology to regression, the quantity $\Sigma(y - y_c)^2$ may be thought of as the sum of squares which measures the unsystematic random disturbances and may be denoted as SSE. The quantity $\Sigma(y_c - \bar{y})^2$ is thought to be the sum of squares which measures the systematic (explained) variability due to regression and denoted as SSR. These sums of squares, together with their respective numbers of df and mean squares (variances), can then be summarized in an analysis of variance table as below:

ANALYSIS OF VARIANCE TABLE FOR REGRESSION

Source	SS	δ	MS
Regression	SSR $= \Sigma(y_c - \bar{y})^2$	$c = 1$	MSR = SSR/1
Error (residual)	SSE $= \Sigma(y - y_c)^2$	$n - 2$	MSE = SSE/$(n - 2)$
Total	SST $= \Sigma(y - \bar{y})^2$	$n - 1$	MST = SST/$(n - 1)$

From the analysis of variance table, we see that

$$\hat{\sigma}_{yx}{}^2 = \frac{\Sigma(y - y_c)^2}{n - 2} \doteq \text{MSE} = \frac{\text{SSE}}{n - 2} \tag{19.33}$$

and that

$$r^2 = \frac{\Sigma(y_c - \bar{y})^2}{\Sigma(y - \bar{y})^2} = \frac{\text{SSR}}{\text{SST}} \tag{19.34}$$

When n is small, r^2 tends to be positively biased. To correct this bias, we compute the corrected coefficient of determination as follows:

$$\dot{r}^2 = 1 - \frac{\text{MSE}}{\text{MST}} \tag{19.35}$$

where

$$\frac{\text{MSE}}{\text{MST}} = \frac{\Sigma(y - y_c)^2/(n - 2)}{\Sigma(y - \bar{y})^2/(n - 1)} = \frac{\Sigma(y - y_c)^2}{\Sigma(y - \bar{y})^2}\left(\frac{n - 1}{n - 2}\right)$$

This result indicates that the correction factor for $\dot{r}^2$ is $(n-1)/(n-2)$, which approaches 1 when n becomes large.

We note also that the ratio

$$F_{1,n-2} = \frac{\text{MSR}}{\text{MSE}} \tag{19.36}$$

can be used to test $B = 0$ against $B > 0$ or $B < 0$.

As can be shown by the answer to Problem 19.12, the sums of squares in the analysis of variance table can be computed as follows:

$$\text{SST} = \sum (y - \bar{y})^2 = \sum y^2 - n(\bar{y})^2, \tag{19.37}$$

$$\text{SSR} = \sum (y_c - \bar{y})^2 = b\left(\sum xy - \frac{1}{n}\sum x \sum y\right), \text{ and} \tag{19.38}$$

$$\text{SSE} = \text{SST} - \text{SSR} \tag{19.39}$$

For the tree data, we have

$$\begin{aligned} \text{SST} &= 58 - 5(3.2)^2 = 6.8, \\ \text{SSR} &= 0.7[55 - \tfrac{1}{5}(15)(16)] = 4.9, \text{ and} \\ \text{SSE} &= 6.8 - 4.9 = 1.9 \end{aligned}$$

These values check with those obtained before. Now, summarizing these sums of squares, together with their df and mean squares, in an analysis of variance table for further computations, we have

ANALYSIS OF VARIANCE FOR TREE DATA

Source	SS	δ	MS
Regression	4.9	1	MSR = 4.9/1 = 4.90
Error	1.9	3	MSE = 1.9/3 = 0.63
Total	6.8	4	MST = 6.8/4 = 1.70

From data in the preceding table, we obtain the following statistics:

1. $\hat{\sigma}_{yx}^2 = \text{MSE} = 0.63$
2. $\hat{\sigma}_{yx} = \sqrt{0.63} = 0.79$
3. $\hat{\sigma}_b^2 = \hat{\sigma}_{yx}^2/\Sigma(x - \bar{x})^2 = \dfrac{0.63}{10} = 0.063$
4. $\hat{\sigma}_b = \sqrt{0.063} = 0.251$
5. $r^2 = 1 - \dfrac{1.9}{6.8} = 0.72$
6. $\dot{r}^2 = 1 - \dfrac{0.63}{1.7} = 0.63$

To test $B = 0$ against $B > 0$ for the tree example, we compute

$$F_{1,3} = \frac{4.90}{0.63} = 7.78$$

However, $P(F_{1,3} > 10.13) = 0.05$, Therefore, $B = 0$ cannot be rejected at $\alpha = 0.05$. The lack of significant dependence of the height of the tree upon the age of the tree is most likely due to the smallness of the sample.

Results of regression analysis are usually stated together in the following form:

$$y_c = a + bx, \quad (r^2 \text{ or } \dot{r}^2)$$
$$(\hat{\sigma}_b)$$

For example, our analysis of the tree data can be presented as

$$y_c = 1.10 + 0.7x \quad (\dot{r}^2 = 0.63)$$
$$(\hat{\sigma}_b = 0.251)$$

19.14

Rank-Correlation Coefficient

When a random sample of n pairs is drawn from a bivariate population whose distribution is unknown, the degree of relationship between X and Y cannot be determined by the technique of correlation coefficient presented before. In such a situation, the covariability or the lack of it between X and Y is established by what is called the *rank-correlation coefficient* developed by C. Spearman in 1904. This measure is especially useful when quantitative measures for certain factors (such as in the evaluation of leadership ability or the judgment of female beauty) cannot be fixed, but the individuals in the group can be arranged in order, thereby obtaining for each individual a number indicating his (her) rank in the group. In any event, the rank-correlation coefficient is applied to a set of ordinal rank numbers, with 1 for the individual ranked first in quantity or quality, and so on, to n for the individual ranked last in a group of n individuals (or n pairs of individuals).

Let us symbolize the rank of individual i on variable X by x_i and the rank of i on variable Y by y_i; then we define d_i as $(x_i - y_i)$. Furthermore, the x_i and y_i may represent two set of ranks assigned to the same set of objects by two different judges. With these notations, the Spearman's rank-correlation coefficient is defined as

$$r_s = 1 - \frac{6\Sigma d_i^2}{n(n^2 - 1)} \tag{19.40}$$

From this definition we see that $-1 \leq r_s < +1$. If the rank orders agree perfectly, the ranks assigned to individuals should have perfect positive correlation. If the ranks assigned to one set, x_i, are exactly the opposite of the ranks in the other set, y_i, then r_s should reflect perfect negative correlation. A zero r_s clearly represents the absence of any relationship between the two sets of ranks.

It is important to note that formula (19.40) has been derived under the assumption that there are no ties in either or both rankings. When ties do occur, they may be treated in any one of the following three ways: First, we may assign mean ranks to sets of tied individuals; namely, when two or more individuals are tied in order, each is assigned the mean of the ranks they would otherwise take. Second, we may compute an ordinary correlation coefficient, r, with the ranks as if they are simply numerical scores; the result may be regarded as a Spearman's correlation adjusted for ties. Finally, if the main aim of our analysis is to test the significance of r_s, we should adopt the conservative procedure of breaking the ties in such a way that the absolute value of r_s would be as small as possible.

When no ties in rank exist, the exact sampling distribution of r_s can be worked out. Tables for distributions of r_s are available for small n. When $n \geq 20$, we may test the hypothesis $r_s = 0$ against an appropriate alternative by using the statistic

$$t = \frac{r_s \sqrt{n-2}}{\sqrt{1 - r_s^2}} \tag{19.41}$$

which is approximately distributed as Student's t with $n - 2$ degrees of freedom. For values $8 \leq n < 20$, the t test can still be employed to yield satisfactory results if the t ratio is computed with an r_s that is calculated with a correction for continuity. Let us denote the corrected rank-correlation coefficient as $\dot{r}_s$; then

$$\dot{r}_s = 1 - \frac{\Sigma d_i^2}{\frac{1}{6}(n)(n^2 - 1) + 1} \tag{19.42}$$

With this measure,

$$t' = \frac{\dot{r}_s \sqrt{n-2}}{\sqrt{1 - \dot{r}_s^2}} \tag{19.43}$$

is again approximately distributed at t_{n-2}.

Let us now consider an example. In a beauty contest, two judges each ranked ten contestants in the order of how they rated "beautiful features of the girls." The results are shown in Table 19.6. Did the judges tend to agree at $\alpha = 0.05$?

$$r_s = \frac{6(24)}{10(10^2 - 1)} = 0.8545$$

Since n is rather small in our illustration, let us compute

$$\dot{r}_s = 1 - \frac{24}{\frac{1}{6}(10)(10^2 - 1) + 1} = 0.8554$$

To determine whether this result is significant at $\alpha = 0.05$, we obtain

$$t' = \frac{0.8554 \sqrt{10 - 2}}{\sqrt{1 - (0.8554)^2}} = 9.03$$

TABLE 19.6

Contestant	Judge I	Judge II	d_i	d_i^2
1	8	9	1	1
2	3	5	2	4
3	9	10	1	1
4	2	1	1	1
5	7	8	1	1
6	10	7	3	9
7	4	3	1	1
8	6	4	2	4
9	1	2	1	1
10	5	6	1	1
				24

Since $P(t_8 > 1.860) = 0.05$, the observed value of rank-correlation is significant. The null hypothesis $r_s = 0$ is rejected in favor of the alternative, $r_s > 0$. The two judges have a high degree of agreement.

19.15 Concluding Remarks

We have in this chapter discussed, in a rather detailed fashion, the simple linear regression and correlation models by presenting the measure of coefficient of determination as the link between the two. It is important to observe that because these two models are constructed under different assumptions, they furnish different types of information. It is, however, not always clear to a beginning student which measure should be used in given problem situation. As an aid in making a selection, it would perhaps be useful to point out, once more, the relationship as well as the differences between the two models.

First, the same type of relationship holds for both the regression and the correlation analysis. This is indicated by the fact that r takes the same sign as b. In this connection, it is also interesting to observe that if the value of b is significant at a given level of significance, r is also significant at that level of significance.

Second, r, via r^2, may be considered as a measure of the closeness-of-fit of the regression line. In general, the greater the value of r, the better is the fit and the more useful is the regression equation as a predictive device.

Third, given the first two observations, we must realize that the main objective of regression analysis is to establish a functional relationship between the dependent and independent variables so that the former can be predicted by the latter; however, the correlation coefficient is a measure of the closeness-of-fit of the regression line and/or a measure of the degree of covariability between X and Y.

Fourth, a given value of correlation coefficient is consistent with an infinite number of regression lines. For instance, in the case of perfect correlation, when all points are shifted upward or downward by an equal amount, the value of correlation coefficient would remain as 1 but a different Y intercept would result. Similarly, if all points are rotated around a given point so that they fall precisely on different straight lines, the slope, and perhaps the Y intercept, will change while the correlation coefficient would still assume the value of 1.

Finally, the main difference in assumptions between the two models is that Y is assumed to be a random variable and X is assumed to take fixed values in the regression analysis, while both X and Y are hypothesized as random variables in the correlation analysis.

Theoretically, therefore, regression is a directional method but correlation is not. The former should be used if one variable is clearly dependent upon the other, or one is measured subsequent to the other. The latter is appropriate when the two variables have a joint distribution, neither of which can be considered as being subsequent to, or as being a consequence of, the other.

Practically, on the other hand, the purpose of the investigation is perhaps the most important consideration in selecting between regression and correlation techniques. The choice depends, in other words, on whether we want to have a predictive equation or merely wish to determine the degree of association. As a matter of fact, in practical situations we are often more interested in finding out the nature of the relationship for the objective of prediction. In this connection, the reason for our interest in the degree of relationship may be only to select the most useful independent variable, among several possible independent variables, for regression analysis with the dependent variable that has been chosen.

Association analysis, while a very useful device, often opens up the possibility of making every kind of misinterpretation. The most unwarranted interpretation stems from the confusion between association and causation. The main source of this confusion arises from the conventional designation of dependent and independent variables. Many people take the independent variable to be the cause and the dependent variable to be the effect in a given situation, but this may not necessarily be correct.

The link between association and causation can be stated thus: The presence of association does not imply causation; but the existence of causation always implies association. Statistical evidence can only establish the presence or absence of association between variables. Whether causation exists or not depends purely on reasoning. For example, there is reason to believe that high family income "causes" high family consumption expenditures; hence these two variables must be positively associated. There is also reason to support the premise that the high price of prime cuts of beef "causes" low quantities of this beef to be sold; hence these two variables must be negatively associated. That good weather is associated with high death rates in areas favored by retired people clearly does not mean that good weather "causes" high death rate.

Why is it that association need not imply causation? One reason is that the association between two variables may be the result of pure chance, such as soil erosion in Alaska and the amount of alcohol consumed in South America. Another reason is that association between two variables may be due to the influence of a third common factor. Since 1945, for example, it has been found that there is a close relationship between teachers' salaries and liquor consumption in this country. Can we say that the increase in liquor consumption is the effect of the increase in teachers' salaries? Certainly not. Indeed, a more plausible explanation is that both variables have been influenced by the continuous increase in national income during the same period of time. A third possibility is that a relationship is real, but it is impossible to determine which variable is the cause and which is the effect; we often find that both variables may be the cause and the effect at the same time. It may also happen that cause and effect change places from time to time. The association between per capita income and per capita expenditures for education, for example, may be of this kind. The higher the per capita income is in a state, the more money can be spent for each student; the more money the state spends for education, the higher will be the per capita income in that state; we cannot be sure, however, that one has produced the other or vice versa.

Associations observed between variables that could not conceivably be causally related are called *spurious* or *nonsense correlations.* More appropriately, however, we should remember that it is the *interpretation* of the degree of association that is spurious, not the degree of association itself. A last word of warning: Errors in association analysis include not only reading causation into spurious associations but also interpreting spuriously a perfectly valid association.

Glossary of Formulas

(19.1) $\mu_{yx} = A + Bx$ **(19.2)** $y = E(Y|X) + e$ **(19.3)** $\sigma_{yx}{}^2 = \dfrac{\Sigma(y - \mu_{yx})^2}{N}$

This set of three equations defines the simple linear regression model. The first is the population regression equation of Y on X. It states that if X, the independent variable, is fixed to vary unit by unit, the means of Y_s, the dependent variable, on X will differ from one another by B—the slope of the population regression equation. A here is the value of Y when $X = 0$. It is called the Y intercept. The second equation expresses an individual value of a subpopulation Y associated with a fixed X value as a deviation from μ_{yx}. Here e is the error term. The last equation is the common variance of the regression. For this model, it is assumed that each and every subpopulation Y has the same variance. The positive square root of this variance gives us the standard deviation of the regression.

(19.4) $y_c = a + bx$ **(19.5)** $b = \dfrac{n\Sigma xy - \Sigma x \Sigma y}{n\Sigma x^2 - (\Sigma x)^2}$ **(19.6)** $a = \bar{y} - b\bar{x}$

The first of these expressions is the sample regression, or prediction equation, where $(a + bx)$ is the linear estimator of $(A + Bx)$. This estimator is a minimum variance estimator by the least squares method if the subpopulation, Ys, do not possess any specific distributions. It is a maximum likelihood estimator if the Ys are normally distributed. In either case, a and b are obtained from solving the same set of two normal equations. Direct solution of these normal equations gives us formulas (19.5) and (19.6), which are used to obtain the sample regression coefficients a and b.

(19.7) $b = b'$ **(19.8)** $a = a' + (k - bc)$ **(19.9)** $a = a'g$

These formulas refer to efficient procedures of computing the sample regression coefficients. The first two are results of shifting the origin, where we define $X' = X - c$ and $Y' = Y - k$, and b' and a' are computed with these transformed variables according to (19.5) and (19.6). The shifting of the origin does not affect the slope and, therefore, $b = b'$. This procedure does affect a, however, and thus $a = a' + (k - bc)$. Equation (19.9) is the result of reducing the scale, where xs and ys are divided by g. Again, b is not affected but a must be converted back to the original scale as dictated by (19.9).

$$(\mathbf{19.10})\ \hat{\sigma}_{yx}{}^2 = \frac{\Sigma(y - y_c)^2}{n - 2} \quad \text{and} \quad (\mathbf{19.11})\ \sigma_{yx}{}^2 = \frac{\Sigma y^2 - a\Sigma y - b\Sigma xy}{n - 2}$$

These are the theoretical and computational equations for sample variance of regression, respectively. They provide an unbiased estimator of the population variance of regression. The positive square root of each of these two expressions is an unbiased estimate of the population standard deviation of regression. (*Note:* The denominator refers to $n - 2$ df due to the fact that two estimators, a and b, are employed to compute it.)

$$(\mathbf{19.12})\ \hat{\sigma}_b{}^2 = \frac{\hat{\sigma}_{yx}{}^2}{\Sigma x^2 - (\Sigma x)^2/n} \qquad (\mathbf{19.13})\ Z = \frac{b - B}{\hat{\sigma}_b} \qquad (\mathbf{19.14})\ t = \frac{b - B}{\hat{\sigma}_b}$$

The sampling distribution of b is normal if Y (or e) is normally distributed, and if $n \geq 30$ with $E(b) = B$, and with variance as estimated by (19.12). To test $B = 0$ against $B \neq 0$, (19.13) is used as the testing statistic when samples are large, and (19.14) is used when samples are small. (*Note:* The standard error of b is the positive square root of the variance of b. When samples are small, the confidence interval for B is also constructed with t distributions for $\delta = n - 2$.)

$$(\mathbf{19.15})\ \hat{\sigma}_{y_c}{}^2 = \hat{\sigma}_{yx}{}^2\left[\frac{1}{n} + \frac{(x - \bar{x})^2}{\Sigma(x - \bar{x})^2}\right]$$

The y_c, as the estimator of μ_{yx}, is distributed as t_{n-2} with $E(y_c) = \mu_{yx}$ and estimated variance as defined here. The positive square root of this estimate is an unbiased estimate of the standard error of y_c, which is used to construct confidence intervals, or prediction band, for μ_{yx}. Note $\Sigma(x - \bar{x})^2 = \Sigma x^2 - (\Sigma x)^2/n$.

$$(\mathbf{19.16})\ \hat{\sigma}^2{}_{y_c - y} = \hat{\sigma}_{yx}{}^2\left[1 + \frac{1}{n} + \frac{(x - \bar{x})^2}{\Sigma(x - \bar{x})^2}\right]$$

The sampling distribution of the difference $y_c - y$ has as its expectation $E(y_c - y) = 0$, and estimated variance as defined here. The positive square root of this estimate is the estimated standard error of $y_c - y$, which can be employed to construct a confidence interval for an individual observation, y, by noting that the ratio $(y_c - y)/(\hat{\sigma}_{y_c - y})$ is distributed as t_{n-2}. The resulting confidence interval estimate of y takes the form: $y_c - t_{n-2;\alpha/2}\hat{\sigma}_{y_c - y} < y < y_c + t_{n-2;\alpha/2}\hat{\sigma}_{y_c - y}$.

$$(\mathbf{19.17})\ \rho = \frac{E(X - \mu_x)(Y - \mu_y)}{\sqrt{E(X - \mu_x)^2}\sqrt{E(Y - \mu_y)^2}}$$

The correlation coefficient for a bivariate normal population is defined in terms of the covariation between X and Y. It is a pure number and ranges from -1 through 0 to $+1$.

$$(\mathbf{19.18})\ r = \frac{n\Sigma xy - \Sigma x\Sigma y}{\sqrt{[n\Sigma x^2 - (\Sigma x)^2][n\Sigma y^2 - (\Sigma y)^2]}} \qquad (\mathbf{19.19})\ r = \frac{\Sigma x'y'}{\sqrt{(\Sigma x'^2)(\Sigma y'^2)}}$$

Equation (19.18) is the sample correlation coefficient defined as the covariation between X and Y in the sample. This statistic, as the corresponding parameter, ranges from -1 through 0 to $+1$. Because r is a pure number, shifting of origin and reduction in the scale will not affect its value. Equation (19.19) is called the product movement formula for r which is computed from $X' = X - \bar{x}$ and $Y' = Y - \bar{y}$.

(19.20) $t = r\sqrt{\dfrac{n-2}{1-r^2}}$ To test $\rho = 0$ against $\rho \neq 0$, we use this ratio as the testing statistic, which is distributed as t_{n-2}. In general, the result of this test should be the same as the t test for population slope.

(19.21) $z_r = \dfrac{1}{2}\,2.3026 \log \dfrac{1+r}{1-r}$ **(19.22)** $\hat{\sigma}_z = \dfrac{1}{\sqrt{n-3}}$ **(19.23)** $Z = \dfrac{z_r - z_\rho}{\hat{\sigma}_z}$

(19.24) $P(z_r - Z_{\alpha/2}\hat{\sigma}_z < z_\rho < z_r + Z_{\alpha/2}\hat{\sigma}_z) = 1 - \alpha$

Here, z_r stands for the normal transformation of r. It is approximately normal with $E(z_r) = z_\rho$, and the estimated standard error is defined by (19.22). To test the hypothesis that the population correlation coefficient assumes a given value, we use (19.23) in the usual manner. Equation (19.24) gives the confidence interval for z_ρ. Values of z_r corresponding to various values of r are given in Table A-XI, so that we do not have to do computations with logarithms. (*Note:* Result with (19.24) should be converted into values of r via the entries in Table A-XI.)

(19.25) $Z = \dfrac{z_{r1} - z_{r2}}{\sqrt{\hat{\sigma}_{z_1}{}^2 + \hat{\sigma}_{z_2}{}^2}}$ This ratio is used to test the significance of a difference between two rs computed from two independent random samples. In the denominator of this expression, $\hat{\sigma}_{z_1} = 1/\sqrt{n_1 - 3}$ and $\hat{\sigma}_{z_2} = 1/\sqrt{n_2 - 3}$.

(19.26) $\Sigma(y - \mu_y)^2 = \Sigma(y - \mu_{yx})^2 + \Sigma(\mu_{yx} - \mu_y)^2$ **(19.27)** $\rho^2 = \dfrac{\Sigma(\mu_{yx} - \mu_y)^2}{\Sigma(y - \mu_y)^2}$

The total error $\Sigma(y - \mu_y)^2$ in regression analysis can be partitioned into two independent parts: the explained error $\Sigma(\mu_{yx} - \mu_y)^2$ and the unexplained error $\Sigma(y - \mu_{yx})^2$. The purpose of the decomposition of the total sum of squares is to derive a definition for the population coefficient of determination, (19.27), which measures mainly the percentage of the total error that is explained by the regression and has a value ranging from 0 to 1.

(19.28) $\sigma_{yx}{}^2 = (1 - \rho^2)\sigma_y{}^2$ This expresses the relationship among the variance of regression, the coefficient of determination, and the variance of Y. It says that the variance of Y has been reduced by $100\rho^2$ percent after the regression line has been fitted. There remains, in other words, a residual of $(1 - \rho^2)$ percent of $\sigma_y{}^2$ that is unexplained by the regression. (*Note:* When $\rho^2 = 0$, $\sigma_{yx}{}^2 = \sigma_y{}^2$, implying that there is no improvement with the regression analysis.)

(19.29) $\sqrt{\rho^2} = \pm\rho$ The square root of the coefficient of determination is the coefficient of correlation, whose sign is the same as that for B.

(19.30) $\Sigma(y - \bar{y})^2 = \Sigma(y - y_c)^2 + \Sigma(y_c - \bar{y})^2$

(19.31) $r^2 = \dfrac{\Sigma(y_c - \bar{y})^2}{\Sigma(y - \bar{y})^2}$ **(19.32)** $\sqrt{r^2} = \pm r$

The first of this group of equations gives the composition of total sum of squares in a sample, from which we define the sample coefficient of determination r^2 as given in (19.31). The square root of r^2 gives the sample correlation coefficient, for which the sign is the same as b.

(19.33) $\hat{\sigma}_{yx}{}^2 = \dfrac{\text{SSE}}{n-2} = \text{MSE}$ **(19.34)** $r^2 = \dfrac{\text{SSR}}{\text{SST}}$

(19.35) $\dot{r}^2 = 1 - \dfrac{\text{MSE}}{\text{MST}}$ **(19.36)** $F_{1,n-2} = \dfrac{\text{MSR}}{\text{MSE}}$

The equations in this group result from the application of the analysis of variance techniques to regression analysis. In terms of sums of squares, together with their respective df and mean squares (variances), various association statistics are computed. These definitions, it must be noted, are comparable to those given before. Two things we may note, however, are: (19.35) gives what is called the corrected r^2 where the correction factor is $(n-1)/(n-2)$; (19.36) is designed to test $B = 0$. The sums of squares are computed with the next three formulas.

(**19.37**) $\text{SST} = \sum (y - \bar{y})^2 = \sum y^2 - n(\bar{y})^2$

(**19.38**) $\text{SSR} = \sum (y_c - \bar{y})^2 = b\left(\sum xy - \frac{1}{n}\sum x \sum y\right)$

(**19.39**) $\text{SSE} = \text{SST} - \text{SSR}$

These are the computational formulas for the sums of squares in the formula for the total error in regression analysis. Each of these sums divided by the appropriate number of df gives a mean square, or variance, value. We note that for SST, $\delta = n - 1$; for SSR, $\delta = 1$; and for SSE, $\delta = n - 2$. Just as for the sums of squares, these numbers of df are also additive. Namely, $n - 1 = (n - 2) + 1$.

(**19.40**) $r_s = 1 - \dfrac{6\Sigma d_i^2}{n(n^2 - 1)}$

(**19.41**) $t = \dfrac{r_s\sqrt{n-2}}{\sqrt{1 - r_s^2}}$

(**19.42**) $\dot{r}_s = 1 - \dfrac{\Sigma d_i^2}{\frac{1}{6}(n)(n^2 - 1) + 1}$

(**19.43**) $t' = \dfrac{\dot{r}_s\sqrt{n-2}}{\sqrt{1 - \dot{r}_s^2}}$

The equations in this group are designed for the anaysis of bivariate data when the population distribution is unspecified or when data are originally given in ranks. The first equation defines what is called the rank-correlation coefficient in which d stands for the difference between pairs of ranks. The $\dot{r}_s$ is the rank-correlation coefficient computed with correction for continuity. When $n > 20$, t is used to test $r_s = 0$; when $8 \leq n \leq 20$, t' is employed to test $\dot{r}_s = 0$. Both t and t' are distributed as the Student's t with $n - 2$ degrees of freedom.

Problems

19.1 Think of as many economic and business problems as you can that can be profitably evaluated by association analysis.

19.2 In your answer to the preceding problem, determine for each case whether regression or correlation analysis is appropriate.

19.3 What does r^2 measure? Why is this measure a link between regression and correlation analysis?

19.4 How do you interpret the following values of r: $+1$, -1, -0.9, and $+0.81$?

19.5 In studying the correlation between lung cancer and cigarette smoking, a student obtained this result: $r = +1.25$. Discuss.

19.6 Why can we not take the presence of association as positive proof of causation between two variables?

19.7 If the correlation coefficient between Y and X_1 is 0.85 and that between Y and X_2 is -0.85, and if a prediction equation is needed, which variable, X_1 or X_2, should be chosen as the independent variable? Explain the reason for your decision.

19.8 Sample standard deviations of regression for two different regression problems are 100 pounds and $50, respectively. From these values, can you conclude that the degree of relationship in the first is weaker than that in the second?

19.9 Why are the signs for b and r the same?

19.10 Given that b and b' are the slopes of two sample regression lines, show that $r^2 = bb'$.

19.11 Show that r defined in terms of covariances and defined as the square root of the coefficient of determination would always yield the same numerical result.

19.12 Show that

$$\text{SSR} = b\left(\sum xy - \frac{1}{n}\sum x \sum y\right)$$

holds.

For each of the next six questions, 19.13–19.18 inclusive, do the following:
a. Construct a scatter diagram.
b. Estimate the population regression equation.
c. Fit a regression line to the data.
d. Test the significance of b at $\alpha = 0.05$ and $\alpha = 0.01$, and construct a 95 percent confidence interval for B.
e. Construct a 95 percent confidence interval for μ_{yx} at fixed value of the independent variable that is underlined below the data tables. Do the same for a new individual observation y.
f. Give an interpretation to the value of a, if it is meaningful, and interpret b carefully.

19.13 The number of defective items produced per unit of time, Y, by a certain machine is thought to vary directly with the speed of machine X, measured in rpm. Observations for twelve time periods yield the following results:

X	Y	X	Y
13.2	9.4	10.9	5.7
14.9	12.2	17.4	12.3
8.1	6.0	13.8	9.2
16.4	11.4	10.2	7.0
13.1	9.6	15.8	9.0
10.8	7.5	12.0	7.0

$\underline{x = 10}$

19.14 The following sample contains the price and the quantity exchanged of a commodity from the sellers' side. Use quantity as the dependent variable in this case.

Price	Quantity	Price	Quantity
25	60	15	40
20	85	20	55
35	110	30	90
40	95	40	115
60	140	50	120
55	160	70	180
45	80	45	95

$\underline{p = 42}$

19.15 The following sample contains the price and the quantity exchanged of a commodity from the buyers' side. Use Q as the dependent variable.

Price	Quantity	Price	Quantity
$6.10	65	$6.80	53
7.40	25	2.50	44
6.40	41	7.00	48
3.00	66	2.30	87
1.00	78	2.00	60
4.00	60	6.00	18
7.10	32	3.80	34
4.50	28	8.00	10

$\underline{p = 4.2}$

19.16 The annual disposable income and consumption expenditure of twelve families selected at random from a certain section of New York City are as follows:

Income	Consumption	Income	Consumption
$ 8,000	$ 7,000	$ 8,000	$ 8,000
15,000	12,000	13,000	11,000
20,000	18,000	7,000	8,000
35,000	30,000	6,000	6,000
28,000	20,000	12,000	10,000
25,000	24,000	15,000	18,000

$\underline{y = \$11{,}500}$

(*Hint:* Use consumption expenditures, C, as a dependent, and disposable income, Y, as the independent, variable. Do your calculations with $Y' = Y/1000$ and $C' = C/1000$.)

19.17 The president of a chain retailing organization believes that there is a positive relationship between the sales of his company's product and per capita income in the year just past. He decides to associate 1967 sales with 1966 per capita income in the 15 cities where his company has branches. The data are as follows:

X: 1966 per capita income in thousands of dollars

Y: 1967 per capita sales in dollars

X	Y	X	Y	X	Y
2.0	15	2.2	23	1.8	12
2.1	25	2.5	25	2.3	22
2.0	17	2.2	16	2.2	18
2.3	23	2.1	18	2.0	18
2.7	12	1.6	11	1.5	10

$\bar{x} = 2.105$

19.18 The following random sample gives the number of hours of study for, X, and grades of, Y, an examination in elementary statistics for twelve students:

X:	3	3	3	4	4	5	5	5	6	6	7	8
Y:	45	60	55	60	75	70	80	75	90	80	75	85

$\bar{x} = 4.5$

For the next four problems, 19.19–19.22 inclusive, do the following:

a. Compute the sample correlation coefficient.
b. Test $\rho = 0$ against $\rho \neq 0$ at $\alpha = 0.05$.
c. Test whether ρ is equal to the value given below each data table at $\alpha = 0.05$ and construct a corresponding confidence interval.
d. Explain if the correlation is real or spurious.

19.19 The following data refer to the entrance test scores, X, and the labor efficiency test, Y, of 20 machine operators:

19.20 Use data in Problem 19.13.

19.21 A researcher has gathered data for annual per capita consumption of chewing tobacco, X, and the number of thefts reported in a sample of urban areas, Y, for the period 1948 to 1955 as shown below:

Year	X (pounds)	Y (thousands)
1948	0.52	73
1949	0.49	71
1950	0.49	77
1951	0.45	89
1952	0.43	97
1953	0.42	102
1954	0.41	97
1955	0.40	104

$\rho = 0.95$

19.22 Use data in Problem 19.16. (Calculate r from Y' and C' as indicated previously.)

19.23 How is the coefficient of determination, r^2, defined? What are the three possible interpretations of r^2?

19.24 From the viewpoint of regression analysis, how is the correlation coefficient defined?

19.25 A sample of 5 adult men for whom heights and weights are observed gives the

Worker	X	Y	Worker	X	Y
1	35	75	11	20	30
2	50	100	12	50	60
3	55	95	13	60	105
4	50	80	14	30	60
5	60	80	15	50	90
6	40	45	16	35	75
7	60	80	17	35	80
8	40	80	18	55	60
9	45	75	19	60	80
10	55	90	20	35	60

$\rho = 0.90$

following results:

Subject	Height: X	Weight: Y
0	5′4″	130 pounds
1	5′5″	145 "
2	5′6″	150 "
3	5′7″	165 "
4	5′8″	170 "

a. Use these data to show that

Total error = Unexplained error + Explained error

b. Compute r^2 and $\hat{r}^2$.

19.26 Do the same as in the preceding problem with data in Problem 19.18.

19.27 Use data in Problem 19.13 to demonstrate that $r = \sqrt{bb'}$.

19.28 Given the following sample data on the amount of fertilizer, X, and yield of potatoes, Y, compute, via the analysis of variance table for sums of squares, the following statistics, (a) $\hat{\sigma}_{yx}$, (b) $\hat{\sigma}_b$, (c) r^2 and $\hat{r}^2$, and test $B = 0$ against $B > 0$ at $\alpha = 0.05$.

X (lb.):	1.5	2.0	2.5	3.0	3.5	4.0	4.5	5.0	5.5
Y (lb.):	10	9	12	14	13	15	17	14	14

19.29 Do the same as in the preceding problem with the following sample data:

IQ, X:	100	95	90	115	120	105	130	140
Test score, Y:	70	75	75	85	90	80	90	95

19.30 Compute r^2 with data in 19.28 with result suggested in the solution for Problem 19.11. (*Hint:* compute r and then square it.)

19.31 Use results in Problems 19.28 and 19.29 to demonstrate that

$$\text{SSE} = \sum (y - y_c)^2 = \frac{\Sigma(x - \bar{x})^2\Sigma(y - \bar{y})^2 - (\Sigma(x - \bar{x})(y - \bar{x}))^2}{\Sigma(x - \bar{x})^2}$$

19.32 The quantities $\Sigma(x - \bar{x})(y - \bar{y})$ and $\Sigma(x - \bar{x})^2$ appear over and over again in the formulas for association analysis. Show that

a. $\sum (x - \bar{x})(y - \bar{y}) = \Sigma xy - \dfrac{\Sigma x \Sigma y}{n}$

b. $\sum (x - \bar{x})^2 = \Sigma x^2 - \dfrac{(\Sigma x)^2}{n}$

19.33 The following are two sets of ranks assigned by two professors of management to 10 characteristics for business leadership:

Characteristic	Prof. I	Prof. II
A	10	6
B	1	9
C	2	3
D	9	10
E	8	2
F	6	7
G	7	5
H	5	1
I	4	8
J	3	4

a. Compute r_s and $\hat{r}_s$.
b. In view of these results, can we conclude that there is agreement between the two professors?

19.34 Data below are costs and sales, in thousands of dollars, for 12 drug stores:

Store	Costs	Sales
1	11	19
2	10	15
3	14	20
4	13	14
5	12	16
6	20	33
7	21	32
8	15	18
9	22	29
10	18	22
11	19	23
12	16	20

a. Compute the rank-correlation coefficient for the above data.
b. Is the result significant at $\alpha = 0.05$? (State your null and alternative hypotheses.)
c. Assuming a normal bivariate population, compute the correlation coefficient r.
d. Is r significant at $\alpha = 0.05$?

20

ASSOCIATION ANALYSIS: MULTIPLE AND NONLINEAR MODELS

20.1

The Multiple Linear Regression Model

In multiple linear regression analysis, it is assumed that the dependent variable Y is related to k independent variables, $X_1, X_2, \cdots, X_k$, by a linear equation of the form

$$1. \quad E(Y|X_1, \cdots, X_k) = A + B_1X_1 + B_2X_2 + \cdots + B_kX_k$$

for which an individual value of Y is defined as

$$2. \quad y = E(Y|X_1, \cdots, X_k) + e$$

The first of these two equations gives us the population regression equation whose graph is a hyper-regression plane in the $k + 1$ dimensional space. In this expression, A and B_i are population regression coefficients. The second expression above gives us the relationship between a single observation of Y in the hyper-regression plane. The e, as before, is the error term, which is a random variable and measures the deviation of an individual y from $E(Y|X_1, \cdots, X_k)$—the mean value of Y associated with a set of fixed values of X_i.

As in the simple association analysis, we have a regression problem if Xs are fixed quantities and a correlation problem if Xs are random variables. Furthermore, if the successive values of e are independent and normally distributed, whether Xs are fixed in advance or possess a multivariate normal distribution, both the Markoff theorem and the maximum likelihood method lead to the least squares procedure and yield the same estimates of the multiple regression coefficients. In other words, we attempt to estimate A and B_i by a and b_i such that

$$\Sigma\hat{e}^2 = \Sigma(y - a - b_1x_1 - \cdots - b_kx_k)^2 = \text{minimum}$$

To achieve this, we need a system of normal equations which can be easily derived from the basic relationship

$$3. \quad a + b_1x_1 + b_2x_2 + \cdots + b_kx_k = y$$

By summing the above equation for all the n observations in the sample, we obtain, as the first normal equation required,

$$na + b_1\Sigma x_1 + b_2\Sigma x_2 + \cdots + b_k\Sigma x_k = \Sigma y$$

The second normal equation is found by multiplying both sides of equation 3 by values of X_1 and summing,

$$a\Sigma x_1 + b_1\Sigma x_1^2 + b_2\Sigma x_1x_2 + \cdots + b_k\Sigma x_1x_k = \Sigma x_1y$$

The third normal equation is found by multiplying both sides of equation 3 by values of X_2 and summing,

$$a\Sigma x_2 + b_1\Sigma x_1x_2 + b_2x_2^2 + \cdots + b_k\Sigma x_2x_k = \Sigma x_2y$$

This process is repeated for each independent variable. When all normal equations are found, there is one for each of the coefficients to be estimated. The last equation for the system, obtained by multiplying both sides of equation 3 by values of X_k is:

$$a\Sigma x_k + b_1\Sigma x_1x_k + b_2\Sigma x_2x_k + \cdots + b_k\Sigma x_k^2 = \Sigma x_ky$$

The first of these normal equations is called the A normal equation, since it is obtained by minimizing $\Sigma\hat{e}^2$ with respect to a. Similarly, the second normal equation is called the B_1 equation, the third the B_2 equation, and so on.

Estimates of the multiple regression coefficients are computed by solving the above system of normal equations—a very tedious procedure if the number of independent variables is at all large. Efficient methods of solving simultaneous equations require a knowledge of matrix algebra, which is not assumed for the reader of this text. Thus, in our discussion that follows, we shall confine ourselves to the two independent variable case which, of course, can be extended to cover cases with three or more independent variables.

20.2

The Linear Regression Model with Two Independent Variables

Suppose that we have three associated variables, which we shall denote as Y, X_2, and X_3. For example, Y might represent pounds of pork purchased by households during a period of time, X_2 prices of pork, and X_3 household incomes during the same period of time. Furthermore, if we know from economic theory or from actual observations that Y is linearly related to X_2 and X_3, then we have, as the linear regression model for the population,

$$E(Y|X_2,X_3) = \mu_{1.23} = A + B_2X_2 + B_3X_3 \tag{20.1}$$

and

$$y_i = \mu_{1.23} + e \tag{20.2}$$

Let us denote the least squares estimate of the relationship (20.1) as

$$\bar{y}_{1.23} = a_{1.23} + b_{12.3}x_2 + b_{13.2}x_3 \tag{20.3}$$

where we also define

$$y_i = \bar{y}_{1.23} + \hat{e} \tag{20.4}$$

To obtain these least squares estimates, we must satisfy the condition that

$$\Sigma\hat{e}^2 = \Sigma(\bar{y}_{1.23} - a_{1.23} - b_{12.3}x_2 - b_{13.2}x_3)^2 = \text{minimum}$$

This condition is satisfied if the regression coefficients are obtained from the following set of normal equations:

$$\begin{aligned} na_{1.23} + b_{12.3}\Sigma x_2 + b_{13.2}\Sigma x_3 &= \Sigma y \\ a_{1.23}\Sigma x_2 + b_{12.3}\Sigma x_2{}^2 + b_{13.2}\Sigma x_2x_3 &= \Sigma x_2y \\ a_{1.23}\Sigma x_3 + b_{12.3}\Sigma x_2x_3 + b_{13.2}\Sigma x_3{}^2 &= \Sigma x_3y \end{aligned}$$

It may now be noted that relation (20.3) is referred to as the equation of the regression of Y on X_2 and X_3. It represents the best fitting plane to the egg-shaped cluster of points in a three-dimensional space as shown by Figure 20.1. The coefficient $a_{1.23}$ is the intercept made by the regression plane on the Y axis, or the elevation of the fitted plane, and is an estimate of A. The b constants are called *partial regression coefficients;* $b_{12.3}$ is an estimate of B_2 and it measures the amount by which a unit change in X_2 is expected to affect Y when X_3 is held constant; $b_{13.2}$ is an estimate of B_3 and it measures the amount of changes in Y per unit change in X_3 when X_2 is held constant. The subscript notation in the coefficients identifies

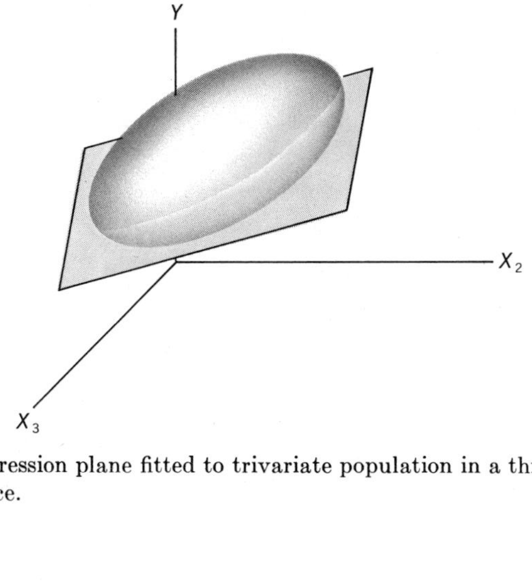

Figure 20.1 Regression plane fitted to trivariate population in a three-dimensional space.

the number of variables in the analysis. The figure 1 in the subscript refers to the variable Y, the figure 2 to X_2, and 3 to X_3. The subscript 1.23 on a reveals that this is the Y intercept in the regression of Y on X_1 and X_2, since 1 comes before and 2 and 3 come after the point. In the subscripts to the partial coefficients, the first figure refers to the dependent variable, the second figure indicates the variable to which this b coefficient is attached, and figures after the point identify which other variables have also been taken into account in estimating the relationship.

Now, turning to the practical problem of solving the system of normal equations that preceded the above paragraph, we note that

$$\begin{aligned}\Sigma(y - \bar{y}) &= 0,\\ \Sigma(x_2 - \bar{x}_2) &= 0, \text{ and}\\ \Sigma(x_3 - \bar{x}_3) &= 0\end{aligned}$$

Thus, if we let

$$\begin{aligned}y' &= y - \bar{y},\\ x_2' &= x_2 - \bar{x}_2, \text{ and}\\ x_3' &= x_3 - \bar{x}_3\end{aligned}$$

we would have then shifted the origin of the normal equations from (0,0,0) to $(\bar{y},\bar{x}_2,\bar{x}_3)$, and the three normal equations stated before would be reduced to two in terms of deviations around the means:

$$\begin{aligned}b_{12.3}\Sigma x_2'^2 + b_{13.2}\Sigma x_2'x_3' &= \Sigma y'x_2'\\ b_{12.3}\Sigma x_2'x_3' + b_{13.2}\Sigma x_3'^2 &= \Sigma y'x_3'\end{aligned} \tag{20.5}$$

Solving these two equations simultaneously, we obtain the partial regression coefficients $b_{12.3}$ and $b_{13.2}$. With these estimates, we can then compute the estimate of A by the relationship,

$$a_{1.23} = \bar{y} - b_{12.3}\bar{x}_2 - b_{13.2}\bar{x}_3 \tag{20.6}$$

In multiple association analysis, the variates may be given in different units of measurements. To facilitate interpretation and comparison of the partial coefficients, we may transform the original data into percentages before calculations are made.

The ABC Cigarette Company is just beginning its eleventh year of operation and is considered a rather successful firm in the industry. In order to schedule production for the coming year, a forecast of the total sales is required. Historical sales data for the past decade are available, as are data for the annual expenditures for advertising. Furthermore, the ABC management believes that its sales are also affected by the price of its product compared to the average price of other brands of cigarettes; that is, sales are influenced by the comparative price index $P = (p/\bar{p})(100)$, where p is the price of ABC's product, and $\bar{p}$ is the mean price of the product in the industry as a whole. Table 20.1 supplies the data upon which the sales forecast is to be made; original data are given in columns 2 to 4 inclusive, and the last three columns contain the percentage values of these data. Also, for

TABLE 20.1 ABC COMPANY HISTORICAL DATA

	Original Data			Original Data Expressed as Percentages		
Year	Y	X_2	X_3	Y	X_2	X_3
1	24	4	80	6.80	6.06	8.25
2	27	4	80	7.65	6.06	8.25
3	31	5	90	8.78	7.58	9.28
4	29	5	100	8.22	7.58	10.31
5	33	6	100	9.35	9.09	10.31
6	38	7	110	10.76	10.61	11.34
7	37	8	120	10.48	12.12	12.37
8	40	8	100	11.33	12.12	10.31
9	45	9	90	12.75	13.64	9.28
10	49	10	100	13.88	15.15	10.31
Total	353	66	970	100.00	100.00	100.00

the original data, we have

Y = annual sales in millions of dollars,
X_2 = advertising expenditure in millions of dollars per year, and
$X_3 = P$, or the comparative price index defined as $P = (p/\bar{p})\,100$

From the relative data above, we first determine

$$\bar{y} = \bar{x}_2 = \bar{x}_3 = {}^{100}\!/_{10} = 10 \text{ percent}$$

Next, we compute

$$\begin{aligned} \Sigma y^2 &= 1046.03 & \Sigma x_2^2 &= 1092.91 & \Sigma x_3^2 &= 1015.18 \\ \Sigma x_2 y &= 1064.11 & \Sigma y x_3 &= 1011.73 & \Sigma x_2 x_3 &= 1020.28 \end{aligned}$$

From these quantities, we construct the following values in terms of deviations around the means:

$$\begin{aligned} \Sigma y'^2 &= \Sigma(y - \bar{y})^2 = \Sigma y^2 - n(\bar{y})^2 = 1046.03 - 10(10)^2 = 46.03 \\ \Sigma x_2'^2 &= 1092.91 - 10(10)^2 = 92.91 \\ \Sigma x_3'^2 &= 1015.18 - 10(10)^2 = 15.18 \\ \Sigma y' x_2' &= \Sigma y x_2 - n(\bar{y})(\bar{x}_2) = 1054.11 - 10(10)(10) = 64.11 \\ \Sigma y' x_3' &= 1011.73 - 10(10)(10) = 11.73 \\ \Sigma x_2' x_3' &= 1020.28 - 10(10)(10) = 20.28 \end{aligned}$$

Substituting these quantities in the system of normal equations (20.5), we have

$$\begin{aligned} 92.91 b_{12.3} + 20.28 b_{13.2} &= 64.11 \\ 20.28 b_{12.3} + 15.18 b_{13.2} &= 11.73 \end{aligned}$$

and solving for the partial coefficients, we find

$$\begin{aligned} b_{12.3} &= 0.73595 \\ b_{13.2} &= -0.21047 \end{aligned}$$

The value of the Y intercept or the elevation is

$$a_{1.23} = 10 - (0.73597)(10) - (-0.21507)(10) = 0.4896$$

Hence, the estimated regression equation becomes

$$\bar{y}_{1.23} = 0.4896 + 0.73597x_2 - 0.21507x_3$$

The preceding result substantiates economic theory or our intuitive belief that advertising affects sales positively, and that increases in relative price cause decreases in sales. More precisely, the value of 0.73597 indicates that if there is a 1 percent increase in advertising expenditures, sales will increase by about 0.74 percent; the value of -0.21507 reveals that if there is an increase in the relative price by 1 percent, there will be a more than 0.21 percent decrease in sales. Furthermore, intuitively, we see that the partial effects of advertisement is about three times as strong as the partial effects of relative price on sales. The relatively small influence of price on sales in this case may be explained by the fact that cigarettes are low priced and are practically a necessity to many smokers; that is, the demand for cigarettes is highly inelastic with respect to price.

20.3 Significance of Partial Regression Coefficients

Having found the sample regression equation, we would like to know two things: (1) Are the values of the partial coefficients significant? Namely, if we find that $b_{12.3} \neq 0$ and that $b_{13.2} \neq 0$, can we also conclude that the corresponding population regression coefficients assume nonzero values? (2) Is there a relative difference between the effects of the independent variables upon the value of the dependent variable? Here, we are interested in determining the net contribution of each independent variable to the dependent variable. These two questions are answered with statistical tests based on the analysis of variance, a method with which we are by now quite familiar.

Given the sample regression equation and thereby the regression plane, we can think of the total deviations of the Y values from the estimated mean as the vertical (parallel to the Y axis) deviations from the fitted regression plane. Furthermore, we can partition this total variation, as in the bivariate case, into two independent parts: one part measures the amount of variations in Y that has been explained by the regression; the other measures the amount of variations in Y that has not been explained by the regression. This second part constitutes, therefore, the residual or error variations in the sampling experiment. Clearly, we see that the following relationship holds:

$$\Sigma(y - \bar{y})^2 = \Sigma(y - \bar{y}_{1.23})^2 + \Sigma(\bar{y}_{1.23} - \bar{y})^2 \qquad \textbf{(20.7)}$$

where $\Sigma(y - \bar{y})^2$ is the total deviation with $\delta = n - 1$. This sum of squares can

TABLE 20.2 ANALYSIS OF VARIANCE TABLE FOR TRIVARIATE REGRESSION

Source	SS	δ	MS
Regression	SSR	$k = 2$	MSR = SSR/k
Residual	SSE	$n - k - 1$	MSE = SSE/$(n - k - 1)$
Total	SST	$n - 1$	MST = SST/$(n - 1)$

be written as

$$\text{SST} = \Sigma(y - \bar{y})^2 = \Sigma y^2 - n(\bar{y})^2 = \Sigma y'^2 \tag{20.8}$$

The $\Sigma(\bar{y}_{1.23} - \bar{y})^2$ is the sum of squares due to regression and, as it can be shown,

$$\text{SSR} = \Sigma(\bar{y}_{1.23} - \bar{y})^2 = b_{12.3}\Sigma x_2'y' + b_{13.2}\Sigma x_3'y' \tag{20.9}$$

For SSR, $\delta = n - k$, where k is the number of partial regression coefficients in the sample regression equation. In our present trivariate analysis, $k = 2$. Having computed SST and SSR for the identity (20.7), we have

$$\text{SSE} = \Sigma(y - \bar{y}_{1.23})^2 = \text{SST} - \text{SSR} \tag{20.10}$$

We note that SSE has $\delta = n - k - 1$, since the estimated relationship

$$y_{1.23} = a_{1.23} + b_{12.3}x_2 + b_{13.2}x_3$$

has $k + 1$ restrictions. Here, in addition to $k = 2b$ coefficients, a is restricted by $\bar{y}$.

The above results can be summarized by the usual analysis of variance table—see Table 20.2.

The mean square of error,

$$\text{MSE} = \frac{\text{SSE}}{n - k - 1} = \frac{\Sigma(y - y_{1.23})^2}{n - k - 1} = \hat{\sigma}^2{}_{1.23} \tag{20.11}$$

is the sample variance of Y (that is, sample variance of the fitted regression plane), and it is an unbiased estimate of the population variance $V(Y) = E(Y - E(Y|X_2,X_3))^2 = \sigma^2{}_{1.23}$. If the subpopulations, Ys, are normally distributed, just as in the simple regression case, MSE measures the closeness-of-fit. The symbol used for the variance here is $\hat{\sigma}^2{}_{1.23}$ for two independent variables; and, under the same system of nomenclature, the sample variance of the regression in the simple case would be $\hat{\sigma}^2{}_{1.2}$ or $\hat{\sigma}^2{}_{1.3}$, depending on which of the independent variables is employed in the simple regression analysis. It is easy to see that if the additional independent variable makes any significant contribution to the dependent variable, the variance computed for the multiple case would be smaller than that obtained from either simple regression equation.

The ratio of MSR to MSE

$$F = \frac{\text{MSR}}{\text{MSE}}$$

is distributed as $F_{k,n-k-1}$ and can be employed to conduct an over-all test that $B_2 = B_3 = 0$. If this null hypothesis is false—that is, if there exists significant

TABLE 20.3

Source	SS	δ	MS
Regression: X_2, X_3	44.71	2	MSR = 44.71/2 = 22.3550
Residual	1.32	7	MSE = 1.32/7 = 0.1886
Total	46.03	9	MST = 46.03/9 = 5.1144

regression—the $\bar{y}_{1.23}$ values will differ significantly from $\bar{y}$, and SSR will be large. As a result, the residuals, SSE, will tend to be small. These imply that the value of F will be large, indicating a significant regression. When the residuals are relatively large, or the improvement brought about by the regression plane is relatively small, a small value of F will result, and the null hypothesis that there exists no significant regression would then be accepted.

Now, returning to our illustrative example, we have

$$\begin{aligned}\text{SST} &= \Sigma y'^2 = 46.03\\ \text{SSR} &= b_{12.3}\Sigma x_2'y' + b_{13.2}\Sigma x_3'y'\\ &= (0.73595)(64.11) - (0.21047)(11.73) = 44.71, \text{ and}\\ \text{SSE} &= \text{SST} - \text{SSR} = 46.03 - 44.71 = 1.32\end{aligned}$$

These sums of squares give the analysis of variance shown in Table 20.3.

From the results in Table 20.3, we see that

$$\text{MSE} = \hat{\sigma}^2{}_{1.23} = 0.1886$$
$$\hat{\sigma}_{1.23} = \sqrt{0.1886} = 0.4347 \text{ percent}$$

Also, the resultant F value is

$$F_{2,7} = \frac{22.3550}{0.1886} = 118.80$$

which is highly significant. There is a highly significant association or regression between sales, advertising, and the relative price index.

To evaluate the separate contributions of each independent variable, we may conduct a step-by-step analysis as follows:

Let

$$b_{12} = \text{coefficient of } X_2 \text{ in simple regression of } Y \text{ on } X_2$$
$$= \frac{\Sigma x_2'y'}{\Sigma x_2'^2} = \frac{64.11}{92.91} = 0.690$$

$$b_{13} = \text{coefficient of } X_3 \text{ in simple regression of } Y \text{ on } X_3$$
$$= \frac{\Sigma x_3'y'}{\Sigma x_3'^2} = \frac{11.73}{15.18} = 0.773$$

With these coefficients in simple regressions, we have then the explained sum of squares due to X_2 alone as

$$\text{SSR}(X_2) = b_{12}\Sigma x_2'y' = (0.690)(64.11) = 44.24$$

TABLE 20.4

Source	SS	δ	MS
X_2	44.24	$k = 1$	44.24
Addition of X_3	0.47	$k = 1$	0.47
X_1 and X_2	44.71	$k = 2$	
Residual	1.32	7	0.1886
Total	46.03	9	—

and the explained sum of squares due to X_3 alone as

$$\text{SSR}(X_3) = b_{13}\Sigma x_3'y' = (0.773)(11.73) = 9.07$$

Now, from $\text{SSR}(X_2)$ we may set up Table 20.4. Note that the total sum of squares due to X_2 and X_3 is known, from Table 20.3, as 44.71. Since the sum of squares due to X_2 alone is 44.23, the additional effect due to X_3 is then found by subtraction: $44.71 - 44.24 = 0.47$.

To test the significance of X_2 alone, we compute the residual sum of squares after X_2 as $\text{SSE}(X_2) = \text{SST} - \text{SSR}(X_2) = 46.03 - 44.24 = 1.79$ with $\delta = 10 - 1 - 1 = 8$, giving a mean square of error of $\text{MSE}(X_2) = 0.224$. The appropriate F ratio is then

$$F_{1,8} = \frac{\text{MSR}(X_2)}{\text{MSE}(X_2)} = \frac{44.24}{0.224} = 197.5$$

which is highly significant.

The additional effect of X_3 on Y can be tested simply by the statistic

$$F_{1,7} = \frac{\text{MSR}(X_3)}{\text{MSE}} = \frac{0.4700}{0.1886} = 2.492$$

which is insignificant at $\alpha = 0.05$.

Alternatively, we may construct Table 20.5 with $\text{SSR}(X_3)$. Again, as we should expect from previous results, the direct effect of X_3 is insignificant and the additional effect of X_2 is highly significant.

These testing results are clearly consistent with our previous interpretation of the same partial regression coefficients.

TABLE 20.5

Source	SS	δ	MS
X_3	9.07	1	9.07
Addition of X_2	35.64	1	35.64
X_2 and X_3	44.71	2	
Residual	1.32	7	0.1886
Total	46.03	9	—

20.4

Coefficient of Multiple Determination

While the closeness-of-fit of the regression plane to the actual points in the three-dimensional space can be measured by the variance or standard deviation of the regression plane, as mentioned in the previous section, this statistic is rather ambiguous and difficult to interpret. A more obvious and clear-cut measure for this purpose is what is called *coefficient of multiple determination*. Recall that the fit of a straight line to the two-variable scatter was measured by the coefficient of determination r^2, defined as the explained sum of squares to the total sum of squares. In exactly the same fashion, we can define the coefficient of multiple determination, denoted as $R^2{}_{1.23}$ for the trivariate case, in terms of residual deviations about the regression plane. That is,

$$R^2{}_{1.23} = 1 - \frac{\text{SSE}}{\text{SST}} = \frac{\text{SSR}}{\text{SST}} \tag{20.12}$$

Note that the total sum of squares, SST $= \Sigma(y - \bar{y})^2$, contains the deviations of the observed points Y from a plane that is fitted to these points and that is horizontal, passing through the point $(\bar{y}, \bar{x}_2, \bar{x}_3)$, and from which the improvement brought about by the regression is measured. Thus, $R^2{}_{1.23}$ may be thought of as a measure of closeness-of-fit of the regression plane to the actual points relative to the point of the means of the variables. Or, just as does r^2, $R^2{}_{1.23}$ measures the percentage of total error that is accounted for by the regression. Thus, the greater the value of $R^2{}_{1.23}$, the smaller is the scatter and the better is the fit.

Now, for our illustrative data, we have

$$R^2{}_{1.23} = \frac{44.71}{46.03} = 0.9713$$

Thus, more than 97 percent of the variations in sales have been explained by the variations in advertising and price index. There remains less than 3 percent of variations in Y that can be explained only by factors which have not been taken into consideration in our analysis.

The coefficient of multiple determination, just as is the coefficient of determination, is a positively biased statistic. To correct this bias downward, we may compute the adjusted coefficient of multiple determination, $\dot{R}^2{}_{1.23}$, as defined here:

$$\dot{R}^2{}_{1.23} = 1 - \frac{\text{MSE}}{\text{MST}} \tag{20.13}$$

(*Note:* Since MSE/MST $=$ (SSE/SST)$((n - 1)/(n - k - 1))$, the correction factor in (20.13) is $(n - 1)/(n - k - 1)$, which approaches 1 from below when n becomes large.)

For the illustrative data,

$$\dot{R}^2_{1.23} = 1 - \frac{0.1886}{5.1144} = 0.9631$$

The square root of the coefficient of multiple determination is called the *coefficient of multiple correlation*, denoted as $R_{1.23}$. The $R_{1.23}$ is always given as positive. Why?

It is important to note that we have considered $R_{1.23}$ here as a measure of closeness-of-fit, not as a measure of covariability between the variables. As such, $R_{1.23}$ is meaningful whether the independent variables are fixed or stochastic. However, when all the variables are random, $R_{1.23}$ may also be considered as a measure of covariability between Y and its best linear estimate, $\bar{y}_{1.23}$; that is, between Y and the group of independent variables, (X_2, X_3). Also, if Y is the only variable, we can only have $R_{1.23}$. If Y, X_2, and X_3 are all random variables, we can have $R_{1.23}$, $R_{1.32}$, and $R_{2.13}$. What measures do these notations represent?

20.5 Partial Correlation Coefficient

Up to this point, our consideration of the trivariate regression analysis has been mainly an extension of the bivariate case. We shall introduce now a new concept called *partial correlation coefficients*, which exists only when there are two or more independent variables.

When we evaluate a multivariate normal distribution, we often wish to measure the covariability between the dependent variable and a particular independent variable by holding all other variables constant; that is, by removing the effects of other variables under the specification: "other things being equal." Such a measure would certainly help us to answer questions like this: Is the correlation between, say, Y and X_2 merely due to the fact that both are affected by X_3, or is there a net covariation between Y and X_2 over and above the association due to the common influence of X_3? Thus, in determining a partial correlation coefficient between Y and X_2, we attempt to remove the influence of X_3 from each of the two variables and to ascertain what net relationship exists between the "unexplained" residuals that remain. It can be shown that partial correlation coefficients are closely related to the partial regression coefficients and can be readily defined in terms of simple correlation coefficients. By the product-movement formula developed in the preceding chapter, we find that

$$r_{12} = \text{simple correlation between } Y \text{ and } X_2 = \frac{\Sigma x_2' y'}{\sqrt{\Sigma x_2'^2 \Sigma y'^2}}$$

$$r_{13} = \text{simple correlation between } Y \text{ and } X_3 = \frac{\Sigma x_3' y'}{\sqrt{\Sigma x_3'^2 \Sigma y'^2}}$$

$$r_{23} = \text{simple correlation between } X_2 \text{ and } X_3 = \frac{\Sigma x_2' x_3'}{\sqrt{\Sigma x_2'^2 \Sigma x_3'^2}}$$

With the values of simple correlation coefficients determined, we may define partial correlation coefficients for the trivariate case in terms of these values as follows:

$$r_{12.3} = \text{partial correlation between } Y \text{ and } X_2 \text{ when } X_3 \text{ is held constant}$$

$$= \frac{r_{12} - r_{13}r_{23}}{\sqrt{(1 - r_{13}^2)(1 - r_{23}^2)}} \tag{20.14}$$

$$r_{13.2} = \text{partial correlation between } Y \text{ and } X_3 \text{ when } X_2 \text{ is held constant}$$

$$= \frac{r_{13} - r_{12}r_{23}}{\sqrt{(1 - r_{12}^2)(1 - r_{23}^2)}} \tag{20.15}$$

$$r_{23.1} = \text{partial correlation between } X_2 \text{ and } X_3 \text{ when } Y \text{ is held constant}$$

$$= \frac{r_{23} - r_{12}r_{13}}{\sqrt{(1 - r_{12}^2)(1 - r_{13}^2)}} \tag{20.16}$$

The simple correlation coefficients are called *zero-order* coefficients. Other zero-order coefficients are simple standard deviations and the regression coefficients in the simple linear case, such as b_{12} and b_{13}. The zero-order coefficients possess no secondary subscripts, that is, no subscripts after the point. The partial correlation and regression coefficients for the trivariate case are called the *first-order* coefficients since there is one secondary subscript after the point. Another example of a first-order coefficient is the standard deviation of Y measured about the simple linear regression equation, since we denote it as $\hat{\sigma}_{y.x} = \hat{\sigma}_{1.2}$. As a matter of fact, we have a hierarchy of coefficients, all identified as the number of secondary subscripts after the point. Coefficients of a given order can generally be expressed in terms of the next lower order, such as expressing partial correlations for the trivariate case in terms of simple correlations. This possibility often provides the simplest computational scheme for three or four independent variables. For example, our trivariate analysis may be built up to the four-variable case, with coefficients of second and third order, which may in turn be defined in terms of coefficients of lower order. This procedure, however, becomes excessively cumbersome and time-consuming, since the number of formulas grows exponentially and the formulas themselves become extremely complicated. These difficulties can be removed to a great extent if matrix algebra is employed, because then we need not be bothered with the subscripts, and the summation signs can be handled efficiently. (The reader is advised to take a course in matrix algebra if he wishes to learn the general model of association analysis.)

Returning now to our illustration, we find that

$$r_{12} = \frac{64.11}{\sqrt{(46.03)(92.91)}} = 0.98028$$

$$r_{13} = \frac{11.73}{\sqrt{(46.03)(15.18)}} = 0.44368$$

$$r_{23} = \frac{20.28}{\sqrt{(92.91)(15.18)}} = 0.54008$$

With these quantities, the partial correlation coefficients are then determined:

$$r_{12.3} = \frac{0.98028 - (0.44368)(0.54008)}{\sqrt{(1 - (0.44368)^2)(1 - (0.54008)^2)}} = 0.98110$$

$$r_{13.2} = \frac{0.44368 - (0.98028)(0.54008)}{\sqrt{(1 - (0.98028)^2)(1 - (0.54008)^2)}} = -0.51719$$

$$r_{23.1} = \frac{0.54008 - (0.98028)(0.44308)}{\sqrt{(1 - (0.98208)^2)(1 - (0.44368)^2)}} = 0.58889$$

In concluding this section, let us make three additional comments on partial correlation coefficients. First, the partial correlations can also be defined in terms of "partial covariance" between two variables, or in terms of the amount of improvement brought about by the addition of another variable. However, these two alternative interpretations lead to the same results as the definition made under the assumption of "other things being equal."

Second, the value of a partial correlation coefficient is usually interpreted via the corresponding *coefficient of partial determination*, which is merely the square of the former. For example, in our illustration, $r_{12.3} = 0.98110$. This gives $r^2_{12.3} = 0.96256$. This means that more than 96 percent of the variation in Y (sales) that is not associated with X_3 (price index) is incrementally associated with X_2 (advertising). Or, alternatively, we say that the errors made in estimating Y from X_3 are reduced by more than 96 percent when X_2 is added to others as an additional independent variable. What does $r_{13.2} = -0.51719$ mean in our illustration?

Third, the t test employed to test the significance of a simple correlation can be employed to test the significance of a partial correlation when the number of degrees of freedom is reduced by the number of variables eliminated. Thus, to test the significance of $r_{12.3}$ in our illustrative example, we have

$$t = r_{12.3}\sqrt{\frac{n - 3}{1 - r^2_{12.3}}} = 0.98028\sqrt{\frac{10 - 3}{1 - (0.98028)^2}} = 4.15$$

which is highly significant. Similar tests on $r_{13.2}$ and $r_{23.1}$ in our illustration indicate insignificant correlations.

20.6

Curvilinear Association Models

Having introduced simple and multiple linear association analysis, we consider a natural extension of a study on relationships between variables to be the coverage of curvilinear association models.

From theory, we may know in advance that the relationship between two variables can be adequately represented only by some curvilinear mathematical functions. For example, the liquidity preference theory suggests that the demand for money to hold as cash balances is a decreasing and nonlinear function of the interest rate. Again, the law of diminishing returns dictates that total production is positively and nonlinearly related to the variable input or inputs. Occasions may also arise in which theoretical considerations enable us to judge only whether an association between two variables should be positive or negative but not whether it is linear or otherwise. The laws of supply and demand are illustrations of this possibility. This being the case the usual procedure is to inspect the scatter diagram constructed with sample data; the general pattern of the scatter may suggest an appropriate form of regression, with linear or some type of nonlinear functions.

We shall, in this section, introduce a few frequently encountered curvilinear functions between economic and business variables.

POLYNOMIAL FUNCTIONS

Very often we find that the most obvious nonlinear relationship between two variables is one in which the dependent variable Y can be approximated by way of a simple polynomial in the independent variable X. A polynomial function is defined by the equation

$$1. \quad E(Y|X) = \mu_{yx} = A + Bx + Cx_2^2 + \cdots + Nx^n$$

where the term involving the highest power of X is known as the *leading term*, and its coefficient is called the *leading coefficient*. The *degree* of the equation is the index of the highest power of X in it. While a relationship between two variables may assume a polynomial function of any degree, we generally seek to have as low a degree as possible. The commonly employed polynomial function in regression analysis is of degree 2; that is, a *parabola*. The sample regression equation is given by the form

$$y_c = a + bx + cx^2 \tag{20.17}$$

where y_c is the computed or predicted value of Y and where the residual sum of squares which we attempt to minimize is

$$\Sigma(y_c - a - bx - cx^2)^2$$

Applying the principle of least squares, we obtain the following set of normal equations to be solved simultaneously for the sample regression coefficients a, b, and c:

$$2. \quad \begin{aligned} \Sigma y &= na + b\Sigma x + c\Sigma x^2 \\ \Sigma xy &= a\Sigma x + b\Sigma x^2 + c\Sigma x^2 \\ \Sigma x^2 y &= a\Sigma x^2 + b\Sigma x^3 + c\Sigma x^4 \end{aligned}$$

RECIPROCAL TRANSFORMATION

Scatter diagrams for, say, sample data with reference to the demand for and supply of a commodity, Y, may indicate that the appropriate regression equations should be respectively,

$$y_c = a + \frac{b}{x} \tag{20.18}$$

and

$$y_c = a - \frac{b}{x} \tag{20.19}$$

where X may stand for the price of Y. We note that in case (20.18), the slope of the function is everywhere negative and decreases in absolute value as X increases. As X approaches zero, Y approaches ∞, and as X approaches ∞, Y approaches a. For case (20.19), the function is everywhere positive and Y approaches a, as X approaches ∞. See Figure 20.2.

For the above functions, *reciprocal transformation* would make the determination of a and b quite simple. Namely, if we define a new variable $Z = 1/X$, then (20.18) and (20.19) would become

$$y_c = a + bZ \tag{20.18a}$$

and

$$y_c = a - bZ \tag{20.19a}$$

respectively. These are now simple linear regressions of Y on Z. Procedures developed in the previous chapter can then be applied to obtain a and b. After a and b are found, the regression of Y on Z can then be easily converted back to the original forms by noting that $Z = 1/X$.

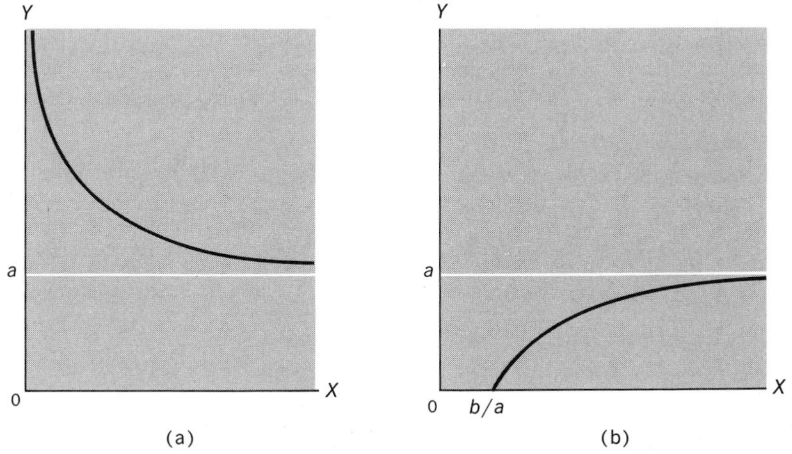

Figure 20.2 Graphs for (a) $y_c = a + b/x$ and (b) $y_c = a - b/x$.

THE EXPONENTIAL FUNCTION AND SEMILOGARITHMIC TRANSFORMATION

A highly useful curvilinear function that applies in situations which are incompatible with polynomial functions is the *exponential function*. When sample data can be fitted with an exponential curve, then the sample regression equation can be written as

$$y_c = ab^x \tag{20.20}$$

For this function, if $a > 0$, then Y will increase indefinitely as X increases, and will approach zero asymptotically as X decreases when $b > 1$. When $b < 1$, the function approaches zero asymptotically as X increases, and increases indefinitely as X decreases. When $b = 1$, we have Y = constant = a. This function, therefore, has no extreme point (maximum or minimum). If $a < 0$, the function is always negative and its behavior is the mirror image of the function for $a > 0$.

For the exponential function, the sum of squares we wish to minimize is

$$\Sigma(y - y_c)^2 = \Sigma(y - ab^x)^2$$

which leads to, as can be shown with aid of calculus, the set of normal equations that follows:

$$\begin{aligned} \Sigma b^x y &= a\Sigma b^{2x} \\ \Sigma x b^{x-1} y &= a\Sigma x b^{2x-1} \end{aligned}$$

This set of equations is clearly inoperative since all the sums involve the unknown b. One way to overcome this difficulty is to take logarithms of both sides of (20.20), yielding

$$\log y_c = \log a + x(\log b) \tag{20.20a}$$

If we let $\log a = c$ and $\log b = d$, this expression becomes

$$\log y_c = c + dx$$

which is now in the form of a linear equation; it is called a semilog function for the obvious reason that in it one variable appears as its logarithm while the other appears as itself. For further simplification, we may put $\log y_c$ as z_c and $\log b$ as d, and rewrite (20.20a) as

$$z_c = c + dx \tag{20.20b}$$

which is a simple linear regression of Z on X. Hence, c and d can be found in the usual manner. To transform $z_c = c + dx$ back to the original form, we merely follow the equations,

$$\begin{aligned} a &= \text{antilog } c \\ b &= \text{antilog } d \\ y_c &= \text{antilog } z_c \end{aligned}$$

Note that in the employment of semilog transformation, all we need to do is to take the logarithms of the Y values before making calculations.

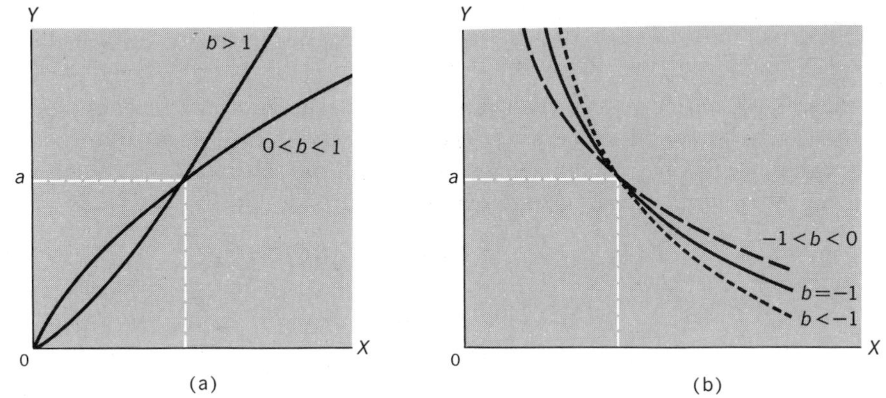

Figure 20.3 Power regression curves.

THE POWER FUNCTION AND DOUBLE-LOG TRANSFORMATION

Occasions may arise in which a power function such as

$$y_c = ax^b \tag{20.21}$$

may represent the relationship between Y and X in the sample. Here, we would like to find the power b that is unknown. For this function, if $b > 1$, Y increases at an increasing rate when X increases. If $0 < b < 1$, the rate of increase of this function decreases continuously. For either value of b, Y approaches ∞ as X approaches ∞. When $b = -1$, this function is represented by a rectangular hyperbola—a curve for which the locus of a point is such that the product XY is a constant. See Figure 20.3.

To facilitate the fit of a power regression curve, we take logarithms of both sides of (20.21), obtaining

$$\log y_c = \log a + b(\log x) \tag{20.21a}$$

This is the result of double-log transformation since both X and Y are given in logarithms. Now, if we let

$$\log y_c = z_c$$
$$\log a = c$$
$$\log x = w$$

we can rewrite the preceeding expression as

$$z_c = c + bw \tag{20.21b}$$

which is a simple linear regression equation of Z on W. Hence, for actual computations, we take logarithms of both X and Y values and proceed to determine the regression coefficients c and b in the usual manner.

MULTIPLE CURVILINEAR REGRESSION

Just as with the linear regression model, nonlinear relationships may be simple as well as multiple in nature. Either theoretical considerations or empirical observations may, for example, suggest that the sample regression should be, say, of the form

$$y_c = a{x_2}^b{x_3}^c \tag{20.22}$$

With this regression equation, the sample partial regression coefficients may be computed again by applying the principle of least squares to the logarithmic values of the variables. Namely, taking logarithms of both sides of (20.22), we may transform the original multiplicative expression into an additive model as

$$\log y_c = \log a + b(\log x_2) + c(\log x_3) \tag{20.22a}$$

If we let

$$\begin{aligned} \log y_c &= z_c \\ \log x_2 &= z_2 \\ \log x_3 &= z_3 \\ \log a &= d \end{aligned}$$

then (20.22a) can be written as follows:

$$z_c = d + bz_2 + cz_3 \tag{20.22b}$$

which is linear in all variables. As a consequence, we can apply the linear methods developed in Section 20.3 to the logarithms of Y, X_2, and X_3 to find the values of the constants in (20.22b). Once d, b, and c are calculated, the sample regression equation in the form of (20.22b) can be converted back to the original form of (20.22) by noting the definitions of Zs and that $d = \log a$.

Double-log transformation is frequently used by economists because it corresponds to the assumption of constant elasticity between the dependent and independent variables, and the simple application of linear methods to logarithms of the variables gives a direct estimate or estimates of elasticities. For example, the production function of the United States as a whole, estimated by Paul H. Douglas and C. W. Cobb, is

$$P = 1.10L^{0.75}C^{0.25}$$

where P is the production index of total output per year, L is the index of labor inputs, and C is the index of capital inputs. Sample indexes for this estimate ran from 1900 to 1922, with 1899 = 100. We are interested to note here that the estimated exponents are the elasticities of production with respect to labor and capital. The first tells us that if there is a 1 percent increase in labor inputs while capital holds constant, there will be 0.75 percent increase in production. Similarly, the second indicates that there will be 0.25 percent increase in production if there is a 1 percent increase in capital with labor held constant.

FURTHER REMARKS

We see that in treating curvilinear relationships we usually search for an initial transformation of data, by way of either reciprocals or logarithms, in such a manner that the relationship between the transformed variables appears to be approximately linear. Linear transformation is made mainly for two reasons. First, as the reader may have already guessed, this procedure is to make calculations easier, since computational problems in nonlinear systems are quite formidable even for versed mathematicians. Second, linear transformation is desirable because statistical theory is primarily developed under the assumption of linearity.

To complete our discussion on curvilinear relationships, let us note the following three facts:

(1) Nonlinear regression coefficients, with or without linear transformations, are least-square estimates of the corresponding population coefficients.

(2) The closeness-of-fit is measured by r^2 or $R^2_{1.23}$, depending upon whether the relationship is simple or multiple. We can measure the residual variation of the dependent variable about the fitted regression function, and then define the appropriate coefficient as 1 minus the ratio of the residual variation to the total variation.

(3) Simple or partial correlation coefficients in the curvilinear models are computed and interpreted in the same manner as in the linear models. However, with nonlinear relationships, simple as well as partial correlation coefficients are always given as positive values because the slope of a curvilinear regression line or plane can change signs at various points.

20.7 Additive Graphic Regression

Instead of finding a regression equation mathematically as we have done so far, we may obtain it graphically. The graphic method can be employed for additive, or linear, models as well as for multiplicative regression functions. The advantage of this procedure is its simplicity; a great deal of labor and time can be saved, especially when it is applied to multiplicative models. The main drawback of the graphic method is the sacrifice in accuracy. However, if great care is taken in its application and some refined procedures are applied, the degree of error can be reduced to such an extent that it will hardly impair the practical utility of results. Thus, graphic regression is often employed to establish economic and business relationships. In this section, we shall discuss the procedure of graphic regression for the multiple linear case. We shall then apply the procedure to simple and multiple nonlinear regression models in the following section.

Let us suppose that a merchandising firm has branches in more than 100 cities in the United States. The advertising of all branches is done exclusively by spot announcements on local radio stations. The firm keeps an efficiency index for each and every branch's management based on such factors as the number of customers' complaints, speed of delivery of merchandise, bad debts of credit sales, and so on. This efficiency index ranges from 0, for the least efficient, to 10, for the most efficient management. The firm wishes to forecast branch-store sales, Y, on the basis of advertising, X_2, and efficiency index, X_3. Eight branch stores have been selected at random for analysis. Data obtained for the selected branches with reference to annual sales in millions of dollars, number of advertisements on

TABLE 20.6

Observation	Y	X_2	X_3
1	12	1	3
2	28	8	6
3	18	2	5
4	18	3	3
5	24	7	4
6	25	6	5
7	24	5	7
8	19	4	2
Total	168	36	35

$\bar{y} = 21.000$, $\bar{x}_2 = 4.500$, $\bar{x}_3 = 4.375$

radio spot announcements per day, and management efficiency index for 1968 are as given in Table 20.6.

Now, if we wish to establish a regression equation of Y on X_2 and X_3 graphically for the above data, we would follow through three simple but main steps.

The first step is to determine the regression equation of Y on X_2, which we shall identify as

$$1. \quad y_{12} = a_{1.2} + b_{12}x_2$$

Next, we would determine the regression of $(y - y_{12})$ on X_3, which we shall identify as

$$2. \quad y_{13} = a_{1.3} + b_{13}x_3$$

Finally, we would combine equations 1 and 2 to obtain the multiple regression equation of Y on X_2 and X_3 by assuming that $b_{12} = b_{12.3}$ and $b_{13} = b_{13.2}$. Thus,

$$\begin{aligned} 3. \quad \bar{y}_{1.23} = y_{12} + y_{13} &= (a_{1.2} + a_{1.3}) + b_{12}x_2 + b_{13}x_3 \\ &= a_{1.23} + b_{12.3}x_2 + b_{13.2}x_3 \end{aligned}$$

We note that the assumption of equality between simple and partial coefficients is the same as assuming that the correlation between the independent variables is zero. Can you explain why?

DETERMINING $y_{12} = a_{1.2} + b_{12}x_2$ GRAPHICALLY

In general, the procedure of determining the regression equation for two variables graphically contains the following three steps:

(1) Draw a scatter diagram of the dependent variable on the independent variable. Since dots in the diagram must later be identified, it is advisable to label them by placing beside them the number representing the corresponding observation. The scatter diagram of sales on advertising is given by Figure 20.4, which shows that the relationship between the two variables in our illustrative problem is linear.

(2) Draw a freehand line through the dots in the scatter diagram in such a way that it will best represent the average pattern of the scatter. Two guideposts in

drawing this freehand line are: (1) If a straight line is employed, it should be made to pass through the mean values of the dependent and independent variables. For our example, $\bar{y} = 21$ and $\bar{x}_2 = 4.5$, which are plotted against each other and identified by the filled-in black square in Figure 20.4. As can be seen, the line passes through this point. This guidepost is based the fact that, according to the principle of least squares, all linear regressions intersect the mean values of the variables. In the case of curvilinear regression, however, the best fit is not the one which passes through the means of the two variables. (2) The line should be so drawn that there are about the same number of dots above and below it. In our illustrative diagram, there are three dots above and three below the line, with the other two dots falling on the line. This guidepost is applicable for both the linear and the nonlinear cases.

(3) Estimates of the Y intercept and the slope of the regression equation are then made from the freehand regression line. This is usually done by utilizing the two-point formula for a straight line. That is, given two points on a straight line, (x_1,y_1) and (x_2,y_2), the equation of the line becomes

$$y = y_1 + \frac{y_2 - y_1}{x_2 - x_1}(x - x_1)$$

While any two points on the line can be used to establish the equation for the line, the convention is to select one point near each end of this line in regression analysis. In our illustration, the two points selected are shown by small circles. The first point, reading off from the scatter diagram, has the coordinates (0,12); the second point happens to be identical with our second observation, whose coordinates are (8,28). Thus, applying the two points, we have as our estimated regression equation of Y on X_2 the following:

$$y = 12 + \frac{28 - 12}{8 - 0}(x - 0) = 12 + 2x$$

or,

$$y_{12} = 12 + 2x_2$$

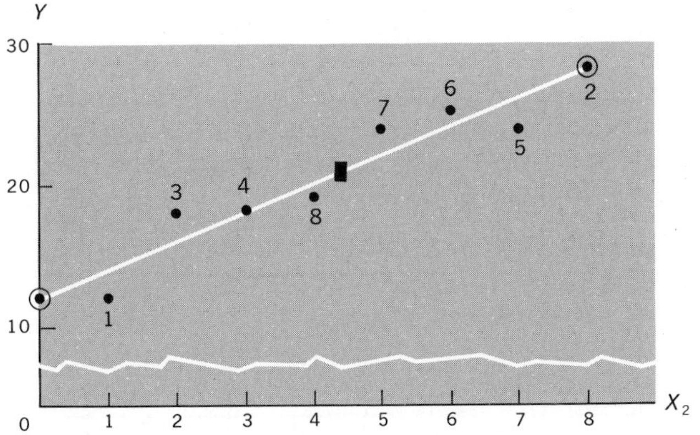

Figure 20.4 Graphic regression of sales y on advertisment x_2.

TABLE 20.7

Observation	Sales y	Advertisement x_2	y_{12}	$y - y_{12}$
1	12	1	14	−2
2	28	8	28	0
3	24	7	26	−2
4	18	2	16	+2
5	18	3	18	0
6	24	5	22	+2
7	19	4	20	−1
8	25	6	24	+1

which says that, other things being equal, the sales will be 12 million dollars per annum if there is no advertisement, and that sales will increase by 2 million dollars per year on the average with each increase in radio spot announcements per day.

The predicted (or computed) values of sales, y_{12}, on the basis of daily radio spot announcements are given in Table 20.7. The last column of this table shows the deviations of predicted values from the actual annual sales, $y - y_{12}$. These are the residual values and can be considered as that part of total variation in sales unaccounted for by variations in advertisement.

DETERMINING $y_{13} = a_{1.3} + b_{13}x_3$ GRAPHICALLY

As stated before, the residual variations, $y - y_{12}$, are variations in sales which have not been explained by the regression of sales on advertisement. Now, we turn to determine the regression of these residuals on the second independent variable, efficiency index, hoping that the former can be totally or partly explained by the latter. This regression can be obtained in exactly the same manner we used before. We first construct a scatter diagram of $y - y_{12}$ on X_3. Next, we draw a freehand line through the zero line at approximately $\bar{x}_2 = 4.375$ (see Figure 20.5). Finally, from the two points (2, −1.8) and (7,2), we obtain the regression equation desired:

$$y_{13} = -1.8 + \frac{2 - (-1.8)}{7 - 2}(x_3 - 2) = -3.32 + 0.76x_3$$

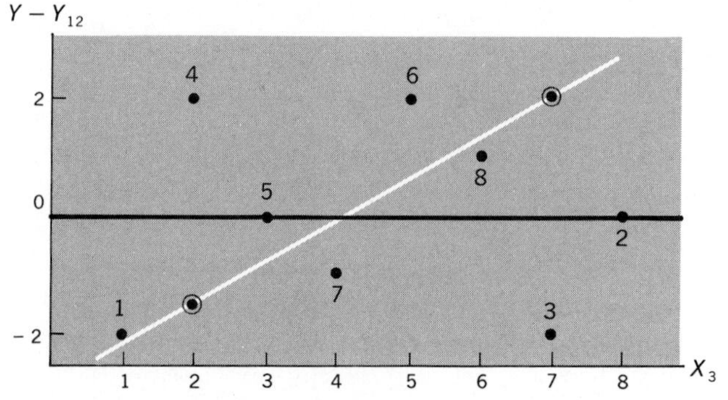

Figure 20.5 Graphic regression of sales on efficiency index.

TABLE 20.8

Observation	$y - y_{12}$	x_3	y_{13}	$(y - y_{12}) - y_{13}$
1	−2	3	−1.04	−0.96
2	0	6	+1.24	−1.24
3	−2	4	−0.28	−1.72
4	+2	5	+0.48	+1.52
5	0	3	−1.04	+1.04
6	+2	7	+2.00	0.00
7	−1	2	−1.80	+0.80
8	+1	5	+0.48	+0.52

What does this regression equation indicate? Since it is the regression of $y - y_{12}$ on X_3, it represents the average effect of efficiency index on annual sales after allowance has been made for the effect of advertisement on sales. In other words, it shows the average relationship between sales and efficiency index, net of advertisement. The values of y_{13} are given in Table 20.8. In the last column of this table, we have the residuals about this regression line, which are the vertical deviations from the regression line of $y - y_{12}$ on X_3; that is, $(y - y_{12}) - y_{13}$.

FINDING THE MULTIPLE REGRESSION EQUATION

Having completed the analysis of the separate parts, we are now ready to combine the results in a meaningful whole. This we do simply by combining the two separate regression equations to obtain the multiple regression equation. Thus, for our illustration, we have

$$\bar{y}_{1.23} = 12 + 2x_2 - 3.32 + 0.76x_3 = 8.68 + 2x_2 + 0.76x_3$$

This equation can now be used to predict the dollar value of annual sales for any branch store on the basis of the two independent variables. For example, if, for a branch store, $X_2 = 1$ and $X_3 = 3$, then the annual sales of that store would be

$$\bar{y}_{1.23} = 8.68 + 2(1) + 0.76(3) = 12.96 \text{ million dollars}$$

The computed, or predicted, values of annual sales, $\bar{y}_{1.23}$, are given in Table 20.9. In the last column of the same table, we have the ratios of the original values to the computed values $y/\bar{y}_{1.23}$. These ratios may be considered as a rough measure of the goodness-of-fit. Since all these ratios are close to 1, we consider the result quite satisfactory.

FURTHER OBSERVATIONS

The method of graphic regression is sometimes called the *method of successive approximations* because it furnishes a framework by which the first approximation to the regression line can be successively improved. In our illustration we have obtained only the first approximations, the result of which we judge to be satisfactory. Had we wished to have more accurate approximations, we would have

TABLE 20.9

Observation	Y	X_2	X_3	$y_{1.23}$	$y/\bar{y}_{1.23}$
1	12	1	3	12.96	0.93
2	28	8	6	29.24	0.96
3	24	7	4	25.72	0.93
4	18	2	5	16.46	1.09
5	18	3	3	17.96	1.00
6	24	5	7	24.00	1.00
7	19	4	2	18.20	1.04
8	25	6	5	24.48	1.02

continued our analysis in the following manner:

First, having obtained the first approximations of the partials $Y-X_2$ and $Y-X_3$, we would proceed to find graphically the regression of the residuals about y_{13}, that is, $(y - y_{12}) - y_{13}$, on X_2. This new regression equation would represent an improved approximation on the partial $Y-X_2$.

Next, we would determine the vertical deviations around the new regression line of Y on X_2, and make a second approximation of the partial $Y-X_3$ by establishing graphically a regression equation from the scatter diagram of these residuals against the values of X_3.

The above two steps can be repeated, using the latest deviations about the latest regression approximations, until no further corrections seem necessary. In general, the lower the covariability between the independent variables, the closer each approximation moves to the mathematically calculated least-square values, at a more rapid rate, since in such a case the simple regressions of Y on X_2 and of Y on X_3 would be very close to their partial relationships in the multiple case. In any event, the last step, naturally, would be the combination of the latest regression equations into the final multiple predictive equation.

It is also possible to extend the method of graphic regression to include three or more independent variables. For example, in our illustration, it would be highly desirable to include population and/or per capita income in each sales district as explanatory variables in the analysis. Suppose that we have population figures available; we may include them as the values of the third independent variable, X_4. Then the residuals about the regression equation of Y on X_3 would be plotted against the values of X_4 and a new set of estimates would be obtained. The regression equation so obtained would represent the partial relationship between annual sales and population when advertisement and efficiency index are held constant. Needless to say, the final predictive equation should now be obtained by combining the three estimated simple regression equations.

Finally, we may also note that in business or economic forecasting, we often employ times series data. Namely, we may try to forecast the value of one series on the basis of other series. The regression of time series data is frequently obtained graphically in the manner discussed in this section. There are, however, a few points worth mentioning with respect to correlation between time series. First, regression with time series is usually estimated on the basis of index numbers. Second, whenever monetary series is involved, it should be deflated by an appropriate price index. Third, the relationship between time series data can often be improved and, therefore, the utility of using regression as a forecasting device increases, by introducing "time lags." Finally, when residuals are still quite large due to the omission of other minor explanatory variables, these residuals can be accounted for by the use of a *time trend*. This is done by taking the final set of residuals of the last regression line and plotting them against time on a scatter diagram. The observations in time, such as months or years, in other words, are regarded as the last independent variable in the analysis, representing the influence of all the ignored minor variables. Of course, the regression equation that has "time" as the independent variable is obtained in the usual manner from the scatter diagram. The introduction of a time trend can often improve the final fit considerably since "time" has been utilized as a *catchall* variable for those influences unspecified.

20.8 Nonlinear Graphic Regression

In Section 20.6, it was pointed out that nonlinear regression models can be converted into linear ones by way of semilog or double-log transformations. Such procedures, however, still require rather laborious calculations in logarithms. We can now simplify our work even more with the graphic method. The application of this method to power and exponential functions has the double benefits of transforming nonlinear relationships into linear ones and the simplicity of finding the regression equations directly from original data. Procedures of the graphic method for nonlinear relationships are the same as those developed in the preceding section with but two exceptions. The first is that we now plot sample data on logarithmic paper, and the second is that the freehand line need not go through the means of variables as in the linear case. We may also note that to plot data on logarithmic paper has the same effects as to scale the logarithms of the variables on arithmetic paper.

When semilog paper is used (that is, when the vertical axis is logarithmic and the horizontal scale is arithmetic), the two-point formula for a straight line becomes

$$1.\quad \log y = \log y_1 + \frac{\log y_2 - \log y_1}{x_2 - x_1}(x - x_1)$$

If double-log paper is employed, the two-point formula for a straight line is given as

$$2.\quad \log y = \log y_1 + \frac{\log y_2 - \log y_1}{\log x_2 - \log x_1}(\log x - \log x_1)$$

We shall now turn to apply the graphic method to nonlinear regressions by taking up a simple and a multiple case, respectively.

TABLE 20.10

Observation	y	x	y_c
1	15	1	15.00
2	18	2	18.69
3	25	3	23.29
4	30	4	29.02
5	35	5	36.17
6	45	6	45.00
7	53	7	56.16
8	70	8	70.00

FITTING $y_c = ab^x$ GRAPHICALLY

Eight trainees learning telegraph code, selected at random, are found capable of receiving a certain number of letters per minute, Y, after the indicated number of months' training, X, as shown in the first three columns in Table 20.10. These data seem to suggest that the average relationship between Y and X can be adequately described by the function

$$y_c = ab^x$$

which, as noted before, can be converted into the linear form by the semilog

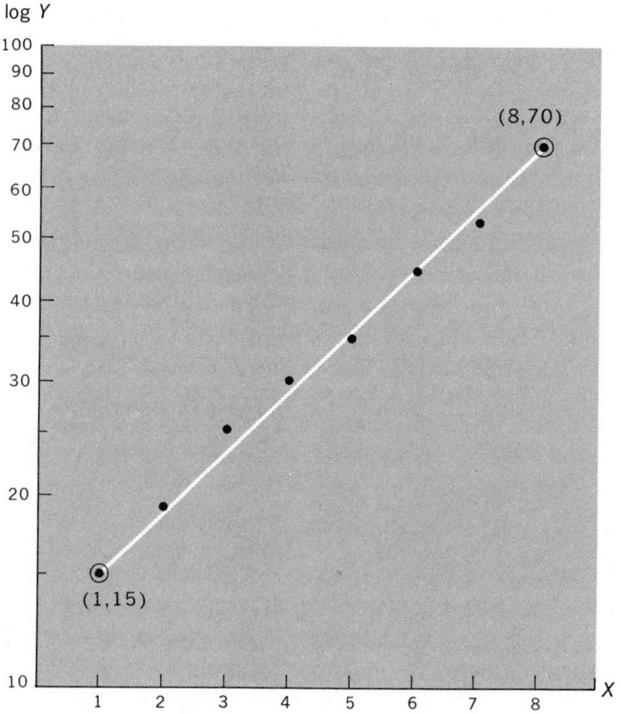

Figure 20.6 Graphic regression for $y_c = ab^x$.

transformation,

$$\log y_c = \log a + x(\log b)$$

To find the above linear regression of the semilogarithmic form, we first plot the sample data on a semilog paper as shown by Figure 20.6. Next, we draw a freehand straight line through the dots which we judge to represent the average relationship between the variables well. For our example, this line is drawn through the two points (1,15) and (8,70). Algebraically, this line may be given as

$$\log y_c = \log y_1 + \frac{\log y_2 - \log y_1}{x_2 - x_1}(x - x_1) = \log 15 + \frac{\log 70 - \log 15}{8 - 1}(x - 1)$$

$$= 1.17609 + \frac{0.66892}{7}(x - 1) = 1.08053 + 0.09556x$$

That is,

$$1.\quad \log y_c = 1.08053 + 0.09556x$$

or,

$$2.\quad y_c = 12.04(1.246^x)$$

as the estimated regression equation.

To obtain the computed values of Y, y_cs, we employ expression 1. For example, if $X = 1$, we have

$$\log y_c = 1.08053 + 0.09556(1) = 1.17609$$

and

$$y_c = 15$$

If $X = 2$,

$$\log y_c = 1.08053 + 0.09556(2) = 1.27165$$

$$y_c = 18.69$$

And so on for other values of X. The y_cs are given in the last column of Table 20.10.

FITTING $X_c = aL^bK^c$ GRAPHICALLY

The production function of a firm can be adequately described by the relationship

$$X_c = aL^bK^c$$

where

X = Quantity of a commodity produced
L = Quantity of labor inputs
K = Quantity of capital inputs

Sample data for the firm's production function are given in the first three columns of Table 20.11. To find this regression function graphically, we first obtain the regression of X on L,

$$X_L = aL^b$$

which, in logarithms, is

$$\log X_L = \log a + b(\log L)$$

TABLE 20.11

X	L	K	$L^{0.49}$	$\frac{X}{L^{0.49}}$	$K^{0.12062}$	$\frac{X}{K^{0.12062}}$	X_c
120	1	2.5	1.00	120	1.11	108	123
180	2	3.0	1.40	129	1.14	158	177
190	3	1.0	1.71	111	1.00	190	190
235	4	1.5	1.98	119	1.05	224	233
285	5	4.0	2.20	130	1.18	257	288
290	6	2.0	2.41	120	1.09	266	292

The graphic expression of the preceding equation is obtained when values of X are plotted against those of L on the double-log paper, as shown by Figure 20.7, and when a freehand line is drawn through the points in the scatter diagram. It may be noted that, for our sample data, the freehand line is drawn through the two points (1,120) and (6,290). Mathematically, this line is given as

$$\log X_L = \log x_1 + \frac{\log x_2 - \log x_1}{\log l_2 - \log l_1}(\log L - \log l_1)$$

$$= \log 120 + \frac{\log 290 - \log 120}{\log 6 - \log 1}(\log L - \log 1)$$

$$= \log 120 + 0.49(\log L)$$

From this result, we have

$$1. \quad X_L = 120L^{0.49}$$

the regression of X on L which we seek.

Here, $L^{0.49}$ is the average amount of variations in production due to the varia-

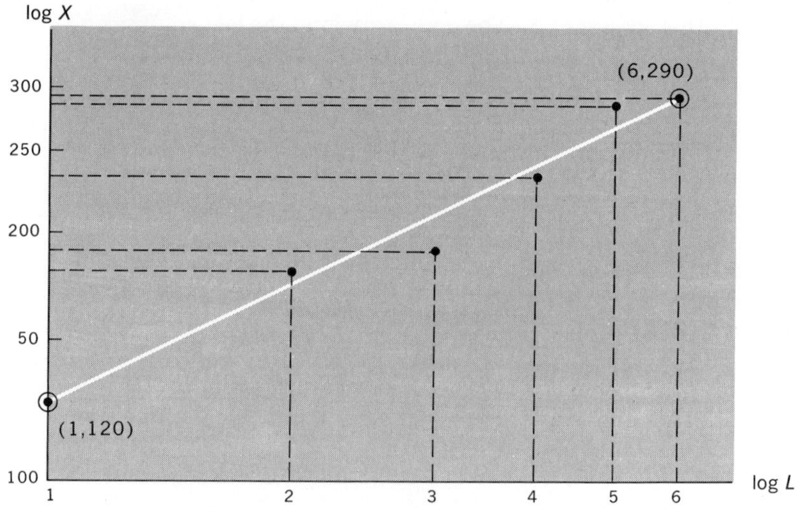

Figure 20.7 Graphic regression of $X_c = aL^b$.

tions in the quantities of labor inputs. To find the values of $L^{0.49}$, we set

$$\log N = \log L^{0.49} = 0.49(\log L)$$

For example, if $L = 1$,

$$\log N = 0.49(\log 1) = 0$$
$$N = 1$$

If $L = 2$,

$$\log N = 0.49(\log 2) = 0.14750$$
$$N = 1.40$$

Similarly for other values of L.

Now we note that the residuals of X after the effects of L have been removed are measured by the differences between X and $L^{0.49}$. In terms of logarithms, these differences are obtained by dividing X by $L^{0.49}$. We shall denote this process as

$$X_L = X/L^{0.49}$$

The values of $L^{0.49}$ and X_L are given in the fourth and fifth columns of Table 20.11, respectively.

Since the residuals of X are the values of X after the effects of L on X have been removed, we may then measure the effects of K on X by finding the regression of these residuals, X_L, on K. The scatter diagram of X_L on K is given by Figure 20.8. The freehand line drawn in this diagram, through (1,111) and (4,130), yields the following expression:

$$\log X_K = \log x_1 + \frac{\log x_2 - \log x_1}{\log k_2 - \log k_1}(\log K - \log k_1)$$

$$= \log 111 + \frac{\log 130 - \log 111}{\log 4 - \log 1}(\log K - \log 1)$$

$$= \log 111 + 0.12062(\log K)$$

Thus,

$$2. \quad X_K = 111K^{0.12062}$$

which shows the average influence of capital on production. The values of $K^{0.12062}$ and $X_K = X/K^{0.12062}$ are given in the sixth and seventh columns of Table 20.11, respectively.

Finally, since $X_L = X/L^{0.49}$, we may incorporate this in equation 2 to obtain

$$3. \quad X_c = 111L^{0.49}K^{0.12062}$$

which is the multiplicative regression equation we seek to find. The values of X_c are determined by multiplying 111 by the product of $L^{0.49}$ and $K^{0.12062}$. For example,

$$111(1.00)(1.11) = 108$$

which is the estimated production volume when $L = 1$ and $K = 2.5$. The X_cs are given in the last column of Table 20.11.

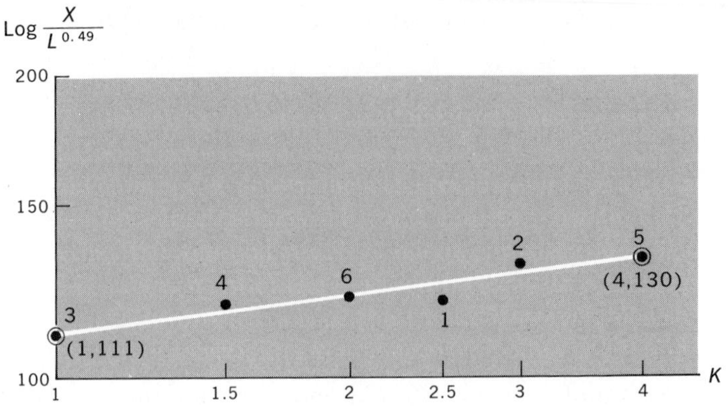

Figure 20.8 Graphic regression of $X_k = aK^c$.

Glossary of Formulas

(20.1) $E(Y|X_2,X_3) = \mu_{1.23} = A + B_2X_2 + B_3X_3$ and **(20.2)** $y_i = \mu_{1.23} + e$

These two equations define the population regression linear model for the trivariate case. The first says that the dependent variable Y is linearly related to the two independent variables X_2 and X_3. The second gives the relationship between a single observation of Y and the three-dimensional regression plane; in this expression e is the random error which measures the deviation of an individual y from the mean value of Y associated with a set of fixed values of X_2 and X_3.

(20.3) $\bar{y}_{1.23} = a_{1.23} + b_{12.3}x_2 + b_{13.2}x_3$ and **(20.4)** $y_i = \bar{y}_{1.23} + \hat{e}$

(20.3) is the least squares estimate of (20.1), and (20.4) defines the relationship between an individual value of Y in the sample and the sample regression plane. Note that in (20.3) a stands as the estimate of A, and is the Y intercept or elevation of the sample regression plane; $b_{12.3}$ and $b_{13.2}$ are called partial regression coefficients and are estimates of B_2 and B_3 respectively.

(20.5) $b_{12.3}\Sigma x_2'^2 + b_{13.2}\Sigma x_2'x_3' = \Sigma y'x_2'$ $\qquad b_{12.3}\Sigma x_2'x_3' + b_{13.2}\Sigma x_3'^2 = \Sigma y'x_3'$

(20.6) $a_{1.23} = \bar{y} - b_{12.3}\bar{x}_2 - b_{13.2}\bar{x}_3$

The first set of equations are the normal equations for the trivariate regression analysis, given in terms of deviations from the means of the variables. This set of normal equations would give us the least squares estimates of the partial regression coefficients. With $b_{12.3}$ and $b_{13.2}$ found, then the value of $a_{1.23}$ can be determined by (20.6).

(20.7) $\Sigma(y - \bar{y})^2 = \Sigma(y - \bar{y}_{1.23})^2 + \Sigma(\bar{y}_{1.23} - \bar{y})^2$

(20.8) SST $= \Sigma(y - \bar{y})^2 = \Sigma y'^2$

(20.9) SSR $= \Sigma(\bar{y}_{1.23} - \bar{y})^2 = b_{12.3}\Sigma x_2'y' + b_{13.2}\Sigma x_3'y'$

(20.10) SSE $= \Sigma(y - \bar{y}_{1.23})^2 =$ SST $-$ SSR

The first of these equations gives the analysis of variance of the trivariate regression case. The left hand side of (20.7) defines the total variations in Y, SST, which has $n - 1$ df. SST is decomposed into two parts: SSR which measures that part of variation Y which has been explained by the regression, and SSE which measures the residual variation. Note also that SSR has

$n - k$ df, where k is the number of partial regression coefficients, and SSE has $n - k - 1$ df. The last three equations in this group are computational formulas for these three sources of variability.

(**20.11**) $\text{MSE} = \dfrac{\text{SSE}}{n - k - 1} = \hat{\sigma}^2_{1.23}$ The mean square of error is an unbiased estimate of the population variance $V(Y) = E(Y - E(Y|X_2,X_2))^2$. The square root of MSE is the sample standard deviation of the regression plane. If Ys are normally distributed, MSE measures the closeness-of-fit. Also note that $F_{k,n-k-1} = \text{MSR}/\text{MSE}$ can be used to conduct an over-all test that $B_2 = B_3 = 0$ in the trivariate analysis.

(**20.12**) $R^2_{1.23} = 1 - \text{SSE}/\text{SST} = \text{SSR}/\text{SST}$ and (**20.13**) $\dot{R}^2_{1.23} = 1 - \text{MSE}/\text{MST}$

The coefficient of multiple determination defined by (20.12) is a positively biased estimate of the corresponding population parameter. This positive bias is corrected by introducing $(n - 1)/(n - k - 1)$ as the correction factor; (20.13) results. In general, the coefficient of multiple determination measures the closeness-of-fit by the regression plane; it measures the percentage in total variation that has been accounted for the regression. The square root of $R^2_{1.23}$, always positive, is the coefficient of multiple correlation.

(**20.14**) $$r_{12.3} = \frac{r_{12} - r_{13}r_{23}}{\sqrt{(1 - r_{13}^2)(1 - r_{23}^2)}}$$

(**20.15**) $$r_{13.2} = \frac{r_{13} - r_{12}r_{23}}{\sqrt{(1 - r_{12}^2)(1 - r_{23}^2)}}$$

(**20.16**) $$r_{23.1} = \frac{r_{23} - r_{12}r_{13}}{\sqrt{(1 - r_{12}^2)(1 - r_{13}^2)}}$$

Partial correlation coefficients in multiple association analysis are here defined in terms of simple correlation coefficients. A partial correlation coefficient measures the net covariation between two of the variables under consideration. It is interpreted in terms of its squared value—coefficient of partial determination. For instance, if $r_{12.3} = 0.90$ and $r^2_{12.3} = 0.81$, then we say that the errors made in estimating Y from X_3 are reduced by 81 percent when X_2 is employed as an additional explanatory variable.

(**20.17**) $y_c = a + bx + cx^2$ The sample regression function of the form of a second-degree polynomial, a parabola, is given by this expression.

(**20.18**) $y_c = a + b/x$ (**20.19**) $y_c = a - b/x$

(**20.18a**) $y_c = a + bZ$ (**20.19a**) $y_c = a - bZ$

When sample regression functions can be adequately described by expressions (20.18) and (20.19), then these nonlinear regressions can be converted into linear forms by the reciprocal transformation. That is, if we define $Z = 1/X$, then the first two regression functions can be rewritten as (20.18a) and (20.19a), respectively; both are now linear regressions of Y on Z.

(**20.20**) $y_c = ab^x$

(**20.20a**) $\log y_c = \log a + x(\log b)$

(**20.20b**) $z_c = c + dx$

The exponential regression function, (20.20), can be converted into a linear one by semilog transformation, (20.20a). Furthermore, if we let $\log y_c = z_c$, $\log b = d$, and $\log a = c$, (20.20a) becomes (20.20b), which is a linear regression of Z on X. The coefficients c and d are found in the usual manner by working the logarithms of Y and the original values of X.

(20.21) $y_c = ax^b$

(20.21a) $\log y_c = \log a + b(\log x)$

(20.21b) $z_c = c + bw$

A power sample regression such as (20.21) is converted into a linear relationship by the double-log transformation as shown by (20.21a). Furthermore, if we let $\log y_c = z_c$, $\log a = c$, and $\log x = w$, then the linear relationship in logarithms can be written as (20.21b), which is a linear function of Z on w. The coefficients c and b are then found in the usual manner by working with the logarithms of Y and logarithms of X.

(20.22) $y_c = ax_2^b x_3^c$

(20.22a) $\log y_c = \log a + b(\log x_2) + c(\log x_3)$

(20.22b) $z_c = d + bz_2 + cz_3$

Given a sample regression in the form of (20.22), a linear relationship such as (20.22b) would result via the double-log transformation. Explain how (20.22) is related to (20.22a).

Problems

20.1 Show that in the trivariate regression analysis

$$a_{1.23} = \bar{y} - b_{12.3}\bar{x}_2 - b_{13.2}\bar{x}_3$$

20.2 Show that $\text{SSR} = \Sigma(\bar{y}_{1.23} - \bar{y})^2 = b_{12.3}\Sigma x_2'y' + b_{13.2}\Sigma x_3'y'$.

20.3 Given the sample regression equation,

$$y_c = a_{1.234} + b_{12.34}x_2 + b_{13.24}x_3 + b_{14.23}x_4$$

derive the set of normal equations required for estimating the partial regression coefficients by the least squares method.

20.4 Write out the set of normal equations in the answer to the preceding problem in terms of deviations from the means of the variables.

20.5 When the partial regression coefficients are found in the four-variable regression analysis, how can $a_{1.234}$ be determined?

20.6 Can you explain why the following formula for the coefficient of multiple determination is valid?

$$R^2_{1.234} = 1 - \frac{\sigma^2_{1.234}}{\sigma_y^2} = \frac{\text{SST} - \text{SSE}}{\text{SST}} = \frac{\text{SSR}}{\text{SST}}$$

20.7 The partial correlation coefficients in the four-variable case are given as

1. $$r_{12.34} = \frac{r_{12.4} - r_{13.4}r_{23.4}}{((1 - r^2_{13.2})(1 - r^2_{23.4}))^{1/2}}$$

2. $$r_{13.24} = \frac{r_{13.4} - r_{12.3}r_{23.4}}{((1 - r^2_{12.4})(1 - r^2_{23.4}))^{1/2}}$$

3. $$r_{14.23} = \frac{r_{14.3} - r_{12.3}r_{24.3}}{((1 - r^2_{12.3})(1 - r^2_{24.3}))^{1/2}}$$

But can you explain why the following identity holds?

$$\frac{r_{12.4} - r_{13.4}}{((1 - r^2_{13.4})(1 - r^2_{23.4}))^{1/2}} \equiv \frac{r_{12.3} - r_{14.3}r_{24.3}}{((1 - r^2_{14.3})(1 - r^2_{24.3}))^{1/2}}$$

20.8 To evaluate the ability of predicting volume of imports by Formosa (the Republic of China) on the basis of the gross national product and of general prices in Formosa in relation to import prices, historical data for 1959 to 1967 were gathered and analyzed. Letting

Y = index of imports of goods and services to Formosa at constant (1959) prices

X_2 = index of gross national product of Formosa at 1959 prices

X_3 = ratio of indices of general Formosa price indices to import price indices the investigators made the following calculations:

$$n = 9$$
$$\Sigma y = 1052 \quad \Sigma x_2 = 1017 \quad \Sigma x_3 = 954$$
$$\bar{y}x = 116.9 \quad \bar{x}_2 = 113 \quad \bar{x}_3 = 106$$
$$\Sigma y^2 = 124{,}228 \quad \Sigma x_2^2 = 115{,}571 \quad \Sigma x_3^2 = 101{,}772$$
$$\Sigma y x_2 = 119{,}750 \quad \Sigma y x_3 = 111{,}433 \quad \Sigma x_2 x_3 = 107{,}690$$

a. Find the linear regression line of Y on X_2 and X_3 by the least squares method.
b. Give the calculated values of Y, $\bar{y}_{1.23}$, with the estimated regression equation in a.

20.9 Give verbal interpretations to the partial regression coefficients obtained in the preceding problem and compare intuitively the relative influence of X_2 and X_3 on Y.

20.10 Ten branch stores of a national retailing company have been selected at random in order to evaluate the effects of population and income on sales in each sales district. The following data have been obtained for this purpose:

Sales (in thousands of dollars)	Population (in thousands)	Income (in millions of dollars)
7	6	15
5	7	12
11	9	18
9	10	16
6	11	10
10	12	15
13	13	17
9	14	10
12	15	12
10	16	11

a. Convert the above data into percentage form.
b. Find the least squares regression equation of sales on population and income.
c. Give verbal interpretations to the values of partial regression coefficients obtained in b.
d. Find the predicted values of Y.

20.11 With data in Problem 20.8,
a. Construct an analysis of variance table.
b. Compute the estimated variance and standard deviation of the regression plane.
c. Conduct an over-all test that $B_1 = B_2 = 0$.
d. Conduct a step-by-step analysis of the separate contributions of each of the independent variables to the dependent variable.

20.12 With data in Problem 20.10, follow the instructions for Problem 20.11.

20.13 Compute $R^2_{1.23}$ and $\dot{R}^2_{1.23}$ for data in Problem 20.8 and interpret results.

20.14 Compute the multiple coefficient of determination and the corresponding adjusted statistic for data in Problem 20.10 and interpret results.

20.15 Compute $r_{12.3}$, $r_{13.2}$, and $r_{23.1}$ for data in Problem 20.8. What does each of these values measure?

20.16 Compute the partial correlation coefficients for data in Problem 20.10 and interpret the results.

20.17 Conduct a t test for each of the partial correlation coefficient values obtained in Problem 20.15 or in Problem 20.16.

20.18 Use results obtained for data in Problem 20.8 to verify the following relationships:

a. $r^2_{13.2} = \dfrac{R^2_{1.23} - r_{12}^2}{1 - r_{12}^2}$

b. $R^2_{1.23} = \dfrac{b_{12.3}\Sigma y'x_2' + b_{13.2}\Sigma y'x_3'}{\Sigma y'^2}$

c. $b_{12.3} = r_{12.3}\left(\dfrac{\hat{\sigma}_{1.3}}{\hat{\sigma}_{2.3}}\right)$ and $b_{13.2} = r_{13.2}\left(\dfrac{\hat{\sigma}_{1.2}}{\hat{\sigma}_{3.2}}\right)$

d. $R^2_{1.23} = 1 - \dfrac{s^2_{1.23}}{s_1^2} = \dfrac{r_{12}^2 + r_{13}^2 - 2r_{12}r_{13}r_{23}}{1 - r_{23}^2}$

e. $1 - R^2_{1.23} = (1 - r_{12}^2)(1 - r^2_{13.2})$

20.19 A study was made of the relationships among grades earned in different subjects by 200 business students selected at random. Using these variables

Y = grades in business statistics
X_2 = grade in freshman mathematics
X_3 = grade in logic
X_4 = grade in general business subjects

the investigator obtained the following statistics:

$$\bar{y} = 75; \quad s_1^2 = 100$$
$$\bar{x}_2 = 24; \quad s_2^2 = 25$$
$$\bar{x}_3 = 15; \quad s_3^2 = 9$$
$$\bar{x}_4 = 36; \quad s_4^2 = 36$$

$r_{12} = 0.90$; $r_{13} = 0.75$; $r_{23} = 0.70$; $r_{24} = 0.70$; $r_{34} = 0.85$

Assuming that Y is linearly related to X_i, find the least squares regression equation of Y on X_2, X_3, and X_4.

20.20 Find the partial correlation coefficients $r_{12.34}$, $r_{13.24}$, and $r_{14.23}$ for data in the preceding problem.

20.21 Compute and interpret the multiple coefficient of determination and the variance of Y about the regression hyper plane for data in Problem 20.19.

20.22 For data below, the variables under consideration are

Y = the breaking strength of steel cable in hundred-weight
X_2 = the diameter of cable in millimeters
X_3 = the percentage of alloy in the steel used
X_4 = the number of separate strands of which each piece of cable is composed

a. Find the regression equation

$$y_c = a_{1.234} + b_{12.34}x_2 + b_{13.24}x_3 + b_{14.23}x_4$$

b. Compute the partial correlation coefficients, the multiple coefficient of determination, and the variance of the regression.

Observation	Y	X_1	X_2	X_3
1	3	2	8	10
2	5	4	5	12
3	7	6	8	10
4	6	5	10	11
5	3	1	5	13
6	7	5	8	16
7	9	6	9	10
8	4	4	6	14
9	8	7	5	13
10	1	1	3	11

20.23 Find the sample regression equation for the data in the previous problem by the graphic method and compare the result with that obtained mathematically.

20.24 The total net investment function for the economy as a whole is assumed to be of the form

$$I = ai^b$$

where I stands for net investment, i for rate of interest, and a and b are the unknown constants. The following sample is obtained:

I:	10.0	6.5	9.5	5.0	4.5	2.5	4.0	2.5	2.2	2.8
i:	3	4	3	5	6	8	7	7	9	8

Determine the regression equation graphically.

20.25 According to the liquidity preference theory of the interest rate, the demand for money to hold as cash balances, for a given level of national income, is a decreasing function of the rate of interest. If we let M stand for the quantity of money demanded and i for interest rate, then an equation of the form

$$M = a + \frac{b}{i}$$

gives a rather close approximation. Given the following sample data, plot a scatter diagram and fit a regression line to it by the least squares method.

M:	13	12	11	17	18	14	13	11	13	10	12	13	10	11	14	23
i:	7	5	3	2	1	3	2	5	6	7	6	4	6	4	2	1

where M are measured in millions of dollars and i are annual interest rates in percentages. (*Hint:* Define a new variable $Z = 1/i$, and find the regression of M on Z from the equation $M = a + bz$.)

20.26 An economist collected data concerning four variables X, W, Y, and Z. He believed that an equation of the form $X_c = aY^bW^cZ^d$, where a, b, c, and d are the unknown constants, could be found from which he could predict X by knowing Y, W, and Z. Outline clearly a procedure by which this aim can be accomplished.

21

DECISION AND GAME THEORIES

21.1

Operations Research

During the past few decades, amidst the growing size and complexity of business, there has been a continuous effort to develop and apply modern scientific techniques to problems involving the operations of a system as a whole. This endeavor is what is now called *operations research*—an omnifarious effort which encompasses methods of inventory control, programming, statistical inference, forecasting, decision and game theories, queuing problems, and Monte Carlo simulation. Some of these topics have already been introduced in this text. Now, in the last two chapters, we shall present some recent methods of operations research which involve probabilistic considerations. Precisely, we shall cover the main aspects of decision and game theories here, and follow up in the next chapter with a discussion on waiting line problems and Monte Carlo simulation.

21.2

The Structure of Decision

The term "decision" has several fundamental implications that must be clearly understood. To begin with, decision is needed in a problem situation where two or more alternative courses of action are available, and where only one of these actions can be taken. Obviously, if there is only one course of action available, no decision is required since that action must be taken in order to solve the problem. The restriction in the second part of this last statement does not in any way limit the handling of practical situations, inasmuch as it is quite permissible to consider any combination of actions as a single action in a decision procedure. What we should do is to reformulate any combination of actions so that this restriction is satisfied. Consequently, the first consideration is to work out what is called a "list of mutually exclusive actions" to meet this restriction in order to avoid later confusion or ambiguity.

Suppose that the research department of a firm has designed a new product which can be put into immediate production. The perfection of this new product creates many problems for managerial decision. Should the new product be introduced to the market? If so, how should it be produced? distributed? advertised? In each of these problem situations there are alternative courses of action to be selected; for instance, the problem of what advertising media to use may include radio, television, newspapers, magazines, direct mailings to potential buyers, or a combination of two or more of these media.

Consider now the federal policy for economic stabilization. If there are, say, ten million workers unemployed today, should the government adopt monetary or fiscal measures to improve the situation? If it decides to employ monetary devices, which instruments are to be used? a reduction in reserve requirements against demand deposits? a decrease in the rediscount rate? purchase of government securities in the open market? an easing of terms for installment purchases? or a combination of some or all of these actions?

In the second place, decision procedure involves selecting, among the alternatives, a single course of action that can be actually carried out. This point is so self-evident that only a brief comment is needed. It is surely a waste of time and resources to think about a problem over which there is no control, or to select a course of action that can never be carried out.

Third, the objective of decision is to select an act which will accomplish some predesignated purpose. To judge whether or not a decision is satisfactory depends upon the outcome or consequence of the act selected by the decision procedure. If the outcome of the act taken is favorable, the decision may be considered as satisfactory. But how do we evaluate whether the outcome is favorable? This question can be answered only if we have specified beforehand just what outcome is to be regarded as favorable. For instance, in deciding which media to use in advertising the new product mentioned earlier, management will be influenced by various objectives. The management may be primarily concerned with reaching the largest possible audience; hence it may decide to use television. It may have the lowest cost as the main objective; if so, it may choose to use direct mailings to its potential buyers. Or it may wish to reach a relatively large audience at a relatively low cost; it may then decide to employ radio or newspapers. In any case, a decision procedure involves the selection of a possible course of action which will lead to an outcome judged to be favorable or not, in accordance with its success or failure to fulfill some predesigned objective. The significance of decision lies exactly in this condition.

Fourth, decision, as a process, has a time element. Decision leads to action which, in turn, leads to outcome. Decision is made at the present, outcome emerges in the future, and, in between, action is being carried out. The time ingredient reveals that decision is predictive: to make a decision, we must predict the outcome or outcomes of that decision. Future outcomes are sometimes called *states of nature.*

The anticipation of states of nature is a very simple matter if the problem at hand is one of the so-called *strict causal chain* variety. For such a chain, events do not keep branching out: each event will lead to another simple event, and so on, through the chain of events. For instance, turning off the switch "causes" the electrical circuit to break which, in turn, "causes" the light to go out. Thus, in a strict causal chain, the prediction is "certain" and no difficulty is present in decision making.

In many problems situations, each course of action may lead to a number of outcomes. For instance, when a coin is tossed, either a head or a tail may appear. Again, if you elect to drive your car this weekend to visit a friend, you may arrive at your destination as planned, or you may find yourself lying in a hospital bed with your arm in a plaster cast, or your car may break down on an unfamiliar road miles away from a garage, and so forth. Clearly, when there are many possible outcomes of an act, it is no longer possible to predict with certainty what will actually happen.

Nevertheless, we are not completely helpless here. We can still predict the possible outcomes of an act in terms of probabilities. To do this, we must distinguish between two cases—cases of risk and cases of partial ignorance. *Risk* refers to a situation where the probability is known for each of a set of outcomes resulting from a specific course of action. For instance, there is risk in predicting the appearance of a head when a coin is tossed once, for it may turn up a tail. Nevertheless, if a coin is tossed over and over again, we can predict that the proportion of heads will approach one half. *Partial ignorance* characterizes a true state of nature where outcomes of an action do not possess known probabilities in the classical or relative frequency sense, but a decision maker can still express various "degrees of belief" with respect to occurrences of possible outcomes. Very often, such initial degrees of belief may be modified with additional information to describe more accurately the "true" state of nature.

Finally, a decision process must take into consideration what is called the evaluation of *conflicting values.* A given outcome may be associated with desirable as well as undesirable aspects, that is, have conflicting values. The *desirability* of an action is the difference between the desirable and undesirable aspects of the outcome of the action itself. But what do we actually mean by desirability and how can we measure it? A moment's reflection reveals that "desirability" is a relative term. It is relative to the objective of a decision maker in a specific problem situation. For business and economic problems, fortunately, we have available only two standards, or units, of measuring desirability: money and utility.

Dollars and cents can be used to measure not only the direct monetary costs (undesirability) and gains (desirability) of an outcome, but also intangibles, such as convenience, goodwill, self-satisfaction, and peace of mind. One way to measure an intangible variable is to assign an arbitrary monetary value to it. Thus, taking the reputation of a firm, the loyalty of customers, and some other abstract valuations into consideration, an accountant may estimate that the goodwill of a firm is worth $100,000.

Another way to convert an intangible into monetary values is to ask an introspective question like this: Would I be willlng to stand in a line that is ten blocks long to buy a ticket at Radio City Music Hall, or would I prefer to give up a quarter, half a dollar, one dollar, or two dollars to avoid standing in the line? In answering this question, one can eventually determine the monetary value of the undesirability of standing in the line.

It is also possible to convert time to monetary values, since each individual consciously or unconsciously values his time at so much money per hour or per day. For example, suppose that it costs you $10 to have your apartment cleaned by a maid, or it costs four hours of your time if you clean it yourself. And suppose that you decide to have a maid's service; then you must have valued your time at more than $2.50 per hour. We have, in this example, ignored the undesirability (or maybe desirability) of manual labor (or maybe exercise) involved in cleaning.

When the conflicting values of an outcome are converted into monetary units, its desirability may be thought of as the difference between total monetary gains and total monetary losses. If an action has only one outcome, the desirability of the outcome is also the desirability of the action. When alternative outcomes are associated with an action, the desirability of the action may be regarded as the average desirability, or expected value, of all possible outcomes. When the desirabilities of alternative courses of action have been measured, we then proceed to select the action that has the highest expected desirability. This may be called the *decision rule*, or *criterion of expected values.*

While the expected monetary value has been used for many decision problems, it is not the decision criterion for all situations. Very often, we may find that a more plausible and appropriate standard of measuring desirability is what is called "utility."

A student of economics is undoubtedly familiar with the fact that the neoclassical economists constructed a utility theory to explain consumer behavior. This theory considers utility as a measure of the psychological state of the consumer—the intensity of his desire, the psychic gains and losses resulting from alternative courses of action available to him. This theory also treats utility as a measure of introspective pleasure—intrinsic satisfaction provided by a commodity. Thus, the neoclassical concept of utility is fundamentally "ordinalistic" in the sense that decision processes are explained merely in terms of an individual's preferences. It involves, then, only ranking or ordering things without assigning proper weights to them.

The concept of utility employed for our later discussion is "cardinal" in nature, in the tradition of Von Neumann and Morgenstern. In the N–M sense, utility is associated with the ordering of choices involving risks that can be evaluated in terms of probability. As we shall see in Section 21.6, the N–M "utility index" is constructed with the aims of computation and prediction in view. In other words, it explains how and why a decision maker should rank risky or uncertain strategic decisions on the basis of "expected utility."

There is, however, one thing in common between the neoclassical and N–M concepts of utility; that is, the utility scale is a personal-value scale, not a dollar-and-cents scale.

21.3 Statistical Decision Theory

Given a problem situation where there are available alternative courses of action, each of which may lead to a set of mutually exclusive outcomes associated with certain probabilities, which course of action should a decision maker take? This is a problem of statistical decision theory and is evaluated in terms of expected monetary outcome.

STATING THE DECISION PROBLEM

Let us consider this example: A baker produces a certain type of fancy cake at a total average cost of \$10, and sells it at a price of \$15. This cake is produced over the weekend and is sold during the following week; such cakes produced but not sold during a week's time are worthless and must be thrown away. According to past experience, the weekly demand for these cakes is never less than 18 or greater than 20; that is, the weekly demand constitutes three possible values: 18, 19, and 20. The question is: How many cakes should the baker produce for each week's sales: 18? 19? 20?

For this example, we clearly have three possible acts:

$$A_1 = \text{produce 18 units}$$
$$A_2 = \text{produce 19 units}$$
$$A_3 = \text{produce 20 units}$$

Whichever act the baker decides to take, the weekly sales may be any one of these three sales volumes (possible outcomes of each one of these three actions).

With the above analysis, we have already taken the first step of statistical decision theory—namely, stating the decision problem: listing all the possible courses of action and the possible outcomes (events) of each action. In general, the listing of events in a given problem situation must be mutually exclusive and exhaustive. Note, also, that the possible outcomes of an act are numerical values here, and as such they are in effect values of a random variable.

PAYOFF TABLES

The next step of statistical decision theory is the construction of a payoff table listing the alternative acts and their possible events (outcomes). A payoff table represents the economics of the problem—a problem of revenue and costs. A payoff may be thought of as a *contingency*, a *conditional value*, or *conditional profits (losses)*. A payoff is a "conditional value" in the sense that associated with each course of action there is a certain profit (or loss), *given* that a specific event (outcome) has occurred. A *payoff* or *contingency table* is then a table which contains all conditional values of all possible combinations of acts and outcomes.

The calculation of payoffs depends on the problem. Sometimes a bit of algebraic reasoning is required. Very often, however, it is a relatively easy matter. For our illustrative example, we may let D be the number of units demanded each week and Q be the quantity produced. Since $18 \leq D \leq 20$, the baker will never consider a level of output which is greater than 20 or less than 18.

Next, payoffs or conditional profits, π, for demand less than quantity produced, $(D < Q)$, can be considered as the difference between total revenue of sales and total costs of production, $(15D - 10Q)$. Total profits, π, for a total demand equal to or greater than the quantity produced, $(D \geq Q)$, can be computed as total revenues from sales of all that was produced $(15Q)$ minus total costs of producing that quantity, $(10Q)$. Symbolically, this becomes

$$\pi = 15D - 10Q, \quad D < Q$$
$$\pi = 15Q - 10Q, \quad D \geq Q$$

With these equations, payoffs associated with each act $(Q = 18, 19, 20)$, given that a specific event $(D = 18, 19, 20)$ has occurred, are given in Table 21.1.

PROBABILITY ASSIGNMENT

With the payoff table, a decision maker may be able to reach the optimal solution of a problem if he has a knowledge of what event is going to occur. For instance, if the baker knew in advance that next week's demand would be 20, he would

TABLE 21.1 PAYOFF TABLE FOR PRODUCTION DECISION

Event	Possible Act A_i		
D	A_1: 18	A_2: 19	A_3: 20
18	\$90.00	\$80.00	\$70.00
19	90.00	95.00	85.00
20	90.00	95.00	100.00

obviously produce 20 units now and make a profit of \$100. However, as is usually the case, such information is not available to the baker. Indeed, if he produces 20 units, the weekly demand may turn out to 18 and his conditional profit will be \$70, not \$100. In the face of uncertainty, a decision maker must make some prediction or forecast, usually in terms of the probability of occurrences of events, and then select a course of action consistent with the prediction. Thus, the third step of statistical decision theory is to assign probabilities to possible events. Suppose that our baker has been producing this type of cake for a number of months; he may then use the relative frequency of past sales data as probabilities. Suppose, also, that the results are as follows:

Event: Demand	Probability
18	0.2
19	0.7
20	0.1

These values may be considered as initial probabilities. If the baker, for one reason or another, believes that next week's demand will be somewhat different, he may modify these probability assignments in accordance with his judgment.

EXPECTED PAYOFF

With event probabilities assigned, the fourth and last step of statistical decision theory is to analyze these probabilities by calculating *expected payoff* (EP) or *expected monetary value* for each course of action. The decision criterion here is to choose as the *optimal act*, OA, the act that yields the highest EP.

As the reader should know from our earlier definition of the expectation of a random variable, the EP of a given act is the sum of products of payoffs for each event and their corresponding probabilities of occurrences. Table 21.2 contains calculations of EP for our illustrative example.

TABLE 21.2 EXPECTED PAYOFF TABLE

		Possible Act					
		A_1:		A_2:		A_3:	
Event: Demand	Probability	Payoff	Expected Value	Payoff	Expected Value	Payoff	Expected Value
18	0.2	\$90	\$18.00	\$80	\$16.00	\$70	\$14.00
19	0.7	90	63.00	95	66.50	85	59.50
20	0.1	90	9.00	95	9.50	100	10.00
Expected Payoff		EP(A_1) =	90.00	EP(A_2) =	92.00	EP(A_3) =	83.50

Thus, the EP of each act is as follows:

Possible Acts	EP
A_1 = produce 18 cakes	\$90.00
A_2 = produce 19 cakes	92.00
A_3 = produce 20 cakes	83.50

From this we see that A_2, produce 19 cakes, is the *optimal act* for the baker because it has the highest expected payoff.

EXPECTED OPPORTUNITY LOSS

An alternative decision criterion of statistical decision theory is what is called *expected opportunity loss*, EOL. We shall see that this criterion leads to the same conclusion as EP.

First, we note that EOL is defined as the amount of payoff foregone by not taking the course of action which will give the highest payoff for each possible event. Opportunity loss for our illustrative problem is the amount of profit lost by not producing the number of units which will give the highest profit for each demand level.

Thus, for example, if the demand is 20 units, then the *optimal act* is to produce 20 units and obtain a profit of \$100. However, if the demand is 20 and the decision is to produce 19 units, the profit is \$95. This act entails an opportunity loss of \$5 (= \$100 − \$95) due to failure to take the best action, that is, produce 20 units given a demand for 20 units. Again, if the demand is 20 and the baker acts by producing 18 units, then this would be a conditional opportunity loss of \$10 (=\$100 − \$90). With similar calculations, we obtain the *conditional opportunity losses* (COL) for our example given in Table 21.3.

Note that the COL of the optimal act is zero; COL of any act other than the OA is positive and is the difference between the payoffs of the OA and the act taken.

Next, it follows logically that if conditional opportunity loss is used as the decision criterion, then the best act is the one which minimizes expected opportunity losses. Calculations of EOL are the same as those for EP, except COL are

TABLE 21.3 CONDITIONAL OPPORTUNITY LOSSES

	OA For Each	Profit of	Conditional Opportunity loss (COL)		
Event: D	Event	OA	A_1	A_2	A_3
18	18	\$ 90	\$ 0	\$10	\$20
19	19	95	5	0	10
20	20	100	10	5	0

TABLE 21.4 EXPECTED OPPORTUNITY LOSSES

		Alternative Act					
		A_1: 18		A_2: 19		A_3: 20	
Event: Demand	Probability	OL	EOL	OL	EOL	OL	EOL
18	0.2	0	\$0.00	10	\$2.00	20	\$4.00
19	0.7	5	3.50	0	0.00	10	7.00
20	0.1	10	1.00	5	0.50	0	0.00
Totals	1.0	—	4.50	—	2.50	—	11.00

used instead of payoffs. Table 21.4 contains computations of EOL for our illustration.

Thus, the optimum act is again A_2, which has the lowest EOL, \$2.50.

VALUES OF PERFECT INFORMATION

If we assume that the payoffs and probabilities shown in Table 21.2 represent a frequently recurring problem, what is the expected value of one occurrence if the decision maker, the baker, knows each time which of the three possible outcomes (weekly demand) is going to occur? How much is such added information worth?

When the baker knows that next week's demand will be 18 units, he takes the action of producing 18 units and obtains a profit of \$90. This occurs 20 percent of the time, and the expected profit is \$18. When the baker knows that next week's demand will be 19 units, he produces 19 units and obtains a profit of \$95. This occurs 70 percent of the time, yielding an expectation of \$66.50. Finally, when he knows the weekly demand will be 20 units, he produces 20 units and gains a profit of \$100. This occurs 10 percent of the time and results in an expected profit of \$10. Hence, if the baker can in some way find a "perfect predictor," whereby each week's demand can be known in advance and the optimal number of cakes can be produced for each week, then the average value of each such occurrence, called the *expected payoff of perfect information* (EPPI), is \$18 + \$66.50 + \$10 = \$94.50.

The EPPI calculated above is the highest expected profit the baker can make if perfect information is available. The value of such information is the difference between EPPI and the EP of the optimal act when there is no such information. Thus, for our example, the value of information is \$94.50 − \$92.00 = \$2.50.

The value of information represents the maximum amount of money which a decision maker could spend to obtain additional information—a perfect predictor. It is also interesting to note that this value is always equal to the EOL of the optimal act. This result should not be surprising, since perfect prediction should reduce the opportunity loss due to uncertainty to zero. This property also provides a check on computations because, for any act, the identity EP + EOL = EPPI must hold.

21.4

The Bayesian Decision Rule

The Bayesian decision rule is but an extension of the statistical decision theory just introduced. In contrast to the latter, the Bayesian approach selects the optimal act for solving a problem by using as its criterion the expected payoffs calculated with posterior probabilities. We shall begin our discussion on this decision rule by way of an example.

DECISION STRATEGIES

Suppose that a firm is completing the research and development phases of a new product. A decision must be made now as to whether or not the new product should be commercialized.

It has been estimated that the production of this new product requires a total fixed cost of \$45 million and a unit variable cost of \$1.50, and that a selling price of \$2.50 may be charged. Furthermore, the marketing research department of this firm has estimated that the possible annual sales volume—that is, states of nature together with their respective probabilities—is as follows:

State of Nature	Probability
H_1: Sales of 20 million units	0.3
H_2: Sales of 15 million units	0.5
H_3: Sales of 2 million units	0.2

With the above information, there are two strategies open to the management:

Strategy I: Make the decision as to whether to introduce the new product on the basis of prior information only.

Strategy II: Conduct a market operations research and make the decision as to whether to introduce the new product on the basis of both prior information and information obtained from research.

What we must do now is to determine which strategy the management should adopt; namely, we must evaluate the relative merits of these two strategies in terms of expected payoffs before research and expected payoffs after research. The evaluation of Strategy I is called *prior analysis*, and that of Strategy II is called *posterior analysis*.

PRIOR ANALYSIS

To evaluate the expected payoff before research, let us assume, for the sake of simplicity, that the fixed investment has a life span of five years and that a total of \$9 million is charged to each year's operation. Under these assumptions, the payoff of each state of nature resulting from the action of introducing the product,

TABLE 21.5 PAYOFF TABLE IN MILLIONS OF DOLLARS

	State of Nature		
Alternatives	H_1	H_2	H_3
A_1	\$11	\$6	−\$7
A_2	0	0	0

A_1, is the difference between total revenue from anticipated sales (units multiplied by price) and total costs (the sum of annual total fixed costs and total variable costs, the latter being the product of sales volume in units times unit variable costs). For example, if it is decided to introduce the new product, and if H_2 = sales of 15 million units is the true state of nature, then the payoff is

$$(15 \text{ million units} \times \$2.50) - (\$9 \text{ million} + (15 \text{ million units} \times \$1.50)) = \$6 \text{ million}$$

If, however, it is decided to take A_2—not to introduce the new product—the payoff is zero irrespective of the state of nature. Results of payoff calculations for this problem are summarized in Table 21.5, where A_1 stands for the action of introducing the new product and A_2 stands for the action of not introducing the new product.

Recall that prior probabilities attached to these states of nature are 0.3 for H_1, 0.5 for H_2, and 0.2 for H_3, respectively, so that the expected payoff for A_1 (introducing the new product before research) is

$$(0.3)(\$11) + (0.5)(\$6) + (0.2)(-\$7) = \$4.9 \text{ million}$$

The expected value for A_2 is, of course, zero.

The first step in posterior analysis is to determine the value of information. This task would be highly simplified if we could assume that the additional information obtained from marketing research is perfect, since then we could identify with certainty the optimal act under Strategy II. In other words, if research results were capable of aiding us to conclude with certainty whether A_1 or A_2 is the optimal act, then the expected payoff would simply be

$$(0.3)(\$11\text{M}) + (0.5)(\$6\text{M}) + (0.2)(0) = \$6.3 \text{ million}$$

Thus, the value of perfect information is \$6.3 − \$4.9 = \$1.4 million. If we were assured of obtaining perfect information, we would then be well advised to spend up to \$1.4 million to get it by conducting marketing research.

However, marketing research seldom, if ever, yields perfect information. There is always a certain degree of uncertainty in the selection of the optimal act even in the presence of research results. Granting that the information obtained is imperfect, how are we then to determine expected payoff after research?

PREPOSTERIOR ANALYSIS

As a matter of fact, a marketing operation test is often expensive. Consequently, the Bayesian approach to decision making suggests that a *preposterior analysis* be made in evaluating a strategy that involves a choice of conducting or not conducting research before a terminal decision is made.

The preposterior analysis begins with a natural query: If research were conducted, what relevant outcomes could we expect to obtain, given the possible relevant states of nature in the prior analysis?

The states of nature considered, for example, in the new product illustration are H_1 = an annual sale of 20 million units; H_2 = an annual sale of 15 million units; and H_3 = an annual sale of 2 million units. Suppose that the possible outcomes of marketing research which are relevant to the evaluation of the true state of nature are as follows:

E_1: Research result indicates an annual sale of 15 million units or more
E_2: Research result indicates an annual sale of between 5 and 15 million units
E_3: Research result indicates an annual sale less than 5 million units

Having hypothesized these possible relevant outcomes, we would evidently be interested in knowing the likelihoods of these outcomes; that is, the conditional probability of a specific outcome given as specific state of nature, $P(E_j|H_i)$. After thorough consideration, suppose we conclude that the weights shown in Table 21.6 can be assigned as the conditional probabilities. Note, that the conditional probabilities assigned to all possible outcomes (E_j), given a specific state of nature as the true state, add up to unity. Also, the sum of likelihoods of a specific outcome, given each and every state of nature, is unity in our example, but this need not be a requirement.

With likelihoods assigned, the second step, as a prerequisite to obtaining posterior probabilities, is to calculate the compound or joint probability that a given state of nature is the true state and that it is identified as such by a given outcome of research. The compound probability of each state of nature and each research outcome is given in Table 21.7. We note, for example, that the probability of H_3 being the true state and being identified by outcome E_2 is

$$P(H_3 \cap E_2) = P(H_3)P(E_2|H_3) = (0.2)(0.1) = 0.02$$

TABLE 21.6 ASSIGNMENTS OF PROBABILITIES TO POSSIBLE OUTCOMES FROM MARKETING RESEARCH, GIVEN H_i AS THE TRUE STATE OF NATURE

States of Nature	Possible Research Outcomes: E_j		
H_i	E_1	E_2	E_3
H_1	0.5	0.4	0.1
H_2	0.4	0.5	0.1
H_3	0.1	0.1	0.8

TABLE 21.7 JOINT PROBABILITIES OF STATE OF NATURE AND RESEARCH OUTCOMES

	Joint Probability			Total
State of Nature	$P(H_i \cap E_1)$	$P(H_i \cap E_2)$	$P(H_i \cap E_3)$	$P(H_i)$
H_1	0.15	0.12	0.03	0.30
H_2	0.20	0.25	0.05	0.50
H_3	0.02	0.02	0.16	0.20
Total: $P(E_j)$	0.37	0.39	0.24	1.00

Several observations about Table 21.7 are of interest. First, probabilities $P(H_i)$ and $P(E_j)$ are called marginal probabilities for the obvious reason that they are located at the margins of the joint probability table. Second, the total of each row is the marginal probability of that state of nature being the true state and is identical with the prior probability initially assigned. Third, the total of each column is the marginal probability of obtaining each specific outcome from marketing research given all alternative states of nature. Given the three states of nature (original hypotheses in Bayes' theorem), for example, the probability of obtaining outcome E_1 is 0.37.

Now, as the third step, we incorporate the additional information that could have been obtained from research into our analysis to revise the prior probabilities—that is, to calculate the posterior probabilities that a given state of nature is the true state, given a specific outcome of research. These probabilities are entered in Table 21.8. Note that each entry is a direct result of applying the formula for Bayes' theorem. For example, the probability that the state of nature is H_2, given outcome E_1, is

$$P(H_2|E_1) = \frac{P(H_2 \cap E_1)}{P(E_1)} = \frac{0.20}{0.37} = 0.541$$

The fourth step in the preposterior analysis is to calculate the *conditional terminal expected payoffs* of each action, given a specific result. Each expected payoff is obtained by using the revised or posterior probabilities. Again, as before

TABLE 21.8 POSTERIOR PROBABILITIES OF STATES OF NATURE AND RESEARCH OUTCOMES

	Posterior Probabilities		
States of Nature	$P(H_i|E_1)$	$P(H_i|E_2)$	$P(H_i|E_3)$
H_1	0.405	0.308	0.125
H_2	0.541	0.641	0.208
H_3	0.054	0.051	0.667
Total	1.000	1.000	1.000

possible research, the conditional expected terminal payoff is zero for A_2 irrespective of the research outcome. The conditional expected terminal payoff for A_1, introducing the new product, is calculated as follows.

1. Given E_1 – an annual sale of 15 million units or more is

$$(0.405)(\$11) + (0.541)(\$6) + (0.054)(-\$7) = \$7.323 \text{ million}$$

2. Given E_2 – an annual sale of between 5 and 15 million units is

$$(0.308)(\$11) + (0.641)(\$6) + (0.051)(-\$7) = \$6.877 \text{ million}$$

3. Given E_3 – an annual sale of less than 5 million units is

$$(0.125)(\$11) + (0.208)(\$6) + (0.667)(-\$7) = -\$2.046 \text{ million}$$

Thus, if research outcome is either E_1 or E_2, the optimal act is A_1: introduce the new product. If research result indicates E_3, the optimal act is A_2: do not introduce the new product, since the conditional expected terminal payoff for A_1 is negative.

With conditional terminal expected payoffs for each action determined, the last step is to compute the *unconditional terminal,* or *over-all expected payoff* after research. This expected value is obtained by multiplying each of the conditional expected payoffs by its marginal probability of occurrence. For our example, we have $(0.37)(\$7.323) + (0.39)(\$6.877) + (0.24)(0) = \$5.39154$ million. Hence, if research were conducted, the expected payoff might be as large as \$5.39154 million in contrast to the expected payoff of \$4.9 million before research. Note, also, that the last term in the calculation is (0.24)(0) because we have already decided that if E_3 results, A_2 is the optimal act.

NET EXPECTED PAYOFF OF RESEARCH

Earlier, the value of perfect information was calculated to be \$1.4 million. It was pointed out that perfect information is a rare commodity, and very often a decision maker must act with partial ignorance. Thus, we define *value of imperfect additional information* as the difference between the best expected payoff of the optimal act before research and that of the optimal act after research. We know that the best act before research is to introduce the new product and that its expected payoff is \$4.9 million. The best act, given possible research, is to introduce the product under outcomes of E_1 and E_2, but not to introduce the new product under outcome E_3. The expected payoff after research is \$5.39154 million. The expected value of additional information is therefore \$0.49154 million, or \$491,540—nearly half a million dollars.

Now, the Bayesian decision rule would be this: Collect additional information for those problem situations where the *net expected value of research* is positive. The net expected value of information is simply the difference between expected value of additional information and the expected cost of conducting research. Thus, if we assume that the expected cost is \$200 thousand in our example, then the net expected payoff of research is \$491,540 – \$200,000 = \$291,540. Clearly, additional information should be obtained before management makes a decision as to whether or not the new product should be introduced.

21.5

Bayesian Inference: Posterior Analysis

We have just demonstrated that when research is expensive, the Bayesian decision rule suggests that a preposterior analysis should be made in order to decide whether additional information should be collected in reaching the final decision. When it is concluded that research should be conducted, how is additional information to be incorporated into the Bayesian decision framework? Or, if sampling is relatively inexpensive and, therefore, a preposterior analysis is unnecessary, how is the Bayesian posterior analysis to be carried out? The answer here is actually a simple one. The posterior analysis is made in exactly the same manner as the preposterior analysis, except that posterior probabilities are now computed with the actual outcomes of sampling instead of with hypothetical values. Furthermore, in the Bayesian posterior analysis, just as in the classical decision procedures, we are also concerned with the optimal sample size in making a terminal decision.

THE ILLUSTRATIVE PROBLEM

Mr. Smith, the owner of a city-wide express and van company in Chicago, is offered the five trucks of another line that is going out of business in exchange for a 25 percent interest in his company. Available to him are the following two acts:

A_1: Accept the offer
A_2: Reject the offer outright

PRIOR ANALYSIS: NO-SAMPLE CASE

For this problem, if A_1 is taken, the key determinant of Mr. Smith's payoff is the number of trucks which may need major reconditioning. Mr. Smith does not possess any reliable information on the condition of the five trucks, but he is able to assign prior probabilities and estimate his payoffs, under different possible states of nature, in accordance with his experience and judgment as shown in Table 21.9.

If Mr. Smith is to make a terminal decision on the basis of prior probabilities alone, his optimal act is A_1, since $\text{EP}(A_2) = 0$, while

$$\begin{aligned}\text{EP}(A_1) &= (10{,}000)(0.35) + (6{,}000)(0.15) + (2{,}000)(0.10) + (0)(0.05) \\ &\quad + (-2{,}000)(0.10) + (-5{,}000)(0.25) = \$3{,}150.00\end{aligned}$$

For convenience, we shall call the prior analysis the no-sample case; that is, $n = 0$.

TABLE 21.9

Number of trucks requiring major reconditioning	Prior Probability	Payoff of Act	
H_i	$P(H_i)$	A_1	A_2
0	0.35	$10,000	0
1	0.15	6,000	0
2	0.10	2,000	0
3	0.05	0,000	0
4	0.10	−2,000	0
5	0.25	−5,000	0

POSTERIOR ANALYSIS

Now, suppose that Mr. Smith is able to make a sample inspection of the trucks before his terminal decision. Furthermore, if we assume that he arbitrarily decides to inspect three trucks at random, then there would be the following four possible sampling results:

E_0: No truck in the sample needs major reconditioning
E_1: One truck in the sample needs major reconditioning
E_2: Two trucks in the sample need major reconditioning
E_3: Three trucks in the sample need major reconditioning.

Given the possible results of observation, use of the complete Bayesian decision rule involves, as in the case of preposterior analysis, the following three steps:

(1) Revise the prior probabilities, or compute the posterior probabilities, on the basis of each of the possible sample outcomes.
(2) Determine the conditional terminal expected payoffs of each act, given a specific result.
(3) Determine the unconditional terminal expected payoffs.

The posterior probabilities are computed with the Bayesian theorem,

$$P(H_i|E_j) = \frac{P(E_j \cap H_i)}{P(E_j)} = \frac{P(E_j|H_i)P(H_i)}{\Sigma P(E_j|H_i)P(H_i)}$$

For example, if one truck in the conditional random sample (that is, selected without replacement) of three is found in need of major reconditioning, then the probability (likelihood) for E_1 to occur given that, say, H_1 is the true state of nature, is expressed by the hypergeometric law as

$$P(E_1|H_1) = \frac{\binom{k}{x}\binom{N-k}{n-x}}{\binom{N}{n}} = \frac{\binom{1}{1}\binom{5-1}{3-1}}{\binom{5}{3}} = \frac{(1)\dfrac{4!}{3!1!}}{\dfrac{5!}{3!2!}} = 0.4$$

TABLE 21.10

H_i	$P(H_i)$	$P(E_1 \mid H_i)$	$P(H_i \cap E_1)$	$P(H_i \mid E_1)$
0	0.35	$\binom{0}{1}\binom{5}{2}/\binom{5}{3} = 0.0$	0.000	0.0000
1	0.15	$\binom{1}{1}\binom{4}{2}/\binom{5}{3} = 0.4$	0.060	0.4444
2	0.10	$\binom{2}{1}\binom{3}{2}/\binom{5}{3} = 0.6$	0.060	0.4444
3	0.05	$\binom{3}{1}\binom{2}{2}/\binom{5}{3} = 0.3$	0.015	0.1111
4	0.10	$\binom{4}{1}\binom{1}{2}/\binom{5}{3} = 0.0$	0.000	0.0000
5	0.25	$\binom{5}{1}\binom{0}{2}/\binom{5}{3} = 0.0$	0.000	0.0000
	1.00		$P(E_1) = 0.135$	1.0000

Other likelihoods of E_1 given other states of nature, H_i, are computed in the same fashion and are entered in the third column of Table 21.10. Note that $P(E_1|H_0) = 0$, since we can never find a truck that needs major reconditioning if no truck in the population needs such service. Similarly, $P(E_1|H_5) = 0$, because if all the trucks in the population need major reconditioning, then we can never have one truck in the sample that needs no such service. With $P(H_i)$ and $P(E_1|H_i)$ given, $P(E_1)$ and $P(H_i|E_1)$ can be readily obtained. All these values are given in Table 21.10.

In terms of the preceding posterior probabilities, the conditional terminal expected payoffs for the particular sample outcome E_1 are

$$\begin{aligned}\text{CTEP}(A_1) &= (10{,}000)(0) + (6{,}000)(0.4444) + (2{,}000)(0.4444) \\ &\qquad + (0)(0.1111) + (-2{,}000)(0) + (-5{,}000)(0) \\ &= \$3{,}555.20 \\ \text{CTEP}(A_2) &= 0\end{aligned}$$

The conditional terminal expected payoffs for other sample outcomes, E_0, E_2, and E_3, are obtained in exactly the same way and are given in Table 21.11.

A glance at Table 21.11 indicates that A_1 is the optimal act if $X \leq 2$, and A_2 is the optimal act if $X = 3$, where X stands for the number of trucks in the sample which require major reconditioning. For this decision rule, the over-all terminal

TABLE 21.11

		CTEP	
E_j	$P(E_j)$	A_1	A_2
E_0	0.420	+\$9,238.00	0
E_1	0.135	+ 3,555.20	0
E_2	0.060	+ 1,000.00	0
E_3	0.300	−\$1,330.00	0

expected payoff for our example becomes

$$\begin{aligned}\text{OTEP} &= (9238.00)(0.42) + (3555.20)(0.135) + (1000.00)(0.060) + (0)(0.300) \\ &= \$4419.91\end{aligned}$$

THE OPTIMAL SAMPLE SIZE

The merit of the Bayesian decision rule is that it maximizes the terminal expected payoffs for a given set of prior probabilities and a particular sample size. Thus, while there exist other decision rules under the same conditions—such as "Take A_1 when there is no truck or one truck needing major reconditioning in a sample of three; otherwise take A_2"—none of these can yield a higher expected payoff than the Bayesian rule. Given the prior probabilities, however, the Bayesian decision rule would yield different terminal net payoffs for different sample sizes. Thus, the complete Bayesian decision framework also includes the determination of the optimal sample size that would lead to the highest possible terminal expected payoff.

To find the optimal sample size, we ascertain the Bayesian rule for each possible sample size and compute the net over-all terminal expected payoff (net of sampling costs). We have already worked out the unconditional terminal expected payoffs for $n = 0$ and $n = 3$. The expected payoffs, sampling costs, and net over-all terminal expected payoffs, NOTEP, for both these cases are summarized in Table 21.12, in which we have also given the corresponding results for $n = 1$ and $n = 2$. Calculations for these are not shown since they are exactly the same as for $n = 3$. Furthermore, we have assumed that the cost of sampling a truck, which includes thorough inspection of parts and a trial run of 50 miles, is \$100.

Results in Table 21.12 indicate that the optimal sample size is $n = 2$, since it yields the highest NOTEP. Now, one additional observation must be made before ending our discussion; that is, why have we stopped with $n = 3$? In other words, why do we not need to show the results for $n = 4$ and $n = N = 5$? The answer here clearly is related to our previous discussion on the value of information. We know that in a given decision problem, decision rules based on incomplete information cannot have expected payoffs higher than the expected payoff with perfect information, which in our case is \$4600.00. (Why?) Had we selected $n = 4$, the sampling costs would have been \$400.00, and then the net terminal expected payoff could not be expected to exceed \$4600.00 − \$400.00 = \$4200.00, which is less

TABLE 21.12

Sample size	Decision rule A_1	Decision rule A_2	OTEP	Sample cost	NOTEP
0	Always	Never	\$3150.00	0	\$3150.00
1	$X = 0$	$X = 1$	3917.54	\$100.00	3817.54
2	$X \leq 1$	$X = 2$	4499.41	200.00	4299.41
3	$X \leq 2$	$X = 3$	4419.91	300.00	4119.91

than the NOTEP for $n = 2$. Similarly, with $n = 5$, we can never expect to have an expected payoff greater than \$4600.00 − \$500.00 = \$4100.00, which is less the NOTEP for $n = 3$.

21.6 Utility Theory

Both the statistical decision theory and the Bayesian decision rule employ monetary value systems as the decision criterion. Such a criterion, however, may not be appropriate when the decision maker is also concerned with the potential expected loss that he cannot afford, together with the possible monetary gain that is uncertain.

Consider, for example, a small businessman with a total capital of, say, \$50,000. Available to him are two investment opportunities, A and B, both of which would require his entire capital to handle. Opportunity A offers him a 100 percent profit with a probability of $\frac{3}{5}$, and a loss of his total capital with a probability of $\frac{2}{5}$. The expected payoff of A is then

$$\text{EP(A)} = (\tfrac{3}{5})(\$100{,}000) + (\tfrac{2}{5})(-\$50{,}000) = \$40{,}000$$

Opportunity B offers him a much smaller but a certain profit of \$10,000. So, the expected payoff of B is

$$\text{EP(B)} = (1)(\$10{,}000) + (0)(\$50{,}000) = \$10{,}000$$

Clearly, if a monetary value system is employed, it would lead to the selection of Opportunity A as the optimal act. However, to our small businessman, A may not necessarily be the best act for him because this act involves the potential loss of his entire capital—a loss that could deprive him of making profits in the future. He may then actually prefer a certain, though smaller, profit of \$10,000 rather than a chance to make four times that amount with possible loss of his entire capital.

The fact that many people prefer a modest but certain gain to a windfall reward to which is attached a possible risk they perhaps cannot afford is the essence of the well-known St. Petersburg Paradox, and this fact inspired Daniel Bernoulli to suggest first that a utility scale rather than a money scale should be used as the decision criterion. Since Bernoulli's initiation, several procedures have been proposed to deal with a utility value system. We shall discuss, however, only one such procedure, that is, the Von Neumann–Morgenstern utility index.

THE N–M UTILITY INDEX

The N–M utility index is designed to predict which of two risky alternatives a person will prefer. If a person exhibits any degree of consistency in his preferences, it is possible to assign utility numbers to alternative outcomes of a risky proposition in accordance with an individual's ranking of the possible outcomes and the probability of each occurring.

A *utility number*, or *index*, represents, then, an individual's value scale, not a consensus, and must have as its foundation some psychological premises. Among them, according to Von Neumann and Morgenstern, are transitivity, desire for higher probability of success, and continuity of preference, properties. *Transitivity* refers to the assumption that if an individual is indifferent in choosing between two alternatives, A and B, then A and B must possess the same utility for him. Also, if he happens to be indifferent about the choice of B and C, then he must be indifferent to A and C.

Next, given two alternatives with identical desirability or reward, an individual would always *desire* the alternative with the *higher probability of success.*

Finally, the *continuity of preference* assumption refers to this psychological property: Let there be made available an arbitrary standard alternative, A, with a most desirable outcome to occur at a probability of α and a most undesirable outcome to occur at a probability $(1 - \alpha)$, and then compare A with any other alternative, say, B. If A is preferred to B when $\alpha = 1$, and if B is preferred to A when $\alpha = 0$, then there must be some in-between value of α, called the "indifference probability," at which an individual is indifferent to A and B.

According to the N–M framework, a decision is said to be made in accordance with the expected utility. *Expected utility* (EU) of a given risky alternative is defined as the sum of utility assignments to all possible outcomes weighted by appropriate probabilities. Thus, if an alternative A can lead to two outcomes, O_1 with probability α and O_2 with probability $(1 - \alpha)$, and if their respective utilities are $U(O_1)$ and $U(O_2)$, then the expected utility of A, denoted as EU(A), is

$$\text{EU}(A) = \alpha U(O_1) + (1 - \alpha)U(O_2) \tag{21.1}$$

EXPECTED UTILITY AS THE DECISION CRITERION

While the simple calculations implied by (21.1) are all there is to the N–M evaluation of the expected utility of an alternative, once the person's preferences of its outcomes are known, there still remains this question: How do we find the utility numbers for all possible outcomes of risky alternative propositions?

It is possible to generalize about an individual's utility function for any commodity, most often money, by the extension of equation (21.1) and the continuity preferences property. We shall present an example to show how a utility number can be assigned to any outcome in the form of money, and to show how expected utility can be employed as a decision criterion.

Consider three investment alternatives, A_1, A_2, and A_3, with different outcomes and probabilities as below:

$$\begin{aligned}
A_1\text{:}\quad & O_1 = \$100{,}000 \text{ profits with } P(O_1) = 0.5 \\
& O_2 = -\$60{,}000 \text{ loss with } P(O_2) = 0.5 \\
A_2\text{:}\quad & O_3 = \$50{,}000 \text{ profits with } P(O_3) = 0.5 \\
& O_4 = -\$20{,}000 \text{ loss with } P(O_4) = 0.5 \\
A_3\text{:}\quad & O_5 = \$10{,}000 \text{ profits with } P(O_5) = 0.5 \\
& O_6 = -\$1{,}000 \text{ loss with } P(O_6) = 0.5
\end{aligned}$$

First, according to the N–M decision framework, we would ask the decision maker to rank these outcomes according to his preferences, suppose that he gives us the following rather obvious ordering:

Rank	
1	$O_1 =$ \$100,000
2	$O_3 =$ 50,000
3	$O_5 =$ 10,000
4	$O_6 =$ −1,000
5	$O_4 =$ −20,000
6	$O_2 =$ −60,000

That is, he prefers O_1 to O_3 to O_5 to O_6 to O_4 to O_2.

To assign utility numbers to the outcomes in such a schedule of preferences, we begin with (by the continuity of preferences premise) a combination of the most preferred and the least preferred events. They are O_1 and O_2, respectively, in our example. Next, we assign two *arbitrary* utility numbers to them. Any two numbers can be used, as long as the number for the most preferred is greater than that for the least preferred outcome. Let us assign, say, a utility number of 5 to O_1 and −15 to O_2; then,

$$U(O_1) = U(\$100{,}000) = 5 \text{ utiles}$$
$$U(O_2) = U(-\$60{,}000) = -15 \text{ utiles}$$

Now, we move on to assign a utility to another outcome, say O_3. To do this, we ask the decision maker to imagine a selection between a sure gain of O_3 versus a one-time gamble associated with a gain of O_1 with probability α, and a gain of O_2 with probability $(1 - \alpha)$. In this connection, we see that if $\alpha = 1$, then he would prefer the gamble of O_1 to O_3, if $\alpha = 0$, rather than the gamble of O_3 to O_1. Why? Clearly, there must be some in-between value of α that would make the decision maker indifferent to a certain gain of O_3 and the gamble with O_1 and O_2 as the only possible outcomes. Once we have found this in-between probability value, then the utility value of O_3 is given by equation (21.1).

Suppose that the decision maker believes that such an "indifference" probability is $\alpha = 0.9$; then his utility index for $O_3 = \$50{,}000$, is

$$U(O_3) = \alpha U(O_1) + (1 - \alpha)U(O_2) = 0.9(5) + 0.1(-15) = 3.0 \text{ utiles}$$

We note that the utility of \$100,000 is much less than twice the utility of \$50,000.

So far we have indicated three points through which the decision maker's utility function passes. Additional utility numbers for other outcomes can be obtained similarly by using a combination of two outcomes, whose utility indexes are known, as the arbitrary standard risky alternative. Consider now the determination of $U(O_5)$. Let us pose this question: At what value of α for, say, O_3, would the decision maker be indifferent to the two alternatives of a sure reward of $O_5 = \$10{,}000$ and the risky combination of O_3 and O_2, with respective probabilities of α and $(1 - \alpha)$, to occur? Suppose he replies that α must be 0.89, then

$$U(O_5) = \alpha U(O_3) + (1 - \alpha)U(O_2) = 0.89(3) + (0.11)(-15) = 1.02 \text{ utiles}$$

Now, for the evaluation of O_6, consider the following equation:

$$U(-1{,}000) = \alpha(-\$60{,}000) + (1 - \alpha)(\$100{,}000)$$

or

$$U(O_6) = \alpha U(O_2) + (1 - \alpha)\, U(O_1)$$

Again, what is the "indifference" probability for O_6, given that the risk alternative is the combination of O_2 and O_1? If α is thought to be 0.26, then

$$U(O_6) = (0.26)(-15) + (0.74)(5) = -0.2 \text{ utiles}$$

There remains one more outcome, $O_4 = -\$20{,}000$, to be assigned a utility number. Let O_4 be evaluated with, say, O_3 and O_2; then we have

$$U(O_4) = \alpha U(O_3) + (1 - \alpha) U(O_2)$$

and if $\alpha = 0.65$, then

$$U(O_4) = 0.65(3) + 0.35(-15) = -3.3 \text{ utiles}$$

Needless to say, the utility of any sum of money between $-\$60{,}000$ and \$100,000 can be calculated in exactly the same way because the utility function is continuous. But can the utility of money greater than \$100,000 or less than $-\$60{,}000$ be calculated in our example? Why or why not?

Let us tabulate the utility function for money calculated before, as follows:

Expected Monetary Value	Utility
\$100,000	5.00
50,000	3.00
10,000	1.02
−1,000	−0.20
−20,000	−3.30
−60,000	−15.00

Returning to the three investment alternatives, we have the following expected utilities:

1. $\text{EU}(A_1) = (0.5)(5) + (0.5)(-15) = -5$ utiles
2. $\text{EU}(A_2) = (0.5)(3) + (0.5)(-3.3) = -0.15$ utiles
3. $\text{EU}(A_3) = (0.5)(1.02) + (0.5)(-0.2) = 0.41$ utiles

From these calculations, we see that investment alternative A_3 is the optimal act since it has the highest expected utility. This is not a surprising result, for the expected profit of A_3 is as large as 10 percent of A_1, and the expected loss of A_3 is as small as $\frac{1}{60}$ of that of A_1. Similar observations between A_3 and A_2 can also be made.

PROPERTIES OF UTILITY FUNCTIONS

Money is thought of as a desirable commodity by nearly everyone. Each individual prefers a larger to a smaller sum of money. The total utility of money then increases with the increase of the amount of money. However, in general, the total utility

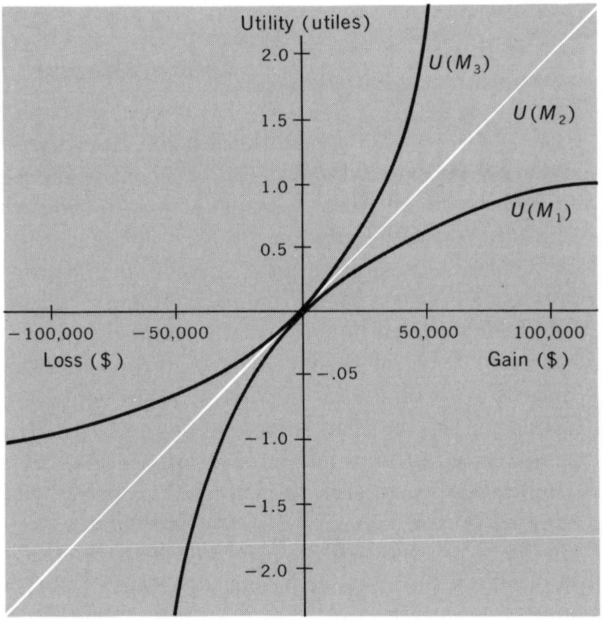

Figure 21.1 Possible utility functions.

of money increases with a decreasing rate of increase per unit increase of money. Namely, the utility function for money is curvilinear as shown by $U(M_1)$ in Figure 21.1.

If, for an individual, the increase in total utility is constant per unit increase in money, then the individual is said to have a linear utility function for money, as is shown by $U(M_2)$ in Figure 21.1. With a linear utility function, the decision criterion of maximizing expected utility also maximizes expected payoff. Thus, the utility theory leads to a different optimal act only if the decision maker's utility function for money is curvilinear.

It is also possible, though unusual, for a person's utility function to be such that his total utility increases with an increasing rate of increase, as shown by $U(M_3)$ in Figure 7.1. Imagine that the individual's happiness is more than doubled when his money is doubled, no matter how much money he has to start with!

On the average, however, it is reasonable to assume that an individual's utility function curve is concave upward, as shown by $U(M_1)$. At extremes, varying from person to person, the utility function is almost sure to approach upper and lower limits; that is, the utility curves flatten out as the monetary gains increase, and goes down sharply as losses mount up. Other things being equal, an individual always feels that his absolute utility index of gaining a certain amount of money is less than the absolute utility of losing the same amount of money. The greater the amount of money involved, the more vivid the sensation of this difference becomes. These observations are consistent with the neoclassical theory of diminishing marginal utility and the findings of psychologists in recent times.

21.7

Game Theory

Both decision and game theories deal with problems of choice involving risks. A decision theory is like a game theory in the sense that the decision maker plays a game of maximizing expected payoffs or utility against nature—possible outcomes of each alternative course of action. Thus, though both theories have quite a bit of apparatus in common, there is one important basic difference between them. In decision theory, nature is, so to speak, not an opponent. The player need not, therefore, count on nature to oppose him by doing anything special. In game theory, at least in the simple two-person zero-sum model (the only one we shall consider here), there is always some positive behavior of the second player that can somehow be predicted. He is determined to do everything possible within his command to oppose the first player; if there is any way at all to increase his payoff, he will surely employ it.

Another distinction, though obvious, between decision and game theories is also of interest. The former is relevant mainly to problems that would be internal to the firm or the household; the latter deals with relationships among independent competing individuals or organizations with conflicting interests.

Our discussion of game theory will be modest—presented merely to introduce the reader to some basic concepts of this fascinating field via the simplest model of game theory.

TWO-PERSON ZERO-SUM GAME

Two bus companies, A and B, serve the same route between two cities and are engaged in a struggle for a larger share of the market. Since the total market share is a fixed 100 percent, every percentage point gained by one must be lost by the other. Such a situation is said to be a *two-person zero-sum* game for the obvious reasons that the game is played by two diametrically opposed players and that the sum of gains and losses is always zero.

Let us now assume that both Company A and B are considering the same three strategies for gaining a larger relative share of the market, as follows:

(1) a_1 or b_1: Serve refreshments during the drive
(2) a_2 or b_2: Introduce air-conditioned buses
(3) a_3 or b_3: Advertise daily on TV stations in the two cities

For convenience, we shall assume that before the game commences, both lines are making no special effort and are sharing the market equally—50 percent each. Furthermore, we shall also assume that each line cannot employ more than one of these moves or strategies at the same time.

Under these assumptions, there are a total of $3 \times 3 = 9$ possible combinations of moves, and each is capable of affecting the market share in a specific way. For instance, if both A and B serve refreshments during the drive, let us say that A would lose 10 percent of the market share to B, which may indicate that B's refreshments are more to the customers' tastes. Also, if A advertises and B, say, serves refreshments, let us assume that A would gain 20 percent of the market from B; apparently, advertisement on TV seems to be more effective than serving refreshments.

Now, for each of the nine combinations we can determine market gains or losses for A as shown in the following table, called the *payoff matrix.*

		B's Strategy		
		b_1	b_2	b_3
A's Strategy	a_1	−10	−11	−1
	a_2	9	−8	−6
	a_3	20	−10	−13

Payoffs in our illustration are given in terms of percentage points of the market share. In practice, they can be stated in anything—a commodity, marbles, money, utility, or even life and death. Also, if the payoff is positive, it means that A has gained at the expense of B; if negative, it means that B has gained at the expense of A.

Now the question is this: Under the assumption that both A and B are acquainted with the information in the payoff matrix and that each is ignorant of the countermoves the other will make, what is the best strategy for A? for B?

MAXIMIN AND MINIMAX STRATEGIES

The conservative approach to the selection of the best strategy is to assume the worst and act accordingly. Thus, according to this approach and with reference to the payoff matrix, if A decides on strategy a_1, it would assume that B will select strategy b_2, thereby reducing A's payoff for a_1 to a *minimum* or *security value* of -11. Similarly the security values for a_2 and a_3 are -8 and -13, respectively.

Note that the security values for the various moves that A can make are the row minimums. Given these minimum values, A is well advised to employ that strategy which yields the maximum of these minimum security values. In our example, A should adopt a_2, aiming at a payoff of -8 to B. This decision rule, which leads to the selection of the largest of the lowest values in which each strategy can result, is called the *maximin strategy.*

Company B, according to this conservative attitude, would assume that for each of its actions, A's countermove will be such that A's market share gain is the highest possible. For example, if B employs strategy b_1, it would assume that A will take strategy a_3, which will give the worst possible loss to B. Similarly the worst payoffs for b_2 and b_3 are -8 and -1, the highest values in columns 2 and 3, respectively. Thus, we see that the maximum in each column is the worst payoff for a corresponding move made by B. The best of these worst payoffs is clearly the minimum value of these highest figures. This figure is -8 in column 2, corresponding to strategy b_2 and the countermove a_2. Therefore, the optimal choice, called B's *minimax strategy*, is b_2.

Note that according to both A's maximin rule and B's minimax rule the payoff is -8. This amount is called the *value of the game*. If the value of the game is positive, the game is said to favor A; if negative, it favors B; and if zero, the game is said to be equitable. The solution of our problem yields a payoff of -8, indicating that the game favors B since B gains 8 percent of the market at the expense of A.

SADDLE POINT

We have now reached a point where if A adopts maximin strategy a_2 then its payoff is exactly the same as what B expects A to have if B employs minimax strategy b_2. A reader may question the wisdom of such decision rules. For example, why does not A strive to gain 20 percent of the market share by employing a_3 instead of losing 8 percent to B by employing a_2? The answer is, if A did so, B might take b_2 or b_3 so that A might lose 10 or 13 percent of the market to B instead of losing only 8 percent. Similarly, one may argue that B should adopt b_3 in order to aim at a gain of 13 percent of the market share from A. However, this payoff is possible only if A makes the moves of a_3. Otherwise, B's gain could be smaller than 8. Similar arguments based on "caution" dictate that a_2 and b_2 are the best strategies for A and B, respectively, because this combination offers both A and B a measure of security. This is so because A's maximin decision criterion gives A the "largest" share of market, which B can be prevented from reducing any further, and B's minimax rule offers B the "lowest" share of market, which A can be stopped from increasing any further.

In other words, maximin and minimax strategies lead the two players of the game to situations in which neither player has any reason or incentive to change his position. A does not want to change because when B plays b_2, he is better off playing a_2 than either a_1 or a_3. B does not want to change because when A plays a_2, he is better off playing b_2 than either b_1 or b_3. We have apparently reached an equilibrium situation.

The payoff at such an equilibrium point is the minimax solution and is known as the *saddle point* of the payoff matrix in the sense that it is the minimum of its row and maximum of its column. Consider the solution of the pair of decisions in our example a_2 and b_2. When A adopts a_2, the payoff decreases from 9 to -8 and then increases from -8 to -6. When B selects b_2, its payoff decreases from -11 to -8 and then increases from -8 to -10. The number -8 in the middle forms a trough when it is viewed from the second row, and it forms a ridge when looked at from the second column. The minimax solution looks exactly like a saddle; hence the name "saddle point," which is at once a minimum, as a trough, and a maximum, as a ridge.

It is possible that there may be more than one saddle point in the payoff matrix of a game. If so, the payoffs corresponding to the saddle points must be equal, and it makes no difference which saddle point is employed to determine optimal moves for the two players. Consider, for example, the following game:

		B's Strategy b_1	b_2	b_3	Row Minimum
A's Strategy	a_1	2	−3	7	−3
	a_2	5	5	6	5*
	a_3	1	4	−4	−4
Column Maximum		5*	5*	7	

Here, we have two saddle points: one corresponds to a_2 and b_1 and the other corresponds to a_2 and b_2. According to minimax criterion, Player A would take move a_2. When he does so, it makes no difference whether Player B employs strategy b_1 or b_2, since in each case B must pay A an amount of 5, say, utiles. Also, since the value of the game in this example is positive, the game is said to favor player A.

A two-person zero-sum game is said to be *strictly determined* if there exists a saddle point, since the saddle point is an accepted solution to the game of finding the best strategy for each of the two players.

DOMINATING STRATEGY

A player is said to possess a *dominating strategy* if a particular strategy is preferred to any other strategies available to him. It is possible for each of the two players to have a dominating strategy. The payoff matrix below illustrates this.

		B's Strategy b_1	b_2
A's Strategy	a_1	3	5
	a_2	−10	1

For such a game it would be stupid for A to adopt a_2, irrespective of B's countermoves, since A can lose as much as 10 while he may gain as little as 1; A therefore can gain more by employing a_1, which dominates a_2. Now, if we eliminate the second row, the move for B also becomes clear. Since B would rather lose 3 than 5, his strategy b_1 dominates his move b_2. Again the solution of this problem is a saddle point because the number 3 is the maximum of the row minimums and the minimum of the column maximums.

It is also possible that there exists no dominating strategy for either player.

Consider, for example, a game which is defined by the following payoff matrix:

		B's Strategy	
		b_1	b_2
A's Strategy	a_1	30	45
	a_2	60	25

This game is clearly to A's advantage since no matter what he does, B must pay him a certain amount. You may wonder why B would play such a game. This is actually not so strange. Imagine that A stands for a union and B stands for a management. During negotiation of a new labor contract, the management usually has to agree to pay the union a certain wage increase no matter what move it may take. Our intention here, however, is to use this game to show that there is no dominating strategy for each player, since A's actual gain from each of its strategies depends very much upon what B would do, and vice versa. Note that there is no saddle point in the payoff matrix. Under these conditions, how can each player select the best strategy for himself?

We may, to start with, employ the conservative approach to see what will happen. Clearly, if in our problem above, A adopts a_1, its minimum gain is 30; if it employs a_2, its minimum gain is 25. Thus, it would seem advisable for A to select a_1, in order to obtain 30—the bigger of the two minimum values. On the other hand, if B adopts a cautious approach and employs b_1, the worst that can happen to him is to lose 60; and if he chooses b_2, the worst that can happen to him is to lose 45. Thus, as a conservative move, he should select b_2 in order to minimize his maximum losses.

When both sides play the game according to the previous analysis, the value of the game is 45. While this is what B expects to lose, it is much more than A expects to gain. Thus, in the absence of a saddle point, a smooth solution is impossible by the pure minimax rule.

It is easy to see, also, that if A should in any way discover what B was about to do next, it could improve its payoff by outguessing B. Suppose, for example, that A observes that B tends to prefer b_1 two out of three times. Then A can aim at the big price of 60 two out of three times by adopting a_2. Needless to say B could also benefit from knowing the possible preferences of A. Clearly, regularity is a dangerous habit in game theory. A player must prevent his opponent from second-guessing his next move. This is achieved by employing a strategy such that even he himself does not know what move he is going to make next. This leads us to consider what is called the mixed strategy.

MIXED STRATEGY

A *mixed strategy* is a combination of two strategies that are picked at random, one at a time according to predetermined probabilities, in contrast to a pure strategy which involves no such chance elements. Let us now develop a mixed strategy for A in our last example.

Suppose that A plays a_1, with probability α; and plays a_2 with probability $1 - \alpha$. Then if B plays b_1, A's expected payoff can be computed with figures in column 1 of the payoff matrix, as follows:

$$1. \quad EP(A) = \alpha(30) + (1 - \alpha)(60) = 60 - 30\alpha$$

The expected payoff for A if B employs strategy b_2 can be derived from the values in column 2 of the payoff matrix:

$$2. \quad EP(A) = \alpha(45) + (1 - \alpha)(25) = 25 + 20\alpha$$

Note that both equations 1 and 2 are linear functions of α.

Now, we wish to determine a value of α such that the expected payoff for A is the same regardless of B's strategy. Such an α value can easily be found by equating equations 1 and 2. That is,

$$3. \quad 60 - 30\alpha = 25 + 20\alpha$$

and

$$\alpha = {}^{35}\!/_{50}$$

With α (the probability that A will employ a_1) determined, we may randomize the selection of strategies by putting 50 identical chips, 35 marked as strategy a_1 and 15 marked as strategy a_2, in a bag. Selecting a chip blindly from the bag for each move, A can expect, from equation 3, to gain

$$60 - 30({}^{35}\!/_{50}) = 25 + 20({}^{35}\!/_{50}) = 39$$

per play of the game. Note that the value is greater than 30, the lowest payoff A originally expected to gain by the unsatisfactory pure strategy.

We might point out that the expected winnings per game can also be calculated directly from the column values in the payoff matrix as follows:

$$30({}^{35}\!/_{50}) + 60({}^{15}\!/_{50}) = 39$$

or

$$45({}^{35}\!/_{50}) + 25({}^{15}\!/_{50}) = 39$$

as before.

Now, we wish to show whether the mixed strategy discussed above is the best move for A by way of a graph for equations 1 and 2 (see Figure 21.2). In this figure, the horizontal scale measures α, which ranges from 0 to 1. The vertical scale measures payoffs.

Note that the line for equation 1, $EP(A) = 60 - 30\alpha$, measures A's payoff when B selects b_1. When $\alpha = 0$, $EP(A) = 60$, and when $\alpha = 1$, $EP(A) = 30$.

Likewise, the line for equation 2, $EP(A) = 25 + 20\alpha$, measures the payoff for A when B employs b_2. Here, when $\alpha = 0$, $EP(A) = 25$; when $\alpha = 1$, $EP(A) = 45$.

We see that the two straight lines intersect at a point G, immediately above $\alpha = {}^{35}\!/_{50} = 0.7$, S, and A's expected payoff is 39. This point may be considered the graphic solution to the best strategy for A under the assumptions that A is faced with the most unfavorable situation and that B plays the game as well as possible. The logic is clearly indicated by the graph. As it can be inferred from the figure, at any point at which $\alpha < 0.7$, A's expected payoff would be less than

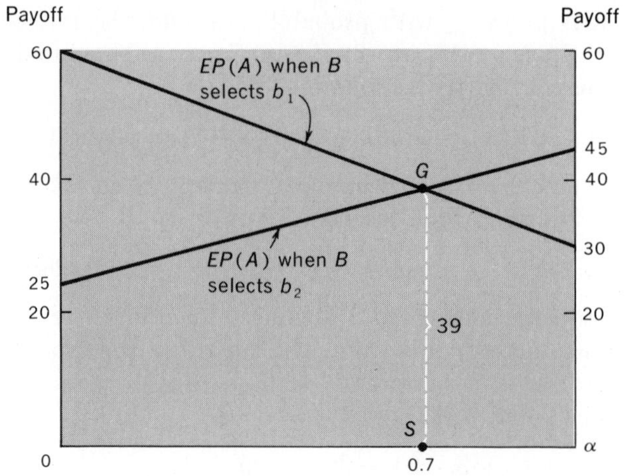

Figure 21.2 Graphic solution of the best move for A in a mixed strategy.

GS if B selects strategy b_2. Similarly, at any point where $\alpha > 0.7$, A's expected payoff would be smaller than *GS* if B employs strategy b_1. Thus, to avoid these possible unfavorable results, A should consider *S* (which corresponds to the intersection of the two lines) the best solution.

The mixed strategy for B can be established in the same manner. Let B play b_1, with a probability of β, and b_2, with $1 - \beta$; then his expected losses when A employs a_1 are

$$1'. \quad \text{EP(B)} = \beta(30) + (1 - \beta)45 = 45 - 15\beta$$

If A selects a_2, B's expected losses become

$$2'. \quad \text{EP(B)} = \beta(60) + (1 - \beta)25 = 25 + 35\beta$$

Equating $1'$ and $2'$,

$$3'. \quad 45 - 15\beta = 25 + 35\beta, \quad \text{we obtain} \quad \beta = 2/5 \text{ or } 0.4$$

After marking 2 chips b_1 and 3 chips b_2, B makes each of his moves by drawing a chip at random; then his expected losses, with reference to $3'$, are

$$45 - 15(0.4) = 25 + 35(0.4) = 39$$

per game. This is less than 45, the original expected loss by selecting a pure strategy b_2. Furthermore, B's expected losses now are equal to A's expected gain per randomized game. Thus we have, through adopting a mixed strategy, produced a value for the game which is satisfactory to both players.

In concluding, we note that one of Von Neumann's great contributions to game theory was the verification of the fact that every zero-sum game, irrespective of the number of moves available to the players, has a unique equilibrium value. As in the case where a saddle point exists, the solution may involve pure strategies. In other cases, as for some pairs of mixed strategies, an equilibrium can still be found.

Problems

21.1 What are the important aspects of a decision process?

21.2 What is the most difficult aspect of decision making? Why does this difficulty arise?

21.3 In what sense do we say that a decision process is predictive?

21.4 Uncertainty has been classified as cases of risks and cases of partial ignorance. What is the difference between these classifications?

21.5 What are the two decision criteria discussed in this chapter? In what way are they similar? How do they differ from each other?

21.6 Give a verbal statement, in your words, of the essence of each of the three decision theories introduced in this chapter.

21.7 Under what condition do the monetary and utility value systems lead to the same result in selecting the optimal act?

21.8 Give the similarities and differences between decision theories and the theory of games.

21.9 Define the following terms in your own words: (a) two-person zero-sum game; (b) payoff matrix; (c) maximin strategy; (d) minimax strategy; (e) saddle point; (f) dominating strategy; (g) mixed strategy.

21.10 Why is the two-person zero-sum game unrealistic? Support your answer with practical illustrations of possible economic and business games that you can think of.

21.11 A baker makes a certain kind of French pastry at night and sells it the next day. This pastry is perishable and must be thrown away if not sold during the day. According to past experience, the daily demand (sales) and respective probabilities are as follows:

D	0	1	2	3	4
Probability	0	0.1	0.3	0.4	0.2

Furthermore, the unit cost and the price of the pastry are \$1 and \$2, respectively. Compute the conditional and expected profits for each act and indicate the optimal act.

21.12 A certain output is manufactured at \$10 and sold at \$16 per unit. This product is such that if it is produced but not sold during a week's time it becomes worthless. The weekly sales records in the past are as follows:

Demand per Week	Number of Weeks Each Sales Level Was Recorded
10	100
11	200
12	600
13	100
	1,000

a. Calculate the expected sales of the month.
b. Prepare a table of payoffs for different possible acts.
c. Prepare a table of expected payoff and select the optimal act.

21.13 With data in the preceding problem, (a) prepare a table of conditional opportunity losses; (b) prepare a table of expected opportunity losses and indicate the optimal act.

21.14 Continuing with data in Problem 12, (a) present the payoff (conditional value) table under the assumption of certainty (that is, a perfect predicting device is available); (b) compute the expected value of perfect information.

21.15 Each day a grocery shop purchases for \$2 a product which it sells for \$4. Every unit sold costs the shop \$0.20 for wrapping. Because the product is perishable, units remaining unsold at the end of the day are returned, without wrapping, to the supplier for reprocessing, and a refund of \$1 per unit is received. The probability distribution of daily demand is estimated as below:

Number of Units Sold Daily D	Probability $P(D)$
0	0.10
1	0.40
2	0.30
3	0.10
4	0.10
	1.00

What is the optimum standing order the grocery shop should place each day?

21.16 If we assume that there are no wrapping costs and no refunds for the previous problem, what is the optimal act?

21.17 A newsstand at a certain subway station sells for 10 cents a copy a daily newspaper for which it pays 5 cents. Unsold newspapers are returned for a refund of 2 cents a copy. The daily sales and corresponding probabilities are as follows:

Daily Sales	Probability
1,000 copies	0.5
2,000 copies	0.4
4,000 copies	0.1

a. How many copies should it order each day?
b. If unsold copies cannot be returned, that is, unsold copies are useless, what is its optimal order each day?

21.18 A man owns a piece of land which can be used as parking lot. A garage operator offers the owner two alternatives for the use of his land: one, a rental-free arrangement based on profit sharing; two, an outright purchase of the land for $500,000. It has been estimated, furthermore, that the first alternative will yield a present value of future rentals of $800,000 if the venture is successful. If it fails, the present value of future rentals is only $200,000. What probability must be attached to "success" in order to make the owner indifferent toward the two alternatives if he is to decide on the basis of expected payoffs?

21.19 A bicycle manufacturer who has sold his product exclusively in the United States until now is contemplating introducing his product in the world market. According to his research staff's report the following possible states of nature—percentages of world market—and their corresponding probabilities, and the payoffs, are as follows.

(1) States of Nature and Probability:
a. H_1: 10% market, $P(H_1) = 0.3$
b. H_2: 5% market, $P(H_2) = 0.7$
c. H_3: 1% market, $P(H_3) = 0.2$

(2) Payoffs:
a. If H_1 is the true state of nature, payoff will be $20 million.
b. If H_2 is the true state of nature, payoff will be $5 million.
c. If H_3 is the true state of nature, payoff will be −$10 million.

Now, if the manufacturer decides to make the decision on this initial information,
a. What is his expected payoff for A_1, introducing his bicycles to the world market?
b. What is his expected payoff for A_2, not introducing his bicycles to the world market?
c. What is the expected payoff under perfect information?
d. What is the value of information?

21.20 In the preceding problem, assume that the bicycle manufacturer wants to make the terminal decision on the basis of both prior and additional information that could have been obtained by marketing research; and suppose that his research staff indicates that the likelihoods of survey results E_1 (relevant to H_1), E_2 (relevant to H_2), and E_3 (relevant to H_3) are

$$\begin{aligned} P(E_1|H_1) &= 0.6 & P(E_2|H_3) &= 0.2 \\ P(E_1|H_2) &= 0.3 & P(E_3|H_1) &= 0.1 \\ P(E_1|H_3) &= 0.1 & P(E_3|H_2) &= 0.2 \\ P(E_2|H_1) &= 0.3 & P(E_3|H_3) &= 0.7 \\ P(E_2|H_2) &= 0.5 \end{aligned}$$

a. Compute joint probabilities, $P(H_i \cap E_j)$.
b. Compute posterior probabilities, $P(H_i|E_j)$.
c. Compute the terminal expected payoff for each act, conditional upon the outcome of market research.
d. What is the optimal act for each possible survey result, E_j?
e. Compute the unconditional (over-all) expected payoff of this strategy.
f. What is the net expected payoff of research if the cost of research is $100,000?

21.21 An agent has agreed to receive, periodically, shipments of four large machines from one of the manufacturing firms he represents. The machines must be moved from the railroad yard upon arrival and installed in good working order in the customers' plants. If the agent accepts the responsibility for the shipment, A_1, he receives a flat fee of $600 each time from the manufacturer, but he must reset, at his own expense, any machine that is damaged during the shipment. The cost of resetting is about $250 per machine on the average. If the agent rejects responsibility for the shipment, A_2, he receives no fee and, of course, the manufacturer pays for the cost of resetting. According to the agent's experience, prior probabilities for the number of damaged machines per shipment may be assigned as follows:

Number of damaged machines:	0	1	2	3	4
Prior probability:	0.45	0.25	0.10	0.05	0.15

a. Determine the payoff functions for A_1 and A_2.
b. If the agent makes the final decision on prior probabilities alone, which is the optimal act? Why?
c. What is the agent's unconditional terminal expected payoff if he is to make the terminal decision on the basis of sample information with $n = 2$?
d. Assuming it costs $30 to inspect one machine, is $n = 2$ the optimal sample size? Verify.

21.22 An investor has a utility index of 1 for a gain of $50,000 and of 0 for a loss of $10,000. He says he is indifferent to a certain gain of $10,000 and a combination of a $50,000 gain, with a probability of 0.8, and a loss of $10,000. What is his utility index of a gain of $10,000?

21.23 For a small businessman $U(-\$1{,}000) = -5$ and $U(5{,}000) = 20$. If he is indifferent to $100 for certain and a chance of 0.4 at $5,000 profit, and an 0.6 chance at a loss of $1000, what is his utility index for $100?

21.24 For the management of a firm, $U(-\$100{,}000) = -100$ and $U(\$100{,}000) = 200$. It is indifferent to a profit of $50,000 for certain and a combination of $100,000 profit and $100,000 loss, with a probability of 0.7 for the former. Give its $U(\$50{,}000)$.

21.25 A college professor has a utility index of 10 for $10,000, of 7 for $5,000, and 0 for $10. What probability combination of $0 and $10,000 would make him indifferent to $5,000 for certain?

21.26 An enterpriser's utility function of money is as follows:

M	$50,000	−$500	0	$50,000
$U(M)$	−100	−3	0	300

For this enterpriser,
a. Should he insure his office building for a fire loss of $50,000 at an annual premium of $500, if the probability of fire on his property is 1/100?
b. Should he invest $500 in a highly speculative venture which may yield a profit of $50,000 with a probability of 0.01?

21.27 Assume that the utility function of a firm can be approximated by $U(M) = 2M - 0.1(M^2)$ where M is in millions of dollars. The firm is faced with the following alternative investment opportunities, both of which require the same amount of capital, as described below:

Alternatives	Outcomes	Payoff	Probability
A_1	O_1	$5	0.6
	O_2	$3	0.4
A_2	O_3	$8	0.4
	O_4	$2	0.6

Which is the best act for the purpose of maximizing expected utility?

21.28 For the previous problem, what is the optimal act, if the firm's utility function is $U(M) = 5 + 0.2M$? If A_2 is selected, can you explain why?

21.29 If the utility function of an individual is represented by the $U(M_1)$ curve in Figure 21.1, which is the optimal act for him with the following two ventures?

A_1: Receive $O_1 = \$50{,}000$ with $P(O_1) = 0.6$
Receive $O_2 = -\$30{,}000$ with $P(O_2) = 0.4$
A_2: Receive $O_3 = \$50{,}000$ with $P(O_3) = 0.9$
Receive $O_4 = -\$100$ with $P(O_3) = 0.1$

21.30 Each payoff matrix below represents a two-person zero-sum game, and the entries are made as payoffs from B to A:

(1)

	b_1	b_2
a_1	9	2
a_2	4	3

(2)

	b_1	b_2
a_1	1	−1
a_2	−1	1

(3)

	b_1	b_2
a_1	2	−2
a_2	−2	3

(4)

	b_1	b_2	b_3
a_1	0	−100	−350
a_2	100	0	−125
a_3	400	150	60

(5)

	b_1	b_2	b_3	b_4
a_1	10	1	7	8
a_2	3	4	5	10
a_3	6	7	4	2

For each payoff matrix determine if there is (a) a dominating strategy for A or B or for both; (b) a saddle point.

21.31 Suppose that the following payoff matrix contains utility gains by A from B:

	b_1	b_2
a_1	5	3
a_2	4	1

a. What strategy will A follow?
b. What strategy will B follow?
c. What is the value of the game?

21.32 In a coin-matching game, A wins $2 if both show tails, and wins $1 if both show heads. If the coins do not match, A loses $1 to B.
a. Present the payoff matrix for the game.
b. Is there a dominating strategy for either player?
c. If A and B follow mixed strategies, what is the value of the game?
d. What is the mixed strategy for A?
e. What is the mixed strategy for B?

21.33 Two toy stores, X and Y, are located on either corner of the same block. The owner of Store X knows that if neither store offers discounted prices, he can expect to have a net profit of $50 on any day. If he offers discounts while the other owner does not, his net profit will increase by $20 a day. If he does not offer discounts while the owner of Y does, his daily net profit will be reduced by $15. If both stores offer discounted prices, he can expect to have a daily net profit of $45. Both act independently.
a. Should the owner of X offer discounted prices, or vice versa, if he wants to maximize his minimum profits?
b. If the above daily net profits of various acts are also applicable to Y, how might the owners conclude that each could expect a net daily profit of $55?
c. Is this game zero-sum?

21.34 Let A be the union and B the management of a firm. During a collective bargaining session for a new contract, each side has two strategies, and payoffs from B to A are as follows:

	b_1	b_2
a_1	15	5
a_2	8	12

a. What is the maximum minimum gain that A can make sure of by following a pure strategy?
b. What is the minimum maximum loss that B can incur for sure by following a pure strategy?
c. If A and B follow mixed strategies, what is the value of the game?
d. What is the mixed strategy for A? for B?
e. Plot the mixed strategies for A and B.

22

QUEUING THEORY AND MONTE CARLO SIMULATION

22.1

Queuing Problems

A. K. Erlang's effort to analyze telephone traffic congestion, with the aim of finding a solution to the randomly arising demand for services provided by the Copenhagen automatic telephone system in 1909, produced a new theory that has become known as the *queuing*, or *waiting-line, theory*. This theory is one of the most valuable tools in management science because many managerial problems can be characterized as "arrival and departure" problems.

We are all too familiar with waiting. We have to wait for our "turn" at a barbershop, a post office, a restaurant, a bank, and so on before we receive attention. Similarly, broken-down machines have to wait to be repaired, ocean liners have to wait for harbor pilots to reach dock, planes must wait for the runways to be clear, or repair stations must wait for machines to break down.

In queuing problems, we often speak about *customers*—such as people waiting for clear telephone lines, machines waiting to be repaired, and planes waiting to land—and *service stations*—such as tables at a restaurant, repairmen in a shop, runways in an airfield, and so on. Each queuing problem involves a variable rate of arrival of customers requiring some type of service, and a variable rate of completing the service performed at the service station.

When we talk about waiting lines we may refer either to those created by customers or by the service stations. Customers may wait in a line simply because existing facilities are inadequate to meet the demand for service; in this case, the waiting line tends to be explosive. Service stations may be waiting because the existing facilities are excessive relative to customers' demand; in this case, service stations could remain idle much of the time. Customers may be waiting temporarily, even when service facilities are adequate, because previously arrived customers are being attended to. Service stations may encounter temporary waiting when, although facilities are adequate in the long run, there is an occasional shortage of demand due to happenstance. These last two cases typify a balanced situation constantly tending toward equilibrium, or a stable situation.

It is the aim of queuing theory to provide models that are capable of influencing the arrival patterns of customers, or to determine the most appropriate amount of service or the number of service stations. Now, we shall call a group of physical units, integrated in such a way that they can operate in unison with a series of organized operations, a *system*. Queuing theory seeks a solution to the problem of waiting by first predicting the behavior of the system of waiting. A solution to a waiting problem, however, is concerned not only with the minimization of the time the customers spend in the system but also with the minimization of costs for waiting time; that is, a waiting problem must be solved in terms of reducing the total costs of those who request, and those who render, the service.

The waiting-line theory offers two basic solutions to queuing problems: the analytical or mathematical approach, and the simulation or the Monte Carlo method. We shall introduce both in this chapter under some simplified conditions.

22.2 The Analytical Single Service-Station Queuing Model

To illustrate the queuing theory that employs mathematical models, we shall introduce first those models which satisfy the following conditions:

(1) *Arrival of customers,* or *inputs:* We shall assume that the arrivals occur at random and that the probability of an arrival at any time remains constant, regardless of what has happened previously and the length of the waiting line. The reader may recall that, under these assumptions, the arrivals would obey the Poisson probability law. Being a Poisson variable, the *mean arrival rate,* or the *average number of arrivals,* per unit of time is λ, which is independent of time. Furthermore, the reciprocal, $1/\lambda$, is the *mean length of time intervals between two successive arrivals.* In other words, we postulate a Poisson arrival pattern and an exponential pattern of interarrival time.

(2) *Queue discipline,* or *Priority rule:* When a customer arrives at the system, he usually has to wait before he receives service. His departure is influenced,

among other things, by the *queue discipline*, which is the established rule by which customers waiting in line are served. We shall, for the time being, assume that the rule is the time-honored one (not always the best) of *first come, first served*. Our rule also includes the requirement that no customer in the system will depart before he receives service.

(3) *Output:* This criterion refers to the number of service stations available, the capacity of each service station, and the distribution of service time. In the discussion that follows, we shall first assume that there is only one service station; that is, one repairman, one tunnel, one information clerk, and so on.

With respect to service time, we note that when the servicing of one arrival takes place between t and $t + \Delta t$, for very small Δt we would have an exponential distribution for the service time delivered and a Poisson distribution for the rate at which services are performed. The exponential postulate for service time holds for a majority of cases, even when the assumption is rather questionable that the probability that servicing will end at any instant is independent of what has happened previously. When the servicing time is exponentially distributed, the *mean service rate*, or the *expected number of services performed* per unit of time, and the standard deviation of servicing times are identical and may be denoted as μ. Note, also, that the reciprocal $1/\mu$ is the *mean servicing time*, or the *expected time per service*.

When the preceding conditions are satisfied, we have what is called "the mathematical model for single service-station, with first come, first served, queuing problems." Governing equations for this model can then be derived or established as given in the following section.

22.3 Equations Governing Queues for the Single Service-Station Model

There are a number of important properties that we must know in evaluating a queuing system. The first question to come to mind is whether or not the service station can handle the customers' demand. Obviously, the answer here can be given by comparing λ (the mean number of servicings per unit of time) with μ (the average number of services per unit of time). It is easy to see that if more customers arrive than can be serviced per unit of time, the waiting line will increase continuously and the system will break down. When $\lambda > \mu$, we are headed for trouble and drastic measures are called for without further analysis. Thus, in our system, we must have $\lambda < \mu$ in order to be functional. The ratio

$$U = \frac{\lambda}{\mu} \tag{22.1}$$

is called the *utilization parameter*, since it measures the degree of the capacity of the service station that is utilized. For example, if customers of a repair shop arrive at a rate of 7 per hour and receive service at an average rate of 10 per hour, then $U = 0.7$. It means that the repair shop is kept busy 70 percent of the time. The utilization parameter can also be interpreted as the *proportion of arrivals who must wait.* If the service station is kept busy 70 percent of the time when $U = 0.7$, then the probability that a customer must wait for service upon his random arrival must be 0.7.

The second question we may ask is this: What is the proportion of time that the service station is idle? The answer to this is very simple. If U is the proportion of time that the service station is kept busy, then the *proportion of time it remains idle* must be

$$I = P_0 = 1 - U \tag{22.2}$$

As I, the percentage of time the service station remains idle, is defined, it is clearly also the *probability that a customer,* upon his random arrival, *will not have wait* at all for service. This probability is denoted as P_0 in (22.2).

From (22.2), we can derive a formula for the *probability of having* n *customers in the system* as follows:

$$P_n = (1 - U)(U)^n \tag{22.3}$$

which includes those in the waiting line and the one being serviced. We sometimes also call P_n the *probability of* n *customers in the system,* for $n \geq 0$. Note, as has already been indicated, that this equation is valid only if $\lambda < \mu$.

In addition to $1/\lambda$ and $1/\mu$, there are various expected values of a waiting line, presenting different aspects, of importance for solving queuing problems. For our present model, they are

(1) *The mean number of customers in the system,* $E(n)$, both waiting and in service:

$$E(n) = \frac{\lambda/\mu}{1 - \lambda/\mu} = \frac{\lambda}{\mu - \lambda} \tag{22.4}$$

(2) *The mean length of the waiting line,* $E(L)$, excluding the customer in service:

$$E(L) = \left(\frac{\lambda}{\mu}\right)\left(\frac{\lambda}{\mu - \lambda}\right) = \frac{\lambda^2}{\mu(\mu - \lambda)} \tag{22.5}$$

(3) *The average waiting time of an arrival in the queue,* $E(W)$:

$$E(W) = \frac{\lambda}{\mu(\mu - \lambda)} \tag{22.6}$$

(4) *The average time an arrival spends in the system,* $E(T)$:

$$E(T) = \frac{1}{\mu - \lambda} \tag{22.7}$$

With these seven basic equations, many questions concerning waiting lines can be answered and solutions to many queuing problems can be obtained. Let us now turn to some illustrative applications of this simple queuing model.

22.4

Applications of the Single Service-Station Queuing Model

As the first illustration, let us consider the problem of determining the optimum number of phone booths. The owner of a restaurant observed that, in the long run, customers used his single phone booth every five minutes during rush hours and that the average length of conversation was four minutes. Suppose that the arrivals follow the Poisson pattern and that the service time is exponentially distributed; then we can outline the following relevant values of this system:

(1) The average number of arrivals and the average number of servicings, respectively, are

$$\lambda = 0.20 \text{ arrival per minute}$$
$$\mu = 0.25 \text{ servicing per minute}$$

(2) With the values of λ and μ determined, we see that the phone booth is kept busy

$$U = \frac{0.20}{0.25} = 0.80$$

or 80 percent of the time.

(3) The probability of forming a queue of length n with $n = 0, 1, 2, \cdots$ becomes

$$\begin{aligned} P_0 &= (1 - U)(U)^0 = (1 - 0.8)(0.8)^0 = 0.2 \\ P_1 &= (1 - 0.8)(0.8) = 0.16 \\ P_2 &= (1 - 0.8)(0.8)^2 = 0.128 \\ P_3 &= (0.2)(0.8)^3 = 0.1024 \\ &\cdots \end{aligned}$$

Thus, when a customer arrives at the booth he has a probability of 0.2 of making the phone call immediately, a probability of 0.16 of standing in the queue alone, a probability of 0.128 of being second in the waiting line, and so on. In general, the probability that a customer will be in the nth position decreases when n increases.

(4) The expected number of customers in the system is

$$E(n) = \frac{0.2}{0.25 - 0.2} = 4 \text{ persons}$$

In general, $E(n)$ varies directly with U, since, as it can be seen, $E(n) = U/(1 - U) = \lambda/(\mu - \lambda)$. When U approaches unity, $E(n)$ grows rapidly and approaches infinity.

(5) The average length of the queue is

$$E(L) = \frac{(0.2)^2}{0.25(0.25 - 0.2)} = 3.2 \text{ persons}$$

Note that $E(n)$ differs from $E(L)$ in that the former includes, but the latter does not include, the customer in service. We note also that $E(L) = E(n)U$.

(6) The average time a customer spends in the system is

$$E(T) = \frac{1}{0.25 - 0.2} = 20 \text{ minutes}$$

(7) The average waiting time a customer spends in the queue is

$$E(W) = \frac{0.2}{0.25(0.25 - 0.2)} = 16 \text{ minutes}$$

It follows from definitions that $E(W) = E(T)U$. We can also state, in view of our present illustration, that the average time a customer spends in the system is the sum of the average waiting time in the queue and the average time a customer spends in service.

What conclusions can we make from the foregoing calculations? We observe that the value of U is rather high and, as a consequence $E(n)$ and $E(T)$ are quite large. In order to reduce the traffic intensity and waiting time, the owner perhaps should consider the installation of a second telephone booth in the restaurant if space is available.

In our second illustration, a manager has to decide which of the two repairmen, X and Y, is to be hired. The frequency of machine repairs in his shop is known to follow the Poisson law, with an average of one breakdown per hour. The nonproductive cost of each idle machine is estimated as \$25 per hour. Repairman X asks for \$20, and Y asks for \$12, per hour respectively. It is also known that the average repair rates of X and Y are 2 and 1.2 machines per hour, respectively. The manager would like to know if it is wise for him to hire the more expensive but faster X.

To solve this problem, we only need to compare the total expected daily cost for both repairmen. This cost is clearly the total wage per day plus the cost of those machines waiting to be, and being, repaired. The expected nonproductive time per hour is the mean repair time multiplied by the expected number of machines in the system, both those being repaired and those in the queue. We note also that the mean number of machine arrivals is $\lambda = 1$ and that the service rate per hour is $\mu_x = 2$ for X, and $\mu_y = 1.2$ for Y. Thus, the expected numbers of machines in the system for X and Y, respectively, are

$$E(n) = \frac{1}{2 - 1} = 1 \qquad \text{for X}$$

and

$$E(n) = \frac{1}{1.2 - 1} = 5 \qquad \text{for Y}$$

Thus, if X is hired, the expected nonproductive machine time in an 8-hour day is $(1)(8) = 8$. The total daily cost in this case becomes

Labor costs + cost of nonproductive machine time

$$= (8)(\$20) + (8)(\$25) = \$360.00$$

If Y is hired, the nonproductive machine time per day is (5)(8) = 40, and the total daily cost becomes

$$(8)(\$12) + (40)(\$25) = \$1096.00$$

It is obvious that the cheaper repairman would cost almost three times as much as the more expensive repairman. The right decision is for the manager to hire X.

As our last example, let us consider a problem of investment decision. A manufacturing firm which produces a high-quality product has an average of 24 pieces of equipment to be repaired each week. The probability of the breakdown of a piece of equipment is approximately constant, so that the arrivals are distributed as under the Poisson law. The repair time is close to an exponential distribution. The opportunity cost of having an equipment breakdown is estimated as \$150 per day.

The firm has two repair facilities, A and B, from which to choose. Facility A requires \$20,000 to install, but Facility B costs \$60,000. Both facilities require an annual labor cost of \$8000 to operate. Furthermore, A can repair machines at a rate of 30 per week, and B at a rate of 60. Finally, both investments are estimated to have economic lives of four years. For the sake of simplicity, we shall ignore interest charges and returns to investment. Under these circumstances, which facility should the firm decide to purchase?

The solution to this problem can be obtained by comparing the total annual costs of the two facilities. For each facility, the total annual cost consists of three parts: (1) annual capital recovery cost, which is ¼ of the total investment expenditure, (2) annual labor cost, and (3) annual cost of lost production-equipment time. Component (3) may be thought to be based on the sum of time waiting to be repaired and time being repaired. The former is to be computed on the basis of $E(W)$—the mean waiting time of an arrival, and the latter is to be evaluated on the basis of the average service time per repair, which is the reciprocal of the mean service rate per unit of time, $1/\mu$.

Now, for Facility A, we have

(1) Annual capital recovery cost = \$5,000.00
(2) Annual labor cost = \$8000.00
(3) Cost of lost production-equipment time
  a. Time waiting to be repaired
    i. Average waiting time of an arrival:

$$E(W) = \frac{24}{30(30 - 24)} = 0.133333 \text{ week}$$

    ii. Total waiting time per year = (24)(52)(0.133333) = 166.4 weeks
  b. Time being repaired:

    i. Mean time per repair $= \dfrac{1}{\mu} = \dfrac{1}{30}$ week

ii. Total repair time per year = (24)(52)($\frac{1}{30}$) = 41.6 week
c. Total lost time per year = 166.4 + 41.6 = 208 weeks
d. Total cost of lost production-equipment time = (5)($150)(208) = $156,000.00

(4) Total annual cost for Facility A = 5000 + 8000 + 156,000 = $169,000.00

For Facility B, we have

(1) Annual capital recovery cost = $15,000.00
(2) Annual labor cost = $8000.00
(3) Cost of lost production-equipment time
a. Time waiting to be repaired

i. $$E(W) = \frac{24}{60(60 - 24)} = 0.011111 \text{ week}$$

ii. Total waiting time per year = (24)(52)(0.011111) = 13.87 weeks
b. Time being repaired
i. Mean time per repair = $\frac{1}{60}$ week
ii. Total repair time per year = (24)(52)($\frac{1}{60}$) = 20.8 weeks
c. Total lost time per year = 13.87 + 20.8 = 34.67 week
d. Total cost of lost production-equipment time = (5)($150)(34.67) = $26,002.50

(4) Total annual cost for Facility B = 15,000 + 8000 + 26,002.50 = $49,002.50

Thus, the more expensive investment proves to be less costly in the long run in this case. The firm should purchase Facility B in order to minimize total cost incidental to repair of production equipment.

22.5

A Note on Priority Rules

So far we have been discussing the single service-station queuing model under the assumption that customers are served in order of arrival. This rule is quite appropriate if "unfairness" will be resented, or if customers are of equal importance and require the same amount of service on the average. In many situations, however, there may be strong reasons for the practice of priority rules. For the sake of courtesy, women and children may be served first. Whenever there are customers for whom the waiting costs are high, for example, where waiting represents a "bottleneck," we may adopt the rule of "most important, first." There are also such rules as "random order" or "nearest calling unit, first." We shall be concerned, however, with this rule: Quickest returned to operation, first. Usually, servicing first the customer needing the least service minimizes the length of the waiting line. If the cost of waiting time for all units is the same, this rule also minimizes total costs of waiting and servicing.

To apply the rule of "quickest returned to operation, first," we may classify customers into two groups, referred to as *fast* and *slow customers*, each having an average servicing time. According to this rule, as soon as a customer is serviced, his place is taken by the customer who arrived first. When there are no fast customers, the place is taken by the slow customer who arrived first. When there are fast customers, the place is taken by the fast customer who arrived first. All fast customers are serviced in order of their arrivals before the slow customers are serviced. Under this priority rule, we have, in effect, two queues: one for the fast, and another for the slow, customers. Of course, once a slow customer is being serviced, he is not interrupted even though a fast customer may have arrived in the mean time.

A moment's thought would lead us to see why this rule would tend to reduce the expected waiting time. For example, if there are two customers, fast A and slow B, waiting to be serviced, and if A requires only 5 minutes and B requires 10 minutes, then B has to wait for only five minutes if A is serviced first, but A has to wait for ten minutes if B is serviced first. The saving in waiting time, when fast A is serviced first, is 10 − 5 = 5 minutes. Be sure to note, however, that there is no gain for slow B, who actually has to wait five minutes longer. The gain—ten minutes—is for fast A only. Also, the total time required to service both is fifteen minutes, regardless of who is served first. Namely, the service time is not affected by changes in the queuing discipline. Or, the utilization parameter remains the same whatever priority rule is employed.

The relative reduction of the average waiting time by our new queue discipline depends upon three factors:

(1) The utilization parameter.

(2) The ratio of the average servicing time of the fast customers to that of the slow customers. Let the average service time for fast customers be $\bar{s}_1$ and that for the slow customers be $\bar{s}_2$; then we may denote this ratio as $R = \bar{s}_1/\bar{s}_2$ which obviously ranges from 0 to 1.

(3) The fraction F of the total number of fast customers, f, to the total number of customers, n; namely, $F = f/n$. Again F ranges from 0 to 1.

With these three quantities determined, the percentage reduction of the expected waiting time, denoted as D, becomes

$$D = \frac{100F(1 - F)(1 - R)U}{1 - F + FR(1 - U)} \tag{22.8}$$

For example, if $U = 0.5$, $\bar{s}_1 = 5$, $\bar{s}_2 = 10$, and $F = 0.5$, then

$$D = \frac{100(0.5)(1 - 0.5)(1 - (5/10))(0.5)}{1 - 0.5 + (0.5)(5/10)(1 - 0.5)} = 10 \text{ percent}$$

Similarly, if $U = 0.95$, $\bar{s}_1 = 3$, $\bar{s}_2 = 10$, and $F = 0.8$, then

$$D = \frac{100(0.7)(1 - 0.7)(1 - 3/10)(0.95)}{1 - 0.7 + (0.7)(3/10)(1 - 0.95)} = 44.9 \text{ percent}$$

Generally, D increases with increases in F. However, when F approaches 1, the gain decreases. Also, other things being equal, the greater U is the greater

D is. This is a very important property, since when U is large, the queue tends to be very long. The reduction of the average waiting time is therefore of supreme importance in the case of great U. In many situations the reduction in average time has led to proportional reduction in costs.

22.6

The Analytical Multiple Service-Station Queuing Model

A multiple service-station queuing model exists when customers in a single waiting line can be serviced by more than one service station equally well. Here, all the k, $k \geq 2$, service stations have identical servicing capacity, and the queue is single in the sense that each customer, without any external pressure, decides on one station to serve him.

When there are k parallel stations and n customers, there will be no waiting line if $n \leq k$. When $n > k$ (the number of arrivals in the system, waiting and/or in service, being more than the number of stations) the system is said to be in state E_n. In such a state, a waiting line is created with $n - k$ as the length of the queue.

For this model, given the mean number of arrival rate, λ, and the mean service rate per unit of time for each station μ, the utilization parameter of the system is defined as

$$U_k = \frac{\lambda}{k\mu} \qquad \text{for } k\mu > \lambda \tag{22.9}$$

Again, if we let the queuing discipline be first come, first served, the *probability that there are n customers in the system*, either being serviced or waiting, becomes

$$P_n = \frac{1}{n!}\left(\frac{\lambda}{\mu}\right)^n P_0 \qquad \text{for } n < k \tag{22.10}$$

or,

$$P_n = \frac{1}{k!k^{n-k}}\left(\frac{\lambda}{\mu}\right)^n P_0 \qquad \text{for } n \geq k \tag{22.11}$$

In these two equations, P_0 stands for the *probability that there is no customer in the system* or the *probability that all the k service stations are idle*. It is defined, for $U_k < 1$ or for $k\mu > \lambda$, as

$$P_0 = \frac{1}{\left[\sum_{i=0}^{k-1} \frac{1}{i!}\left(\frac{\lambda}{\mu}\right)^i\right] + \left[\frac{1}{k!}\left(\frac{\lambda}{\mu}\right)^k \left(\frac{k\mu}{k\mu - \lambda}\right)\right]} \tag{22.12}$$

Now, the *probability that an arrival at random has to wait for service* is the same as the *probability that there are k or more customers in the system.* Denoting the number of customers between k and n as m, we have this probability:

$$P_m = \frac{\mu(\lambda/\mu)^k}{(k-1)!(k\mu - \lambda)} P_0 \tag{22.13}$$

If we multiply P_m by the ratio $\lambda/(k\mu - \lambda)$, we obtain the *mean length of the queue*, excluding customers being serviced, as

$$E(L) = \frac{\lambda\mu(\lambda/\mu)^k}{(k-1)!(k\mu - \lambda)^2} P_0 \tag{22.14}$$

The *mean number of customers in the system*, given $E(L)$, becomes

$$E(n) = \frac{\lambda}{\mu} + E(L) \tag{22.15}$$

The *average waiting time of an arrival in the queue* is:

$$E(W) = \frac{\mu(\lambda/\mu)^k}{(k-1)!(k\mu - \lambda)^2} P_0 \tag{22.16}$$

The *average time a customer spends in the system*, given $E(W)$, is defined as

$$E(T) = \frac{1}{\mu} + E(W) \tag{22.17}$$

Equations (22.9) through (22.17) govern the behaviors of waiting lines in the multiple service-station queuing model.

22.7 Applications of the Multiple Service-Station Queuing Model

A textile manufacturing firm has a large number of identical machines whose breakdown rate is estimated at 60 per day. It has now three service stations for repair, each having a mean service rate of 25 per day. For this system, answers to these questions constitute an evaluation:

(a) How many hours, for an 8-hour day, is a service station kept busy?

(b) What is the probability that all three stations are idle at any instant? two are idle? one is idle?

(c) What is the expected number of idle service stations from time to time? The mean idle time per station?

(d) What is the probability that a machine arrives and has to wait to be serviced?

(e) What is the mean length of the queue from time to time?
(f) What is the average waiting time of machine arrivals?
(g) What over-all conclusion can we reach about this system?

Answers are easily obtained with the equations governing multiple service-station models.

(a) Since for this system, $U_k = 60/3(25) = 0.8$, the expected number of hours a service station is kept busy each day becomes $8(0.8) = 6.4$ hours.

(b) The probability that all three stations are idle at any instant is the probability that there is no customer in the system. It is

$$P_0 = \frac{1}{\left[\sum_0^2 \frac{1}{i!}\left(\frac{60}{25}\right)^i\right] + \left[\frac{1}{3!}\left(\frac{60}{25}\right)^3\left(\frac{3(25)}{3(25) - 60}\right)\right]}$$

where

$$\sum_0^2 \frac{1}{i!}\left(\frac{60}{25}\right)^i = 1 + 2.4 + \frac{1}{2}(2.4)^2 = 6.28$$

and

$$\frac{1}{3!}\left(\frac{60}{25}\right)^3\left(\frac{3(25)}{3(25) - 60}\right) = \frac{1}{6}(2.4)^3(5) = 11.52$$

Therefore

$$P_0 = \frac{1}{6.28 + 11.52} = 0.0562$$

This indicates that about less than 6 percent of the time three or fewer breakdowns can be repaired immediately, without waiting in the queue.

The probability that two stations are idle is the probability that there is only one customer in the system. Therefore,

$$P_1 = \frac{1}{1!}\left(\frac{60}{25}\right)(0.0562) = 0.13488$$

The probability that one station is idle or that there are two customers in the system is

$$P_2 = \frac{1}{2!}\left(\frac{60}{25}\right)^2(0.0562) = 0.16186$$

(c) The expected number of idle stations can be computed in the usual way for expectations. Namely, denoting this quantity as $E(I)$, we have

$$\begin{aligned} E(I) &= kP_0 + (k-1)P_1 + (k-2)P_2 + \cdots + (1)P_{n-1} \\ &= 3(0.0562) + 2(0.13486) + 0.16186 = 0.6 \text{ station} \end{aligned}$$

The mean idle time per station is $E(I)/k = 0.6/3 = 0.2$, or 20 percent of the time. This checks with the fact that the average idle time per station is $1 - U_k = 1 - 0.8 = 0.2$.

(d) The probability that a machine arrival will have to wait is

$$P_m = \frac{25(60/25)^3}{(3-1)![3(25)-60]}(0.0562) = 0.65$$

(e) The mean length of the queue is

$$E(L) = \frac{(60)(25)(60/25)^3}{(3-1)![3(25)-60]^2}(0.0562) = 2.59 \text{ machines}$$

With this value given, we can also determine the mean number of machines in the system as $(60/25) + 2.59 = 4.99$ machines.

(f) The average waiting time of a machine which has arrived in the system is

$$E(W) = \frac{25(60/25)^3}{(3-1)![3(25)-60]^2}(0.0562) = 0.04316 \text{ day}$$

Given this, the average time a machine spends in the system is $E(T) = 1/25 + 0.04316 = 0.08316$ day.

(g) At first glance, this system seems quite satisfactory, since the mean number of idle service stations is only 0.6 station, or the average idle time per station is as small as 0.2 (or 20 percent of the time). In other words, the service stations are kept quite busy most of the time, and we seem to have got our money's worth. However, looking at the system from the point of view of waiting time, we notice these aspects: a machine waits to be repaired 65 percent of the time; the mean number of machines in the system is nearly as large as five; and the average time a machine spends in the system is greater than 0.04 of an 8-hour workday. These values are unusually high if the opportunity cost of lost production time is high relative to total daily operating costs of service stations.

Now, suppose that the lost production time is estimated at $200 per day per machine; then the total cost of lost production time becomes

$$\lambda(E(T))(100) = 60(0.08316)(200) = \$997.92$$

In addition, let us assume that the total operating cost of each station is $100 per day, then the total cost of this system is

$$3(100) + 997.20 = \$1297.92 \text{ per day}$$

At this point, the management might consider adding another service station similar to those in operation. This would be the right move to make, since with four identical service stations, we would have

$$U_k = \frac{60}{4(25)} = 0.6$$

which indicates that the proportion of time during which the service facilities are idle is $1 - 0.6 = 0.4$, and that the expected number of idle service stations is

$4(0.4) = 1.6$ stations. However, we would also have

$$P_0 = \frac{1}{\left[\sum_0^3 \frac{1}{i!}\left(\frac{60}{25}\right)^i\right] + \left[\frac{1}{4!}\left(\frac{60}{25}\right)^4\left(\frac{4(25)}{4(25) - 60}\right)\right]}$$

$$= \frac{1}{[(1 + 2.4 + (1/2)(2.4)^2 + (1/6)(2.4)^3] + [(1/24)(2.4)^4(100/40)]} = 0.0831$$

$$E(W) = \frac{25(60/25)^4}{(4-1)![4(25-60)]^2}(0.0831) = 0.008 \text{ day}$$

and

$$E(T) = 1/25 + 0.008 = 0.048 \text{ day}$$

The total cost with four service stations becomes, then,

$$4(100) + 60(0.048)(200) = \$976.00 \text{ per day}$$

Thus we see that with four stations, while the proportion of time of idle facilities, or the expected number of idle stations, has increased considerably, the average time a nonproductive machine spends in the system has been correspondingly reduced. The reduction of $E(T)$ from 0.08316 to 0.048 has made it possible, despite an increase of $E(I)$ from 0.2 to 0.4, to reduce total cost incidental to repair by

$$1{,}297.92 - 976.00 = \$321.92 \text{ per day}$$

This is clearly an economy of large-scale production in the operation!

22.8

Monte Carlo Simulation

We have now learned what the essence of queuing problems is, namely, that when customers need services at arbitrary moments, there can arise two unpleasant situations: either a customer has to wait to be served, or a service station is idle. Both circumstances cost money. The problem is to find the optimum size of service capacity, or the optimum number of service stations, that will minimize costs, given certain probabilities of arrivals and servicing time.

Solutions have been given to a number of waiting problems, mathematically, under the assumptions that arrivals follow the Poisson probability law and that the service time is exponentially distributed. With such a judiciously selected set of premises, we have presented computations that might have given the impression that queuing problems are quite simple. However, we must remind the reader that the appearance of simplicity is due to the fact that our assumptions were deliberately chosen to make calculations amenable to relatively simple algebraic manipulations. When models become more complex and realistic, the level of mathematics required becomes quite involved. It is even possible that sometimes a queuing problem may be so complicated that it frustrates a competent mathematician. Again, even when an analytical solution is possible, that solution may be objectionable—on the grounds that it is difficult or impossible to understand—to people who are in the position of making final decisions. In such situations, the best tactic is to run a "simulation" study of the problem. We shall call such a study the *Monte Carlo simulation* because simulation is a statistical method related to probabilities of chance.

Monte Carlo simulation is the friend of unsophisticated mathematicians. To understand and to use it, little mathematical training is needed. It can be easily adapted to any situation, provided that alternatives can be specified quantitatively and that data required can be predicted with acceptable confidence.

Monte Carlo is a process of solving a problem by simulating original data with random-number generators. Its application requires only two basic things. First, we must have a *model* that represents an image of reality as we see it. The model here is nothing more than the probability distribution of the variable under consideration. The significant merit of simulation is that it can be applied even when the probability distributions cannot be expressed explicitly in any of the theoretical forms which have been introduced in this text. All that is required is a table or a graph of a cumulative probability distribution of a variable, which may be obtained either by direct observation (such as an appropriate survey to determine the actual behavior of, say, arrivals) or, indirectly, by the use of past records.

The second thing needed is a *mechanism* to simulate the model. The mechanism can be any such random-number generator as a pair of dice, a spinning pointer, a roulette wheel, a table of random digits, or a high-speed computer properly instructed. In short, the Monte Carlo method is to simulate, via random devices, real-world situations that are probabilistic in nature.

The use of cumulative probability distributions and random-number generators enables us to obtain almost the same probability distribution that occurs, or would occur, in the real world. For example, suppose that, on the basis of past experience, the cumulative probability distribution of assembling time for a certain mechanism is as shown by Figure 22.1; then we may simulate this process by using, say, a table of random numbers. Each random number selected is considered a probability value, and we enter it on the vertical scale of our diagram. We then obtain the corresponding value of the variable, X, by drawing a line perpendicular to the horizontal axis, from the point where the curve meets the horizontal line that we used to indicate the probability given by the selected random number. If we proceed in this fashion for a large number of times, we would have many X values and their corresponding probabilities. If we organize these sample values into a cumulative distribution, we would expect it to represent the actual future behavior of the assembling process. It is easy to see that the representation, or prediction, of the real-world event—assembling time, becomes more accurate the larger the number of trials in our simulation. We also note that the larger the sample from which the original empirical distribution, Figure 22.1, has been secured, the more

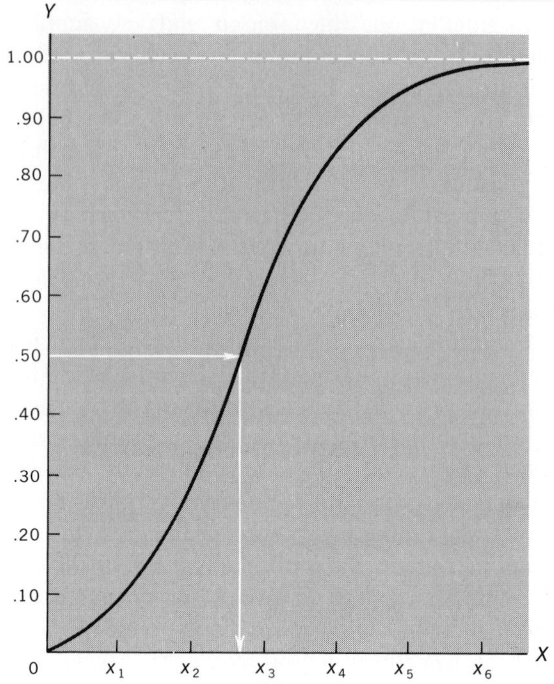

Figure 22.1 Cumulative probabilities from an experimental distribution.

reliable is its representation of the true state of nature and, therefore, the better is the result of simulated sampling.

Let us consider another simple example of simulation. Suppose it is known that a machine is out of order and requires repair $\frac{3}{37}$ of the time, and that the breakdowns are independent. We may, in this case, conveniently simulate the future breakdowns of the machine by spinning a roulette wheel. We may use any 3 of the 37 numbers, such as 0, 1, and 2, to represent breakdowns. When any one of these numbers comes up, we say that the machine is out of order, and when any of the other 37 numbers turns up, we say that the machine is functioning properly. Furthermore, we may arbitrarily prescribe that each spin of the wheel represents the state of the machine for, say, 30 minutes. We could then proceed to simulate the machine breakdowns by examining a sample of three time periods ($1\frac{1}{2}$ hours) by spinning the roulette wheel three times. The binomial probabilities, $b(x;3,\frac{3}{37})$, can then be approximated by observing relative frequencies, with $x = 0, 1, 2$, and 3 breakdowns in a large number of such trials. The fraction of cases in which a given number of breakdowns—say, in 10,000 such trials, or in the next 5000 hours of operation—will differ negligibly from the number expected in the long run according to the law of large numbers.

The Monte Carlo technique can be applied in many economic and business situations. It can be used to forecast sales demand, cost of production, or which team is going to win in the World Series; to evaluate production schedules, inventory policies, or plans for pricing, advertising, distribution, and installment; and to solve complicated queuing problems. In all these, and many other, situations, the general procedure of simulation is the same. We shall, in next two sections, present two illustrative applications of Monte Carlo methods to waiting-line problems.

22.9 Monte Carlo Solution: The Optimum Number of Trucks

Due to a rise in its sales, a manufacturing firm needs more factory space. One alternative solution being considered is the purchase of a warehouse, ten miles from the manufacturing site, for storing finished goods. This plan requires the moving of finished items from the plant to the warehouse at the end of each workday. Management must determine the optimum number of trucks to be hired or bought. In order to do this, it must estimate, assuming various numbers of trucks, the average amount of finished goods which will remain unmoved each day and the daily average of truck capacity which will remain unused.

In this case, clearly, a knowledge of the weight of the output produced and the weight of the output that can be moved each day is essential. Daily production volumes in pounds, according to past data, are distributed as shown in Table 22.1.

The type of truck which should be used for moving this output is known to have an average loading capacity of 4,500 pounds. The number of trips which can be made each 8-hour day depends on loading and unloading time as well as on the time required to travel the ten miles between the plant and the warehouse. The distribution of the number of pounds of product a fully loaded truck can move from the plant to the warehouse during an 8-hour period is estimated as shown in Table 22.2.

From data given in Table 22.1, we find that the expected number of pounds of product produced per day is

$$410{,}000(0.08) + 430{,}000(0.12) + 450{,}000(0.25) + 470{,}000(0.40) + 490{,}000(0.10) + 510{,}000(0.05) = 459{,}400 \text{ pounds}$$

TABLE 22.1 DISTRIBUTION OF DAILY PRODUCTION

Pounds Produced Daily	Relative Frequency	Cumulative Frequency
400,000–419,999	0.08	0.08
420,000–439,999	0.12	0.20
440,000–459,999	0.25	0.45
460,000–479,999	0.40	0.85
480,000–499,999	0.10	0.95
500,000–519,999	0.05	1.00

TABLE 22.2 DISTRIBUTION OF POUNDS OF PRODUCT MOVED PER 8-HOUR DAY PER TRUCK

Pounds Moved	Relative Frequency	Cumulative Frequency
38,000–39,999	0.15	0.15
40,000–41,999	0.25	0.40
42,000–43,999	0.35	0.75
44,000–45,999	0.15	0.90
46,000–47,999	0.06	0.96
48,000–49,999	0.04	1.00

From data in Table 22.2, we find that the expected number of pounds of product moved per day per truck is

$$39{,}000(0.15) + 41{,}000(0.25) + 43{,}000(0.35) + 45{,}000(0.15) + 47{,}000(0.06) + 49{,}000(0.04) = 42{,}680 \text{ pounds}$$

Suppose that we wish to use ten identical trucks; then the expected total number of pounds to be moved per day would be $42{,}680(10) = 426{,}800$ pounds, which is smaller than the expected number of pounds of product produced per day.

We can now simulate what will happen on any number of days we desire to sample. Of course, the more days we sample, the more reliable our results will be. However, merely for the purpose of illustration, we shall take a sample of only 20 days.

The first step in our simulation is to construct a cumulative probability graph for each of the two variables now under consideration. The graph for production data is given by Figure 22.2, and that for data on quantities moved is given by Figure 22.3. We note that in both diagrams, class marks are used on the horizontal scales. Also, the horizontal scale for Figure 22.3 is 10 times the class marks in Table 22.2, since 10 identical trucks are being simulated.

To simulate production, we enter the Table of Random Digits (A-XVI) and select 20 three-digit numbers, each for a day's production. These numbers are given in column (1) of Table 22.3. Note that the first number selected is 095. We

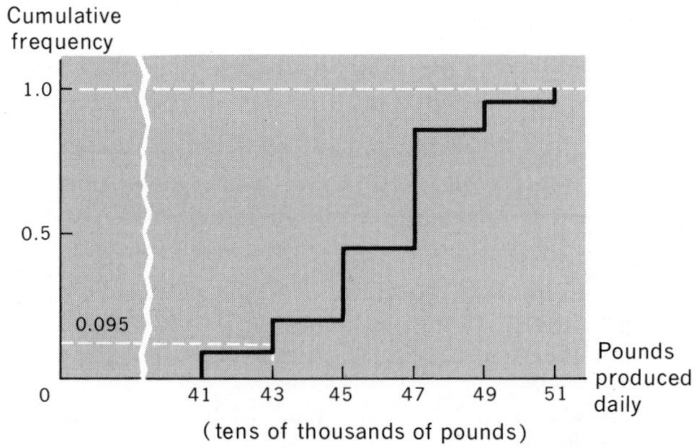

Figure 22.2 Cumulative distribution of daily production.

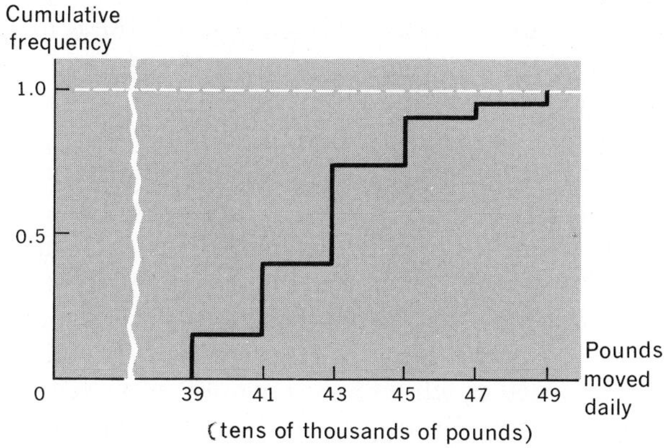

Figure 22.3 Cumulative distribution of product moved.

then enter the cumulative distribution in Figure 22.2 on the vertical axis at 0.095 and draw a horizontal line across until it meets the cumulative probability curve. Finally, we drop a vertical line from this point to the abscissa, and read off the sample value, which is 430,000 pounds. This value is our first predicted daily production figure. All the production figures in column (2) are so determined with the corresponding random numbers in column (1).

TABLE 22.3 SIMULATED DAILY PRODUCTION AND MOVED QUANTITY

Random Number (1)	Production (2)	Random Number (3)	Quantity Moved (4)	Quantity Unmoved (5)	Idle truck Capacity (6)
095	430,000	601	430,000	—	—
973	510,000	889	450,000	60,000	—
410	450,000	882	450,000	—	—
176	430,000	245	410,000	20,000	—
030	410,000	230	410,000	—	—
164	430,000	842	450,000	—	20,000
676	470,000	839	450,000	20,000	—
879	490,000	443	430,000	60,000	—
954	510,000	996	490,000	20,000	—
328	450,000	080	390,000	60,000	—
505	470,000	552	430,000	40,000	—
871	490,000	109	390,000	100,000	—
075	410,000	938	470,000	—	60,000
453	470,000	866	450,000	20,000	—
839	470,000	358	410,000	60,000	—
930	490,000	077	390,000	100,000	—
480	470,000	977	490,000	—	20,000
190	430,000	947	470,000	—	40,000
514	470,000	992	490,000	—	20,000
180	430,000	386	410,000	20,000	—
Total	—	—	—	580,000	160,000
Average	—	—	—	29,000	8,000

Following the procedure described above, we obtain the values in columns (3) and (4), which are our simulated quantities of products moved daily for 20 days.

Using these simulated quantities of output produced and output moved, we can compare readily these two sets of values, thus obtaining values for quantities produced that are unmoved, and for truck capacities that are unused. These are given in columns (5) and (6). The results of these calculations reveal that the daily average unmoved finished product amounts to 29,000 pounds, and the daily average idle truck capacity is 8000 pounds. How can we use these summary values?

In answering the question just posed, we note that these figures take on new dimensions if they are translated into monetary values. The daily cost of unmoved items, due to inconvenience or other reasons, can be estimated. The daily cost of using ten trucks can be determined quite easily. The total daily combined costs of using ten trucks can then be calculated. Employing the same simulation procedure and calculations, we can estimate the total daily combined costs of using, say, 8, 9, 11, or 12 trucks. By comparing total costs per day at various levels of truck utilization, we can finally ascertain the optimum number of trucks to be hired or purchased.

The example just presented is capable of simple mathematical solution. It has been employed merely for the purpose of making clear the basic methodology of Monte Carlo simulation. In actual practice, Monte Carlo models are much more complicated in order to be realistic, and simulated samples are much greater in order to be reliable. In our illustration, for example, more reliable results would be obtained if we had simulated 100 days, or even 1000 days. When large simulated samples are used, even a table of random digits becomes a slow and clumsy aid. The efficient way is to turn to high-speed computers which, when properly instructed, can generate 5000 three-digit random numbers in less than 2 minutes. Recently there has been a growing interest in simulation methods, mainly due to the development and popularization of computer technology.

22.10

Monte Carlo Solution: The Optimum Size of a Service Team

A new manufacturing firm, having been in business for 3000 operating hours, has a large number of identical machines in use. It has a single service station with a team of repairmen. The repair of any machine is a joint effort by the team. When the service station is occupied by a machine and other breakdowns occur, a waiting line is created. Data concerning the number of breakdowns per hour and the number of repairs requiring various time periods for servicing have been carefully recorded for the past 3000 hours. Probabilities (relative frequencies) of these two sets of events based on past data are shown in Tables 22.4 and 22.5, respectively.

The manager now wishes to evaluate whether the existing repair facility is of

TABLE 22.4 DATA ON BREAKDOWNS

Breakdowns per Hour	Probability	Cumulative Probability	Interval of Three-Digit Random Numbers
0	0.900	0.900	000–899
1	0.090	0.990	900–989
2	0.008	0.998	990–997
3	0.002	1.000	998–999

the optimum size, or whether it actually minimizes both the waiting machine cost and the idle time of the service team. We can easily do this by predicting, by Monte Carlo simulation, what can be expected in the future. In order to have reliable results, it would be best to simulate the next 250 (or even more) breakdowns. However, due to the limitation of manual facility, we shall simulate only the first 24 breakdowns as an illustration of the procedure involved. We should, of course, bear in mind that our conclusion may be somewhat distorted due to the smallness of the simulated sample.

The first thing we should do is to assign intervals of random numbers to machine breakdowns per hour and to numbers of hours required for repairing the machines. These intervals are given in the last columns of Table 22.4 and 22.5. Observe that in both cases the intervals of random numbers assigned are proportional to probabilities of occurrences of the respective events.

Simulated data and various summary values are shown in Table 22.6. The meanings of the symbols in the columns and the calculations are explained as follows:

Column (1) contains numbers of machine breakdowns, B_i. We plan to simulate a total of 24 breakdowns.

Column (2) contains random numbers selected to simulate breakdowns, RNB. These numbers are chosen from Table A-XVI. We have started from the first row and the first three columns (three-digit numbers). These numbers are used to simulate two types of events as given in columns (3) and (4). Note that random numbers in the interval 000–899 are ignored since they indicate 0 breakdown.

Column (3) contains simulated occurrence time of breakdowns, O_i. Each figure here corresponds to the position of RNB in Table A-XVI. For example, the first RNB 977 is in the second row (or the second place) in Table A-XVI, and thus we say that the first breakdown occurs at the second hour. Similarly, the second RNB

TABLE 22.5 DATA ON TIME REQUIRED FOR REPAIR

Hours of Repair	Probability	Cumulative Probability	Interval of Three-Digit Random Numbers
1	0.251	0.251	000–250
2	0.375	0.626	251–625
3	0.213	0.839	626–838
4	0.124	0.963	839–962
5	0.037	1.000	963–999

994 is in the 18th place in the table, and we say that the second breakdown occurs at the 18th hour. And so on.

Column (4) gives the randomly selected breakdowns per hour, N_i. These values are simulated by RNB in (2) with reference to Table 22.4. For instance, the first random number in (2) is 977, and from Table 22.4 we find that this number corresponds to one breakdown since 977 falls in the interval 900–989.

The first 24 breakdowns have thus been simulated as shown in columns (1) through (4). It is interesting to note that the total number of hours simulated is 207—206 extractions of random numbers in (2), and that the number of 2 breakdowns per hour occurs four times in total of 24 even though it has a probability of only 0.008 to happen. Also, we note that there is not a single occurrence of 3 breakdowns per hour although this is quite possible. Such distortions can obviously be corrected to a great extent by the use of larger samples.

Column (5) gives idle machine time, due to waiting, W_i. Each idle machine time due to waiting is obtained by taking the difference between the ending time of the repair of the previous breakdown, E_{i-1}, (10) and the current time that a breakdown occurs, O_i, in (3); that is, $W_i = E_{i-1} - O_i$.

Column (6) comprises three-digit random numbers which are employed to simulate repair time, RNR. These numbers are selected by starting from the first row and the eighth column in Table A-XVI.

Column (7) consists of randomly selected repair time, R_i. These values are determined by RNR in (6) with reference to Table 22.5. For instance, the first RNR 386 in (6) refers to a 2-hour repair time in Table 22.5.

Column (8) presents the repair team's idle time, I_i, and $I_i = O_i - E_{i-1}$.

Column (9) shows starting time of repair, S_i. The S_i is either O_i or E_i, depending on whichever is smaller.

Column (10) contains ending time of repair, E_i. It is the sum of starting time of repair and randomly selected repair time; that is, $E_i = S_i + R_i$.

Column (11) gives total idle machine time for repairing, T_i, which is simply the sum of W_i and R_i.

To summarize the relationships among O_i, E_i, W_i, I_i, and S_i, we may note three different cases:

(a) If $E_{i-1} < O_i$, then $S_i = O_i$. In this case, $W_i = 0$ and $I_i = O_i - E_{i-1}$.

(b) If $E_{i-1} > O_i$, then $S_i = E_{i-1}$. In this case, $W_i = E_{i-1} - O_i$ and $I_i = 0$.

(c) If $E_{i-1} = O_i$, then $S_i = E_{i-1}$. In this case, $W_i = I_i = 0$ and $E_i = S_i + R_i$.

These three cases are identified by heavy vertical bars on the left side of Table 22.6.

From simulated data and quantities already computed in Table 22.6, we can now determine various mean values of interest. They are

(1) *Average repair time:*

$$\bar{R} = \frac{\text{Total repair time}}{\text{Total number of breakdowns}} = \frac{\Sigma R_i}{n} = \frac{57}{24} = 2.375 \text{ hours}$$

(2) *Average idle machine time due to waiting:*

$$\bar{W} = \frac{\text{Total waiting time}}{\text{Total number of breakdowns}} = \frac{\Sigma W_i}{n} = \frac{33}{24} = 1.375 \text{ hours}$$

TABLE 22.6 SIMULATION TABLE FOR 24 BREAKDOWNS

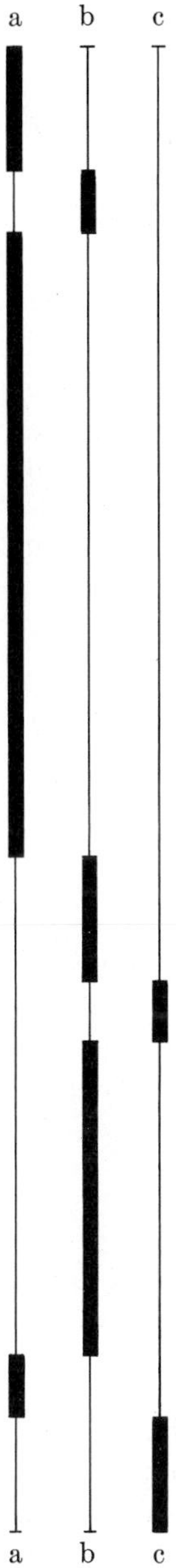

B_i (1)	RNB (2)	O_i (3)	N_i (4)	W_i (5)	RNR (6)	R_i (7)	I_i (8)	S_i (9)	E_i (10)	T_i (11)
—	—	0	—	—	—	—	—	—	—	—
1	977	2	1	—	386	2	2	2	4	2
2	994	18	2	—	762	3	14	18	21	3
3				3	926	4	—	21	25	7
4	969	44	1	—	564	2	19	44	46	2
5	968	57	1	—	439	2	11	57	59	2
6	964	70	1	—	331	2	11	70	72	2
7	975	74	1	—	859	4	2	74	78	4
8	981	88	1	—	429	2	10	88	90	2
9	971	97	1	—	244	1	7	97	98	1
10	965	103	1	—	246	1	5	103	104	1
11	920	116	1	—	775	3	12	116	119	3
12	905	130	1	—	474	2	11	130	132	2
13	996	169	2	—	995	5	37	169	174	5
14				5	512	2	—	174	176	7
15	938	173	1	3	213	1	—	176	177	4
16	977	177	2	—	735	3	—	177	180	3
17				3	277	2	—	180	182	5
18	947	178	1	4	472	2	—	182	184	6
19	992	179	2	5	243	1	—	184	185	6
20				6	370	2	—	185	187	8
21	976	183	1	4	576	2	—	187	189	6
22	944	199	1	—	653	3	10	199	202	3
23	973	202	1	—	961	4	—	202	206	4
24	916	206	1	—	424	2	—	206	208	2
Total	—	—	24	33	—	57	154	—	—	90

(3) *Total average idle machine time:*

$$\bar{T} = \frac{\Sigma T_i}{n} = \bar{R} + \bar{W} = \frac{90}{24} = 2.375 + 1.375 = 3.750 \text{ hours}$$

(4) *Average idle time of the repair team,* or *average waiting time between breakdowns:*

$$\bar{I} = \frac{\Sigma I_i}{n} = \frac{154}{24} = 6.417 \text{ hours}$$

All the preceding average values may be interpreted as mean values per breakdown viewed from different angles.

Our calculations reveal that the average idle time of the repairing team is almost twice as large as the total average idle machine time. Do these results indicate that the service capacity is too large for the demand for service? To answer this question, we must compare the cost of the wasted time for machines and that of the wasted time of the repair team. Now, suppose we assume that the cost of lost production due to idle machines is estimated at \$25 per hour, and that the repair team consists of 5 repairmen, each being paid an hourly wage rate of \$8; then we would have

(1) Pure repair cost $= \Sigma R_i(25 + 5(8)) = 57(65) = \$3705.00$

(2) Inefficiency cost due to

a. Cost of idle machine time $= \Sigma W_i(25) = (33)(25) = \$825.00$

b. Cost of repair team idle time $= \Sigma I_i(40) = 154(40) = \$6160.00$

These calculations may be considered as expected future expenses. Since the cost of waste time of the repairmen for a 207-hour simulation is extremely high compared to the cost of waste time of the machines, the manager should probably reduce the number of the repairmen on the service team in order to reduce the idle time of the repair team. The reader must be reminded again that this decision is based on 24 simulated breakdowns. In practice, as we mentioned earlier, computation for problems of this kind should perhaps be based on 250 or more breakdowns. With the use of high-speed computers, such large samples can be simulated and summary values computed in less than 10 minutes!

22.11

Concluding Remarks

We have in this chapter presented some of the essential aspects of the queuing theory. This theory has the objective of finding optimum solutions to waiting-line problems. The analysis of queues has as its focus the utilization parameter. The proportion of time that service facilities remain completely idle decreases as the utilization factor increases. The average waiting time of arrivals increases sharply as the arrival rate approaches the total capacity of the service facility. The mean length of the waiting line increases sharply, too, as the utilization factor approaches unity.

We often find that more repairmen, or more expensive equipment for repair, are actually cheaper in the long run in terms of total service and waiting costs. These observations lead us to see economies of large-scale production in the operation. In this respect, it is also easy to see why waiting lines and capacity requirements can be favorably affected by centralization of certain service operations—such as filing, reporting, stenographic services, and many functions in selling and manufacturing. This is so because, if the efficiency of each service station remains the same after centralization, we would have a single queue in contrast to a waiting line for each service station as before, which should make improvements in service possible. Why?

As to the comparative merits of analytical and Monte Carlo solutions to waiting-line problems, we note that the use of the former requires theoretical models whose strict assumptions must be approximately met by real-life situations. The Monte Carlo method for simulating risk has the advantage of flexibility, since it can be applied to theoretical distributions, as well as to empirical distributions, which may not satisfy the properties of any theoretical model. In addition, the Monte Carlo technique is capable of solving many problems for which analytical methods have not yet been developed, or for which available mathematical solutions are far too involved. Above all, it is a powerful tool available to research workers who are not mathematicians.

We must realize, however, that with Monte Carlo simulation we have the same problems of statistical variations and experimental design as with real-life sampling or direct experimentation. Frequently, very large samples are required and simulation must continue for a long time to reach stable conditions. Therefore, when a mathematical model is realistic and can be manipulated with simple algebra to obtain the optimum solution, it possesses a definite advantage over Monte Carlo simulation.

In conclusion, we may also note that while both simulation and direct experimentation are subject to sampling error, the former is more desirable than the latter on many accounts. With direct experimentation, we must hold the same conditions, a thing quite difficult to achieve, in repeated sampling; the method may also disrupt actual operations and be very expensive. Monte Carlo simulation, however, can be conducted continuously "on paper" with little cost. It permits us, therefore, to evaluate many alternatives which we could not attempt to do in real-life experimentation.

Glossary of Formulas

Equations (22.1) through (22.7) govern the single service-station queuing model. For these equations to hold it is required that $\lambda < \mu$. Also, arrivals and service time are assumed to follow the Poisson and exponential probability laws, respectively.

(22.1) $U = \dfrac{\lambda}{\mu}$ This is the definition of the utilization parameter. It is also called the *traffic intensity* or *utilization factor*. Here, λ is the mean number of arrivals per unit of time, and μ is the mean number servicings per unit of time. The U measures the proportion of time that the service station is kept busy or the probability that a random arrival has to wait.

(**22.2**) $I = P_0 = 1 - U$ This equation gives the proportion of time the service station remains idle. As such, it is also the probability that a randomly arrived customer does not have to wait; here, P_0 indicates the probability that a customer arrives and finds no other customer in the system.

(**22.3**) $P_n = (1 - U)(U)^n$ This gives the probability of n customers in the system, $n \geq 0$. It includes those in the queue and the one being serviced.

(**22.4**) $E(n) = \dfrac{\lambda}{\mu - \lambda}$ This expression is the definition of the mean number of customers in the system at any instant. It includes both those customers in the queue and those in service.

(**22.5**) $E(L) = \dfrac{\lambda^2}{\mu(\mu - \lambda)}$ This is the expected length of the waiting line. The reader should note the relationship between $E(n)$ and $E(L)$.

(**22.6**) $E(W) = \dfrac{\lambda}{\mu(\mu - \lambda)}$ This defines the average waiting time of a customer in the queue. Note that $E(W) = (\lambda/\mu)E(T)$. What does it mean?

(**22.7**) $E(T) = \dfrac{1}{\mu - \lambda}$ This gives the average time a customer spends in the system. Again, the reader should observe the relationship between $E(W)$ and $E(T)$. What is it?

(**22.8**) $D = \dfrac{100F(1 - F)(1 - R)U}{1 - F + FR(1 - U)}$ This equation gives the percentage reduction in the expected waiting time when the priority rule of "quickest returned to operation, first" is applied. Here, R is ratio of the average servicing time of the fast customers to that of the slow customers, and F is the fraction of fast customers to the total number of customers. With this priority rule, the reduction of average waiting time often leads to proportional reduction in costs. It has been observed that one can obtain savings of 25 percent or more in many a case.

Equations (22.9) through (22.17) govern the behaviors of waiting lines in multiple service-station queuing models. For these equations, Poisson arrivals and exponential service time are also assumed. Furthermore, these expressions hold only if $k\mu > \lambda$.

(**22.9**) $U_k = \dfrac{\lambda}{k\mu}$ The utilization parameter of k parallel service stations is so defined. It is interpreted in the same way as U. Here k is the number of service stations, λ is the mean number of arrivals per unit of time, and μ is the mean number of servicings per unit of time per station.

(**22.10**) $P_n = \dfrac{1}{n!}\left(\dfrac{\lambda}{\mu}\right)^n P_0$ (**22.11**) $P_n = \dfrac{1}{k!k^{n-k}}\left(\dfrac{\lambda}{\mu}\right)^n P_0$

These two equations give the probability that there are n customers in the system. (22.10) is used when $n < k$ but (22.11) should be employed when $n > k$. In these expressions, P_0 stands for the probability that there is no customer in the system and is defined by (22.12).

(**22.12**) $P_0 = \dfrac{1}{\left[\sum_{i=0}^{k-1} \dfrac{1}{i!}\left(\dfrac{\lambda}{\mu}\right)^i\right] + \left[\dfrac{1}{k!}\left(\dfrac{\lambda}{\mu}\right)^k \left(\dfrac{k\mu}{k\mu - \lambda}\right)\right]}$ Being the probability that there is no customer in the system, P_0 also stands for the probability that all the k service stations are idle.

(22.13) $P_m = \dfrac{\mu(\lambda/\mu)^k}{(k-1)!(k\mu - \lambda)} P_0$ The probability that a customer arriving at random in a multiple station model has to wait for service is the probability that there are k or more customers in the system. This probability is defined by (22.13), where the subscript m is the number of arrivals from k up to n.

(22.14) $E(L) = \dfrac{\lambda\mu(\lambda/\mu)^k}{(k-1)!(k\mu - \lambda)^2} P_0$ This expression gives the mean length of the queue. Since $E(L)$ excludes those customers in service, it is equal to $[\lambda/(k\mu - \lambda)]P_m$. Can you explain why verbally?

(22.15) $E(n) = \dfrac{\lambda}{\mu} + E(L)$ This defines the mean number of customers in the system. You may wish to explain in your own way the relationship between $E(n)$ and $E(L)$.

(22.16) $E(W) = \dfrac{\mu(\lambda/\mu)^k}{(k-1)!(k\mu - \lambda)^2} P_0$ The average waiting time of an arrival in the queue is given by this equation. Observe the differences among the expressions for $E(L)$, P_m, and $E(W)$. How are they related to each other?

(22.17) $E(T) = \dfrac{1}{\mu} + E(W)$ The average time a customer or element spends in the system is defined here. Give a verbal statement of the relationship between $E(W)$ and $E(T)$.

Problems

22.1 Do you recall how the exponential distribution is related to the Poisson distribution?

22.2 Why do we require that the mean arrival rate must be less than the mean service rate per unit of time in queuing problems?

22.3 Under what condition or conditions is a queuing problem considered as multichannel?

22.4 In the single service-station queuing model it is required that $\lambda < \mu$. Discuss why, and in what form, this requirement is present in the multiple service-station model.

22.5 What are the queue disciplines mentioned in the text? Indicate situations in which each of these disciplines is appropriate.

22.6 What do we mean by Monte Carlo simulation? What are two basic things required for simulation?

22.7 What are the comparative merits or demerits of analytical and Monte Carlo solutions to queuing problems?

22.8 In what do we say that simulated sampling is superior to direct experimentation?

22.9 The mean arrival rate of customers at a repair shop is 20 per hour, and the shop has a mean service rate of 40 per hour.

a. What is the utilization factor of this system?
b. What is the probability of forming a waiting line of length n in this system, for $n = 0, 1, 2, 3$, and 4?
c. What is the average number of customers in the system at any instance?
d. What is the mean time between arrivals?
e. What is the mean service time per servicing?

22.10 The mean arrival rate of machine breakdowns in a factory is 80 per week, and the factory's service station can repair 120 machines per week. For this system,

a. What is the proportion of time that the service station remains idle?
b. What is the probability that there are n down machines in the queue, for $n = 0, 1, 2, 3$, and 4?
c. What is the expected number of broken-down machines in the system?
d. What is the mean waiting time of an arrival?
e. What is the average time a broken-down machine spends in the system?

22.11 Arrivals at an information desk that has one information clerk are considered to be Poisson, with an average time of 10 minutes between one arrival and the next. The time that customers spend in seeking information of one kind or another is assumed to be exponential, with a mean of 3 minutes.
a. What is the probability that a person arriving at the desk will have to wait?
b. What is the average number of elements in the system?
c. The manager of the organization will hire a second clerk if convinced that an arrival would expect to have to wait at least 3 minutes for receiving information. By how much must the flow of arrivals be increased in order to justify a second clerk?

22.12 The owner of a gas station expects a customer every 5 minutes. Services takes, on the average, 4 minutes.
a. What are the mean arrival rate and mean service time, respectively?
b. What is the probability that a customer will not have to wait?
c. What is the probability that there will be a waiting line?
d. The owner would hire an additional attendant if he found that a customer had to wait for 2 minutes or longer before being serviced. Would the current arrival rate justify such an addition?

22.13 In a manufacturing firm, machines break down at a rate of 0.8 per hour, and the nonproductive cost of each machine waiting to be repaired and being repaired is estimated at $25 per hour. There are two repairing teams available to the firm. Team A asks for $25 per hour and guarantees that it can easily repair 1.5 machines per hour. Team B asks for $20 per hour and informs the firm that it can fix 1 machine per hour. Given that the service (time) rates of the two teams are reliable, which team should the firm employ?

22.14 In the preceding problem, if Team A asks for $30 per hour and Team B asks for $15 per hour, which team should be employed?

22.15 A clothing factory has a large number of identical machines; the mean breakdown rate is 1 per hour and the cost of idle time is $10 per hour. The manager of this firm is considering two organizations, X and Y, for the job of providing maintenance of his machines. Organization X is known to have a repairing rate of 2 machines per hour and a charge of $20 per hour; Organization Y is known to have a repairing rate of 1.5 machines per hour and a charge of $12 per hour. Which organization should be given the contract?

22.16 Arrivals at a watch repair shop occur at a rate of 10 per 8-hour day, and the watch repairman has mean service time of 30 minutes per repair. If he repairs watches in the order of arrivals, what is the repairman's expected idle time per day? How many watches are ahead of the average number of watches just arrived?

22.17 Assuming the rule of "quickest return to operation, first" and assuming $R = 0.3$, calculate the percentage reduction in the average waiting time as a function of F and U as indicated by the following table:

F	U			
	0.5	*0.7*	*0.85*	*0.99*
0.1	3	?	?	7
0.2	?	?	?	14
0.3	10	14	?	?
0.4	?	?	?	?
0.5	?	?	?	?
0.6				
0.7				
0.8				
0.9				

22.18 Comment on D in relation to F and U as revealed by calculations in the preceding problem.

22.19 The arrival rate at a repair shop is 180 per day. The shop has three service stations, each having a service rate of 100 per day.
a. What is the probability that the repair shop is empty of customers?
b. What is the probability that an arrival must wait?
c. What is the probability that two stations are idle? one station is idle?
d. What is the average proportion of time any station is idle?
e. What is the average waiting time of arrivals?
f. What is the average length of waiting line?

22.20 There are four officials on the main floor of a bank to receive customers who may inquire about financial operations and other matters. The average arrival is 60 persons in a 6-hour service day. Each station, that is, official, spends an irregular amount of time in serving customers, with an average service time of 20 minutes. Customers are served in order of their arrivals. The manager responsible for this service function wishes answers to these questions:

a. How many hours, for each 30-hour service week, does an official spend in performing his job?

b. What is the average time a customer is kept in the system?

c. What is the probability that an official is waiting for a customer?

d. What is the expected number of idle officials at any instant?

22.21 Suppose that, in the preceding problem, each official is paid $5 an hour, and the manager believes that the goodwill and business lost because customers have to wait for service is worth at least 25 cents per minute. In view of these cost data, should the number of officials be reduced to three or increased to five?

22.22 A bank has two tellers during slack periods. One handles demand deposit accounts and the other takes care of savings accounts. The service rate of each teller is 25 customers per hour. The arrival rate at the demand deposit window is 20 per hour, while that at the savings window is only 15 per hour. What are the average waiting times for the two kinds of customers? Suppose that each teller handles both types of customers while the service rates remain as before; what effect could be expected on the average waiting time of customers? What would be the effect if the integraton can be achieved only by reducing each teller's mean service rate to 22 per hour?

22.23 In the children's clothing department of a department store, there are one salesclerk, one wrapper, and one cashier. The salesclerk serves a customer in an average of 1.5 minutes. The wrapper takes 1 minute to wrap, on the average. The mean time of the cashier to receive payment is 0.5 minute. Customers arrive at a rate of 30 per hour.

a. What is the mean number of customers being served by the three clerks?

b. What is the mean number of customers in the waiting line of the whole system?

c. What is the total number of customers in the whole system, including those waiting and those being serviced?

d. What is the probability of having 3 customers waiting to be served by the salesclerk, 2 customers waiting for the wrapper, and 1 customer waiting for the cashier, at any instant?

22.24 Consider the preceding problem again.

a. What is total average time a customer spends in the complete system?

b. What would be the effects on the average lengths of waiting lines and average waiting times if each of the three clerks were to perform all the three functions of serving, wrapping, and receiving payments (i) under the assumption that the average total time for servicing, wrapping, and receiving payment is 3 minutes? (ii) under the assumption that the total average time for servicing has increased to 3.5 minutes?

22.25 Write down the numbers 0, 1, · · · , 8, and 9 on ten identical physical objects and put them in an urn. Generate a series of 50 four-digit random numbers.

22.26 Calculate relative frequencies and cumulative relative frequencies for the following data:

Class on monthly orders in 100-month period	No. of months in which corresponding quantity of orders has been received
000–199	10
200–399	21
400–599	32
600–799	15
800–999	10
1000–1199	8
1200–1399	4
	100

22.27 Simulate a possible distribution of orders for the next 20 months for data in Problem 22.26 with random numbers generated in Problem 22.25.

22.28 Simulate production data and quantity moved for 15 days with data in Tables 22.1 and 22.2, assuming that 11 trucks are used. Also calculate the daily averages of quantities unmoved and idle truck capacities.

22.29 Use the random-number generator in Problem 22.25 to simulate 14 more breakdowns for the illustration in Section 22.10. Combine your simulated data with those in Table 22.6. Do the computations again, as in the illustration, with the combined data.

APPENDIXES

A

The Summation Sign

A.1

Single, Double, and Triple Summation Signs

The Greek capital letter sigma, Σ, is a shorthand notation for designating a sum. For instance, the sum

$$x_1 + x_2 + \cdots + x_n$$

may be conveniently written with the *single summation sign* as

$$\sum_{i=1}^{n} x_i$$

which is read as "the summation of x_i from $i = 1$ to $i = n$." Here, i is called the *index of summation* and is a variable ranging over the integers $1, 2, \cdots, n$. The symbol $i = 1$ below the Σ sign indicates that 1 is the initial value taken on by i, and the n above the Σ sign indicates the terminal value of i. The symbol x_i is called the *summand*; it is a function of i which takes on values $x_1, x_2, \cdots, x_n$, respectively, as i takes on values $1, 2, \cdots, n$. The sign Σ says that the values taken by the summand are to be added.

The symbol used for the index of summation is sometimes called a *dummy index* for the reason that it is completely arbitrary. For example,

$$x_1 + x_2 + \cdots + x_n = \sum_{i=1}^{n} x_i = \sum_{j=1}^{n} x_j = \sum_{k=1}^{n} x_k = \cdots$$

Another arbitrariness of the summation index is that the initial value of the index can be altered in a compensating way so that the net sum remains unchanged. Namely,

1. $$\sum_{i=1}^{n} x_i = \sum_{i=0}^{n-1} x_{i+1} = \sum_{i=2}^{n+1} x_{i-1} = \cdots$$

2. $$\sum_{n=0}^{\infty} \frac{x^n}{n!} = \sum_{n=1}^{\infty} \frac{x^{n-1}}{(n-1)!} = \sum_{n=-1}^{\infty} \frac{x^{n+1}}{(n+1)!} = \cdots$$

As illustrations, we see that

$$\sum_{i=1}^{3} i = 1 + 2 + 3 = 6$$

$$\sum_{i=1}^{3} i^2 = 1^2 + 2^2 + 3^2 = 14$$

$$\sum_{i=0}^{r} \frac{1}{2^i} = \frac{1}{2^0} + \frac{1}{2^1} + \frac{1}{2^2} + \cdots + \frac{1}{2^r}$$

$$\sum_{i=1}^{5} x_i = x_1 + x_2 + \cdots + x_5$$

$$\sum_{i=6}^{9} x_i = x_6 + x_7 + x_8 + x_9$$

and from the last two examples, we can readily appreciate that

$$x_1 + x_2 + \cdots + x_9 = \sum_{i=1}^{5} x_i + \sum_{i=6}^{9} x_i = \sum_{i=1}^{9} x_i$$

Before proceeding any further, we may point out that if the range of summation is completely clear by the context, the index of summation can be omitted. Thus, we can use Σx to replace $\sum_{i=1}^{n} x_i$. Similarly, $\Sigma x^2 - (\Sigma x)^2$ can be used in place of $\sum_{i=1}^{n} x_i^2 - \left(\sum_{i=1}^{n} x_i\right)^2$.

When the values of a variable X, the summand, are arranged as a two-way table; that is, when there are cj groups, each of which contains n_j individual values, we can indicate the sum of x_{ij} over all groups by a *double summation sign*, which is regarded as a single summation on one subscript followed by a single summation on a second summation. Namely, we can evaluate a double summation notation by applying the single summation twice, as follows:

$$\begin{aligned}\sum_{j=1}^{c} \sum_{i=1}^{n_j} x_{ij} &= \sum_{j=1}^{j} (x_{1j} + x_{2j} + x_{3j} + \cdots + x_{n_j j}) \\ &= (x_{11} + x_{21} + \cdots + x_{n_1 1}) \\ &\quad + (x_{12} + x_{22} + \cdots + x_{n_2 2}) + \cdots \\ &\quad + (x_{1c} + x_{2c} + \cdots + x_{n_c c})\end{aligned}$$

The above double summation is interpreted as: "Take a particular group c and sum the values over all the n_j different observations in that group; having done this for each of the c groups, sum the results over all the groups."

Consider now the case of the *triple summation sign*, the largest such symbol used. If the values of a variable X are arranged in a table which consists of r rows, c columns, and rc cells, where each individual value belongs to one and only one row, one and only one column, and one and only one cell in the table, then the ith observation in the jkth cell is denoted as x_{ijk}. Also, we denote the number of observations in cell jk as n_{jk}. The notation of the sum of the observations in such a table is symbolized as

$$\sum_{j=1}^{c} \sum_{k=1}^{r} \sum_{i=1}^{n_{jk}} x_{ijk}$$

which states: "Take a particular cell, jk, and sum all the individual values in that cell; then, still in column j, sum the results of doing this for each row k over all the rows; finally, sum the results for each column j over all the columns." A little reflection will reveal that this is simply the sum of all the individual values in the table.

A.2 Theorems of Summation

THEOREM I

If a is some constant value over the n different observations i, then

$$\sum_{i=1}^{n} a = na$$

since

$$\sum_{i=1}^{n} a = \underbrace{a + a + \cdots + a}_{n \text{ terms}} = na$$

Extending this rule to double and triple summations, we have

$$\sum_{j=1}^{c} \sum_{i=1}^{n_j} a = \sum_{j=1}^{c} (n_j)a$$

and

$$\sum_{j=1}^{c} \sum_{k=1}^{r} \sum_{i=1}^{n_{jk}} a = \sum_{j=1}^{c} \sum_{k=1}^{r} (n_{jk})a$$

THEOREM II

Given the value a, which is a constant over all individual values which enter into the summation, then

$$\sum_{i=1}^{n} ax_i = a \sum_{i=1}^{n} x_i$$

since

$$\begin{aligned}\sum_{i=1}^{n} ax_i &= ax_1 + ax_2 + \cdots + ax_n \\ &= a(x_1 + x_2 + \cdots x_n) \\ &= a \sum_{i=1}^{n} x_i\end{aligned}$$

This rule can also be applied to several sums; namely,

$$\sum_j \sum_i ax_{ij} = a \sum_j \sum_i x_{ij}$$

and so forth.

THEOREM III

If the only operation to be carried out before a sum is taken is itself a sum (or difference), then the summation may be distributed; that is,

$$\sum_i (x_i + y_i - z_i) = \sum_i x_i + \sum_i y_i - \sum_i z_i$$

since

$$\begin{aligned}\sum_i (x_i + y_i - z_i) &= (x_1 + y_1 - z_1) + (x_2 + y_2 - z_2) \\ &\quad + \cdots + (x_n + y_n - z_n) \\ &= (x_1 + x_2 + \cdots + x_n) + (y_1 + y_2 + \cdots + y_n) \\ &\quad - (z_1 + z_2 + \cdots + z_n) \\ &= \sum_i x_i + \sum_i y_i - \sum_i z_i\end{aligned}$$

Theorem III is a very handy tool for algebraic manipulations on summation signs. For example,

$$\begin{aligned}\sum_j \sum_i (x_{ij} - a)^2 &= \sum_j \sum_i (x_{ij}^2 - 2ax_{ij} + a^2) \\ &= \sum_j \sum_i x_{ij}^2 - \sum_j \sum_i 2ax_{ij} + \sum_j \sum_i a^2 \\ &= \sum_j \sum_i x_{ij}^2 - 2a \sum_j \sum_i x_{ij} + a^2 \sum_j n_j\end{aligned}$$

THEOREM IV

If some operation is to be carried out on the individual values of X before the summation, this is indicated by mathematical punctuation, or symbols having the force of punctuation. Unless the summation notation is included within this punctuation, the summation is to be made after the other operation. For instance,

$$\Sigma(x)^2 = \Sigma(x_1^2 + x_2^2 + \cdots + x_n^2)$$

whereas

$$(\Sigma x)^2 = (x_1 + x_2 + \cdots + x_n)^2$$

Also,

$$\Sigma \sqrt{x_i} = \sqrt{x_1} + \sqrt{x_2} + \cdots + \sqrt{x_n}$$

whereas

$$\sqrt{\Sigma x} = \sqrt{x_1 + x_2 + \cdots + x_n}$$

This rule is especially important where multiple summation signs are involved. For example,

$$\sum_j \Big(\sum_i x_{ij}\Big)^2 = \sum_j (x_{1j} + x_{2j} + \cdots + x_{nj})^2$$

THEOREM V

If both limits on the summation signs are constants, then the order of summation may be interchanged; that is,

$$\sum_{j=1}^{b} \sum_{i=1}^{a} x_{ij} = \sum_{i=1}^{a} \sum_{j=1}^{b} x_{ij}$$

THEOREM VI

$$\Sigma x_i y_i = x_1 y_1 + x_2 y_2 + \cdots + x_n y_n;\ \Sigma x_i y_i \neq \Sigma x_i(\Sigma y_i)$$

THEOREM VII

$$\Sigma\Sigma a x_i y_i = a\Sigma\Sigma x_i y_i$$

THEOREM VIII

$$\Sigma(a x_i + b y_i) = a\Sigma x_i + b\Sigma y_i$$

where a and b are both constants.

THEOREM IX

$$\sum_j^{c} \sum_i^{n_j} y_j x_{ij} = \sum^{c} y_i \Big(\sum_i^{n_j} x_{ij}\Big)$$

Problems

Let

$$t_{.1} = \sum_{i}^{n_1} x_{i1} \text{ be sum of sample values drawn from population 1;}$$

$$t_{.2} = \sum_{i}^{n_2} x_{i2} \text{ be sum of sample values drawn from population 2;}$$

$$t_{.j} = \sum_{i}^{n_j} x_{ij} \text{ be sum of sample values drawn from population } j;$$

$$t_{..} = \sum_{j}\sum_{i} x_{ij} = \sum_{j} t_{.j};$$

$$\bar{x}_{.1} = t_{.1}/n_1,\ \bar{x}_{.2} = t_{.2}/n_2,\ \bar{x}_{.j} = t_{.j}/n_j, \text{ and } \bar{x}_{..} = t_{..}/\Sigma n_i = t_{..}/N.$$

Prove that

$$1.\quad \sum_{j=1}^{r}\sum_{i=1}^{n_j} (x_{ij} - \bar{x}_{.j}) = 0$$

and

$$2.\quad \sum_{j=1}^{r}\sum_{i=1}^{n_j} (x_{ij} - \bar{x}_{..})^2 = \sum_{j}\sum_{i} x_{ij}^2 - \frac{t_{..}^2}{N}$$

ANSWERS:

$$1.\quad \sum_{j}\sum_{i} (x_{ij} - \bar{x}_{.j}) = \sum_{j}\sum_{i} x_{ij} - \sum_{j}\sum_{i} \bar{x}_{.j}$$

but

$$\sum_{j}\sum_{i} \bar{x}_{.j} = \sum_{j} n_j\bar{x}_{.j} = \sum_{j}\sum_{i} x_{ij}$$

Therefore,

$$\sum_{j}\sum_{i} (x_{ij} - \bar{x}_{.j}) = 0$$

$$2.\quad \sum_{j}\sum_{i} (x_{ij} - \bar{x}_{.j})^2 = \sum_{j}\sum_{i} (x_{ij}^2 - 2\bar{x}_{..}x_{ij} + \bar{x}_{..}^2)$$

$$= \sum_{j}\sum_{i} x_{ij}^2 - \sum_{j}\sum_{i} 2\bar{x}_{..}x_{ij} + \sum_{j}\sum_{i} \bar{x}_{..}^2$$

but

$$\sum_{j}\sum_{i} 2\bar{x}_{..}x_{ij} = 2\bar{x}_{..}\sum_{j}\sum_{i} x_{ij} = 2(t_{..}/N)t_{..} = 2(t_{..}^2/N)$$

and

$$\sum_{j}\sum_{i} \bar{x}_{..} = N\bar{x}_{..}^2 = N(t_{..}/N)^2 = t_{..}^2/N$$

Therefore,

$$\sum_j \sum_i (x_{ij} - \bar{x}_{..})^2 = \sum_j \sum_i x_{ij}^2 - 2\frac{t_{..}^2}{N} + \frac{t_{..}^2}{N}$$

$$= \sum_j \sum_i x_{ij}^2 - \frac{t_{..}^2}{N}$$

B

Suggestions for Further Reading

This book is an introduction to business and economic statistics. Yet, if the reader has fully mastered the contents in this text, he will discover that the more advanced statistical methods he might find in further courses of statistics will not present him with any main difficulties except those of a purely technical nature.

This appendix lists a number of books suggested to the reader interested in pursuing his studies of statistics further.

The first group of books recommended is at the introductory level. These books will, therefore, benefit the reader who wishes to supplement his study in the present text. This group may be divided into two categories: (1) general textbooks and (2) textbooks in economic and business statistics.

General Textbooks

Arkin, Herbert, and Raymond R. Colton. *Statistical Methods*, 4th ed. New York: Barnes & Noble, Inc., 1956.

Chernoff, Herman, and Lincoln E. Moses. *Elementary Decision Theory*. New York: John Wiley & Sons, Inc., 1959.

Croxton, Frederick E., and Dudley J. Cowden. *Applied General Statistics,* 2d ed. Englewood Cliffs, N.J.: Prentice-Hall, Inc., 1967.
Freund, John E. *Modern Elementary Statistics.* Englewood Cliffs, N.J.: Prentice-Hall, Inc., 1952.
Griffin, John I. *Statistics: Methods and Applications.* New York: Holt, Rinehart and Winston, Inc., 1962.
Guenther, W. C. *Concepts of Statistical Inference.* New York: McGraw-Hill Book Company, Inc., 1965.
McMarthy, Philip J. *Introduction to Statistical Reasoning.* New York: McGraw-Hill Book Company, Inc., 1957.
Rosander, A. C. *Elementary Principles of Statistics.* Princeton, N.J.: D. Van Nostrand Company, Inc., 1951.
Yamane, Toro. *Statistics: An Introductory Analysis,* 2d ed. New York: Harper & Row, Publishers, 1967.
Snedecor, G. W. *Statistical Methods,* 5th ed. Ames, Iowa: Collegiate Press, Inc., 1956.
Tippett, L. H. C. *Statistics.* New York: Oxford University Press, 1953.
Walker, Helen M., and Joseph Lev. *Elementary Statistical Methods.* New York: Holt, Rinehart and Winston, Inc., 1958.
Wallis, W. Allen, and Harry V. Roberts. *Statistics: A New Approach.* New York: The Free Press, 1956.
Yule, G. Udny, and M. G. Kendall. *An Introduction to the Theory of Statistics,* 14th ed. London: Charles Griffin & Co., Ltd., 1950.

Textbooks in Economic and Business Statistics

Allen, R. G. D. *Statistics for Economists.* London: Hutchinson & Co., Ltd., 1949.
Bryant, Edward C. *Statistical Analysis,* 2d ed. New York: McGraw-Hill Book Company, Inc., 1966.
Croxton, F. E., and D. J. Cowden. *Practical Business Statistics,* 3d ed. Englewood Cliffs, N.J.: Prentice-Hall, Inc., 1960.
Davis, George R., and Dale Yoder. *Business Statistics.* New York: John Wiley & Sons, Inc., 1950.
Freund, J. E., and F. J. Williams. *Elementary Business Statistics: The Modern Approach.* Englewood Cliffs, N.J.: Prentice-Hall, Inc., 1964.
Kurnow, E., G. J. Glasser, and F. R. Ottman. *Statistics for Business Decisions.* Homewood, Ill.: Richard D. Irwin, Inc., 1959.
Lewis, Edward E. *Methods of Statistical Analysis in Economics and Business.* Boston: Houghton Mifflin Company, 1953.
Mills, Frederick C. *Statistical Methods,* 3d ed. New York: Holt, Rinehart and Winston, Inc., 1955.
Neiswanger, William A. *Elementary Statistical Methods,* 3d ed. New York: The Macmillan Company, 1960.
Neter, John, and William Wassermann. *Fundamental Statistics for Business and Economics,* 3d ed. Boston: Allyn and Bacon, 1966.

Paden, Donald W., and E. F. Lindquist. *Statistics for Economics and Business*, 2d ed. New York: McGraw-Hill Book Company, Inc., 1956.

Stockton, John R. *Business Statistics*, 2d ed. Cincinnati: South-Western Publishing Co., 1962.

A few standard reference books on statistics are useful to keep at hand.

Arkin, Herbert, and Raymond R. Colton. *Tables for Statisticians*. New York: Barnes & Noble, Inc., 1950.

Burington, Richard S., and D. C. May. *Handbook of Probability and Statistics*. New York: McGraw-Hill Book Company, Inc., 1953.

Cox, E. B. *Basic Tables in Business and Economics*. New York: McGraw-Hill Book Company, Inc., 1966.

Hauser, Philip M., and William R. Leonard. *Government Statistics for Business Use*, 2d ed. New York: John Wiley & Sons, Inc., 1956.

Kendall, M. G., and W. R. Buckland. *A Dictionary of Statistical Terms*. New York: Hafner Publishing Co., 1957.

Writings in statistics at the intermediate level are essentially concerned with special topics. References that follow are grouped for convenience; in some instances a reference pertains to several topics but is shown only once under a part division.

SURVEY TECHNIQUES AND DATA FLOW These references discuss methods of data collection, organization of the collection procedure, and methods of processing and presenting data, with particular stress on studies of human population.

Blankenship, Albert B. *Consumer and Opinion Research: The Questionnaire Technique*. New York: Harper & Row, Publishers, 1943.

Goode, W., and P. Hatt. *Methods in Social Research*. New York: McGraw-Hill Book Company, Inc., 1952.

Jenkinson, Bruce L. *Bureau of the Census Manual of Tabular Presentation*. Washington, D.C.: Government Printing Office, 1949.

Modley, Rudolph. *How to Use Pictorial Statistics*. New York: Harper & Row, Publishers, 1937.

Myers, John H. *Statistical Presentation*. Patterson, N.J.: Littlefield, Adams & Co., 1956.

Parten, Mildred B. *Surveys, Polls and Samples*. New York: Harper & Row, Publishers, 1950.

INDUCTIVE STATISTICS For most of the references in this group, in addition to a knowledge of basic statistical methods, some acquaintance with beginning calculus is assumed.

Cochran, William G. *Sampling Techniques*. New York: John Wiley & Sons, Inc., 1953.

Deming, William E. *Some Theory of Sampling*. New York: John Wiley & Sons, Inc., 1950.

Deming, William E. *Sample Design in Business Research.* New York: John Wiley & Sons, Inc., 1960.

Hansen, Morris H., William N. Hurwitz, and William G. Madow. *Sample Survey Methods and Theory.* New York: John Wiley & Sons, Inc., Vol. I, 1953.

Jastram, Roy W. *Elements of Statistical Inference.* Berkeley: California Book Co., Ltd., 1947.

Vance, Lawrence L., and John Neter. *Statistical Sampling for Auditors and Accountants.* New York: John Wiley & Sons, Inc., 1956.

Walker, Helen M., and Joseph Lev. *Statistical Inference.* New York: Holt, Rinehart and Winston, Inc., 1953.

Weiss, Lionel. *Statistical Decision Theory.* New York: McGraw-Hill Book Company, Inc., 1961.

Yates, Frank. *Sampling Methods for Censuses and Surveys.* New York: Hafner Publishing Co., 1953.

QUALITY CONTROL Clear discussion on the construction of control charts and their use can be found in the following.

A.S.T.M. Manual on Quality Control Materials. Philadelphia: American Society for Testing and Materials, 1951.

American Standard, Z1.3—1958, Control Chart Method for Controlling Quality during Production. New York: American Standards Association, Inc., 1958.

Cowden, Dudley J. *Statistical Methods in Quality Control.* New York: McGraw-Hill Book Company, Inc., 1957.

Duncan, Acheson J. *Quality Control and Industrial Statistics.* Homewood, Ill.: Richard D. Irwin, Inc., 1959.

Feigenbaum, A. V. *Quality Control—Principles, Practice, and Administration.* New York: McGraw-Hill Book Company, Inc., 1951.

Grant, Eugene L. *Statistical Quality Control.* New York: McGraw-Hill Book Company, Inc., 1952.

Shewhart, W. A. *Economic Control of Quality of Manufactured Product.* Princeton, N.J.: D. Van Nostrand Company, Inc., 1931.

Association Analysis

Anderson, T. W. *Introduction to Multivariate Statistical Analysis.* New York: John Wiley and Sons, Inc., 1958.

Deming, William E. *Statistical Adjustment of Data.* New York: John Wiley & Sons, Inc., 1941.

Ezekiel, Mordecai. *Methods of Correlation Analysis.* New York: John Wiley & Sons, Inc., 1941.

Ferber, Robert. *Statistical Techniques in Market Research.* New York: McGraw-Hill Book Company, Inc., 1949.

Schultz, Henry. *The Theory and Measurement of Demand.* Chicago: The University of Chicago Press, 1938.

Operation's Research

Brabb, George. *Introduction to Quantitative Management.* New York: Holt, Rinehart and Winston, Inc. 1968.

Churchman, C. West, and others. *Introduction to Operations Research.* New York: John Wiley & Sons, 1957.

Goetz, Billy E. *Quantitative Methods.* New York: McGraw-Hill Book Co., 1965.

Horowitz, Ira. *An Introduction to Quantitative Business Analysis.* New York: McGraw-Hill, 1965.

Sasieni, Maurice, and others. *Operations Research.* New York: John Wiley & Sons, 1964.

Index Numbers

Fisher, Irving. *The Making of Index Numbers.* Boston: Houghton Mifflin Company, 1927.

Joint Economic Committee, 87th Congress, 1st Session. *Government Price Statistics: Hearings before the Subcommittee on Economic Statistics.* Washington, D.C.: Government Printing Office, 1961.

Mitchell, Wesley C. *The Making and Using of Index Numbers.* Washington, D.C.: Government Printing Office, 1938.

Mudgett, Bruce D. *Index Numbers.* New York: John Wiley & Sons, Inc., 1951.

Persons, Warren M. *The Construction of Index Numbers.* Boston: Houghton Mifflin Company, 1928.

Time Series Analysis and Forecasting

Abramson, Adolph G., and Russell H. Mack (Eds.). *Business Forecasting in Practice; Principles and Cases.* New York: John Wiley & Sons, Inc., 1956.

Burns, Arthur F. *Production Trends in the United States Since 1870.* New York: National Bureau of Economic Research, 1934.

Burns, Arthur F., and Wesley C. Mitchell. *Measuring Business Cycles.* New York: National Bureau of Economic Research, 1946.

Kuznets, Simon. *Seasonal Variations in Industry and Trade.* New York: National Bureau of Economic Research, 1933.

National Bureau of Economic Research, Conference on Research in Income and Wealth. *Short-Term Economic Forecasting* (Studies in Income and Wealth). Princeton, N.J.: Princeton University Press, Vol. XVII, 1955.

National Bureau of Economic Research, Conference on Research in Income and Wealth. *Long-Range Economic Projections* (Studies in Income and Wealth). Princeton, N.J.: Princeton University Press, Vol. XVI, 1954.

Newbury, Frank D. *Business Forecasting: Principles and Practices.* New York: McGraw-Hill Book Company, Inc., 1952.

Snyder, Richard M. *Measuring Business Changes.* New York: John Wiley & Sons, Inc., 1955.

Spencer, Milton H., Colin G. Clark, and Peter W. Hoguet. *Business and Economic Forecasting*. Homewood, Ill.: Richard D. Irwin, Inc., 1961.

Wolfe, Henry Dean. *Business Forecasting Methods*. New York: Holt, Rinehart and Winston, Inc., 1966.

PROBABILITY AND MATHEMATICAL STATISTICS Although studies of more advanced theorems of probability and mathematical theory of statistics usually require a considerable amount of higher mathematics, references on these topics given below pre-suppose beginning calculus with only one exception. Seven of these references—Cramér, Feller, Fry, Goldberg, Kolmogorov, Parzen, and Munroe—deal primarily with probability theory and only secondarily with statistics. Of these, Cramér, Goldberg, and Munroe are designed for beginning students of probability. Feller's is a definitive treatment for probability theory in the discrete case. Fry is a pioneering work on applications of probability to engineering problems, Kolmogorov is concerned with the postulational or axiomatic approach to probability and is a standard text in this area. It is, however, extremely compact and the entire book is suitable only for the advanced student.

Books by Alexander, Brunk, Fraser, Hoel, and Hogg and Craig are introductions to mathematical statistics. Both Cramér and Wilks are thorough treatises in mathematical statistics and basic references for the reader who wishes to pursue the subject. Mood demands more mathematical background than elementary calculus, but the student should find his persistence and effort quite rewarding. Cramér includes, in the first section of his book, a comprehensive survey of the mathematical tools necessary for the pursuit of advanced mathematical statistics, which includes set theory, measure theory, Lebesgue integration, matrix theory, and a variety of special integrals and formulas.

Dixon and Massey is an elementary textbook on statistical methods but exceptionally broad in its coverage and comprehensive for its collection of tables.

Alexander, Howard W. *Elements of Mathematical Statistics*. New York: John Wiley & Sons, Inc., 1961.

Brunk, H. D. *An Introduction to Mathematical Statistics*. Englewood Cliffs, N.J.: Prentice-Hall, Inc., 1960.

Cramér, H. *Mathematical Methods of Statistics*. Princeton, N.J.: Princeton University Press, 1946.

Cramér, H. *The Elements of Probability Theory and Some of Its Applications*. New York: John Wiley & Sons, Inc., 1955.

Dixon, W. J., and F. J. Massey. *Introduction to Statistical Analysis*, 2d ed. New York: McGraw-Hill Book Company, Inc., 1957.

Feller, W. *An Introduction to Probability Theory and Its Applications*, 2d ed. New York: John Wiley & Sons, Inc., 1957.

Fraser, D. A. S. *Statistics: An Introduction*. New York: John Wiley & Sons, Inc., 1958.

Fry, T. C. *Probability and Its Engineering Uses*. Princeton, N.J.: D. Van Nostrand Company, 1929.

Goldberg, S. *Probability: An Introduction*. Englewood Cliffs, N.J.: Prentice-Hall, Inc., 1960.

Hoel, P. G. *Introduction to Mathematical Statistics,* 2d ed. New York: John Wiley & Sons, Inc., 1954.

Hogg, R. V., and A. T. Craig. *Introduction to Mathematical Statistics.* New York: The Macmillan Company, 1959.

Kolmogorov, A. N. *Foundations of the Theory of Probability.* New York: Chelsea Publishing Co., 1950.

Mack, S. F. *Elementary Statistics.* New York: Holt, Rinehart and Winston, Inc., 1960.

Mood, A. M. *Introduction to the Theory of Statistics.* New York: McGraw-Hill Book Company, Inc., 1950.

Munroe, M. E. *Theory of Probability.* New York: McGraw-Hill Book Company, Inc., 1951.

Parzen, Emanuel. *Modern Probability Theory and Its Applications.* New York: John Wiley & Sons, Inc., 1960.

Wilks, S. S. *Mathematical Statistics.* Princeton, N.J.: Princeton University Press, 1944.

Interpretation

The books listed below, mainly written for the general public, stress the interpretation of statistical results. They point out a number of misuses of statistics and can be read with profit by all students of statistics.

Huff, Darrell. *How to Lie With Statistics.* New York: W. W. Norton & Company, Inc., 1954.

Levinson, Horace C. *Chance, Luck, and Statistics,* 2d ed., New York: Dover Publications, 1963. (First edition published under the title *The Science of Chance.*)

C

Statistical Tables

TABLE A-I* SQUARES AND SQUARE ROOTS

n	n^2	$\sqrt{n}$	$\sqrt{10n}$
1.00	1.0000	1.00000	3.16228
1.01	1.0201	1.00499	3.17805
1.02	1.0404	1.00995	3.19374
1.03	1.0609	1.01489	3.20936
1.04	1.0816	1.01980	3.22490
1.05	1.1025	1.02470	3.24037
1.06	1.1236	1.02956	3.25576
1.07	1.1449	1.03441	3.27109
1.08	1.1664	1.03923	3.28634
1.09	1.1881	1.04403	3.30151
1.10	1.2100	1.04881	3.31662
1.11	1.2321	1.05357	3.33167
1.12	1.2544	1.05830	3.34664
1.13	1.2769	1.06301	3.36155
1.14	1.2996	1.06771	3.37639
1.15	1.3225	1.07238	3.39116
1.16	1.3456	1.07703	3.40588
1.17	1.3689	1.08167	3.42053
1.18	1.3924	1.08628	3.43511
1.19	1.4161	1.09087	3.44964
1.20	1.4400	1.09545	3.46410
1.21	1.4641	1.10000	3.47851
1.22	1.4884	1.10454	3.49285
1.23	1.5129	1.10905	3.50714
1.24	1.5376	1.11355	3.52136
1.25	1.5625	1.11803	3.53553
1.26	1.5876	1.12250	3.54965
1.27	1.6129	1.12694	3.56371
1.28	1.6384	1.13137	3.57771
1.29	1.6641	1.13578	3.59166
1.30	1.6900	1.14018	3.60555
1.31	1.7161	1.14455	3.61939
1.32	1.7424	1.14891	3.63318
1.33	1.7689	1.15326	3.64692
1.34	1.7956	1.15758	3.66060
1.35	1.8225	1.16190	3.67423
1.36	1.8496	1.16619	3.68782
1.37	1.8769	1.17047	3.70135
1.38	1.9044	1.17473	3.71484
1.39	1.9321	1.17898	3.72827
1.40	1.9600	1.18322	3.74166
1.41	1.9881	1.18743	3.75500
1.42	2.0164	1.19164	3.76829
1.43	2.0449	1.19583	3.78153
1.44	2.0736	1.20000	3.79473
1.45	2.1025	1.20416	3.80789
1.46	2.1316	1.20830	3.82099
1.47	2.1609	1.21244	3.83406
1.48	2.1904	1.21655	3.84708
1.49	2.2201	1.22066	3.86005
1.50	2.2500	1.22474	3.87298
1.51	2.2801	1.22882	3.88587
1.52	2.3104	1.23288	3.89872
1.53	2.3409	1.23693	3.91152
1.54	2.3716	1.24097	3.92428
1.55	2.4025	1.24499	3.93700
1.56	2.4336	1.24900	3.94968
1.57	2.4649	1.25300	3.96232
1.58	2.4964	1.25698	3.97492
1.59	2.5281	1.26095	3.98748
1.60	2.5600	1.26491	4.00000
1.61	2.5921	1.26886	4.01248
1.62	2.6244	1.27279	4.02492
1.63	2.6569	1.27671	4.03733
1.64	2.6896	1.28062	4.04969
1.65	2.7225	1.28452	4.06202
1.66	2.7556	1.28841	4.07431
1.67	2.7889	1.29228	4.08656
1.68	2.8224	1.29615	4.09878
1.69	2.8561	1.30000	4.11096
1.70	2.8900	1.30384	4.12311
1.71	2.9241	1.30767	4.13521
1.72	2.9584	1.31149	4.14729
1.73	2.9929	1.31529	4.15933
1.74	3.0276	1.31909	4.17133
1.75	3.0625	1.32288	4.18330
1.76	3.0976	1.32665	4.19524
1.77	3.1329	1.33041	4.20714
1.78	3.1684	1.33417	4.21900
1.79	3.2041	1.33791	4.23084
1.80	3.2400	1.34164	4.24264
1.81	3.2761	1.34536	4.25441
1.82	3.3124	1.34907	4.26615
1.83	3.3489	1.35277	4.27785
1.84	3.3856	1.35647	4.28952
1.85	3.4225	1.36015	4.30116
1.86	3.4596	1.36382	4.31277
1.87	3.4969	1.36748	4.32435
1.88	3.5344	1.37113	4.33590
1.89	3.5721	1.37477	4.34741
1.90	3.6100	1.37840	4.35890
1.91	3.6481	1.38203	4.37035
1.92	3.6864	1.38564	4.38178
1.93	3.7249	1.38924	4.39318
1.94	3.7636	1.39284	4.40454
1.95	3.8025	1.39642	4.41588
1.96	3.8416	1.40000	4.42719
1.97	3.8809	1.40357	4.43847
1.98	3.9204	1.40712	4.44972
1.99	3.9601	1.41067	4.46094

** Reprinted with permission of The Macmillan Company from Macmillan Selected Mathematics Tables by E. R. Hedrick. Copyright © 1936 by The Macmillan Company, renewed by Dorothy H. McWilliams, Clyde L. Hedrick and Elisabeth B. Miller.*

TABLE A-I SQUARES AND SQUARE ROOTS (Continued)

n	n^2	$\sqrt{n}$	$\sqrt{10n}$
2.00	4.0000	1.41421	4.47214
2.01	4.0401	1.41774	4.48330
2.02	4.0804	1.42127	4.49444
2.03	4.1209	1.42478	4.50555
2.04	4.1616	1.42829	4.51664
2.05	4.2025	1.43178	4.52769
2.06	4.2436	1.43527	4.53872
2.07	4.2849	1.43875	4.54973
2.08	4.3264	1.44222	4.56070
2.09	4.3681	1.44568	4.57165
2.10	4.4100	1.44914	4.58258
2.11	4.4521	1.45258	4.59347
2.12	4.4944	1.45602	4.60435
2.13	4.5369	1.45945	4.61519
2.14	4.5796	1.46287	4.62601
2.15	4.6225	1.46629	4.63681
2.16	4.6656	1.46969	4.64758
2.17	4.7089	1.47309	4.65833
2.18	4.7524	1.47648	4.66905
2.19	4.7961	1.47986	4.67974
2.20	4.8400	1.48324	4.69042
2.21	4.8841	1.48661	4.70106
2.22	4.9284	1.48997	4.71169
2.23	4.9729	1.49332	4.72229
2.24	5.0176	1.49666	4.73286
2.25	5.0625	1.50000	4.74342
2.26	5.1076	1.50333	4.75395
2.27	5.1529	1.50665	4.76445
2.28	5.1984	1.50997	4.77493
2.29	5.2441	1.51327	4.78539
2.30	5.2900	1.51658	4.79583
2.31	5.3361	1.51987	4.80625
2.32	5.3824	1.52315	4.81664
2.33	5.4289	1.52643	4.82701
2.34	5.4756	1.52971	4.83735
2.35	5.5225	1.53297	4.84768
2.36	5.5696	1.53623	4.85798
2.37	5.6169	1.53948	4.86826
2.38	5.6644	1.54272	4.87852
2.39	5.7121	1.54596	4.88876
2.40	5.7600	1.54919	4.89898
2.41	5.8081	1.55242	4.90918
2.42	5.8564	1.55563	4.91935
2.43	5.9049	1.55885	4.92950
2.44	5.9536	1.56205	4.93964
2.45	6.0025	1.56525	4.94975
2.46	6.0516	1.56844	4.95984
2.47	6.1009	1.57162	4.96991
2.48	6.1504	1.57480	4.97996
2.49	6.2001	1.57797	4.98999

n	n^2	$\sqrt{n}$	$\sqrt{10n}$
2.50	6.2500	1.58114	5.00000
2.51	6.3001	1.58430	5.00999
2.52	6.3504	1.58745	5.01996
2.53	6.4009	1.59060	5.02991
2.54	6.4516	1.59374	5.03984
2.55	6.5025	1.59687	5.04975
2.56	6.5536	1.60000	5.05964
2.57	6.6049	1.60312	5.06952
2.58	6.6564	1.60624	5.07937
2.59	6.7081	1.60935	5.08920
2.60	6.7600	1.61245	5.09902
2.61	6.8121	1.61555	5.10882
2.62	6.8644	1.61864	5.11859
2.63	6.9169	1.62173	5.12835
2.64	6.9696	1.62481	5.13809
2.65	7.0225	1.62788	5.14782
2.66	7.0756	1.63095	5.15752
2.67	7.1289	1.63401	5.16720
2.68	7.1824	1.63707	5.17687
2.69	7.2361	1.64012	5.18652
2.70	7.2900	1.64317	5.19615
2.71	7.3441	1.64621	5.20577
2.72	7.3984	1.64924	5.21536
2.73	7.4529	1.65227	5.22494
2.74	7.5076	1.65529	5.23450
2.75	7.5625	1.65831	5.24404
2.76	7.6176	1.66132	5.25357
2.77	7.6729	1.66433	5.26308
2.78	7.7284	1.66733	5.27257
2.79	7.7841	1.67033	5.28205
2.80	7.8400	1.67332	5.29150
2.81	7.8961	1.67631	5.30094
2.82	7.9524	1.67929	5.31037
2.83	8.0089	1.68226	5.31977
2.84	8.0656	1.68523	5.32917
2.85	8.1225	1.68819	5.33854
2.86	8.1796	1.69115	5.34790
2.87	8.2369	1.69411	5.35724
2.88	8.2944	1.69706	5.36656
2.89	8.3521	1.70000	5.37587
2.90	8.4100	1.70294	5.38516
2.91	8.4681	1.70587	5.39444
2.92	8.5264	1.70880	5.40370
2.93	8.5849	1.71172	5.41295
2.94	8.6436	1.71464	5.42218
2.95	8.7025	1.71756	5.43139
2.96	8.7616	1.72047	5.44059
2.97	8.8209	1.72337	5.44977
2.98	8.8804	1.72627	5.45894
2.99	8.9401	1.72916	5.46809

TABLE A-I SQUARES AND SQUARE ROOTS (Continued)

n	n^2	$\sqrt{n}$	$\sqrt{10n}$
3.00	9.0000	1.73205	5.47723
3.01	9.0601	1.73494	5.48635
3.02	9.1204	1.73781	5.49545
3.03	9.1809	1.74069	5.50454
3.04	9.2416	1.74356	5.51362
3.05	9.3025	1.74642	5.52268
3.06	9.3636	1.74929	5.53173
3.07	9.4249	1.75214	5.54076
3.08	9.4864	1.75499	5.54977
3.09	9.5481	1.75784	5.55878
3.10	9.6100	1.76068	5.56776
3.11	9.6721	1.76352	5.57674
3.12	9.7344	1.76635	5.58570
3.13	9.7969	1.76918	5.59464
3.14	9.8596	1.77200	5.60357
3.15	9.9225	1.77482	5.61249
3.16	9.9856	1.77764	5.62139
3.17	10.0489	1.78045	5.63028
3.18	10.1124	1.78326	5.63915
3.19	10.1761	1.78606	5.64801
3.20	10.2400	1.78885	5.65685
3.21	10.3041	1.79165	5.66569
3.22	10.3684	1.79444	5.67450
3.23	10.4329	1.79722	5.68331
3.24	10.4976	1.80000	5.69210
3.25	10.5625	1.80278	5.70088
3.26	10.6276	1.80555	5.70964
3.27	10.6929	1.80831	5.71839
3.28	10.7584	1.81108	5.72713
3.29	10.8241	1.81384	5.73585
3.30	10.8900	1.81659	5.74456
3.31	10.9561	1.81934	5.75326
3.32	11.0224	1.82209	5.76194
3.33	11.0889	1.82483	5.77062
3.34	11.1556	1.82757	5.77927
3.35	11.2225	1.83030	5.78792
3.36	11.2896	1.83303	5.79655
3.37	11.3569	1.83576	5.80517
3.38	11.4244	1.83848	5.81378
3.39	11.4921	1.84120	5.82237
3.40	11.5600	1.84391	5.83095
3.41	11.6281	1.84662	5.83952
3.42	11.6964	1.84932	5.84808
3.43	11.7649	1.85203	5.85662
3.44	11.8336	1.85472	5.86515
3.45	11.9025	1.85742	5.87367
3.46	11.9716	1.86011	5.88218
3.47	12.0409	1.86279	5.89067
3.48	12.1104	1.86548	5.89915
3.49	12.1801	1.86815	5.90762

n	n^2	$\sqrt{n}$	$\sqrt{10n}$
3.50	12.2500	1.87083	5.91608
3.51	12.3201	1.87350	5.92453
3.52	12.3904	1.87617	5.93296
3.53	12.4609	1.87883	5.94138
3.54	12.5316	1.88149	5.94979
3.55	12.6025	1.88414	5.95819
3.56	12.6736	1.88680	5.96657
3.57	12.7449	1.88944	5.97495
3.58	12.8164	1.89209	5.98331
3.59	12.8881	1.89473	5.99166
3.60	12.9600	1.89737	6.00000
3.61	13.0321	1.90000	6.00833
3.62	13.1044	1.90263	6.01664
3.63	13.1769	1.90526	6.02495
3.64	13.2496	1.90788	6.03324
3.65	13.3225	1.91050	6.04152
3.66	13.3956	1.91311	6.04979
3.67	13.4689	1.91572	6.05805
3.68	13.5424	1.91833	6.06630
3.69	13.6161	1.92094	6.07454
3.70	13.6900	1.92354	6.08276
3.71	13.7641	1.92614	6.09098
3.72	13.8384	1.92873	6.09918
3.73	13.9129	1.93132	6.10737
3.74	13.9876	1.93391	6.11555
3.75	14.0625	1.93649	6.12372
3.76	14.1376	1.93907	6.13188
3.77	14.2129	1.94165	6.14003
3.78	14.2884	1.94422	6.14817
3.79	14.3641	1.94679	6.15630
3.80	14.4400	1.94936	6.16441
3.81	14.5161	1.95192	6.17252
3.82	14.5924	1.95448	6.18061
3.83	14.6689	1.95704	6.18870
3.84	14.7456	1.95959	6.19677
3.85	14.8225	1.96214	6.20484
3.86	14.8996	1.96469	6.21289
3.87	14.9769	1.96723	6.22093
3.88	15.0544	1.96977	6.22896
3.89	15.1321	1.97231	6.23699
3.90	15.2100	1.97484	6.24500
3.91	15.2881	1.97737	6.25300
3.92	15.3664	1.97990	6.26099
3.93	15.4449	1.98242	6.26897
3.94	15.5236	1.98494	6.27694
3.95	15.6025	1.98746	6.28490
3.96	15.6816	1.98997	6.29285
3.97	15.7609	1.99249	6.30079
3.98	15.8404	1.99499	6.30872
3.99	15.9201	1.99750	6.31664

TABLE A-I SQUARES AND SQUARE ROOTS (Continued)

n	n^2	$\sqrt{n}$	$\sqrt{10n}$
4.00	16.0000	2.00000	6.32456
4.01	16.0801	2.00250	6.33246
4.02	16.1604	2.00499	6.34035
4.03	16.2409	2.00749	6.34823
4.04	16.3216	2.00998	6.35610
4.05	16.4025	2.01246	6.36396
4.06	16.4836	2.01494	6.37181
4.07	16.5649	2.01742	6.37966
4.08	16.6464	2.01990	6.38749
4.09	16.7281	2.02237	6.39531
4.10	16.8100	2.02485	6.40312
4.11	16.8921	2.02731	6.41093
4.12	16.9744	2.02978	6.41872
4.13	17.0569	2.03224	6.42651
4.14	17.1396	2.03470	6.43428
4.15	17.2225	2.03715	6.44205
4.16	17.3056	2.03961	6.44981
4.17	17.3889	2.04206	6.45755
4.18	17.4724	2.04450	6.46529
4.19	17.5561	2.04695	6.47302
4.20	17.6400	2.04939	6.48074
4.21	17.7241	2.05183	6.48845
4.22	17.8084	2.05426	6.49615
4.23	17.8929	2.05670	6.50384
4.24	17.9776	2.05913	6.51153
4.25	18.0625	2.06155	6.51920
4.26	18.1476	2.06398	6.52687
4.27	18.2329	2.06640	6.53452
4.28	18.3184	2.06882	6.54217
4.29	18.4041	2.07123	6.54981
4.30	18.4900	2.07364	6.55744
4.31	18.5761	2.07605	6.56506
4.32	18.6624	2.07846	6.57267
4.33	18.7489	2.08087	6.58027
4.34	18.8356	2.08327	6.58787
4.35	18.9225	2.08567	6.59545
4.36	19.0096	2.08806	6.60303
4.37	19.0969	2.09045	6.61060
4.38	19.1844	2.09284	6.61816
4.39	19.2721	2.09523	6.62571
4.40	19.3600	2.09762	6.63325
4.41	19.4481	2.10000	6.64078
4.42	19.5364	2.10238	6.64831
4.43	19.6249	2.10476	6.65582
4.44	19.7136	2.10713	6.66333
4.45	19.8025	2.10950	6.67083
4.46	19.8916	2.11187	6.67832
4.47	19.9809	2.11424	6.68581
4.48	20.0704	2.11660	6.69328
4.49	20.1601	2.11896	6.70075
4.50	20.2500	2.12132	6.70820
4.51	20.3401	2.12368	6.71565
4.52	20.4304	2.12603	6.72309
4.53	20.5209	2.12838	6.73053
4.54	20.6116	2.13073	6.73795
4.55	20.7025	2.13307	6.74537
4.56	20.7936	2.13542	6.75278
4.57	20.8849	2.13776	6.76018
4.58	20.9764	2.14009	6.76757
4.59	21.0681	2.14243	6.77495
4.60	21.1600	2.14476	6.78233
4.61	21.2521	2.14709	6.78970
4.62	21.3444	2.14942	6.79706
4.63	21.4369	2.15174	6.80441
4.64	21.5296	2.15407	6.81175
4.65	21.6225	2.15639	6.81909
4.66	21.7156	2.15870	6.82642
4.67	21.8089	2.16102	6.83374
4.68	21.9024	2.16333	6.84105
4.69	21.9961	2.16564	6.84836
4.70	22.0900	2.16795	6.85565
4.71	22.1841	2.17025	6.86294
4.72	22.2784	2.17256	6.87023
4.73	22.3729	2.17486	6.87750
4.74	22.4676	2.17715	6.88477
4.75	22.5625	2.17945	6.89202
4.76	22.6576	2.18174	6.89928
4.77	22.7529	2.18403	6.90652
4.78	22.8484	2.18632	6.91375
4.79	22.9441	2.18861	6.92098
4.80	23.0400	2.19089	6.92820
4.81	23.1361	2.19317	6.93542
4.82	23.2324	2.19545	6.94262
4.83	23.3289	2.19773	6.94982
4.84	23.4256	2.20000	6.95701
4.85	23.5225	2.20227	6.96419
4.86	23.6196	2.20454	6.97137
4.87	23.7169	2.20681	6.97854
4.88	23.8144	2.20907	6.98570
4.89	23.9121	2.21133	6.99285
4.90	24.0100	2.21359	7.00000
4.91	24.1081	2.21585	7.00714
4.92	24.2064	2.21811	7.01427
4.93	24.3049	2.22036	7.02140
4.94	24.4036	2.22261	7.02851
4.95	24.5025	2.22486	7.03562
4.96	24.6016	2.22711	7.04273
4.97	24.7009	2.22935	7.04982
4.98	24.8004	2.23159	7.05691
4.99	24.9001	2.23383	7.06399

TABLE A-I SQUARES AND SQUARE ROOTS (Continued)

n	n^2	$\sqrt{n}$	$\sqrt{10n}$
5.00	25.0000	2.23607	7.07107
5.01	25.1001	2.23830	7.07814
5.02	25.2004	2.24054	7.08520
5.03	25.3009	2.24277	7.09225
5.04	25.4016	2.24499	7.09930
5.05	25.5025	2.24722	7.10634
5.06	25.6036	2.24944	7.11337
5.07	25.7049	2.25167	7.12039
5.08	25.8064	2.25389	7.12741
5.09	25.9081	2.25610	7.13442
5.10	26.0100	2.25832	7.14143
5.11	26.1121	2.26053	7.14843
5.12	26.2144	2.26274	7.15542
5.13	26.3169	2.26495	7.16240
5.14	26.4196	2.26716	7.16938
5.15	26.5225	2.26936	7.17635
5.16	26.6256	2.27156	7.18331
5.17	26.7289	2.27376	7.19027
5.18	26.8324	2.27596	7.19722
5.19	26.9361	2.27816	7.20417
5.20	27.0400	2.28035	7.21110
5.21	27.1441	2.28254	7.21803
5.22	27.2484	2.28473	7.22496
5.23	27.3529	2.28692	7.23187
5.24	27.4576	2.28910	7.23878
5.25	27.5625	2.29129	7.24569
5.26	27.6676	2.29347	7.25259
5.27	27.7729	2.29565	7.25948
5.28	27.8784	2.29783	7.26636
5.29	27.9841	2.30000	7.27324
5.30	28.0900	2.30217	7.28011
5.31	28.1961	2.30434	7.28697
5.32	28.3024	2.30651	7.29383
5.33	28.4089	2.30868	7.30068
5.34	28.5156	2.31084	7.30753
5.35	28.6225	2.31301	7.31437
5.36	28.7296	2.31517	7.32120
5.37	28.8369	2.31733	7.32803
5.38	28.9444	2.31948	7.33485
5.39	29.0521	2.32164	7.34166
5.40	29.1600	2.32379	7.34847
5.41	29.2681	2.32594	7.35527
5.42	29.3764	2.32809	7.36206
5.43	29.4849	2.33024	7.36885
5.44	29.5936	2.33238	7.37564
5.45	29.7025	2.33452	7.38241
5.46	29.8116	2.33666	7.38918
5.47	29.9209	2.33880	7.39594
5.48	30.0304	2.34094	7.40270
5.49	30.1401	2.34307	7.40945

n	n^2	$\sqrt{n}$	$\sqrt{10n}$
5.50	30.2500	2.34521	7.41620
5.51	30.3601	2.34734	7.42294
5.52	30.4704	2.34947	7.42967
5.53	30.5809	2.35160	7.43640
5.54	30.6916	2.35372	7.44312
5.55	30.8025	2.35584	7.44983
5.56	30.9136	2.35797	7.45654
5.57	31.0249	2.36008	7.46324
5.58	31.1364	2.36220	7.46994
5.59	31.2481	2.36432	7.47663
5.60	31.3600	2.36643	7.48331
5.61	31.4721	2.36854	7.48999
5.62	31.5844	2.37065	7.49667
5.63	31.6969	2.37276	7.50333
5.64	31.8096	2.37487	7.50999
5.65	31.9225	2.37697	7.51665
5.66	32.0356	2.37908	7.52330
5.67	32.1489	2.38118	7.52994
5.68	32.2624	2.38238	7.53658
5.69	32.3761	2.38537	7.54321
5.70	32.4900	2.38747	7.54983
5.71	32.6041	2.38956	7.55645
5.72	32.7184	2.39165	7.56307
5.73	32.8329	2.39374	7.56968
5.74	32.9476	2.39583	7.57628
5.75	33.0625	2.39792	7.58288
5.76	33.1776	2.40000	7.58947
5.77	33.2929	2.40208	7.59605
5.78	33.4084	2.40416	7.60263
5.79	33.5241	2.40624	7.60920
5.80	33.6400	2.40832	7.61577
5.81	33.7561	2.41039	7.62234
5.82	33.8724	2.41247	7.62889
5.83	33.9889	2.41454	7.63544
5.84	34.1056	2.41661	7.64199
5.85	34.2225	2.41868	7.64853
5.86	34.3396	2.42074	7.65506
5.87	34.4569	2.42281	7.66159
5.88	34.5744	2.42487	7.66812
5.89	34.6921	2.42693	7.67463
5.90	34.8100	2.42899	7.68115
5.91	34.9281	2.43105	7.68765
5.92	35.0464	2.43311	7.69415
5.93	35.1649	2.43516	7.70065
5.94	35.2836	2.43721	7.70714
5.95	35.4025	2.43926	7.71362
5.96	35.5216	2.44131	7.72010
5.97	35.6409	2.44336	7.72658
5.98	35.7604	2.44540	7.73305
5.99	35.8801	2.44745	7.73951

TABLE A-I SQUARES AND SQUARE ROOTS (Continued)

n	n^2	$\sqrt{n}$	$\sqrt{10n}$
6.00	36.0000	2.44949	7.74597
6.01	36.1201	2.45153	7.75242
6.02	36.2404	2.45357	7.75887
6.03	36.3609	2.45561	7.76531
6.04	36.4816	2.45764	7.77174
6.05	36.6025	2.45967	7.77817
6.06	36.7236	2.46171	7.78460
6.07	36.8449	2.46374	7.79102
6.08	36.9664	2.46577	7.79744
6.09	37.0881	2.46779	7.80385
6.10	37.2100	2.46982	7.81025
6.11	37.3321	2.47184	7.81665
6.12	37.4544	2.47386	7.82304
6.13	37.5769	2.47588	7.82943
6.14	37.6996	2.47790	7.83582
6.15	37.8225	2.47992	7.84219
6.16	37.9456	2.48193	7.84857
6.17	38.0689	2.48395	7.85493
6.18	38.1924	2.48596	7.86130
6.19	38.3161	2.48797	7.86766
6.20	38.4400	2.48998	7.87401
6.21	38.5641	2.49199	7.88036
6.22	38.6884	2.49399	7.88670
6.23	38.8129	2.49600	7.89303
6.24	38.9376	2.49800	7.89937
6.25	39.0625	2.50000	7.90569
6.26	39.1876	2.50200	7.91202
6.27	39.3129	2.50400	7.91833
6.28	39.4384	2.50599	7.92465
6.29	39.5641	2.50799	7.93095
6.30	39.6900	2.50998	7.93725
6.31	39.8161	2.51197	7.94355
6.32	39.9424	2.51396	7.94984
6.33	40.0689	2.51595	7.95613
6.34	40.1956	2.51794	7.96241
6.35	40.3225	2.51992	7.96869
6.36	40.4496	2.52190	7.97496
6.37	40.5769	2.52389	7.98123
6.38	40.7044	2.52587	7.98749
6.39	40.8321	2.52784	7.99375
6.40	40.9600	2.52982	8.00000
6.41	41.0881	2.53180	8.00625
6.42	41.2164	2.53377	8.01249
6.43	41.3449	2.53574	8.01873
6.44	41.4736	2.53772	8.02496
6.45	41.6025	2.53969	8.03119
6.46	41.7316	2.54165	8.03741
6.47	41.8609	2.54362	8.04363
6.48	41.9904	2.54558	8.04984
6.49	42.1201	2.54755	8.05605

n	n^2	$\sqrt{n}$	$\sqrt{10n}$
6.50	42.2500	2.54951	8.06226
6.51	42.3801	2.55147	8.06846
6.52	42.5104	2.55343	8.07465
6.53	42.6409	2.55539	8.08084
6.54	42.7716	2.55734	8.08703
6.55	42.9025	2.55930	8.09321
6.56	43.0336	2.56125	8.09938
6.57	43.1649	2.56320	8.10555
6.58	43.2964	2.56515	8.11172
6.59	43.4281	2.56710	8.11788
6.60	43.5600	2.56905	8.12404
6.61	43.6921	2.57099	8.13019
6.62	43.8244	2.57294	8.13634
6.63	43.9569	2.57488	8.14248
6.64	44.0896	2.57682	8.14862
6.65	44.2225	2.57876	8.15475
6.66	44.3556	2.58070	8.16088
6.67	44.4889	2.58263	8.16701
6.68	44.6224	2.58457	8.17313
6.69	44.7561	2.58650	8.17924
6.70	44.8900	2.58844	8.18535
6.71	45.0241	2.59037	8.19146
6.72	45.1584	2.59230	8.19756
6.73	45.2929	2.59422	8.20366
6.74	45.4276	2.59615	8.20975
6.75	45.5625	2.59808	8.21584
6.76	45.6976	2.60000	8.22192
6.77	45.8329	2.60192	8.22800
6.78	45.9684	2.60384	8.23408
6.79	46.1041	2.60576	8.24015
6.80	46.2400	2.60768	8.24621
6.81	46.3761	2.60960	8.25227
6.82	46.5124	2.61151	8.25833
6.83	46.6489	2.61343	8.26438
6.84	46.7856	2.61534	8.27043
6.85	46.9225	2.61725	8.27647
6.86	47.0596	2.61916	8.28251
6.87	47.1969	2.62107	8.28855
6.88	47.3344	2.62298	8.29458
6.89	47.4721	2.62488	8.30060
6.90	47.6100	2.62679	8.30662
6.91	47.7481	2.62869	8.31264
6.92	47.8864	2.63059	8.31865
6.93	48.0249	2.63249	8.32466
6.94	48.1636	2.63439	8.33067
6.95	48.3025	2.63629	8.33667
6.96	48.4416	2.63818	8.34266
6.97	48.5809	2.64008	8.34865
6.98	48.7204	2.64197	8.35464
6.99	48.8601	2.64386	8.36062

TABLE A-I SQUARES AND SQUARE ROOTS (Continued)

n	n^2	$\sqrt{n}$	$\sqrt{10n}$
7.00	49.0000	2.64575	8.36660
7.01	49.1401	2.64764	8.37257
7.02	49.2804	2.64953	8.37854
7.03	49.4209	2.65141	8.38451
7.04	49.5616	2.65330	8.39047
7.05	49.7025	2.65518	8.39643
7.06	49.8436	2.65707	8.40238
7.07	49.9849	2.65895	8.40833
7.08	50.1264	2.66083	8.41427
7.09	50.2681	2.66271	8.42021
7.10	50.4100	2.66458	8.42615
7.11	50.5521	2.66646	8.43208
7.12	50.6944	2.66833	8.43801
7.13	50.8369	2.67021	8.44393
7.14	50.9796	2.67208	8.44985
7.15	51.1225	2.67395	8.45577
7.16	51.2656	2.67582	8.46168
7.17	51.4089	2.67769	8.46759
7.18	51.5524	2.67955	8.47349
7.19	51.6961	2.68142	8.47939
7.20	51.8400	2.68328	8.48528
7.21	51.9841	2.68514	8.49117
7.22	52.1284	2.68701	8.49706
7.23	52.2729	2.68887	8.50294
7.24	52.4176	2.69072	8.50882
7.25	52.5625	2.69258	8.51469
7.26	52.7076	2.69444	8.52056
7.27	52.8529	2.69629	8.52643
7.28	52.9984	2.69815	8.53229
7.29	53.1441	2.70000	8.53815
7.30	53.2900	2.70185	8.54400
7.31	53.4361	2.70370	8.54985
7.32	53.5824	2.70555	8.55570
7.33	53.7289	2.70740	8.56154
7.34	53.8756	2.70924	8.56738
7.35	54.0225	2.71109	8.57321
7.36	54.1696	2.71293	8.57904
7.37	54.3169	2.71477	8.58487
7.38	54.4644	2.71662	8.59069
7.39	54.6121	2.71846	8.59651
7.40	54.7600	2.72029	8.60233
7.41	54.9081	2.72213	8.60814
7.42	55.0564	2.72397	8.61394
7.43	55.2049	2.72580	8.61974
7.44	55.3536	2.72764	8.62554
7.45	55.5025	2.72947	8.63134
7.46	55.6516	2.73130	8.63713
7.47	55.8009	2.73313	8.64292
7.48	55.9504	2.73496	8.64870
7.49	56.1001	2.73679	8.65448

n	n^2	$\sqrt{n}$	$\sqrt{10n}$
7.50	56.2500	2.73861	8.66025
7.51	56.4001	2.74044	8.66603
7.52	56.5504	2.74226	8.67179
7.53	56.7009	2.74408	8.67756
7.54	56.8516	2.74591	8.68332
7.55	57.0025	2.74773	8.68907
7.56	57.1536	2.74955	8.69483
7.57	57.3049	2.75136	8.70057
7.58	57.4564	2.75318	8.70632
7.59	57.6081	2.75500	8.71206
7.60	57.7600	2.75681	8.71780
7.61	57.9121	2.75862	8.72353
7.62	58.0644	2.76043	8.72926
7.63	58.2169	2.76225	8.73499
7.64	58.3696	2.76405	8.74071
7.65	58.5225	2.76586	8.74643
7.66	58.6756	2.76767	8.75214
7.67	58.8289	2.76948	8,75785
7.68	58.9824	2.77128	8.76356
7.69	59.1361	2.77308	8.76926
7.70	59.2900	2.77489	8.77496
7.71	59.4441	2.77669	8.78066
7.72	59.5984	2.77849	8.78635
7.73	59.7529	2.78029	8.79204
7.74	59.9076	2.78209	8.79773
7.75	60.0625	2.78388	8.80341
7.76	60.2176	2.78568	8.80909
7.77	60.3729	2.78747	8.81476
7.78	60.5284	2.78927	8.82043
7.79	60.6841	2.79106	8.82610
7.80	60.8400	2.79285	8.83176
7.81	60.9961	2.79464	8.83742
7.82	61.1524	2.79643	8.84308
7.83	61.3089	2.79821	8.84873
7.84	61.4656	2.80000	8.85438
7.85	61.6225	2.80179	8.86002
7.86	61.7796	2.80357	8.86566
7.87	61.9369	2.80535	8.87130
7.88	62.0944	2.80713	8.87694
7.89	62.2521	2.80891	8.88257
7.90	62.4100	2.81069	8.88819
7.91	62.5681	2.81247	8.89382
7.92	62.7264	2.81425	8.89944
7.93	62.8849	2.81603	8.90505
7.94	63.0436	2.81780	8.91067
7.95	63.2025	2.81957	8.91628
7.96	63.3616	2.82135	8.92188
7.97	63.5209	2.82312	8.92749
7.98	63.6804	2.82489	8.93308
7.99	63.8401	2.82666	8.93868

TABLE A-I SQUARES AND SQUARE ROOTS (Continued)

n	n^2	$\sqrt{n}$	$\sqrt{10n}$
8.00	64.0000	2.82843	8.94427
8.01	64.1601	2.83019	8.94986
8.02	64.3204	2.83196	8.95545
8.03	64.4809	2.83373	8.96103
8.04	64.6416	2.83549	8.96660
8.05	64.8025	2.83725	8.97218
8.06	64.9636	2.83901	8.97775
8.07	65.1249	2.84077	8.98332
8.08	65.2864	2.84253	8.98888
8.09	65.4481	2.84429	8.99444
8.10	65.6100	2.84605	9.00000
8.11	65.7721	2.84781	9.00555
8.12	65.9344	2.84956	9.01110
8.13	66.0969	2.85132	9.01665
8.14	66.2596	2.85307	9.02219
8.15	66.4225	2.85482	9.02774
8.16	66.5856	2.85657	9.03327
8.17	66.7489	2.85832	9.03881
8.18	66.9124	2.86007	9.04434
8.19	67.0761	2.86182	9.04986
8.20	67.2400	2.86356	9.05539
8.21	67.4041	2.86531	9.06091
8.22	67.5684	2.86705	9.06642
8.23	67.7329	2.86880	9.07193
8.24	67.8976	2.87054	9.07744
8.25	68.0625	2.87228	9.08295
8.26	68.2276	2.87402	9.08845
8.27	68.3929	2.87576	9.09395
8.28	68.5584	2.87750	9.09945
8.29	68.7241	2.87924	9.10494
8.30	68.8900	2.88097	9.11043
8.31	69.0561	2.88271	9.11592
8.32	69.2224	2.88444	9.12140
8.33	69.3889	2.88617	9.12688
8.34	69.5556	2.88791	9.13236
8.35	69.7225	2.88964	9.13783
8.36	69.8896	2.89137	9.14330
8.37	70.0569	2.89310	9.14877
8.38	70.2244	2.89482	9.15423
8.39	70.3921	2.89655	9.15969
8.40	70.5600	2.89828	9.16515
8.41	70.7281	2.90000	9.17061
8.42	70.8964	2.90172	9.17606
8.43	71.0649	2.90345	9.18150
8.44	71.2336	2.90517	9.18695
8.45	71.4025	2.90689	9.19239
8.46	71.5716	2.90861	9.19783
8.47	71.7409	2.91033	9.20326
8.48	71.9104	2.91204	9.20869
8.49	72.0801	2.91376	9.21412

n	n^2	$\sqrt{n}$	$\sqrt{10n}$
8.50	72.2500	2.91548	9.21954
8.51	72.4201	2.91719	9.22497
8.52	72.5904	2.91890	9.23038
8.53	72.7609	2.92062	9.23580
8.54	72.9316	2.92233	9.24121
8.55	73.1025	2.92404	9.24662
8.56	73.2736	2.92575	9.25203
8.57	73.4449	2.92746	9.25743
8.58	73.6164	2.92916	9.26283
8.59	73.7881	2.93087	9.26823
8.60	73.9600	2.93258	9.27362
8.61	74.1321	2.93428	9.27901
8.62	74.3044	2.93598	9.28440
8.63	74.4769	2.93769	9.28978
8.64	74.6496	2.93939	9.29516
8.65	74.8225	2.94109	9.30054
8.66	74.9956	2.94279	9.30591
8.67	75.1689	2.94449	9.31128
8.68	75.3424	2.94618	9.31665
8.69	75.5161	2.94788	9.32202
8.70	75.6900	2.94958	9.32738
8.71	75.8641	2.95127	9.33274
8.72	76.0384	2.95296	9.33809
8.73	76.2129	2.95466	9.34345
8.74	76.3876	2.95635	9.34880
8.75	76.5625	2.95804	9.35414
8.76	76.7376	2.95973	9.35949
8.77	76.9129	2.96142	9.36483
8.78	77.0884	2.96311	9.37017
8.79	77.2641	2.96479	9.37550
8.80	77.4400	2.96648	9.38083
8.81	77.6161	2.96816	9.38616
8.82	77.7924	2.96985	9.39149
8.83	77.9689	2.97153	9.39681
8.84	78.1456	2.97321	9.40213
8.85	78.3225	2.97489	9.40744
8.86	78.4996	2.97658	9.41276
8.87	78.6769	2.97825	9.41807
8.88	78.8544	2.97993	9.42338
8.89	79.0321	2.98161	9.42868
8.90	79.2100	2.98329	9.43398
8.91	79.3881	2.98496	9.43928
8.92	79.5664	2.98664	9.44458
8.93	79.7449	2.98831	9.44987
8.94	79.9236	2.98998	9.45516
8.95	80.1025	2.99166	9.46044
8.96	80.2816	2.99333	9.46573
8.97	80.4609	2.99500	9.47101
8.98	80.6404	2.99666	9.47629
8.99	80.8201	2.99833	9.48156

TABLE A-I SQUARES AND SQUARE ROOTS (Concluded)

n	n^2	$\sqrt{n}$	$\sqrt{10n}$	n	n^2	$\sqrt{n}$	$\sqrt{10n}$
9.00	81.0000	3.00000	9.48683	**9.50**	90.2500	3.08221	9.74679
9.01	81.1801	3.00167	9.49210	9.51	90.4401	3.08383	9.75192
9.02	81.3604	3.00333	9.49737	9.52	90.6304	3.08545	9.75705
9.03	81.5409	3.00500	9.50263	9.53	90.8209	3.08707	9.76217
9.04	81.7216	3.00666	9.50789	9.54	91.0116	3.08869	9.76729
9.05	81.9025	3.00832	9.51315	9.55	91.2025	3.09031	9.77241
9.06	82.0836	3.00998	9.51840	9.56	91.3936	3.09192	9.77753
9.07	82.2649	3.01164	9.52365	9.57	91.5849	3.09354	9.78264
9.08	82.4464	3.01330	9.52890	9.58	91.7764	3.09516	9.78775
9.09	82.6281	3.01496	9.53415	9.59	91.9681	3.09677	9.79285
9.10	82.8100	3.01662	9.53939	**9.60**	92.1600	3.09839	9.79796
9.11	82.9921	3.01828	9.54463	9.61	92.3521	3.10000	9.80306
9.12	83.1744	3.01993	9.54987	9.62	92.5444	3.10161	9.80816
9.13	83.3569	3.02159	9.55510	9.63	92.7369	3.10322	9.81326
9.14	83.5396	3.02324	9.56033	9.64	92.9296	3.10483	9.81835
9.15	83.7225	3.02490	9.56556	9.65	93.1225	3.10644	9.82344
9.16	83.9056	3.02655	9.57079	9.66	93.3156	3.10805	9.82853
9.17	84.0889	3.02820	9.57601	9.67	93.5089	3.10966	9.83362
9.18	84.2724	3.02985	9.58123	9.68	93.7024	3.11127	9.83870
9.19	84.4561	3.03150	9.58645	9.69	93.8961	3.11288	9.84378
9.20	84.6400	3.03315	9.59166	**9.70**	94.0900	3.11448	9.84886
9.21	84.8241	3.03480	9.59687	9.71	94.2841	3.11609	9.85393
9.22	85.0084	3.03645	9.60208	9.72	94.4784	3.11769	9.85901
9.23	85.1929	3.03809	9.60729	9.73	94.6729	3.11929	9.86408
9.24	85.3776	3.03974	9.61249	9.74	94.8676	3.12090	9.86914
9.25	85.5625	3.04138	9.61769	9.75	95.0625	3.12250	9.87421
9.26	85.7476	3.04302	9.62289	9.76	95.2576	3.12410	9.87927
9.27	85.9329	3.04467	9.62808	9.77	95.4529	3.12570	9.88433
9.28	86.1184	3.04631	9.63328	9.78	95.6484	3.12730	9.88939
9.29	86.3041	3.04795	9.63846	9.79	95.8441	3.12890	9.89444
9.30	86.4900	3.04959	9.64365	**9.80**	96.0400	3.13050	9.89949
9.31	86.6761	3.05123	9.64883	9.81	96.2361	3.13209	9.90454
9.32	86.8624	3.05287	9.65401	9.82	96.4324	3.13369	9.90959
9.33	87.0489	3.05450	9.65919	9.83	96.6289	3.13528	9.91464
9.34	87.2356	3.05614	9.66437	9.84	96.8256	3.13688	9.91968
9.35	87.4225	3.05778	9.66954	9.85	97.0225	3.13847	9.92472
9.36	87.6096	3.05941	9.67471	9.86	97.2196	3.14006	9.92975
9.37	87.7969	3.06105	9.67988	9.87	97.4169	3.14166	9.93479
9.38	87.9844	3.06268	9.68504	9.88	97.6144	3.14325	9.93982
9.39	88.1721	3.06431	9.69020	9.89	97.8121	3.14484	9.94485
9.40	88.3600	3.06594	9.69536	**9.90**	98.0100	3.14643	9.94987
9.41	88.5481	3.06757	9.70052	9.91	98.2081	3.14802	9.95490
9.42	88.7364	3.06920	9.70567	9.92	98.4064	3.14960	9.95992
9.43	88.9249	3.07083	9.71082	9.93	98.6049	3.15119	9.96494
9.44	89.1136	3.07246	9.71597	9.94	98.8036	3.15278	9.96995
9.45	89.3025	3.07409	9.72111	9.95	99.0025	3.15436	9.97497
9.46	89.4916	3.07571	9.72625	9.96	99.2016	3.15595	9.97998
9.47	89.6809	3.07734	9.73139	9.97	99.4009	3.15753	9.98499
9.48	89.8704	3.07896	9.73653	9.98	99.6004	3.15911	9.98999
9.49	90.0601	3.08058	9.74166	9.99	99.8001	3.16070	9.99500

TABLE A-II COMMON LOGARITHMS

N	0	1	2	3	4	5	6	7	8	9
10	0000	0043	0086	0128	0170	0212	0253	0294	0334	0374
11	0414	0453	0492	0531	0569	0607	0645	0682	0719	0755
12	0792	0828	0864	0899	0934	0969	1004	1038	1072	1106
13	1139	1173	1206	1239	1271	1303	1335	1367	1399	1430
14	1461	1492	1523	1553	1584	1614	1644	1673	1703	1732
15	1761	1790	1818	1847	1875	1903	1931	1959	1987	2014
16	2041	2068	2095	2122	2148	2175	2201	2227	2253	2279
17	2304	2330	2355	2380	2405	2430	2455	2480	2504	2529
18	2553	2577	2601	2625	2648	2672	2695	2718	2742	2765
19	2788	2810	2833	2856	2878	2900	2923	2945	2967	2989
20	3010	3032	3054	3075	3096	3118	3139	3160	3181	3201
21	3222	3243	3263	3284	3304	3324	3345	3365	3385	3404
22	3424	3444	3464	3483	3502	3522	3541	3560	3579	3598
23	3617	3636	3655	3674	3692	3711	3729	3747	3766	3784
24	3802	3820	3838	3856	3874	3892	3909	3927	3945	3962
25	3979	3997	4014	4031	4048	4065	4082	4099	4116	4133
26	4150	4166	4183	4200	4216	4232	4249	4265	4281	4298
27	4314	4330	4346	4362	4378	4393	4409	4425	4440	4456
28	4472	4487	4502	4518	4533	4548	4564	4579	4594	4609
29	4624	4639	4654	4669	4683	4698	4713	4728	4742	4757
30	4771	4786	4800	4814	4829	4843	4857	4871	4886	4900
31	4914	4928	4942	4955	4969	4983	4997	5011	5024	5038
32	5051	5065	5079	5092	5105	5119	5132	5145	5159	5172
33	5185	5198	5211	5224	5237	5250	5263	5276	5289	5302
34	5315	5328	5340	5353	5366	5378	5391	5403	5416	5428
35	5441	5453	5465	5478	5490	5502	5514	5527	5539	5551
36	5563	5575	5587	5599	5611	5623	5635	5647	5658	5670
37	5682	5694	5705	5717	5729	5740	5752	5763	5775	5786
38	5798	5809	5821	5832	5843	5855	5866	5877	5888	5899
39	5911	5922	5933	5944	5955	5966	5977	5988	5999	6010
40	6021	6031	6042	6053	6064	6075	6085	6096	6107	6117
41	6128	6138	6149	6160	6170	6180	6191	6201	6212	6222
42	6232	6243	6253	6263	6274	6284	6294	6304	6314	6325
43	6335	6345	6355	6365	6375	6385	6395	6405	6415	6425
44	6435	6444	6454	6464	6474	6484	6493	6503	6513	6522
45	6532	6542	6551	6561	6571	6580	6590	6599	6609	6618
46	6628	6637	6646	6656	6665	6675	6684	6693	6702	6712
47	6721	6730	6739	6749	6758	6767	6776	6785	6794	6803
48	6812	6821	6830	6839	6848	6857	6866	6875	6884	6893
49	6902	6911	6920	6928	6937	6946	6955	6964	6972	6981
50	6990	6998	7007	7016	7024	7033	7042	7050	7059	7067
51	7076	7084	7093	7101	7110	7118	7126	7135	7143	7152
52	7160	7168	7177	7185	7193	7202	7210	7218	7226	7235
53	7243	7251	7259	7267	7275	7284	7292	7300	7308	7316
54	7324	7332	7340	7348	7356	7364	7372	7380	7388	7396
N	0	1	2	3	4	5	6	7	8	9

TABLE A-II COMMON LOGARITHMS (Concluded)

N	0	1	2	3	4	5	6	7	8	9
55	7404	7412	7419	7427	7435	7443	7451	7459	7466	7474
56	7482	7490	7497	7505	7513	7520	7528	7536	7543	7551
57	7559	7566	7574	7582	7589	7597	7604	7612	7619	7627
58	7634	7642	7649	7657	7664	7672	7679	7686	7694	7701
59	7709	7716	7723	7731	7738	7745	7752	7760	7767	7774
60	7782	7789	7796	7803	7810	7818	7825	7832	7839	7846
61	7853	7860	7868	7875	7882	7889	7896	7903	7910	7917
62	7924	7931	7938	7945	7952	7959	7966	7973	7980	7987
63	7993	8000	8007	8014	8021	8028	8035	8041	8048	8055
64	8062	8069	8075	8082	8089	8096	8102	8109	8116	8122
65	8129	8136	8142	8149	8156	8162	8169	8176	8182	8189
66	8195	8202	8209	8215	8222	8228	8235	8241	8248	8254
67	8261	8267	8274	8280	8287	8293	8299	8306	8312	8319
68	8325	8331	8338	8344	8351	8357	8363	8370	8376	8382
69	8388	8395	8401	8407	8414	8420	8426	8432	8439	8445
70	8451	8457	8463	8470	8476	8482	8488	8494	8500	8506
71	8513	8519	8525	8531	8537	8543	8549	8555	8561	8567
72	8573	8579	8585	8591	8597	8603	8609	8615	8621	8627
73	8633	8639	8645	8651	8657	8663	8669	8675	8681	8686
74	8692	8698	8704	8710	8716	8722	8727	8733	8739	8745
75	8751	8756	8762	8768	8774	8779	8785	8791	8797	8802
76	8808	8814	8820	8825	8831	8837	8842	8848	8854	8859
77	8865	8871	8876	8882	8887	8893	8899	8904	8910	8915
78	8921	8927	8932	8938	8943	8949	8954	8960	8965	8971
79	8976	8982	8987	8993	8998	9004	9009	9015	9020	9025
80	9031	9036	9042	9047	9053	9058	9063	9069	9074	9079
81	9085	9090	9096	9101	9106	9112	9117	9122	9128	9133
82	9138	9143	9149	9154	9159	9165	9170	9175	9180	9186
83	9191	9196	9201	9206	9212	9217	9222	9227	9232	9238
84	9243	9248	9253	9258	9263	9269	9274	9279	9284	9289
85	9294	9299	9304	9309	9315	9320	9325	9330	9335	9340
86	9345	9350	9355	9360	9365	9370	9375	9380	9385	9390
87	9395	9400	9405	9410	9415	9420	9425	9430	9435	9440
88	9445	9450	9455	9460	9465	9469	9474	9479	9484	9489
89	9494	9499	9504	9509	9513	9518	9523	9528	9533	9538
90	9542	9547	9552	9557	9562	9566	9571	9576	9581	9586
91	9590	9595	9600	9605	9609	9614	9619	9624	9628	9633
92	9638	9643	9647	9652	9657	9661	9666	9671	9675	9680
93	9685	9689	9694	9699	9703	9708	9713	9717	9722	9727
94	9731	9736	9741	9745	9750	9754	9759	9763	9768	9773
95	9777	9782	9786	9791	9795	9800	9805	9809	9814	9818
96	9823	9827	9832	9836	9841	9845	9850	9854	9859	9863
97	9868	9872	9877	9881	9886	9890	9894	9899	9903	9908
98	9912	9917	9921	9926	9930	9934	9939	9943	9948	9952
99	9956	9961	9965	9969	9974	9978	9983	9987	9991	9996
N	0	1	2	3	4	5	6	7	8	9

TABLE A-III THE CUMULATIVE BINOMIAL DISTRIBUTION*

n	r	$p = .10$	$p = .20$	$p = .25$	$p = .30$	$p = .40$	$p = .50$
5	0	.59049	.32768	.23730	.16807	.07776	.03125
	1	.91854	.73728	.63281	.52822	.33696	.18750
	2	.99144	.94208	.89648	.83692	.68256	.50000
	3	.99954	.99328	.98437	.96922	.91296	.81250
	4	.99999	.99968	.99902	.99757	.98976	.96875
	5	1.00000	1.00000	1.00000	1.00000	1.00000	1.00000
10	0	.34868	.10737	.05631	.02825	.00605	.00098
	1	.73610	.37581	.24403	.14931	.04636	.01074
	2	.92981	.67780	.52559	.38278	.16729	.05469
	3	.98720	.87913	.77588	.64961	.38228	.17187
	4	.99837	.96721	.92187	.84973	.63310	.37695
	5	.99985	.99363	.98027	.95265	.83376	.62305
	6	.99999	.99914	.99649	.98941	.94524	.82812
	7	1.00000	.99992	.99958	.99841	.98771	.94531
	8		1.00000	.99997	.99986	.99832	.98926
	9			1.00000	.99999	.99990	.99902
	10				1.00000	1.00000	1.00000
15	0	.20589	.03518	.01336	.00475	.00047	.00003
	1	.54904	.16713	.08018	.03527	.00517	.00049
	2	.81594	.39802	.23609	.12683	.02711	.00369
	3	.94444	.64816	.46129	.29687	.09050	.01758
	4	.98728	.83577	.68649	.51549	.21728	.05923
	5	.99775	.93895	.85163	.72162	.40322	.15088
	6	.99969	.98194	.94338	.86886	.60981	.30362
	7	.99997	.99576	.98270	.94999	.78690	.50000
	8	1.00000	.99921	.99581	.98476	.90495	.69638
	9		.99989	.99921	.99635	.96617	.84912
	10		.99999	.99988	.99933	.99065	.94077
	11		1.00000	.99999	.99991	.99807	.98242
	12			1.00000	.99999	.99972	.99631
	13				1.00000	.99997	.99951
	14					1.00000	.99997
	15						1.00000
20	0	.12158	.01153	.00317	.00080	.00004	.00000
	1	.39175	.06918	.02431	.00764	.00052	.00002
	2	.67693	.20608	.09126	.03548	.00361	.00020
	3	.86705	.41145	.22516	.10709	.01596	.00129
	4	.95683	.62965	.41484	.23751	.05095	.00591
	5	.98875	.80421	.61717	.41637	.12560	.02069
	6	.99761	.91331	.78578	.60801	.25001	.05766
	7	.99958	.96786	.89819	.77227	.41589	.13159
	8	.99994	.99002	.95907	.88667	.59560	.25172
	9	.99999	.99741	.98614	.95204	.75534	.41190
	10	1.00000	.99944	.99606	.98286	.87248	.58810
	11		.99990	.99906	.99486	.94347	.74828
	12		.99998	.99982	.99872	.97897	.86841
	13		1.00000	.99997	.99974	.99353	.94234

** E. C. Molina Poisson's Binomial Exponential Limit (Princeton, N. J.: D. Van Nostrand Company, Inc., 1949), pp. 276–280. Courtesy of D. Van Nostrand Company, Inc.*

TABLE A-III THE CUMULATIVE BINOMIAL DISTRIBUTION (Continued)

n	r	$p = .10$	$p = .20$	$p = .25$	$p = .30$	$p = .40$	$p = .50$
20	14			1.00000	.99996	.99839	.97931
	15				.99999	.99968	.99409
	16				1.00000	.99995	.99871
	17					.99999	.99980
	18					1.00000	.99998
	19						1.00000
25	0	.07179	.00378	.00075	.00013	.00000	.00000
	1	.27121	.02739	.00702	.00157	.00005	.00000
	2	.53709	.09823	.03211	.00896	.00043	.00001
	3	.76359	.23399	.09621	.03324	.00237	.00008
	4	.90201	.42067	.21374	.09047	.00947	.00046
	5	.96660	.61669	.37828	.19349	.02936	.00204
	6	.99052	.78004	.56110	.34065	.07357	.00732
	7	.99774	.89088	.72651	.51185	.15355	.02164
	8	.99954	.95323	.85056	.67693	.27353	.05388
	9	.99992	.98267	.92867	.81056	.42462	.11476
	10	.99999	.99445	.97033	.90220	.58577	.21218
	11	1.00000	.99846	.98027	.95575	.73228	.34502
	12		.99963	.99663	.98253	.84623	.50000
	13		.99992	.99908	.99401	.92220	.65498
	14		.99999	.99979	.99822	.96561	.78782
	15		1.00000	.99996	.99955	.98683	.88524
	16			.99999	.99990	.99567	.94612
	17			1.00000	.99998	.99879	.97836
	18				1.00000	.99972	.99268
	19					.99995	.99796
	20					.99999	.99954
	21					1.00000	.99992
	22						.99999
	23						1.00000
50	0	.00515	.00001	.00000	.00000		
	1	.03379	.00019	.00001	.00000		
	2	.11173	.00129	.00009	.00000		
	3	.25029	.00566	.00050	.00003		
	4	.43120	.01850	.00211	.00017		
	5	.61612	.04803	.00705	.00072	.00000	
	6	.77023	.10340	.01939	.00249	.00001	
	7	.87785	.19041	.04526	.00726	.00006	
	8	.94213	.30733	.09160	.01825	.00023	
	9	.97546	.44374	.16368	.04023	.00076	.00000
	10	.99065	.58356	.26220	.07885	.00220	.00001
	11	.99678	.71067	.38162	.13904	.00569	.00005
	12	.99900	.81394	.51099	.22287	.01325	.00015
	13	.99971	.88941	.63704	.32788	.02799	.00047
	14	.99993	.93928	.74808	.44683	.05396	.00130
	15	.99998	.96920	.83692	.56918	.09550	.00330
	16	1.00000	.98556	.90169	.68388	.15609	.00767
	17		.99374	.94488	.78219	.23688	.01642

TABLE A-III THE CUMULATIVE BINOMIAL DISTRIBUTION (Continued)

n	r	p = .10	p = .20	p = .25	p = .30	p = .40	p = .50
50	18		.99749	.97127	.85944	.33561	.03245
	19		.99907	.98608	.91520	.44648	.05946
	20		.99968	.99374	.95224	.56103	.10132
	21		.99990	.99738	.97491	.67014	.16112
	22		.99997	.99898	.98772	.76602	.23994
	23		.99999	.99963	.99441	.84383	.33591
	24		1.00000	.99988	.99763	.90219	.44386
	25			.99996	.99907	.94266	.55614
	26			.99999	.99966	.96859	.66409
	27			1.00000	.99988	.98397	.76006
	28				.99996	.99238	.83888
	29				.99999	.99664	.89868
	30				1.00000	.99863	.94054
	31					.99948	.96755
	32					.99982	.98358
	33					.99994	.99233
	34					.99998	.99670
	35					1.00000	.99870
	36						.99953
	37						.99985
	38						.99995
	39						.99999
	40						1.00000
100	0	.00003					
	1	.00032					
	2	.00194					
	3	.00784					
	4	.02371	.00000				
	5	.05758	.00002				
	6	.11716	.00008				
	7	.20605	.00028	.00000			
	8	.32087	.00086	.00001			
	9	.45129	.00233	.00004			
	10	.58316	.00570	.00014	.00000		
	11	.70303	.01257	.00039	.00001		
	12	.80182	.02533	.00103	.00002		
	13	.87612	.04691	.00246	.00006		
	14	.92743	.08044	.00542	.00016		
	15	.96011	.12851	.01108	.00040		
	16	.97940	.19234	.02111	.00097		
	17	.98999	.27119	.03763	.00216		
	18	.99542	.36209	.06301	.00452	.00000	
	19	.99802	.46016	.09953	.00889	.00001	
	20	.99919	.55946	.14883	.01646	.00002	
	21	.99969	.65403	.21144	.02883	.00004	
	22	.99989	.73893	.28637	.04787	.00011	
	23	.99996	.81091	.37018	.07553	.00025	
	24	.99999	.86865	.46167	.11357	.00056	
	25	1.00000	.91252	.55347	.16313	.00119	

TABLE A-III THE CUMULATIVE BINOMIAL DISTRIBUTION (Concluded)

n	r	$p = .10$	$p = .20$	$p = .25$	$p = .30$	$p = .40$	$p = .50$
100	26		.94417	.64174	.22440	.00240	
	27		.96585	.72238	.29637	.00460	.00000
	28		.97998	.79246	.37678	.00843	.00001
	29		.98875	.85046	.46234	.01478	.00002
	30		.99394	.89621	.54912	.02478	.00004
	31		.99687	.93065	.63311	.03985	.00009
	32		.99845	.95540	.71072	.06150	.00020
	33		.99926	.97241	.77926	.09125	.00044
	34		.99966	.98357	.83714	.13034	.00089
	35		.99985	.99059	.88392	.17947	.00176
	36		.99994	.99482	.92012	.23861	.00332
	37		.99998	.99725	.94695	.30681	.00602
	38		.99999	.99860	.96602	.38219	.01049
	39		1.00000	.99931	.97901	.46208	.01760
	40			.99968	.98750	.54329	.02844
	41			.99985	.99283	.62253	.04431
	42			.99994	.99603	.69674	.06661
	43			.99997	.99789	.76347	.09667
	44			.99999	.99891	.82110	.13563
	45			1.00000	.99946	.86891	.18410
	46				.99974	.90702	.24206
	47				.99988	.93621	.30865
	48				.99995	.95770	.38218
	49				.99998	.97290	.46021
	50				.99999	.98324	.53979
	51				1.00000	.98999	.61782
	52					.99424	.69135
	53					.99680	.79794
	54					.99829	.81590
	55					.99912	.86437
	56					.99956	.90333
	57					.99979	.93339
	58					.99990	.95569
	59					.99996	.97156
	60					.99998	.98240
	61					.99999	.98951
	62					1.00000	.99398
	63						.99668
	64						.99824
	65						.99911
	66						.99956
	67						.99980
	68						.99991
	69						.99996
	70						.99998
	71						.99999
	72						1.00000

TABLE A-IV THE HYPERGEOMETRIC DISTRIBUTION*

N	n	k	r or x	P(r)	p(x)
10	1	1	0	0.900000	0.900000
10	1	1	1	1.000000	0.100000
10	2	1	0	0.800000	0.800000
10	2	1	1	1.000000	0.200000
10	2	2	0	0.622222	0.622222
10	2	2	1	0.977778	0.355556
10	2	2	2	1.000000	0.022222
10	3	1	0	0.700000	0.700000
10	3	2	1	1.000000	0.300000
10	3	2	0	0.466667	0.466667
10	3	2	1	0.933333	0.466667
10	3	2	2	1.000000	0.066667
10	3	3	0	0.291667	0.291667
10	3	3	1	0.816667	0.525000
10	3	3	2	0.991667	0.175000
10	3	3	3	1.000000	0.008333
10	4	1	0	0.600000	0.600000
10	4	1	1	1.000000	0.400000
10	4	2	0	0.333333	0.333333
10	4	2	1	0.866667	0.533333
10	4	2	2	1.000000	0.133333
10	4	3	0	0.166667	0.166667
10	4	3	1	0.666667	0.500000
10	4	3	2	0.966667	0.300000
10	4	3	3	1.000000	0.033333
10	4	4	0	0.071429	0.071429
10	4	4	1	0.452381	0.380952
10	4	4	2	0.880952	0.428571
10	4	4	3	0.995238	0.114286
10	4	4	4	1.000000	0.004762
10	5	1	0	0.500000	0.500000
10	5	1	1	1.000000	0.500000
10	5	2	0	0.222222	0.222222
10	5	2	1	0.777778	0.555556
10	5	2	2	1.000000	0.222222
10	5	3	0	0.083333	0.083333
10	5	3	1	0.500000	0.416667
10	5	3	2	0.916667	0.416667
10	5	3	3	1.000000	0.083333
10	5	4	0	0.023810	0.023810
10	5	4	1	0.261905	0.238095
10	5	4	2	0.738095	0.476190
10	5	4	3	0.976190	0.238095
10	5	4	4	1.000000	0.023810
10	5	5	0	0.003968	0.003968
10	5	5	1	0.103175	0.099206
10	5	5	2	0.500000	0.396825
10	5	5	3	0.896825	0.396825
10	5	5	4	0.996032	0.099206
10	5	5	5	1.000000	0.003968
10	6	1	0	0.400000	0.400000
10	6	1	1	1.000000	0.600000
10	6	2	0	0.133333	0.133333
10	6	2	1	0.666667	0.533333
10	6	2	2	1.000000	0.333333
10	6	3	0	0.033333	0.033333
10	6	3	1	0.333333	0.300000
10	6	3	2	0.833333	0.500000
10	6	3	3	1.000000	0.166667
10	6	4	0	0.004762	0.004762
10	6	4	1	0.119048	0.114286
10	6	4	2	0.547619	0.428571
10	6	4	3	0.928571	0.380952
10	6	4	4	1.000000	0.071429
10	6	5	1	0.023810	0.023810
10	6	5	2	0.261905	0.238095
10	6	5	3	0.738095	0.476190
10	6	5	4	0.976190	0.238095
10	6	5	5	1.000000	0.023810
10	6	6	2	0.071429	0.071429

Extracted with permission from Gerald J. Lieberman and Donald B. Owen, "Tables of the Hypergeometric Probability Distribution," Stanford University Press, Stanford, Calif., 1961.

TABLE A-IV THE HYPERGEOMETRIC DISTRIBUTION (Concluded)

N	n	k	r or x	P(r)	p(x)
10	6	6	3	0.452381	0.380952
10	6	6	4	0.880952	0.428571
10	6	6	5	0.995238	0.114286
10	6	6	6	1.000000	0.004762
10	7	1	0	0.300000	0.300000
10	7	1	1	1.000000	0.700000
10	7	2	0	0.066667	0.066667
10	7	2	1	0.533333	0.466667
10	7	2	2	1.000000	0.466667
10	7	3	0	0.008333	0.008333
10	7	3	1	0.183333	0.175000
10	7	3	2	0.708333	0.525000
10	7	3	3	1.000000	0.291667
10	7	4	1	0.033333	0.033333
10	7	4	2	0.333333	0.300000
10	7	4	3	0.833333	0.500000
10	7	4	4	1.000000	0.166667
10	7	5	2	0.083333	0.083333
10	7	5	3	0.500000	0.416667
10	7	5	4	0.916667	0.416667
10	7	5	5	1.000000	0.083333
10	7	6	3	0.166667	0.166667
10	7	6	4	0.666667	0.500000
10	7	6	5	0.966667	0.300000
10	7	6	6	1.000000	0.033333
10	7	7	4	0.291667	0.291667
10	7	7	5	0.816667	0.525000
10	7	7	6	0.991667	0.175000
10	7	7	7	1.000000	0.008333
10	8	1	0	0.200000	0.200000
10	8	1	1	1.000000	0.800000
10	8	2	0	0.022222	0.022222
10	8	2	1	0.377778	0.355556
10	8	2	2	1.000000	0.622222
10	8	3	1	0.066667	0.066667
10	8	3	2	0.533333	0.466667
10	8	3	3	1.000000	0.466667
10	8	4	2	0.133333	0.133333
10	8	4	3	0.666667	0.533333
10	8	4	4	1.000000	0.333333
10	8	5	3	0.222222	0.222222
10	8	5	4	0.777778	0.555556
10	8	5	5	1.000000	0.222222
10	8	6	4	0.333333	0.333333
10	8	6	5	0.866667	0.533333
10	8	6	6	1.000000	0.133333
10	8	7	5	0.466667	0.466667
10	8	7	6	0.933333	0.466667
10	8	7	7	1.000000	0.066667
10	8	8	6	0.622222	0.622222
10	8	8	7	0.977778	0.355556
10	8	8	8	1.000000	0.022222
10	9	1	0	0.100000	0.100000
10	9	1	1	1.000000	0.900000
10	9	2	1	0.200000	0.200000
10	9	2	2	1.000000	0.800000
10	9	3	2	0.300000	0.300000
10	9	3	3	1.000000	0.700000
10	9	4	3	0.400000	0.400000
10	9	4	4	1.000000	0.600000
10	9	5	4	0.500000	0.500000
10	9	5	5	1.000000	0.500000
10	9	6	5	0.600000	0.600000
10	9	6	6	1.000000	0.400000
10	9	7	6	0.700000	0.700000
10	9	7	7	1.000000	0.300000
10	9	8	7	0.800000	0.800000
10	9	8	8	1.000000	0.200000
10	9	9	8	0.900000	0.900000
10	9	9	9	1.000000	0.100000

TABLE A-V THE CUMULATIVE POISSON DISTRIBUTION*

r	$\mu = .1$	$\mu = .2$	$\mu = .3$	$\mu = .4$	$\mu = .5$
0	.90484	.81873	.74082	.67302	.60653
1	.99532	.98248	.96306	.93845	.90980
2	.99985	.99885	.99640	.99207	.98561
3	1.00000	.99994	.99973	.99922	.99825
4		1.00000	.99998	.99994	.99983
5			1.00000	1.00000	.99999
6					1.00000

r	$\mu = .6$	$\mu = .7$	$\mu = .8$	$\mu = .9$	$\mu = 1.0$
0	.54881	.49658	.44933	.40657	.36788
1	.87810	.84419	.80879	.77248	.73576
2	.97688	.96586	.95258	.93714	.91970
3	.99664	.99425	.99092	.98654	.98101
4	.99961	.99921	.99859	.99766	.99634
5	.99996	.99991	.99982	.99966	.99941
6	1.00000	.99999	.99998	.99996	.99992
7		1.00000	1.00000	1.00000	.99999
8					1.00000

r	$\mu = 2$	$\mu = 3$	$\mu = 4$	$\mu = 5$	$\mu = 6$
0	.13534	.04979	.01832	.00674	.00248
1	.40601	.19915	.09158	.04043	.01735
2	.67668	.42319	.23810	.12465	.06197
3	.85712	.64723	.43347	.26503	.15120
4	.94735	.81526	.62884	.44049	.28506
5	.98344	.91608	.78513	.61596	.44568
6	.99547	.96649	.88933	.76218	.60630
7	.99890	.98810	.94887	.86663	.74398
8	.99976	.99620	.97864	.93191	.84724
9	.99995	.99890	.99187	.96817	.91608
10	.99999	.99971	.99716	.98630	.95738
11	1.00000	.99993	.99908	.99455	.97991
12		.99998	.99973	.99798	.99117
13		1.00000	.99992	.99930	.99637
14			.99998	.99977	.99860
15			1.00000	.99993	.99949
16				.99998	.99982
17				1.00000	.99994
18					.99998
19					1.00000

** Extracted from E. C. Molina, Poisson's Binomial Exponential Limit, (Princeton, N. J.: D. Van Nostrand Company, Inc., 1949), pp. 289–291. Courtesy of D. Van Nostrand Company, Inc.*

TABLE A-V THE CUMULATIVE POISSON DISTRIBUTION (Concluded)

r	$\mu = 7$	$\mu = 8$	$\mu = 9$	$\mu = 10$
0	.00091	.00033	.00012	.00004
1	.00730	.00302	.00123	.00050
2	.02964	.01375	.00623	.00277
3	.08176	.04238	.02123	.01034
4	.17299	.09963	.05496	.02925
5	.30071	.19124	.11569	.06709
6	.44971	.31337	.20678	.13014
7	.59871	.45296	.32390	.22022
8	.72909	.59255	.45565	.33282
9	.83050	.71662	.58741	.45793
10	.90148	.81589	.70599	.58304
11	.94665	.88808	.80301	.69678
12	.97300	.93620	.87577	.79156
13	.98719	.96582	.92615	.86446
14	.99428	.98274	.95853	.91654
15	.99759	.99177	.97796	.95126
16	.99904	.99628	.98889	.97296
17	.99964	.99841	.99468	.98572
18	.99987	.99935	.99757	.99281
19	.99996	.99975	.99894	.99655
20	.99999	.99991	.99956	.99841
21	1.00000	.99997	.99982	.99930
22		.99999	.99993	.99970
23		1.00000	.99998	.99988
24			.99999	.99995
25			1.00000	.99998
26				.99999
27				1.00000

TABLE A-VI EXPONENTIAL FUNCTIONS

x	e^{-x}	e^{x}
.01	.9900	1.0101
.02	.9802	1.0202
.03	.9704	1.0305
.04	.9608	1.0408
.05	.9512	1.0513
.06	.9418	1.0618
.07	.9324	1.0725
.08	.9231	1.0833
.09	.9139	1.0942
.10	.9048	1.1052
.20	.8187	1.2214
.30	.7408	1.3499
.40	.6703	1.4918
.50	.6065	1.6487
.60	.5488	1.8221
.70	.4966	2.0138
.80	.4493	2.2255
.90	.4066	2.4596
1.00	.3679	2.7183
2.00	.1353	7.3891
3.00	.04979	20.0886
4.00	.01832	54.598
5.00	.00674	148.41
6.00	.00248	403.43
7.00	.000912	1096.6
8.00	.000335	2981.0
9.00	.000123	8103.1
10.00	.000045	22026.0

TABLE A-VII VALUES OF THE STANDARD NORMAL DISTRIBUTION FUNCTION

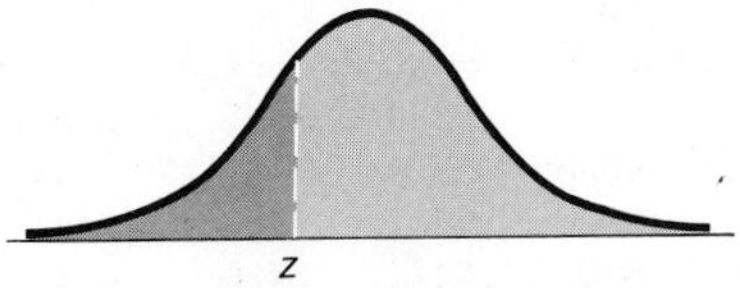

z	0	1	2	3	4	5	6	7	8	9
−3.	.0013	.0010	.0007	.0005	.0003	.0002	.0002	.0001	.0001	.0000
−2.9	.0019	.0018	.0017	.0017	.0016	.0016	.0015	.0015	.0014	.0014
−2.8	.0026	.0025	.0024	.0023	.0023	.0022	.0021	.0021	.0020	.0019
−2.7	.0035	.0034	.0033	.0032	.0031	.0030	.0029	.0028	.0027	.0026
−2.6	.0047	.0045	.0044	.0043	.0041	.0040	.0039	.0038	.0037	.0036
−2.5	.0062	.0060	.0059	.0057	.0055	.0054	.0052	.0051	.0049	.0048
−2.4	.0082	.0080	.0078	.0075	.0073	.0071	.0069	.0068	.0066	.0064
−2.3	.0107	.0104	.0102	.0099	.0096	.0094	.0091	.0089	.0087	.0084
−2.2	.0139	.0136	.0132	.0129	.0126	.0122	.0119	.0116	.0113	.0110
−2.1	.0179	.0174	.0170	.0166	.0162	.0158	.0154	.0150	.0146	.0143
−2.0	.0228	.0222	.0217	.0212	.0207	.0202	.0197	.0192	.0188	.0183
−1.9	.0287	.0281	.0274	.0268	.0262	.0256	.0250	.0244	.0238	.0233
−1.8	.0359	.0352	.0344	.0336	.0329	.0322	.0314	.0307	.0300	.0294
−1.7	.0446	.0436	.0427	.0418	.0409	.0401	.0392	.0384	.0375	.0367
−1.6	.0548	.0537	.0526	.0516	.0505	.0495	.0485	.0475	.0465	.0455
−1.5	.0668	.0655	.0643	.0630	.0618	.0606	.0594	.0582	.0570	.0559
−1.4	.0808	.0793	.0778	.0764	.0749	.0735	.0722	.0708	.0694	.0681
−1.3	.0968	.0951	.0934	.0918	.0901	.0885	.0869	.0853	.0838	.0823
−1.2	.1151	.1131	.1112	.1093	.1075	.1056	.1038	.1020	.1003	.0985
−1.1	.1357	.1335	.1314	.1292	.1271	.1251	.1230	.1210	.1190	.1170
−1.0	.1587	.1562	.1539	.1515	.1492	.1469	.1446	.1423	.1401	.1379
− .9	.1841	.1814	.1788	.1762	.1736	.1711	.1685	.1660	.1635	.1611
− .8	.2119	.2090	.2061	.2033	.2005	.1977	.1949	.1922	.1894	.1867
− .7	.2420	.2389	.2358	.2327	.2297	.2266	.2236	.2206	.2177	.2148
− .6	.2743	.2709	.2676	.2643	.2611	.2578	.2546	.2514	.2483	.2451
− .5	.3085	.3050	.3015	.2981	.2946	.2912	.2877	.2843	.2810	.2776
− .4	.3446	.3409	.3372	.3336	.3300	.3264	.3228	.3192	.3156	.3121
− .3	.3821	.3783	.3745	.3707	.3669	.3632	.3594	.3557	.3520	.3483
− .2	.4207	.4168	.4129	.4090	.4052	.4013	.3974	.3936	.3897	.3859
− .1	.4602	.4562	.4522	.4483	.4443	.4404	.4364	.4325	.4286	.4247
− .0	.5000	.4960	.4920	.4880	.4840	.4801	.4761	.4721	.4681	.4641

(*Table concluded on next page.*)

TABLE A-VII VALUES OF THE STANDARD NORMAL DISTRIBUTION FUNCTION (Concluded)

z	0	1	2	3	4	5	6	7	8	9
.0	.5000	.5040	.5080	.5120	.5160	.5199	.5239	.5279	.5319	.5359
.1	.5398	.5438	.5478	.5517	.5557	.5596	.5636	.5675	.5714	.5753
.2	.5793	.5832	.5871	.5910	.5948	.5987	.6026	.6064	.6103	.6141
.3	.6179	.6217	.6255	.6293	.6331	.6368	.6406	.6443	.6480	.6517
.4	.6554	.6591	.6628	.6664	.6700	.6736	.6772	.6808	.6844	.6879
.5	.6915	.6950	.6985	.7019	.7054	.7088	.7123	.7157	.7190	.7224
.6	.7257	.7291	.7324	.7357	.7389	.7422	.7454	.7486	.7517	.7549
.7	.7580	.7611	.7642	.7673	.7703	.7734	.7764	.7794	.7823	.7852
.8	.7881	.7910	.7939	.7967	.7995	.8023	.8051	.8078	.8106	.8133
.9	.8159	.8186	.8212	.8238	.8264	.8289	.8315	.8340	.8365	.8389
1.0	.8413	.8438	.8461	.8485	.8508	.8531	.8554	.8577	.8599	.8621
1.1	.8643	.8665	.8686	.8708	.8729	.8749	.8770	.8790	.8810	.8830
1.2	.8849	.8869	.8888	.8907	.8925	.8944	.8962	.8980	.8997	.9015
1.3	.9032	.9049	.9066	.9082	.9099	.9115	.9131	.9147	.9162	.9177
1.4	.9192	.9207	.9222	.9236	.9251	.9265	.9278	.9292	.9306	.9319
1.5	.9332	.9345	.9357	.9370	.9382	.9394	.9406	.9418	.9430	.9441
1.6	.9452	.9463	.9474	.9484	.9495	.9505	.9515	.9525	.9535	.9545
1.7	.9554	.9564	.9573	.9582	.9591	.9599	.9608	.9616	.9625	.9633
1.8	.9641	.9648	.9656	.9664	.9671	.9678	.9686	.9693	.9700	.9706
1.9	.9713	.9719	.9726	.9732	.9738	.9744	.9750	.9756	.9762	.9767
2.0	.9772	.9778	.9783	.9788	.9793	.9798	.9803	.9808	.9812	.9817
2.1	.9821	.9826	.9830	.9834	.9838	.9842	.9846	.9850	.9854	.9857
2.2	.9861	.9864	.9868	.9871	.9874	.9878	.9881	.9884	.9887	.9890
2.3	.9893	.9896	.9898	.9901	.9904	.9906	.9909	.9911	.9913	.9916
2.4	.9918	.9920	.9922	.9925	.9927	.9929	.9931	.9932	.9934	.9936
2.5	.9938	.9940	.9941	.9943	.9945	.9946	.9948	.9949	.9951	.9952
2.6	.9953	.9955	.9956	.9957	.9959	.9960	.9961	.9962	.9963	.9964
2.7	.9965	.9966	.9967	.9968	.9969	.9970	.9971	.9972	.9973	.9974
2.8	.9974	.9975	.9976	.9977	.9977	.9978	.9979	.9979	.9980	.9981
2.9	.9981	.9982	.9982	.9983	.9984	.9984	.9985	.9985	.9986	.9986
3.	.9987	.9990	.9993	.9995	.9997	.9998	.9998	.9999	.9999	1.0000

Note 1: If a random variable X is not "standard," its values must be "standardized": $Z = (X - \mu)/\sigma$. That is,

$$P(X \leq x) = N\left(\frac{x - \mu}{\sigma}\right).$$

Note 2: For $z \geq 4$, $N(z) = 1$ to four decimal places; for $z \leq -4$, $N(z) = 0$ to four decimal places.

TABLE A-VIII THE CHI-SQUARE DISTRIBUTION

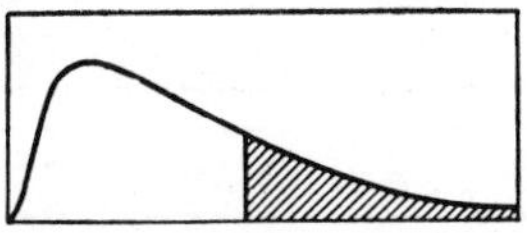

df	Probability that chi-square value will be exceeded							
	.995	.990	.975	.950	.050	.025	.010	.005
1	---	---	---	.004	3.84	5.02	6.63	7.88
2	.01	.02	.05	.10	5.99	7.38	9.21	10.60
3	.07	.11	.22	.35	7.81	9.35	11.34	12.84
4	.21	.30	.48	.71	9.49	11.14	13.28	14.86
5	.41	.55	.83	1.15	11.07	12.83	15.09	16.75
6	.68	.87	1.24	1.64	12.59	14.45	16.81	18.55
7	.99	1.24	1.69	2.17	14.07	16.01	18.48	20.28
8	1.34	1.65	2.18	2.73	15.51	17.53	20.09	21.96
9	1.73	2.09	2.70	3.33	16.92	19.02	21.67	23.59
10	2.16	2.56	3.25	3.94	18.31	20.48	23.21	25.19
11	2.60	3.05	3.82	4.57	19.68	21.92	24.72	26.76
12	3.07	3.57	4.40	5.23	21.03	23.34	26.22	28.30
13	3.57	4.11	5.01	5.89	22.36	24.74	27.69	29.82
14	4.07	4.66	5.63	6.57	23.68	26.12	29.14	31.32
15	4.60	5.23	6.26	7.26	25.00	27.49	30.58	32.80
16	5.14	5.81	6.91	7.96	26.30	28.85	32.00	34.27
17	5.70	6.41	7.56	8.67	27.59	30.19	33.41	35.72
18	6.26	7.01	8.23	9.39	28.87	31.53	34.81	37.16
19	6.84	7.63	8.91	10.12	30.14	32.85	36.19	38.58
20	7.43	8.26	9.59	10.85	31.41	34.17	37.57	40.00
21	8.03	8.90	10.28	11.59	32.67	35.48	38.93	41.40
22	8.64	9.54	10.98	12.34	33.92	36.78	40.29	42.80
23	9.26	10.20	11.69	13.09	35.17	38.08	41.64	44.18
24	9.89	10.86	12.40	13.85	36.42	39.36	42.98	45.56
25	10.52	11.52	13.12	14.61	37.65	40.65	44.31	46.93
26	11.16	12.20	13.84	15.38	38.89	41.92	45.64	48.29
27	11.81	12.88	14.57	16.15	40.11	43.19	46.96	49.64
28	12.46	13.56	15.31	16.93	41.34	44.46	48.28	50.99
29	13.12	14.26	16.05	17.71	42.56	45.72	49.59	52.34
30	13.79	14.95	16.79	18.49	43.77	46.98	50.89	53.67
40	20.71	22.16	24.43	26.51	55.76	59.34	63.69	66.77
50	27.99	29.71	32.36	34.76	67.50	71.42	76.15	79.49
60	35.53	37.48	40.48	43.19	79.08	83.30	88.38	91.95
70	43.28	45.44	48.76	51.74	90.53	95.02	100.43	104.22
80	51.17	53.54	57.15	60.39	101.88	106.63	112.33	116.32
90	59.20	61.75	65.65	69.13	113.14	118.14	124.12	128.30
100	67.33	70.06	74.22	77.93	124.34	129.56	135.81	140.17

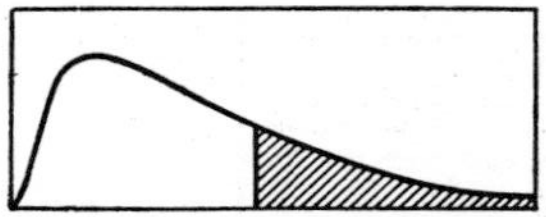

TABLE A-IX THE F DISTRIBUTION

Upper 10% points

δ_2 \ δ_1	1	2	3	4	5	6	7	8	9	10	12	15	20	24	30	40	60	120	∞
1	39·86	49·50	53·59	55·83	57·24	58·20	58·91	59·44	59·86	60·19	60·71	61·22	61·74	62·00	62·26	62·53	62·79	63·06	63·33
2	8·53	9·00	9·16	9·24	9·29	9·33	9·35	9·37	9·38	9·39	9·41	9·42	9·44	9·45	9·46	9·47	9·47	9·48	9·49
3	5·54	5·46	5·39	5·34	5·31	5·28	5·27	5·25	5·24	5·23	5·22	5·20	5·18	5·18	5·17	5·16	5·15	5·14	5·13
4	4·54	4·32	4·19	4·11	4·05	4·01	3·98	3·95	3·94	3·92	3·90	3·87	3·84	3·83	3·82	3·80	3·79	3·78	3·76
5	4·06	3·78	3·62	3·52	3·45	3·40	3·37	3·34	3·32	3·30	3·27	3·24	3·21	3·19	3·17	3·16	3·14	3·12	3·10
6	3·78	3·46	3·29	3·18	3·11	3·05	3·01	2·98	2·96	2·94	2·90	2·87	2·84	2·82	2·80	2·78	2·76	2·74	2·72
7	3·59	3·26	3·07	2·96	2·88	2·83	2·78	2·75	2·72	2·70	2·67	2·63	2·59	2·58	2·56	2·54	2·51	2·49	2·47
8	3·46	3·11	2·92	2·81	2·73	2·67	2·62	2·59	2·56	2·54	2·50	2·46	2·42	2·40	2·38	2·36	2·34	2·32	2·29
9	3·36	3·01	2·81	2·69	2·61	2·55	2·51	2·47	2·44	2·42	2·38	2·34	2·30	2·28	2·25	2·23	2·21	2·18	2·16
10	3·29	2·92	2·73	2·61	2·52	2·46	2·41	2·38	2·35	2·32	2·28	2·24	2·20	2·18	2·16	2·13	2·11	2·08	2·06
11	3·23	2·86	2·66	2·54	2·45	2·39	2·34	2·30	2·27	2·25	2·21	2·17	2·12	2·10	2·08	2·05	2·03	2·00	1·97
12	3·18	2·81	2·61	2·48	2·39	2·33	2·28	2·24	2·21	2·19	2·15	2·10	2·06	2·04	2·01	1·99	1·96	1·93	1·90
13	3·14	2·76	2·56	2·43	2·35	2·28	2·23	2·20	2·16	2·14	2·10	2·05	2·01	1·98	1·96	1·93	1·90	1·88	1·85
14	3·10	2·73	2·52	2·39	2·31	2·24	2·19	2·15	2·12	2·10	2·05	2·01	1·96	1·94	1·91	1·89	1·86	1·83	1·80
15	3·07	2·70	2·49	2·36	2·27	2·21	2·16	2·12	2·09	2·06	2·02	1·97	1·92	1·90	1·87	1·85	1·82	1·79	1·76
16	3·05	2·67	2·46	2·33	2·24	2·18	2·13	2·09	2·06	2·03	1·99	1·94	1·89	1·87	1·84	1·81	1·78	1·75	1·72
17	3·03	2·64	2·44	2·31	2·22	2·15	2·10	2·06	2·03	2·00	1·96	1·91	1·86	1·84	1·81	1·78	1·75	1·72	1·69
18	3·01	2·62	2·42	2·29	2·20	2·13	2·08	2·04	2·00	1·98	1·93	1·89	1·84	1·81	1·78	1·75	1·72	1·69	1·66
19	2·99	2·61	2·40	2·27	2·18	2·11	2·06	2·02	1·98	1·96	1·91	1·86	1·81	1·79	1·76	1·73	1·70	1·67	1·63
20	2·97	2·59	2·38	2·25	2·16	2·09	2·04	2·00	1·96	1·94	1·89	1·84	1·79	1·77	1·74	1·71	1·68	1·64	1·61
21	2·96	2·57	2·36	2·23	2·14	2·08	2·02	1·98	1·95	1·92	1·87	1·83	1·78	1·75	1·72	1·69	1·66	1·62	1·59
22	2·95	2·56	2·35	2·22	2·13	2·06	2·01	1·97	1·93	1·90	1·86	1·81	1·76	1·73	1·70	1·67	1·64	1·60	1·57
23	2·94	2·55	2·34	2·21	2·11	2·05	1·99	1·95	1·92	1·89	1·84	1·80	1·74	1·72	1·69	1·66	1·62	1·59	1·55
24	2·93	2·54	2·33	2·19	2·10	2·04	1·98	1·94	1·91	1·88	1·83	1·78	1·73	1·70	1·67	1·64	1·61	1·57	1·53
25	2·92	2·53	2·32	2·18	2·09	2·02	1·97	1·93	1·89	1·87	1·82	1·77	1·72	1·69	1·66	1·63	1·59	1·56	1·52
26	2·91	2·52	2·31	2·17	2·08	2·01	1·96	1·92	1·88	1·86	1·81	1·76	1·71	1·68	1·65	1·61	1·58	1·54	1·50
27	2·90	2·51	2·30	2·17	2·07	2·00	1·95	1·91	1·87	1·85	1·80	1·75	1·70	1·67	1·64	1·60	1·57	1·53	1·49
28	2·89	2·50	2·29	2·16	2·06	2·00	1·94	1·90	1·87	1·84	1·79	1·74	1·69	1·66	1·63	1·59	1·56	1·52	1·48
29	2·89	2·50	2·28	2·15	2·06	1·99	1·93	1·89	1·86	1·83	1·78	1·73	1·68	1·65	1·62	1·58	1·55	1·51	1·47
30	2·88	2·49	2·28	2·14	2·05	1·98	1·93	1·88	1·85	1·82	1·77	1·72	1·67	1·64	1·61	1·57	1·54	1·50	1·46
40	2·84	2·44	2·23	2·09	2·00	1·93	1·87	1·83	1·79	1·76	1·71	1·66	1·61	1·57	1·54	1·51	1·47	1·42	1·38
60	2·79	2·39	2·18	2·04	1·95	1·87	1·82	1·77	1·74	1·71	1·66	1·60	1·54	1·51	1·48	1·44	1·40	1·35	1·29
120	2·75	2·35	2·13	1·99	1·90	1·82	1·77	1·72	1·68	1·65	1·60	1·55	1·48	1·45	1·41	1·37	1·32	1·26	1·19
∞	2·71	2·30	2·08	1·94	1·85	1·77	1·72	1·67	1·63	1·60	1·55	1·49	1·42	1·38	1·34	1·30	1·24	1·17	1·00

TABLE A-IX THE F DISTRIBUTION (Continued)

Upper 5% points

δ_2 \ δ_1	1	2	3	4	5	6	7	8	9	10	12	15	20	24	30	40	60	120	∞
1	161·4	199·5	215·7	224·6	230·2	234·0	236·8	238·9	240·5	241·9	243·9	245·9	248·0	249·1	250·1	251·1	252·2	253·3	254·3
2	18·51	19·00	19·16	19·25	19·30	19·33	19·35	19·37	19·38	19·40	19·41	19·43	19·45	19·45	19·46	19·47	19·48	19·49	19·50
3	10·13	9·55	9·28	9·12	9·01	8·94	8·89	8·85	8·81	8·79	8·74	8·70	8·66	8·64	8·62	8·59	8·57	8·55	8·53
4	7·71	6·94	6·59	6·39	6·26	6·16	6·09	6·04	6·00	5·96	5·91	5·86	5·80	5·77	5·75	5·72	5·69	5·66	5·63
5	6·61	5·79	5·41	5·19	5·05	4·95	4·88	4·82	4·77	4·74	4·68	4·62	4·56	4·53	4·50	4·46	4·43	4·40	4·36
6	5·99	5·14	4·76	4·53	4·39	4·28	4·21	4·15	4·10	4·06	4·00	3·94	3·87	3·84	3·81	3·77	3·74	3·70	3·67
7	5·59	4·74	4·35	4·12	3·97	3·87	3·79	3·73	3·68	3·64	3·57	3·51	3·44	3·41	3·38	3·34	3·30	3·27	3·23
8	5·32	4·46	4·07	3·84	3·69	3·58	3·50	3·44	3·39	3·35	3·28	3·22	3·15	3·12	3·08	3·04	3·01	2·97	2·93
9	5·12	4·26	3·86	3·63	3·48	3·37	3·29	3·23	3·18	3·14	3·07	3·01	2·94	2·90	2·86	2·83	2·79	2·75	2·71
10	4·96	4·10	3·71	3·48	3·33	3·22	3·14	3·07	3·02	2·98	2·91	2·85	2·77	2·74	2·70	2·66	2·62	2·58	2·54
11	4·84	3·98	3·59	3·36	3·20	3·09	3·01	2·95	2·90	2·85	2·79	2·72	2·65	2·61	2·57	2·53	2·49	2·45	2·40
12	4·75	3·89	3·49	3·26	3·11	3·00	2·91	2·85	2·80	2·75	2·69	2·62	2·54	2·51	2·47	2·43	2·38	2·34	2·30
13	4·67	3·81	3·41	3·18	3·03	2·92	2·83	2·77	2·71	2·67	2·60	2·53	2·46	2·42	2·38	2·34	2·30	2·25	2·21
14	4·60	3·74	3·34	3·11	2·96	2·85	2·76	2·70	2·65	2·60	2·53	2·46	2·39	2·35	2·31	2·27	2·22	2·18	2·13
15	4·54	3·68	3·29	3·06	2·90	2·79	2·71	2·64	2·59	2·54	2·48	2·40	2·33	2·29	2·25	2·20	2·16	2·11	2·07
16	4·49	3·63	3·24	3·01	2·85	2·74	2·66	2·59	2·54	2·49	2·42	2·35	2·28	2·24	2·19	2·15	2·11	2·06	2·01
17	4·45	3·59	3·20	2·96	2·81	2·70	2·61	2·55	2·49	2·45	2·38	2·31	2·23	2·19	2·15	2·10	2·06	2·01	1·96
18	4·41	3·55	3·16	2·93	2·77	2·66	2·58	2·51	2·46	2·41	2·34	2·27	2·19	2·15	2·11	2·06	2·02	1·97	1·92
19	4·38	3·52	3·13	2·90	2·74	2·63	2·54	2·48	2·42	2·38	2·31	2·23	2·16	2·11	2·07	2·03	1·98	1·93	1·88
20	4·35	3·49	3·10	2·87	2·71	2·60	2·51	2·45	2·39	2·35	2·28	2·20	2·12	2·08	2·04	1·99	1·95	1·90	1·84
21	4·32	3·47	3·07	2·84	2·68	2·57	2·49	2·42	2·37	2·32	2·25	2·18	2·10	2·05	2·01	1·96	1·92	1·87	1·81
22	4·30	3·44	3·05	2·82	2·66	2·55	2·46	2·40	2·34	2·30	2·23	2·15	2·07	2·03	1·98	1·94	1·89	1·84	1·78
23	4·28	3·42	3·03	2·80	2·64	2·53	2·44	2·37	2·32	2·27	2·20	2·13	2·05	2·01	1·96	1·91	1·86	1·81	1·76
24	4·26	3·40	3·01	2·78	2·62	2·51	2·42	2·36	2·30	2·25	2·18	2·11	2·03	1·98	1·94	1·89	1·84	1·79	1·73
25	4·24	3·39	2·99	2·76	2·60	2·49	2·40	2·34	2·28	2·24	2·16	2·09	2·01	1·96	1·92	1·87	1·82	1·77	1·71
26	4·23	3·37	2·98	2·74	2·59	2·47	2·39	2·32	2·27	2·22	2·15	2·07	1·99	1·95	1·90	1·85	1·80	1·75	1·69
27	4·21	3·35	2·96	2·73	2·57	2·46	2·37	2·31	2·25	2·20	2·13	2·06	1·97	1·93	1·88	1·84	1·79	1·73	1·67
28	4·20	3·34	2·95	2·71	2·56	2·45	2·36	2·29	2·24	2·19	2·12	2·04	1·96	1·91	1·87	1·82	1·77	1·71	1·65
29	4·18	3·33	2·93	2·70	2·55	2·43	2·35	2·28	2·22	2·18	2·10	2·03	1·94	1·90	1·85	1·81	1·75	1·70	1·64
30	4·17	3·32	2·92	2·69	2·53	2·42	2·33	2·27	2·21	2·16	2·09	2·01	1·93	1·89	1·84	1·79	1·74	1·68	1·62
40	4·08	3·23	2·84	2·61	2·45	2·34	2·25	2·18	2·12	2·08	2·00	1·92	1·84	1·79	1·74	1·69	1·64	1·58	1·51
60	4·00	3·15	2·76	2·53	2·37	2·25	2·17	2·10	2·04	1·99	1·92	1·84	1·75	1·70	1·65	1·59	1·53	1·47	1·39
120	3·92	3·07	2·68	2·45	2·29	2·17	2·09	2·02	1·96	1·91	1·83	1·75	1·66	1·61	1·55	1·50	1·43	1·35	1·25
∞	3·84	3·00	2·60	2·37	2·21	2·10	2·01	1·94	1·88	1·83	1·75	1·67	1·57	1·52	1·46	1·39	1·32	1·22	1·00

TABLE A-IX THE F DISTRIBUTION (Concluded)

Upper 1% points

δ_2 \ δ_1	1	2	3	4	5	6	7	8	9	10	12	15	20	24	30	40	60	120	∞
1	4052	4999·5	5403	5625	5764	5859	5928	5982	6022	6056	6106	6157	6209	6235	6261	6287	6313	6339	6366
2	98·50	99·00	99·17	99·25	99·30	99·33	99·36	99·37	99·39	99·40	99·42	99·43	99·45	99·46	99·47	99·47	99·48	99·49	99·50
3	34·12	30·82	29·46	28·71	28·24	27·91	27·67	27·49	27·35	27·23	27·05	26·87	26·69	26·60	26·50	26·41	26·32	26·22	26·13
4	21·20	18·00	16·69	15·98	15·52	15·21	14·98	14·80	14·66	14·55	14·37	14·20	14·02	13·93	13·84	13·75	13·65	13·56	13·46
5	16·26	13·27	12·06	11·39	10·97	10·67	10·46	10·29	10·16	10·05	9·89	9·72	9·55	9·47	9·38	9·29	9·20	9·11	9·02
6	13·75	10·92	9·78	9·15	8·75	8·47	8·26	8·10	7·98	7·87	7·72	7·56	7·40	7·31	7·23	7·14	7·06	6·97	6·88
7	12·25	9·55	8·45	7·85	7·46	7·19	6·99	6·84	6·72	6·62	6·47	6·31	6·16	6·07	5·99	5·91	5·82	5·74	5·65
8	11·26	8·65	7·59	7·01	6·63	6·37	6·18	6·03	5·91	5·81	5·67	5·52	5·36	5·28	5·20	5·12	5·03	4·95	4·86
9	10·56	8·02	6·99	6·42	6·06	5·80	5·61	5·47	5·35	5·26	5·11	4·96	4·81	4·73	4·65	4·57	4·48	4·40	4·31
10	10·04	7·56	6·55	5·99	5·64	5·39	5·20	5·06	4·94	4·85	4·71	4·56	4·41	4·33	4·25	4·17	4·08	4·00	3·91
11	9·65	7·21	6·22	5·67	5·32	5·07	4·89	4·74	4·63	4·54	4·40	4·25	4·10	4·02	3·94	3·86	3·78	3·69	3·60
12	9·33	6·93	5·95	5·41	5·06	4·82	4·64	4·50	4·39	4·30	4·16	4·01	3·86	3·78	3·70	3·62	3·54	3·45	3·36
13	9·07	6·70	5·74	5·21	4·86	4·62	4·44	4·30	4·19	4·10	3·96	3·82	3·66	3·59	3·51	3·43	3·34	3·25	3·17
14	8·86	6·51	5·56	5·04	4·69	4·46	4·28	4·14	4·03	3·94	3·80	3·66	3·51	3·43	3·35	3·27	3·18	3·09	3·00
15	8·68	6·36	5·42	4·89	4·56	4·32	4·14	4·00	3·89	3·80	3·67	3·52	3·37	3·29	3·21	3·13	3·05	2·96	2·87
16	8·53	6·23	5·29	4·77	4·44	4·20	4·03	3·89	3·78	3·69	3·55	3·41	3·26	3·18	3·10	3·02	2·93	2·84	2·75
17	8·40	6·11	5·18	4·67	4·34	4·10	3·93	3·79	3·68	3·59	3·46	3·31	3·16	3·08	3·00	2·92	2·83	2·75	2·65
18	8·29	6·01	5·09	4·58	4·25	4·01	3·84	3·71	3·60	3·51	3·37	3·23	3·08	3·00	2·92	2·84	2·75	2·66	2·57
19	8·18	5·93	5·01	4·50	4·17	3·94	3·77	3·63	3·52	3·43	3·30	3·15	3·00	2·92	2·84	2·76	2·67	2·58	2·49
20	8·10	5·85	4·94	4·43	4·10	3·87	3·70	3·56	3·46	3·37	3·23	3·09	2·94	2·86	2·78	2·69	2·61	2·52	2·42
21	8·02	5·78	4·87	4·37	4·04	3·81	3·64	3·51	3·40	3·31	3·17	3·03	2·88	2·80	2·72	2·64	2·55	2·46	2·36
22	7·95	5·72	4·82	4·31	3·99	3·76	3·59	3·45	3·35	3·26	3·12	2·98	2·83	2·75	2·67	2·58	2·50	2·40	2·31
23	7·88	5·66	4·76	4·26	3·94	3·71	3·54	3·41	3·30	3·21	3·07	2·93	2·78	2·70	2·62	2·54	2·45	2·35	2·26
24	7·82	5·61	4·72	4·22	3·90	3·67	3·50	3·36	3·26	3·17	3·03	2·89	2·74	2·66	2·58	2·49	2·40	2·31	2·21
25	7·77	5·57	4·68	4·18	3·85	3·63	3·46	3·32	3·22	3·13	2·99	2·85	2·70	2·62	2·54	2·45	2·36	2·27	2·17
26	7·72	5·53	4·64	4·14	3·82	3·59	3·42	3·29	3·18	3·09	2·96	2·81	2·66	2·58	2·50	2·42	2·33	2·23	2·13
27	7·68	5·49	4·60	4·11	3·78	3·56	3·39	3·26	3·15	3·06	2·93	2·78	2·63	2·55	2·47	2·38	2·29	2·20	2·10
28	7·64	5·45	4·57	4·07	3·75	3·53	3·36	3·23	3·12	3·03	2·90	2·75	2·60	2·52	2·44	2·35	2·26	2·17	2·06
29	7·60	5·42	4·54	4·04	3·73	3·50	3·33	3·20	3·09	3·00	2·87	2·73	2·57	2·49	2·41	2·33	2·23	2·14	2·03
30	7·56	5·39	4·51	4·02	3·70	3·47	3·30	3·17	3·07	2·98	2·84	2·70	2·55	2·47	2·39	2·30	2·21	2·11	2·01
40	7·31	5·18	4·31	3·83	3·51	3·29	3·12	2·99	2·89	2·80	2·66	2·52	2·37	2·29	2·20	2·11	2·02	1·92	1·80
60	7·08	4·98	4·13	3·65	3·34	3·12	2·95	2·82	2·72	2·63	2·50	2·35	2·20	2·12	2·03	1·94	1·84	1·73	1·60
120	6·85	4·79	3·95	3·48	3·17	2·96	2·79	2·66	2·56	2·47	2·34	2·19	2·03	1·95	1·86	1·76	1·66	1·53	1·38
∞	6·63	4·61	3·78	3·32	3·02	2·80	2·64	2·51	2·41	2·32	2·18	2·04	1·88	1·79	1·70	1·59	1·47	1·32	1·00

Source: E. S. Pearson and H. O. Hartley, Biometrika Tables for Statisticians, Volume I. Table 18.

TABLE A-X TABLE OF "STUDENT'S" DISTRIBUTION: VALUE OF *t*

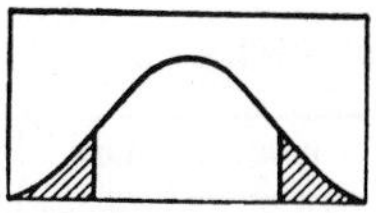

Degrees of freedom	Probability												
	0.9	0.8	0.7	0.6	0.5	0.4	0.3	0.2	0.1	0.05	0.02	0.01	0.001
1	0.158	0.325	0.510	0.727	1.000	1.376	1.963	3.078	6.314	12.706	31.821	63.657	636.619
2	0.142	0.289	0.445	0.617	0.816	1.061	1.386	1.886	2.920	4.303	6.965	9.925	31.598
3	0.137	0.277	0.424	0.584	0.765	0.978	1.250	1.638	2.353	3.182	4.541	5.841	12.924
4	0.134	0.271	0.414	0.569	0.741	0.941	1.190	1.533	2.132	2.776	3.747	4.604	8.610
5	0.132	0.267	0.408	0.559	0.727	0.920	1.156	1.476	2.015	2.571	3.365	4.032	6.869
6	0.131	0.265	0.404	0.553	0.718	0.906	1.134	1.440	1.943	2.447	3.143	3.707	5.959
7	0.130	0.263	0.402	0.549	0.711	0.896	1.119	1.415	1.895	2.365	2.998	3.499	5.408
8	0.130	0.262	0.399	0.546	0.706	0.889	1.108	1.397	1.860	2.306	2.896	3.355	5.041
9	0.129	0.261	0.398	0.543	0.703	0.883	1.100	1.383	1.833	2.262	2.821	3.250	4.781
10	0.129	0.260	0.397	0.542	0.700	0.879	1.093	1.372	1.812	2.228	2.764	3.169	4.587
11	0.129	0.260	0.396	0.540	0.697	0.876	1.088	1.363	1.796	2.201	2.718	3.106	4.437
12	0.128	0.259	0.395	0.539	0.695	0.873	1.083	1.356	1.782	2.179	2.681	3.055	4.318
13	0.128	0.259	0.394	0.538	0.694	0.870	1.079	1.350	1.771	2.160	2.650	3.012	4.221
14	0.128	0.258	0.393	0.537	0.692	0.868	1.076	1.345	1.761	2.145	2.624	2.977	4.140
15	0.128	0.258	0.393	0.536	0.691	0.866	1.074	1.341	1.753	2.131	2.602	2.947	4.073
16	0.128	0.258	0.392	0.535	0.690	0.865	1.071	1.337	1.746	2.120	2.583	2.921	4.015
17	0.128	0.257	0.392	0.534	0.689	0.863	1.069	1.333	1.740	2.110	2.567	2.898	3.965
18	0.127	0.257	0.392	0.534	0.688	0.862	1.067	1.330	1.734	2.101	2.552	2.878	3.922
19	0.127	0.257	0.391	0.533	0.688	0.861	1.066	1.328	1.729	2.093	2.539	2.861	3.883
20	0.127	0.257	0.391	0.533	0.687	0.860	1.064	1.325	1.725	2.086	2.528	2.845	3.850
21	0.127	0.257	0.391	0.532	0.686	0.859	1.063	1.323	1.721	2.080	2.518	2.831	3.819
22	0.127	0.256	0.390	0.532	0.686	0.858	1.061	1.321	1.717	2.074	2.508	2.819	3.792
23	0.127	0.256	0.390	0.532	0.685	0.858	1.060	1.319	1.714	2.069	2.500	2.807	3.767
24	0.127	0.256	0.390	0.531	0.685	0.857	1.059	1.318	1.711	2.064	2.492	2.797	3.745
25	0.127	0.256	0.390	0.531	0.684	0.856	1.058	1.316	1.708	2.060	2.485	2.787	3.725
26	0.127	0.256	0.390	0.531	0.684	0.856	1.058	1.315	1.706	2.056	2.479	2.779	3.707
27	0.127	0.256	0.389	0.531	0.684	0.855	1.057	1.314	1.703	2.052	2.473	2.771	3.690
28	0.127	0.256	0.389	0.530	0.683	0.855	1.056	1.313	1.701	2.048	2.467	2.763	3.674
29	0.127	0.256	0.389	0.530	0.683	0.854	1.055	1.311	1.699	2.045	2.462	2.756	3.659
30	0.127	0.256	0.389	0.530	0.683	0.854	1.055	1.310	1.697	2.042	2.457	2.750	3.646
40	0.126	0.255	0.388	0.529	0.681	0.851	1.050	1.303	1.684	2.021	2.423	2.704	3.551
60	0.126	0.254	0.387	0.527	0.679	0.848	1.046	1.296	1.671	2.000	2.390	2.660	3.460
120	0.126	0.254	0.386	0.526	0.677	0.845	1.041	1.289	1.658	1.980	2.358	2.617	3.373
∞	0.126	0.253	0.385	0.524	0.674	0.842	1.036	1.282	1.645	1.960	2.326	2.576	3.291

for 1 tail test divide prob headings by 2

This table is abridged from Table II of Fisher & Yates: Statistical Tables for Biological, Agricultural and Medical Research, published by Oliver & Boyd Ltd., Edinburgh, and by permission of the authors and publishers.

TABLE A-XI

Values of $z = \left(\frac{1}{2}\right) \ln \left(\frac{1+r}{1-r}\right)^*$

(For negative values of r put a minus sign in front of the tabled numbers.)

r	0.00	0.01	0.02	0.03	0.04	0.05	0.06	0.07	0.08	0.09
0.0	.00000	.01000	.02000	.03001	.04002	.05004	.06007	.07012	.08017	.09024
0.1	.10034	.11045	.12058	.13074	.14093	.15114	.16139	.17167	.18198	.19234
0.2	.20273	.21317	.22366	.23419	.24477	.25541	.26611	.27686	.28768	.29857
0.3	.30952	.32055	.33165	.34283	.35409	.36544	.37689	.38842	.40006	.41180
0.4	.42365	.43561	.44769	.45990	.47223	.48470	.49731	.51007	.52298	.53606
0.5	.54931	.56273	.57634	.59014	.60415	.61838	.63283	.64752	.66246	.67767
0.6	.69315	.70892	.72500	.74142	.75817	.77530	.79281	.81074	.82911	.84795
0.7	.86730	.88718	.90764	.92873	.95048	.97295	.99621	1.02033	1.04537	1.07143
0.8	1.09861	1.12703	1.15682	1.18813	1.22117	1.25615	1.29334	1.33308	1.37577	1.42192
0.9	1.47222	1.52752	1.58902	1.65839	1.73805	1.83178	1.94591	2.09229	2.29756	2.64665

** Extracted by permission from Wilfrid J. Dixon and Frank J. Massey, Jr., Introduction to Statistical Analysis, 2d ed., p. 468. McGraw-Hill Book Company, New York, 1957.*

TABLE A-XII 5% AND 1% SIGNIFICANCE POINTS FOR THE COEFFICIENT OF SERIAL CORRELATION (CIRCULAR DEFINITION)*

	Positive tail		Negative tail	
n	5%	1%	5%	1%
5	0.253	0.297	−0.753	−0.798
6	0.345	0.447	0.708	0.863
7	0.370	0.510	0.674	0.799
8	0.371	0.531	0.625	0.764
9	0.366	0.533	0.593	0.737
10	0.360	0.525	0.564	0.705
11	0.353	0.515	0.539	0.679
12	0.348	0.505	0.516	0.655
13	0.341	0.495	0.497	0.634
14	0.335	0.485	0.479	0.615
15	0.328	0.475	0.462	0.597
20	0.299	0.432	0.399	0.524
25	0.276	0.398	0.356	0.473
30	0.257	0.370	0.325	0.433
35	0.242	0.347	0.300	0.401
40	0.229	0.329	0.279	0.376
45	0.218	0.314	0.262	0.356
50	0.208	0.301	0.248	0.339
55	0.199	0.289	0.236	0.324
60	0.191	0.278	0.225	0.310
65	0.184	0.268	0.216	0.298
70	0.178	0.259	0.207	0.287
75	0.173	0.250	−0.199	−0.276

For values of n above 75, use the following formulas to determine the significant points:

For the 5% significance level: $\dfrac{-1 \pm 1.645\sqrt{n-2}}{n-1}$

For the 1% significance level: $\dfrac{-1 \pm 2.326\sqrt{n-2}}{n-1}$

R. L. Anderson, "Distribution of the serial correlation coefficient," Annals of Mathematical Statistics, ***13****, No. 1, 1942, pp. 1–13.*

** Reproduced by permission of the editors.*

TABLE A-XIII 5% AND 1% SIGNIFICANCE POINTS FOR THE RATIO OF THE MEAN SQUARE SUCCESSIVE DIFFERENCE TO THE VARIANCE*

Values of $\frac{\delta^2}{s^2}$ for Different Levels of Significance

	Values of k		*Values of k′*	
n	$P = .01$	$P = .05$	$P = .95$	$P = .99$
4	.8341	1.0406	4.2927	4.4992
5	.6724	1.0255	3.9745	4.3276
6	.6738	1.0682	3.7318	4.1262
7	.7163	1.0919	3.5748	3.9504
8	.7575	1.1228	3.4486	3.8139
9	.7974	1.1524	3.3476	3.7025
10	.8353	1.1803	3.2642	3.6091
11	.8706	1.2062	3.1938	3.5294
12	.9033	1.2301	3.1335	3.4603
13	.9336	1.2521	3.0812	3.3996
14	.9618	1.2725	3.0352	3.3458
15	.9880	1.2914	2.9943	3.2977
16	1.0124	1.3090	2.9577	3.2543
17	1.0352	1.3253	2.9247	3.2148
18	1.0566	1.3405	2.8948	3.1787
19	1.0766	1.3547	2.8675	3.1456
20	1.0954	1.3680	2.8425	3.1151
21	1.1131	1.3805	2.8195	3.0869
22	1.1298	1.3923	2.7982	3.0607
23	1.1456	1.4035	2.7784	3.0362
24	1.1606	1.4141	2.7599	3.0133
25	1.1748	1.4241	2.7426	2.9919
26	1.1883	1.4336	2.7264	2.9718
27	1.2012	1.4426	2.7112	2.9528
28	1.2135	1.4512	2.6969	2.9348
29	1.2252	1.4594	2.6834	2.9177
30	1.2363	1.4672	2.6707	2.9016
31	1.2469	1.4746	2.6587	2.8864
32	1.2570	1.4817	2.6473	2.8720
33	1.2667	1.4885	2.6365	2.8583
34	1.2761	1.4951	2.6262	2.8451
35	1.2852	1.5014	2.6163	2.8324
36	1.2940	1.5075	2.6068	2.8202
37	1.3025	1.5135	2.5977	2.8085
38	1.3108	1.5193	2.5889	2.7973
39	1.3188	1.5249	2.5804	2.7865
40	1.3266	1.5304	2.5722	2.7760
41	1.3342	1.5357	2.5643	2.7658
42	1.3415	1.5408	2.5567	2.7560
43	1.3486	1.5458	2.5494	2.7466
44	1.3554	1.5506	2.5424	2.7376
45	1.3620	1.5552	2.5357	2.7289
46	1.3684	1.5596	2.5293	2.7205
47	1.3745	1.5638	2.5232	2.7125
48	1.3802	1.5678	2.5173	2.7049
49	1.3856	1.5716	2.5117	2.6977
50	1.3907	1.5752	2.5064	2.6908
51	1.3957	1.5787	2.5013	2.6842
52	1.4007	1.5822	2.4963	2.6777
53	1.4057	1.5856	2.4914	2.6712
54	1.4107	1.5890	2.4866	2.6648
55	1.4156	1.5923	2.4819	2.6585
56	1.4203	1.5955	2.4773	2.6524
57	1.4249	1.5987	2.4728	2.6465
58	1.4294	1.6019	2.4684	2.6407
59	1.4339	1.6051	2.4640	2.6350
60	1.4384	1.6082	2.4596	2.6294

* *B. I. Hart, "Significance levels for the ratio of the mean square successive difference to the variance," Annals of Mathematical Statistics, 13, No. 4, 1942, pp. 445–447. Reproduced by permission of the editors.*

TABLE A-XIV CONFIDENCE INTERVALS FOR THE MEDIAN*

N	Largest k	$\alpha \leq .05$	Largest k	$\alpha \leq .01$	N	Largest k	$\alpha \leq .05$	Largest k	$\alpha \leq .01$
6	1	.031			36	12	.029	10	.004
7	1	.016			37	13	.047	11	.008
8	1	.008	1	.008	38	13	.034	11	.005
9	2	.039	1	.004	39	13	.024	12	.009
10	2	.021	1	.002	40	14	.038	12	.006
11	2	.012	1	.001	41	14	.028	12	.004
12	3	.039	2	.006	42	15	.044	13	.008
13	3	.022	2	.003	43	15	.032	13	.005
14	3	.013	2	.002	44	16	.049	14	.010
15	4	.035	3	.007	45	16	.036	14	.007
16	4	.021	3	.004	46	16	.026	14	.005
17	5	.049	3	.002	47	17	.040	15	.008
18	5	.031	4	.008	48	17	.029	15	.006
19	5	.019	4	.004	49	18	.044	16	.009
20	6	.041	4	.003	50	18	.033	16	.007
21	6	.027	5	.007	51	19	.049	16	.005
22	6	.017	5	.004	52	19	.036	17	.008
23	7	.035	5	.003	53	19	.027	17	.005
24	7	.023	6	.007	54	20	.040	18	.009
25	8	.043	6	.004	55	20	.030	18	.006
26	8	.029	7	.009	56	21	.044	18	.005
27	8	.019	7	.006	57	21	.033	19	.008
28	9	.036	7	.004	58	22	.048	19	.005
29	9	.024	8	.008	59	22	.036	20	.009
30	10	.043	8	.005	60	22	.027	20	.006
31	10	.029	8	.003	61	23	.040	21	.010
32	10	.020	9	.007	62	23	.030	21	.007
33	11	.035	9	.005	63	24	.043	21	.005
34	11	.024	10	.009	64	24	.033	22	.008
35	12	.041	10	.006	65	25	.046	22	.006

If the observations are arranged in order of size $X_1 < X_2 < X_3 < \cdots < X_N$, then we are 100(1 — α)% confident of the population median being between X_k and X_{N-k+1}, where k and α are given above.

** Reproduced with permission from K. R. Nair, "Table of confidence interval for the median in samples from any continuous population," Sankhya, vol. 4 (1940), Pergamon Press, pp. 551–558.*

TABLE A-XV DISTRIBUTION OF THE TOTAL NUMBER OF RUNS R IN SAMPLES OF SIZE (n_1, n_2)*

(n_1,n_2) \ R	2	3	4	5	6	7	8	9	10	11	12	13	14	15	16	17	18	19	20
(2,3)	.200	.500	.900	1.000															
(2,4)	.133	.400	.800	1.000															
(2,5)	.095	.333	.714	1.000															
(2,6)	.071	.286	.643	1.000															
(2,7)	.056	.250	.583	1.000															
(2,8)	.044	.222	.533	1.000															
(2,9)	.036	.200	.491	1.000															
(2,10)	.030	.182	.455	1.000															
(3,3)	.100	.300	.700	.900	1.000														
(3,4)	.057	.200	.543	.800	.971	1.000													
(3,5)	.036	.143	.429	.714	.929	1.000													
(3,6)	.024	.107	.345	.643	.881	1.000													
(3,7)	.017	.083	.283	.583	.833	1.000													
(3,8)	.012	.067	.236	.533	.788	1.000													
(3,9)	.009	.055	.200	.491	.745	1.000													
(3,10)	.007	.045	.171	.455	.706	1.000													
(4,4)	.029	.114	.371	.629	.886	.971	1.000												
(4,5)	.016	.071	.262	.500	.786	.929	.992	1.000											
(4,6)	.010	.048	.190	.405	.690	.881	.976	1.000											
(4,7)	.006	.033	.142	.333	.606	.833	.954	1.000											
(4,8)	.004	.024	.109	.279	.533	.788	.929	1.000											
(4,9)	.003	.018	.085	.236	.471	.745	.902	1.000											
(4,10)	.002	.014	.068	.203	.419	.706	.874	1.000											
(5,5)	.008	.040	.167	.357	.643	.833	.960	.992	1.000										
(5,6)	.004	.024	.110	.262	.522	.738	.911	.976	.998	1.000									
(5,7)	.003	.015	.076	.197	.424	.652	.854	.955	.992	1.000									
(5,8)	.002	.010	.054	.152	.347	.576	.793	.929	.984	1.000									
(5,9)	.001	.007	.039	.119	.287	.510	.734	.902	.972	1.000									
(5,10)	.001	.005	.029	.095	.239	.455	.678	.874	.958	1.000									
(6,6)	.002	.013	.067	.175	.392	.608	.825	.933	.987	.998	1.000								
(6,7)	.001	.008	.043	.121	.296	.500	.733	.879	.966	.992	.999	1.000							
(6,8)	.001	.005	.028	.086	.226	.413	.646	.821	.937	.984	.998	1.000							
(6,9)	.000	.003	.019	.063	.175	.343	.566	.762	.902	.972	.994	1.000							
(6,10)	.000	.002	.013	.047	.137	.288	.497	.706	.864	.958	.990	1.000							
(7,7)	.001	.004	.025	.078	.209	.383	.617	.791	.922	.975	.996	.999	1.000						
(7,8)	.000	.002	.015	.051	.149	.296	.514	.704	.867	.949	.988	.998	1.000	1.000					
(7,9)	.000	.001	.010	.035	.108	.231	.427	.622	.806	.916	.975	.994	.999	1.000					
(7,10)	.000	.001	.006	.024	.080	.182	.355	.549	.743	.879	.957	.990	.998	1.000					
(8,8)	.000	.001	.009	.032	.100	.214	.405	.595	.786	.900	.968	.991	.999	1.000	1.000				
(8,9)	.000	.001	.005	.020	.069	.157	.319	.500	.702	.843	.939	.980	.996	.999	1.000	1.000			
(8,10)	.000	.000	.003	.013	.048	.117	.251	.419	.621	.782	.903	.964	.990	.998	1.000	1.000			
(9,9)	.000	.000	.003	.012	.044	.109	.238	.399	.601	.762	.891	.956	.988	.997	1.000	1.000	1.000		
(9,10)	.000	.000	.002	.008	.029	.077	.179	.319	.510	.681	.834	.923	.974	.992	.999	1.000	1.000	1.000	
(10,10)	.000	.000	.001	.004	.019	.051	.128	.242	.414	.586	.758	.872	.949	.981	.996	.999	1.000	1.000	1.000

*C. Eisenhart and F. Swed, "Tables for testing randomness of grouping in a sequence of alternatives," *Annals of Mathematical Statistics*, 14 (1943), p. 66. Reproduced by permission of the editors.

TABLE A-XV DISTRIBUTION OF THE TOTAL NUMBER OF RUNS (Concluded)

The values listed on the previous page give the change that R or fewer runs will occur. For example, for two samples of size 4, the chance of three or fewer runs is .114. For sample sizes $n_1 = n_2$ larger than 10 the following table can be used. The columns headed 0.5, 1, 2.5, 5 give values of R such that R or fewer runs occur with chance less than the indicated percentage. For example, for $n_1 = n_2 = 12$ the chance of 8 or fewer runs is about .05. The columns headed 95, 97.5, 99, 99.5 give values of R for which the chance of R or more runs is less than 5, 2.5, 1, 0.5 per cent.

$n_1 = n_2$	0.5	1	2.5	5	95	97.5	99	99.5	Mean	Var.	s.d.
11	5	6	7	7	16	16	17	18	12	5.24	2.29
12	6	7	7	8	17	18	18	19	13	5.74	2.40
13	7	7	8	9	18	19	20	20	14	6.24	2.50
14	7	8	9	10	19	20	21	22	15	6.74	2.60
15	8	9	10	11	20	21	22	23	16	7.24	2.69
16	9	10	11	11	22	22	23	24	17	7.74	2.78
17	10	10	11	12	23	24	25	25	18	8.24	2.87
18	10	11	12	13	24	25	26	27	19	8.74	2.96
19	11	12	13	14	25	26	27	28	20	9.24	3.04
20	12	13	14	15	26	27	28	29	21	9.74	3.12
25	16	17	18	19	32	33	34	35	26	12.24	3.50
30	20	21	22	24	37	39	40	41	31	14.75	3.84
35	24	25	27	28	43	44	46	47	36	17.25	4.15
40	29	30	31	33	48	50	51	52	41	19.75	4.44
45	33	34	36	37	54	55	57	58	46	22.25	4.72
50	37	38	40	42	59	61	63	64	51	24.75	4.97
55	42	43	45	46	65	66	68	69	56	27.25	5.22
60	46	47	49	51	70	72	74	75	61	29.75	5.45
65	50	52	54	56	75	77	79	81	66	32.25	5.68
70	55	56	58	60	81	83	85	86	71	34.75	5.89
75	59	61	63	65	86	88	90	92	76	37.25	6.10
80	64	65	68	70	91	93	96	97	81	39.75	6.30
85	68	70	72	74	97	99	101	103	86	42.25	6.50
90	73	74	77	79	102	104	107	108	91	44.75	6.69
95	77	79	82	84	107	109	112	114	96	47.25	6.87
100	82	84	86	88	113	115	117	119	101	49.75	7.05

For large values of n_1 and n_2, particularly for $n_1 = n_2$ greater than 10, a normal approximation may be used. The mean and variance are

$$\frac{2n_1n_2}{n_1 + n_2} + 1 \quad \text{and} \quad \frac{2n_1n_2(2n_1n_2 - n_1 - n_2)}{(n_1 + n_2)^2(n_1 + n_2 - 1)}$$

respectively. For example, for $n_1 = n_2 = 20$ the mean is 21, and the variance is 9.74. The 97.5 and 2.5 percentiles are $21 + 1.96\sqrt{9.74}$ and $21 - 1.96\sqrt{9.74}$ or 27.1 and 14.9. The approximation is improved if $\frac{1}{2}$ is subtracted from the computed values. The resulting percentiles would then be 26.6 and 14.4. For two samples each of size n the mean is $n + 1$, and the variance is $n(n - 1)/(2n - 1)$.

TABLE A-XVI EQUIDISTIRIBUTED RANDOM NUMBERS

03 47 43 73 86	36 96 47 36 61	46 98 63 71 62	33 26 16 80 45	60 11 14 10 95
97 74 24 67 62	42 81 14 57 20	42 53 32 37 32	27 07 36 07 51	24 51 79 89 73
16 76 62 27 66	56 50 26 71 07	32 90 79 78 53	13 55 38 58 59	88 97 54 14 10
12 56 85 99 26	96 96 68 27 31	05 03 72 93 15	57 12 10 14 21	88 26 49 81 76
55 59 56 35 64	38 54 82 46 22	31 62 43 09 90	06 18 44 32 53	23 83 01 30 30
16 22 77 94 39	49 54 43 54 82	17 37 93 23 78	87 35 20 96 43	84 26 34 91 64
84 42 17 53 31	57 24 55 06 88	77 04 74 47 67	21 76 33 50 25	83 92 12 96 76
63 01 63 78 59	16 95 55 67 19	98 10 50 71 75	12 86 73 58 07	44 39 52 38 79
33 21 12 34 29	78 64 56 07 82	52 42 07 44 38	15 51 00 13 42	99 66 02 79 54
57 60 86 32 44	09 47 27 96 54	49 17 46 09 62	90 52 84 77 27	08 02 73 43 28
18 18 07 92 46	44 17 16 58 09	79 83 86 19 62	06 76 50 03 10	55 23 64 05 05
25 62 38 97 75	84 16 07 44 99	83 11 46 32 24	20 14 85 88 45	10 93 72 88 71
23 42 40 64 74	82 97 77 77 81	07 45 32 14 08	32 98 94 07 72	93 85 79 10 75
52 36 28 19 95	50 92 26 11 97	00 56 76 31 38	80 22 02 53 53	86 60 42 04 53
37 85 94 35 12	83 39 50 08 30	42 34 07 96 88	54 42 06 87 98	35 85 29 48 39
70 29 17 12 13	40 33 20 38 26	13 89 51 03 74	17 76 37 13 04	07 74 21 19 30
56 62 18 37 35	96 83 50 87 75	97 12 25 93 47	70 33 24 03 54	97 77 46 44 80
99 49 55 22 77	88 42 95 45 72	16 64 36 16 00	04 43 18 66 79	94 77 24 21 90
16 08 15 04 72	33 27 14 34 09	45 59 34 68 49	12 72 07 34 45	99 27 72 95 14
31 16 93 32 43	50 27 89 87 19	20 15 37 00 49	52 85 66 60 44	38 68 88 11 80
68 34 30 13 70	55 74 30 77 40	44 22 78 84 26	04 33 46 09 52	68 07 97 06 57
74 57 25 65 76	59 29 97 68 60	71 91 38 67 54	13 58 18 24 76	15 54 55 95 52
27 42 37 86 53	48 55 90 65 72	96 57 69 36 10	96 46 92 42 45	97 60 49 04 91
00 39 68 29 61	66 37 32 20 30	77 84 57 03 29	10 45 65 04 26	11 04 96 67 24
29 94 98 94 24	68 49 69 10 82	53 75 91 93 30	34 25 20 57 27	40 48 73 51 92
16 90 82 66 59	83 62 64 11 12	67 19 00 71 74	60 47 21 29 68	02 02 37 03 31
11 27 94 75 06	06 09 19 74 66	02 94 37 34 02	76 70 90 30 86	38 45 94 30 38
35 24 10 16 20	33 32 51 26 38	79 78 45 04 91	16 92 53 56 16	02 75 50 95 98
38 23 16 86 38	42 38 97 01 50	87 75 66 81 41	40 01 74 91 62	48 51 84 08 32
31 96 25 91 47	96 44 33 49 13	34 86 82 53 91	00 52 43 48 85	27 55 26 89 62
66 67 40 67 14	64 05 71 95 86	11 05 65 09 68	76 83 20 37 90	57 16 00 11 66
14 90 84 45 11	75 73 88 05 90	52 27 41 14 86	22 98 12 22 08	07 52 74 95 80
68 05 51 18 00	33 96 02 75 19	07 60 62 93 55	59 33 82 43 90	49 37 38 44 59
20 46 78 73 90	97 51 40 14 02	04 02 33 31 08	39 54 16 49 36	47 95 93 13 30
64 19 58 97 79	15 06 15 93 20	01 90 10 75 06	40 78 78 89 62	02 67 74 17 33
05 26 93 70 60	22 35 85 15 13	92 03 51 59 77	59 56 78 06 83	52 91 05 70 74
07 97 10 88 23	09 98 42 99 64	61 71 62 99 15	06 51 29 16 93	58 05 77 09 51
68 71 86 85 85	54 87 66 47 54	73 32 08 11 12	44 95 92 63 16	29 56 24 29 48
26 99 61 65 53	58 37 78 80 70	42 10 50 67 42	32 17 55 85 74	94 44 67 16 94
14 65 52 68 75	87 59 36 22 41	26 78 63 06 55	13 08 27 01 50	15 29 39 39 43

Source: From Table XXXIII of Fisher and Yates: Statistical Tables for Biological, Agricultural, and Medical Research, published by Oliver & Boyd, Ltd., Edinburgh and London, by permission of the authors and publishers.

TABLE A-XVI. EQUIDISTRIBUTED RANDOM NUMBERS

INDEX